Introduction to Aircraft Flight Mechanics: Performance, Static Stability, Dynamic Stability, and Classical Feedback Control

Introduction to Aircraft Flight Mechanics: Performance, Static Stability, Dynamic Stability, and Classical Feedback Control

Thomas R. Yechout
with
Steven L. Morris
David E. Bossert
Wayne F. Hallgren

EDUCATION SERIES
Joseph A. Schetz
Series Editor-in-Chief
Virginia Polytechnic Institute and State University
Blacksburg, Virginia

Published by
American Institute of Aeronautics and Astronautics, Inc.
1801 Alexander Bell Drive, Reston, VA 20191-4344

American Institute of Aeronautics and Astronautics, Inc., Reston, Virginia

5

Library of Congress Cataloging-in-Publication Data

Yechout, Thomas R.
Introduction to aircraft flight mechanics: performance, static stability, dynamic stability, and classical feedback control/Thomas R. Yechout with Steven L. Morris, David E. Bossert, Wayne F. Hallgren.
p. cm.—(Education Series)
Includes bibliographical references and index.
1. Airplanes—Performance. 2. Aerodynamics. 3. Flight. 4. Flight control.
I. Morris, Steven L., 1952 II. Title III. Series: AIAA education series.
TL671.4.T43 2003 629.132–dc21 2003007339
ISBN 1-56347-577-4

Foreword

Introduction to Aircraft Flight Mechanics: Performance, Static Stability, Dynamic Stability, and Classical Feedback Control by Thomas R. Yechout with Steven L. Morris, David E. Bossert, and Wayne F. Hallgren as contributors, all from the Department of Aeronautics of the U.S. Air Force Academy, is an outstanding textbook for use in undergraduate aeronautical engineering curricula. The text evolved from lecture notes at the Academy and it incorporates many suggestions literally from hundreds of cadets to improve its pedagogical value. The text reflects a wealth of experience by the authors. It covers all the essential topics needed to teach performance, static and dynamic stability, and classical feedback control of the aircraft at the introductory level.

The ten chapters of this text cover the following topics: (1) Review of Basic Aerodynamics, (2) Review of Basic Propulsion, (3) Aircraft Performance, (4) Aircraft Equations of Motion, (5) Aircraft Static Stability, (6) Linearizing Equations of Motion, (7) Aircraft Dynamic Stability, (8) Classical Feedback Control, (9) Aircraft Stability and Control Augmentation, and (10) Special Topics (mainly additional analysis techniques for feedback control and the various types of aircraft flight control systems). This text should contribute greatly to the learning of the fundamental principles of flight mechanics that is the crucial requirement in any aeronautical engineering curricula.

The AIAA Education Series of textbooks and monographs, inaugurated in 1984, embraces a broad spectrum of theory and application of different disciplines in aeronautics and astronautics, including aerospace design practice. The series also includes texts on defense science, engineering, and management. These texts serve as teaching tools as well as reference materials for practicing engineers, scientists, and managers. The complete list of textbooks published in the series can be found on the end pages of this volume.

J. S. PRZEMIENIECKI
Editor-in-Chief (Retired)
AIAA Education Series

Table of Contents

Preface

This textbook was created as a resource for teaching aircraft performance, static stability, dynamic stability, and classical feedback control as part of an undergraduate aeronautical engineering curriculum. Chapters 1 through 5 are intended for a one-semester course in performance and static stability, while Chapters 6 through 9 are intended for a sequential one-semester course in dynamic stability and feedback control. The text is intended to provide an understandable first exposure to these topics as well as a logical progression of subject matter. These courses are normally taken during the junior year following a fundamental course in aeronautics. This text in draft form was used as the course text for the first two courses in aircraft flight mechanics at the U.S. Air Force Academy during a four-year period preceding publication. The experience and student feedback obtained was used to improve and expand the text. The text was also used at the Air Force Academy for an undergraduate aeronautical engineering elective course in aircraft feedback control systems, normally taken after completion of the aircraft dynamic stability and feedback control course. Chapters 6 through 9 were covered at a fairly rapid pace and Chapter 10 provided new material and additional depth. This text may also serve as a reference for the practicing engineer.

Thomas R. Yechout
January 2003

Acknowledgments

Many individuals contributed to the development of this textbook. First, I would like to thank Col. Michael L. Smith and Col. D. Neal Barlow, the past and present heads of the Department of Aeronautics at the U. S. Air Force Academy, for their encouragement and support throughout the five years required to bring this effort from concept to reality. Special thanks are due to Lt. Col. (Ret.) Steven L. Morris for his contributions to Chapters 2 and 4 and for his many years of contributing to the development of flight mechanics courses at the Air Force Academy. Thanks also to Col. (Ret.) Wayne F. Hallgren for his contributions to Chapters 1 and 3 and to Lt. Col. David E. Bossert for his contributions to Chapter 10. Excellent contributions and review were provided by Bill Blake and Dave Leggett of the Air Force Research Laboratory, Flight Vehicles Directorate, Dr. Jeff Ashworth of Embry Riddle Aeronautical University, Dr. Dennis Bernstein of the University of Michigan, and Meredith Cawley of the AIAA staff. Thanks are also due to Lt. Col. Dave Bossert, Lt. Col. Steve Pluntze, Col. (Ret.) Gene Rose, Capt. Alex Sansone, Lt. Col. Scott Wells, and Dr. Tom Cunningham, all of the Air Force Academy Department of Aeronautics, for the time devoted to providing detailed review comments. In addition, AIAA reviewers and numerous Air Force Academy cadets majoring in aeronautical engineering provided review comments. Finally, I would like to thank my wife, Kathy, for her continued support throughout the development of this textbook and the career that I love.

Thomas R. Yechout

1
A Review of Basic Aerodynamics

Lift, drag, thrust, and weight are the four primary forces acting on an aircraft in flight (refer to Fig. 1.1).

Lift and drag are "aerodynamic forces" arising because of the relative motion between the aircraft and the surrounding air. Thrust is provided by the propulsive system, and the force due to gravity is called "weight."

Ultimately, we want to adequately predict an aircraft's motion. An understanding of lift, drag, and thrust is essential to this end. This chapter provides the basics of lift and drag, while Chapter 2 introduces propulsion. These chapters are not designed to replace an aerodynamics or propulsion course, but do provide a baseline we can build on.

1.1 Fundamental Concepts and Relationships

We will begin our discussion with a review of fundamental aerodynamic concepts and relationships. A sound understanding of these concepts is necessary to establish a solid foundation for the study of aircraft flight mechanics.

1.1.1 Properties of a Flowfield and a Discussion of Units

The study of aerodynamics deals with the flow of air. As a body moves through air, or any fluid (liquid or gas) for that matter, the surrounding air is disturbed. The term "flowfield" is common in the language of aerodynamics and is used to refer to the air in the vicinity of the body.

Pressure, density, temperature, and velocity are the key physical properties of aerodynamics. A goal of the aeronautical engineer is to quantify these properties at every point in the flowfield. We will begin by defining each of these properties:

1) Pressure (p) "is the normal force per unit area exerted on a surface due to the time rate of change of momentum of the gas molecules impacting the surface" (Ref. 1). At sea level, atmospheric pressure is approximately 2116 psf. Pressure distributions on an aircraft, caused by the same physical mechanism (namely an exchange of momentum between air molecules and a body) will be discussed elsewhere in this book.
2) The density (ρ) of air is its mass (weight/acceleration due to gravity) per unit volume. A high-density flow implies closely compacted air molecules.
3) Temperature (T) is a measure of the average kinetic energy of the air molecules. A high temperature indicates that the air molecules are moving randomly at relatively high speeds.

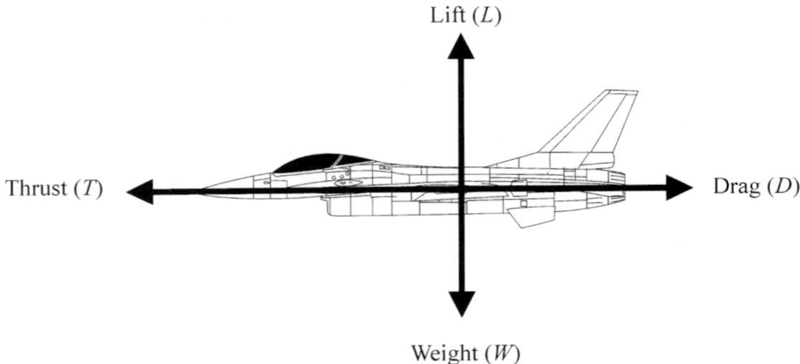

Fig. 1.1 Simplified illustration of the four forces acting on an aircraft.

4) Velocity (V) is a vector quantity; it has both magnitude and direction. The velocity at any point in the flowfield is the velocity of an infinitesimally small fluid element (differential "chunk" of air) as it sweeps through that point.

The English Engineering System is used in this text. Based on a consistent set of units (from Newton's 2nd law), this system is typically chosen in the study of flight mechanics. Assuming constant mass, Newton's 2nd law is:

$$F = ma$$

A pound force (lb) is defined as the force necessary to accelerate one slug (our unit of mass) one foot, per second squared. Table 1.1 displays the dimensions and units used for our fundamental properties.

Consider a flowfield as shown in Fig. 1.2. Our four properties are called "point properties." In general, they vary from point to point within the flowfield. Additionally, these properties can be a function of time; this is called "unsteady" flow. Pressure can be a function of not only location, but also of time, for example $p = p(x, y, z, t)$.

A "steady flow" assumption removes the time dependency; therefore, $p = p(x, y, z)$. Obviously, this makes our analysis more simple. For the case of

Table 1.1 Dimensions and units used in this book

PROPERTY	DIMENSIONS	UNITS
Pressure (p)	force/area	lb/ft^2 (psf)
Density (ρ)	mass/volume	slug/ft^3
Temperature (T)	n/a	deg Rankine (°R)
Velocity (V)	length/time	ft/s (fps)

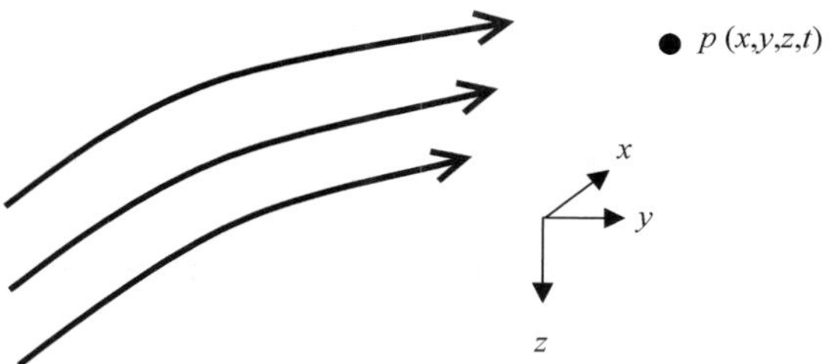

Fig. 1.2 Point in a flowfield.

straight, level, and unaccelerated flight, steady flow is a reasonable assumption. It would be unreasonable to assume steady flow for a rapid pitch-up maneuver.

Another important concept, which relates to velocity, is the definition of a "streamline." A streamline is a curve that is tangent to the velocity vectors in a flow. For example, refer to Fig. 1.3.

Consider two arbitrary streamlines in the flow, as shown in Fig. 1.4. A consequence of the definition of a streamline in steady flow is that the mass flow (slug/s) passing through cross section 1 must be the same as that passing through 2. By definition, there is no mechanism for mass to cross a streamline—the mass flow rate must be conserved between the two streamlines. Shortly, the significance of this will become more clear.

For flow over an airfoil, the distance between streamlines is decreased as they pass over and above the airfoil. As we will see, this indicates an increase in velocity.

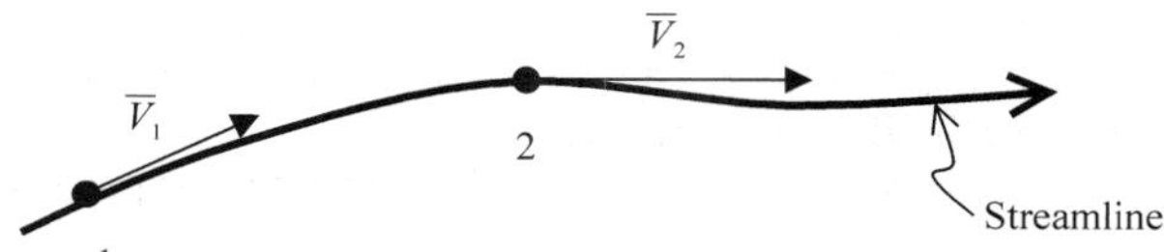

Fig. 1.3 Streamline in a flowfield.

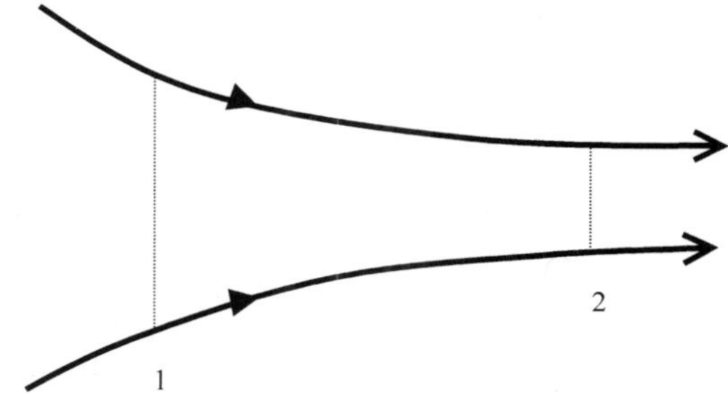

Fig. 1.4 Two streamlines in a flowfield.

1.1.2 Equation of State for a Perfect Gas

A perfect gas assumes that intermolecular forces are negligible. For the pressures and densities characteristic of flight mechanics applications, this assumption is extremely reasonable. The equation governing a perfect gas is:

$$p = \rho RT$$

where R is the specific gas constant, a function of the gas considered. For example, its value for air is different than for argon. For normal air (not, for example, chemically reacting air) the value of R is:

$$R = 1716 \frac{\text{ft-lb}}{(\text{slug})(^\circ\text{R})} [\text{English Units}] = 287 \frac{\text{J}}{(\text{kg})(\text{K})} [\text{Metric Units}]$$

Looking at the units of R, temperature must be in degrees Rankine [English Units] or Kelvin [Metric Units] to properly use the equation of state.

Example 1.1

An aircraft is flying at an air pressure of 10 psi and a temperature of −20°F. What is the air density for these conditions?

Using the equation of state and solving for density, we have:

$$\rho = \frac{p}{RT}$$

We must next convert to consistent units.

$$p = 10 \text{ psi} = (10 \text{ psi})(144 \text{ psf/psi}) = 1440 \text{ psf}$$
$$T = -20^\circ\text{F} = 460 + (-20^\circ) = 440^\circ\text{R}$$

Finally,

$$\rho = \frac{1440}{(1716)(440)} = 0.00191 \text{slug/ft}^3$$

1.1.3 Hydrostatic Equation

Consider a differential fluid element of air shown in Fig. 1.5. Its mass is dm, and it has dimensions as shown below. In the vertical direction there are two forces—weight and the forces due to pressure acting on the top and bottom surface areas (dA).

Consider a force balance in the vertical, or z, direction,

$$\Sigma \text{F}_\text{z} = p\text{d}A - (p + \text{d}p)\text{d}A - (\rho \text{d}A\text{d}h)g$$

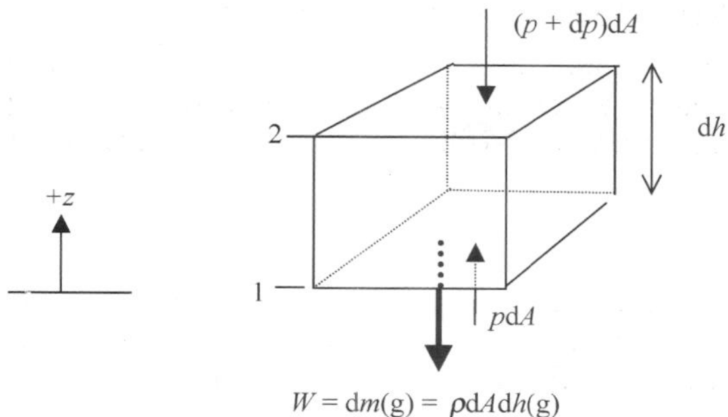

Fig. 1.5 Differential fluid element of air.

or

$$dp = -\rho g dh$$

Called the hydrostatic equation because it was originally derived for water, this differential equation relates a change in h, or altitude, with a change in pressure. Note, a positive increase in altitude corresponds to a negative change in pressure. As altitude increases, pressure decreases. Integrating between heights 1 and 2 in the vertical direction yields

$$\int_1^2 dp = -\int_1^2 \rho g dh$$

Assuming constant ρ and g,

$$p_2 - p_1 = -\rho g(h_2 - h_1) \tag{1.1}$$

This relationship is known as the manometry equation and is valid for a fluid (typically a liquid) of constant density in a uniform gravitational field.

Example 1.2

A lake is 50 ft deep. What is the difference in pressure between the bottom of the lake and the surface given that the density of water is 1.94 slug/ft^3?

Using the manometry equation [Eq. (1.1)], we have

$$p_2 - p_1 = -\rho g(h_2 - h_1)$$

We will designate position 1 as the surface of the water ($h_1 = 0$) and position 2 as the bottom of the lake ($h_2 = -50$ ft). We then have

$$p_2 - p_1 = -(1.94)(32.2)(-50 - 0) = 3123.4\,\text{psf}$$

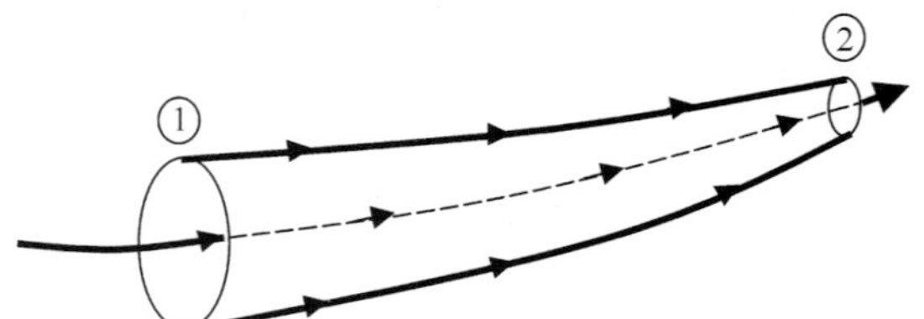

Fig. 1.6 Three-dimensional stream tube.

Thus, the pressure at the bottom of the lake is 3123.4 psf higher than at the surface.

1.1.4 Continuity Equation

The laws of aerodynamics are governed by physical principles. When these principles are applied to an appropriate model, useful equations can be derived. The continuity equation is based on the physical law that mass is conserved. Consider the "stream tube" in Fig. 1.6, which can be thought of as a bundle of streamlines. Let us also recall that mass cannot cross a streamline.

If we assume a steady flow, such that the properties everywhere in the flowfield are time independent, then the mass flow rate across 1 and 2 must be the same. Now, we will define a **one-dimensional flow**, which is a flow in which the properties are assumed constant at each cross section (perpendicular to the flow's velocity) of the flow. To help your understanding, consider Fig. 1.7 in which 1 and 2 are arbitrary points on a cross section of the flow.

By assuming one-dimensional flow, we neglect any variation in the velocity across a specific cross section. The amount of incremental mass, dm, that enters the stream tube during the incremental time, dt, can be defined as

$$\mathrm{d}m = \rho A \mathrm{d}x = \rho A V \mathrm{d}t$$

where A is the cross-sectional area of the stream tube. The following expression then follows for the mass flow rate through the stream tube:

$$\frac{\mathrm{d}m}{\mathrm{d}t} = \dot{m} = \rho A V$$

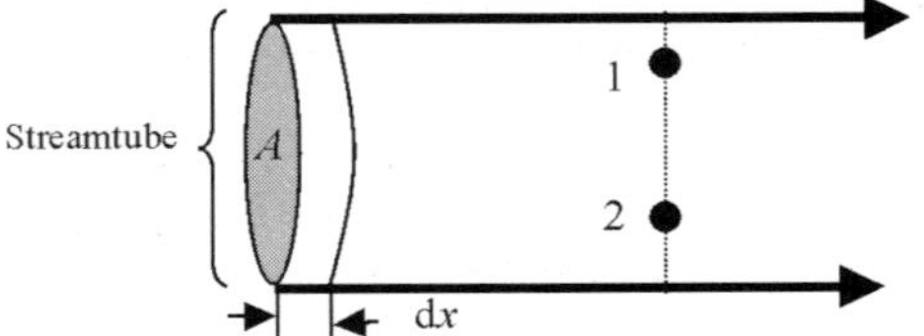

$$p_1 = p_2$$
$$\rho_1 = \rho_2$$
$$\overline{V}_1 = \overline{V}_2$$
$$T_1 = T_2$$

Fig. 1.7 Illustration of one-dimensional flow.

Because mass is conserved, the mass flow rate is the same at any cross section in the stream tube and the continuity equation reduces to the following simple result:

$$\rho_1 A_1 V_1 = \rho_2 A_2 V_2 \tag{1.2}$$

or, $\rho AV =$ constant.

The dimensions are mass per time. It should make sense that the mass flow rate is a function of density, velocity, and cross-sectional area.

Example 1.3

Consider the following nozzle.

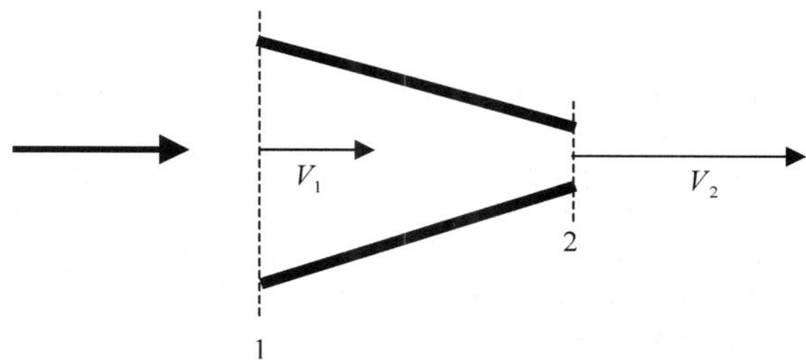

Find the velocity V_2 at the nozzle exit given that

$$V_1 = 35 \text{ ft/s} \qquad \rho_1 = 0.002 \text{ slug/ft}^3 \qquad A_1 = 0.5 \text{ ft}^2$$
$$\rho_2 = 0.0015 \text{ slug/ft}^3 \qquad A_2 = 0.05 \text{ ft}^2$$

Using the continuity equation [Eq. (1.2)] and solving for V_2,

$$V_2 = \frac{\rho_1 A_1 V_1}{\rho_2 A_2} = \frac{(0.002)(0.5)(35)}{(0.0015)(0.05)} = 466.7 \text{ ft/s}$$

1.1.5 Incompressible and Compressible Flow

Under certain conditions, it is reasonable to assume the flowfield is essentially incompressible, or constant density flow. This assumption is typically made for low-speed flowfields, where velocities everywhere (all x, y, z locations) are less than 330 ft/s. Later we will define Mach number (M) and note that this threshold corresponds to a Mach number of 0.3 at sea-level, standard-day conditions.

Note that the continuity equation reduces to the following for the case of a one-dimensional, steady, and incompressible flow. The dimensions have, of course, changed—they are now ft^3/s, or volumetric flow rate.

$$AV = \text{Constant}$$

As an aside, water (another fluid) is virtually incompressible. For this reason, flowfield density variations are typically ignored for water and other liquids.

Example 1.4

For the nozzle of Example 1.3, assume the fluid is incompressible water and V_1 remains at 35 ft/s. Find V_2 and the volumetric flow rate.

Using the incompressible form of the continuity equation and solving for V_2, we have

$$V_2 = \frac{V_1 A_1}{A_2} = \frac{(35)(0.5)}{0.05} = 350 \text{ ft/s}$$

The volumetric flow rate would be,

$$\text{volumetric flow rate} = V_1 A_1 = V_2 A_2 = (35)(0.5) = 17.5 \text{ ft}^3/\text{s}$$

1.1.6 Bernoulli's Equation

Newton's 2nd law is used again (refer to the hydrostatic equation) to derive another extremely useful equation. Consider a differential fluid element moving along a streamline, as shown in Fig. 1.8.

In general, the forces acting on the fluid element are (refer to Fig. 1.9)

1) Weight, or force due to gravity
2) Normal forces (pressure times surface area) acting on all six sides
3) Tangential forces due to the friction between adjacent fluid elements

For the purpose of this discussion, only the forces in the streamline direction are shown. In fact, forces due to pressure and friction act on all six surfaces.

Assume a steady flow and neglect the weight of the fluid element—in essence, we are assuming the fluid element's weight is small (for air) in comparison to the pressure forces "pushing" the element along the streamline. Furthermore, neglect the effects of friction. By applying Newton's 2nd law,

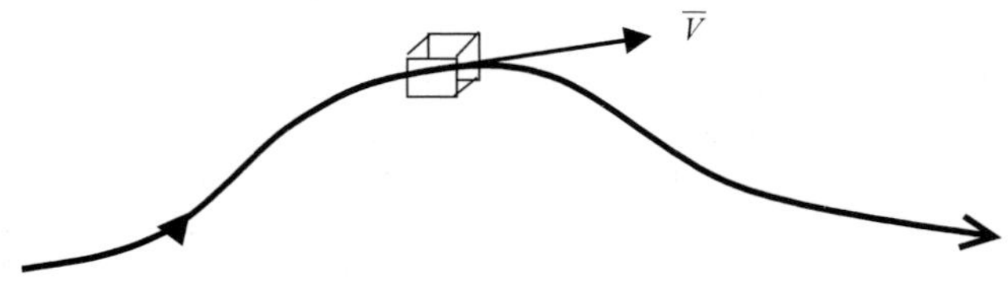

Fig. 1.8 Differential fluid element moving along a streamline.

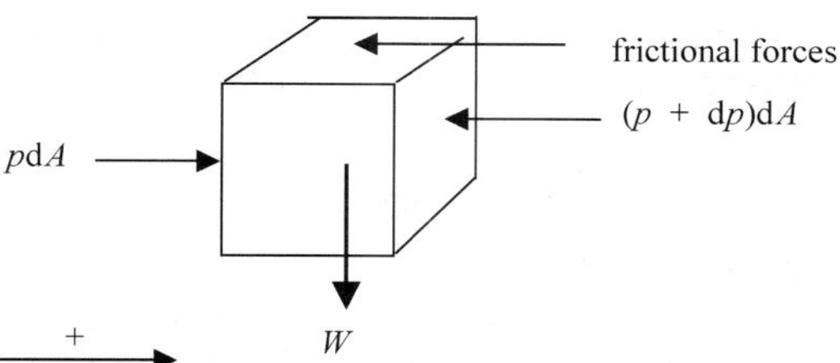

Fig. 1.9 Forces acting on a fluid element.

$F = ma$ reduces to the following differential equation (for details, refer to Ref. 2), called Euler's equation:

$$dp = -\rho V dV$$

Euler's equation is not convenient for easily solving problems. If we make one more assumption, incompressible flow, density is assumed constant, and the equation is easily integrated between two points along a streamline.

$$\int_1^2 dp = -\int_1^2 \rho V dV$$

$$p_2 - p_1 = -\rho\left[\frac{V_2^2}{2} - \frac{V_1^2}{2}\right]$$

$$p_1 + \frac{1}{2}\rho V_1^2 = p_2 + \frac{1}{2}\rho V_2^2 \tag{1.3}$$

or

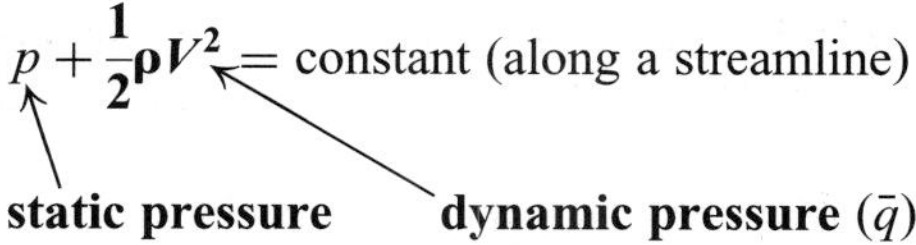

As long as the assumptions are valid, the summation of static pressure (p) and dynamic pressure ($\bar{q} = \frac{1}{2}\rho V^2$) remains constant along the streamline. Frequently, "total pressure" (p_0) is used to identify the constant, or

$$p_0 = p + \bar{q} = \textbf{total pressure}$$

Equation (1.3) is called Bernoulli's equation or the momentum equation and is one of the classics of aerodynamics. The equation is algebraic. Remember, this only applies for incompressible flow. Additionally, the four assumptions behind Euler's equation are still buried in the result—the equation is applied along a streamline, steady flow is assumed, and forces due to weight and friction are neglected.

Let's pause for a moment. Note that both the continuity and the momentum equations relate properties (pressure, density, and velocity) between points in a flow, say A and B. On the other hand, the equation of state can only be applied at a single point; it says nothing about how the properties at point B relate to the properties at point A.

Example 1.5

An F-15 on approach to Tyndall Air Force Base is flying at 120 kn. The atmospheric pressure and density are 2116 psf and 0.00238 slug/ft^3, respectively. At a point on the upper surface of the wing, the pressure is measured as 2060 psf. Find the velocity of the flow at this point on the wing and the total pressure acting on the aircraft.

We can use Bernoulli's equation [(Eq. (1.3)] since the flow is incompressible:

$$p_1 + \frac{1}{2}\rho V_1^2 = p_2 + \frac{1}{2}\rho V_2^2$$

We will designate a point out in front of the aircraft as position 1 and the point on the wing as position 2. First find the total pressure based on position 1 conditions. We will convert the airspeed to consistent units.

$$V_1 = 120 \text{ kn} = (120 \text{ kn})\left(1.69\frac{\text{ft/s}}{\text{kn}}\right) = 202.8 \text{ ft/s}$$

and then find the total pressure.

$$p_0 = p_1 + \frac{1}{2}\rho V_1^2 = 2116 + \frac{1}{2}(0.00238)(202.8)^2 = 2165 \text{ psf}$$

Because total pressure is constant,

$$p_0 = p_2 + \frac{1}{2}\rho V_2^2$$

we can solve for V_2:

$$V_2 = \sqrt{\frac{2(p_0 - p_2)}{\rho}} = \sqrt{\frac{2(2165 - 2060)}{0.00238}} = 297 \text{ ft/s}$$

1.1.7 *The Speed of Sound*

For a perfect gas, the speed of sound (a) is calculated from the following equation (refer to Ref. 3 for a detailed derivation):

$$a = \sqrt{\gamma RT} \tag{1.4}$$

In this equation γ is the ratio of specific heats. For most aerodynamic applications, it is assumed to be a constant equal to 1.4 for air. Note that the speed of sound for a perfect gas is only a function of temperature. The propagation of a sound wave takes place through molecular collisions. If the air molecules are moving faster, because they are excited by high temperatures, then the speed of the sound wave is faster. Temperature must be in °R to obtain a speed of sound in ft/s.

Example 1.6

What is the speed of sound if the air temperature is 70°F?
First, we convert to absolute temperature.

$$T = 70°\text{F} = 70 + 460 = 530°\text{R}$$

Using Eq. (1.4),

$$a = \sqrt{\gamma RT} = \sqrt{(1.4)(1716)(530)} = 1128 \text{ ft/s}$$

1.1.8 *Mach Number and Aerodynamic Flight Regimes*

Figure 1.10 presents a representative streamline around an aircraft. At any point in the flowfield, the local Mach number (M) can be defined as the ratio of the local flow velocity to the local speed of sound,

$$M = \frac{V}{a} \tag{1.5}$$

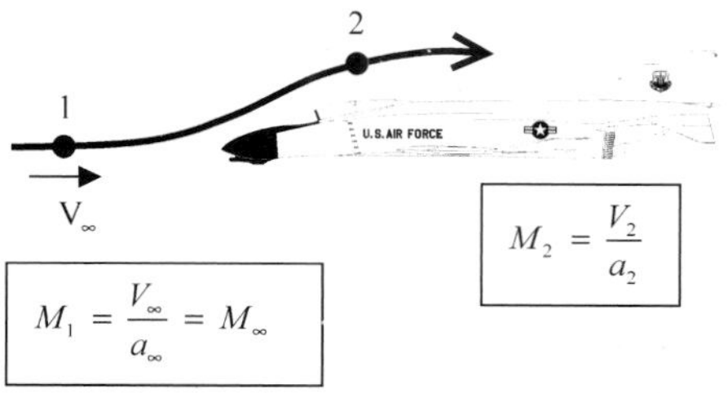

Fig. 1.10 Local Mach numbers for a streamline around an aircraft.

The Mach number at point 2 is probably larger than the Mach number at point 1 because of the acceleration of the flow velocity around the contours of the aircraft. Point 1 is sufficiently far ahead of the aircraft such that the properties at that point have not been disturbed by the presence of the aircraft. With the aircraft as the reference frame, V_1 becomes the aircraft's true airspeed. The conditions at point 1 are called freestream conditions and denoted by a subscript ∞.

If the local speed of sound is the same as the local velocity, the Mach number is 1.0 (or "sonic") at that point in the flow.

Figure 1.11 defines aerodynamic flight regimes based on freestream Mach number. Four regimes are defined:

1) When the local Mach number is less than 1.0 everywhere in the flowfield, the flow is "subsonic."
2) When the local Mach number is greater than 1.0 everywhere in the flowfield, the flow is "supersonic."
3) When the flowfield has regions of both subsonic and supersonic flow, the flowfield is "transonic." Depending on airspeed and geometry, transonic flow typically occurs at "freestream" Mach numbers between approximately 0.8 and 1.2.
4) A flow is called "hypersonic" when certain physical phenomena become important that were not important at lower speeds. These include, for example, high temperature effects and relatively thin shock layers. Typically, Mach 5 is used as the hypersonic threshold, but this value is greatly dependent on the shape of the body of interest. Refer to Ref. 4 for more detail.

The dynamic pressure, $\bar{q}$, may easily be defined in terms of Mach number using Eqs. (1.4), (1.5), and the equation of state for a perfect gas.

$$\bar{q} = \frac{1}{2}\rho V^2 = \frac{1}{2}\rho(\mathrm{Ma})^2 = \frac{1}{2}\rho M^2 \gamma RT$$

From the equation of state,

$$\rho = \frac{p}{RT}$$

We thus have an alternate form for $\bar{q}$:

$$\bar{q} = \frac{1}{2}\gamma p M^2 \tag{1.6}$$

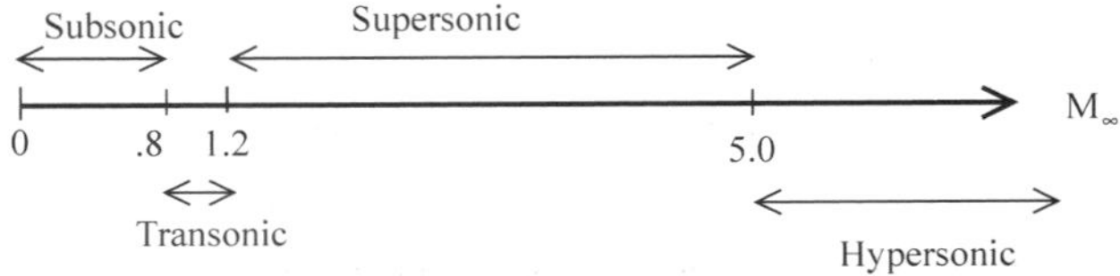

Fig. 1.11 Aerodynamic flight regimes.

1.2 The Standard Atmosphere

A discussion of aerodynamics would not be complete without introducing the concept of a standard atmosphere. Air pressure, temperature, and density (also viscosity) are functions of altitude. The standard atmosphere, typically presented in a tabular form, assigns values to these properties as they change with altitude. It provides a common reference for Department of Defense (DoD), academia, and industry.

For example, suppose the Air Force wants to purchase an interceptor. To compare climb performance (critical to an interceptor mission) between competing aircraft, manufacturers present data based on their aircraft operating on a standard day, which is an imaginary day when the pressure, temperature, and density behave exactly as defined in the standard atmosphere. Otherwise, it would be nearly impossible to accurately assess how one aircraft performs against another.

The standard atmosphere was generated by starting from an assumed (easiest property to measure) temperature distribution. Figure 1.12 shows an ideal variation of temperature with altitude based on many experimental samplings. The temperature is assumed to remain constant between approximately 36,000 and 82,000 ft; this is called the isothermal region.

With a known temperature profile, two laws of physics (hydrostatic equation and the equation of state) were used to mathematically "build" the standard atmosphere. A current version of the standard atmosphere is presented in Appendix B, and a more detailed discussion of the standard atmosphere development is presented in Ref. 5. The standard atmosphere properties of temperature, density, and pressure may be presented in the form of ratios, as defined in Eq. (1.7). Note: θ, σ, and δ all have the value of 1.0 at sea level conditions:

$$\theta = \frac{T}{T_{\mathrm{SL}}} \qquad \sigma = \frac{\rho}{\rho_{\mathrm{SL}}} \qquad \delta = \frac{p}{p_{\mathrm{SL}}} \tag{1.7}$$

Empirical equations have been developed to predict the temperature and pressure ratios as a function of altitude. These predictions are aligned with the

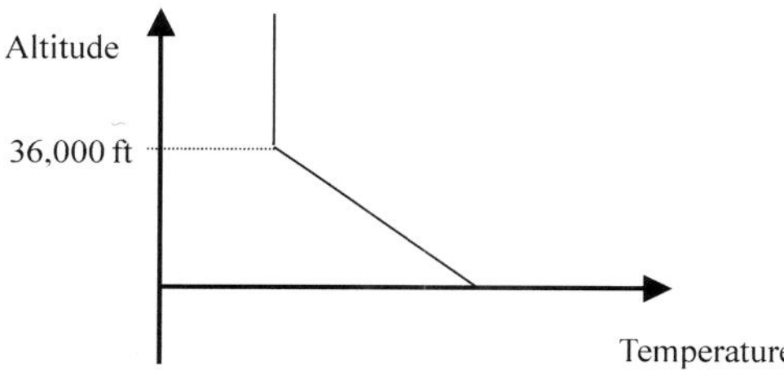

Fig. 1.12 Standard atmosphere temperature variation.

1962 U.S. Standard Atmosphere. They are divided into the altitude regions below and above approximately 36,000 ft (the troposphere region where temperature decreases at a linear rate, and the isothermal stratosphere region).

For altitudes (h) less than or equal to 36,089 ft, we have

$$\left.\begin{aligned} \theta &= 1 - 6.875 \times 10^{-6}\, h \\ \delta &= (1 - 6.875 \times 10^{-6}\, h)^{5.2561} \\ \sigma &= \frac{\delta}{\theta} \end{aligned}\right\} h \leq 36{,}089 \text{ ft} \tag{1.8}$$

For altitudes from 36,000 ft to approximately 65,600 ft, we have

$$\left.\begin{aligned} \theta &= 0.75189 \\ \delta &= 0.2234 e^{(4.806 \times 10^{-5}(36{,}089 - h))} \\ \sigma &= \frac{\delta}{\theta} \end{aligned}\right\} 36{,}089 \text{ ft} < h < 65{,}600 \text{ ft} \tag{1.9}$$

The altitude (h) must be input in feet in the previous equations. The relationship shown for density ratio can be derived using the equation of state for a perfect gas.

Frequently, in the language of flight and aeronautical engineering, the terms **pressure**, **temperature**, and **density altitudes** are used. Consider an aircraft flying at 10,000 ft above sea level, as shown in Fig. 1.13.

For the ambient pressure and temperatures shown, we use the standard atmosphere table to find these values. The standard atmosphere altitude corresponding to a pressure of 1484 psf is 9500 ft, and the aircraft is said to be flying at a **pressure altitude** (h_p) of 9500 ft. Pressure altitude says nothing about how high the aircraft is above the ground. Rather, the aircraft is "seeing" an air pressure as though it were flying at 9500 ft on a standard day.

Similarly, a **temperature altitude** (h_T) can be defined. For example, with an ambient temperature of 479.5°R, the aircraft is said to be at an 11,000-ft temperature altitude because 479.5°R is the standard atmosphere temperature for 11,000 ft.

Density altitude (h_ρ) follows the same approach. Pressure, temperature, and density altitude relate pressure, temperature, and density, respectively, to the standard atmosphere model. Simply stated, density altitude is the standard

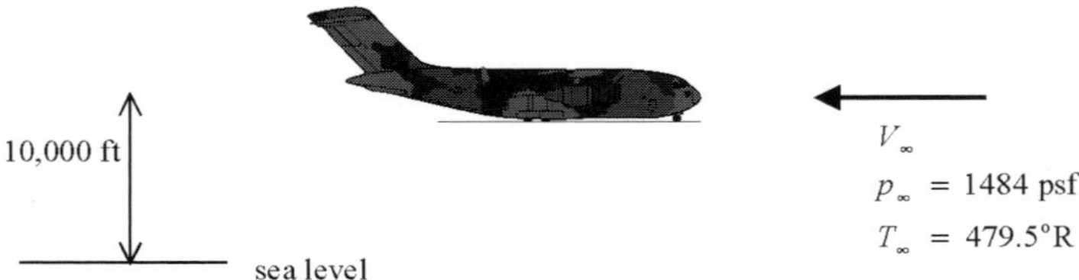

Fig. 1.13 Aircraft at specified flight conditions.

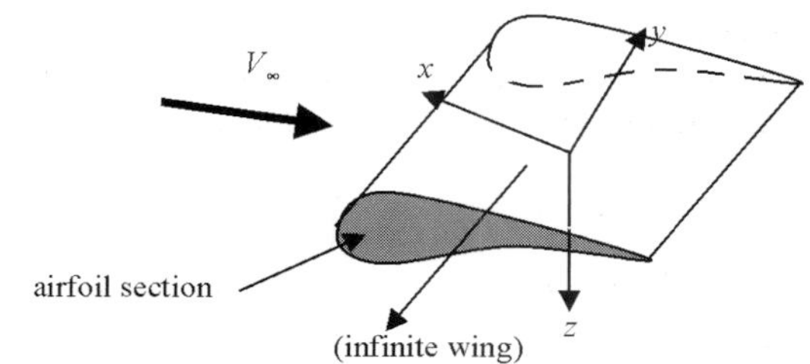

Fig. 1.14 Airfoil section.

atmosphere altitude that has the same value of density as the conditions under consideration.

1.3 Airfoil Fundamentals

Figure 1.14 presents a sketch of an airfoil, or the cross section of a wing. By definition, the flow over the airfoil is assumed to vary only in the x and z (z is perpendicular to the surface) direction. The span (in the y-direction) is assumed to approach infinity. Frequently, airfoils are also called two-dimensional wings or infinite wings because wing-tip effects (refer to Sec. 1.4) are ignored.

1.3.1 Source of Aerodynamic Forces and Relative Motion

In Fig. 1.15, an airfoil is shown in a flowfield. The only way nature can transmit an aerodynamic force is through pressure and shear stress distributions acting on the airfoil surface. Pressure and shear stress act at every point on the body, for example, points 1 and 2 in Fig. 1.15. Pressure always acts perpendicular to the surface. Shear stress (τ_w) acts tangentially to the surface (or "wall")—like pressure, it has the dimensions of force per unit area (refer to Sec. 1.3.3.1). The net effect is an aerodynamic force (F_{aero}). Later, we will break F_{aero} into lift and drag components and consider the moment created by the pressure and shear stress distributions.

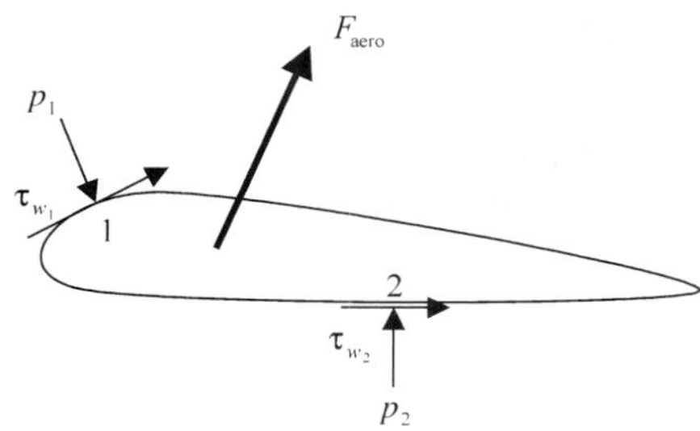

Fig. 1.15 Pressure and shear stress vectors on an airfoil.

Fig. 1.16 Airfoil on a test stand and an airfoil in flight.

Incidentally, the same is true of any body in a flowfield, be it an automobile, a ski jumper, or a cyclist—pressure and shear stress are the only physical mechanisms that generate an aerodynamic force.

As you might expect, aerodynamic forces depend on the relative velocity between the body and the air. Consider two cases, as shown in Fig. 1.16. Shown is an airfoil on a test stand with air blowing over it at 300 ft/s and an identical airfoil flying at 300 ft/s through still air.

The two airfoils have the same aerodynamic force, which is why wind tunnels work.

1.3.2 Lift

As shown in Fig. 1.17, lift (L) and drag (D) are the components of the aerodynamic force perpendicular and parallel, respectively, to the relative wind (V_∞). In this section, we will focus on lift.

Recall that nature transmits an aerodynamic force through pressure and shear stress distributions. The pressure distribution over an airfoil, or wing for that matter, is primarily responsible for lift. Consider a pressure distribution as shown in Fig. 1.18. The longer arrows denote pressures higher than freestream pressure; shorter arrows are lower pressures.

Simply stated, lift is generated by creating a net pressure difference between the upper and lower surfaces. As we will see later in this chapter, an airfoil's geometry is one of the keys to efficiently generating lift.

By referring to Fig. 1.19 and looking at continuity and Bernoulli's equation, we can gain some insight into how lift is generated. Although these two equations have several assumptions buried in them, they very nicely capture the basic physics to explain lift. Air must speed up to get over the curved upper surface of an airfoil. This can be viewed as an area constriction or nozzle effect, with the continuity equation predicting an increase in velocity. As the

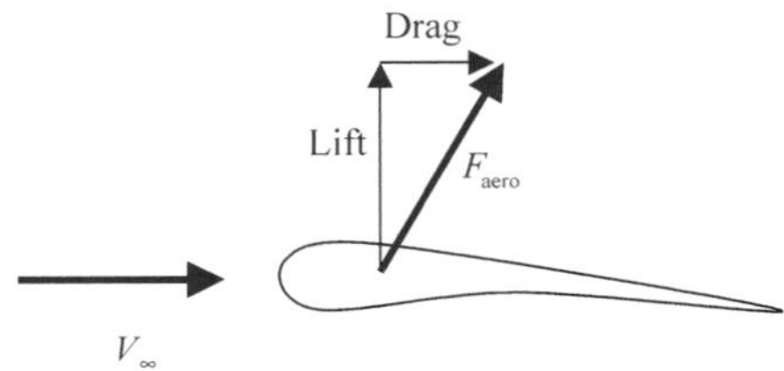

Fig. 1.17 Lift and drag components of F_{aero}.

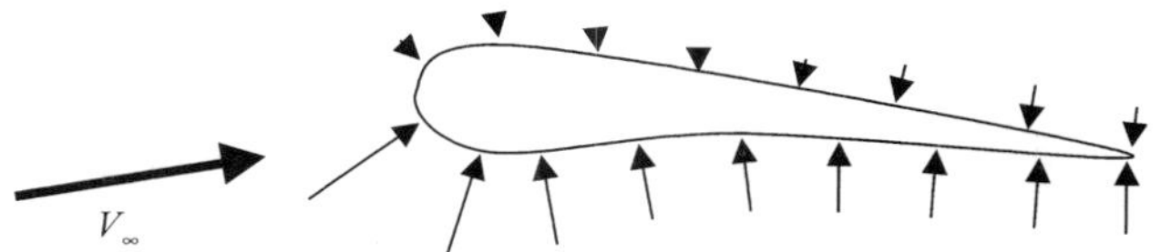

Fig. 1.18 Pressure distribution over an airfoil.

air speeds up, the pressure goes down, as predicted by Bernoulli's equation. The reduced pressure on the upper surface, relative to the higher pressure on the lower surface, creates a lift force in the upward direction.

Note that the streamlines get closer (cross-sectional area goes down) as they pass over the upper surface. From continuity, we know velocity must increase to pass the mass between the streamlines; this is no different than putting your thumb over the end of a garden hose to speed up the water. From Bernoulli's equation, we know that if velocity increases, pressure decreases. Therefore, a high velocity implies low pressure, low velocity implies high pressure, and a pressure differential between the upper and lower surfaces leads to lift. As you might expect, the amount of lift depends on a number of parameters, for example flow velocity and airfoil shape. This is discussed in subsequent sections.

1.3.3 Drag

Refer again to Fig. 1.17 in which drag is the component of the aerodynamic force parallel to V_∞. To gain an appreciation for drag, where it comes from and its consequences, we need some more tools. We will start with the concept of a viscous flow.

1.3.3.1 Introduction to viscous flow. A viscous flow is one in which the effects of viscosity, thermal conduction, and/or mass diffusion are important. As a particle moves about in space, it carries with it its momentum, energy, and mass (its identity). Viscosity is due to the transport of momentum—it is important if a flowfield has large velocity gradients. Thermal conduction results from the

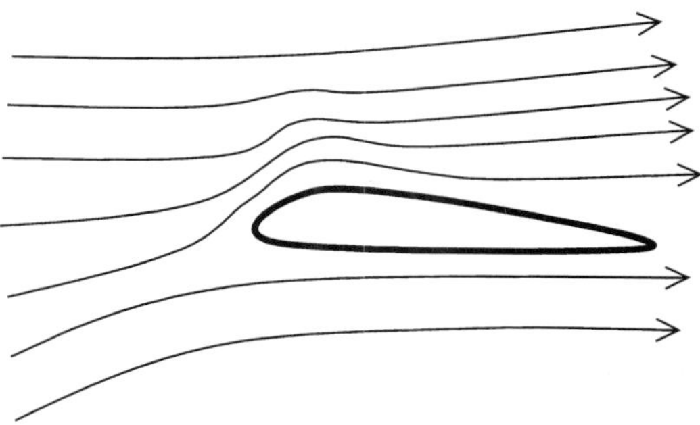

Fig. 1.19 Flow streamlines over an airfoil.

transport of energy and similarly is significant in regions of strong temperature gradients. Mass diffusion is due to the transport of mass—it is important in regions of strong concentration gradients, for example, in a chemically reacting flowfield.

For the purpose of this book, we will ignore the effects of thermal conduction and mass diffusion. The airspeeds we are concerned with do not yield flowfields where these effects are important. Therefore, in this book, a viscous flow implies regions in which there are strong velocity gradients. Velocity gradients cause shear stress. Remember from Sec. 1.3.1 that pressure and shear stress distributions generate aerodynamic forces.

To understand why shear stresses exist, consider a shear flow as sketched in Fig. 1.20.

The streamlines are horizontal; however, velocity varies in the y direction. Therefore, a velocity gradient, in the y-direction, exists. Because there is a velocity gradient, a fluid element above the plane a–b is moving faster than a fluid element below the plane. There is a rubbing action, or friction, between the fluid elements because of the different velocities. Because of an exchange of momentum (mass times velocity) between the fluid elements, a force is exerted on the plane a–b. We give it a special name, shear stress (τ_{a-b}), where the subscript denotes the stress is acting on the plane a–b.

As you might expect, shear stress (τ) is proportional to the strength of the velocity gradient. The constant of proportionality is called the coefficient of viscosity and is given the symbol μ.

$$\tau \propto \left.\frac{\mathrm{d}V}{\mathrm{d}y}\right)_{a-b} \qquad \tau = \mu \left.\frac{\mathrm{d}V}{\mathrm{d}y}\right)_{a-b} \tag{1.10}$$

Physically, viscosity is a measure of a fluid's resistance to shear and has the dimensions of mass/(length · time). The standard day sea level value for viscosity is

$$\mu = 3.7373 \times 10^{-7} \mathrm{slug/(ft \cdot s)}$$

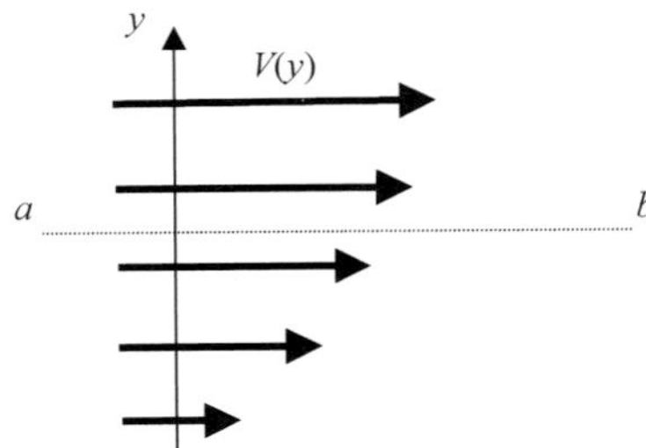

Fig. 1.20 Illustration of a shear flow.

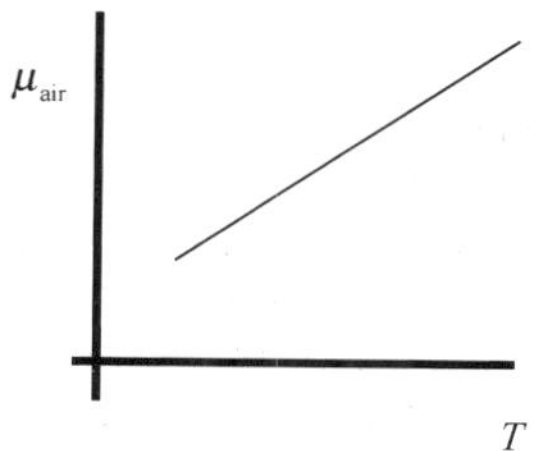

Fig. 1.21 Coefficient of viscosity as a function of temperature.

The coefficient of viscosity is a function of temperature. Qualitatively, its value changes as shown in Fig. 1.21.

From the figure, it is apparent that the viscosity of air increases as temperature increases. Recall, viscosity is due to the transport of momentum. Air molecules have more velocity and, hence, more momentum in high-temperature flowfields.

1.3.3.2 The concept of a boundary layer. Consider the flow over an airfoil, as shown in Fig. 1.22.

There is a relatively small region adjacent to the airfoil where the effects of viscosity (or friction) must be taken into account. Called a boundary layer, the concept was introduced by Ludwig Prandtl in 1904. Outside this region, the flowfield is assumed to be inviscid or frictionless.

Within a boundary layer, velocity gradients are severe—the flow is retarded because of the presence of the body. From our previous discussion, this implies that viscous effects are important. As presented in Fig. 1.23, a boundary-layer profile describes how the flow velocity changes in a direction normal (in the y direction as shown) to the surface of a wing, fuselage, or any solid surface exposed to an air stream.

Note that the velocity is zero at the surface, which is the so-called "no slip boundary condition." Frequently, the subscript w (for "wall") is used to denote the surface boundary conditions. The effect of friction, between the air and the body, diminishes in the y direction. Hence, the velocity increases through the boundary layer (in the y direction) until eventually the presence of friction is

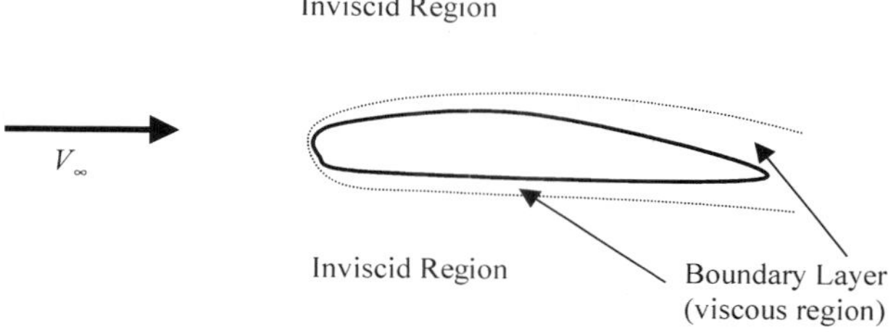

Fig. 1.22 Flow over an airfoil with boundary layer region.

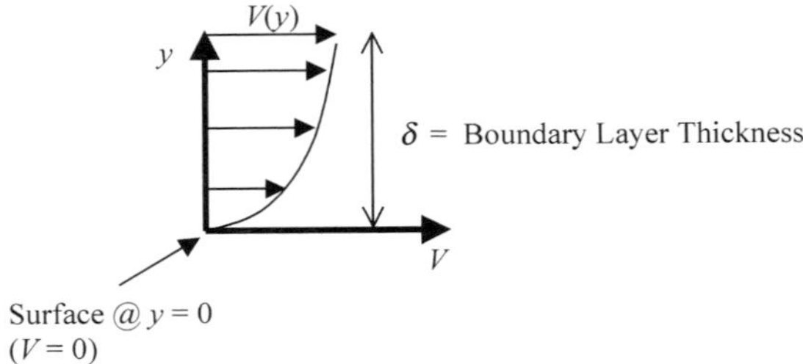

Fig. 1.23 Boundary-layer profile.

no longer felt, and the velocity gradient approaches zero. The boundary layer thickness is denoted by δ.

Two types of boundary layers exist: laminar and turbulent. A laminar boundary layer is characterized by smooth and regular streamlines, or smoothly layered flow. In contrast, a turbulent boundary layer is "random, irregular, and tortuous."[7] Laminar and turbulent velocity profiles are different. To gain some insight into the differences, consider the flow over a flat plate. At some streamwise distance, two representative velocity profiles are shown in Fig. 1.24:

The turbulent profile does not imply a nice, orderly boundary layer. Rather, what is really shown is how the average velocity changes in the y direction. From the figure, note the following

1) A turbulent boundary layer is thicker than a laminar boundary layer ($\delta_{\text{turb}} > \delta_{\text{lam}}$). For a flat plate, the freestream velocity (V_∞) is 99.9% recovered at the edge of the boundary layer.
2) The velocity gradient at the wall is greater for a turbulent boundary layer. For a given y distance, the turbulent velocity is greater than the laminar velocity.

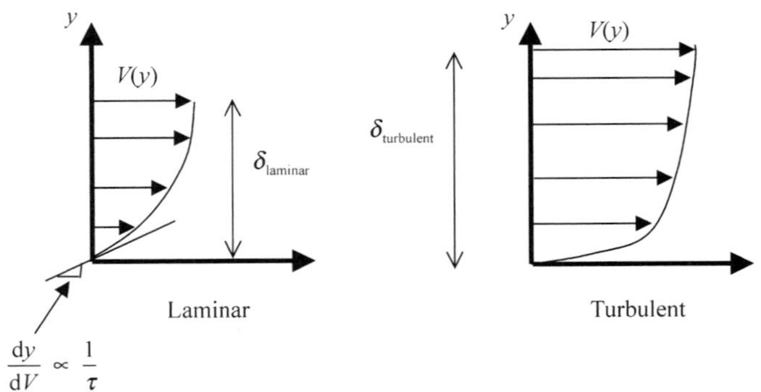

Fig. 1.24 Laminar and turbulent boundary layer profiles.

3) The shear stress at the wall is higher for a turbulent boundary layer $[\tau_w)_{\text{turb}} > \tau_w)_{\text{lam}}]$. This implies a higher skin friction drag for a turbulent boundary layer.

1.3.3.3 Reynolds number and transition. Consider the flow over a sharp flat plate as shown in Fig. 1.25; the distance "x" is measured from the leading edge.

The Reynolds number, based on a characteristic length (in this case x), is defined as

$$Re_x = \frac{\rho_\infty V_\infty x}{\mu_\infty} \tag{1.11}$$

Physically, the Reynolds number is a ratio of inertia forces to viscous forces—it is nondimensional. For nearly all our applications, the Reynolds number is relatively high (on the order of 10^4-10^8) and inertia forces dominate.

The Reynolds number is useful in predicting if a boundary layer is laminar or turbulent. Transition is defined as the point (in reality a relatively small region) where the boundary layer changes from laminar to turbulent. Once again, consider the flow over a sharp flat plate as shown in Fig. 1.26.

Typically, a boundary layer starts as laminar. Eventually, for a variety of reasons, it will transition to being turbulent. The distance x_{crit} locates the transition point. In reality, there will be a transition region, where the boundary layer has pockets of both laminar and turbulent flow. A Reynolds number, based on x_{crit}, is defined as

$$Re_{x_{\text{crit}}} = \frac{\rho_\infty V_\infty x_{\text{crit}}}{\mu_\infty} \tag{1.12}$$

or

$$x_{\text{crit}} = \frac{Re_{x_{\text{crit}}} \mu_\infty}{\rho_\infty V_\infty} \tag{1.13}$$

For the case of flow over a smooth flat plate, the critical Reynolds number is approximately 500,000. Therefore, given the freestream conditions, the critical

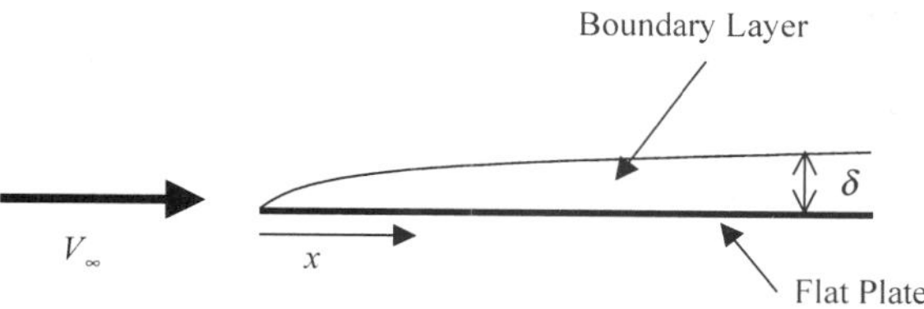

Fig. 1.25 Flow over a flat plate.

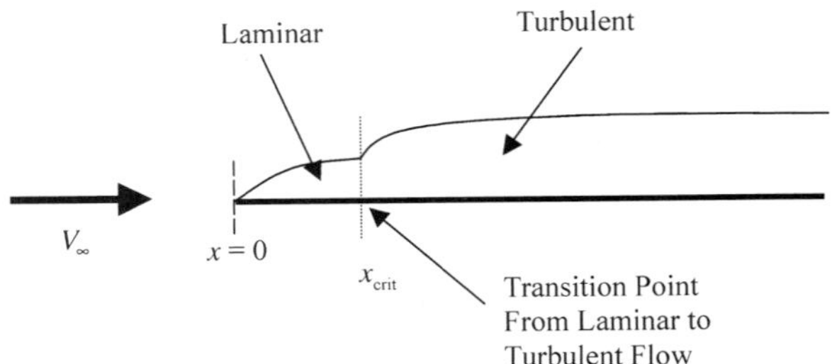

Fig. 1.26 Typical boundary layer transition from laminar to turbulent flow.

Reynolds number provides a means to predict where transition will occur, which is very handy, but very arbitrary.

Various factors influence where transition takes place. For example, surface roughness will trip a boundary layer and cause it to go turbulent earlier than expected. Additionally, surface temperature, Mach number, and pressure gradients all affect transition. Because of its significant ramifications, researchers will continue working to find better ways to predict transition.

Example 1.7

A flat plate with a 1-ft length in a wind tunnel test section is being tested at 150 and 300 ft/s at standard sea-level conditions. If the critical Reynolds number is 500,000, find the location where the flow transitions from laminar to turbulent for each velocity.

Using Eq. (1.13), we have at 150 ft/s:

$$x_{\text{crit}} = \frac{Re_{x_{\text{crit}}}\mu_\infty}{\rho_\infty V_\infty} = \frac{(500{,}000)(3.737 \times 10^{-7})}{(0.00238)(150)} = 0.523 \text{ ft}$$

At 300 ft/s we have:

$$x_{\text{crit}} = \frac{Re_{x_{\text{crit}}}\mu_\infty}{\rho_\infty V_\infty} = \frac{(500{,}000)(3.737 \times 10^{-7})}{(0.00238)(300)} = 0.262 \text{ ft}$$

Thus, we can see that the transition point moves forward as the velocity is increased. The 300 ft/s case is illustrated next.

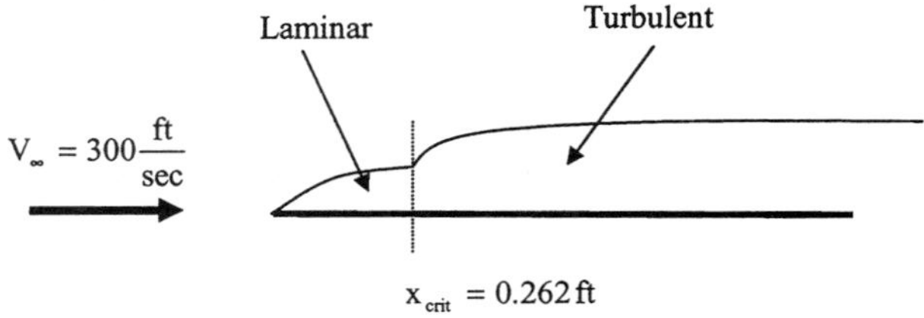

1.3.3.4 Skin friction drag. Because of the presence of friction between an aerodynamic body and the flowfield, two forms of drag are created: skin friction drag and pressure drag. We will discuss skin friction drag first.

Consider the flowfield in Fig. 1.27.

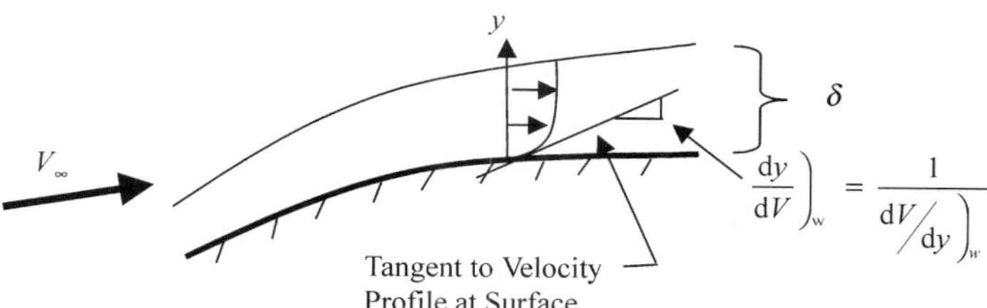

Fig. 1.27 Boundary layer velocity profile.

Within the boundary layer, the velocity increases from zero at the surface. This velocity gradient causes a shear stress (τ_w) at the surface, or wall—recall Eq. (1.10) for shear stress, which has the dimensions of force per unit area. When integrated over the entire surface area, the result is called skin friction drag and given the notation D_f.

The skin friction coefficient, C_f, is defined as

$$C_f = \frac{D_f}{\bar{q}_\infty S_{\text{wet}}} \tag{1.14}$$

S_{wet} is the so-called "wetted area," which is a surface area that would get wet if the aerodynamic body were in water. C_f is obviously going to depend on freestream conditions and boundary layer type. For the simple case of flow over a flat plate, experimental/theoretical values of skin friction coefficient are presented in the following equations:

$$C_f = \frac{1.328}{\sqrt{Re_L}} \quad \text{(laminar)} \qquad C_f = \frac{0.074}{(Re_L)^{1/5}} \quad \text{(turbulent)}$$

The equations assume a fully laminar or turbulent boundary layer from the leading edge (that is, with no transition).

Although simple relationships, the results give some valuable physical insight. Note the Reynolds number dependence, where L is the characteristic length (the total running length of the flat plate in this case). For a given Reynolds number, C_f is greater for the turbulent case. Because there is such a close relationship between Reynolds number and boundary layer type (hence shear stress), this result is not surprising. Regardless of the aerodynamic shape, a laminar boundary layer will cause less skin friction drag than a turbulent boundary layer.

$$D_f)_{\text{turbulent}} > D_f)_{\text{laminar}}$$

Example 1.8

A rectangular wing has a 5-ft chord and a 40-ft span. Estimate the skin friction drag acting on the wing at a velocity of 100 ft/s and sea-level conditions assuming the critical Reynolds number is 600,000.

We will first calculate the transition location using Eq. (1.13).

$$x_{\text{crit}} = \frac{Re_{x_{\text{crit}}}\mu_\infty}{\rho_\infty V_\infty} = \frac{(600{,}000)(3.737 \times 10^{-7})}{(0.00238)(100)} = 0.942 \text{ ft}$$

We will next assume turbulent flow for the entire wing and compute D_f using the second equation from Table 1.2.

$$D_f = C_f\left(\frac{1}{2}\right)\rho V^2 S$$

$$C_f = \frac{0.074}{Re_L^{0.2}}$$

For the entire wing, $L = 5$ ft, so that

$$Re_L = \frac{\rho VL}{\mu} = \frac{(0.00238)(100)(5)}{3.737 \times 10^{-7}} = 3.18 \times 10^6$$

and

$$S \approx (5 \text{ ft})(40 \text{ ft}) = 200 \text{ ft}^2 \text{ for the upper surface of the wing}$$

$$C_f = \frac{0.074}{(3.18 \times 10^6)^{0.2}} = 0.0037$$

$$D_{f_{\substack{\text{turbulent}\\\text{wing}}}} = (0.0037)\left(\frac{1}{2}\right)(0.00238)(100)^2(200) = 8.81 \text{ lb (on one wing surface)}$$

However, the flow is not turbulent over the entire wing; therefore, we next calculate the turbulent skin friction drag associated with the laminar region so it can be subtracted out from the previous result. For the laminar region

$$Re_L = \frac{\rho VL}{\mu} = \frac{(0.00238)(100)(0.942)}{3.737 \times 10^{-7}} = 6.0 \times 10^5$$

$$C_f = \frac{0.074}{(6 \times 10^5)^{0.2}} = 0.00517$$

$$D_{f_{\substack{\text{turbulent}\\\text{forward}\\\text{wing}}}} = (0.00517)\left(\frac{1}{2}\right)(0.00238)(100)^2(0.942 \times 40) = 2.32 \text{ lb (one surface)}$$

The turbulent skin friction drag on the portion of the wing aft of x_{crit} is thus:

$$D_{f_{\substack{\text{turbulent}\\ \text{aft wing}}}} = D_{f_{\substack{\text{turbulent}\\ \text{wing}}}} - D_{f_{\substack{\text{turbulent}\\ \text{forward}\\ \text{wing}}}} = 8.81 - 2.32 = 6.49 \text{ lb (one surface)}$$

We must next calculate the skin friction drag on the forward portion of the wing which is in laminar flow. Using the first equation from Table 1.2,

$$C_f = \frac{1.328}{\sqrt{Re_L}} = \frac{1.328}{\sqrt{6.0 \times 10^5}} = 0.00171$$

where

$$Re_L = \frac{\rho VL}{\mu} = \frac{(0.00238)(100)(0.942)}{3.737 \times 10^{-7}} = 6.0 \times 10^5$$

The laminar skin friction drag on the upper wing surface is then,

$$D_{f_{\substack{\text{laminar}\\ \text{forward}\\ \text{wing}}}} = C_f\left(\frac{1}{2}\right)\rho V^2 S = 0.00171\left(\frac{1}{2}\right)(0.00238)(100)^2(0.94 \times 40) = 0.765 \text{ lb}$$

and the total skin friction drag on the upper surface is

$$D_f = D_{f_{\substack{\text{laminar}\\ \text{forward}\\ \text{wing}}}} + D_{f_{\substack{\text{turbulent}\\ \text{aft wing}}}} = 0.765 + 6.49 = 7.255 \text{ lb}$$

For the upper and lower surface of the wing, we simply multiply by two.

$$D_{f_{\text{wing}}} = 2(7.255) = 14.53 \text{ lb}$$

A sketch (not to scale) of the wing is shown:

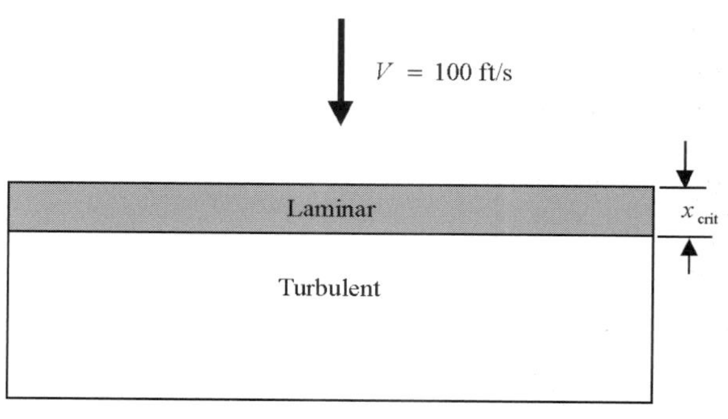

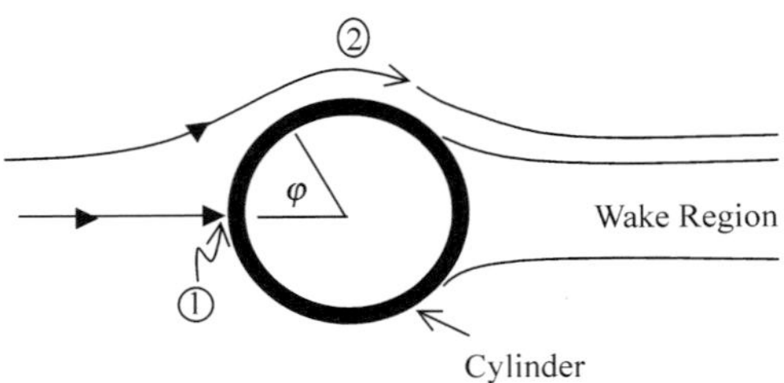

Fig. 1.28 Flow over a cylinder.

1.3.3.5 Pressure drag. As discussed, the presence of friction leads to skin friction drag. Additionally, friction also causes another form of drag; this is called drag due to the pressure field or simply pressure drag. To understand the concept of pressure drag, consider the viscous flow over a cylinder, as sketched in Fig. 1.28—two representative streamlines are shown.

Because the velocity is zero at point 1, it is called a stagnation point. Pressure is a maximum here—recall the inverse relationship between velocity and static pressure. Between 1 and 2, the flow accelerates to its maximum velocity and achieves a minimum pressure. Aft of point 2, the pressure begins to increase. In the meantime, friction has sufficiently reduced the flow's energy such that it cannot overcome the increasing pressure aft of point 2. The flow separates and a wake is formed on the aft side of the cylinder. Pressure drag is created.

Qualitatively, the pressure (p) on the cylinder's surface is shown in Fig. 1.29—φ (see Fig. 1.28) is 0 deg at the stagnation point, 90 deg at Point 2. Note the difference between the high pressure on the front (pushing the cylinder to the right) and relatively lower pressure on the back (pushing to the left). This leads to pressure drag.

Physically, the same thing happens for an airfoil. Consider Fig. 1.30. Once again, the pressure is highest at the stagnation point. Over the upper surface, the velocity increases to a maximum—a minimum pressure is reached. Aft of this point, the pressure increases. When pressure increases in the streamwise direction, the pressure gradient is called adverse. In contrast, if the pressure gradient decreases in the streamwise direction, the gradient is favorable. When the flowfield has insufficient momentum (or energy) to overcome the adverse

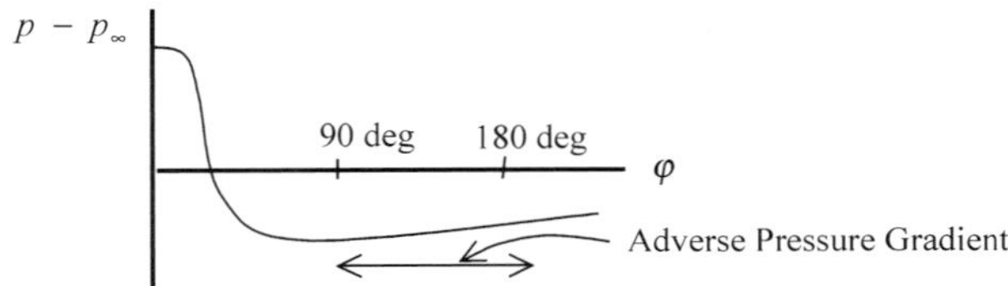

Fig. 1.29 Pressure distribution on a cylinder's surface.

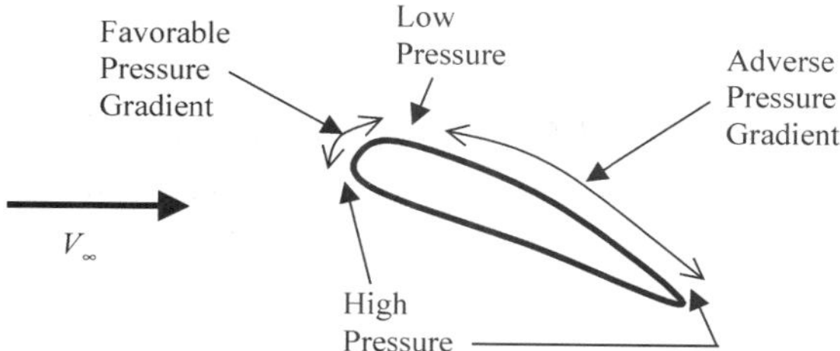

Fig. 1.30 Pressure regions on an airfoil.

pressure gradient, it separates from the upper surface and the airfoil stalls. Drag goes up; lift goes down.

Because the physics of flow separation is so important, we will look at it again but from the perspective of the velocity profiles inside the boundary layer. Consider the boundary layer on the upper surface of an airfoil as shown in Fig. 1.31—x is a running length along the airfoil's surface.

We will assume the boundary layer transitions to turbulent at x_{crit}. Friction takes its toll, and eventually the fluid particles near the surface have insufficient energy to overcome the adverse pressure. The fluid particles slow down, and then stop—this is the separation point. The velocity gradient at the wall is zero. Downstream of this point, the fluid particles may actually backup (called flow reversal) because of an adverse pressure gradient.

How can separation be delayed? We can energize the boundary layer by increasing the momentum, or kinetic energy, of the fluid elements within it. This is most easily accomplished by tripping the boundary layer to make it turbulent. Recall, a turbulent boundary layer has a larger (than laminar) velocity gradient near the wall. Therefore, a turbulent boundary layer has more momentum and thus is able to withstand an adverse pressure gradient longer before separating. Vortex generators, and various other boundary layer control (BLC) devices, are all designed to delay separation and consequently reduce the penalties associated with pressure drag.

Obviously, the stronger the adverse pressure gradient, the more susceptible an airfoil will be to flow separation and a stalled condition. Therefore, airfoil

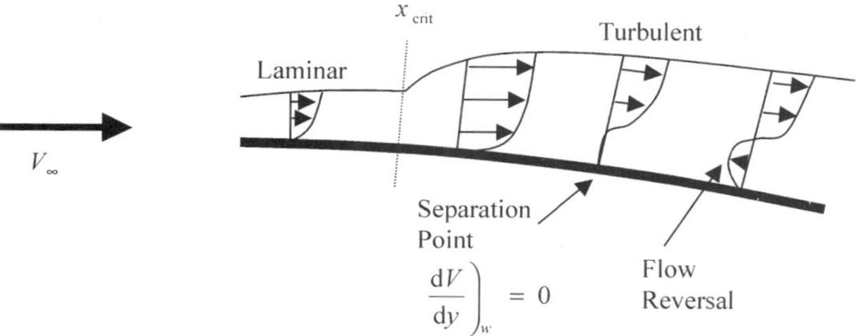

Fig. 1.31 Boundary layer velocity profiles.

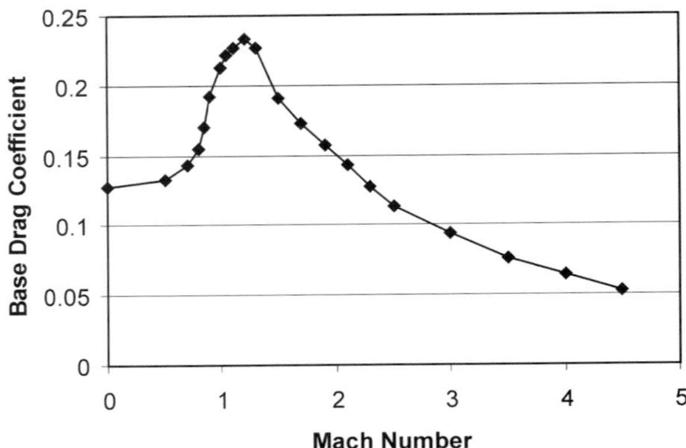

Fig. 1.32 Variation of base drag coefficient with Mach number for a missile shape with 7.2 fineness ratio (see Ref. 6).

design and orientation to the freestream velocity are critical to an efficient lifting surface.

Another type of pressure drag is referred to as base drag. Base drag is typically associated with long, slender shapes, such as missiles and fuselages, which have relatively blunt aft ends. For example, a cylindrical missile of constant diameter may simply end at the rear of the missile without tapering to a point. A separated flow region will exist at the base or aft area of the missile because the flow will not be able to stay attached around the sharp corner of this base region. Base drag can account for up to 50% of the total drag on a missile or projectile. Fig. 1.32 presents a variation of base drag with Mach number for a missile with a length to diameter ratio of 7.2.

1.3.3.6 Profile drag. The combined drag because of skin friction and flowfield separation (pressure drag) is called profile drag.

$$D = D_f + D_p$$

We have arrived at one of the great compromises of aerodynamics. A laminar boundary layer decreases skin friction drag but very likely will increase pressure drag. A turbulent boundary layer will typically reduce pressure drag, but will increase skin friction drag. Ultimately, the shape/orientation of the body will dictate which type of drag is dominant. As you might expect, skin friction is dominant for slender bodies, while pressure drag dominates blunt or bluff bodies.

Consider the flow over a sphere. Qualitatively, a graph of profile drag vs Reynolds number is shown in Fig. 1.33. Incidentally, the easiest way to change Reynolds number is to change velocity. Note the dramatic decrease in drag when the Reynolds number is sufficiently large enough to cause the laminar boundary layer to transition to turbulent. Although skin friction drag increases

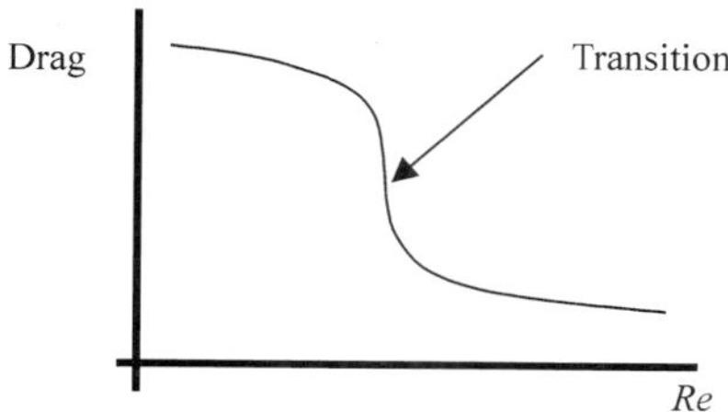

Fig. 1.33 Drag variation with Reynolds number.

with a turbulent boundary layer, pressure drag is clearly dominant for a sphere (bluff body). A turbulent boundary layer, with its higher energy flow, can better overcome the strong adverse pressure gradient, thereby reducing the overall profile drag.

Example 1.9

An illustration of two spheres in flow at the same velocity is shown below. There's only one difference between the two cases—the one on the left has a smooth surface and the one on the right has a dimpled surface to trip a turbulent boundary layer. Which ball has lower separation drag?

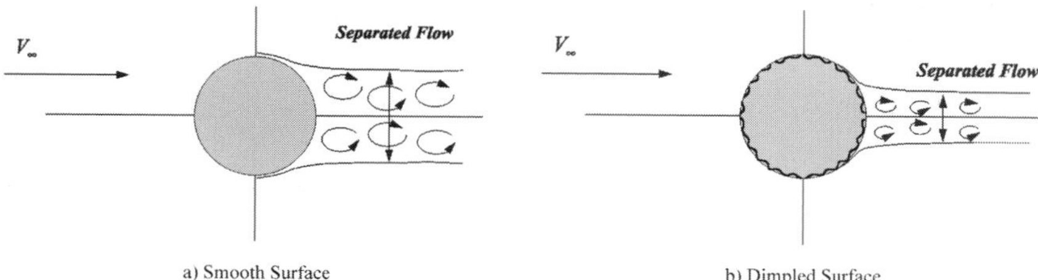

Separation is significantly delayed on the ball on the right. As a result, the overall profile drag is reduced because separation drag is significantly larger than skin friction drag. For this same reason, golf balls have dimples so that the ball will travel farther in the air. In Sec. 1.3.5, we'll examine airfoil data and reinforce the impact a boundary layer has on profile drag.

1.3.4 Airfoil Terminology

Airfoils are the fundamental building block for aircraft wing and tail surface design. In this section we will discuss the definition of airfoil configurations and aerodynamic characteristics.

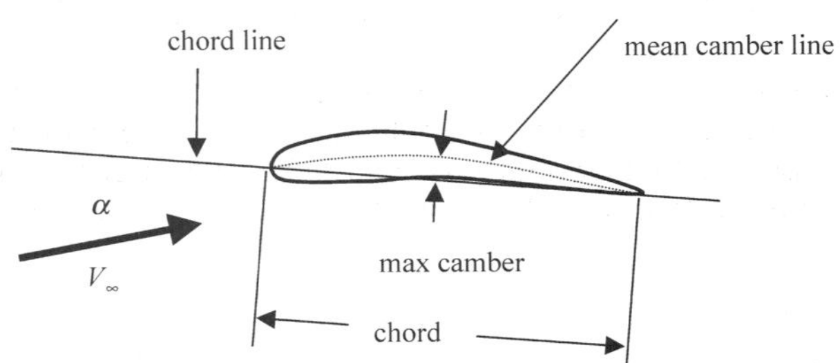

Fig. 1.34 Airfoil geometry.

1.3.4.1 Geometry and nomenclature. The concept of an airfoil (wing cross section) was introduced at the beginning of Sec. 1.3. Defining the geometry of an airfoil can get complicated (refer to Ref. 7). In this book, we will use a few common definitions to grasp the basics of airfoil geometry. Consider Fig. 1.34.

Leading edge and trailing edge are self-explanatory. Other definitions follow:

1) A straight line, passing through the leading and trailing edge, is called the chord line. The straight-line distance between the leading and trailing edge is the chord.
2) The mean camber line is the locus of points halfway between the upper and lower surfaces, as measured perpendicular to the mean camber line itself—positive camber is shown (typical for wing sections).
3) The max camber (sometimes called simply camber) is the maximum distance between the mean camber line and the chord line, as measured perpendicular to the chord line.
4) The thickness is the distance between the upper surface and lower surface, as measured perpendicular to the mean camber line.
5) The angle of attack (α) is the angle between the chord line and the freestream velocity (V_∞), or relative wind.
6) The airfoil in Fig. 1.34 is cambered. For the case of an uncambered, or symmetric, airfoil the top and bottom surface are identical—the mean camber line is the same as the chord line.

For convenience, government and industry have devised numerous ways to geometrically describe, with numbers and letters, various airfoil shapes. One extremely common airfoil designation is the National Advisory Committee for Aeronautics (NACA), a precursor to NASA, four-digit series. Four digits define an airfoil shape: the first is the amount of maximum camber in percent of chord, the second is the location of the maximum camber in tenths of chord, the last two digits are the maximum thickness of the airfoil in percent of chord. For example, a NACA 2412 airfoil would have a two percent ($0.02 \times$chord) maximum camber, the maximum camber would occur at the 40% chord location ($x = 0.4 \times$chord), and the maximum thickness would be 12% of the chord length ($0.12 \times$chord). There are several other NACA designations to describe airfoil shapes. A good reference is *Theory of Wing Sections* by Abbot and Von Doenhoff.

Data for the four-digit series was published by NACA in the 1920s and 1930s and is still widely used to verify computer code and experimental results. We'll use NACA four-digit graphs in Sec. 1.3.5 to demonstrate how to read and interpret airfoil data. Appendix C presents data for selected NACA airfoils.

1.3.4.2 Lift, drag, and moment coefficients. Consider an airfoil, at some angle of attack, as shown in Fig. 1.35. Again, lift and drag are the components of the aerodynamic force perpendicular and parallel to the freestream velocity, respectively. In general, the pressure and shear stress distributions also cause a moment (M), where pitch up (as shown) is considered positive. When comparing airfoil (or for that matter, aircraft) performance, values of lift, drag, and moments are somewhat meaningless. For example, one aircraft might generate twice the lift, but do so very inefficiently, in terms of its design and airspeed.

Therefore, dimensionless coefficients are used. Lift, drag, and moment coefficients lend themselves beautifully when comparing aerodynamic performance. Their definition and significance stem from a principle called dynamic similarity. Consider the flow over two bodies. By definition, the flows are dynamically similar if

1) Geometric similarity exists (the bodies look alike, scale models) and
2) Similarity parameters are the same.

If the flows are dynamically similar, then the force and moment coefficients will be equal and the streamline pattern over each body will be geometrically similar.

The key is determining the governing similarity parameters. Dimensional analysis (for example, the Buckingham Pi theorem) provides a mechanism. By applying dimensional analysis to an aircraft,[8] the following force/moment coefficients are defined:

$$C_L = \frac{L}{\bar{q}_\infty S} \tag{1.15}$$

$$C_D = \frac{D}{\bar{q}_\infty S} \tag{1.16}$$

$$C_M = \frac{M}{\bar{q}_\infty S\bar{c}} \tag{1.17}$$

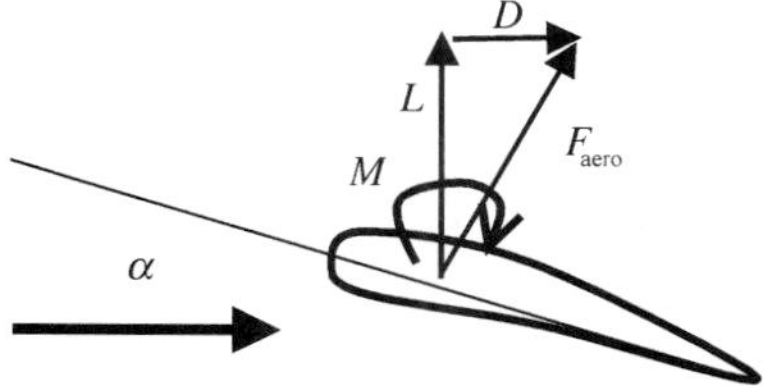

Fig. 1.35 Aerodynamic forces and moments on an airfoil.

S is a reference area—planform (top view) area of the aircraft's wing, $\bar{q}_\infty$ is the freestream dynamic pressure $(\frac{1}{2}\rho V_\infty^2)$, and $\bar{c}$ is the aircraft's mean aerodynamic chord (MAC) as defined in Sec. 1.4.1.

When discussing airfoils, these forms of the coefficients are inconvenient. Recall from Sec. 1.3, wing tips are considered to approach infinity, thus making the planform area meaningless. To avoid this, airfoil data are presented in terms of lift, drag, and moment, per unit span. Refer to the airfoil section in Fig. 1.36.

The distance between wing tips is called the span (b). When collecting airfoil data, this is the width of the wind tunnel's test section, ensuring that wing tip effects are not included in the force and moment results. Because airfoil sections are not tapered, the mean chord is just the airfoil's chord. Therefore, the planform area is simply

$$S = b \times c$$

Substituting the above for S, and manipulating the equations, leads to the following form of the coefficients. Note that the numerator is now lift per unit span, for example. A lower case is also used to denote airfoil (not aircraft) data.

$$C_l = \frac{L/b}{\bar{q}_\infty c} \tag{1.18}$$

$$C_d = \frac{D/b}{\bar{q}_\infty c} \tag{1.19}$$

$$C_m = \frac{M/b}{\bar{q}_\infty c^2} \tag{1.20}$$

These coefficients, regardless of which form, are a function of three similarity parameters: Mach number, Reynolds number, and angle of attack. Therefore, if a scale model is tested in a wind tunnel, with Reynolds number, Mach number, and angle of attack equal to those in a flight test, the coefficients should accurately predict the forces and moments in flight, which is an extremely powerful experimental tool!

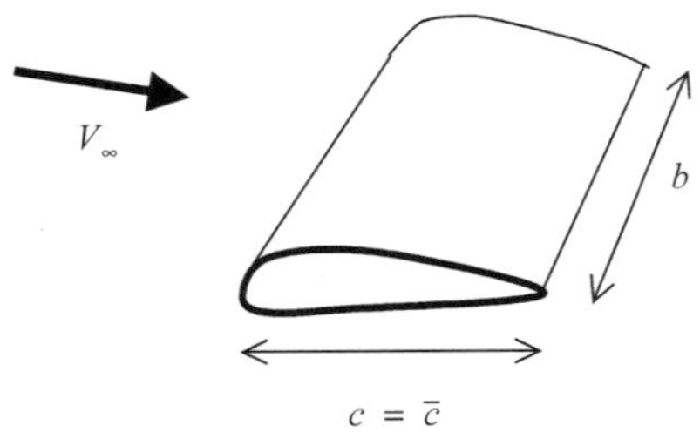

Fig. 1.36 Airfoil characteristics.

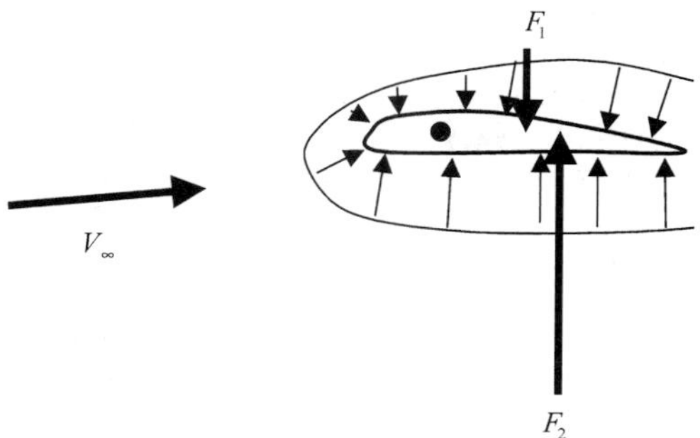

Fig. 1.37 Pressure distribution about an airfoil.

How do lift, drag, and moment coefficients change with these similarity parameters? The answer is discussed in later sections.

1.3.4.3 Center of pressure and aerodynamic center. Any point can be chosen on an airfoil to represent the aerodynamic forces (lift and drag) and moment. Before defining two special points, the center of pressure and the aerodynamic center, we will discuss how a moment arises. Figure 1.37 shows the pressure distribution about an airfoil (ignoring shear stress contributions). F_1 is the net downward force because of the pressure distribution on the upper surface of the airfoil. Likewise, F_2 is the net upward force. If we choose to support the airfoil about an arbitrary point at the quarter chord location (indicated by the black dot), an aerodynamic moment also results about this point. In this case, the airfoil will tend to pitch down about the quarter chord point because of the aerodynamic pressure distribution. For a given angle of attack and Reynolds number, there is one location, called the center of pressure where the aerodynamic moment about that point is zero. Refer to Fig. 1.38.

The center of pressure is not a very convenient reference point, in that a change in either angle of attack or Reynolds number (visualize as a change in freestream velocity) will cause the center of pressure location to shift.

In contrast, there is one point on the airfoil, called the **aerodynamic center**, where the moment coefficient about that point remains constant with changes

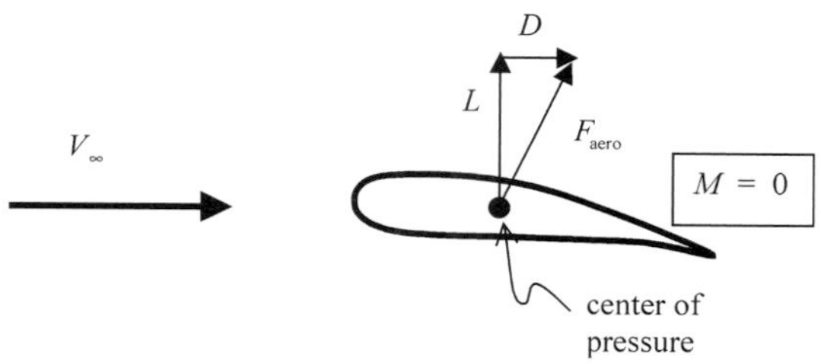

Fig. 1.38 Illustration of center of pressure.

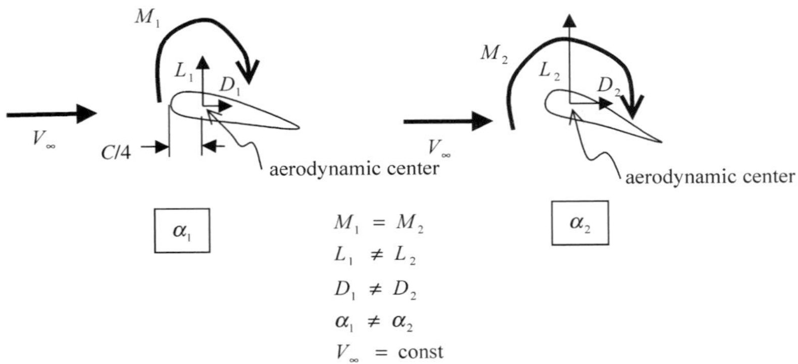

Fig. 1.39 Illustration of aerodynamic center for subsonic flow.

in angle of attack and Reynolds number. It is fixed on the airfoil and located at approximately the quarter chord for subsonic flow. If you supported an airfoil at the aerodynamic center, the aerodynamic moment about the aerodynamic center would stay constant as angle of attack was changed (and velocity was held constant). Likewise, if both angle of attack and velocity were changed, the aerodynamic moment coefficient about the aerodynamic center would stay constant. See Fig. 1.39.

Again, pitch up is positive by our convention. For an airfoil with positive camber in subsonic flow, the moment about the aerodynamic center will be negative (nose down). In transonic and supersonic flow conditions, the location of the aerodynamic center moves aft.

1.3.5 Airfoil Data

Recall that airfoil force and moment coefficients are a function of angle of attack, Reynolds number, and Mach number. In this section, we will focus on how angle of attack and Reynolds number affect these coefficients. Mach effects will be discussed in the next section.

A typical lift curve (graph of lift coefficient vs angle of attack), for a positively cambered airfoil, is presented in Fig. 1.40.

A plot of C_l vs α is one of the classics of aerodynamics. A couple key points about a lift curve follow:

1) At some angle of attack, α_{stall}, lift dramatically decreases and the airfoil stalls because of flow separation.
2) Just before stalling, the airfoil reaches a maximum lift coefficient; this is denoted by $C_{l_{\max}}$.

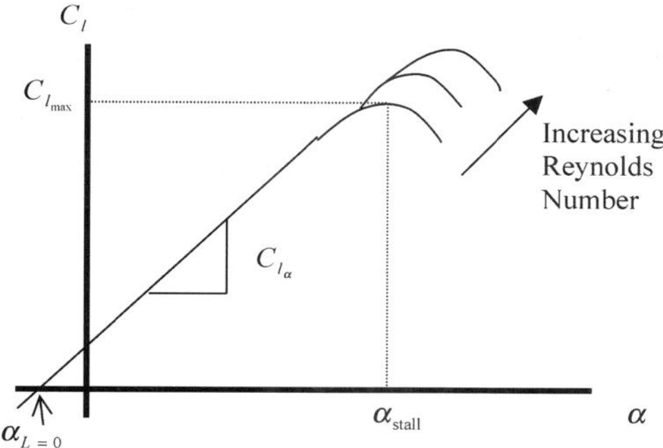

Fig. 1.40 Airfoil lift curve.

3) The lift curve slope (denoted by C_{l_α}) is typically linear below stall; its slope is about 0.11/deg.
4) At some angle of attack, the lift coefficient is zero; this is called the zero lift angle of attack, or $\alpha_{L=0}$. This occurs at a negative angle of attack (chord line oriented below the freestream velocity) for an airfoil with positive camber. At $\alpha_{L=0}$, no lift is generated.
5) Lift is generated when an airfoil with positive camber is at zero degrees angle of attack. This is usually desirable in aircraft design.
6) An increase in Reynolds number tends to increase max lift and delay the onset of stall. Does this make sense? (Hint: think about the physics of stall.)

The lift curve for a symmetric airfoil looks basically the same. However, there is one significant difference: the curve passes through the origin. Convince yourself that this makes sense. The equation for predicting C_l in the linear region is

$$C_l = C_{l_\alpha}(\alpha - \alpha_{L=0}) \tag{1.21}$$

This equation is used to predict C_l as a function of angle of attack when C_{l_α} and $\alpha_{L=0}$ are known.

A typical **drag polar**, or graph of drag coefficient vs lift coefficient, is shown in Fig. 1.41. Because lift coefficient and angle of attack vary linearly (before stall), it is easier to interpret these graphs if lift coefficient is visualized as an angle of attack. High C_l implies high α.

C_d is the airfoil's profile drag coefficient—it includes skin friction and pressure drag. Like the lift curve, it is critical to understand what a drag polar is "saying." Key points:

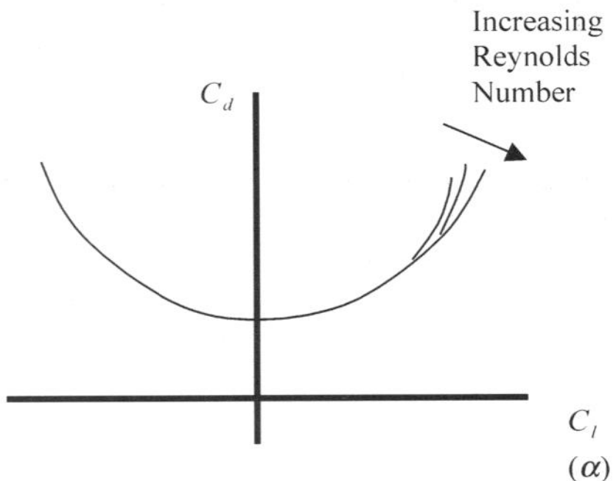

Fig. 1.41 Airfoil drag polar.

1) When C_d is plotted against C_l, the graph is parabolic in nature. As lift increases, drag increases in a parabolic fashion. To explain this, consider what the oncoming flow "sees" at low and high lift coefficients (or α's).
   a) Drag is a minimum at the smaller lift coefficients (small α). Here, the oncoming flow "sees" a slender body—skin friction dominates and there is very little pressure drag.
   b) Drag reaches a maximum at the larger lift coefficients (high α). The airfoil is no longer slender—it is a blunt body! Skin friction drag is still present, but pressure drag becomes increasingly important!
2) This drag polar is for a symmetric airfoil. The data are symmetric about the y axis, with minimum drag at a lift coefficient equal to zero ($\alpha = 0$). For a positively cambered airfoil, the curve shifts to the right.
3) As with the lift curve, Reynolds number has little effect at the low lift coefficients. However, as the angle of attack increases, Re becomes important. Why?

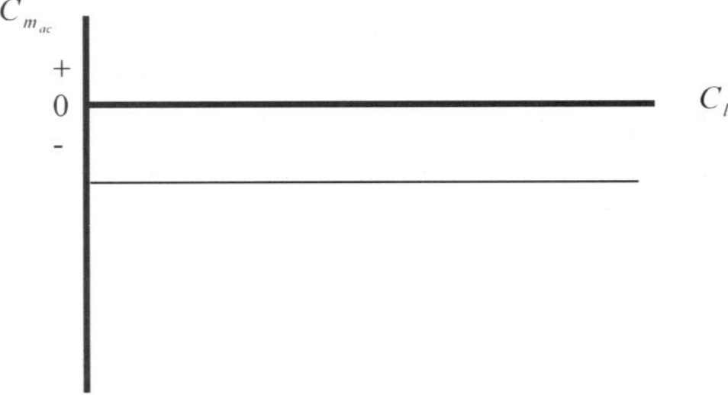

Fig. 1.42 Moment coefficient about the aerodynamic center lift coefficient.

Figure 1.42 is a graph of the moment coefficient about the aerodynamic center vs lift coefficient (or angle of attack) for a positively cambered airfoil. The key points are

1) The moment coefficient about the aerodynamic center remains constant with Reynolds number and angle of attack variation.
2) The moment coefficient about the aerodynamic center for an airfoil with positive camber is negative—the airfoil has a pitch-down tendency. Later, we will see that this has important ramifications in terms of longitudinal stability.

Example 1.10

A NACA 4412 airfoil with a 2-ft chord and a 5-ft span is being tested in a wind tunnel at standard sea-level conditions and a test section velocity of 240 ft/sec and an angle of attack of 8 deg. What is the airfoil's maximum thickness, maximum camber, location of maximum camber, and zero-lift angle of attack? Also, calculate the lift, drag, and pitching moment about the aerodynamic center.

The airfoil maximum thickness, camber, and location of maximum camber depend only on the NACA 4412 airfoil shape and length of the airfoil chord. The first digit of the 4412 designation specifies a maximum camber, which is 4% of the 2-ft chord or 0.08 ft. The second digit indicates the chordwise location of the point of maximum camber which is 0.4 c or 0.8 ft aft of the leading edge. The last two digits specify a 12% thick airfoil, and therefore the maximum thickness is 0.12 c or 0.24 ft. The aerodynamic properties of the airfoil may depend on Reynolds number, which for the given test conditions is

$$Re = \frac{\rho V c}{\mu} = \frac{(0.00238)(240)(2)}{3.737 \times 10^{-7}} = 3.06 \times 10^6$$

We will thus use the airfoil curves for $Re = 3 \times 10^6$. The value of the zero lift angle of attack does not, in fact, vary significantly with Reynolds number as we check the first NACA chart. The C_l at an angle of attack of 8 deg does show some slight variation with Reynolds number. These values are obtained from the NACA 4412 airfoil charts[7] as

$$\alpha_{L=0} = -4^\circ$$
$$C_{l_{\alpha=8^\circ}} = 1.2$$

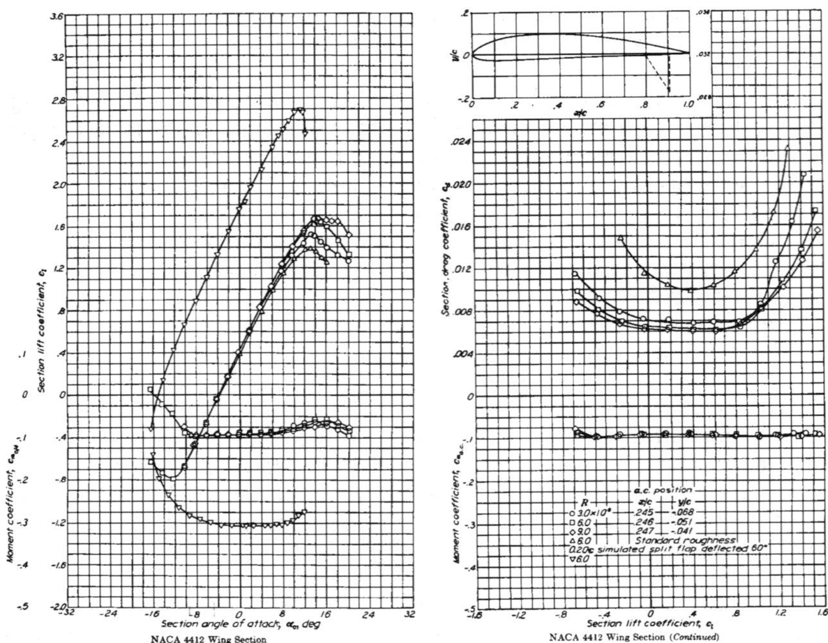

NACA 4412 Wing Section NACA 4412 Wing Section (*Continued*)

The profile drag coefficient and pitching moment coefficient about the aerodynamic center are obtained from the second chart just shown for a C_l of 1.2 and a Reynolds number of 3×10^6.

$$C_d = 0.013 \quad \text{and} \quad C_{m_{AC}} = -0.1$$

We can then determine the lift, drag, and moment about the aerodynamic center given that S is 10 ft^2 (2 ft chord $\times$ 5 ft span).

$$L = C_l\left(\frac{1}{2}\rho V^2\right)S = 1.2\left[\frac{1}{2}(0.00238)(240)^2\right](10) = 822.5 \text{ lb}$$

$$D = C_d\left(\frac{1}{2}\rho V^2\right)S = 0.013\left[\frac{1}{2}(0.00238)(240)^2\right](10) = 8.91 \text{ lb}$$

$$M_{AC} = C_{m_{AC}}\left(\frac{1}{2}\rho V^2\right)Sc = -0.1\left[\frac{1}{2}(0.00238)(240)^2\right](10)(2) = -137.1 \text{ ft} \cdot \text{lb}$$

Note that the second chart also gives the exact location of the aerodynamic center which is very close to the quarter chord, as previously discussed.

1.3.6 Compressibility (Mach) Effects

Earlier we said that lift, drag, and moment coefficients were a function of Reynolds number, angle of attack, and Mach number. The airfoil data we examined in the previous section were for low speed or incompressible flow. The influence of Reynolds number and angle of attack were shown, but no compressibility (or Mach) effects were shown.

At Mach 0.3, our threshold for compressible flow, the density of air changes approximately 5% from its static value. As Mach number increases beyond 0.3, it is no longer reasonable to ignore the effects of compressibility. Depending on the shape of the body, at Mach numbers approaching the speed of sound and beyond, shock waves develop, significantly changing the aerodynamic properties of the flowfield. Figure 1.43 identifies how the flowfield properties change across a normal shock. In this figure, station 1 is ahead of the normal shock, and station 2 is behind.

A shock wave is very thin (on the order of 10^{-5} cm) and very viscous.[9] Velocity and Mach number abruptly decrease across a shock. Total pressure, which is a measure of the flow's energy, decreases. All the static properties increase, including pressure. It is this "shock jump" in pressure that has the most profound effect on the force/moment coefficients.

To understand how compressibility influences force and moment coefficients, we need to introduce the definition of the critical Mach number (M_{crit}). Consider an airfoil as shown in Fig. 1.44 and assume the freestream Mach number is gradually increased.

As the Mach number increases, the properties in the flowfield surrounding the airfoil will naturally change. At some freestream Mach number, called the critical Mach number, sonic flow will first be achieved at a point in the flowfield (usually close to the surface of the airfoil).

Figure 1.45 is a qualitative sketch showing the variation of an airfoil's lift coefficient with Mach number.

As you might expect from the rapid changes in lift coefficient, the flowfield is changing dramatically as the Mach number is increased. Figure 1.46, based on flow visualization, shows changes in shock wave formation for the points labeled *a* through *e* in Fig. 1.45.

The following are the significant points from Figs. 1.45 and 1.46:

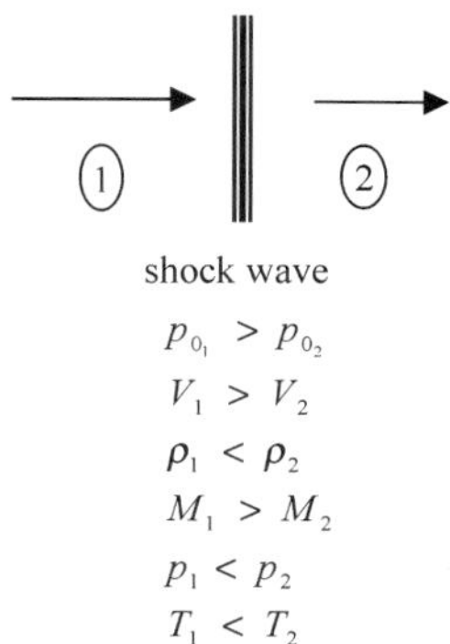

Fig. 1.43 Property changes across a normal shock.

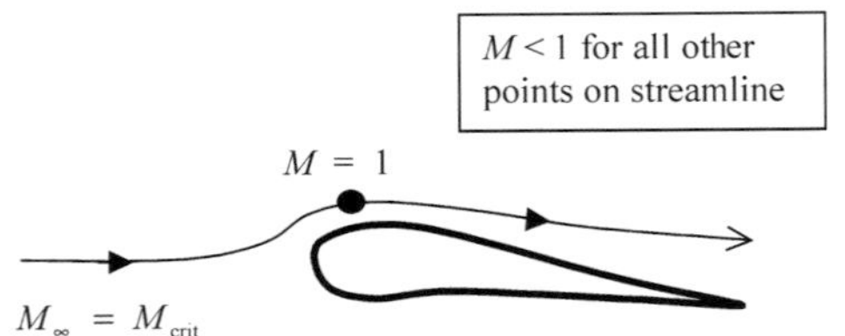

Fig. 1.44 Illustration of critical Mach number.

1) The flowfield is subsonic until point a. Typically, a compressibility correction known as the Prandtl–Glauert rule is used in this region. The equation is shown:

$$C_l = \frac{C_{l_0}}{\sqrt{1 - M^2}} \tag{1.22}$$

C_{l_0} is an incompressible lift coefficient (the subscript 0 signifies Mach = 0.0). This is the lift coefficient found in typical airfoil data as discussed in Sec. 1.3.5. The Prandtl–Glauert rule is a reasonable correction below the critical Mach number, M_{crit}. A rule of thumb is to only use Prandtl–Glauert to Mach 0.7.

2) At point b, the flow is supersonic over most of the upper surface, terminating in a shock wave. Pressure increases across a shock, thus causing increased likelihood of the flow separating from the airfoil's surface. An adverse pressure gradient is created by the presence of the shock wave.

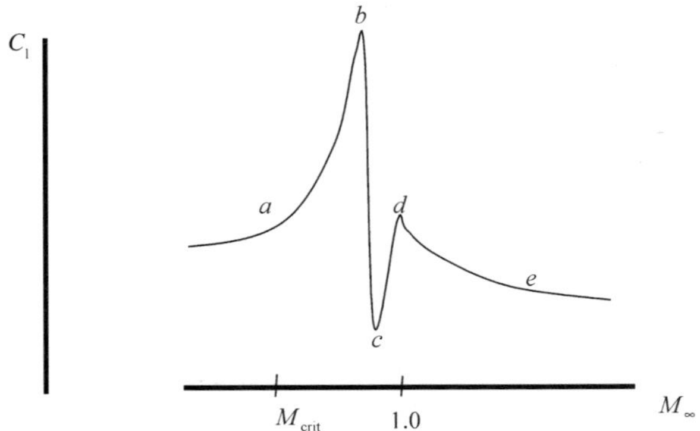

Fig. 1.45 Variation of airfoil lift coefficient with Mach number.

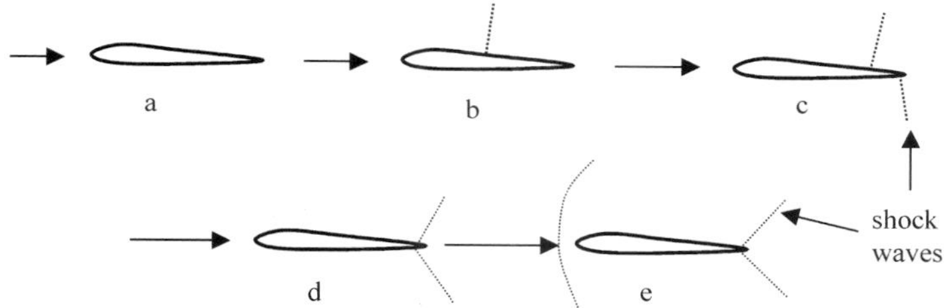

Fig. 1.46 Airfoil shock wave formation.

3) At point c, the flow over the lower surface is essentially all supersonic. The pressure on the lower surface is less than it was at condition b. The upper surface is relatively unchanged. The net effect is a smaller lift coefficient.
4) Between points c and d, the shock waves, on both surfaces, move aft and the lift coefficient increases.
5) As the freestream Mach number increases, a bow shock (or bow wave) forms. Because the velocity decreases across this initial shock, the shocks at the trailing edge are relatively weaker. The pressure differential, between the upper and lower surface, decreases and lift coefficient decreases.

It is important to recognize that the dynamic pressure, $\bar{q}_\infty$ $(1/2\rho_\infty V_\infty^2)$ is increasing with Mach number. Therefore, although the lift coefficient may be decreasing, lift can actually be increasing through the dynamic pressure term.

$$L = C_l \bar{q}_\infty S$$

At high Mach numbers, the lift coefficient is typically an order of magnitude less than its value at low speeds.

A qualitative sketch of how an airfoil's drag coefficient changes with Mach number is shown in Fig. 1.47.

Note the increase in drag preceding Mach 1.0. This is where the term drag barrier initially came from. At some freestream Mach number, beyond M_{crit}, drag increases rapidly. This is called the drag divergence Mach number, M_{DD}. Drag divergence is primarily because of the formation of shock waves on the airfoil's surface. This, in turn, causes drag because of flowfield separation. A typical definition of where the drag divergence Mach number occurs is $\partial C_D/\partial M > 0.1$ (Ref. 10).

High speed flow and the accompanying compressibility have introduced a new form of drag called wave drag, $D_w(C_{d_w}$ in coefficient form). This form of drag is only present at transonic and supersonic speeds. In addition to the drag associated with shock-induced flow separation, drag is created simply by the pressure increase across shocks. For example, consider the supersonic flow over a wedge, as shown in Fig. 1.48.

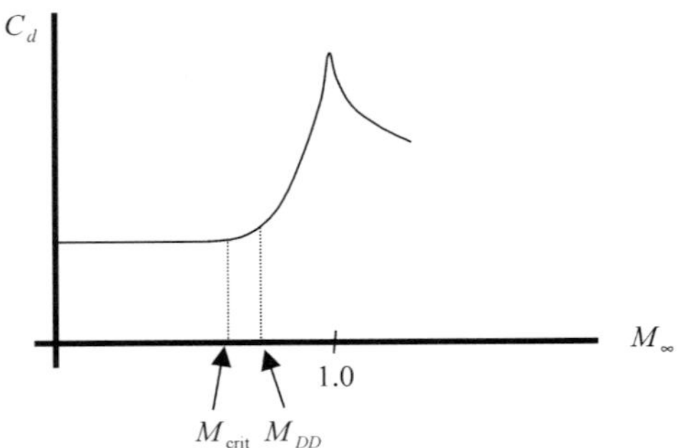

Fig. 1.47 Variation of airfoil drag coefficient with Mach number.

Because the pressure behind the oblique shock wave is higher than the free-stream pressure (p_∞), an adverse pressure gradient exists that can result in flow separation and additive drag (wave drag). In summary, the airfoil drag coefficient now consists of three contributors: skin friction drag (C_{d_f}), pressure drag (C_{d_p}), and wave drag (C_{d_w}).

$$C_d = C_{d_f} + C_{d_p} + C_{d_w} \tag{1.23}$$

Finally, what happens to the moment about the aerodynamic center as Mach number increases? As you might expect, the coefficient will typically change in the transonic region. The most important Mach effect, however, is the fact that the location of the aerodynamic center shifts from roughly the quarter-chord to the mid-chord as supersonic Mach numbers are achieved. As we'll see later, this shift has a tremendous effect on pitch stability.

1.4 Finite Wings

To this point, we have only addressed airfoils, or infinite span wings. Before we attack a complete airplane, this section introduces the aerodynamics associated with wing tips and finite span wings. First, however, we will define some terms used to describe a wing's geometry.

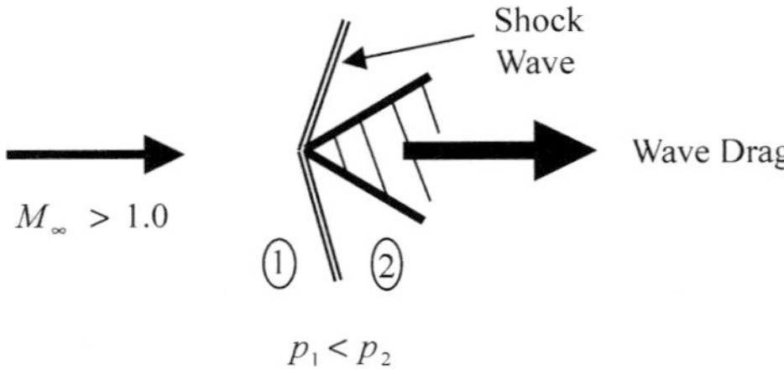

Fig. 1.48 Supersonic flow over a wedge.

1.4.1 Finite Wing Geometry

In our discussion of force and moment coefficients we introduced wing area (S) and span (b). Figure 1.49 (top view of a finite wing) shows these again. The following are some other useful definitions:

1) A wing's aspect ratio (AR) is defined in Eq. (1.24). It is dimensionless.

$$AR = b^2/S \tag{1.24}$$

2) The root chord (c_r) and the tip chord (c_t) are the distances of the chord line at the root and tip, respectively. The ratio of the tip to root chord is called the taper ratio, λ.

$$\lambda = \frac{c_t}{c_r} \tag{1.25}$$

A rectangular wing has no taper; a delta wing approaches a taper ratio of zero.

3) A wing's sweep angle is often defined at the leading edge (Λ_{LE}), the quarter-chord ($\Lambda_{c/4}$), or the mid-chord ($\Lambda_{c/2}$).
4) A wing's mean aerodynamic chord (MAC), or $\bar{c}$, is defined as

$$\bar{c} = \text{MAC} = \frac{2}{S}\int_0^{b/2} c^2 dy$$

where y is as shown in Fig. 1.49 and c is the chord at any y position. The MAC can be interpreted as the representative chord length for the forces and moments acting on a wing. For a straight, tapered wing, the MAC can be shown to be

$$\bar{c} = \frac{2}{3}c_r\left(\frac{\lambda^2 + \lambda + 1}{\lambda + 1}\right)$$

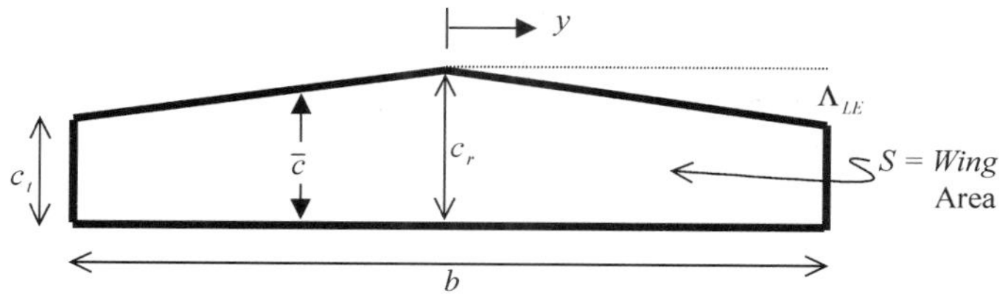

Fig. 1.49 Finite wing geometry.

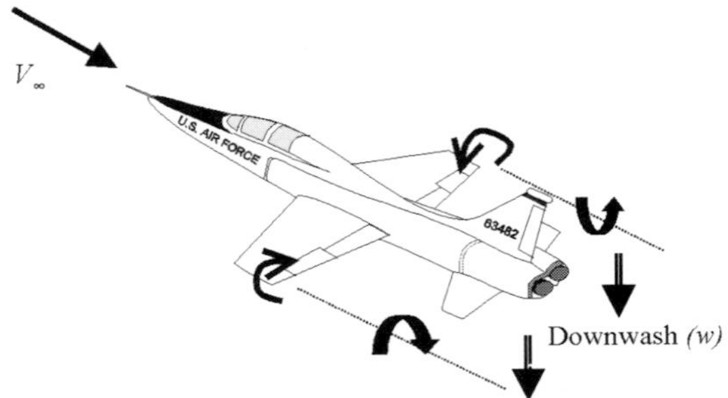

Fig. 1.50 Generation of wing tip vorticies and downwash.

1.4.2 Induced Drag

Induced drag is the penalty paid for generating lift on a finite wing. Consider a finite wing as shown on the T-38 aircraft in Fig. 1.50.

A wing generates lift by creating a pressure differential between the upper and lower surfaces. Wing tip vortices are generated as the high-pressure air on the lower wing surface near the wing tip seeks the relatively lower pressure on the upper surface. These small "tornadoes" induce a downward component of velocity, called downwash (w). The freestream velocity is displaced through the induced angle of attack (α_i), as shown in Fig. 1.51.

The wing "sees" V_{local}. Figure 1.52 shows a wing's cross section. L' is the component of the aerodynamic force perpendicular to the local velocity. In effect, the lift vector has been rotated aft—a new form of drag, induced drag (D_i), is introduced.

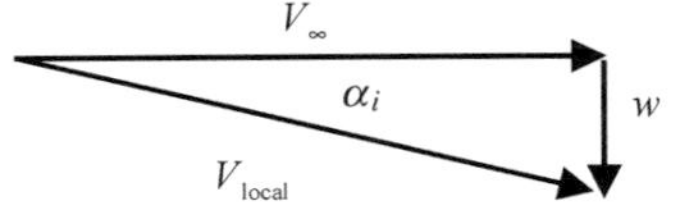

Fig. 1.51 Induced angle of attack.

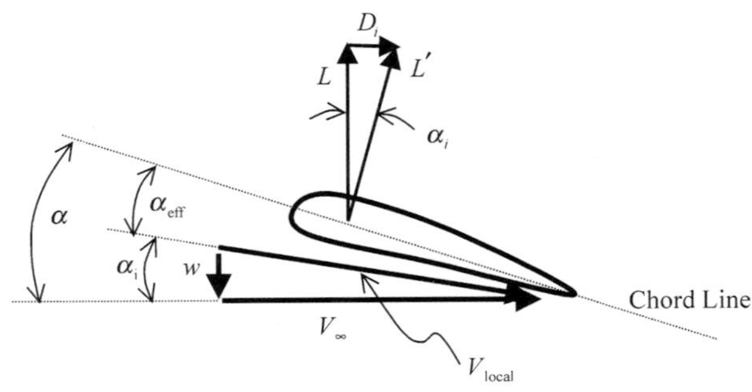

Fig. 1.52 Induced drag description.

The wing effectively "sees" a lower angle of attack, α_{eff}. Therefore, less lift is generated. Deriving an equation for induced drag is relatively easy (for example, see Ref. 11). The result is presented (Note: A D subscript is used to distinguish this from airfoil drag):

$$D_i = C_{D_i}\bar{q}_\infty S \tag{1.26}$$

where

$$C_{D_i} = \frac{C_L^2}{\pi e AR} \tag{1.27}$$

A few important points about [Eqs. (1.26) and (1.27)]

1) The span efficiency factor is e. A value of 1.0 is optimum and applies for the case of constant downwash along the wing's span (obtained from an elliptical lift distribution). For other planforms the efficiency is less, typically ranging between 0.95 and 1.0.
2) A high aspect ratio wing reduces induced drag. For this reason, high aspect ratio wings are used on the U-2 reconnaissance aircraft and gliders.
3) Induced drag is proportional to the lift coefficient, squared. Therefore, at high angles of attack (high lift), induced drag dominates. High lift implies greater pressure differentials, thus stronger vortices, and thus more induced drag.

1.4.3 Drag Summary

Finite wing geometry (wing tips) introduces a fourth form of drag—induced drag. Skin friction drag, pressure drag, and wave drag (if above the drag divergence Mach number) still exist. In summary, the drag coefficient for a finite wing is written as:

$$C_D = C_d + \frac{C_L^2}{\pi e AR} \tag{1.28}$$

where

$$C_d = \underbrace{C_{d_f} + C_{d_p}}_{\text{profile drag}} + C_{d_w}$$

This information is typically presented graphically as a drag polar, as shown in Fig. 1.53 (again, capital subscripts distinguish this from airfoil data). Note the effect of decreasing aspect ratio.

1.4.4 Lift Coefficient for a Finite Wing

We modified airfoil drag (C_d) data to account for the effect of wing tips. Similarly, this section describes how airfoil data are used to predict the lift coefficient for a finite wing. Qualitatively, the effect of aspect ratio on the lift curve slope is shown in Fig. 1.54, where C_L is used to denote a finite wing lift coefficient.

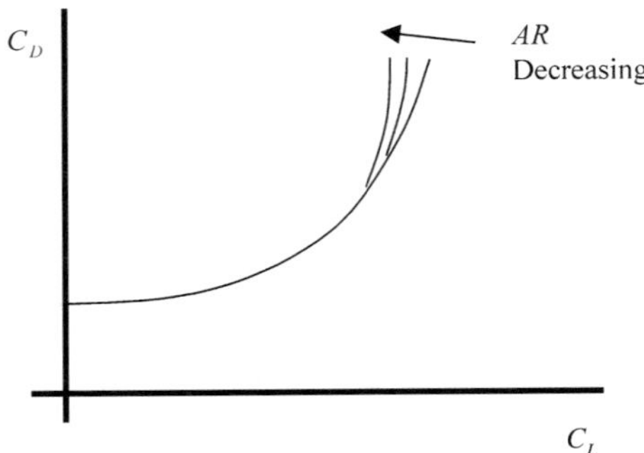

Fig. 1.53 Effect of aspect ratio on the drag polar.

Note the following

1) As aspect ratio decreases, the lift-curve slope (C_{L_α}) decreases. For the same angle of attack, the lift coefficient is smaller for the finite wing.
2) The zero lift angle of attack ($\alpha_{L=0}$) does not change. Because the wing is not generating lift, wing-tip vortices are not formed. In effect, the finite and infinite cases behave the same.
3) Given an angle of attack, the lift coefficient for the finite wing can be calculated from the following equation, which takes into account a reduced lift-curve slope:

$$C_L = C_{L_\alpha}(\alpha - \alpha_{L=0}) \tag{1.29}$$

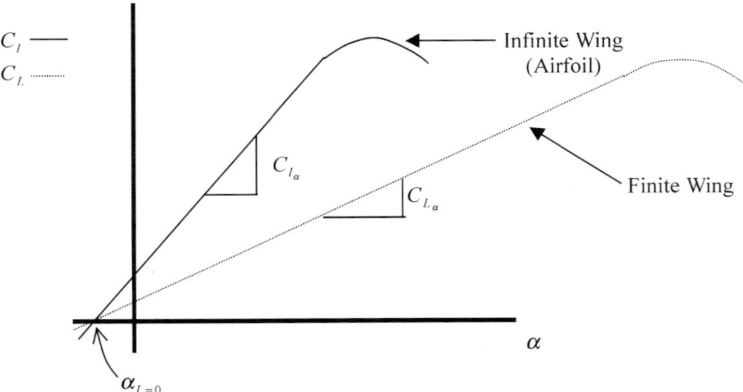

Fig. 1.54 Lift curves for infinite and finite wings.

where

$$C_{L_\alpha} = \frac{C_{l_\alpha}}{1 + \dfrac{57.3C_{l_\alpha}}{\pi eAR}} \tag{1.30}$$

Again, a capital L (C_L) distinguishes this lift coefficient from that of an airfoil. As mentioned, the zero lift angle of attack can be obtained from airfoil data. As we have seen, the lift curve slope for an airfoil (C_{l_α}) is approximately 0.11/deg. Unless otherwise specified, this is a reasonable number to use in Eq. (1.30).

Example 1.11

An unswept flying wing has an aspect ratio of 10 and incorporates a NACA 4412 airfoil (as in Example 1.10). For a Reynolds number of 6×10^6 and a span efficiency factor of 0.95, find C_L and C_D at an angle of attack of 4 deg.

Using the NACA 4412 airfoil charts in Example 1.10, we find

$$C_l = 0.85 \quad \text{and} \quad \alpha_{L=0} = -4 \text{ deg}$$

for the stated angle of attack and Reynolds number. The airfoil drag coefficient is

$$C_d = 0.0065$$

We next find C_{L_α} for the finite wing using Eq. (1.30) and a $C_{l_\alpha} = 0.11/\text{deg}$.

$$C_{L_\alpha} = \frac{C_{l_\alpha}}{1 + \dfrac{57.3C_{l_\alpha}}{\pi eAR}} = \frac{0.11}{1 + \dfrac{57.3(0.11)}{(3.14)(0.95)(10)}} = 0.0908/\text{deg}$$

Equation (1.29) is used to determine C_L.

$$C_L = C_{L_\alpha}(\alpha - \alpha_{L=0}) = 0.0908[4 \text{ deg} - (-4 \text{ deg})] = 0.7264$$

C_D is determined from Eq. (1.28).

$$C_D = C_d + \frac{C_L^2}{\pi eAR} = 0.0065 + \frac{(0.7264)^2}{(3.14)(0.95)(10)} = 0.0242$$

Notice that the lift coefficient decreases for a finite wing and that the drag coefficient increases.

1.5 Aircraft Aerodynamics

We have discussed the aerodynamics of airfoils and finite wings. It is now time to use this essential background to introduce aircraft aerodynamics. Here we will be discussing the aerodynamics of the entire aircraft, including the wing, tail surfaces, and fuselage.

1.5.1 Load Factor, Aerodynamic Coefficients, and Stall Airspeed

To begin, we will define load factor (n), or cockpit g where n is the ratio of an aircraft's lift to its weight, or

$$n = L/W \tag{1.31}$$

For example: at $2\,g$, an aircraft is generating an amount of lift equal to twice its weight.

The lift and drag coefficients for a complete aircraft are defined below, where lift is replaced by nW to keep the equation in its more general form.

$$\begin{aligned} C_L &= \frac{L}{\bar{q}S} = \frac{nW}{\bar{q}S} \\ C_D &= \frac{D}{\bar{q}S} \end{aligned} \tag{1.32}$$

Lift and drag include the contributions from not only the wing, but also the fuselage, horizontal/vertical tail, strakes, and external stores. The reference area, S, now typically includes a portion of the fuselage as shown in Fig. 1.55.

The slowest speed an airplane can fly in straight, level, and unaccelerated flight is called the stall speed (V_{stall}). For these flight conditions, lift is equal to weight ($n = 1$). The equation for stall speed is derived as follows:

$$\begin{aligned} L = W &= C_L\bar{q}S \\ &= C_L\frac{1}{2}\rho V^2 S \end{aligned}$$

Solving for velocity

$$V = \sqrt{\frac{2W}{\rho S C_L}} \tag{1.33}$$

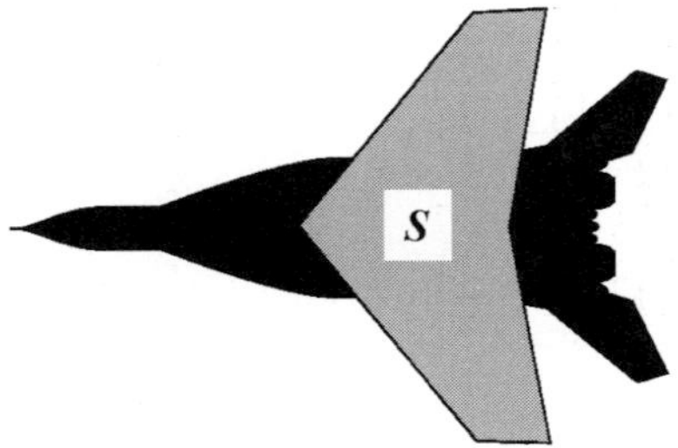

Fig. 1.55 Illustration of aircraft wing reference area.

At stall, C_L becomes equal to $C_{L_{\max}}$.

$$V_{\text{stall}} = \sqrt{\frac{2W}{\rho S C_{L_{\max}}}} \tag{1.34}$$

$C_{L_{\max}}$ is the maximum lift coefficient for the aircraft's configuration. In the above form, the stall speed is a "true airspeed." As altitude increases, density decreases, and an aircraft stalls at a higher true airspeed—not very convenient in terms of flight operations. In Chapter 3, we'll introduce other airspeeds (indicated, calibrated, etc.) and see how to avoid this inconvenience.

Example 1.12

Determine the stall airspeed at sea level for a 10,000-lb T-38 with 20-deg flaps. The wing reference area is 170 ft^2. Also, determine the load factor if the same aircraft is at an angle of attack of 10 deg with the flaps up at sea level with an airspeed of 265 kn. Use the following chart.

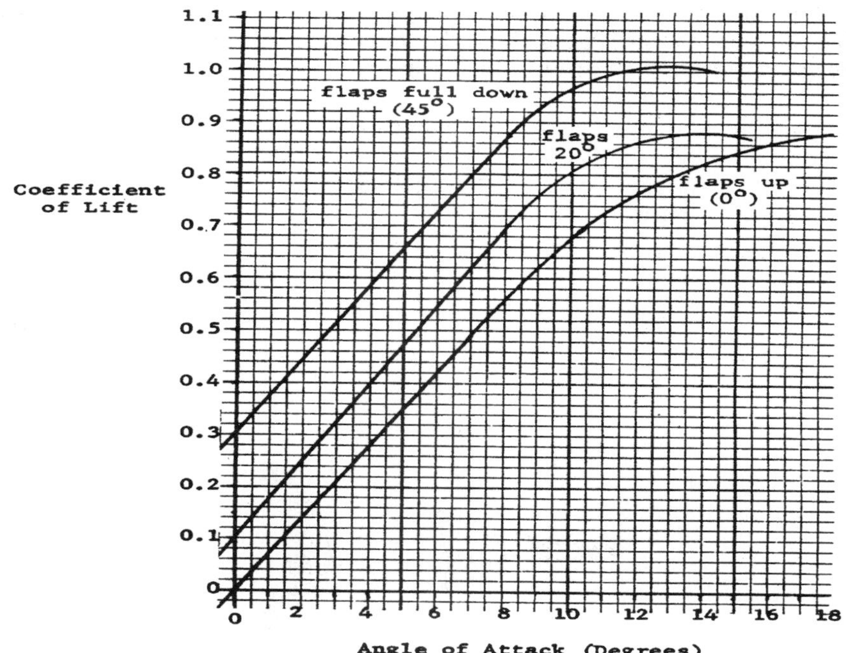

(Source: Department of Aeronautics, USAF Academy)

To determine the stall speed, we first determine $C_{L_{max}}$ for 20-deg flaps from the chart.

$$C_{L_{max}} = 0.88$$

Then, Eq. (1.34) can be used to find the stall speed.

$$V_{stall} = \sqrt{\frac{2W}{\rho S C_{L_{max}}}} = \sqrt{\frac{2(10{,}000)}{(0.00238)(170)(0.88)}} = 237 \text{ ft/s}$$

To find the load factor, the lift coefficient for an angle of attack of 10 deg (flaps up) is found from the chart.

$$C_L = 0.67$$

Then, Eq. (1.32) is solved for load factor.

$$n = \frac{C_L \bar{q} S}{W} = \frac{(0.67)\left(\frac{1}{2}(0.00238)\left(265 \times 1.69 \frac{\text{ft/s}}{\text{kn}}\right)^2\right)(170)}{10{,}000} = 2.72$$

At these conditions, the T-38 is pulling $2.72\,g$.

1.5.2 Aircraft Drag Polar

Typically, the aerodynamics of an airplane are presented as a drag polar—this is in the form of an equation, a graph, or both. Recall from the airfoil and finite wing discussions that a drag polar (by definition) shows the relationship between lift and drag coefficients for a specific aerodynamic body. In equation form, the drag polar of an aircraft is

$$C_D = C_{D_0} + \frac{C_L^2}{\pi e AR} \tag{1.35}$$

or simply

$$C_D = C_{D_0} + K C_L^2 \tag{1.36}$$

C_{D_0} is called the parasite drag coefficient or zero lift drag coefficient. Below the drag divergence Mach number, it is approximated by a constant (independent of lift) for a specific aircraft configuration. Included in this term are profile drag (skin friction and zero lift pressure drag) and interference drag. Interference drag is generated when more than one body (for example, stores on a wing) is placed in the same flowfield, creating eddies, turbulence, and/or restrictions to smooth flow. For example, if an external store is hung on a wing, the combined drag will typically be more than the summation of the individual store and wing drag. Occasionally, the addition of blended surfaces, as in the case of the F-15 conformal fuel tanks, will reduce the interference drag.

The term $C_L^2/\pi eAR$ in Eq. (1.35) is the drag due to lift, or induced drag term and is sometimes referred to as C_{D_i}. The induced drag on all lifting surfaces (wing, strakes, and horizontal tail) and the increment of pressure drag when the aircraft is generating lift are included in this term. The term K in Eq. (1.36) is referred to as the induced drag factor and can be seen to be equal to $1/\pi eAR$. If Mach effects are unimportant, K is a constant for a specific configuration. The term e is called the Oswald efficiency factor. It can be thought of as a "fudge factor," obtained through wind-tunnel and/or flight tests, which takes into account such effects as a nonelliptical lift distribution and the variation of pressure drag with lift. Typically, e is on the order of 0.8, but no greater than 1.0.

It is very convenient to present the drag polar graphically. Typically, this is done in two ways, as presented in Fig. 1.56.

We have previously seen the parabolic presention of the drag polar. The second graph is a linear presentation because C_D is plotted as a function of C_L^2. This form of the drag polar is convenient for determining the induced drag factor, K, which is simply the slope of the line. It is also a convenient format for plotting individual flight test data points when determination of the drag polar is the end objective. In this format, a linear curve fit to the data is supported by theory.

Mach effects are typically defined through the values of C_{D_0} and K. For example, Table 1.2 illustrates how these "constants" change for the F-16.

Another useful measure of drag used by aircrews is the drag count. A **drag count** is defined as one ten thousandth of a drag coefficient, or

$$1 \text{ drag count} \Rightarrow \text{a } C_D \text{ of } 0.0001$$

The drag count is simply a "user friendly" way to express the drag coefficient. A C_D of 0.025 is equivalent to 250 drag counts. Drag counts are especially useful when adding external protuberances to an aircraft. For example, the addition of a forward radome on the AC-130H gunship adds approximately 23 drag counts to the total aircraft drag. This is equivalent to a drag coefficient

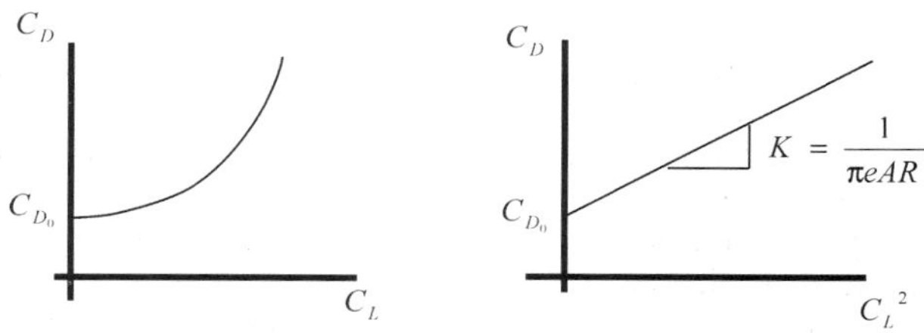

Fig. 1.56 Two ways of presenting the drag polar.

increase of 0.0023, but 23 drag counts proves to be an easier number to remember.

Example 1.13

Using Table 1.2, find the total drag counts for the F-16 at a lift coefficient of 0.2 and Mach 0.86.

From Table 1.2, we have,

$$C_{D_0} = 0.0169 \quad \text{and} \quad K = 0.117$$

Using Eq. (1.36),

$$C_D = C_{D_0} + KC_L^2 = 0.0169 + (0.117)(0.2)^2 = 0.02158$$

For this condition, the aircraft would have 215.8 drag counts.

1.5.3 Total Aircraft Drag

The total drag on an aircraft is simply

$$D = C_D \bar{q} S$$

It is useful to factor out velocity (V_∞) in the previous equation, as shown:

$$D = [C_{D_0} + KC_L^2] \frac{1}{2} \rho V^2 S$$

Substituting in

$$C_L = \frac{nW}{\bar{q}S} = \frac{nW}{\frac{1}{2}\rho V^2 S}$$

Table 1.2 Variation of C_{D_0} and K with Mach number for the F-16 (Ref. 2)

Mach	C_{D_0}	K
0.1	0.0169	0.117
0.86	0.0169	0.117
1.05	0.0430	0.128
1.5	0.0382	0.252
2.0	0.0358	0.367

We have

$$D = \underbrace{\left[C_{D_0}\frac{1}{2}\rho S\right]}_{\text{parasite drag}} V^2 + \underbrace{\left[\frac{2K(nW)^2}{\rho S}\right]}_{\text{induced drag}} \frac{1}{V^2} \tag{1.37}$$

From Eq. (1.37), it is clear that parasite drag is proportional to the square of velocity. Induced drag behaves as the inverse of velocity squared. Qualitatively, this is shown in Fig. 1.57. The total drag is the sum of parasite and induced drag. From Eq. (1.37), it can be seen that changes in altitude (ρ), load factor (n), configuration (C_{D_0}, S, and K), and/or weight will affect an aircraft's total drag. Reference 12 illustrates these effects.

We will see that the drag vs velocity curve is very important in defining and optimizing aircraft performance characteristics such as range and endurance.

1.6 Historical Snapshot—The AC-130H Drag Reduction Effort

In 1998, wind tunnel research was begun at the Air Force Academy Aeronautics Laboratory to investigate drag reduction approaches for the AC-130H Gunship.[13] The AC-130H is a C-130 aircraft modified with a 40-mm Bofors cannon and a 105-mm Howitzer that fire out the left side of the aircraft, along with a full complement of offensive and defensive avionics. The aircraft is capable of performing various missions such as close air support and air interdiction. A picture of an AC-130H is presented in Fig. 1.58.

The numerous external protuberances required by these missions resulted in a high level of drag on the AC-130H when compared to the standard C-130. The primary focus of the drag reduction effort was to increase the aircraft's ceiling, range, and loiter time, and to decrease fuel consumption. The primary objectives of this research included

1) determining the added drag counts of 14 external protuberances on the AC-130H using a 1/48th scale wind tunnel model;

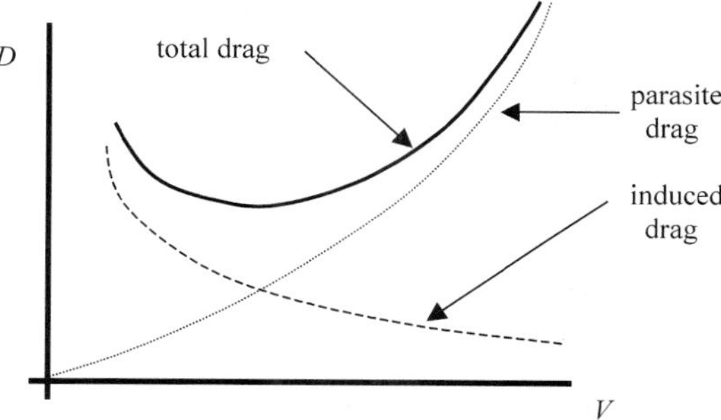

Fig. 1.57 Typical drag vs velocity curve for an aircraft.

Fig. 1.58 AC-130H Gunship.

2) designing and evaluating drag reduction modifications that could be easily installed and would not reduce the aircraft's operational capability; and
3) projecting the operational impact of identified drag reduction modifications in terms of loiter time, range, fuel savings, and ceiling.

To accomplish these objectives, wind tunnel testing was accomplished on a variety of configurations and modifications using the Air Force Academy Subsonic Wind Tunnel. A picture of this closed circuit wind tunnel is presented in Fig. 1.59.

The 1/48th scale test model is shown mounted in the tunnel test section in Fig. 1.60.

Drag coefficients resulting from the research were converted to drag counts to make accurate comparisons and effective operational projections. The program identified a maximum drag reduction potential of 53 drag counts based on incorporation of nine recommended modifications. This translated to a reduction in the parasite drag coefficient, C_{D_0}, of 0.0053. Because the parasite drag coefficient of a clean C-130 was approximately 0.03 as compared to the AC-130H C_{D_0} of 0.045, the identified drag reduction potential represented approximately 35 percent of the added incremental drag (0.015) associated with gunship modifications. To put this in perspective, the parasite drag portion of the aircraft total drag curve as illustrated in Fig. 1.57 would be lowered by 35 percent which has important implications for aircraft performance improvement. A performance simulation program projected a 1730-ft increase in ceiling, and either a fuel savings of 1483 lb, a mission radius increase of 53 nautical miles, or a loiter time increase of 52 min for a 5-h combat mission.

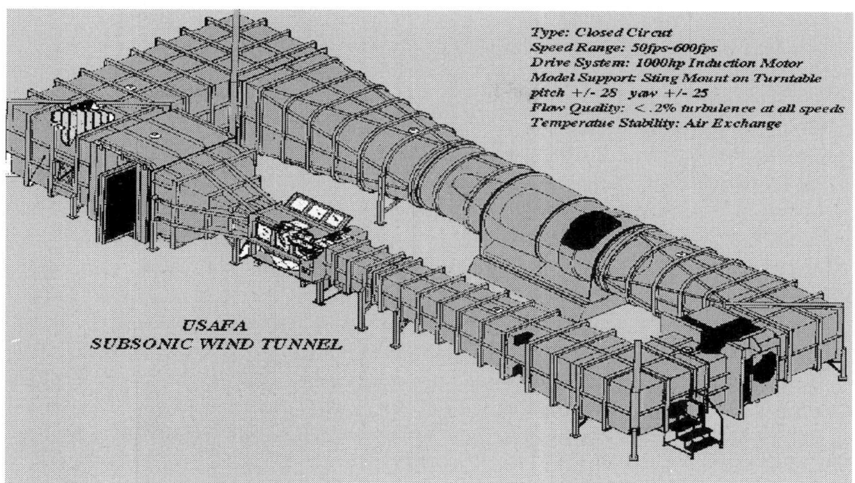

Fig. 1.59 U.S. Air Force Academy subsonic wind tunnel.

Fig. 1.60 AC-130 wind tunnel model.

These were significant improvements in combat capability resulting from a reduction in parasite drag.

References

[1]Anderson, J. D., *Introduction to Flight*, 3rd ed., McGraw-Hill, New York, 1989, p. 51.

[2]Brandt, S. A., Stiles, R. J., Bertin, J. J., and Whitford, R., *Introduction to Aeronautics: A Design Perspective*, AIAA Education Series, AIAA, Reston, VA, 1997, pp. 52–53.

[3]Anderson, J. D., *Fundamentals of Aerodynamics*, McGraw-Hill, New York, 1984, pp. 318–323.

[4]Anderson, J. D, *Hypersonic and High Temperature Gas Dynamics*, McGraw–Hill, New York, pp. 13–30.

[5]Brandt, S. A., Stiles, R. J., Bertin, J. J., and Whitford, R., *Introduction to Aeronautics: A Design Perspective*, AIAA Education Series, AIAA, Reston, VA, 1997, pp. 40–45.

[6]Moore, F. G., Hymer, T., and Wilcox, F. J., Jr., "Improved Empirical Model for Base Drag Prediction on Missile Configurations Based on New Wind Tunnel Data," Naval Surface Warfare Center, Dahlgren Division, TR 92-509, Oct. 1992.

[7]Abbott, I. H., and Von Doenhoff, A. E., *Theory of Wing Sections*, Dover Publications, Inc., New York, 1959.

[8]Anderson, J. D., *Introduction to Flight*, 3rd ed., McGraw–Hill, New York, 1989, pp. 182–185.

[9]Anderson, J. D., *Fundamentals of Aerodynamics*, McGraw–Hill, New York, 1984, pp. 305–375.

[10]Lan, C. E., and Roskam, J., *Airplane Aerodynamics and Performance*, Roskam Aviation and Engineering Corporation, Ottawa, KS, 1988.

[11]Anderson, J. D., *Introduction to Flight*, 3rd ed., McGraw Hill, New York, 1989, pp. 222–225.

[12]Brandt, S. A., Stiles, R. J., Bertin, J. J., and Whitford, R., *Introduction to Aeronautics: A Design Perspective*, AIAA Education Series, AIAA, Reston, VA, 1997, p. 172.

[13]Yechout, T. R., Chadsey, D. S., and Rutgers, N. G., "Identification of AC-130H Drag Reduction Potential Using a 1/48th Scale Wind Tunnel Model and Prediction of Resulting Performance Enhancements," U.S. Air Force Academy Dept. of Aeronautics TR 00-02, Aug. 2000.

Problems

1.1 Define the pressure, density, and temperature ratios. Based on these ratios and the Perfect Gas Law derive an expression for density ratio (σ) in terms of pressure ratio (δ) and temperature ratio (θ).

1.2 A flight test engineer wants to obtain drag data at an ambient pressure of 1000 psf. (Use definitions of σ, δ, and θ. Interpolate to check your answer.)

(a) At what altitude should the aircraft fly?
(b) What type of altitude is this?
(c) The weather report shows the temperature is $-15°$C at this altitude. Is it hotter or colder than standard and by how many degrees?
(d) What is the aircraft density altitude at these conditions?

1.3 An aircraft is flying at 88 ft/s at sea level on a standard day. Find the velocity at a point on the wing where the static pressure is 2070 psf.

1.4 An aircraft is flying at a velocity of 50 m/s at an altitude of 5 km on a standard day. At one point on the wing, the local velocity is 70 m/s. Find the freestream dynamic pressure, the flowfield total pressure, and the local static pressure at the point where the velocity is 70 m/s.

1.5 Sketch a typical C_L vs angle of attack curve and a C_D vs C_L curve for $AR = \infty$. Show how the curves would change for an AR of 10 and 5.

1.6 Given a span efficiency of 0.95 and an aspect ratio of 10, find the three-dimensional lift curve slope (C_{L_α}) and the slope of the C_D vs C_L^2 curve (K).

1.7 The Fairchild Republic A-10 with the following characteristics is in level unaccelerated flight.

$$C_{D_0} = 0.032 \qquad S = 506 \text{ ft}^2 \qquad W = 23{,}200 \text{ lb}$$
$$AR = 6.5 \qquad e = 0.87 \qquad C_{L_{\max}} = 2.0$$
$$\text{Max } T_{SL} = 9060 \text{ lb (each engine)}$$

(a) Write the drag polar equation for the A-10 at this flight condition.
(b) Find the stall velocity in knots at sea level.
(c) What are the lift and drag of the A-10 at 300 knots at sea level?

2
A Review of Basic Propulsion

To sustain flight for any appreciable amount of time, aircraft need some type of propulsion system. The purpose of the propulsion system is to produce a controllable force called thrust, which is made to act in concert with the other forces on the aircraft (lift, drag, and weight) to produce the desired translational motion. In most cases thrust is used to accelerate the aircraft along the flight path and to counteract drag; however, it is certainly possible for thrust to be used in other ways such as to augment lift (for example, V-STOL aircraft like the AV-8A Harrier).

2.1 Types of Propulsion Systems

A number of different propulsion choices exist for the aircraft designer. This section provides a brief overview of different propulsion systems and their advantages and disadvantages.

2.1.1 Piston–Propeller

The piston (or reciprocating) engine–propeller combination is probably the most efficient propulsion system for low-speed aircraft (Fig. 2.1). This is true because the propeller produces thrust by increasing the momentum of a relatively large amount of air. Also, the reciprocating engine is efficient in terms of fuel consumption. Fuel consumption is usually expressed as pounds of fuel per hour per brake horsepower and given the name of brake-specific fuel consumption (BSFC). Typical values of BSFC for aircraft reciprocating engines in use today are 0.4 to 0.5 lb/bhp-h for cruise power. At flight speeds above approximately Mach 0.3, the efficiency of the propeller starts to drop off because of compressibility effects on the blades. The engine size necessary to produce the required thrust makes other propulsion systems more attractive at this point.

2.1.2 Turboprop

A turboprop propulsion system uses a gas turbine engine instead of a reciprocating engine to power the propeller. A typical layout is shown in Fig. 2.2. As air enters the engine, it is compressed somewhat by the ram effect of air hitting the engine and being slowed down in the inlet. Most of the compression is done in the compressor. Fuel is added and the mixture is burned in the combustor. The hot gases are expelled to turn one turbine to drive the compressor, and another turbine to drive the propeller. Because the turbine speed is on the order of 16,000 rpm, a reduction gear is necessary to run the

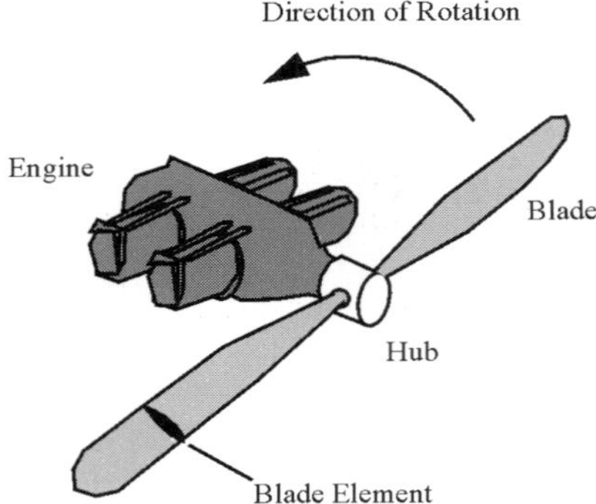

Fig. 2.1 Piston–propeller.

propeller. The thermodynamic cycle on which the engine operates is known as the Brayton cycle. Specific fuel consumption (SFC), similar to BSFC for reciprocating engines, is expressed as pounds of fuel per hour per equivalent shaft horsepower (ESHP). ESHP is the shaft power delivered to the propeller plus effective power of any additional thrust produced by the exhaust gases. Cruise SFC for modern turboprop engines is in the neighborhood of 0.6 lb/eshp-hr. The extremely large airflow through the gas turbine engine, when compared to that of the reciprocating engine, gives it a much larger power-to-weight ratio. Also, the power output capability of a turboprop increases somewhat with flight velocity because of the ram effect experienced by the air. These two factors increase the efficiency of the system up to flight speeds of approximately Mach 0.6.

2.1.3 Turbojet

A turbojet is a gas turbine engine in which all of the thrust is produced by the exit velocity of the exhaust gas. Only the power necessary to drive the compressor (and a small amount to drive aircraft accessories, such as generators and hydraulic pumps) is extracted from the turbine. Figure 2.3 shows a

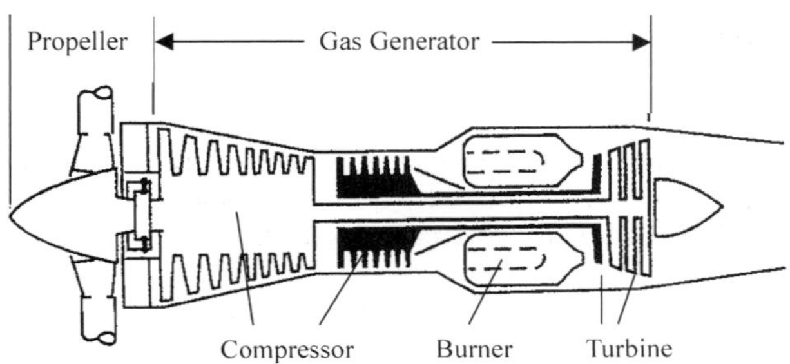

Fig. 2.2 Turboprop.

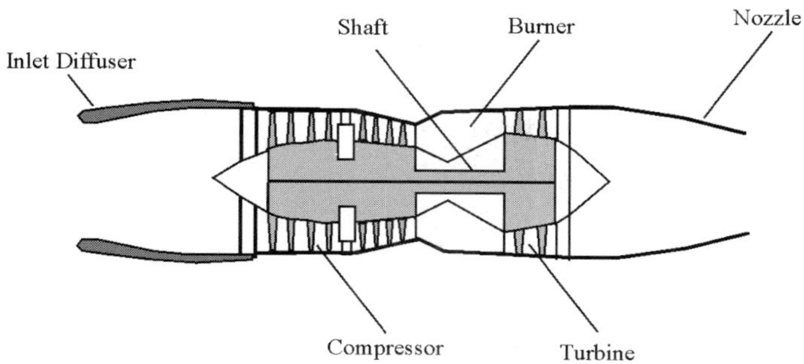

Fig. 2.3 Turbojet engine.

typical turbojet. After leaving the turbine section, the high-pressure exhaust gases are accelerated through the nozzle and exit the engine close to atmospheric pressure. The thermodynamic cycle for a turbojet is a modified Brayton cycle.

The most significant turbojet engine design criterion that limits the thrust output (other than size) is the maximum allowable turbine inlet temperature. The higher this temperature can be, the more energy that can be added to the air, which means more energy is available to provide thrust after driving the turbine. The maximum turbine inlet temperature depends on the material characteristics of the turbine and whether or not blade cooling is used. A typical value for maximum turbine inlet temperature is approximately 2000°F for current high performance turbojets.

Fuel consumption for turbojet engines is referenced to thrust produced instead of power produced and is called thrust-specific fuel consumption (TSFC). It is usually expressed in pounds of fuel per hour per pound of thrust. Typical values for a turbojet are usually in the neighborhood of 1.0 /h. Individual engine component efficiencies and the compressor pressure ratio have a major effect on the overall engine efficiency. Higher values of compressor pressure ratio generally yield a greater efficiency. Typically this ratio varies from 5:1 to 15:1.

2.1.4 Turbojet with Afterburner

In supersonic high-performance aircraft, turbojets are usually augmented with afterburners. Figure 2.4 shows a schematic of this type of engine. Note the only difference between this engine and Fig. 2.3 is the addition of a long afterburner duct. After air passes through the turbine, additional fuel can be selectively added and burned in the afterburner duct. The maximum allowable temperatures are significantly higher than the maximum turbine inlet temperature, and as a result, a thrust increase of more than 50% is common for afterburner use. A variable area nozzle is necessary because of the differences in the volume of airflow, depending on whether or not the afterburner is in operation, as the exhaust gas is expanded to atmospheric pressure. Afterburner

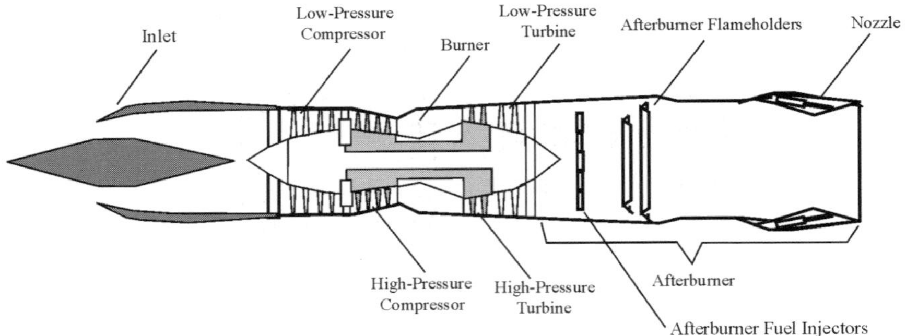

Fig. 2.4 Turbojet with afterburner.

operation causes a loss in efficiency, bringing the TSFC up to approximately 2.0 /hr. The turbojet, augmented as necessary by afterburning, serves most efficiently in the Mach 1.0–2.5 range.

2.1.5 *Turbofan*

Spanning the speed regime between the turboprop and turbojet, the turbofan provides the most efficient available propulsion. As expected, the turbofan can be considered a cross between the ducted turboprop and a turbojet. Figure 2.5 provides a schematic of a typical turbofan engine. Figure 2.5(a) illustrates a low-bypass ratio turbofan and Fig. 2.5(b) shows a high-bypass ratio turbofan.

A percentage of the air passes through the fan only and is expelled, providing some thrust. This is called the secondary or fan airflow. The rest of the air, called the primary or core airflow, passes through the entire engine as in the conventional turbojet. The ratio of the secondary mass flow rate to the primary mass flow rate is called the engine bypass ratio (BPR) and is essentially constant for a given engine.

Turbofans begin to lose their efficiency advantage over turbojets when approaching sonic flight velocity because of their relatively large frontal area. The actual velocity at which this occurs depends largely on the engine BPR, as might be expected. The TF-39 has a BPR of approximately 8:1 and powers the C-5, which cruises around Mach 0.8. The F-100 engine has a BPR of less than one and powers the F-15 and F-16, whose missions require velocity capabilities well into the supersonic region.

Fuel consumption of turbofans is measured by TSFC, just as with turbojets. It generally runs approximately 0.5 /h for cruise conditions, although it can vary significantly depending on the BPR. For example, TSFC for the TF-39 is approximately 0.4 /h for normal cruise, where TSFC for the F-100 is close to 0.7 /h with afterburner.

2.1.6 *Ramjet*

At high flight velocities the air hitting the engine intake produces enough compression for engine operation without having to power an active compres-

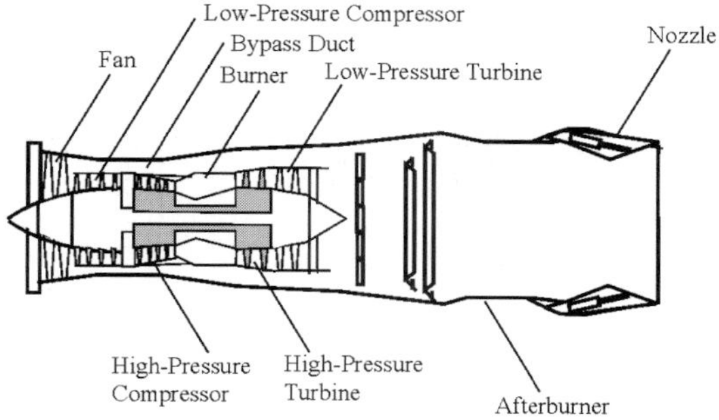

(a) Low-bypass ratio turbofan

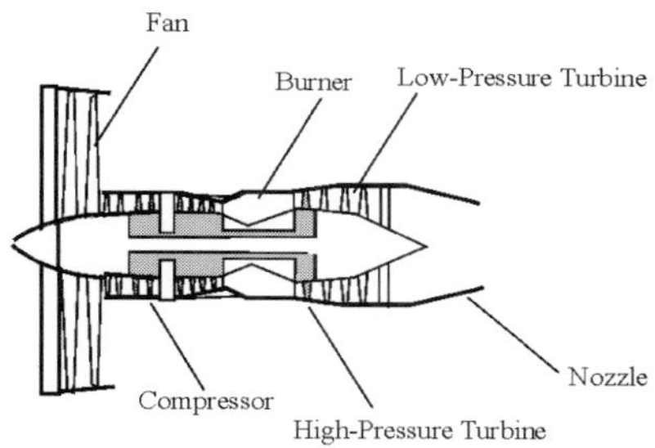

(b) High-bypass ratio turbofan

Fig. 2.5 Turbofan bypass ratio diagrams.

sor. This enables the ramjet to operate without any moving parts. As shown in Fig. 2.6, a turbine is not necessary because there is no compressor. These engines have the advantage of being very simple. The obvious drawback is that they must have sufficient flight velocity before they can produce thrust. For this reason, they are used mainly on missiles augmented with rocket engines. As far as efficiency is concerned, ramjets begin to surpass afterburning turbojets at flight speeds of approximately Mach 3 with TSFCs in the neighborhood of 3 /h.

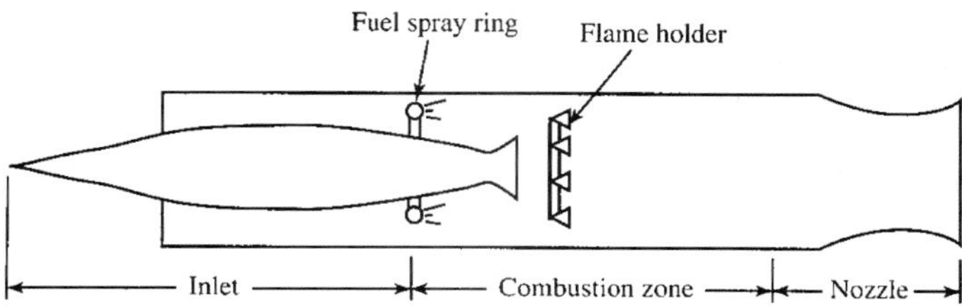

Fig. 2.6 Ramjet.

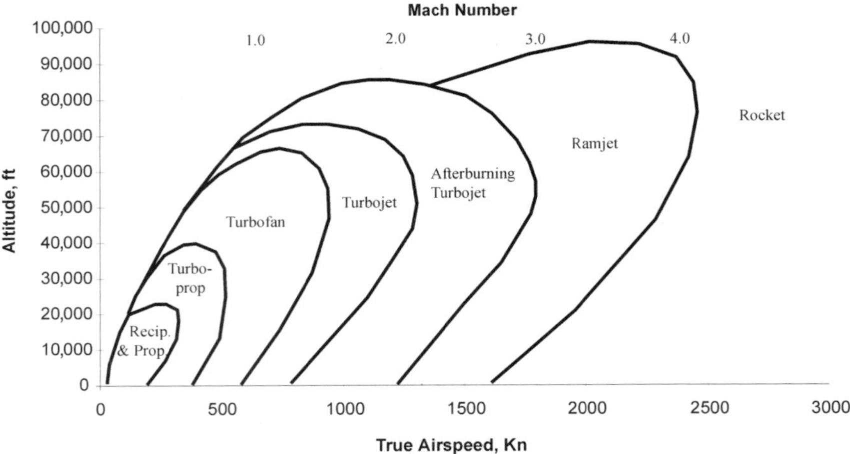

Fig. 2.7 Operating regimes.

Figure 2.7 gives a qualitative summary of propulsion systems discussed in this section.

Notice that there is considerable overlap in the altitude–airspeed envelopes for these different propulsion systems, but Fig. 2.7 gives a general idea of the specific areas of use.

2.2 Propulsion System Characteristics

Propulsion system characteristics can be predicted using a variety of methods. This text will present a simplified approach using the thrust equation and corrected properties. These approaches are usually sufficient for analysis of flight mechanics characteristics. Of course, more sophisticated approaches, such as an engine contractor's prediction program, can be used if these resources are available and more accuracy is needed.

2.2.1 Thrust Equation

In understanding how an aircraft performs it is essential to develop the aircraft thrust equation. To accomplish this, the engine can be modeled as a control volume. The net force acting on the control volume can be thought of as two separate forces. There is a force because of the change of velocity (linear momentum) through the control volume, and there may be a differential pressure force. Figure 2.8 shows a simple schematic of an engine. The thrust force can be quantified by an application of Newton's 2nd law. Newton's 2nd law in vector equation form is

$$\Sigma\bar{F} = \frac{\mathrm{d}(m\bar{V})}{\mathrm{d}t} \tag{2.1}$$

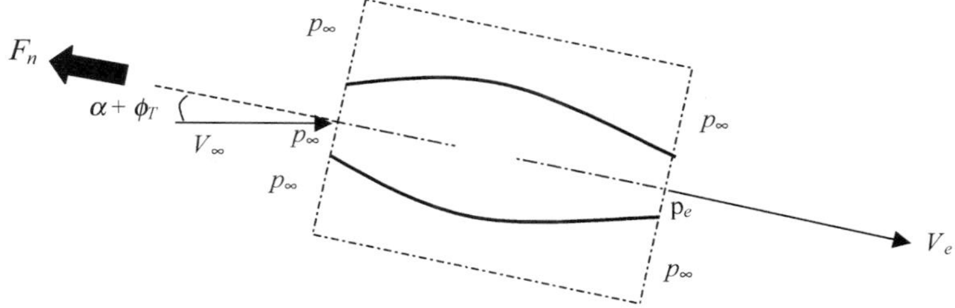

Fig. 2.8 Net thrust schematic.

The force exerted on the flow is in the same direction as the change of momentum. Therefore, the force exerted on the flow in Fig. 2.8 acts to the right assuming that $V_e > V_\infty$. Newton's 3rd law states that for every action there is an equal and opposite reaction. In the analysis of a propulsion system, this reaction is the thrust force. Because the force on the flow is to the right (that is, force is exerted to speed the flow up), the thrust force (the equal and opposite force on the propulsion system) acts to the left in Fig. 2.8. Assuming steady, one-dimensional flow at the entrance and exit of the control volume, the difference in linear momentum at the entrance and exit is represented by the force equation:

$$\Sigma \bar{F} = \frac{\mathrm{d}(m\bar{V})}{\mathrm{d}t} = \dot{m}_e \bar{V}_e - \dot{m}_\infty \bar{V}_\infty = -\bar{F}_n \tag{2.2}$$

where

$\dot{m}_\infty = \rho_\infty A V_\infty$ (mass flow rate of air going into the engine)
$\dot{m}_e = \dot{m}_\infty + \dot{m}_f$ (mass flow rate at the exit)
$\dot{m}_f =$ the mass flow rate of the fuel
$\bar{V}_e =$ the nozzle exit velocity
$\bar{V}_\infty =$ the freestream velocity entering the engine

The second force acting on the propulsion system is because of the differential pressures at the entrance and exit, assuming that the entrance and exit areas are parallel to each other. Therefore, the net pressure force ($\bar{F}_p$) perpendicular to the areas in vector form is

$$\bar{F}_p = p_e \bar{A}_{e_{\mathrm{norm}}} - p_\infty \bar{A}_{i_{\mathrm{norm}}} \tag{2.3}$$

where

$\bar{A}_{e_{\text{norm}}}$ = the area of the nozzle exit perpendicular to the engine centerline
$\bar{A}_{i_{\text{norm}}}$ = the area of the inlet perpendicular to the engine centerline

The total net thrust, F_n, expressed in vector form is therefore

$$\bar{F}_n = -(\dot{m}_e \bar{V}_e - \dot{m}_\infty \bar{V}_\infty) + p_e \bar{A}_{e_{\text{norm}}} - p_\infty \bar{A}_{i_{\text{norm}}} \tag{2.4}$$

or

$$\bar{F}_n = -[(\dot{m}_\infty + \dot{m}_f)\bar{V}_e - \dot{m}_\infty \bar{V}_\infty] + p_e \bar{A}_{e_{\text{norm}}} - p_\infty \bar{A}_{i_{\text{norm}}} \tag{2.5}$$

The equation can be further simplified by making several additional assumptions. First, the mass flow rate of the air is assumed to be significantly greater than the mass flow rate of the fuel (that is, $\dot{m}_\infty \gg \dot{m}_f$). Therefore, the mass flow rate of the fuel will be neglected (that is, $\dot{m}_f \cong 0$). Second, it is also assumed that freestream velocity vector is parallel to the nozzle exit velocity vector (namely, $[\alpha + \phi_{\text{T}}] \approx 0$). Third, the area of the inlet and area of the nozzle are assumed to act perpendicular to the freestream velocity. Finally, the areas of the inlet and nozzle are assumed to be approximately equal ($A_e \approx A_i$). Pictorially, this is shown in Fig. 2.9.

These assumptions make the thrust vector equation into a scalar equation as shown:

$$F_n = \dot{m}_\infty V_e - \dot{m}_\infty V_\infty + p_e A_e - p_\infty A_e \tag{2.6}$$

or

$$F_n = \dot{m}_\infty V_e - \dot{m}_\infty V_\infty + (p_e - p_\infty) A_e \tag{2.7}$$

If the nozzle is assumed to be operating on design, then $p_e = p_\infty$ and the thrust equation simplifies to

$$F_n = \dot{m}_\infty V_e - \dot{m}_\infty V_\infty \tag{2.8}$$

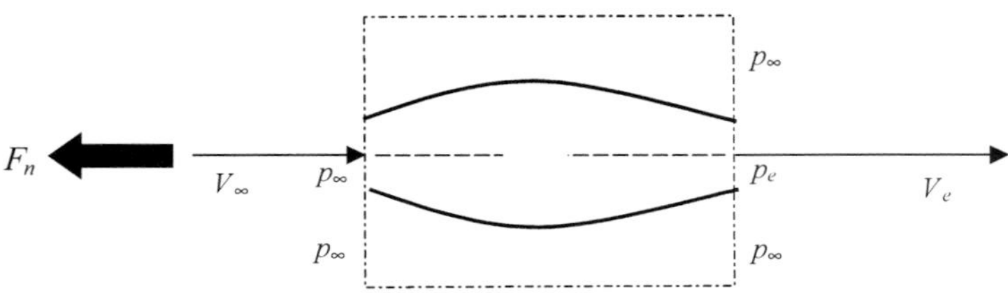

Fig. 2.9 Thrust schematic after assumptions.

This is by no means accurate enough for detailed jet engine analysis and thrust calculations, but it serves as an approximation for determining net thrust provided the engine is operating close to its design point.

The first term, $\dot{m}_\infty V_e$, is the gross thrust, F_g, of the engine. F_g becomes the static thrust at $V_\infty = 0$. Therefore, data collected on a static test stand yields the gross thrust of an engine. Engine manufacturers provide data on an engine in terms of this static thrust, usually presented for standard sea-level conditions.

The second term, $-\dot{m}_\infty V_\infty$, is known as the ram drag, F_e (namely, $F_e = \dot{m}_\infty V_\infty$). Therefore,

$$F_n = F_g - \dot{m}_\infty V_\infty \tag{2.9}$$

Ram drag is not usually discussed in propulsion courses. Ram drag is the force resulting from the change of linear momentum in bringing the flow from the freestream velocity to a velocity near zero in the inlet. Recall the purpose of an inlet to slow the velocity to near zero at the compressor face, which in turn increases the pressure. This force opposes the gross thrust, so it causes the net thrust to be less.

Typically, ram drag dominates in the incompressible flow regime at $M < 0.3$. As the freestream velocity increases in this regime, $\dot{m}_\infty$ increases in both the gross thrust and ram drag term. However, V_∞ is also present in the ram drag term, causing that term to increase more. As a result, the net thrust initially decreases with increasing velocity as shown in Fig. 2.10. In the compressible flow regime ($M > 0.3$) ram effect, or ram recovery, is important. In this regime there is an increase in the density and pressure as the velocity is slowed down in the inlet because of compressibility effects. This higher pressure at the compressor face results in a larger gross thrust that increases faster than the rise in ram drag. As a result, there is an increase in the net thrust with increasing velocities at the higher Mach numbers, as shown in Fig. 2.10.

2.2.2 Functional Relationship of Thrust

The net thrust of the engine depends on a number of parameters. These include ρ_∞, p_∞, T_∞, V_∞, $\dot{m}_\infty$, N, and D, where N is engine rpm based on

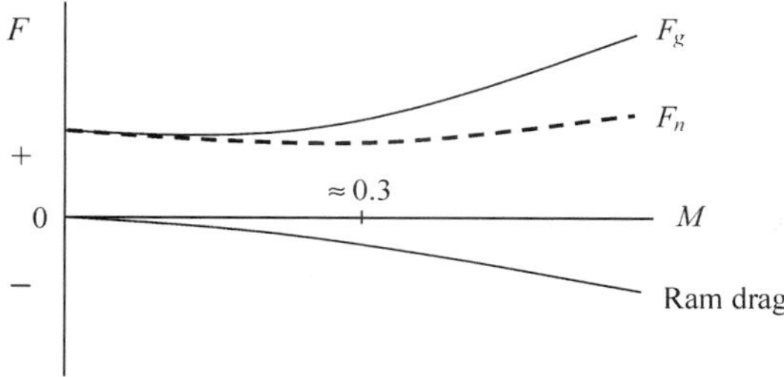

Fig. 2.10 Ram drag and ram effect for nonafterburning turbojet.

throttle setting and D is the engine diameter. The number of variables can be reduced because p_∞ and T_∞ fixes the value of ρ_∞. These parameters also make up mass flow rate. Therefore,

$$F_n = f(p_\infty, T_\infty, V_\infty, N, D) \tag{2.10}$$

As a result, the presentation of thrust data for an engine must reflect the conditions at which the data were obtained. Often, data are presented for an aircraft in thrust curves for a specific rpm setting and a given altitude based on standard conditions, such as the T-38 data shown in Fig. 2.11. Note that the plot yields the net thrust as a function of Mach number for military power (100% rpm) and maximum power (100% rpm with afterburner) at standard sea-level conditions. This provides valuable information but does not tell what happens if the aircraft is operating at any other throttle setting or a flight condition other than standard sea-level values at that altitude. Therefore, an infinite number of plots would be required to characterize the thrust performance at all possible flight conditions. An alternate, and very useful, method of presenting thrust information involves the use of corrected properties in conjunction with the net thrust equation in Eq. (2.10).

2.2.3 Corrected Properties

Dimensional analysis shows, and is verified by experimental results, that the number of variables can be greatly reduced by using corrected properties. The primary corrected properties used are corrected gross thrust, corrected thrust-specific fuel consumption (TSFC), and corrected airflow, which are usually defined as a function of corrected rpm and M. These properties depend on the temperature and pressure ratios (θ and δ) defined in Eq. (1.7) and are summarized as follows:

$N_c = \dfrac{N}{\sqrt{\theta}}$ is **corrected rpm**

$F_{g_c} = \dfrac{F_g}{\delta}$ is **corrected gross thrust**

$\dfrac{\dot{w}_f}{F_g\sqrt{\theta}}$ is **corrected TSFC** based on gross thrust

$\dfrac{\dot{w}_f}{F_n\sqrt{\theta}}$ is corrected TSFC based on net thrust

$\dfrac{\dot{w}_{\text{air}}\sqrt{\theta}}{\delta}$ is **corrected airflow**

Static Case $M_\infty = 0$ (2.11)

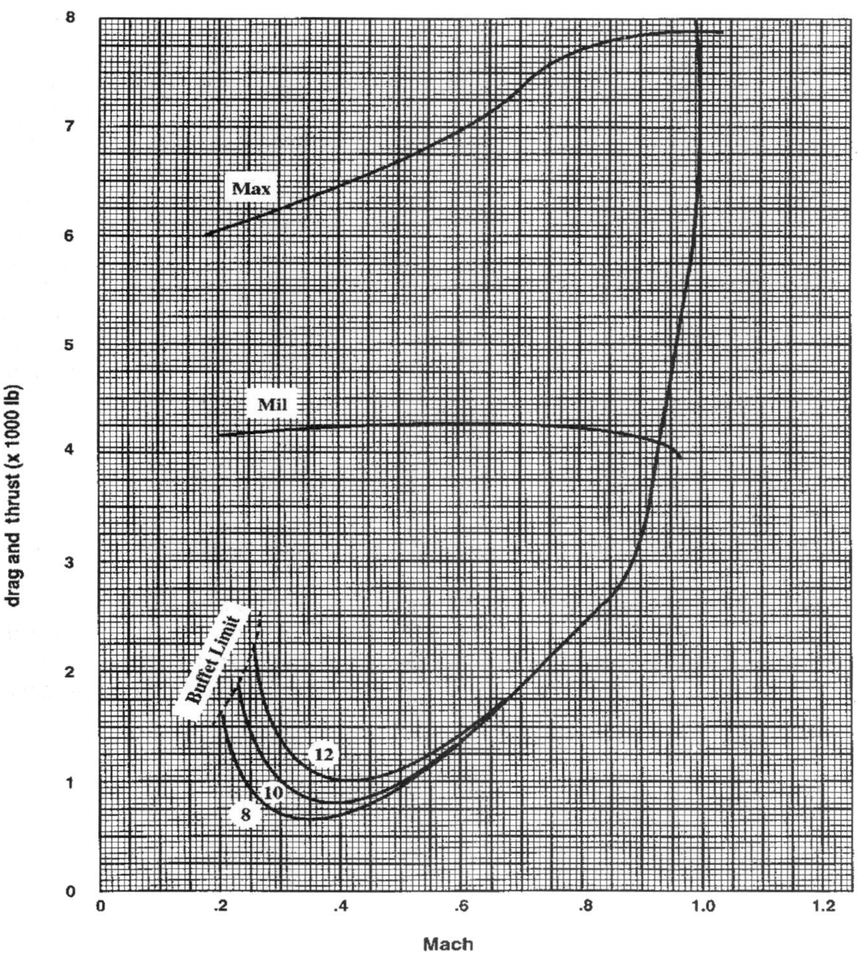

Fig. 2.11 T-38 thrust data.

For the turbojet, the gross thrust, TSFC, and airflow data can be obtained in static tests ($V_\infty = 0$) on a test stand and corrected to standard sea-level conditions for presentation. These corrected data can be plotted on a series of nominal graphs as shown in Fig. 2.12 that can be used to analyze any possible operating condition. These graphs are very powerful and can be used to determine the gross thrust, TSFC, and mass airflow for any atmospheric condition and throttle setting.

Note that the aforementioned quantities are based on the static case ($V_\infty = 0$). The corrected properties are the same for the in-flight case, $V_\infty > 0$, except that θ and δ are replaced in the equations with θ_T and δ_T, respectively, where

$$\theta_T = \frac{T_o}{T_{sl}} \tag{2.12}$$

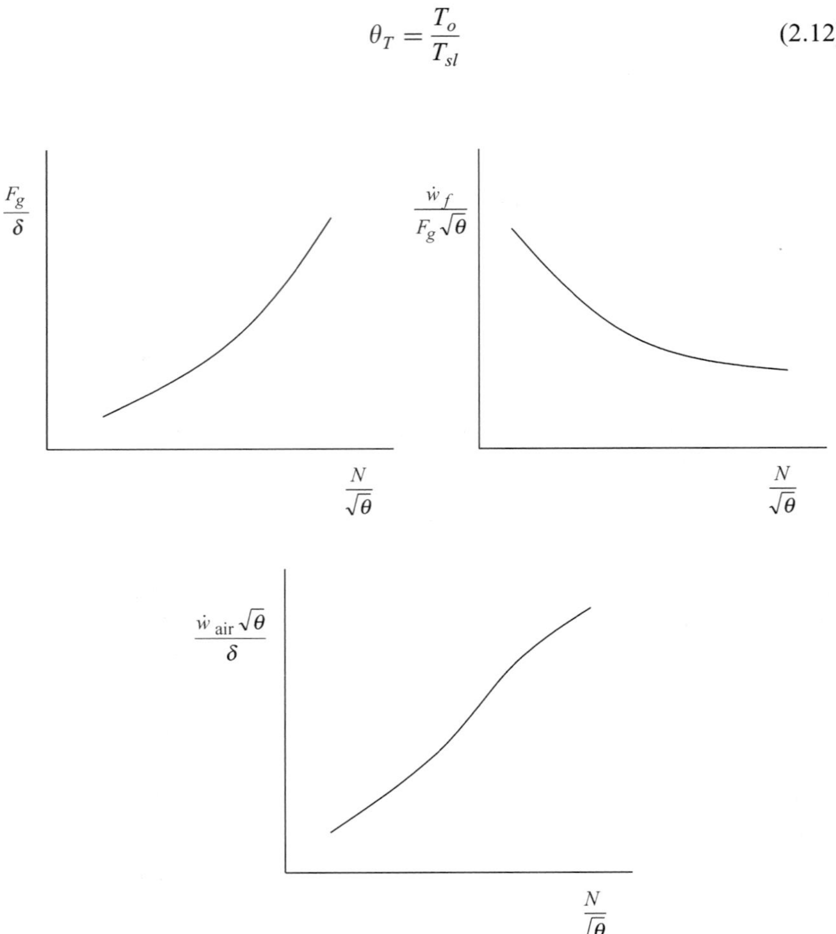

Fig. 2.12 Corrected properties vs corrected rpm.

and

$$\delta_T = \frac{P_o}{p_{sl}} \tag{2.13}$$

Note that the total temperature, T_o, and the total pressure, p_o, are used in θ_T and δ_T, respectively, instead of the static temperature, T, and static pressure, p. From the isentropic relations, θ_T and δ_T are easily found to be

$$\theta_T = \theta\left(1 + \frac{\gamma - 1}{2}M^2\right) \tag{2.14}$$

$$\delta_T = \delta\left(1 + \frac{\gamma - 1}{2}M^2\right)^{\frac{\gamma}{\gamma-1}} \tag{2.15}$$

For the static case ($M = 0$), $\theta = \theta_T$ and $\delta = \delta_T$, which means the static case is just a subset of the in-flight case. Therefore, more general forms of the corrected properties are

$$N_c = \frac{N}{\sqrt{\theta_T}} \quad \text{is corrected rpm}$$

$$F_{g_c} = \frac{F_g}{\delta_T} \quad \text{is corrected gross thrust}$$

$$\frac{\dot{w}_f}{F_g\sqrt{\theta_T}} \quad \text{is corrected TSFC based on gross thrust}$$

$$\frac{\dot{w}_f}{F_n\sqrt{\theta_T}} \quad \text{is corrected TSFC based on net thrust}$$

$$\frac{\dot{w}_{\text{air}}\sqrt{\theta_T}}{\delta_T} \quad \text{is corrected airflow} \tag{2.16}$$

For the in-flight case (namely, $V_\infty > 0$), these total properties, θ_T and δ_T, replace the static properties, θ and δ, respectively, on the axes in Fig. 2.12.

Actual data for the J69 engine used in the T-37B aircraft are shown in Figs. 2.13 through 2.15. Figure 2.13 depicts the corrected gross thrust as a function of corrected rpm. Note that the corrected gross thrust data provided are for a single engine only; therefore, the results must be multiplied by two to reflect the two engine T-37B.

Figure 2.14 shows the corrected TSFC as function of corrected rpm for the J69 engine. Because TSFC is represented in pounds of fuel per hour per pound of thrust produced, it will have the same value for one engine or two engines. It is the actual flow rate of the fuel in pounds per hour, $\dot{w}_f$, which must be doubled for two engines.

Figure 2.15 shows the corrected airflow as function of corrected rpm. Just as in Fig. 2.13, the data presented are for only one engine and must be multiplied by two for the T-37B.

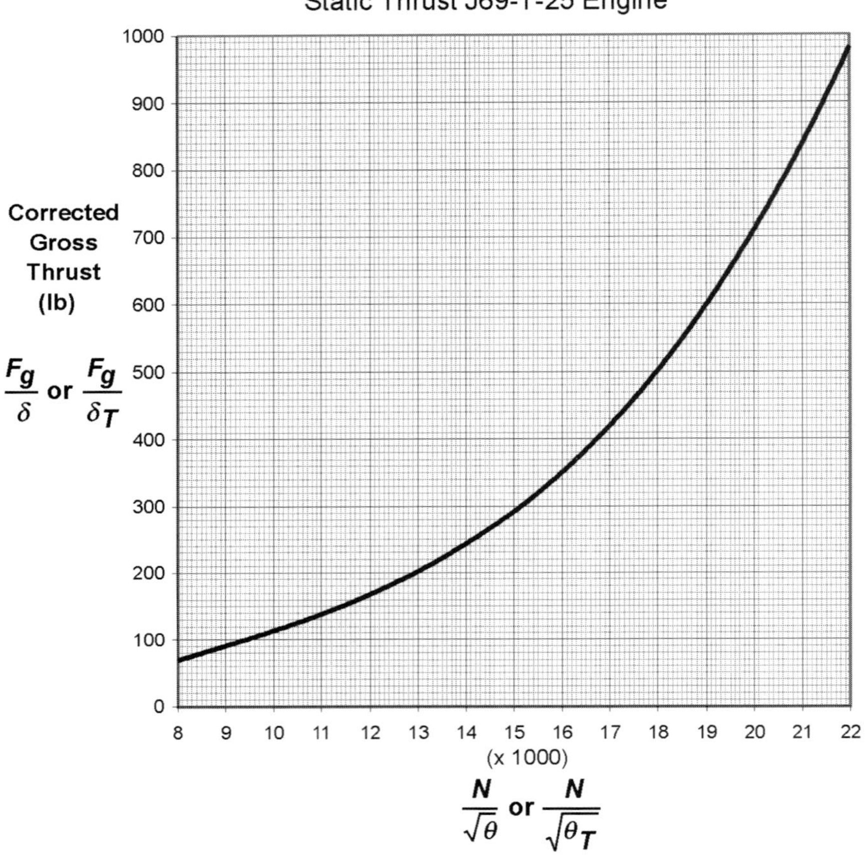

Fig. 2.13 Corrected gross thrust vs corrected rpm for J69.

Note that the data are a function of both corrected rpm and Mach number. For Mach numbers other than those listed, the user must interpolate to estimate the corrected airflow.

The use of the corrected property plots (Figs. 2.13 through 2.15) is best illustrated with the following T-37 example.

Example 2.1

A T-37 is cruising at 20,000 ft pressure altitude at 300 kn true airspeed. The outside air temperature is $-20°$F and the actual engine speed is 18,000 rpm. Assume that the angle between the ram drag and gross thrust is 180 deg. Determine the net thrust and the fuel flow rate.

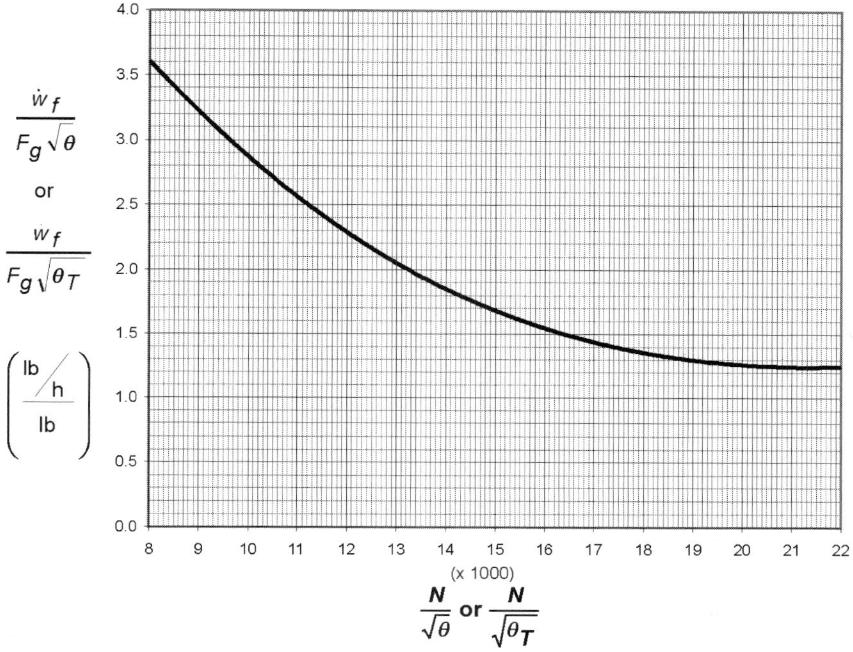

Fig. 2.14 Corrected TSFC vs corrected rpm for J69.

Solution: Because the aircraft is in flight, the corrected properties must be determined based on the total temperature ratio and the total pressure ratio (namely, θ_T and δ_T, respectively). To determine these ratios it is first necessary to determine the Mach number.

$$M = \frac{V}{a} = \frac{V}{\sqrt{\gamma RT}} = \frac{(300\ \text{kn})(1.689\ \text{ft/sec/kn})}{\sqrt{(1.4)(1716\ \text{ft}\cdot\text{lb/slug}\cdot^\circ\text{R})(-20^\circ\text{F} + 460^\circ\text{F})}} = 0.493$$

θ is determined using

$$\theta = \frac{T}{T_{sl}} = \frac{(-20^\circ\text{F} + 460^\circ\text{F})}{518.67^\circ\text{R}} = 0.8483$$

and δ can be read directly from the standard atmosphere table because the aircraft is operating at a pressure altitude of 20,000 ft.

$$\delta = 0.4595$$

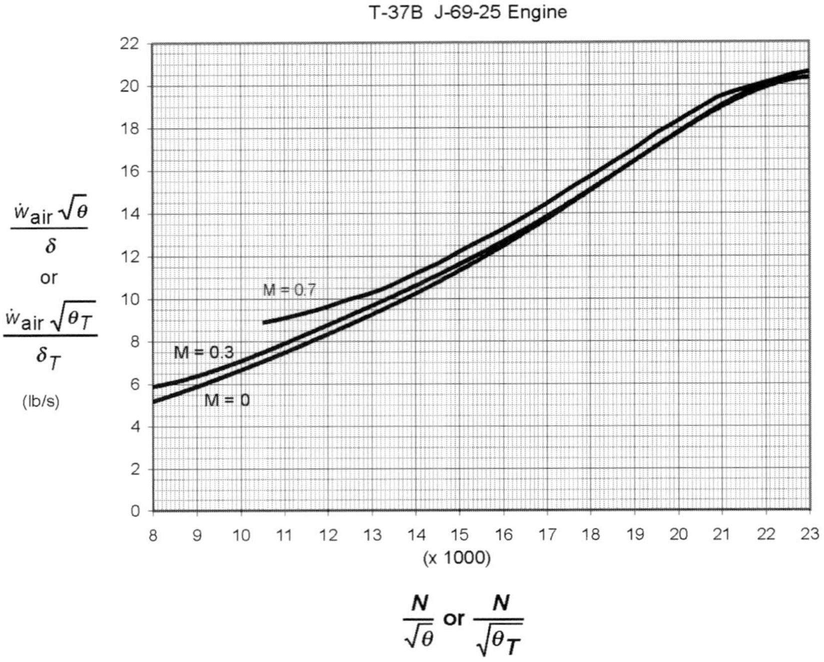

Fig. 2.15 Corrected airflow vs corrected rpm.

The total properties are found using the isentropic relations,

$$\theta_T = \theta\left(1 + \frac{\gamma - 1}{2}M^2\right) = (0.8483)\left[1 + \frac{(1.4 - 1)}{2}(0.493)^2\right] = 0.8895$$

and

$$\delta_T = \delta\left(1 + \frac{\gamma - 1}{2}M^2\right)^{\frac{\gamma}{\gamma-1}} = (0.4595)\left[1 + \frac{(1.4 - 1)}{2}(0.493)^2\right]^{\frac{1.4}{1.4-1}} = 0.5425$$

The corrected rpm, N_c, is

$$N_c = \frac{N}{\sqrt{\theta_T}} = \frac{18{,}000}{\sqrt{0.8895}} = 19{,}085 \text{ rpm}$$

Using Fig. 2.13, the gross corrected thrust, F_{g_c}, is found to be approximately 610 p for one engine. Therefore, the actual gross thrust, F_g, for one engine is found using Eq. (2.11).

$$F_g = F_{g_c}\delta_T = (610\ \text{lbf})(0.5425) = 330.9\ \text{lbf}$$

Therefore, for both engines

$$F_g = (2\ \text{engines})(330.9\ \text{lbf/engine}) = 661.8\ \text{lbf}$$

The net thrust, F_n, is found using Eq. (2.9).

$$F_n = F_g - \dot{m}_\infty V_\infty$$

Therefore, it is necessary to determine the mass flow rate of the air through both engines. The airflow is determined using data in Fig. 2.15. Using the graph

$$\frac{\dot{w}_{\text{air}}\sqrt{\theta_T}}{\delta_T} \approx 16.8\ \text{lb/s}$$

Therefore, the actual airflow for one engine is

$$\dot{w}_{\text{air}} = (16.8\ \text{lb/s})\frac{\delta_T}{\sqrt{\theta_T}} = (16.8\ \text{lb/s})\left(\frac{0.5425}{\sqrt{0.8895}}\right) = 9.66\ \text{lb/s}$$

which means the airflow for both engines is

$$\dot{w}_{\text{air}} = (2\ \text{engines})(9.66\ \text{lb/s/engine}) = 19.32\ \text{lb/s}$$

The mass flow rate is found using

$$\dot{m}_\infty = \frac{\dot{w}_{\text{air}}}{g} = \frac{19.32\ \text{lb/s}}{32.2\ \text{ft/s}^2} = 0.6\ \text{slug/s}$$

Therefore, the net thrust is

$$\begin{aligned} F_n &= F_g - \dot{m}_\infty V_\infty = 661.8\ \text{lb} - (0.6\ \text{slug/s})[(300\ \text{kn})(1.69\ \text{ft/s/kn})] \\ &= 357.8\ \text{lb} \end{aligned}$$

To find the fuel flow rate, $\dot{w}_f$, Fig. 2.14 is used to find the corrected TSFC.

$$\text{TSFC}_c = \frac{\dot{w}_f}{F_g\sqrt{\theta_T}} = 1.3/\text{h}$$

Therefore,

$$\dot{w}_f = (1.3/\text{h})(F_g\sqrt{\theta_T}) = (1.3/\text{h})(661.8\ \text{lb})\sqrt{0.8895} = 811.4\ \text{lb/h}$$

2.2.4 Summary

The combination of the thrust equation and the use of corrected properties provide a relatively straightforward method to estimate the thrust produced for aircraft with simple turbojets, such as the T-37. Similar approaches are available for the more common turbofans. This approach can be extremely useful in quickly predicting aircraft performance and capabilities. Complex engines require more elaborate procedures that are usually supplied by the engine manufacturer.

2.3 Historical Snapshot—Aircraft Performance Modeling and the Learjet Model 35

In the early 1980s, a joint research effort between the National Aeronautics and Space Administration (NASA) Dryden Research Center and the University of Kansas Department of Aerospace Engineering developed an efficient method for defining aircraft and engine characteristics based on a limited amount of flight test data.[1] The overall method was referred to as performance modeling and used a combination of acceleration and deceleration flight test

Fig. 2.16 Learjet Model 35 aircraft.

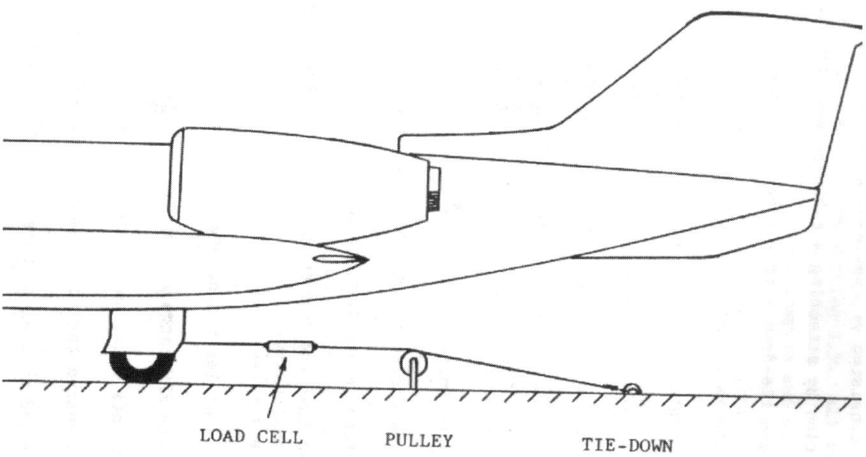

Fig. 2.17 Thrust run load cell/tie down configuration.

maneuvers to define aircraft characteristics such as lift and drag, along with engine characteristics such as corrected thrust and corrected air flow. These characteristics were then input into a performance modeling computer program that would iterate to predict aircraft performance characteristics such as range and fuel flow for any flight condition. A Learjet Model 35 aircraft was used as

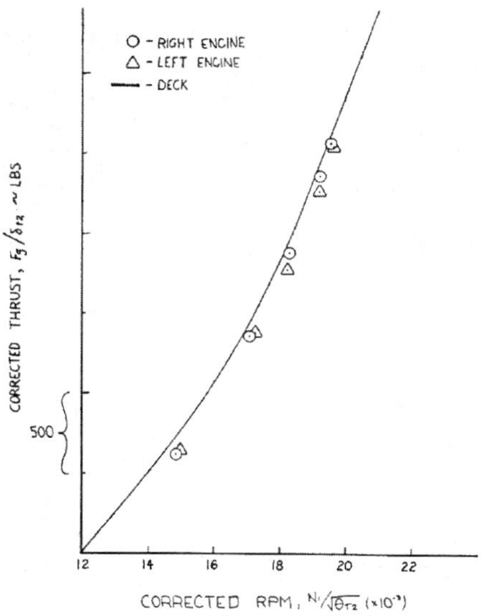

Fig. 2.18 Learjet 35 thrust run corrected thrust data.

the flight test vehicle to develop and evaluate the approach. The aircraft is shown in Fig. 2.16.

Of particular application to the material in this chapter, the individual characteristics of each engine on the aircraft had to be evaluated and compared to the generic engine model (sometimes referred to as the engine deck) supplied by the engine manufacturer. This was accomplished by conducting a ground thrust run where the aircraft was tied down and the static thrust on each engine was measured with a load cell. The thrust run setup is illustrated in Fig. 2.17.

The thrust run data was reduced using corrected property form and compared to the engine deck. The actual graph for corrected thrust vs corrected rpm is presented in Fig. 2.18. Notice that the engines on the test aircraft were producing somewhat less thrust than that predicted by the engine deck.

The final step was to adjust the engine deck to reflect the actual characteristics of each engine based on the thrust run. This is an important step in any flight test program since determination of aircraft drag is directly dependent on the thrust prediction.

Reference

[1]Yechout, T. R., and Braman, K. B., *Development and Evaluation of a Performance Modeling Flight Test Approach Based on Quasi Steady-State Maneuvers*, NASA CR 170414, NASA Dryden Flight Research Facility, April 1984.

Problems

2.1 The mass flow of air through a turbojet engine in flight is 0.5 slug/s. The jet exhaust velocity is 1000 ft/s and the flight velocity is 180 kn TAS. Neglect the mass of the fuel burned.
  (a) Find the magnitude of the gross thrust.
  (b) Find the magnitude of the ram drag.

2.2 What installed static military thrust ($N = 21{,}700$ rpm) would you expect for the T-37 with a pressure altitude of 6200 ft and an ambient temperature of 85°F?

2.3 A T-37 is cruising at 10,000 ft pressure altitude at 300 kn true airspeed. The outside air temperature is −20°F and the engine speed is 18,000 rpm. Find the net thrust and fuel flow. Assume the angle between the ram drag and gross thrust is 180 deg.

3
Aircraft Performance

3.1 Airspeed

An understanding of airspeed is critical to interpreting and discussing aircraft performance. To this point, we have only dealt with true airspeed. This section addresses indicated, calibrated, equivalent, and ground speed.

A pilot needs a direct indication related to the performance of the aircraft. This is done by measuring, and displaying, airspeed on an airspeed indicator. However, we'll see that the airspeed the pilot reads can be quite different than the true velocity (V_∞) of the aircraft.

First, we will introduce some terminology:

1) Static pressure (p) is the pressure because of random molecular motion and is devoid of any contribution from the flow velocity.
2) Total pressure (p_o) is the pressure that exists at a stagnation point, or would exist at any point in the flow, if it were isentropically slowed to zero velocity.
3) Indicated airspeed (V_i) is the airspeed displayed in the cockpit and is obtained from pitot-static instrumentation and "fed" into the airspeed indicator.
4) Calibrated airspeed (V_c) is the indicated airspeed corrected for position error.
5) Equivalent airspeed (V_e) is the calibrated airspeed corrected for nonstandard sea-level pressure.
6) True airspeed (V) is airspeed relative to the air mass. It is equivalent airspeed corrected for non-standard, sea-level density.
7) Ground speed (V_g) is speed relative to the ground. It is true airspeed corrected for wind.
8) Position error (ΔV_p) is obtained from flight test and is a correction between indicated and calibrated airspeed, used to account for error in static port placement and known instrument errors.
9) f Factor (f) is a nonstandard sea-level pressure correction factor and is the ratio between equivalent and calibrated airspeeds.

A pitot-static tube (or a pitot tube with a separate static source) is used to measure an aircraft's airspeed. The static source senses the static pressure (p) of the freestream air and the pitot tube senses the air's total pressure (p_o). These pressures are fed to either a mechanical or digital airspeed measurement system. In a mechanical system, static and total pressures are fed to opposite

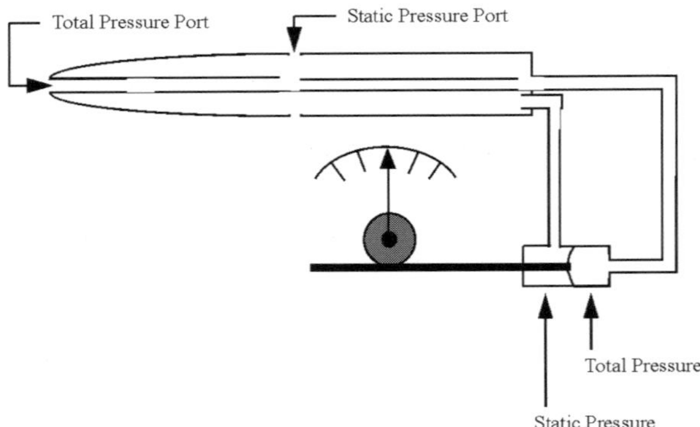

Fig. 3.1 Aircraft pitot–static airspeed system.

sides of a thin, metallic, airspeed diaphragm. The diaphragm is in a sealed case as shown in Fig. 3.1.

The mechanical expansion and contraction of the diaphragm (which is coupled to a system of gears and levers) translates the pressure difference ($p_o - p$, or Δp) into a reading on the aircraft's airspeed indicator. The airspeed the pilot sees on the face of this instrument is called the indicated airspeed, or V_i.

Airspeed indicators must be simple and reliable instruments designed to relate a pressure difference (Δp), from the pitot–static system, to a velocity for all speed regimes, namely both incompressible and compressible flow. Historically, this distinction gave rise to two types of airspeed indicators—those for incompressible flow and those for compressible flow. Aircraft built before approximately 1925 operated exclusively at incompressible airspeeds; they had incompressible airspeed indicators. Because these are inaccurate for high-speed flight, we will concentrate on defining airspeeds using the compressible flow equations.

The following equation (from Ref. 1) defines true airspeed, as a function of density and pressure, and is valid for incompressible and compressible subsonic flow.

$$V = \sqrt{\left\{\left(\frac{1}{\rho}\right)7p\left[\left(\frac{\Delta p}{p}+1\right)^{\frac{1}{3.5}}-1\right]\right\}} \tag{3.1}$$

The velocity (without a subscript) is true airspeed, defined as the speed of the aircraft relative to the air mass in which it is moving. To help distinguish between true airspeed and ground speed (V_g), consider the following example:

Example 3.1

An aircraft flying at 300 kn true airspeed has a 50 kn tailwind. What is its ground speed?

To obtain ground speed, simply use vector addition.

$$\overrightarrow{V_{\text{airplane}} = 300\ \text{kn}} + \overrightarrow{V_{\text{wind}} = 50\ \text{kn}} = \overrightarrow{V_{\text{ground}} = 350\ \text{kn}}$$

It is easy to see that ground speed is simply true airspeed corrected for wind conditions. Referring to Eq. (3.1), note that true airspeed is a function of three variables (Δp, p and ρ), not a simple equation to engineer into a mechanical instrument. In addition, values of ρ, which in turn depend on pressure and temperature, are difficult to determine accurately with instruments. For these reasons, it is difficult to build a simple and reliable airspeed indicator based on Eq. (3.1).

Engineers surmounted this problem by simplifying the equation. In the factory, airspeed indicators are machined with gears calibrated to use sea-level standard atmospheric values of p and ρ. So, in effect, an airspeed indicator is calibrated to solve the expression

$$V_c = \sqrt{\left\{\left(\frac{1}{\rho_{SL}}\right)7p_{SL}\left[\left(\frac{\Delta p}{p_{SL}}+1\right)^{\frac{1}{3.5}}-1\right]\right\}} \tag{3.2}$$

where V_c is called the calibrated airspeed. However, this is still not what is indicated on the airspeed indicator. The static ports on the aircraft may be located such that they do not accurately measure the freestream static pressure under all flight conditions. This is referred to as position or installation error. Additionally, there may be small inaccuracies in the machining of the instrument. To account for these discrepancies, errors are quantified during flight testing and equated to a velocity change (ΔV_p) called position error. Therefore, the relationship between what is displayed on the airspeed indicator (indicated airspeed, V_i) and the calibrated airspeed is given as

$$V_c = V_i + \Delta V_p \tag{3.3}$$

In other words, on a perfect airspeed indicator (zero position error), a pilot reading indicated airspeed would also be reading calibrated airspeed. However, in most cases ΔV_p does not equal zero, and indicated airspeed is slightly different than calibrated airspeed.

To obtain true airspeed [Eq. (3.1)] from calibrated airspeed [Eq. (3.2)], two corrections must be made—one for the actual existing pressure, and the other for the actual density. We'll make these corrections in two steps.

1) First, we will look at the pressure correction. Define equivalent airspeed, V_e, as

$$V_e = \sqrt{\left\{\left(\frac{1}{\rho_{SL}}\right)7p\left[\left(\frac{\Delta p}{p}+1\right)^{\frac{1}{3.5}}-1\right]\right\}} \tag{3.4}$$

Note the values of actual pressure are used here, as opposed to the sea-level values in Eq. (3.2). To relate V_c to V_e, a pressure correction factor (called "f") is used:

$$V_e = f\,V_c \tag{3.5}$$

where

$$f = \frac{V_e}{V_c} = \frac{\sqrt{\left\{\left(\left(\frac{1}{\rho_{SL}}\right)7p\left[\left(\frac{\Delta p}{p}+1\right)^{\frac{1}{3.5}}-1\right]\right)\right\}}}{\sqrt{\left\{\left(\frac{1}{\rho_{SL}}\right)7p_{SL}\left[\left(\frac{\Delta p}{p_{SL}}+1\right)^{\frac{1}{3.5}}-1\right]\right\}}} \tag{3.6}$$

Note that f depends only on Δp and p. All other variables are constant. The value of Δp can be obtained from the calibrated airspeed, and p can be obtained by specifying a pressure altitude. In this manner, Table 3.1 of f factors can be produced—this table is independent of aircraft. Note that in some literature, including undergraduate pilot training texts, this pressure correction factor is referred to as a compressibility correction.

Table 3.1 f Factor table

Pressure altitude, ft	Calibrated airspeed, kn								
	100	125	150	175	200	225	250	275	300
5000	0.999	0.999	0.999	0.998	0.998	0.997	0.997	0.996	0.995
10000	0.999	0.998	0.997	0.996	0.995	0.994	0.992	0.991	0.989
15000	0.998	0.997	0.995	0.994	0.992	0.990	0.987	0.985	0.982
20000	0.997	0.995	0.993	0.990	0.987	0.984	0.981	0.977	0.973
25000	0.995	0.993	0.990	0.986	0.982	0.978	0.973	0.968	0.963
30000	0.993	0.990	0.986	0.981	0.975	0.970	0.963	0.957	0.950
35000	0.991	0.986	0.981	0.974	0.967	0.959	0.951	0.943	0.934
40000	0.988	0.982	0.974	0.966	0.957	0.947	0.937	0.926	0.916
45000	0.984	0.976	0.966	0.956	0.944	0.932	0.920	0.907	0.895
50000	0.979	0.969	0.957	0.944	0.930	0.915	0.901	0.886	0.871

2) Now, we will tackle the density correction. If we multiply Eq. (3.4) by $(\rho_{SL}/\rho)^{0.5}$, Eq. (3.1) (the equation for true velocity) is reobtained, as shown:

$$V = V_e\sqrt{\frac{\rho_{SL}}{\rho}} = \sqrt{\left\{\left(\frac{\rho_{SL}}{\rho}\right)\left(\frac{1}{\rho_{SL}}\right)7p\left[\left(\frac{\Delta p}{p}+1\right)^{\frac{1}{3.5}}-1\right]\right\}}$$

$$= \sqrt{\left\{\left(\frac{1}{\rho}\right)7p\left[\left(\frac{\Delta p}{p}+1\right)^{\frac{1}{3.5}}-1\right]\right\}}$$

Therefore,

$$V = V_e\sqrt{\frac{\rho_{SL}}{\rho}} \tag{3.7}$$

Because the density ratio ρ_{SL}/ρ is usually ≥ 1, V is usually $\geq V_e$. Notice, if we are flying at sea level on a standard day, $\rho_{SL}/\rho = 1.0$, and from the equation above, $V = V_e$. In other words, V_e is the same as true airspeed at sea level on a standard day.

Recall, lift, drag, and moments depend on dynamic pressure, q.

$$\bar{q} = \frac{1}{2}\rho V^2 \tag{3.8}$$

If we replace the true velocity (V), with Eq. (3.7) we obtain

$$\bar{q} = \frac{1}{2}\rho_{SL} V_e^2 \tag{3.9}$$

Notice that the dynamic pressure does not depend on altitude if we define it using equivalent airspeed rather than true. In other words, at a given angle of attack, lift, drag, and moment remain the same for a particular V_e, regardless of altitude. This becomes very useful to both engineers and pilots.

The following summarizes our discussion:

1) indicated to calibrated: apply position error (ΔV_p).

$$V_c = V_i + \Delta V_p \tag{3.10}$$

2) calibrated to equivalent: adjust for actual pressure at altitude.

$$V_e = f\, V_c \tag{3.11}$$

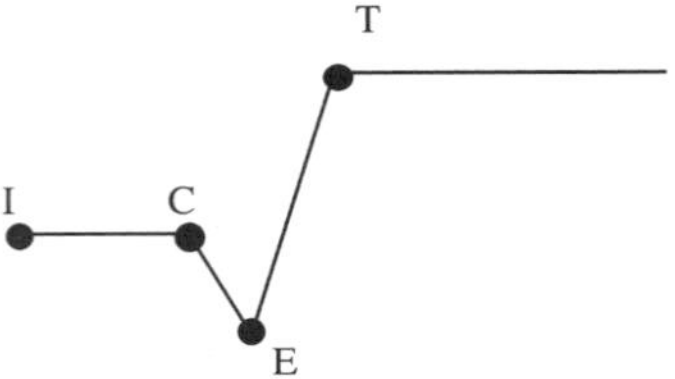

Fig. 3.2 ICE-T convention.

3) equivalent to true: adjust for actual density at altitude.

$$V = V_e\sqrt{\frac{\rho_{SL}}{\rho}} = V_e\sqrt{\frac{1}{\sigma}} \tag{3.12}$$

4) true to ground speed: vector addition of wind to true airspeed.

$$\vec{V}_G = \vec{V} + \vec{V}_{\text{wind}} \tag{3.13}$$

This may seem like a lot to remember. A handy aid used by pilots is the acronym ICE-T (Indicated, Calibrated, Equivalent, and True Airspeed). When used with the radical sign as shown in Fig. 3.2, it relates the magnitudes of all the different velocities to each other.

Notice that indicated and calibrated airspeeds are nearly the same. Equivalent is usually less than calibrated, and true airspeed is usually greater than the others.

Finally, does a pilot need to dig out paper and pencil to calculate true airspeed? The answer is no. Most pilots use a hand-held computer allowing them to carry out the necessary calculations with relative ease. Even better, most modern aircraft use flight data computers. Any airspeed the pilot wants can usually be displayed with the push of a button.

Example 3.2

An aircraft has an indicated airspeed of 200 kn and is flying at an altitude of 25,000 ft (assume standard atmosphere conditions). If the position error is +1 kn, find the true airspeed.

We start by finding the calibrated airspeed with Eq. (3.10).

$$V_c = V_i + \Delta V_p = 200 + 1 = 201 \text{ kn}$$

To convert calibrated airspeed to equivalent airspeed, we use Eq. (3.11) and Table 3.1 to find the f factor.

$$V_e = f\,V_c = (0.982)(201) = 197.4 \text{ kn}$$

The final step involves using Eq. (3.12) to determine true airspeed.

$$V = V_e\sqrt{\frac{\rho_{SL}}{\rho}} = (197.4)\sqrt{\frac{0.00238}{0.00107}} = 294.4 \text{ kn}$$

Notice that we have carried units of knots throughout the problem. Although knots are not consistent units, they can be used for airspeed problems if used for each step in the conversion process.

Example 3.3

An aircraft has a 20-kn headwind and would like a 200-kn ground speed. If the aircraft is flying at 10,000 ft (standard day), what indicated airspeed should it fly if the position error is -1 kn?

We must first determine the true airspeed that the aircraft needs to maintain.

$$V = V_{\text{ground}} + V_{\text{wind}} = 200 + 20 = 220 \text{ kn}$$

Next, we determine the equivalent airspeed using Eq. (3.12).

$$V_e = V\sqrt{\frac{\rho}{\rho_{SL}}} = (220)\sqrt{\frac{0.00176}{0.00238}} = 189.2 \text{ kn}$$

To determine the calibrated airspeed, we must use Table 3.1 in reverse. Because calibrated and equivalent airspeeds are close in value, we will simply enter Table 3.1 with our equivalent airspeed to determine the f factor. To be perfectly correct, we should iterate on the f factor value (go in the reverse direction). With the simple approach and Eq. (3.11), we have,

$$V_c = \frac{V_e}{f} = \frac{189.2}{0.995} = 190.2 \text{ kn}$$

Finally, we determine the indicated airspeed using Eq. (3.10).

$$V_i = V_c - \Delta V_p = 190.2 - (-1) = 191.2 \text{ kn}$$

3.2 Equations of Motion for Straight, Level, and Unaccelerated Flight

Consider an aircraft flying a curvilinear path as shown in Fig. 3.3. To fly a curved path, a net force (F_{net}), composed of force components parallel and perpendicular to the flight path, must exist. In applying Newton's 2nd law, acceleration parallel (linear) to the flight path is dV/dt—centripetal acceleration is V^2/r, where r is the radius of curvature of the curvilinear path.

Lift, drag, thrust, and weight are the four forces acting on an aircraft—these are shown in Fig. 3.4.

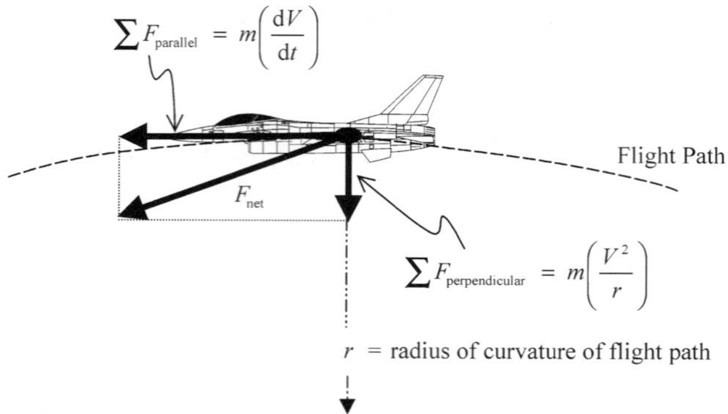

Fig. 3.3 Aircraft flying a curvilinear path.

A few points to remember about the figure:

1) Lift and drag, as discussed in Chapter 1, are the components of the aerodynamic force perpendicular and parallel to the relative wind (V_∞).
2) The x axis is a body-fixed axis, aligned with the "reference line" of the aircraft.
3) The aircraft's angle of attack (α) is between the x axis and the relative wind.
4) In general, the thrust vector is not aligned with the aircraft's reference line. Rather, it is displaced through an angle ϕ_T, called the thrust angle.
5) The flight path angle (γ) is the angle between the horizon and the relative wind.
6) The pitch angle (Θ) is the angle between the aircraft's reference line and the horizon. It is sometimes referred to as pitch attitude or the deck angle (on transports, the cargo deck is usually taken as the aircraft's reference line).

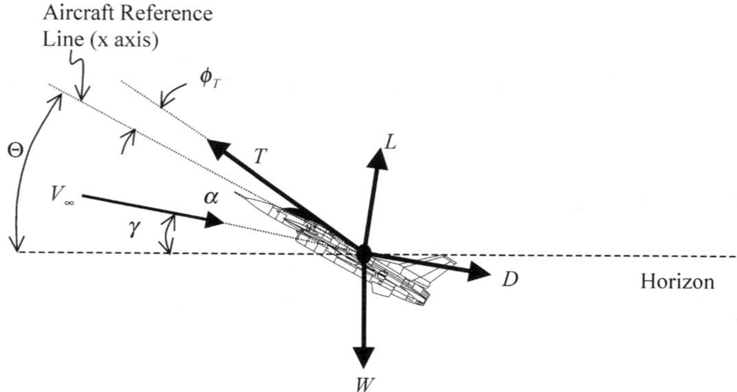

Fig. 3.4 Forces and angles acting on an aircraft in flight.

Summing forces, parallel and perpendicular to the flight path, yields the following.

$$\Sigma F_{\text{parallel}} = m\left(\frac{\mathrm{d}V}{\mathrm{d}t}\right) = T\cos(\phi_T + \alpha) - D - W\sin\gamma \tag{3.14}$$

$$\Sigma F_{\text{perpendicular}} = m\left(\frac{V^2}{r}\right) = W\cos\gamma - L - T\sin(\alpha + \phi_T) \tag{3.15}$$

Because these equations are in a general form, we will return to them frequently. By making appropriate assumptions, we will reduce these equations to predict aircraft performance for a wide variety of situations (for example, glides and climbs).

The first case we will look at is rather simple, yet provides terrific insight. Consider the special case of straight, level, and unaccelerated flight (SLUF). We will apply these assumptions to the general equations of motion (EOM) which are Eqs. (3.14) and (3.15):

1) "Straight" implies that the aircraft's radius of curvature approaches infinity. Therefore, the centripetal acceleration term (V^2/r) goes to zero.
2) The flight path angle, γ, is zero for "level" flight. The $\cos\gamma = 1.0$ and the $\sin\gamma = 0.0$.
3) Linear acceleration ($\mathrm{d}V/\mathrm{d}t$) is zero for "unaccelerated" flight.

Additionally, if we assume the thrust angle and angle of attack are small, then $\cos(\phi_T + \alpha) = 1$ and $\sin(\phi_T + \alpha) = 0$. With these assumptions, the equations of motion reduce to:

$$L = W \text{ (load factor} = 1) \tag{3.16}$$

$$T = D \tag{3.17}$$

A simple, yet useful, result for level unaccelerated flight is that an aircraft's lift balances its weight. The engines must produce sufficient thrust to overcome aerodynamic drag.

3.3 Thrust and Power Curves

Frequently, the term "thrust required" (T_R), is used—it is the amount of thrust necessary to overcome drag, regardless of the maneuver. For SLUF, thrust required is simply equal to the total aircraft drag, as presented in Sec. 1.5.3.

$$T_R = D = [C_{D_0} + KC_L^2]\bar{q}S \tag{3.18}$$

or

$$T_R = \left[C_{D_0}\frac{1}{2}\rho S\right]V^2 + \left[\frac{2K(nW)^2}{\rho S}\right]\frac{1}{V^2} \tag{3.19}$$

Thrust available (T_A), on the other hand, is the amount of thrust available to the pilot and is strictly engine(s) dependent. As we saw in Chapter 2, T_A is a function of altitude, airspeed, and throttle setting. Figure 3.5 is a generic "thrust curve," where T_R and T_A are plotted as a function of velocity (or Mach number).

A great deal of information is captured in an aircraft's thrust curve. Remember that drag (therefore, T_R) changes with altitude, weight/load factor, and/or configuration—again, refer to Sec. 1.5.3. Change any of these, and the thrust required curve changes.

1) Assume for a minute, the pilot's maximum thrust available is represented by the $T_{A)_1}$ curve. Note the following:
   a) At all airspeeds below V_5 thrust available exceeds thrust required. Therefore, by adjusting the throttle, SLUF can be achieved at any point on the T_R curve.
   b) At low airspeeds, the aircraft is "stall limited." Even though there is sufficient thrust available to overcome drag (or T_R), the aircraft stalls below V_1.
   c) V_5 is the maximum velocity achievable in SLUF conditions.
2) Now, assume the pilot adjusts the throttle to the thrust available represented by the $T_{A)_2}$ curve. Note the following:
   a) SLUF can only be achieved at V_2 and V_4.
   b) Below V_2 and beyond V_4, the thrust required exceeds the thrust available. The aircraft will slow down to either the stall limit or to V_4.

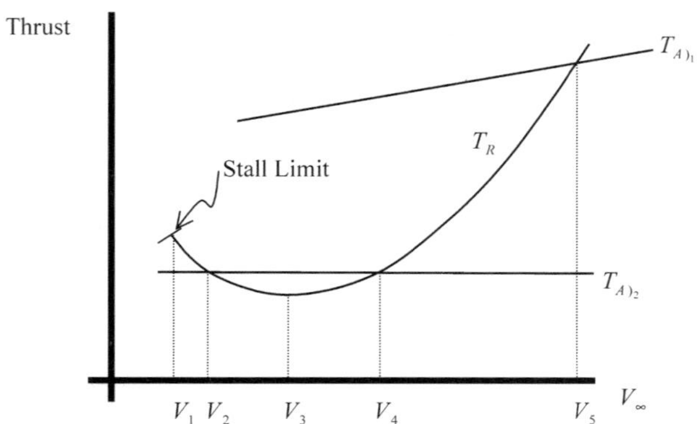

Fig. 3.5 Generic thrust curves.

c) Between V_2 and V_4, thrust available exceeds thrust required and the aircraft will accelerate to V_4. To stabilize the aircraft in SLUF at V_2, the pilot must generally reduce power, increase angle of attack, and then increase power back to $T_{A)_2}$ to achieve this slowflight condition.

3) V_3 is an extremely significant airspeed—total aircraft drag (or T_R) is a minimum. The ratio of lift to drag (L/D) is maximized. Also, parasite and induced drag are equal.

Example 3.4

A 12,000-lb T-38 is at 10,000 ft. Using the T-38 Performance Charts in Appendix D, find

1) $L/D)_{max}$
2) The induced drag at $L/D)_{max}$
3) The true airspeed at $L/D)_{max}$
4) The stall Mach number
5) The maximum Mach number for mil power

We begin by referring to the T-38 10,000-ft performance chart in Appendix D and the 12,000-lb drag curve:

1) $L/D)_{max}$ occurs at the minimum point on the drag curve. For 1-g SLUF we have:

$$\left.\frac{L}{D}\right)_{max} = \frac{W}{D_{min}} = \frac{12{,}000}{1000} = 12$$

2) The induced drag and parasite drag are equal at $L/D)_{max}$. Thus

$$D_{induced} = \frac{1}{2}D_{min} = \frac{1}{2}(1000) = 500 \text{ lb}$$

3) From the chart, the Mach number at $L/D)_{max}$ is 0.5. Using the standard atmosphere tables, the speed of sound at 10,000 ft (standard atmosphere) is 1077.4 ft/s. Thus,

$$V = Ma = (0.5)(1077.4) = 538.7 \text{ ft/s}$$

4) The stall (buffet) Mach number is 0.305 as read from the chart.
5) The max Mach number at mil power is 0.96 as read from the chart.

Power curves typically present power required (P_R) and power available (P_A) as a function of the trim velocity of the aircraft. The power curves evolve directly from the quantities of thrust available and thrust required using the following relationships:

$$
\begin{aligned}
P_R &= T_R V_\infty = D V_\infty \\
P_A &= T_A V_\infty
\end{aligned}
\tag{3.20}
$$

A check of the units shows that a thrust force times a distance divided by time is work per unit time or power. The power curves can be generated from an aircraft's thrust curves by simply multiplying thrust required times the corresponding trim velocity at several points to obtain a series of power required points, which will define the power required curve. A similar approach may be used to obtain the power available curve. Typical thrust and power curves for the same aircraft are presented in Fig. 3.6.

In Fig. 3.6, the scale of the velocity axis is the same for both the thrust and power graphs. Take note of how $V_{\max}$ and $V_{L/D)_{\max}}$ translate to their respective points on each graph. In particular, note that the tangent from the origin to the P_R curve graphically locates the velocity for $T_{R_{\min}}$. The slope of the tangent line is actually $P_R/V = T_R V/V = T_R$, so that the minimum slope, which occurs with the tangent, locates the velocity for $T_{R_{\min}}$, and thus $D_{\min}$ and $L/D)_{\max}$. Both thrust and power curves vary with aircraft weight, configuration,

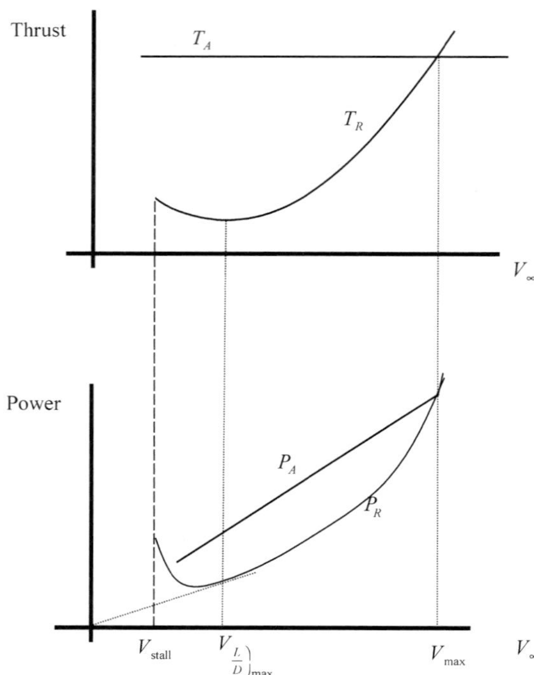

Fig. 3.6 Typical thrust and power curves for the same aircraft.

load factor, and altitude. Both are very useful in aircraft performance predictions.

3.4 Takeoff and Landing Performance

Takeoff is one of the most demanding phases of flight. The aircraft's engines are typically operating at maximum rating. Airspeeds are low, so control is minimal. Combine these considerations with low altitude and there is little room for error. In this section, we will examine the forces during takeoff, apply Newton's 2nd law, and arrive at two ways to predict takeoff performance: 1) "average acceleration method," and 2) numerical method.

3.4.1 Takeoff Phases and Ground Roll Distance

As shown in Fig. 3.7, takeoff performance includes a ground roll, aircraft rotation, transition, and climb-out phase. In our discussion, we will address only the ground-roll distance, S_G. As shown, the velocity corresponding to S_G is V_G.

V_G is the lift-off velocity (V_{LOF}), corrected for winds ($V_G = V_{LOF} +/- V_{\mathrm{WIND}}$)—add for a tail wind, subtract for a headwind. The liftoff velocity is typically expressed as some factor above the aircraft's stall speed. We will use $\boldsymbol{V_{LOF} = 1.1 V_{\mathrm{STALL}}}$ where V_{STALL} is defined in Eq. (1.34).

We will now derive a general expression for the ground-roll distance. Referring to Fig. 3.7, we will assume no wind and integrate between brake release (both S and V are equal to zero) and S_G (where V is V_G). Using the chain rule:

$$\mathrm{d}S = \frac{\mathrm{d}S}{\mathrm{d}t}\frac{\mathrm{d}t}{\mathrm{d}V}\mathrm{dV} = V\frac{1}{a}\,\mathrm{d}V = \frac{1}{a}V\,\mathrm{d}V$$

where a is acceleration or dV/dt. Now integrating both sides from brake release to the ground roll distance,

$$\int_{S=0}^{S=S_G} \mathrm{d}S = \int_{V=0}^{V=V_G} \frac{1}{a} V\,\mathrm{dV} \tag{3.21}$$

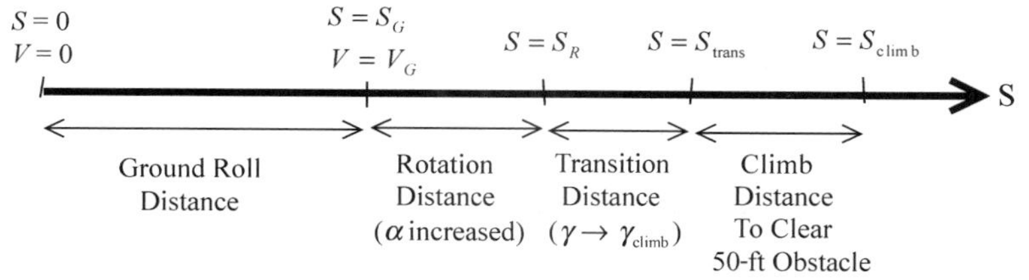

$S = 0$ $V = 0$ | $S = S_G$ $V = V_G$ | $S = S_R$ | $S = S_{\mathrm{trans}}$ | $S = S_{\mathrm{climb}}$ | S

Ground Roll Distance | Rotation Distance (α increased) | Transition Distance ($\gamma \rightarrow \gamma_{\mathrm{climb}}$) | Climb Distance To Clear 50-ft Obstacle

Fig. 3.7 Takeoff phases.

Look at the right-hand side of Eq. (3.21). To integrate it, we will take two approaches during the ground roll: 1) assume acceleration is a constant, and 2) use a numerical approach, and account for acceleration changing. In either case, we will have to evaluate acceleration, which is obtained from Newton's 2nd law ($F = ma$).

3.4.2 Forces During Takeoff

As mentioned previously, we will use Newton's 2nd law and identify the forces during the takeoff roll. Consider Fig. 3.8 below in which the runway slope is greatly exaggerated.

Lift, drag, thrust, and weight are shown. φ is the runway's slope—a positive slope is shown (worst case). Additionally, there is a retarding force (R) because of friction between the wheels and the runway. This is developed below, where μ_r is the rolling friction coefficient.

$$\begin{aligned}\Sigma F_z &= ma_z = 0\\ &= W\cos\varphi - L - N\end{aligned}$$

For small φ, $\cos \approx 1$, and

$$N \approx W - L$$

The retarding force then becomes

$$R \approx \mu_r(W - L) \tag{3.22}$$

Typical values of μ_r range between 0.02 for a dry concrete runway to 0.3 for very soft ground. Returning back to Fig. 3.8, we can apply Newton's 2nd law in the x direction to obtain an expression for acceleration during the ground roll.

$$\Sigma F_x = ma_x = \frac{W}{g}a_x = T - D - W\sin\varphi - R$$

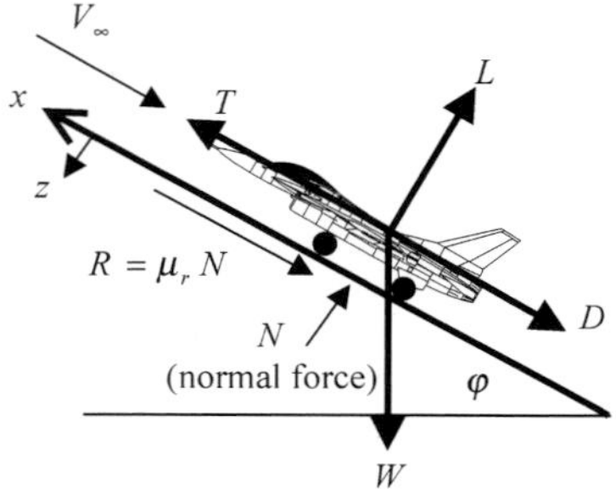

Fig. 3.8 Forces during takeoff roll.

If we assume φ is small, $\sin\varphi \approx \varphi$, and

$$a_x \approx \frac{g}{W}[T - D - W\varphi - \mu_r(W - L)] \tag{3.23}$$

To gain some further insight, assume for a minute that the runway slope is zero. Additionally, ignore any thrust variation with velocity. Figure 3.9 shows how the forces (with these assumptions) change during the takeoff roll.

Note the following:

1) Lift and drag start at zero. They both increase through the takeoff roll; both forces are proportional to V^2.
2) W is constant during the takeoff roll (ignoring fuel consumption).
3) R decreases during the takeoff roll because the normal force, N, decreases as lift increases. R is zero at liftoff (weight is equal to lift).
4) With our assumptions, the difference between thrust and $[D + \mu_r(W - L)]$ is the net acceleration force (again, we have assumed $W\varphi$ is zero). This is a direct indicator of acceleration capability.

Now, the essentials are in place to estimate takeoff roll. Again, we will take two approaches. The easiest will be first.

3.4.3 Average Acceleration Method

If we assume acceleration is constant (denoted by $\bar{a}$) during the takeoff roll, Eq. (3.21) can be integrated as shown:

$$\begin{aligned} S_G &= \frac{1}{\bar{a}}\int_{V=0}^{V=V_G} V\,\mathrm{d}V = \frac{1}{\bar{a}}\frac{V^2}{2}\Big)_0^{V_G} \\ S_G &= \frac{1}{\bar{a}}\frac{V_G^2}{2} = \frac{V_G^2}{2\bar{a}} \end{aligned} \tag{3.24}$$

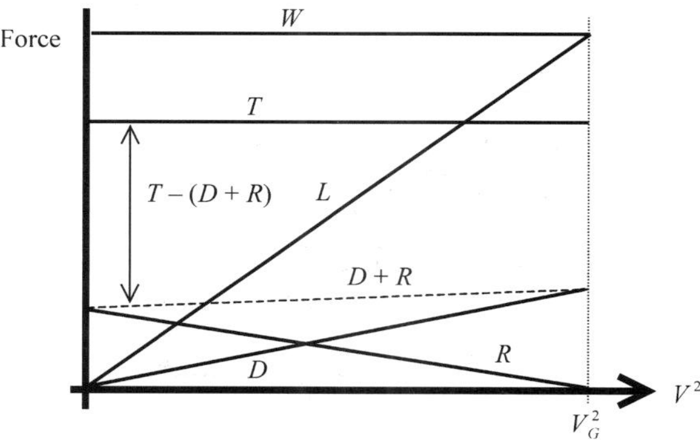

Fig. 3.9 Variation of forces during takeoff roll.

A headwind will reduce V_G by the wind component parallel to the runway and, in a similar manner, a tail wind will increase V_G. Thus, a headwind will decrease S_G and a tail wind will increase S_G. This is why the runway direction for takeoff (and landing) is generally selected based on having a headwind component. Equation (3.24) can be simply modified to account for a head or tail wind:

$$S_G = \frac{1}{\bar{a}}\frac{V_G^2}{2} = \frac{(V_{LOE} \mp V_{\text{wind}})^2}{2\bar{a}}$$

The minus sign is used for a head wind component and the plus sign is used for a tail wind component.

What value of acceleration do we use in Eq. (3.24)? In Sec. 3.4.2, we derived an expression for acceleration [Eq. (3.23)]. It is repeated next, however, in terms of average values:

$$\bar{a} = \frac{g}{W}[T_{\text{avg}} - D_{\text{avg}} - W\varphi - \mu_r(W - L_{\text{avg}})]$$

For this approach, we will assume thrust, aircraft weight, rolling friction coefficient, and runway slope are constant during the takeoff roll. To evaluate lift and drag, we will use an average velocity, as derived next. Let

$$D_{\text{avg}} = C_D\frac{1}{2}\rho V_{\text{avg}}^2 S$$

$$D_{\text{avg}} = \frac{D_{S=0} + D_{S=S_G}}{2}$$

$$C_D\frac{1}{2}\rho V_{\text{avg}}^2 S = \frac{C_D\frac{1}{2}\rho V_{S=0}^2 S + C_D\frac{1}{2}\rho V_{S=S_G}^2 S}{2}$$

Of course, for no wind conditions, $V_{S=0}$. Therefore

$$V_{\text{avg}} = \frac{V_G}{\sqrt{2}} = .707[V_{LOF}] \tag{3.25}$$

With a headwind or tailwind, V_{avg} could be adjusted slightly using this approach but it is normally not, because the average velocity of Eq. (3.25) using $V_{\text{avg}} = V_{LOF}/\sqrt{2}$ provides sufficient accuracy for the average acceleration method. If engine thrust data are available, we should use the thrust at V_{avg}. Additionally, lift and drag are functions of lift coefficient C_L. Again, we will use an optimal value, called "$C_{L_{\text{opt}}}$." To define this value, we will try to maximize $\bar{a}$ with respect to C_L.

$$\frac{\partial\bar{a}}{\partial C_L} = \frac{\partial}{\partial C_L}\left[\frac{g}{W}\{T_{\text{avg}} - D_{\text{avg}} - \mu_r(W - L_{\text{avg}}) - W\varphi\}\right] = 0$$

Recalling that $D = (C_{D_0} + KC_L^2)\bar{q}S$, we can see that only the $D_{\rm avg}$ and $L_{\rm avg}$ terms are a function of C_L. Taking the partial derivative, we then have:

$$0 = -2\frac{g}{W}KC_L\bar{q}S + \frac{g}{W}\mu_r\bar{q}S$$

and solving for C_L

$$C_L = \frac{\mu_r}{2K} = C_{L_{\rm opt}} \tag{3.26}$$

As you can see, we have made several simplifying assumptions to estimate an aircraft's ground roll. In the next section, we'll use a numerical approach to achieve greater accuracy.

Example 3.5

Using the average acceleration method, find the max power takeoff ground roll, with no wind and a zero runway slope, for a 12,000-lb T-38 at sea level and 6000 ft given the following conditions:

$S = 170 \text{ ft}^2 \qquad C_{L_{\max}} = 0.88 \qquad \mu_r = 0.025 \qquad C_{D_0} = 0.02 \qquad K = 0.2$

We will first work the problem for sea level. Using Eq. (1.33), we determine the stall speed.

$$V_{\rm stall} = \sqrt{\frac{2W}{\rho S C_{L_{\max}}}} = \sqrt{\frac{2(12{,}000)}{(0.00238)(170)(0.88)}} = 259.6 \text{ ft/s}$$

We next determine the lift-off velocity,

$$V_{LOF} = 1.1V_{\rm stall} = 1.1(259.6) = 285.6 \text{ ft/s}$$

and the average velocity [using Eq. (3.25)].

$$V_{\rm avg} = 0.707[V_{LOF}] = 0.707(285.6) = 201.9 \text{ ft/s}$$

The average Mach number then becomes

$$M_{\rm avg} = \frac{V_{\rm avg}}{a_{SL}} = \frac{201.9}{1116.4} = 0.181$$

The T-38 sea-level thrust and drag chart at Max power and 0.181 Mach gives us

$$T_{avg} \approx 6000 \text{ lb}$$

Using Eq. (3.26),

$$C_{L_{\rm opt}} = \frac{\mu_r}{2K} = \frac{0.025}{2(0.2)} = 0.0625$$

and

$$C_{D_{\rm avg}} = C_{D_0} + KC_{L_{\rm opt}}^2 = 0.02 + (0.2)(0.0625)^2 = 0.0208$$

The average dynamic pressure is

$$\bar{q}_{\rm avg} = \frac{1}{2}\rho V_{\rm avg}^2 = \frac{1}{2}(0.00238)(201.9)^2 = 48.5 \text{ psf}$$

We next determine the average acceleration

$$\begin{aligned} \bar{a} &= \frac{g}{W}[T_{\rm avg} - D_{\rm avg} - \mu_r(W - L_{\rm avg})] \\ &= \frac{32.2}{12{,}000}[6000 - (0.0208)(48.5)(170) \\ &\quad - 0.025\{12{,}000 - (0.0625)(48.5)(170)\}] \\ \bar{a} &= 14.9 \text{ ft/s}^2 \end{aligned}$$

and use Eq. (3.24) to calculate the ground roll with $V_G = V_{LOF}$ for the no wind case.

$$S_{G_{SL}} = \frac{1}{\bar{a}}\frac{V_G^2}{2} = \frac{V_G^2}{2\bar{a}} = \frac{(285.6)^2}{2(14.9)} = 2737 \text{ ft}$$

The same steps are used to calculate the ground roll at 6000 ft with appropriate changes in air density, the speed of sound, and interpolation of the T-38 thrust curves. Intermediate values for this analysis are:

$V_{LOF} = 312.6$ ft/s $V_{\rm avg} = 220.8$ ft/s $T_{\rm avg} \approx 5100$ lb $\bar{a} = 12.5$ ft/s^2

The ground roll at 6000 ft is

$$S_{G_{6000\text{ ft}}} = 3909 \text{ ft}$$

Notice the 42% increase in ground roll distance with the increase in altitude. This results from a decrease in thrust and an increase in V_{LOF}.

3.4.4 Numerical Method

The last section discussed how to solve the takeoff distance problem assuming that acceleration remains constant and can be defined at $V_{\rm average} =$

$V_{LOF}/\sqrt{2}$. This may not be a good assumption for some airplanes if the acceleration varies significantly during the takeoff roll. For many airplanes, the takeoff acceleration drops off somewhat as V_{LOF} is approached. To obtain a more precise prediction of the takeoff roll, a numerical integration technique, or a numerical method, may be used.

3.4.4.1 Euler method. We will start the analysis using a simple Euler method for numerical integration. Recall Eq. (3.23),

$$a = \frac{g}{W}[T - D - W\varphi - \mu_r(W - L)]$$

Rewriting this in terms of C_L and C_D, we have

$$a = \frac{g}{W}\left[T - W\varphi - \mu_r W - (C_D - \mu C_L)\frac{1}{2}\rho V^2 S\right] \tag{3.27}$$

which gives us the instantaneous acceleration at any point during the ground roll. Assuming the pitch attitude of the aircraft remains constant during the ground roll, we can see that the angle of attack of the aircraft will remain constant along with C_L and C_D. If we also assume that ρ, W, and μ_r remain constant, then determination of the acceleration at any point during the ground roll becomes a function of the thrust (T) and the velocity. Using the Euler method and Eq. (3.27), we can divide the takeoff roll into several small time intervals, assume the acceleration is constant during each interval, and obtain the velocity at the end of each interval. Figure 3.10 illustrates this approach.

The time interval, Δt, is usually a constant and made as small as practical (not larger that 0.1 s).

$$\Delta t = t_1 - t_0 = t_2 - t_1 \Rightarrow \text{etc.}$$

For the first interval, $V_0 = 0$ at $t_0 = 0$ (because this is at brake release), and the initial acceleration is

$$a_0 = \frac{g}{W}[T_{V=0} - W\varphi - \mu_r W]$$

The velocity V_1 at time t_1 can be computed using the Euler method.

$$V_1 = V_0 + a_0(t_1 - t_0) = a_0(\Delta t)$$

The velocity V_2 can be calculated in a similar fashion using the result for V_1 and Eq. (3.27) for acceleration.

$$V_2 = V_1 + a_1(t_2 - t_1) = V_1 + a_1(\Delta t)$$

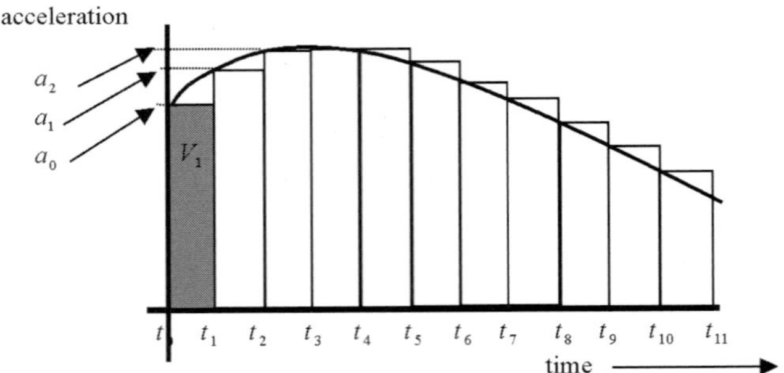

Fig. 3.10 Euler integration to obtain velocity.

with

$$a_1 = \frac{g}{W}\left[T - W\varphi - \mu_r W - (C_D - \mu C_L)\frac{1}{2}\rho V_1^2\right]$$

The general form of the Euler integration for velocity then becomes

$$V_{n+1} = V_n + a_n(\Delta t) \tag{3.28}$$

where

$$t_{n+1} = t_n + \Delta t \tag{3.29}$$

Using this approach and a series of integrations, the ground roll velocity can be determined as a function of time.

Next, we can find the distance traveled during each time interval by using the Euler method on the velocity vs time characteristics. With this method we now assume that V is constant over each time interval (just as we previously assumed acceleration was constant during an interval). Figure 3.11 illustrates this approach.

Remembering that S_0 and V_0 are zero, we have

$$S_1 = S_0 + V_0(t_1 - t_0) = 0$$

and

$$S_2 = S_1 + V_1(t_2 - t_1) = S_1 + V_1(\Delta t)$$

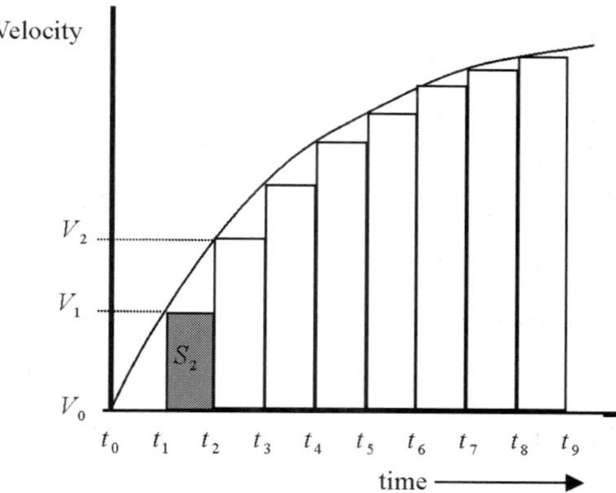

Fig. 3.11 Euler integration to obtain distance.

or in general form

$$S_{n+1} = S_n + V_n(t_{n+1} - t_n) = S_n + V_n \Delta t \tag{3.30}$$

Because the velocity curve is much steeper than the acceleration curve, the Euler method results in considerable error unless Δt is chosen to be very small. Greater accuracy can be achieved using Heun's predictor–corrector method (often referred to as the modified Euler method).

3.4.4.2 Heun's method. Heun's method uses an average of the dependant variable value for each time interval rather than the value at the beginning of the interval. For rapidly changing curves such as takeoff velocity vs time, Heun's method improves the accuracy of the integration over that provided by the Euler method if the same Δt is used. This can be seen in Fig. 3.12. For takeoff distance, the equations become

$$S_1 = S_0 + \frac{V_0 + V_1}{2}(t_1 - t_0) = S_0 + \frac{V_0 + V_1}{2}(\Delta t)$$

$$S_2 = S_1 + \frac{V_1 + V_2}{2}(\Delta t)$$

or in general form,

$$S_{n+1} = S_n + \frac{(V_n + V_{n+1})}{2}(\Delta t) \tag{3.31}$$

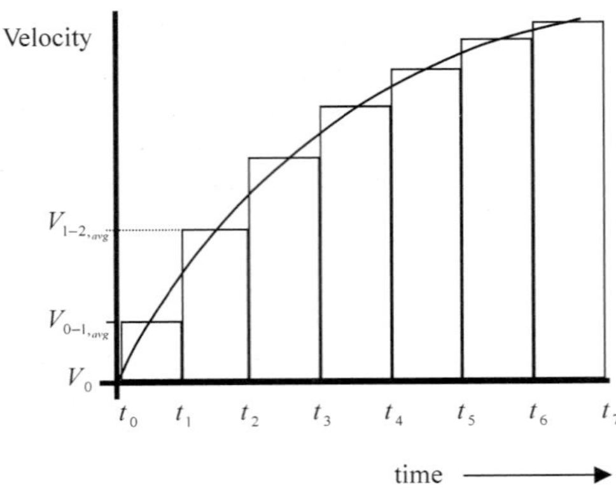

Fig. 3.12 Heun's method integration to obtain distance.

3.4.4.3 Approach. The Euler method works well for relatively flat curves such as acceleration vs time during a takeoff roll. As discussed, Heun's method provides more accuracy when integrating the velocity vs time curve to obtain distance. Several other numerical integration methods may also be used; however, with the speed of modern computer, Δt can be made very small and good accuracy can be obtained using the following approach to numerically determine takeoff distance:

1) Choose a Δt as small as practical but not larger than 0.1 s
2) Using Eq. (3.27) and remembering that $V_0 = S_0 = 0$ at time $t_0 = 0$, compute a_0.
3) Use the Euler method [Eq. (3.28)] to compute V_1.
4) Use Heun's method [Eq. (3.31)] to compute S_1.
5) Increment the time using Eq. (3.29).
6) Repeat Steps 2–5 for succeeding time intervals until $V_n = V_{LOF} \pm V_{\text{wind}}$. Remember that the acceleration term (a_n) in Eq. (3.28) is computed using Eq. (3.27) and V_n for each iteration.

If needed, even more accurate numerical methods such as a Runge–Kutta integration are available in software analysis packages such as MATLAB® (a registered trademark of The MathWorks, Inc.).

3.4.5 Landing Performance

Landing is another very demanding phase of flight. As with takeoff, the aircraft is operating at airspeeds near the stall and, in addition, precise control of the flight path is important to assure touchdown at the desired point on the runway and with a minimum sink rate. Another challenge involved in landing is dissipating the energy of the aircraft once on the ground to bring it to a stop. Landing performance is very dependent on pilot technique.

Landing performance is generally divided into three phases: the approach, the flare, and the ground roll. The approach is typically flown at approximately

1.3 V_{stall} to allow a margin of safety above the stall for turbulence and control input perturbations. The flare is generally initiated within a wingspan of the ground and consists of a gradual power reduction to idle and increase in the angle of attack to slow the aircraft to approximately 1.15 V_{stall} for touchdown (V_{TD}) and to reduce the aircraft's sink rate. The ground-roll phase begins after touchdown with the objective of bringing the aircraft to a stop. To minimize the ground-roll distance, the total retarding force acting on the aircraft must be maximized. The retarding force may consist of brake application, drag from a drag chute and/or speed brakes, and reverse thrust, in addition to the normal drag of the aircraft. Pilot techniques are also important during the ground roll. For example, on some aircraft, holding full aft stick during the ground roll provides a favorable tradeoff between increased aircraft drag and reduced download on the main gear.

The forces during landing ground roll include those discussed for takeoff in Sec. 3.4.2 and presented in Fig. 3.8. For landing, the rolling friction coefficient (μ_r) becomes much larger because of the application of brakes (≈ 0.5 for a dry concrete runway with the brakes fully applied), the drag (D) may become significantly larger because of speed brakes and/or a drag chute, and the aircraft thrust will be a small value or negative (if thrust reversing is used). Referring to Eq. (3.23) with the assumption of a level runway ($\varphi = 0$), we have

$$a_x \approx \frac{g}{W}[T - D - \mu_r(W - L)]$$

Because we will have a deceleration, rather than an acceleration, during the landing ground roll, we will recast the above equation to emphasize deceleration with a negative sign at the beginning of the equation.

$$a_x \approx -\frac{g}{W}[-T + D + \mu_r(W - L)]$$

Again, if thrust reversing is used, T is less than zero ($T < 0$) and the deceleration is larger (a_x more negative). We can estimate the ground roll distance ($S_{G_{\substack{\text{ground}\\\text{roll}}}}$ with the following equation:

$$S_{G_{\text{landing}}} = \int_{V_{TD}}^{0} \frac{V}{a_x}\,\mathrm{d}V$$

which is similar to the approach we took for takeoffs. Recall from Eqs. (1.36) and (1.32) that

$$D = C_D\bar{q}S = (C_{D_0} + KC_L^2)\bar{q}S$$

and

$$L = C_L\bar{q}S$$

where $\bar{q} = \frac{1}{2}\rho V^2$.

Combining, we have

$$S_{G_{\text{landing}}} = \int_{V_{TD}}^{0} \frac{V}{-\dfrac{g}{W}[-T + D + \mu_r(W - L)]} \, \mathrm{d}V$$

$$= \int_{V_{TD}}^{0} \frac{V}{-\dfrac{g}{W}\left[-T + C_D\left(\frac{1}{2}\rho V^2\right)S + \mu_r\left(W - C_L\left[\frac{1}{2}\rho V^2\right]S\right)\right]} \, \mathrm{d}V$$

and integrating,

$$S_{G_{\text{landing}}} = \frac{W}{g\rho S(C_D - \mu_r C_L)} \ln\left[1 + \frac{(C_D - \mu_r C_L)\rho S V_{TD}^2}{2(-T + \mu_r W)}\right] \tag{3.32}$$

Equation (3.32) can be used to estimate the landing ground roll during the period of wheel braking. Remember that C_D must include the effects of speed brakes and/or a drag chute.

The previous analysis assumes that the thrust (or reversed thrust) is acting parallel to the ground roll. The F-15 STOL/MTD test aircraft used two-dimensional nozzles with thrust reversers that allowed a variable thrust angle with upper and lower surface deflections. For landing ground roll, the upper surface provided a maximized reverser angle, while the lower surface reverser angle was reduced with velocity to prevent hot gas injestion.[2] In such a case, the thrust inclination angle relative to the runway must be accounted for in the development of Eq. (3.32).

3.5 Gliding Flight

Consider an aircraft in a poweroff glide, as shown in Fig. 3.13. The forces of lift, drag, and weight are acting on the aircraft.

Recall, the flight path angle (γ) is defined as the angle between the horizon and the relative wind—in this case it is a negative angle (V_∞ is below the horizon). To avoid confusion, we will define $\bar{\gamma}$ as the magnitude of the angle.

$$-\gamma = \bar{\gamma}$$

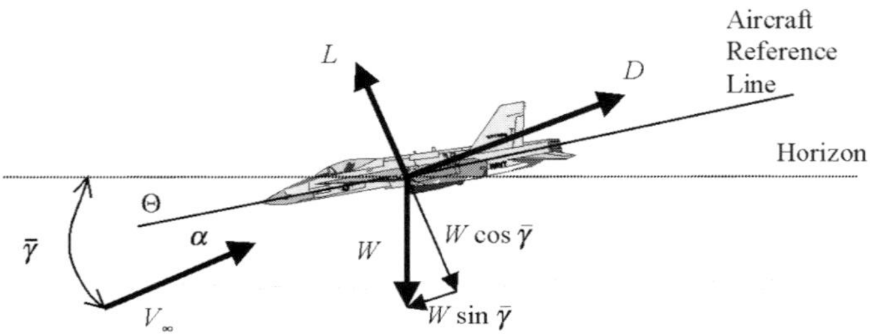

Fig. 3.13 Poweroff gliding flight.

We will assume straight (but not level) and unaccelerated gliding flight. Summing forces parallel and perpendicular to the relative wind or flight path, the equations of motion for gliding flight are obtained. Note that both the linear and centripetal acceleration terms are zero.

$$\Sigma F_{\text{parallel}} = m\frac{\mathrm{d}V}{\mathrm{d}t} = W\sin\bar{\gamma} - D = 0$$

or

$$\sin\bar{\gamma} = \frac{D}{W}$$

$$\Sigma F_{\text{perpendicular}} = m\frac{V^2}{r} = W\cos\bar{\gamma} - L = 0$$

or

$$\cos\bar{\gamma} = \frac{L}{W}$$

Combining the two results, the following is obtained:

$$\tan\bar{\gamma} = \frac{\sin\bar{\gamma}}{\cos\bar{\gamma}} = \frac{D}{L} \tag{3.33}$$

We can derive an expression for glide range using the geometry of a glide as shown in Fig. 3.14.

From the geometry shown,

$$\tan\bar{\gamma} = \frac{H}{R} \tag{3.34}$$

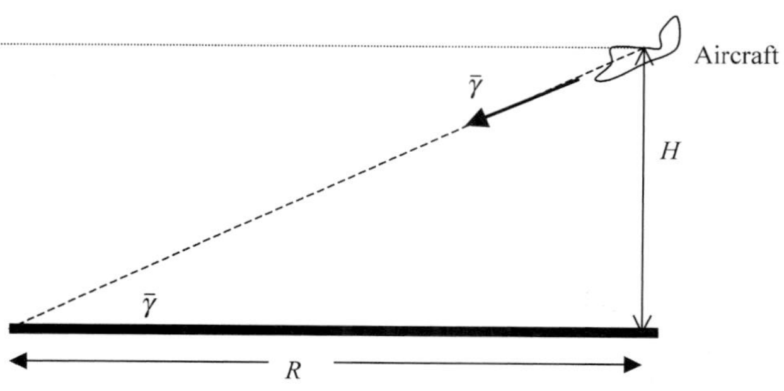

Fig. 3.14 Geometry of a glide.

where H is the altitude at the start of the glide and R is the horizontal glide range. From Eqs. (3.33) and (3.34), we can see that $H/R = D/L$. The glide range then is simply

$$R = H\left(\frac{L}{D}\right) \tag{3.35}$$

Not surprisingly, glide distance is a function of initial altitude. Also, by maximizing the lift-to-drag ratio, glide range is maximized. Therefore, a pilot should fly at the airspeed (or, more correctly, the angle of attack) corresponding to $L/D_{\max}$. From Sec. 3.3, this is the minimum drag point on the thrust required curve.

Glide ratio is defined as R/H. It is horizontal distance covered, per decrease in altitude—units are nondimensional. From the previous discussion,

$$\frac{R}{H} = \frac{L}{D} \tag{3.36}$$

Therefore, an aircraft's $L/D_{\max}$ is also its maximum glide ratio. $L/D_{\max}$ occurs at $D_{\min}$ on the drag vs velocity graph and, as can be seen from Fig. 1.57, parasite and induced drag are equal at this point. Thus, $C_{D_o} = KC_L^2$ at the condition for maximum glide ratio.

Another important aspect of gliding flight is rate of descent. Obviously, if this is minimized, time in the air (or endurance) is maximized. Consider Fig. 3.15.

We are interested in finding a relationship for the vertical velocity (V_v), or rate of descent (also called sink rate). From the geometry shown before, $\sin\bar{\gamma} = V_v/V_\infty$, or $V_v = V_\infty \sin\bar{\gamma}$. Substituting for $\sin\bar{\gamma}$, from the equations of motion developed in this section, gives the following:

$$\text{rate of descent} = V_v = V_\infty \sin\bar{\gamma} = V_\infty(D/W) = P_R/W \tag{3.37}$$

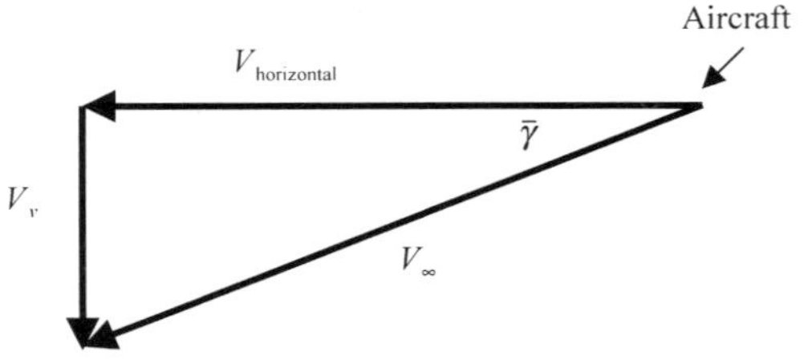

Fig. 3.15 Velocity components for an aircraft in a glide.

Therefore, to minimize sink rate (or vertical velocity), the aircraft should be flown at minimum power required (refer to Sec. 3.3). The max endurance for gliding flight then can be conservatively estimated as

$$E_{\max_{(\text{glide})}} = \frac{H}{(\text{rate of descent})_{\min}} = \frac{WH}{P_{R_{\min}}} \tag{3.38}$$

The C_L and C_D relationship associated with the rate of descent can be easily developed with the assumption of a small glide angle ($\cos\bar{\gamma} \approx 1$ and $L \approx W$) as follows:

$$\text{rate of descent} = \frac{P_r}{W} = V\frac{D}{W} \approx V\frac{D}{L} = V\frac{C_D}{C_L} = \frac{C_D}{C_L}\sqrt{\frac{2W}{\rho S C_L}}$$

$$\text{rate of descent} \approx \frac{C_D}{C_L^{3/2}}\sqrt{\frac{2W}{\rho S}}$$

Thus, to minimize the rate of descent, the aircraft should be flown at an angle of attack (or C_L) where the ratio $C_L^{3/2}/C_D$ is a maximum. In addition, we can use this to determine the relationship between parasite and induced drag at $C_L^{3/2}/C_D)_{\max}$. Because

$$\frac{\partial\left(\dfrac{C_L^{3/2}}{C_D}\right)}{\partial C_L} = 0 \qquad @\left(\frac{C_L^{3/2}}{C_D}\right)_{\max}$$

Using the formula for the derivative of the quotient of two functions,

$$d\left(\frac{u}{v}\right) = \frac{v\,du - u\,dv}{v^2}$$

with $u = C_L^{3/2}$, and $v = C_D$, we have,

$$\frac{\partial\left(\dfrac{C_L^{3/2}}{C_D}\right)}{\partial C_L} = \frac{\frac{3}{2}(C_{D_0} + KC_L^2)C_L^{1/2} - C_L^{3/2}(2KC_L)}{(C_{D_0} + KC_L^2)^2} = 0$$

$$\left(\frac{3}{2}\right)C_{D_0}C_L^{1/2} + \left(\frac{3}{2}\right)KC_L^{5/2} - 2KC_L^{5/2} = 0$$

$$\left(\frac{3}{2}\right)C_L^{1/2}C_{D_0} = \left(\frac{1}{2}\right)KC_L^{5/2} \tag{3.39}$$

$$3C_{D_0} = KC_L^2$$

Thus, the induced drag is equal to three times the parasite drag at $P_{R_{\min}}$ where the rate of descent is a minimum and where endurance is maximized.

Referring to Fig. 3.16, glide performance is easily summarized on a power required curve.

Please note that maximum glide range is achieved at a higher velocity than minimum rate of descent.

Example 3.6

A T-37 has a drag polar $C_D = 0.02 + 0.057C_L^2$. Find its max glide ratio and max glide range from 20,000 ft to sea level.

The max glide ratio occurs at $L/D)_{\max}$ where $C_{D_0} = KC_L^2$. Thus

$$\left.\frac{R}{H}\right)_{\max} = \left.\frac{L}{D}\right)_{\max} = \left.\frac{C_L}{C_D}\right)_{\max} = \frac{\sqrt{\frac{C_{D_0}}{K}}}{2C_{D_0}}$$

$$= \frac{1}{2\sqrt{KC_{D_0}}} = \frac{1}{2\sqrt{0.057(0.02)}} = 14.8$$

The max glide range from 20,000 ft would be

$$\left.\frac{R}{H}\right)_{\max} = 14.8 = \frac{R}{20{,}000}$$

$$R = 14.8(20{,}000) = 296{,}000 \text{ ft} = 56.1 \text{ miles}$$

Example 3.7

What is the maximum glide range a 12,000-lb T-38 can achieve from 20,000 ft? Use the T-38 performance charts.

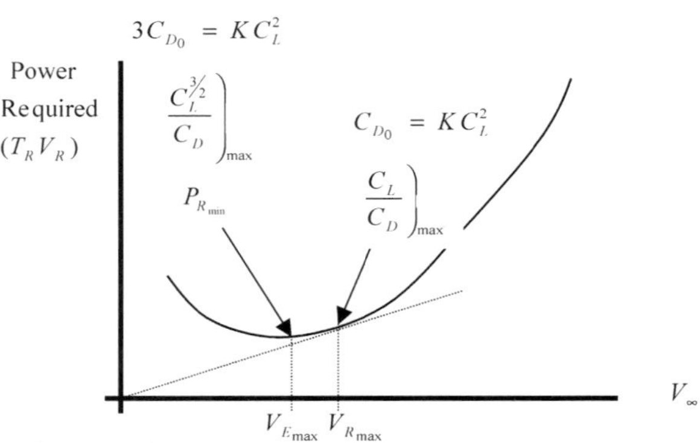

Fig. 3.16 Power required for an aircraft in a glide.

Because $L/D)_{max}$ occurs at D_{min}, we use the 20,000 ft T-38 drag curve to find

$$D_{min} = 1070 \text{ lb}$$

and

$$\left.\frac{R}{H}\right)_{max} = \left.\frac{L}{D}\right)_{max} = \frac{W}{D_{min}} = \frac{12{,}000}{1070} = 11.4$$

the max glide range is

$$R = 11.4(20{,}000) = 228{,}000 \text{ ft} = 43.2 \text{ miles}$$

Notice that this is approximately 13 miles shorter than for the T-37.

Example 3.8

Find the true airspeed for max range and max endurance for an unpowered F-4 at 18,000 ft. $W = 45,000$ lb, $C_D = 0.027 + 0.209C_L^2$, $S = 530$ ft^2.

For max glide range

$$C_{D_0} = KC_L^2$$
$$0.027 = 0.209\ C_L^2$$
$$C_L = \sqrt{\frac{C_{D_0}}{K}} = \sqrt{\frac{0.027}{0.209}} = 0.359$$

The equation for the true airspeed at a given lift coefficient is

$$V = \sqrt{\frac{2W}{\rho SC_L}} = \sqrt{\frac{2(45{,}000)}{(0.00136)(530)(0.359)}} = 589.7 \text{ ft/s} = 348.9 \text{ kn}$$

For max glide endurance

$$3C_{D_0} = KC_L^2$$
$$3(0.027) = 0.209C_L^2$$
$$C_L = \sqrt{\frac{3C_{D_0}}{K}} = \sqrt{\frac{3(0.027)}{0.209}} = 0.623$$

The true airspeed for max endurance becomes

$$V = \sqrt{\frac{2W}{\rho SC_L}} = \sqrt{\frac{2(45{,}000)}{(0.00136)(530)(0.623)}} = 447.7 \text{ ft/s} = 264.9 \text{ kn}$$

3.6 Climbs

We will use the same approach with climb performance that we did with glides. We identify the forces, make some simplifying assumptions, and determine the equations of motion. Consider Fig. 3.17.

The general forms of the equations of motion were derived in Sec. 3.2. They are repeated next with parallel and perpendicular subscripts referring to parallel and perpendicular to the flight path:

$$\Sigma F_{\text{parallel}} \Rightarrow m\left(\frac{\mathrm{d}V}{\mathrm{d}t}\right) = T\cos(\phi_T + \alpha) - D - W\sin\gamma$$

$$\Sigma F_{\text{perpendicular}} \Rightarrow m\left(\frac{V^2}{r}\right) = W\cos\gamma - L - T\sin(\alpha + \phi_T)$$

We will assume a straight and unaccelerated climb. Also, we will assume the following:

$$\alpha + \phi_T \approx 0 \text{ [therefore, } \cos(\alpha + \phi_T) \approx 1.0 \text{ and } \sin(\alpha + \phi_T) \approx 0.0]$$

$$\gamma < 15 \text{ deg [therefore, } \cos\gamma \approx 1.0]$$

With these assumptions, the equations of motion reduce to

$$\Sigma F_{\text{parallel}} \Rightarrow \sin\gamma = \left(\frac{T - D}{W}\right), \text{ or } \gamma = \sin^{-1}\left(\frac{T - D}{W}\right) \tag{3.40}$$

$$\Sigma F_{\text{perpendicular}} \Rightarrow L = W \tag{3.41}$$

We can conclude that to obtain a maximum sustained climb angle ($\gamma_{\max}$), weight should be minimized and the aircraft should operate at maximum excess thrust, $(T - D)_{\max}$, as shown in Fig. 3.18.

An expression for an aircraft's rate of climb (V_v or ROC) can be obtained by considering Fig. 3.19 and the equations of motion.

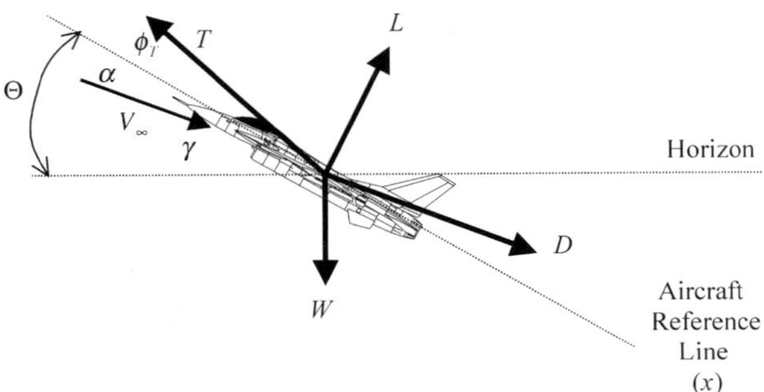

Fig. 3.17 Aircraft in a climb.

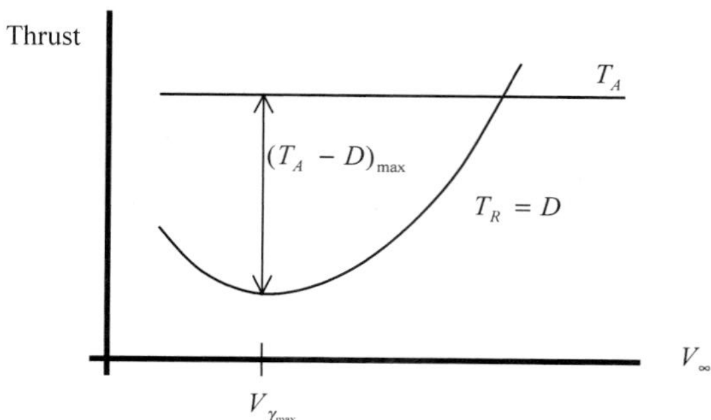

Fig. 3.18 Determination of $V_{\gamma_{max}}$ from thrust vs velocity graph.

From the geometry, $\sin\gamma = V_v/V_\infty$, or $V_v = \text{ROC} = V_\infty \sin\gamma$. Substituting this result into Eq. (3.40), we obtain

$$\text{rate of climb} = V_v = V_\infty\left(\frac{T-D}{W}\right) = \frac{P_A - P_R}{W} = \frac{\text{excess power}}{\text{weight}} \tag{3.42}$$

Therefore, maximum rate of climb occurs at maximum excess power, as illustrated on the power curves in Fig. 3.20. Note the graphical technique for locating this point if P_A is a straight line.

Example 3.9

An 8000-lb T-38 is in a steady climb at 10,000 ft and 0.5 Mach. Determine the rate of climb for military (Mil) and max thrust. Also, determine the climb angle if using max thrust.

Using Eq. (3.42) and the T-38 performance chart for 10,000 ft, mil power and 8000 lb, we have,

$$\text{rate of climb} = V_\infty\left(\frac{T-D}{W}\right) = Ma\left(\frac{3280 - 800}{8000}\right)$$

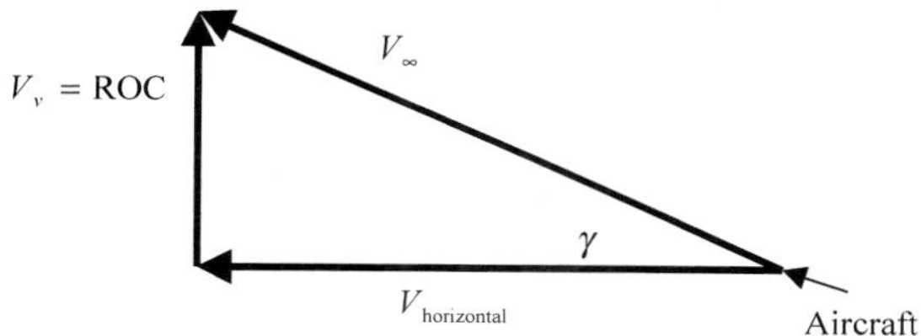

Fig. 3.19 Velocity components for an aircraft in a climb.

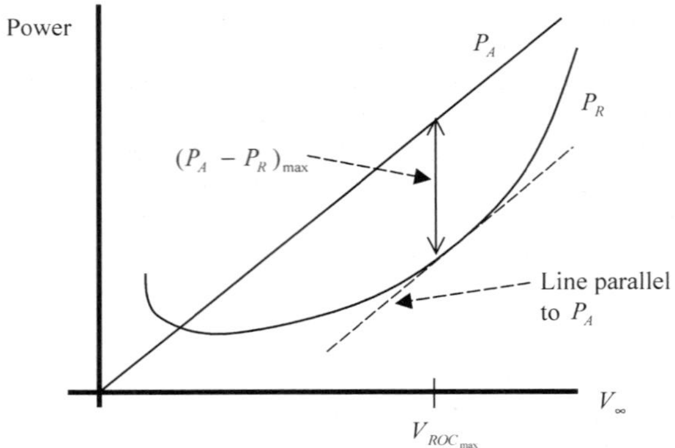

Fig. 3.20 Determination of max rate of climb airspeed.

The standard atmosphere speed of sound at 10,000 feet is 1077.4 ft/s. Thus,

$$ROC_{\substack{\text{Mil}\\ \text{Power}}} = (0.5)(1077.4)\left(\frac{3280 - 800}{8000}\right) = 167.0 \text{ ft/s}$$

For max thrust,

$$\text{ROC}_{\substack{\text{Max}\\ \text{Power}}} = (0.5)(1077.4)\left(\frac{5200 - 800}{8000}\right) = 296.3 \text{ ft/s}$$

We can determine the max thrust climb angle using Eq. (3.40).

$$\gamma = \sin^{-1}\left(\frac{T - D}{W}\right) = \sin^{-1}\left(\frac{5200 - 800}{8000}\right) = 33.4 \text{ deg}$$

Of course, this climb angle violates the assumption of a climb angle less than 15 deg which is used to develop Eq. (3.40), so it is not an exact answer. To obtain an exact answer, the equations should be iterated with $L = W \cos \gamma$.

3.7 Endurance

Aircraft endurance is measured in terms of time, and quantifies how long an aircraft can remain airborne based on a given amount of fuel available. Endurance considerations have a high priority in the design of aircraft that must loiter as part of their mission requirements. The A-10, for example, had a 2-h loiter time requirement in its original design mission profile so that it could remain in a combat area and provide close air support quickly in response to a ground request. Endurance is also important for aircraft such as the E-3A AWACS and KC-135/KC-10 tankers, which must maximize time on station. In simple terms, maximizing endurance simply consists of flying at a throttle setting that minimizes fuel flow and allows the aircraft to maintain altitude. After a few moments of thought, it should become apparent that the minimum fuel flow required to maintain altitude goes down as the weight of the aircraft decreases because the induced drag portion of thrust required decreases. Thus, to really quantify the endurance of an aircraft, we must evaluate the aircraft over the applicable weight excursions as fuel is burned. Fortunately, a simple integration allows us to do this and the equations are fairly straightforward.

3.7.1 Specific Endurance

In quantifying endurance, we begin with the concept of specific endurance (*SE*). Specific endurance is endurance, in terms of time, normalized with respect to fuel burned. It can be thought of, for any short period of time during the mission, as time airborne per pound of fuel consumed or

$$SE = \frac{\text{time airborne}}{\text{fuel consumed}} \tag{3.43}$$

Figure 3.21 presents a simple way of looking at specific endurance. If the endurance of an aircraft is plotted as a function of the fuel consumed, specific endurance can be thought of as the slope of the curve at any point during the mission. It can be seen that, as the weight of the aircraft decreases because of fuel burned, the specific endurance increases.

A more useful form of the specific endurance equation is cast in terms of fuel flow,

$$SE = \frac{1}{\text{fuel flow}} = \frac{1}{\dot{W}_f} \tag{3.44}$$

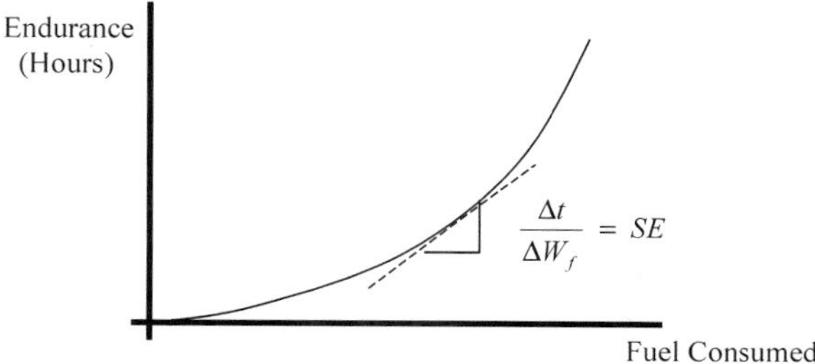

Fig. 3.21 Typical endurance relationship.

where fuel flow rate generally has units of pounds per hour or pounds per second. As we saw in Chapter 2, $\dot{W}_f$ can be expressed in terms of thrust-specific fuel consumption ($TSFC$). Recall $TSFC$ is equal to

$$TSFC = \frac{(\text{lb of fuel})}{(\text{h})(\text{lb of thrust})} = \text{h}^{-1} = (TSFC_{\text{sea level}})\sqrt{\theta} \tag{3.45}$$

$$\dot{W}_f = \frac{\text{lb of fuel}}{\text{h}} = (TSFC)T_A \cong (TSFC)T_R \tag{3.46}$$

and, with the assumption of level flight where thrust available is approximately equal to thrust required, we have

$$SE = \frac{1}{(TSFC)T_R} \tag{3.47}$$

It is important to remember that specific endurance applies to a moment in flight for an aircraft, and will vary with temperature (because $TSFC$ varies with temperature), altitude (density), aircraft velocity, and aircraft weight (because T_R varies with these three parameters). Also notice that specific endurance does not include the amount of fuel available.

Theoretically, the velocity for maximizing specific endurance is at the minimum drag or $L/D_{\max}$ point (as will be shown in the next section). This is based on the level flight simplifying assumptions that $L = W$ and $T = D$. In reality, because the thrust vector is generally inclined slightly above the flight path and helps to overcome some of the weight, the velocity for maximizing specific endurance is slightly lower than the velocity for minimum drag. Thus, control of the aircraft is more challenging because this is in the region of reverse command (stabilizing at a lower velocity requires more thrust). As a matter of practicality, operational procedures generally dictate flying at the velocity for minimum drag and accepting a small penalty in endurance. For the purposes of this text, we will assume maximum endurance occurs at minimum drag.

Evaluation of endurance during flight test is typically accomplished by flying cruise (stabilized) points across the velocity range of the aircraft at constant values of W/δ to standardize weight and altitude variations. Quasi-steady maneuvers (level accelerations and decelerations) may also be used,[3] however, generally stabilized or trim points are still used to spot check the data. To maintain W/δ constant, altitude is increased as fuel is burned. Fuel flow is recorded at each point along with temperature, velocity, and pressure altitude. Corrected fuel flow can then be calculated along with the specific endurance parameter (SEP), which is the reciprocal of corrected fuel flow, $\delta\sqrt{\theta}/\dot{W}_f$. A plot, such as that presented in Fig. 3.22, is prepared for several

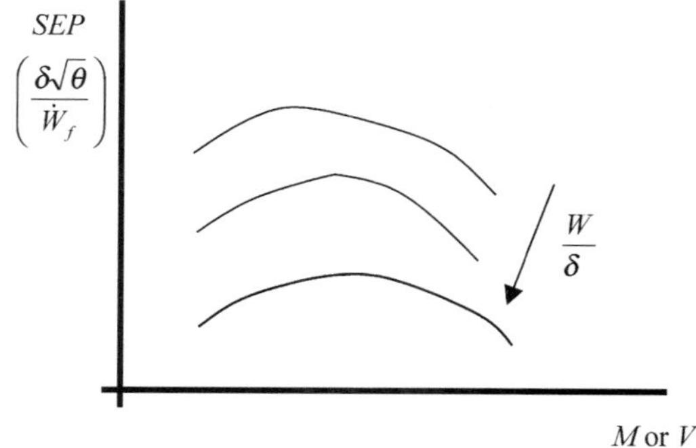

Fig. 3.22 Typical specific endurance parameter plot.

values of W/δ, so that the maximum *SEP* can be determined as a function of Mach number or true airspeed along with W/δ.

Another way to present the data is with the plot presented in Fig. 3.23. Notice that the minimum corrected fuel flow line, which indicates the conditions to maximize specific endurance, falls to the left of the minimum drag point, as discussed earlier. Fortunately, we will see that maximizing endurance from the pilot's standpoint is much easier than this. There is one angle of attack that corresponds to maximum endurance for level flight.

Knowing an aircraft's specific endurance, makes it relatively easy to estimate total endurance. We will take two approaches: 1) an average value equation, and 2) the Breguet endurance equation.

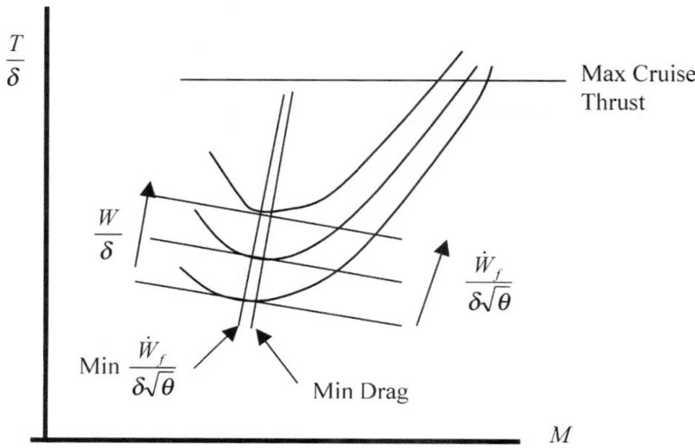

Fig. 3.23 Corrected thrust vs Mach number.

3.7.2 Average Value Endurance Equation

Recall that specific endurance applies to a moment in flight for an aircraft. Knowing the available fuel, an aircraft's endurance can be estimated by the following equation:

$$E_{\text{AVG}} \cong SE(W_0 - W_1) \cong \frac{1}{(TSFC)T_R}(W_0 - W_1) \tag{3.48}$$

Called the "average value endurance equation," W_0 is the aircraft's initial weight and W_1 is the final weight. The difference is the amount of fuel available for the endurance mission. The equation says that endurance is improved by maximizing available fuel and minimizing both thrust-specific fuel consumption and thrust required (or drag). This makes sense. Sometimes the equation is written in the following form, particularly if thrust curves are available.

$$E_{\text{AVG}} \cong \frac{1}{(TSFC)D_{\text{AVG}}}(W_0 - W_1) \cong \frac{\Delta W_f}{(TSFC)D_{\text{AVG}}} \tag{3.49}$$

ΔW_f is the available fuel and D_{AVG} is the drag (or thrust required) at the average weight of the aircraft during the endurance mission. The following example shows how to use these equations to estimate endurance. Because *TSFC* is typically presented in units of h^{-1}, endurance has the units of hours—ΔW_f and D_{AVG} are in units of pounds.

Example 3.10

Determine the average value endurance for a T-38 at 20,000 ft with an initial weight of 10,000 lb and 2000 lb of fuel flying at 0.5 Mach. *TSFC* is 1.01/h.

Using the T-38 performance charts, the drag at 0.5 Mach for a 10,000-lb T-38 is 900 lb. At 8,000 lb, the drag is 675 lb. Using this and Eq. (3.49), we have

$$E_{\text{AVG}} \cong \frac{\Delta W_f}{(TSFC)D_{\text{AVG}}} = \frac{2000}{(1.01)\left(\dfrac{900 + 675}{2}\right)} = 2.51 \text{ h}$$

3.7.3 Breguet Endurance Equation

Mathematically, endurance is simply equal to the following. We are interested in determining the total time an aircraft can stay aloft.

$$E = \int_{t_0}^{t_1} \mathrm{d}t$$

where time t_0 is the start of the endurance mission; t_1 is the end. Recall, that the rate of fuel flow ($\dot{W}_f$), is simply the change in the aircraft's weight per unit time, or

$$\dot{W}_f = \frac{-\,\mathrm{d}W}{\mathrm{d}t}$$

Solving for $\mathrm{d}t$, and using the definition of specific endurance ($SE = 1/\dot{W}_f$), endurance can be recast in the following form:

$$E = \int_{t_0}^{t_1} \mathrm{d}t = \int_{W_0}^{W_1} \frac{-\mathrm{d}W}{\dot{W}_f} = \int_{W_0}^{W_1} -SE\ \mathrm{d}W = \int_{W_0}^{W_1} \frac{-\mathrm{d}W}{(TSFC)T_A} = \int_{W_1}^{W_0} \frac{\mathrm{d}W}{(TSFC)T_A}$$

Let us work on thrust available (T_A). If we assume straight, level, and unaccelerated flight ($SLUF$), then

$$(1)\ T_A = T_R = D$$
$$(2)\ L = W$$

Therefore, $T_A = D = (D/L)W = (C_D/C_L)W$

Let us also assume *TSFC* (fixed altitude and throttle setting) and angle of attack (α) are constant—a constant angle of attack implies a constant ratio of C_D/C_L. With these assumptions, our endurance equation is modified as follows:

$$E = \frac{1}{TSFC}\left(\frac{C_L}{C_D}\right)\int_{W_1}^{W_0} \frac{1}{W}\,\mathrm{d}W = \frac{1}{TSFC}\left(\frac{C_L}{C_D}\right)\ell\mathrm{n}\,\frac{W_0}{W_1} \tag{3.50}$$

This is the Breguet endurance equation. Physically it says the same thing as the average value equation. Aircraft endurance is maximized by minimizing *TSFC* (fly at high altitude) and maximizing the lift-to-drag ratio and available fuel.

Example 3.11

A T-37 at 20,000 ft has a drag polar of $C_D = 0.02 + 0.057C_L^2$. The aircraft has an initial weight of 6000 lb and 500 lb of usable fuel. If the *TSFC* at sea level is 0.9 /h, find the max endurance.

To maximize endurance, the aircraft must fly at $L/D)_{\max}$ which was previously determined from the drag polar to be 14.8 in Example 3.6. We will use Eq. (3.50) to determine the endurance, but first we must find the *TSFC* at 20,000 ft using Eq. (3.45).

$$TSFC = TSFC_{\substack{\text{sea}\\\text{level}}}\sqrt{\theta} = 0.9\sqrt{0.8625} = 0.836/\mathrm{h}$$

We can now find the max endurance using Eq. (3.50).

$$E = \frac{1}{TSFC}\left(\frac{C_L}{C_D}\right)\ell\text{n}\,\frac{W_0}{W_1} = \frac{1}{0.836}(14.8)\,\ell\text{n}\,\frac{6000}{5500} = 1.54\text{ h}$$

Notice that the units for endurance in this problem are hours. If the *TSFC* would have had units of 1/min, the units for endurance would have been minutes.

3.8 Range

Aircraft range is measured in terms of distance, and quantifies how far an aircraft can fly based on a given amount of fuel available. Range considerations will generally be a high priority requirement in the design process for all aircraft. Transport aircraft such as the C-17 and bombers such as the B-2 had very exacting range requirements based on the need to meet global requirements. The range capability of modern aircraft has improved considerably over the years because of significant improvements in engine technology and aerodynamic design. In fundamental terms, maximizing range simply consists of flying at a throttle setting that maximizes the ratio of distance traveled to fuel consumed and that allows the aircraft to maintain altitude. With a little reflection, the ratio of distance traveled to fuel consumed is the same as the ratio of true airspeed to fuel flow ($\dot{W}_f$). Because the weight of the aircraft decreases as fuel is burned, induced drag decreases as a flight progresses and the fuel flow required to maintain altitude also goes down. Thus, to quantify the range of an aircraft, we must evaluate the aircraft over the applicable weight excursions as fuel is burned, similar to the approach taken for endurance.

3.8.1 Specific Range

In quantifying range, we begin with the concept of specific range (*SR*). Specific range is range, in terms of distance, normalized with respect to fuel burned. Like specific endurance, it should be thought of as the ratio of distance traveled to fuel consumed during any short period of time during the mission. Figure 3.24 presents a simple way of looking at specific range. Specific range

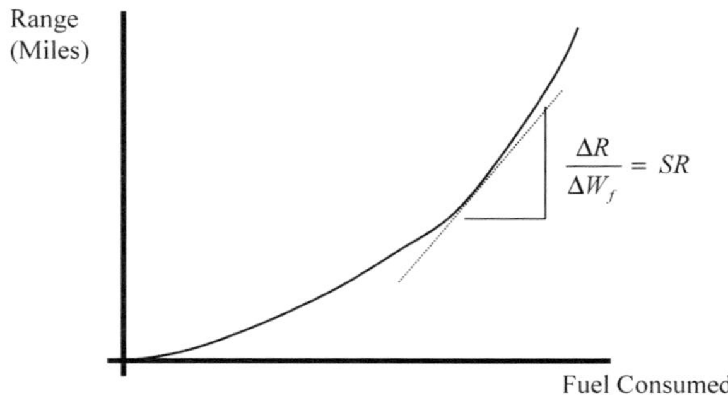

Fig. 3.24 Typical range relationship.

can be thought of as the slope of the curve at any point during the mission. As the weight of the aircraft decreases because of fuel burned, the specific range increases.

A more useful form of the specific range equation is cast in terms of velocity and fuel flow,

$$SR = \frac{\text{distance traveled}}{\text{fuel consumed}} = \frac{V}{\dot{W}_f} \tag{3.51}$$

Because $\dot{W}_f$ can be expressed in terms of thrust-specific fuel consumption as

$$\dot{W}_f = \frac{\text{lb of fuel}}{\text{h}} = (TSFC)T_A \cong (TSFC)T_R$$

we can express specific range as

$$SR = \frac{V}{\dot{W}_f} = \frac{V}{(TSFC)T_R} \tag{3.52}$$

with the assumption of level flight where thrust is approximately equal to thrust required. It is important to note that V in the above relationship is true airspeed and that specific range applies to a moment in flight. Specific range will vary with the same parameters as specific endurance, namely, temperature, altitude, velocity, and weight. Specific range does not include the amount of fuel available.

For level flight with the assumption of $T = D$ and $L = W$, the velocity for maximizing specific range can be determined using a thrust required vs velocity plot as presented in Fig. 3.25. Assuming *TSFC* is a fixed value, *SR* can be maximized by maximizing V/T_R. Notice that the slope of a line originating at the origin and intersecting the thrust required curve is the reciprocal of this, or T_R/V. Thus, the velocity to maximize *SR* can simply be found by drawing the line tangent to the curve as shown. In reality, because inclination of the thrust vector and winds must be considered, the velocity to maximize specific range will be near this point but not exactly at it.

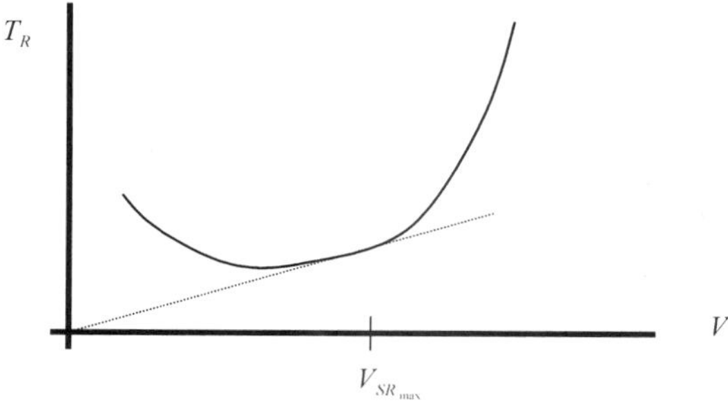

Fig. 3.25 Typical thrust required vs velocity plot.

Evaluation of range during flight test is generally accomplished using the same cruise (stabilized) points as those discussed for endurance. Constant values of W/δ are flown to standardize weight and altitude variations, and the same parameters are recorded at each point. The specific range parameter (*SRP*) is calculated with the following relationship,

$$SRP = \frac{V\delta}{\dot{W}_f} \tag{3.53}$$

and plotted vs Mach number or true airspeed, as presented in Fig. 3.26 so that the maximum SRP can be determined as a function of Mach and W/δ.

As with endurance, we will see that maximizing range from the pilot's standpoint is fairly easy. There is also one angle of attack (lower than that for endurance) that corresponds to the max range condition for level flight.

3.8.2 Average Value Range Equation

Like we did in our discussion of endurance, we can develop an average value equation to estimate an aircraft's range. Range is simply aircraft velocity times endurance, or:

$$R = E \cdot V_\infty \tag{3.54}$$

Note that endurance is typically presented in hours, so be careful with the units of freestream velocity, V_∞. Substituting in our expression for average endurance, the following is obtained—recall, D_{AVG} is the drag (or thrust required), at the average weight of the aircraft, during the mission.

$$R_{\text{AVG}} = E_{\text{AVG}} \cdot V_\infty = \frac{\Delta W_f}{(TSFC)D_{\text{AVG}}} \cdot V_\infty = \frac{\Delta W_f}{(TSFC)(D_{\text{AVG}}/V_\infty)} \tag{3.55}$$

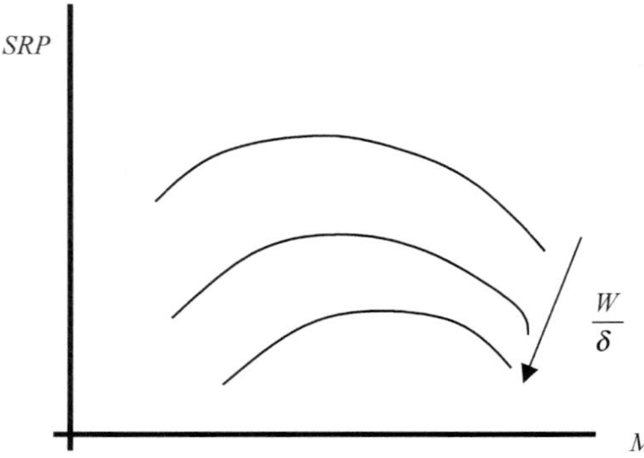

Fig. 3.26 Typical specific range parameter plot.

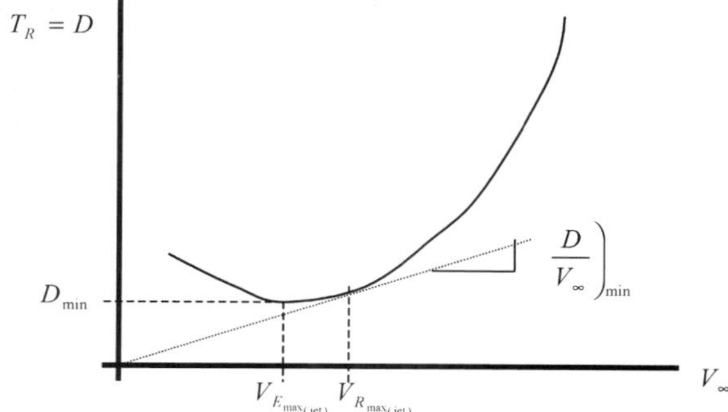

Fig. 3.27 Thrust required vs velocity graph for aircraft average weight.

Like endurance, maximizing available fuel and minimizing *TSFC* will increase range. However, now the ratio of D_{AVG}/V_∞ (not simply D_{AVG}) should be minimized to maximize range. On a thrust curve, the airspeed to fly at $(D_{AVG}/V_\infty)_{MIN}$ can be determined as shown in Fig. 3.27.

Recall, flying at minimum drag (or maximum L/D) will maximize endurance. The airspeed to achieve maximum range is higher than that for maximum endurance.

Example 3.12

Determine the average value range for a T-38 at 20,000 ft with an initial weight of 10,000 lb and 2000 lb of fuel flying at 0.5 Mach. The *TSFC* is 1.01/h.

These are the same conditions as in Example 3.10 except that range must be determined. We will use Eq. (3.55) and the same drag values determined in Example 3.10.

$$V_\infty = Ma = 0.5(1036.8) = 518.4 \text{ ft/s}$$

$$R_{AVG} = E_{AVG} \cdot V_\infty$$

$$= (2.51 \text{ h})(3600 \text{ s/h})(518.4 \text{ ft/s}) = 4{,}684{,}262 \text{ ft} = 887.2 \text{ miles}$$

Notice that with the Mach specified at 0.5, this is not the max range for the aircraft under these altitude and weight conditions because Mach 0.5 is not the max range Mach at the tangent to the drag curve. Notice also that we need to be careful with units when calculating range. Endurance is usually calculated in hours while range usually has units of feet.

3.8.3 Breguet Range Equation

In this section, we will derive a general equation for the range of a jet aircraft in powered cruise—the result will be in an integral form. We will evaluate it for two specific cases—constant altitude cruise and constant speed cruise.

We will start by obtaining a differential expression for distance ($\mathrm{d}s$) traveled.

$$V = \frac{\mathrm{d}s}{\mathrm{d}t} \Rightarrow \mathrm{d}s = V\,\mathrm{d}t$$

We will find an expression for $\mathrm{d}t$ in the previous equation. Recall, aircraft weight is decreasing as fuel is burned—the rate flow of fuel ($\dot{W}_f$) is typically presented in pounds per hour.

$$\dot{W}_f = \frac{-\mathrm{d}W}{\mathrm{d}t} \Rightarrow \mathrm{d}t = \frac{-\mathrm{d}W}{\dot{W}_f} = \frac{-\mathrm{d}W}{(TSFC)T_A}$$

Therefore,

$$\mathrm{d}s = V\,\mathrm{d}t = V\left[\frac{-\mathrm{d}W}{(TSFC)T_A}\right]$$

We will now integrate the previous equation. At the start of the range mission, s is zero and the aircraft weight is W_0. At the end of the mission, s is R (for range), and the aircraft weight is W_1. The difference between W_0 and W_1 is the fuel burned during the mission, or ΔW_f.

$$\int_{S_0}^{S_1=R} ds = -\int_{W_0}^{W_1} \frac{V\,\mathrm{d}W}{(TSFC)T_A} = \int_{W_1}^{W_0} \frac{V\,\mathrm{d}W}{(TSFC)T_A} = \int_{W_1}^{W_0}\left[\frac{V}{(TSFC)T_A}\right]\mathrm{d}W = R$$

The term in the bracket is simply specific range (refer to Sec. 3.8.1). Therefore, the previous equation is equivalent to

$$R = \int_{W_1}^{W_0} (SR)\,\mathrm{d}W$$

This is the general range equation. We will next examine how to maximize cruise range. Assuming *TSFC* is some fixed value (constant altitude and throttle setting), maximizing the ratio of V/T_A will maximize R—this is equivalent to minimizing T_A/V (or D/V, because $T_A \approx T_R = D$). This is the same result we obtained during our discussion of the average value range equation, namely fly at $(D/V)_{\mathrm{MIN}}$ to maximize range.

Now, we will expand the general range equation by using the relationships for straight, level, unaccelerated flight.

$$T_A = D = C_D q_\infty S\left(\frac{W}{L}\right) = \frac{(C_D q_\infty S)W}{C_L q_\infty S} = \left(\frac{C_D}{C_L}\right)W$$

$$L = W = C_L q_\infty S = C_L \frac{1}{2}\rho_\infty V_\infty^2 S \Rightarrow V_\infty = \sqrt{\frac{2W}{\rho_\infty C_L S}}$$

Therefore, the general range equation can be expressed as

$$R = \int_{W_1}^{W_0} \sqrt{\frac{2W}{\rho_\infty C_L S}} \left(\frac{1}{TSFC}\right)\left(\frac{C_L}{C_D}\right)\left(\frac{1}{W}\right) \mathrm{d}W$$
$$R = \int_{W_1}^{W_0} \sqrt{\frac{2W}{\rho_\infty S}} \left(\frac{1}{TSFC}\right)\left(\frac{C_L^{1/2}}{C_D}\right)\left(\frac{1}{W}\right) \mathrm{d}W \tag{3.56}$$

Ignoring all terms except C_L and C_D, we see that range is maximized by flying at the angle of attack (and velocity) to maximize the ratio of $C_L^{1/2}/C_D$, that is $(C_L^{1/2}/C_D)_{\mathrm{MAX}}$. It can be shown that this ratio is maximized at $(D/V)_{\mathrm{MIN}}$, precisely the same point on a thrust required curve as discussed in Sec. 3.8.2. Additionally, refer to Problem 3.9, which states that the relationship between parasite and induced drag at this point is

$$C_{D_0} = 3KC_L^2 \tag{3.57}$$

Because parasite drag is three times induced drag, we know that this is a faster velocity than we would fly for minimum drag. Recall that $L/D_{\max}$ occurs at $D_{\min}$, where parasite and induced drag are equal.

Now we are ready to evaluate Eq. (3.56). Once again, we will take two approaches.

3.8.3.1 Constant altitude cruise.

For convenience, a form of the general range equation is repeated next. We will evaluate it for the special case of a constant altitude cruise.

$$R = \int_{W_1}^{W_0} \sqrt{\frac{2W}{\rho_\infty S}} \left(\frac{1}{TSFC}\right)\left(\frac{C_L^{1/2}}{C_D}\right)\left(\frac{1}{W}\right) \mathrm{d}W$$

Our simplifying assumptions are

1) A constant altitude implies constant density, ρ_∞.
2) Because lift and drag coefficients are functions of angle of attack, C_L and C_D are also constant.
3) $TSFC$ is a constant.

With these assumptions, our range equation is simplified to the following, called the Breguet range equation:

$$R = \sqrt{\frac{2}{\rho_\infty S}}\left(\frac{1}{TSFC}\right)\left(\frac{C_L^{1/2}}{C_D}\right)\int_{W_1}^{W_0}\frac{dW}{\sqrt{W}}$$
$$R = \sqrt{\frac{2}{\rho_\infty S}}\left(\frac{2}{TSFC}\right)\left(\frac{C_L^{1/2}}{C_D}\right)(\sqrt{W_0} - \sqrt{W_1}) \tag{3.58}$$

We can conclude that range is maximized by flying high (low density and low *TSFC*), maximizing available fuel, and maximizing the ratio of $C_L^{1/2}/C_D$. Unfortunately, there are a few limitations to this equation:

1) Flight test data shows that *TSFC* may actually increase (instead of decrease) above 50,000 ft. However, aircraft typically operate below this altitude.
2) As fuel is burned, weight decreases. To maintain a constant angle of attack (or lift coefficient), airspeed must be decreased. Therefore, the pilot will retard the throttle, which may violate our constant *TSFC* assumption.

In the next section, an example is worked demonstrating use of Eq. (3.58). Readers are cautioned to watch units and keep *TSFC* in the consistent units of 1/s.

3.8.3.2 Constant speed cruise (cruise climb). We will start from the general range equation, repeated next. In this section, we will evaluate it for the case of a cruise climb. In this technique, weight is traded (as fuel is burned) for altitude.

$$R = \int_{W_1}^{W_0}\left[\frac{V}{(TSFC)T_A}\right]\mathrm{d}W$$

In this method, we will assume the following:

1) Constant velocity.
2) Constant angle of attack—again, this implies C_L and C_D are constant.
3) *TSFC* is a constant.

With these assumptions, the previous equation reduces to:

$$R = \frac{V}{TSFC}\int_{W_1}^{W_0}\frac{\mathrm{d}W}{T_A}$$

Referring to Sec. 3.8.3, we can once again (assuming straight, level, unaccelerated flight) replace thrust available and velocity with the following:

$$T_A = \left(\frac{C_D}{C_L}\right) W$$

$$V = \sqrt{\frac{2W}{\rho_\infty C_L S}}$$

Continuing, range for a cruise climb mission can be expressed as

$$R = \frac{V}{TSFC}\int_{W_1}^{W_0}\left(\frac{C_L}{C_D}\right)\frac{\mathrm{d}W}{W} = \sqrt{\frac{2W}{\rho_\infty C_L S}}\frac{1}{TSFC}\frac{C_L}{C_D}\int_{W_1}^{W_0}\frac{\mathrm{d}W}{W}$$

Integrating the previous equation yields

$$R = \sqrt{\frac{2W}{\rho_\infty S}}\frac{1}{TSFC}\frac{C_L^{1/2}}{C_D}\ln\frac{W_0}{W_1} = \sqrt{\frac{2}{S}\left(\frac{W}{\rho_\infty}\right)}\frac{1}{TSFC}\frac{C_L^{1/2}}{C_D}\ln\frac{W_0}{W_1} \tag{3.59}$$

For velocity to remain constant, (W/ρ_∞) must be kept constant—the aircraft climbs as weight decreases.

Examining our final equation, range is maximized in a cruise climb by maximizing fuel, flying high (low density and *TSFC*), and maximizing the ratio of $C_L^{1/2}/C_D$. It is interesting to note that these are the same conclusions we reached when flying a constant altitude cruise. As in the first method, assuming a constant *TSFC* introduces some error. Although a constant throttle setting is maintained, *TSFC* will decrease as altitude increases.

Although both range missions (constant altitude and cruise climb) introduce some error, they do provide good insight into how range is maximized. Which method gives the best range performance? The answer is a cruise climb. The following example, using both methods, will demonstrate this.

Example 3.13

Using the information in Example 3.11 for the T-37 and $S = 184$ ft^2, determine the max range for the T-37 at 20,000 ft using a constant altitude cruise and a cruise climb.

We first must determine $C_L^{1/2}/C_D)_{\max}$ using Eq. (3.57). Because

$$C_{D_0} = 3KC_L^2$$

we have

$$C_L = \sqrt{\frac{C_{D_0}}{3K}} = \sqrt{\frac{0.02}{3(0.057)}} = 0.342$$

and

$$C_D = C_{D_0} + KC_L^2 = 0.02 + 0.057(0.342)^2 = 0.0267$$

Thus,

$$\left.\frac{C_L^{1/2}}{C_D}\right)_{\max} = \frac{(0.342)^{1/2}}{0.0267} = 21.9$$

Next, to maintain consistent units, we will convert $TSFC$ to units of 1/s.

$$TSFC = 0.836/\text{h} = 0.836\left(\frac{\text{h}}{3600\ \text{s}}\right) = 0.000232/\text{s}$$

We will evaluate the constant altitude cruise first using Eq. (3.58).

$$\begin{aligned} R &= \sqrt{\frac{2}{\rho_\infty S}}\left(\frac{2}{TSFC}\right)\left(\frac{C_L^{1/2}}{C_D}\right)(\sqrt{W_0} - \sqrt{W_1}) \\ &= \sqrt{\frac{2}{(0.001267)(184)}}\left(\frac{2}{0.000232}\right)(21.9)(\sqrt{6000} - \sqrt{5500}) \\ &= 1{,}823{,}529\ \text{ft} = 345\ \text{miles} \end{aligned}$$

Finally, we will evaluate the cruise climb, which begins at 20,000 ft, using Eq. (3.59) and recalling that W/ρ stays constant as the aircraft is performing the climb.

$$\begin{aligned} R &= \sqrt{\frac{2W}{\rho_\infty S}}\frac{1}{TSFC}\frac{C_L^{1/2}}{C_D}\ln\frac{W_0}{W_1} \\ &= \sqrt{\frac{2(6000)}{(0.001267)(184)}}\frac{1}{0.000232}(21.9)\ln\left(\frac{6000}{5500}\right) \\ &= 1{,}863{,}483.6\ \text{ft} = 352.9\ \text{miles} \end{aligned}$$

Notice that approximately 8 more miles are obtained with the cruise climb.

3.8.4 Range Factor Method

Another method to quantify the range of an aircraft during flight test involves the parameter range factor (RF). RF is defined as

$$RF = (SR)W = \frac{VW}{\dot{W}_f} \tag{3.60}$$

It proves useful because it can be calculated from stabilized point flight test data and used in a direct calculation of the cruise climb range equation developed in the previous section. Recalling that

$$R = -\int_{W_0}^{W_1} (SR)\, \mathrm{d}W$$

We can recast this equation in terms of RF by multiplying through by one, in the form of: W/W.

$$R = -\int_{W_0}^{W_1} (SR) W \frac{\mathrm{d}W}{W}$$

Or,

$$R = -\int_{W_0}^{W_1} RF \frac{\mathrm{d}W}{W} = RF\, \ell\mathrm{n}\left(\frac{W_0}{W_1}\right) \tag{3.61}$$

Notice that the range of the aircraft for a given set of flight conditions is then directly proportional to the RF. Because the RF can be obtained from stabilized point (trimmed level flight) flight test data, we can determine the maximum range for a given fuel load by maximizing the RF. However, we need a logical flight test approach to determine the maximum RF for every altitude, airspeed, and weight combination within the aircraft's capabilities. Using Eq. (3.60) and the fact that

$$V = Ma = M\sqrt{\gamma RT}\frac{\sqrt{T_{SL}}}{\sqrt{T_{SL}}} = M\sqrt{\theta}\sqrt{\gamma RT_{SL}}$$

we see that

$$RF = \frac{M\sqrt{\theta}\sqrt{\gamma RT_{SL}}W}{\dot{W}_f}\left(\frac{\delta}{\delta}\right) = M\left[\frac{1}{\frac{\dot{W}_f}{\sqrt{\theta}\delta}}\right]\frac{W}{\delta}\sqrt{\gamma RT_{SL}} \tag{3.62}$$

This shows that RF is a function of three primary variables:

$$RF = f\left[M, \left(\frac{W}{\delta}\right), \left(\frac{\dot{W}_f}{\sqrt{\theta}\delta}\right)\right]$$

The flight test approach seeks to determine the optimal Mach and W/δ to obtain maximum RF. Several W/δ values are chosen which are representative of the aircraft's capabilities. Stabilized points are flown at each constant value of W/δ across the Mach range of the aircraft. To hold W/δ constant, the

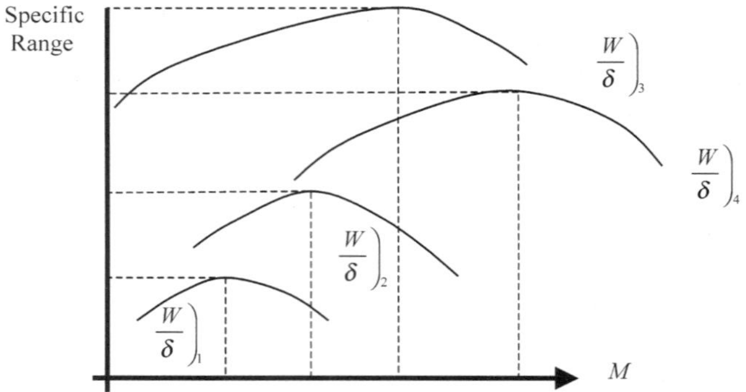

Fig. 3.28 Specific range flight test data.

aircraft increases altitude as fuel is burned off. Specific range data are plotted as a function of Mach for each value of W/δ, as shown in Fig. 3.28.

The maximum specific range is then determined for each value of W/δ as shown. Two plots are next constructed using the maximum specific range points as shown in Fig. 3.29. The optimal range factor plot is based on the maximum specific range data multiplied by the weight of the aircraft for that

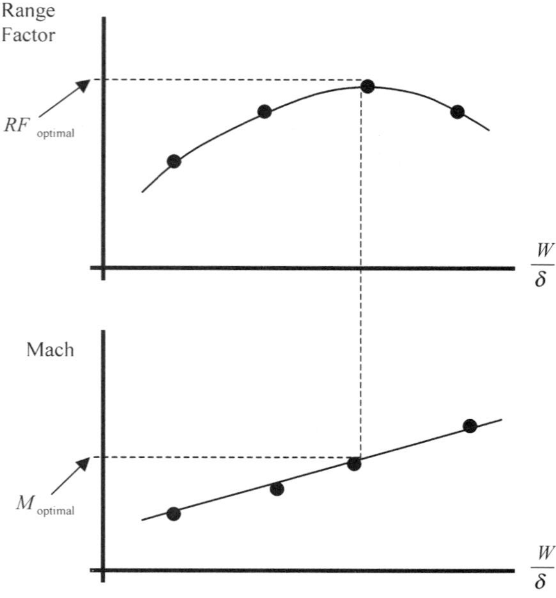

Fig. 3.29 Determination of optimal W/δ and Mach to maximize range.

particular flight condition. The peak of the optimal RF plot is then used to determine the optimal W/δ and Mach as shown.

The range of the aircraft can then be maximized by flying at the optimal W/δ (a cruise climb) and the optimal Mach.

3.9 Turn Performance and *V-n* Diagrams

Until this point, we have assumed straight flight in all our discussions in which the aircraft's radius of curvature approached infinity. This is not the case when examining turning performance. In this section, we will develop equations to estimate turn radius and turn rate (deg/s) for three specific cases: level, pull-up, and pull-down. Additionally, we will introduce the concept of a "V-n diagram"; this diagram describes, in terms of velocity and load factor, an aircraft's operating envelope.

3.9.1 Fundamental Terminology

Consider an aircraft performing a turn, as shown in Fig. 3.30 (rear view). The bank angle (Φ) is the angle between the horizon and the wing (or y body axis; refer to Chapter 4). Typically, ailerons are used to control an aircraft's bank angle. Velocity and angle of attack (or lift coefficient) control the magnitude of the lift vector. Additionally, thrust, weight, and drag are also acting on the airplane.

3.9.2 Level Turns

The aircraft in Fig. 3.31 (rear view) is performing a level turn. To maintain a constant altitude, the velocity vector must act in the horizontal plane; thus,

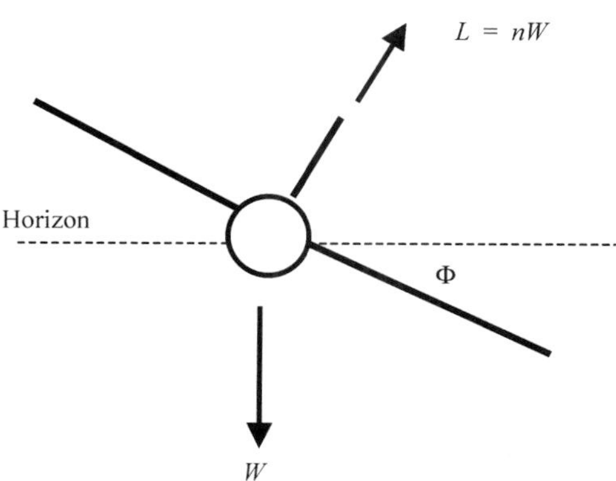

Fig. 3.30 Aircraft in a turn.

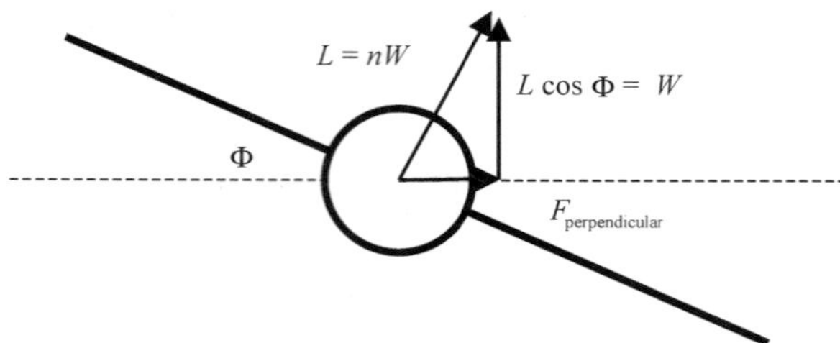

Fig. 3.31 Aircraft in a level turn.

the vertical component of the lift vector must equal the weight. Summing forces in the vertical direction,

$$\begin{aligned} \Sigma F_{\text{vert}} &= 0 = L\cos\Phi - W \\ W &= L\cos\Phi \\ \frac{L}{W} &= n = \frac{1}{\cos\Phi} \end{aligned} \tag{3.63}$$

where n is defined as the load factor. The load factor is generally referred to as the number of gs that the aircraft is pulling. Equation 3.63 defines the relationship between load factor and bank angle for a level turn. Note that the equation is independent of aircraft type. Therefore, any aircraft in a sustained level 60-deg banked turn will be pulling $2g$.

Again referring to Fig. 3.31, we sum forces in the horizontal plane perpendicular to the velocity vector and set them equal to the centripetal force (using Newton's 2nd law). From this geometry, we can derive expressions for turn radius (R) and turn rate (ω). From the Pythagorean theorem:

$$\begin{aligned} \Sigma F_{\text{horiz}} &= \frac{mV^2}{R} = \frac{WV^2}{gR} = L\sin\Phi = \sqrt{L^2 - W^2} = W\sqrt{n^2-1} \\ \frac{V^2}{gR} &= \sqrt{n^2-1} \end{aligned}$$

Solving for the turn radius,

$$R = \frac{V^2}{g\sqrt{n^2-1}} \tag{3.64}$$

Of course, the velocity we are referring to in Eq. (3.64) is the true airspeed of the aircraft or V_∞. Because turn rate is simply equal to V_∞/R, an expression is easily developed, as shown:

$$\omega = \frac{V_\infty}{R} = \frac{V_\infty}{\left(\dfrac{V_\infty^2}{g\sqrt{n^2 - 1}}\right)} = \frac{g\sqrt{n^2 - 1}}{V_\infty} \tag{3.65}$$

Using consistent units in Eq. (3.65), the units of turn rate will fall out to be radians per second. Because this is inconvenient, in terms of a physical interpretation, turn rate is typically presented in degrees per second. If this is desired, simply multiply the result of Eq. (3.65) by 57.3 deg/rad.

It is apparent, looking at the equations for turn radius and rate, that a combination of low airspeed and high g loading will give the best level-turn performance. This is also true for the next two cases: pull-up and pull-down. Equations (3.64) and (3.65) do not tell us anything about the ability of the aircraft to sustain the turn. Sustained turns require that sufficient thrust be available to overcome the drag in the turn. Sustained turning capability can be evaluated by calculating the aircraft drag for a given turning condition and comparing it to the thrust available. Starting with Eq. (1.35), to sustain a turn, the following relationship must be satisfied:

$$T_{R_{\substack{\text{turning}\\\text{flight}}}} = C_D\bar{q}S = \left(C_{D_0} + \frac{C_L^2}{\pi eAR}\right)\bar{q}S = T_{A_{\substack{\text{turning}\\\text{flight}}}}$$

Because the lift coefficient in a turn is directly proportional to load factor,

$$C_L = \frac{nW}{\bar{q}S},$$

it is easy to see that the induced drag significantly increases in a turn and, consequently, the thrust available must be sufficient to match the total drag or the turn will not be sustained. Further analysis of sustained turning performance involves the specific excess power characteristics of the aircraft. This subject is addressed in Refs. 1 and 2.

Example 3.14

Determine the load factor, bank angle, and turn radius for an aircraft in a level turn at a true airspeed of 120 kn and a turn rate of 15 deg/s.

We first convert the airspeed and turn rate to consistent units.

$$V_\infty = 120\ \text{kn} = 120\left(1.69\,\frac{\text{ft/s}}{\text{kn}}\right) = 202.8\ \text{ft/s}$$

$$\omega = 15\ \text{deg/s} = 15\left(\frac{\text{rad}}{57.3\ \text{deg}}\right) = 0.262\ \text{rad/s}$$

From Eq. (3.65),

$$\omega = \frac{g\sqrt{n^2 - 1}}{V_\infty}$$

we can solve for load factor,

$$n = \sqrt{\left(\frac{\omega V_\infty}{g}\right)^2 + 1} = \sqrt{\left(\frac{(0.262)(202.8)}{32.2}\right)^2 + 1} = 1.93$$

Equation (3.64) may be used to find the turn radius,

$$R = \frac{V_\infty^2}{g\sqrt{n^2 - 1}} = \frac{(202.8)^2}{32.2\sqrt{(1.93)^2 - 1}} = 773.8 \text{ ft}$$

An easier alternate approach is to recognize that $V_\infty = R\omega$.

$$R = \frac{V_\infty}{\omega} = \frac{202.8}{0.262} = 774.0 \text{ ft}$$

The bank angle is found from Eq. (3.63).

$$\phi = \cos^{-1}\left(\frac{1}{n}\right) = \cos^{-1}\left(\frac{1}{1.93}\right) = 58.8 \text{ deg}$$

3.9.3 Pull-Ups

Assume an aircraft is in level flight at the bottom of a loop, as shown in Fig. 3.32. At the start of the loop, the pilot pulls back on the stick initiating the loop.

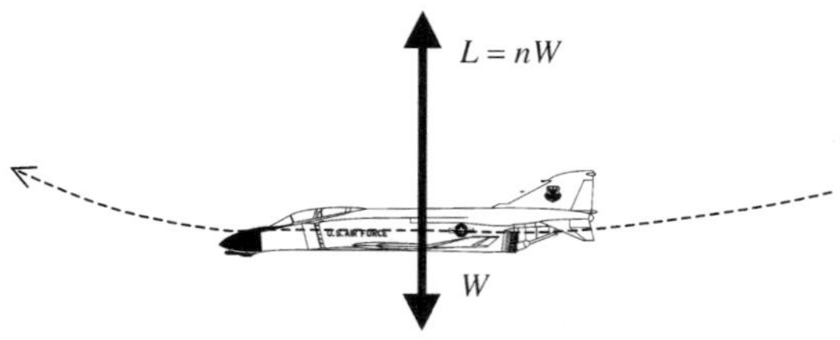

Fig. 3.32 Aircraft in a pull-up.

Again, by applying Newton's 2nd law in the centripetal force direction, expressions for turn rate and radius can be derived as shown, where $F_\perp$ is the force perpendicular to the velocity vector.

$$\Sigma F_\perp = m\left(\frac{V_\infty^2}{R}\right) = \frac{W}{g}\left(\frac{V_\infty^2}{R}\right) = L - W = nW - W = W(n-1)$$

Solving for turn radius and rate, the following is obtained:

$$R = \frac{V_\infty^2}{g(n-1)} \tag{3.66}$$

$$\omega = \frac{V_\infty}{R} = \frac{g(n-1)}{V_\infty} \tag{3.67}$$

Example 3.15

An F-22 is performing a 5-g pull-up at 10,000 ft and 500 kn true airspeed. What is the turn rate and turn radius?

We begin by converting the airspeed to consistent units.

$$V_\infty = 500 \text{ kn} = (500)\left(1.69\frac{\text{ft/s}}{\text{kn}}\right) = 845 \text{ ft/s}$$

We then use Eq. (3.67) to determine the turn rate,

$$\begin{aligned}\omega &= \frac{V_\infty}{R} = \frac{g(n-1)}{V_\infty} = \frac{32.2(5-1)}{845} = 0.152 \text{ rad/s} \\ &= 0.152\left(\frac{57.3 \text{ deg}}{\text{rad}}\right) = 8.71 \text{ deg/s}\end{aligned}$$

and Eq. (3.66) is used to determine the turn radius.

$$R = \frac{V_\infty^2}{g(n-1)} = \frac{(845)^2}{32.2(5-1)} = 5544 \text{ ft}$$

3.9.4 Pull-Downs

The aircraft is now inverted (avoiding negative gs) at the top of a loop as shown in Fig. 3.33. Similar to the pull-up (except that lift is now in the same direction as weight), we can derive expressions for turn rate and radius.

$$\sum F_\perp = m\left(\frac{V_\infty^2}{R}\right) = \frac{W}{g}\left(\frac{V_\infty^2}{R}\right) = L + W = nW + W = W(n+1)$$

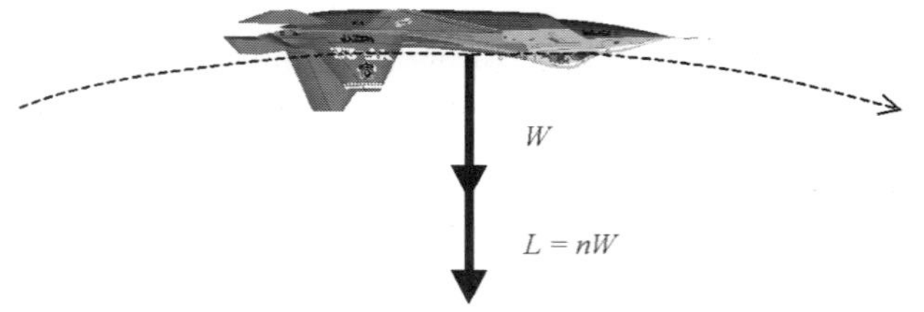

Fig. 3.33 Aircraft in a pull-down.

Solving for turn radius and rate, the following is obtained:

$$R = \frac{V_\infty^2}{g(n+1)} \tag{3.68}$$

$$\omega = \frac{V_\infty}{R} = \frac{g(n+1)}{V_\infty} \tag{3.69}$$

As you might expect, the inverted pull-down gives the best turning performance because the weight of the aircraft is helping perform the turn (sometimes called "God's g").

Example 3.16

An F-22 is at the same conditions (10,000 ft, 500 kn, and pulling 5 g) as in Example 3.14. This time, the aircraft executes a pull-down. Compare the turn rate and turn radius to that achieved for the pull-up.

From Example 3.15, we know the true airspeed in consistent units is 845 ft/s. We use Eq. (3.69) to determine the turn rate,

$$\omega = \frac{V_\infty}{R} = \frac{g(n+1)}{V_\infty} = \frac{32.2(5+1)}{845} = 0.229 \text{ rad/s} = 13.1 \text{ deg/s}$$

and Eq. (3.68) to determine the turn radius.

$$R = \frac{V_\infty^2}{g(n+1)} = \frac{(845)^2}{32.2(5+1)} = 3696 \text{ ft}$$

In comparing these values with those for the pull-up (from Example 3.15), we see that the turn rate is significantly higher and the turn radius is significantly smaller. This leads to the correct conclusion that, with the same load factor and airspeed, a pull-down will result in a quicker, tighter turn than a pull-up.

3.9.5 V-n Diagrams

Most Air Force aircraft have a V-n diagram in the flight manual... the limits are memorized by the pilots. In general, aircraft are limited in their performance capabilities by aerodynamics (for example, stall), structure, and/or engine capabilities. A V-n diagram captures an aircraft's operating envelope. Before presenting the diagram, we will discuss aerodynamic and structural limits.

As we have seen, aircraft are limited aerodynamically by stall. In Sec. 1.5, we developed an equation for stall airspeed—it is repeated in Eq. (3.70), except in a more general form where lift has been replaced by nW (earlier we assumed a load factor of 1, or $L = W$).

$$V_{\text{STALL}} = \sqrt{\frac{2nW}{\rho_\infty S C_{L_{\text{MAX}}}}} \tag{3.70}$$

When pulling gs, an aircraft stalls at a higher airspeed—a V-n diagram will show this trend. Additionally, an aircraft can withstand only a finite amount of g loading until it structurally fails. This g loading is typically denoted by n_{max}. At one airspeed, called the **corner velocity**, an aircraft is operating at the brink of stall, and pulling max g (or maximum load factor), as defined:

$$\text{Corner Velocity} = V^* = \sqrt{\frac{2n_{\text{max}}W}{\rho_\infty S C_{L_{\text{MAX}}}}} \tag{3.71}$$

In our discussion of turns, we said that maximum performance is obtained at a combination of low airspeed and high g loading... this is the corner velocity. It is clear that this equation identifies both an aerodynamic (through the maximum lift coefficient) and a structural (through maximum load factor) limitation for performance.

Figure 3.34 is a typical V-n diagram with velocity (or sometimes Mach number) on the horizontal axis and load factor on the vertical axis.

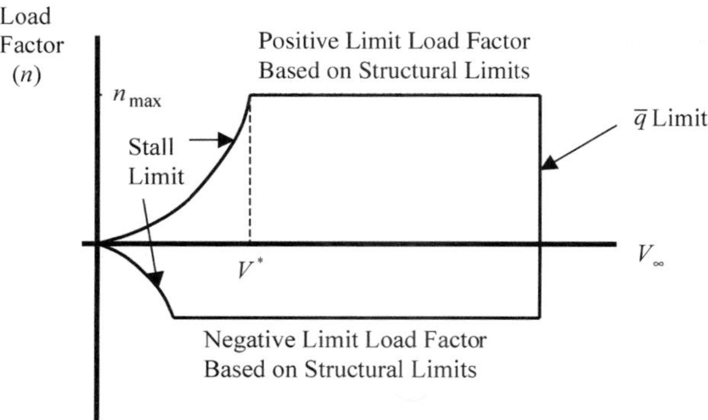

Fig. 3.34 Typical *V-n* diagram.

The following are the key points about an aircraft's V-n diagram:

1) A V-n diagram is good for one weight, altitude, and configuration.
2) The area inside is called the operating envelope in which the aircraft can be safely operated at combination of airspeed and load factor.
3) Refer to the positive and negative stall limits—again, the higher the load factor, the higher the stall airspeed. At a load factor of 1.0 (lift equal to weight), the 1-g stall limit is quickly identified.
4) The corner velocity V^* is shown. It is at the corner of the aircraft's stall limit and structural limit.
5) A positive and negative limit load factor are identified—the aircraft will sustain damage if these are exceeded. Structurally, an aircraft is typically designed to handle more positive than negative gs (as is the pilot).
6) At the high airspeeds the aircraft reaches a "q limit," or "redline." Not as straightforward, this could be an aeroelastic limit or a temperature limit. Something undesirable, in terms of structure and handling qualities, happens if this limit is exceeded.
7) The effects of changes in weight and altitude are shown in Figs. 3.35 and 3.36.
8) If equivalent velocity, V_e, is used on the horizontal axis of the V-n diagram, then the V_e-n diagram will not be affected by altitude changes.

Example 3.17

Determine the velocity and load factor that will result in the highest turn rate and smallest turn radius for a 12,000-lb T-38 at 15,000 ft. Use the T-38 performance charts (V-n diagram).

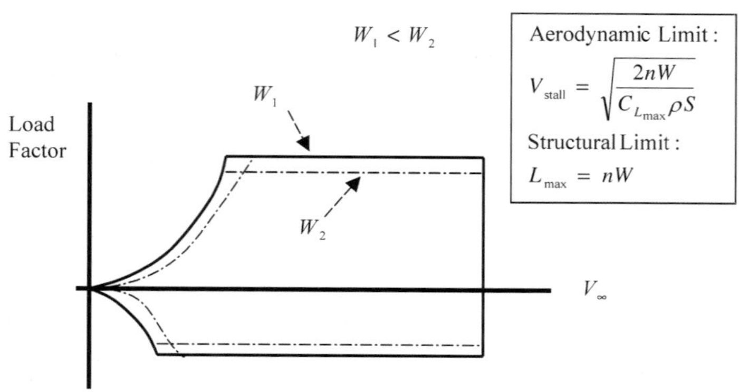

Fig. 3.35 Effect of weight on V-n diagram.

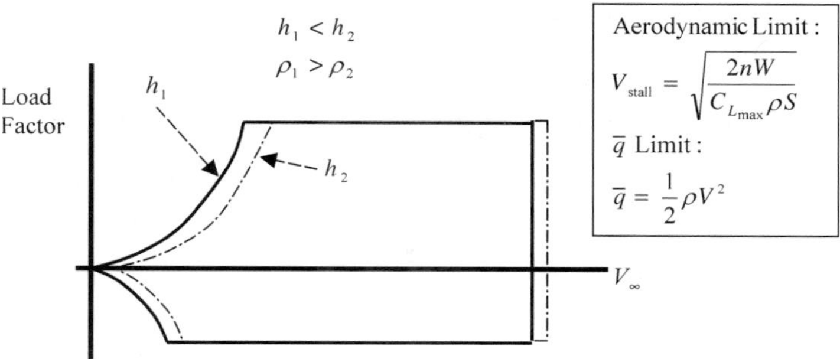

Fig. 3.36 Effect of altitude on *V-n* diagram.

Using the T-38 V-n diagram for 12,000 lb, we see that

$$n_{max} = 6$$

and that the corner velocity for 15,000 ft is 0.85 Mach. We convert the Mach number to ft/s knowing that the speed of sound at 15,000 ft is 1057.4 ft/s.

$$V^* = Ma = 0.85(1057.4 \text{ ft/s}) = 898.8 \text{ ft/s}$$

3.10 Historical Snapshot

3.10.1 Nordic Ski Jumping Aerodynamics

In the early 1990s, the Air Force Academy Aeronautics Department was asked by the U.S. Ski Team to develop an aerodynamic database that would support optimizing the glide ratio of Nordic ski jumpers.[4] The "V" technique, where the jumper holds the tips of the skis apart to form a "V" during the free flight, was just emerging as a preferred method in international competition. A variety of configurations were evaluated in the Air Force Academy Subsonic Wind Tunnel (see Fig. 1.59) using the 1 to 5.5 scale test model presented in Fig. 3.37.

Variable configuration parameters included body camber, arm angle, ski/leg angle, toe out angle, ankle angle, and binding position. Because a ski jumper in flight is subject to the same aerodynamic forces as an aircraft (see Fig. 3.38), the overall objective of the effort was to find a combination of these parameters that would maximize the lift-to-drag ratio of the jumper.

As discussed in Sec. 3.5, the glide ratio is directly proportional to the lift over drag ratio. Wind tunnel results showed that the maximum glide ratio for a jumper with parallel skis (no "V") was approximately 1.1. Figure 3.39 shows that this could be increased to 1.25 with a 22.5 degree "toe out" of the skis, which forms the conventional "V" configuration, as long as the ski/leg angle was 10 deg. This represented a significant performance improvement (approxi-

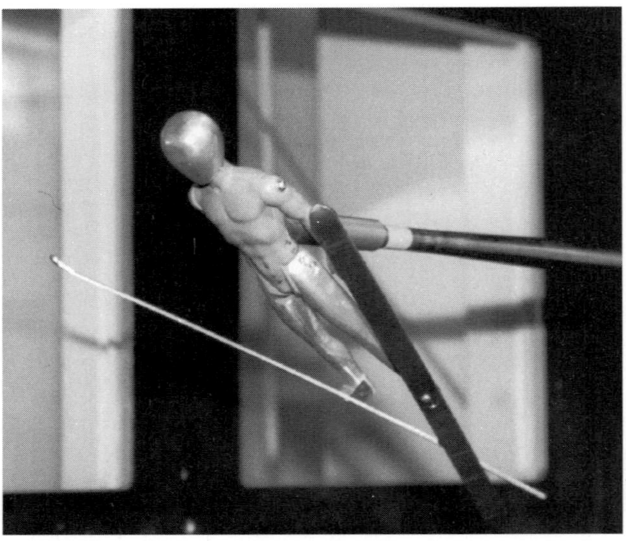

Fig. 3.37 Nordic ski jumper wind tunnel model.

mately 13%) for Olympic competition. Indeed, this is the reason that virtually every Nordic jumper now uses the "V" technique.

Another interesting result found during the testing was the influence of ankle angle, a configuration change where the jumper rolls his ankles "arches up." A 20-deg ankle angle combined with the optimal 10-deg ski/leg angle, and 22.5-deg toe out was found to have the potential to increase L/D to approximately 1.7, which is a huge increase (see Fig. 3.40).

This configuration proved to be difficult physically for the jumpers to maintain because the skis were positioned at a very high value of lift. U.S. Ski Team Olympic athletes were briefed on the results of the wind tunnel testing

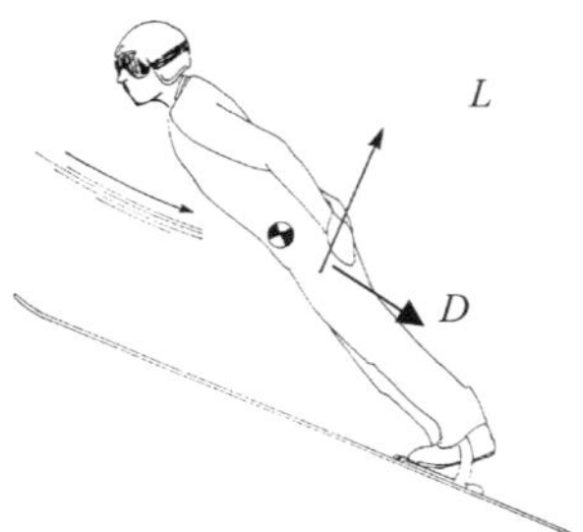

Fig. 3.38 Nordic ski jumper in flight.

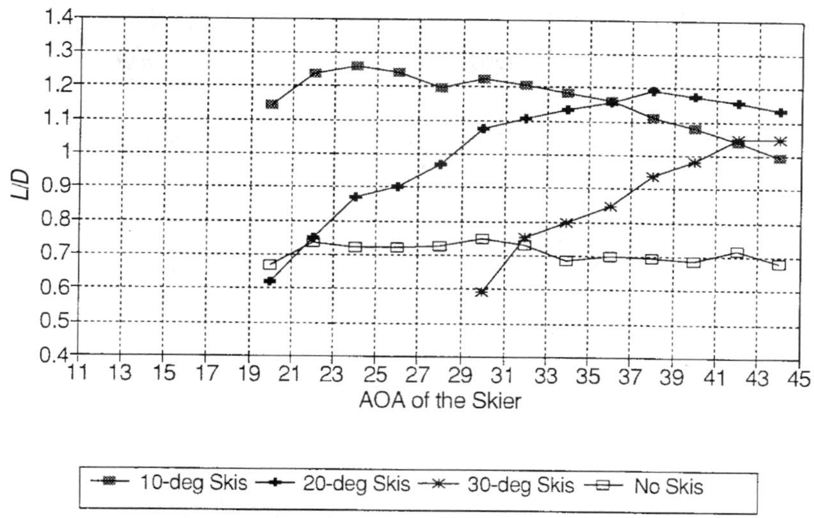

Fig. 3.39 ***L/D* vs Alpha at *M* = 0.2 for 22.5-deg toe out, arms down, no body camber and various ski/leg angles.**

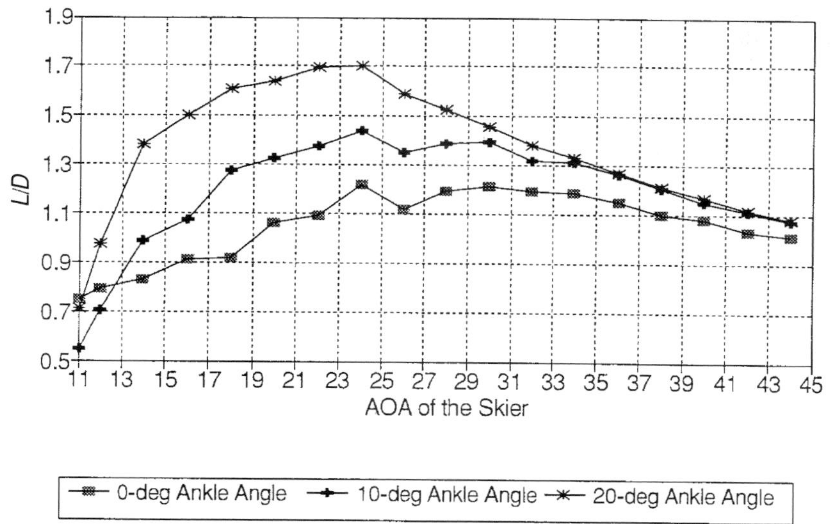

Fig. 3.40 ***L/D* vs Alpha at *M* = 0.2 10-deg skis, 22.5 toe out, arms down for various ankle angles.**

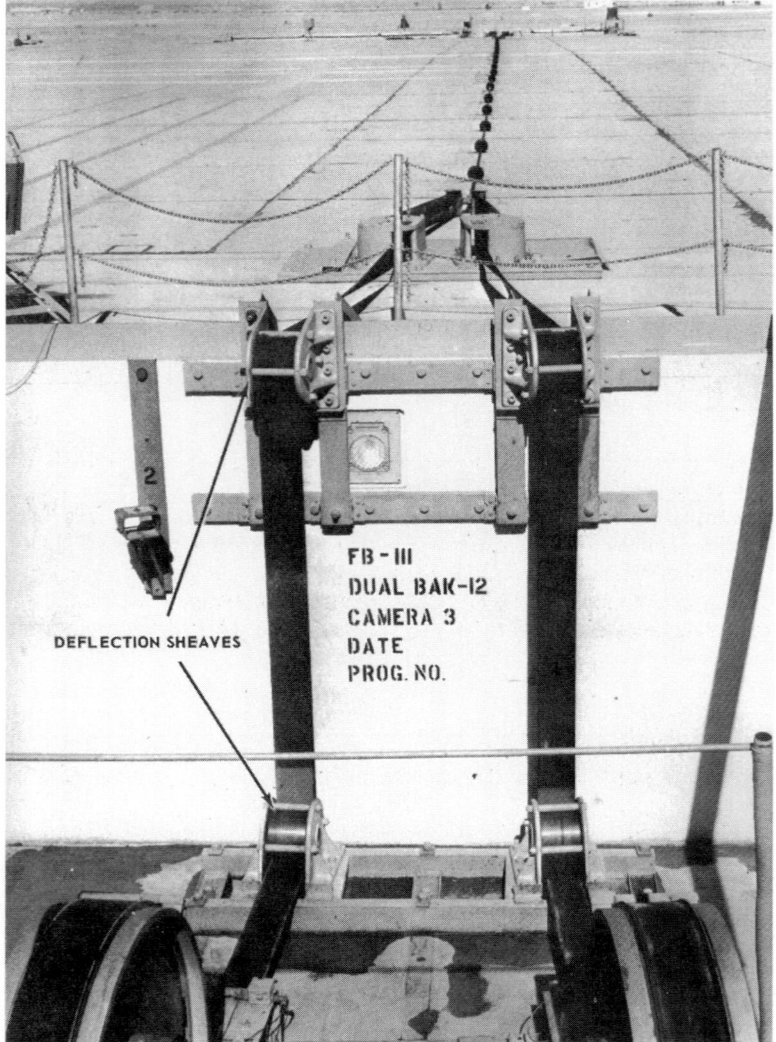

Fig. 3.41 View of Dual BAK-12 from pit looking across the runway.

and the recommended techniques were incorporated into their training in preparation for the 1994 Olympics. The result was the highest finish ever by a U.S. team.

3.10.2 FB-111 Barrier Test Program

Aircraft arresting gear systems (normally called barriers) are common at airports (air bases) with high-performance fighter aircraft. The landing distance

Fig. 3.42 FB-111 test aircraft engaging Dual BAK-12 system.

predicted by Eq. (3.32) is directly dependent on the rolling friction coefficient (μ_r), which is dependent on the condition of the runway surface and the braking capability designed into the aircraft. This, combined with the relatively high landing speeds of fighter aircraft, can result in situations where the aircraft may run out of runway, especially in emergency situations. Most fighter aircraft have tail hooks to engage an arresting gear system typically installed at each end of a runway. An arresting gear system consists of a steel cable stretched

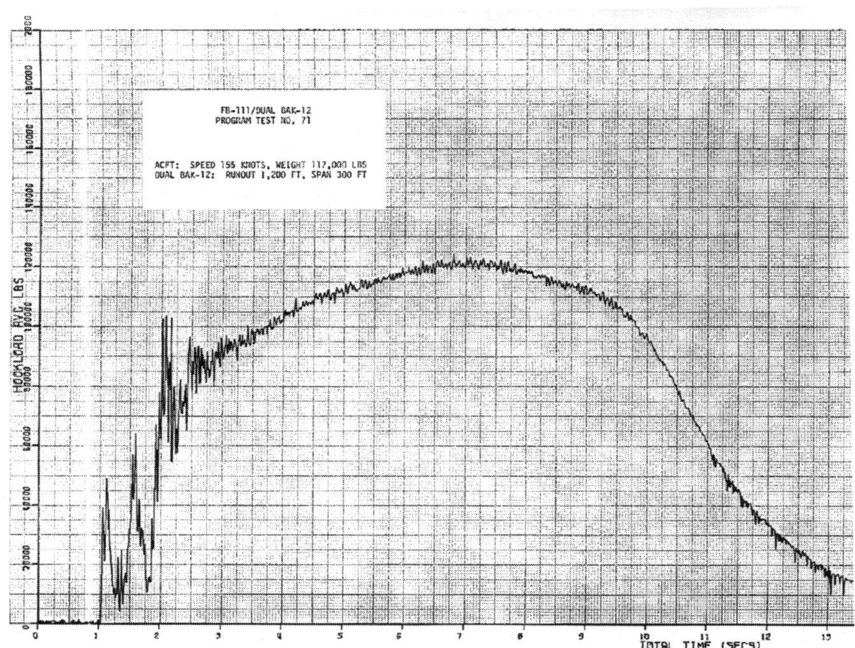

Fig. 3.43 Hookload vs time for FB-111/Dual BAK-12 test.

across the runway and supported approximately 4-in. above the runway surface, which is connected via a nylon tape to large energy absorbers (installed in pits below ground) on each side of the runway. After the aircraft engages the cable with the tail hook, the arresting gear system brings the aircraft to a stop within 900 to 1200 feet (for U.S. Air Force systems).

In the late 1960s, the emergence of the heavy weight FB-111 posed a challenge to existing Air Force arresting gear systems. The FB-111 had a maximum gross weight of 117,000 lb and could conceivably engage the barrier at speeds as high as 170 kn. This energy level exceeded the capacity of existing systems at the time. The Air Force's response was to develop a system (called the Dual BAK-12) with two energy absorbers on each side of the runway to double the energy absorption capability (see Figs. 3.40 and 3.41).

A test program was conducted at the Air Force Flight Test Center, Edwards Air Force Base, in the early 1970s to evaluate the FB-111's compatibility with the Dual BAK-12 system.[5] A total of 73 test engagements were performed, one of which is shown in Fig. 3.42.

Figure 3.43 presents hookload data for one of the heavy weight tests. Notice that the maximum hookload experienced is approximately 120,000 lb, which is below the design limit strength of 147,000 lb.

The Dual BAK-12 system was deployed to FB-111 bases and successfully supported the aircraft until its retirement from active service in the 1990s.

References

[1]Brandt, S. A., Stiles, R. J., Bertin, J. J., and Whitford, R., *Introduction to Aeronautics: A Design Perspective*, AIAA Education Series, AIAA, Reston, VA, 1997.

[2]Blake, W. B., and Laughrey, J. A., "F-15 STOL/MTD Hot Gas Ingestion Wind Tunnel Test Results," AIAA TP 87-1922, June 1987.

[3]Yechout, T. R., and Braman, K. B., "Development and Evaluation of a Performance Modeling Flight Test Approach Based on Quasi Steady-State Maneuvers," NASA Contractor Report 170414, April 1984.

[4]Cutter, D. A., and Yechout, T. R., "Nordic Ski Jumping Aerodynamics," AIAA TP 94-0008, *32nd Aerospace Sciences Meeting*, Reno, NV, Jan. 1994.

[5]Yechout, T. R., "FB-111/Dual BAK-12 Arresting System Compatibility Tests," Air Force Flight Test Center TR TR-71-13, April 1971.

Problems

3.1 A test pilot on the F-16XL test team is told by the contractor's structural engineers to fly at a constant 300 KEAS (knots equivalent airspeed) to get the desired test points for the F-16XL's flutter limits. The position error correction is -10 kn for the F-16 at 300 KIAS.

(a) What indicated airspeed must be flown in knots to get data at 20,000-ft pressure altitude? (319 KIAS) (Assume $V_{cal} = V_e$ and iterate).

(b) What type of velocity will the ground-based radar record when the pilot is at the test point in part a and there is no wind? ($V_t = V_g$) What if there is a wind? ($\bar{V}_g = \bar{V}_t + \bar{V}_w$).

(c) At what true velocity (in knots) must the pilot fly to conduct a test point at 30,000 ft on a standard day? What about at 10,000 ft?

3.2 Dynamic pressure is represented by q in the basic lift and drag equations. Give an expression for q in terms of density (ρ) and true velocity (V). Also, show that q can be expressed in terms of Mach number (M) and pressure (P), and it may be expressed in terms of density at sea level (ρ_{SL}) and equivalent airspeed (V_e).

3.3 On a standard day at 20,000 ft, if V_E is 430 kn, what is the true velocity and Mach number?

3.4 Equation (3.23) is an expression for the acceleration of an aircraft during the ground roll portion of a takeoff. Acceleration (expressed as a function of time) can be integrated twice with respect to time to find the takeoff distance. If the acceleration is assumed constant, the integration can be done by hand. Otherwise, the integration involves calculating the acceleration at specific points in time to do a numeric integration. When the acceleration is assumed constant, it is normally evaluated at $V = V_{LOF}/\sqrt{2}$. For an 11,000 lb_f T-38 with 20 deg of flaps at sea-level standard conditions, a friction coefficient of 0.025, and an optimal rolling C_L, use the maximum thrust available at sea level from the T-38 data to calculate: (Assume $\alpha_T = 0$, zero bank angle, $C_{L_{opt}} = 0.1$).

(a) The ground roll (S_G) for this T-38 by evaluating the acceleration at $V = V_{LOF}/\sqrt{2}$ with no wind.
(b) The ground roll with a 15-mph headwind.
(c) The ground roll with a 15-mph tailwind.

3.5 From Sec. 3.4.3, takeoff roll can be approximated by the following expression—recall that acceleration is assumed constant:

$$S_G = \frac{V_G^2}{2\bar{a}}$$

For a KC-10 with the following characteristics, determine its takeoff ground roll.

$C_L = 0.2083$	$C_D = 0.0543$
$W = 590{,}000\ \text{lb}_\text{f}$	$\mu = 0.025$
$V_{LOF} = 280$ ft/s	$T(@V = 280\ \text{ft/s}) = 39{,}500\ \text{lb}_\text{f}/\text{eng}$
No runway slope	$T(@V = 0) = 49{,}300\ \text{lb}_\text{f}/\text{eng}$
$S = 3647.0\ \text{ft}^2$	sea-level standard day
3 engines	

3.6 Derive the equation for range in unaccelerated gliding flight, $R = H(L/D)$:

(a) Draw a sketch of an aircraft in gliding flight (with flight path angle), and show the three forces acting on it. Sum the forces parallel and perpendicular to the flight path to obtain two equations that are equal to zero (unaccelerated flight),

(b) Obtain a relationship for flight path angle in terms of altitude (H) and range over the ground (R),

(c) Combine these equations to obtain range in terms of the forces acting on the aircraft.

3.7 Use the T-38 data to determine the maximum glide range (in nm) a 12,000-lb T-38 can attain from 10,000 ft to sea level.

3.8 Sketch a thrust required curve and a power required curve and show where $(L/D)_{max}$ occurs on each curve.

3.9 For a jet whose thrust is not a function of velocity, maximum range occurs when $C_L^{1/2}/C_D$ is a maximum. This occurs when

$$\frac{\partial(C_L^{1/2}/C_D)}{\partial C_L} = 0 \qquad \text{Remember: } \mathrm{d}\left(\frac{u}{v}\right) = \frac{v\,\mathrm{d}u - u\,\mathrm{d}v}{v^2}$$

(a) Show that $C_{D_0} = 3KC_L^2$ at $(C_L^{1/2}/C_D)_{max}$. Explain how this point is found on a plot of drag vs velocity. (Hint: When taking the partial derivative indicated, recall the drag polar equation for C_D. Solve the partial derivative for C_{D_0}.)

(b) Show that $C_{D_0} = 1/3KC_L^2$ at $(C_L^{3/2}/C_D)_{max}$. Explain how this point is found on a plot of power vs velocity.

3.10 An F-4 is in level flight at 18,000 ft on a standard day. The aircraft weighs 45,000 lb_f initially and is cruising at 510 kn true airspeed. Assume you can use Eq. (1.30) to find C_{L_α}. (This is not a perfect assumption, as will be seen in Chapter 5).

$$q = 506\ \text{lb}_f/\text{ft}^2 \qquad S = 530\ \text{ft}^2 \qquad AR = 2.78$$
$$C_D = 0.027 + 0.115C_L^2 \qquad \alpha_{0L} = 0^\circ \qquad e = 1$$

(a) What is the F-4's angle of attack?

(b) What will be the F-4's angle of attack if the pilot slows down to cruise at 300 kn true airspeed at 18,000 ft on a standard day?

(c) Why did it increase or decrease?

3.11 Range and endurance. Fill in the blanks:

	$\frac{C_L^{(X)}}{C_D}\|_{max}$	$C_{D_0} = (Y)C_{Di}$	Glider: Max Range or Endurance	Jet: Max Range or Endurance	Velocity
D_{min}					
P_{min}					
$(D/V)_{min}$					
Choose	$X = 1/2$, 1, 3/2	$Y = 1/3$, 1, 3	Emax, Rmax, or N/A	Emax, Rmax, or N/A	Slowest, middle, fastest

3.12 Cruise performance. The following data apply to a turbojet aircraft:

$$C_D = 0.02 + 0.057C_L^2 \quad \text{Initial weight} = 6000 \text{ lb}$$
$$S = 184 \text{ ft}^2 \quad TSFC = c = 1.25/\text{h at sea level}$$

Find the range for this jet at 30,000 ft if the pilot is flying for max range and has 1000 lb of fuel available.

3.13 For the KC-10 with the following characteristics,

$$RF_{opt} = 12{,}100\,\text{NM} \qquad (W/\delta)_{opt} = 1.75 \times 10^6 \text{ lb} \qquad M_{opt} = 0.81$$

determine the following:

(a) the altitude to fly for maximum range at a gross weight of 505,750 lb.

(b) the range of this KC-10 if it burns 100,000 lb of fuel.

(c) the time to fly this range at the Mach number for optimum W/δ at 505,750 lb gross weight.

3.14 For a T-38 at 20,000 ft, answer the following questions:

(a) The subsonic drag polar equation (assuming no variation with Mach) is: $C_D = 0.015 + 0.125C_L^2$. Find the maximum time the aircraft can remain airborne if the pilot flies at maximum endurance. Use the information from Appendix D and assume that the installed military power *TSFC* applies and that the initial and final weights are 10,000 and 8000 lb respectively. At what velocity will this aircraft fly, and what will be its range at this flight condition?

(b) What is L/D_{max} at 30,000 ft? Is it different than L/D_{max} at 20,000 ft? Would there be a difference in maximum endurance time at 30,000 ft compared to 20,000 ft for the weights given in Part (a)? If so, why? (No change in L/D_{max}. Endurance changes).

3.15 The Fairchild A-10 has the following characteristics in flight at sea level.

$$C_{D_0} = 0.032 \quad S = 505.9 \text{ ft}^2 \quad Wt = 28{,}000 \text{ lb}_f$$
$$e = 0.87 \quad AR = 6.5 \quad \text{Max}T_{SL} = 9000 \text{ lb}_f/\text{engine}$$

(a) Find the velocity for maximum climb angle and the climb angle.
(b) Find the climb rate for this climb angle.
(c) Find the velocity for maximum cruise endurance.
(d) Find the velocity for maximum cruise range.

3.16 An F-16 is at 250 kn in a level 60-deg banked turn. Calculate the load factor, turn rate, and turn radius.

3.17 Use the V-n diagrams in the T-38 data to answer the following:
(a) What is the corner velocity for a 9600 lb T-38 at 25,000 ft?
(b) What is n_{max} for a 12,000 lb T-38 at $M = 0.8$ at sea level?
(c) What is n_{max} for a 12,000 lb T-38 at $M = 0.8$ at 35,000 ft?
(d) Why are these two limits different in Parts (b) and (c)?

3.18 The following information is provided for a nonafterburning fighter at sea level, static, standard-day conditions. Answer the following questions:

$$n_{max} = 7.33 \quad T/W = 0.40 \quad C_{L_{max}} = 1.12$$
$$K = 0.12 \quad W/S = 60 \text{ lb}_f/\text{ft}^2 \quad C_{D_0} = 0.015$$

(a) Calculate the corner velocity.
(b) Find the minimum instantaneous turn radius for a level turn.
(c) Find the maximum instantaneous turn rate for a level turn.
(d) What is the aircraft's L/D in this corner velocity turn?
(e) Does this aircraft have sufficient thrust to sustain this turn at corner velocity? (Assume α_t small) What T/W ratio is required to sustain this turn?

4
Aircraft Equations of Motion

To understand how an aircraft behaves, it is essential to develop and understand the aircraft equations of motion (EOM). The EOM consist of the right-hand side of the equations made up of the applied forces and moments, and the left-hand side of the equations providing the aircraft response. The aircraft equations of motion are obtained by applying Newton's 2nd law to a rigid aircraft. Newton's 2nd law states that the summation of the applied forces acting on the aircraft is equal to the time rate of change of linear momentum, and that the summation of the applied moments acting on the aircraft is equal to the time rate of change of angular momentum. To develop these equations of motion, it is necessary to understand the various axis systems. The following section discusses the axis systems used in our development of the EOM.

4.1 Aircraft Axis Systems

In this chapter we will concern ourselves with three axis systems. These include the body axis system fixed to the aircraft, the Earth axis system, which we will assume to be an inertial axis system fixed to the Earth, and the stability axis system, which is defined with respect to the relative wind. Each of these systems is useful in that they provide a convenient system for defining a particular vector, such as, the aerodynamic forces, the weight vector, or the thrust vector.

4.1.1 Body Axis System

The body axis system is fixed to the aircraft with its origin at the aircraft's center of gravity. The x axis is defined out the nose of the aircraft along some reference line. The reference line may be chosen to be the chord line of the aircraft or may be along the floor of the aircraft, as is often the case in large transports. The y axis is defined out the right wing of the aircraft, and the z axis is defined as down through the bottom of the aircraft in accordance with the right-hand rule, as shown in Fig. 4.1. The pilot sits in the body axis system, making it a very useful reference frame. Additionally, it is relatively easy to determine the moments and products of inertia in the body axis system because it is fixed to the aircraft.

4.1.2 Earth Axis System

The Earth axis system is fixed to the Earth with its z axis pointing to the center of the Earth. The x axis and y axis are orthogonal and lie in the local horizontal plane with the origin at the aircraft center of gravity. Often, the x

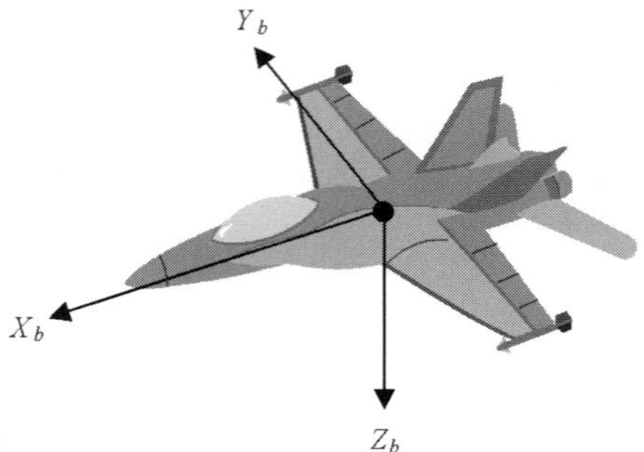

Fig. 4.1 Body axis system.

axis is defined as North and the y axis defined as East. The Earth axis system is assumed to be an inertial axis system for aircraft problems. This is important because Newton's 2nd law is valid only in an inertial system. While this assumption is not totally accurate, it works well for aircraft problems where the aircraft rotation rates are large compared to the rotation rate of the Earth.

4.1.3 Stability Axis System

The stability axis system is rotated relative to the body axis system through the angle of attack. This means that the stability x axis points in the direction of the projection of the relative wind onto the xz plane of the aircraft. The origin of the stability axis system is also at the aircraft center of gravity. The y axis is out the right wing and coincident with the y axis of the body axis system. The z axis is orthogonal and points downward in accordance with the right-hand rule. This is illustrated in Fig. 4.2. The stability axis system is particularly useful in defining the aerodynamic forces of lift and drag.

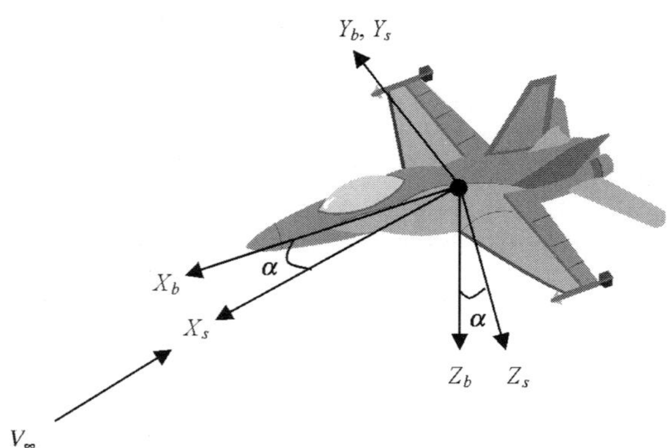

Fig. 4.2 Stability axis system.

4.2 Coordinate Transformations

As stated previously, it is convenient to express certain vectors in a particular coordinate system. For example, the weight vector of the aircraft is conveniently represented in the Earth axis system where there is only a component in the positive z direction because the vector acts toward the center of the Earth, that is,

$$\bar{F}_{\text{Weight}_{\text{Earth}}} = \bar{F}_{\text{Gravity}_{\text{Earth}}} = \bar{F}_{G_E} = \begin{bmatrix} 0 \\ 0 \\ W \end{bmatrix}_{\text{Earth}} = \begin{bmatrix} 0 \\ 0 \\ mg \end{bmatrix}_{\text{Earth}} \tag{4.1}$$

The aerodynamic forces are conveniently displayed in the stability axis, where drag acts in the negative x direction and lift acts in the negative z axis, that is,

$$\bar{F}_{\text{Aero}} = \bar{F}_A = \begin{bmatrix} -D \\ F_{A_y} \\ -L \end{bmatrix}_{\text{Stability}} \tag{4.2}$$

Likewise, the thrust vector can easily be expressed in the body axis as

$$\bar{F}_{\text{Thrust}} = \bar{F}_T = \begin{bmatrix} T\cos(\phi_T) \\ 0 \\ -T\sin(\phi_T) \end{bmatrix}_{\text{Body}} \tag{4.3}$$

where ϕ_T is the angle between the x-body axis and the thrust vector, T. While these equations are conveniently displayed in a particular axis system, they must be all transformed into the same axis system before they can be summed in the equations of motion. As a result, it is very important to understand how to transform a vector from one axis system to another.

4.2.1 Earth Axis to Body Axis Transformation

Transforming a vector from the Earth axis system to the body axis system requires three consecutive rotations about the z axis, y axis, and x axis, respectively. In flight mechanics, the Euler angles are used to rotate the "vehicle carried" Earth axis system into coincidence with the body axis system. The Euler angles are expressed as heading (Ψ), pitch (Θ), and roll (Φ). Euler angles are very useful in describing the orientation of the aircraft with respect to inertial space. The proper order of rotation is illustrated in Fig. 4.3.

The yaw angle, Ψ, is defined as the angle between the projection of the x-body axis onto the horizontal plane and the x axis of the Earth axis system. With the Earth x axis defined as North, the yaw angle is the same as the vehicle heading angle. The pitch angle, Θ, is the angle measured in a vertical plane between the x-body axis and the horizontal plane. The roll angle, Φ, is the angle measured in the yz plane of the vehicle body axis system, between the y-body axis and the horizontal plane. This is the same as the bank angle

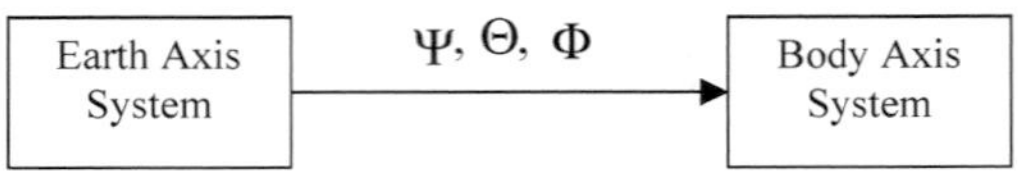

Fig. 4.3 Earth to body transformation.

for a given ψ and θ, and is a measure of the rotation about the x axis to put the aircraft in the desired position from a wing's horizontal condition. The accepted limits on the Euler angles are

$$
\begin{aligned}
0 \leq \Psi &\leq 360 \text{ deg} \\
-90 \text{ deg} \leq \Theta &\leq 90 \text{ deg} \\
-180 \text{ deg} \leq \Phi &\leq 180 \text{ deg}
\end{aligned}
$$

The importance of the sequence of the Euler angle rotations cannot be overemphasized. If the sequence is performed in a different order than ψ, θ, and ϕ, the final result will be incorrect. The following illustrates the transformation of a vector, $\bar{F}_E$, in the Earth axis system into the body axis system, where

$$
\bar{F}_E = X_E \hat{i}_E + Y_E \hat{j}_E + Z_E \hat{k}_E \tag{4.4}
$$

and $\hat{i}_E$, $\hat{j}_E$, and $\hat{k}_E$ are the unit vectors in the Earth axis system. Therefore,

$$
\bar{F}_E = \begin{bmatrix} X_E \\ Y_E \\ Z_E \end{bmatrix}_{\text{Earth}} \tag{4.5}
$$

If we rotate through the yaw angle, ψ, about the z-Earth axis, $\hat{k}_E$, we end up in some intermediate axis system $\hat{i}'$, $\hat{j}'$, and $\hat{k}'$. (See Fig. 4.4.)

The vector in the intermediate axis system ($\hat{i}'$, $\hat{j}'$, and $\hat{k}'$) is:

$$
\bar{F}' = X'\hat{i}' + Y'\hat{j}' + Z'\hat{k}' \tag{4.6}
$$

where

$$
\begin{aligned}
X' &= X_E \cos\Psi + Y_E \sin\Psi \\
Y' &= -X_E \sin\Psi + Y_E \cos\Psi \\
Z' &= Z_E
\end{aligned} \tag{4.7}
$$

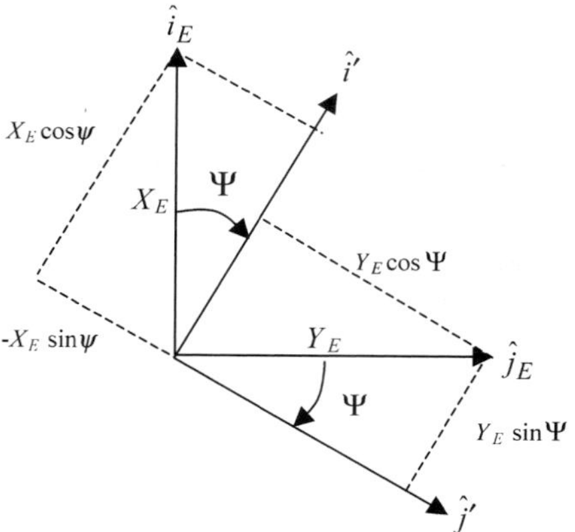

Fig. 4.4 Rotation through Ψ.

so

$$\begin{bmatrix} X' \\ Y' \\ Z' \end{bmatrix} = \underbrace{\begin{bmatrix} \cos\Psi & \sin\Psi & 0 \\ -\sin\Psi & \cos\Psi & 0 \\ 0 & 0 & 1 \end{bmatrix}}_{R_3(\Psi)} \begin{bmatrix} X_E \\ Y_E \\ Z_E \end{bmatrix} \tag{4.8}$$

$$\bar{F}' = R_3(\Psi)\bar{F}_E \tag{4.9}$$

We will now rotate this intermediate vector, $\bar{F}'$, through some pitch angle, Θ, to some other intermediate axis system, $\hat{i}''$, $\hat{j}''$, and $\hat{k}''$ as shown in Fig. 4.5.

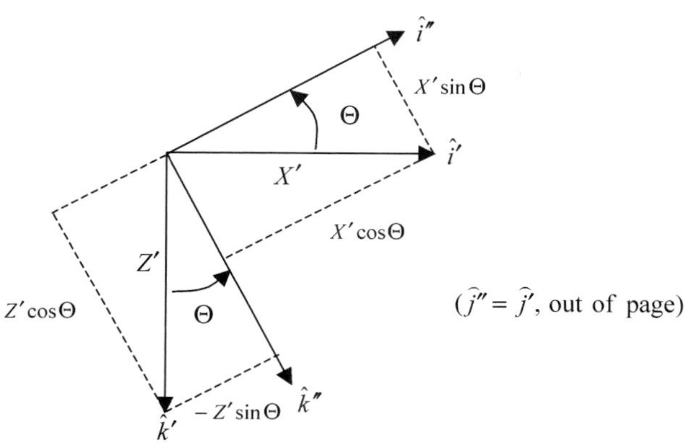

Fig. 4.5 Rotation through Θ.

The vector in the intermediate axis is

$$\bar{F}'' = X''i'' + Y''j'' + Z''k'' \tag{4.10}$$

where

$$\begin{aligned} X'' &= X' \cos\Theta - Z' \sin\Theta \\ Y'' &= Y' \\ Z'' &= X' \sin\Theta + Z' \cos\Theta \end{aligned} \tag{4.11}$$

$$\begin{bmatrix} X'' \\ Y'' \\ Z'' \end{bmatrix} = \underbrace{\begin{bmatrix} \cos\Theta & 0 & -\sin\Theta \\ 0 & 1 & 0 \\ \sin\Theta & 0 & \cos\Theta \end{bmatrix}}_{R_2(\Theta)} \begin{bmatrix} X' \\ Y' \\ Z' \end{bmatrix} \tag{4.12}$$

so,

$$\bar{F}'' = R_2(\Theta)\bar{F}' = R_2(\Theta)R_3(\Psi)\bar{F}_E \tag{4.13}$$

Finally we will rotate the vector, $\bar{F}''$, through some roll angle, Φ, into the body axis system, $\hat{i}$, $\hat{j}$, and $\hat{k}$ as shown in Fig. 4.6.

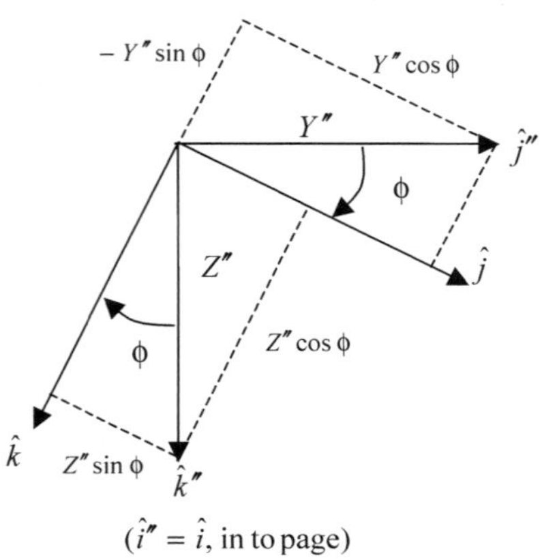

Fig. 4.6 Rotation through Φ.

The vector in the body axis system is

$$\bar{F}_B = X_B\hat{i} + Y_B\hat{j} + Z_B\hat{k} \tag{4.14}$$

where

$$\begin{aligned} X_B &= X'' && (\hat{i}\text{ direction}) \\ Y_B &= Y''\cos\Phi + Z''\sin\Phi && (\hat{j}\text{ direction}) \\ Z_B &= -Y''\sin\Phi + Z''\cos\Phi && (\hat{k}\text{ direction}) \end{aligned} \tag{4.15}$$

$$\begin{bmatrix} X_B \\ Y_B \\ Z_B \end{bmatrix} = \underbrace{\begin{bmatrix} 1 & 0 & 0 \\ 0 & \cos\Phi & \sin\Phi \\ 0 & -\sin\Phi & \cos\Phi \end{bmatrix}}_{R_1(\Phi)} \begin{bmatrix} X'' \\ Y'' \\ Z'' \end{bmatrix} \tag{4.16}$$

so,

$$\bar{F}_B = R_1(\Phi)\bar{F}'' = R_1(\Phi)R_2(\Theta)R_3(\Psi)\bar{F}_E \tag{4.17}$$

Therefore, any vector in the Earth axis system can be transformed into the body axis system using the following transformation.

$$\bar{F}_B = R_1(\Phi)R_2(\Theta)R_3(\Psi)\bar{F}_E \tag{4.18}$$

This transformation is very useful in transforming the weight vector of an aircraft expressed in the Earth axis into the body axis system. As shown earlier in Eq. (4.1), the aircraft's weight vector in the Earth axis system is

$$\bar{F}_{\text{Gravity}_{\text{Earth}}} = \bar{F}_{G_E} = \begin{bmatrix} 0 \\ 0 \\ W \end{bmatrix}_{\text{Earth}} = \begin{bmatrix} 0 \\ 0 \\ mg \end{bmatrix}_{\text{Earth}} \tag{4.19}$$

The vector in the body axis system is easily found using the transformation from Eq. (4.18) as shown in Eq. (4.20).

$$\bar{F}_{\text{Gravity}_B} = \begin{bmatrix} 1 & 0 & 0 \\ 0 & \cos\Phi & \sin\Phi \\ 0 & -\sin\Phi & \cos\Phi \end{bmatrix} \begin{bmatrix} \cos\Theta & 0 & -\sin\Theta \\ 0 & 1 & 0 \\ \sin\Theta & 0 & \cos\Theta \end{bmatrix} \begin{bmatrix} \cos\Psi & \sin\Psi & 0 \\ -\sin\Psi & \cos\Psi & 0 \\ 0 & 0 & 1 \end{bmatrix} \begin{bmatrix} 0 \\ 0 \\ mg \end{bmatrix}_E \tag{4.20}$$

This yields the following weight vector in the body axis system:

$$\bar{F}_{\text{Gravity}_B} = \begin{bmatrix} -mg\sin\Theta \\ mg\sin\Phi\cos\Theta \\ mg\cos\Phi\cos\Theta \end{bmatrix}_B \tag{4.21}$$

In going from the body axis system to the Earth axis system, we go through $-\Phi$, then $-\Theta$, and then $-\Psi$. So the transformation is

$$\bar{F}_E = R_3(-\Psi)R_2(-\Theta)R_1(-\Phi)\bar{F}_B \tag{4.22}$$

For these orthonormal transformation matrices

$$\begin{aligned} R_1(-\Phi) &= R_1^T(\Phi) \\ R_2(-\Theta) &= R_2^T(\Theta) \\ R_3(-\Psi) &= R_3^T(\Psi) \end{aligned} \tag{4.23}$$

where the superscript T indicates the transposition of the matrix.

Therefore,

$$\bar{F}_E = R_3^T(\Psi)R_2^T(\Theta)R_1^T(\Phi)\bar{F}_B \tag{4.24}$$

This is convenient for transforming the acceleration or velocity vector in the body axis system into a vector in the Earth axis system, as might be measured by a radar site tracking the aircraft.

4.2.2 Stability Axis to Body Axis Transformation

It is also important to transform a vector in the stability axis system into the body axis system. This is useful in transforming the aerodynamic forces from their convenient axis system, Eq. (4.7), into the body axis system. This is accomplished by rotating the stability axis system through a positive angle of attack, as shown in Fig. 4.7.

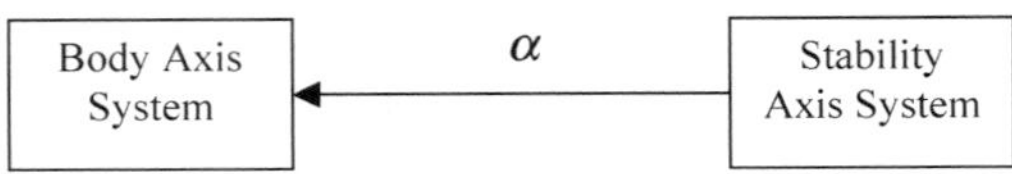

Fig. 4.7 Stability to body axis transformation.

The transformation from stability to body is simply a rotation about the y axis through the angle α, or an $R_2(\alpha)$ transformation, namely,

$$\bar{F}_{\text{Aero}_{\text{Body}}} = R_2(\alpha)\bar{F}_{\text{Aero}_{\text{Stability}}} \tag{4.25}$$

$$\begin{bmatrix} F_{A_x} \\ F_{A_y} \\ F_{A_z} \end{bmatrix}_{\text{Body}} = \underbrace{\begin{bmatrix} \cos\alpha & 0 & -\sin\alpha \\ 0 & 1 & 0 \\ \sin\alpha & 0 & \cos\alpha \end{bmatrix}}_{R_2(\alpha)} \begin{bmatrix} -D \\ F_{A_y} \\ -L \end{bmatrix}_{\text{Stability}} \tag{4.26}$$

Therefore,

$$\begin{aligned} F_{A_{X_B}} &= -D\cos\alpha + L\sin\alpha \\ F_{A_{y_B}} &= F_{A_{y_S}} \\ F_{A_{z_B}} &= -D\sin\alpha - L\cos\alpha \end{aligned} \tag{4.27}$$

4.2.3 *Summary of Axes Transformation*

Figure 4.8 provides a block diagram showing the complete set of transformations from the Earth axis system to the stability axis system. As already stated, the arrow shows a positive transformation from one axis system to another.

4.3 Aircraft Force Equations

This section develops the three aircraft force equations. The force equations consist of aircraft response (in terms of accelerations) on the left-hand side of the equations, and the applied forces on the right-hand side of the equations. Newton's 2nd law states that the time rate of change of linear momentum is equal to the summation of the applied forces acting on the aircraft's center of gravity:

$$\left[\frac{\mathrm{d}(m\bar{V})}{\mathrm{d}t}\right]_{\text{Inertial}} = \bar{F} \tag{4.28}$$

It is extremely important to understand that Newton's 2nd law is only valid in an inertial reference frame. An inertial reference frame is an axis system that is

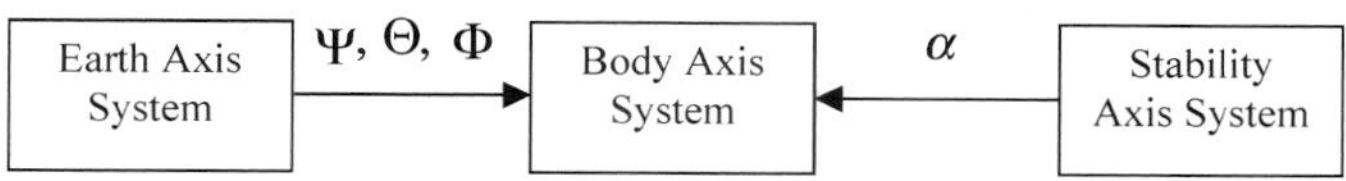

Fig. 4.8 Earth axis system to stability axis system transformation.

fixed in space with no relative motion. For the purposes of analyzing aircraft, it will be assumed that the Earth axis system is an inertial reference frame, even though it does have rotation. It is assumed that the rotation rate of the Earth is small compared to the rotation rates of the aircraft.

4.3.1 Aircraft Response

In developing the response side of the aircraft force equations, several additional assumptions will be made. First, it is assumed that the aircraft is a rigid body. This assumes that the different parts of the aircraft are not moving with respect to each other. The mass of the aircraft is also assumed to be constant, which is reasonable over a relatively short duration of time. This assumption allows Newton's 2nd law to be rewritten as

$$m\left[\frac{\mathrm{d}(\bar{V})}{\mathrm{d}t}\right]_{\text{Inertial}} = m\bar{a}_{\text{Inertial}} = \bar{F} \tag{4.29}$$

While Newton's 2nd law is only valid with respect to an inertial reference frame, the equations can be expressed in the vehicle body axis system. If the equations are expressed in the body axis system, the fact that the system is rotating with respect to an inertial reference frame must be taken into account. This is accomplished using

$$(\bar{a}_{\text{Inertial}})_{Body} = \dot{\bar{V}}_{\text{Body}} + \bar{\omega}_{\text{Body}} \times \bar{V}_{\text{Body}} \tag{4.30}$$

The velocity vector in the body axis system, $\bar{V}_{\text{Body}}$, is defined as

$$\bar{V}_{\text{Body}} = U\hat{i} + V\hat{j} + W\hat{k} \tag{4.31}$$

where U, V, and W are the velocities in the x, y, and z body axes, respectively. The aircraft angular rate in the body axis system, $\bar{\omega}_{\text{Body}}$, is defined as

$$\bar{\omega}_{\text{Body}} = P\hat{i} + Q\hat{j} + R\hat{k} \tag{4.32}$$

where P, Q, and R are the roll, pitch, and yaw rates, respectively, expressed in the body axis. Therefore,

$$(\bar{a}_{\text{Inertial}})_{Body} = \begin{bmatrix} \dot{U} \\ \dot{V} \\ \dot{W} \end{bmatrix}_{\text{Body}} + \begin{vmatrix} \hat{i} & \hat{j} & \hat{k} \\ P & Q & R \\ U & V & W \end{vmatrix}_{\text{Body}} \tag{4.33}$$

This results in

$$(\bar{a}_{\text{Inertial}})_{Body} = \begin{bmatrix} \dot{U} + QW - RV \\ \dot{V} + RU - PW \\ \dot{W} + PV - QU \end{bmatrix}_{\text{Body}} \tag{4.34}$$

Multiplying the inertial acceleration in the body axis system by the mass of the aircraft yields the three force equations

$$m\begin{bmatrix} \dot{U}+QW-RV \\ \dot{V}+RU-PW \\ \dot{W}+PV-QU \end{bmatrix}_{\text{Body}} = \begin{bmatrix} F_x \\ F_y \\ F_z \end{bmatrix}_{\text{Body}} = \bar{F}_{\text{Body}} \tag{4.35}$$

Therefore,

$$\begin{aligned} m(\dot{U}+QW-RV) &= F_x \\ m(\dot{V}+RU-PW) &= F_y \\ m(\dot{W}+PV-QU) &= F_z \end{aligned} \tag{4.36}$$

4.3.2 *Applied Forces*

The previous section developed the left-hand side, or response side, of the force equations. The right-hand side of each equation consists of the applied forces that act on the aircraft. They consist of the gravity forces, the aerodynamic forces, and the thrust forces.

$$\begin{aligned} m(\dot{U}+QW-RV) &= F_{G_x}+F_{A_x}+F_{T_x} \\ m(\dot{V}+RU-PW) &= F_{G_y}+F_{A_y}+F_{T_y} \\ m(\dot{W}+PV-QU) &= F_{G_z}+F_{A_z}+F_{T_z} \end{aligned} \tag{4.37}$$

Because the left-hand sides of the equations were developed in the body axis system, the right-hand side must also be in the body axis system. Therefore, each of the forces must be represented in the body axis system for the previous equations to be valid. The gravity forces, aerodynamic forces, and thrust forces were previously determined in the body axis system in Secs. 4.2.1, 4.2.2, and 4.2, respectively. Therefore, the three force equations in the body axis system are

$$\begin{aligned} m(\dot{U}+QW-RV) &= -mg\sin\Theta+(-D\cos\alpha+L\sin\alpha)+T\cos\Phi_T \\ m(\dot{V}+RU-PW) &= mg\sin\Phi\cos\Theta+F_{A_y}+F_{T_y} \\ m(\dot{W}+PV-QU) &= mg\cos\Phi\cos\Theta+(-D\sin\alpha-L\cos\alpha)-T\sin\Phi_T \end{aligned} \tag{4.38}$$

4.4 Moment Equations

The three moment equations are determined by applying Newton's 2nd law in a manner similar to the three force equations. Newton's 2nd law states that the time rate of change in the angular momentum of the aircraft is equal to the applied moments acting on the aircraft, namely,

$$\left[\frac{\mathrm{d}\bar{H}}{\mathrm{d}t}\right]_{\text{Inertial}} = \bar{M} \tag{4.39}$$

$\bar{H}$ is the angular momentum of the aircraft and is defined as

$$\bar{H} = \bar{r} \times (m\bar{V}) \tag{4.40}$$

Equation (4.39) will be used, along with some simplifying assumptions, to develop the three rotational equations of motion.

4.4.1 Response Side of Moment Equations

A six-step procedure will be used to methodically build up the response side of the three moment equations. This provides both a mathematical and physical insight into the equations.

4.4.1.1 Step 1. The first step is to examine a small elemental mass, dm, of the aircraft that is located at some distance from the aircraft's center of gravity. It will be assumed that the elemental mass is rotating about the aircraft center of gravity with a positive roll rate, pitch rate, and yaw rate (P, Q, and R, respectively). The distance from the center of gravity to the small mass is defined as

$$\bar{r}_{\mathrm{d}m} = x\hat{i} + y\hat{j} + z\hat{k} \tag{4.41}$$

where x, y, and z are the distances in the x, y, and z axes of the body axis system, respectively. This is shown in Fig. 4.9.

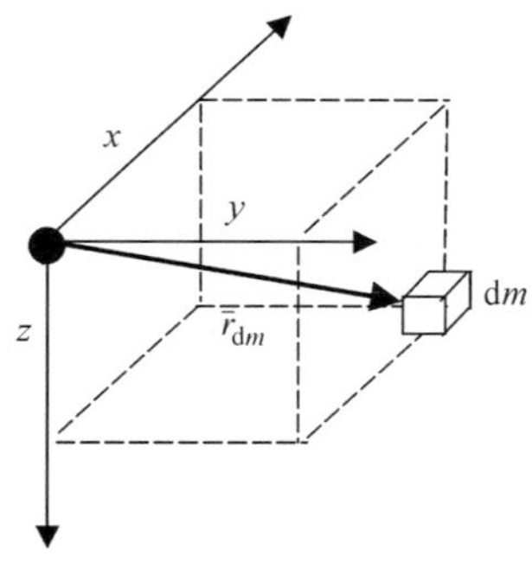

Fig. 4.9 Differential mass in body axis system.

4.4.1.2 Step 2. Next an expression is developed for the velocity of the small mass, dm, solely because of its rotation about the center of gravity. The velocity for the movement of the center of gravity of the aircraft was taken into account in the development of the three force equations. The velocity, $\bar{V}_{dm}$, of the mass relative to the center of gravity is determined using the expression

$$\bar{V}_{dm} = \left[\frac{d\bar{r}_{dm}}{dt}\right]_{Body} + \bar{\omega}_{Body} \times \bar{r}_{dm} \tag{4.42}$$

Because the aircraft was previously assumed to be a rigid body, $\bar{r}_{dm}$ is constant, so

$$\left[\frac{d\bar{r}_{dm}}{dt}\right]_{Body} = 0 \tag{4.43}$$

and therefore,

$$\bar{V}_{dm} = \bar{\omega}_{Body} \times \bar{r}_{dm} \tag{4.44}$$

Mathematically this yields

$$\bar{V}_{dm} = \begin{vmatrix} \hat{i} & \hat{j} & \hat{k} \\ P & Q & R \\ x & y & z \end{vmatrix} \tag{4.45}$$

$$\bar{V}_{dm} = (Qz - Ry)\hat{i} + (Rx - Pz)\hat{j} + (Py - Qx)\hat{k} \tag{4.46}$$

4.4.1.3 Step 3. Next an expression is developed for the linear momentum of dm solely because of its rotation about the center of gravity. The linear momentum is found simply by multiplying the mass times the velocity, namely,

$$\begin{aligned} \text{linear momentum} &= dm\bar{V} \\ &= dm[(Qz - Ry)\hat{i} + (Rx - Pz)\hat{j} + (Py - Qx)\hat{k}] \end{aligned} \tag{4.47}$$

4.4.1.4 Step 4. An expression for the angular momentum of the differential mass, dm, is developed using

$$d\bar{H}_{dm} = \bar{r}_{dm} \times (dm\bar{V}_{dm}) \tag{4.48}$$

Therefore,

$$d\bar{H}_{dm} = \begin{vmatrix} \hat{i} & \hat{j} & \hat{k} \\ x & y & z \\ dm(Qz - Ry) & dm(Rx - Pz) & dm(Py - Qx) \end{vmatrix} \tag{4.49}$$

After carrying out the cross product and regrouping the terms, the three components of the angular momentum are:

$$
\begin{aligned}
\mathrm{d}H_x &= P(y^2 + z^2)\mathrm{d}m - Qxy\,\mathrm{d}m - Rxz\,\mathrm{d}m \\
\mathrm{d}H_y &= Q(x^2 + z^2)\mathrm{d}m - Ryz\,\mathrm{d}m - Pxy\,\mathrm{d}m \\
\mathrm{d}H_z &= R(x^2 + y^2)\mathrm{d}m - Pxz\,\mathrm{d}m - Qyz\,\mathrm{d}m
\end{aligned}
\tag{4.50}
$$

4.4.1.5 Step 5. The next step is to integrate the expressions for the angular momentum of dm over the entire aircraft. Because P, Q, and R are not functions of the mass, they can be taken outside of the integration. Therefore, the three components for the angular momentum of the entire aircraft are

$$
\begin{aligned}
H_x &= \int \mathrm{d}H_x = P\int (y^2 + z^2)\mathrm{d}m - Q\int xy\,\mathrm{d}m - R\int xz\,\mathrm{d}m \\
H_y &= \int \mathrm{d}H_y = Q\int (x^2 + z^2)\mathrm{d}m - R\int yz\,\mathrm{d}m - P\int xy\,\mathrm{d}m \\
H_z &= \int \mathrm{d}H_z = R\int (x^2 + y^2)\mathrm{d}m - P\int xz\,\mathrm{d}m - Q\int yz\,\mathrm{d}m
\end{aligned}
\tag{4.51}
$$

The **moments of inertia** are defined as

$$
\begin{aligned}
I_{xx} &= \int (y^2 + z^2)\mathrm{d}m \\
I_{yy} &= \int (x^2 + z^2)\mathrm{d}m \\
I_{zz} &= \int (x^2 + y^2)\mathrm{d}m
\end{aligned}
\tag{4.52}
$$

The moments of inertia are indications of the resistance to rotation about that axis (that is, I_{xx} indicates the resistance to rotation about the x axis of the aircraft). The **products of inertia** are

$$
\begin{aligned}
I_{xy} &= \int xy\,\mathrm{d}m \\
I_{xz} &= \int xz\,\mathrm{d}m \\
I_{yz} &= \int yz\,\mathrm{d}m
\end{aligned}
\tag{4.53}
$$

The products of inertia are an indication of the symmetry of the aircraft. Substituting the moments and products of inertia into Eq. (4.51) yields

$$
\begin{aligned}
H_x &= PI_{xx} - QI_{xy} - RI_{xz} \\
H_y &= QI_{yy} - RI_{yz} - PI_{xy} \\
H_z &= RI_{zz} - PI_{xz} - QI_{yz}
\end{aligned}
\tag{4.54}
$$

where

$$\bar{H} = H_x\hat{i} + H_y\hat{j} + H_z\hat{k} \tag{4.55}$$

This can also be easily found by applying an expression for angular momentum usually developed in basic physics courses, which is

$$\bar{H} = \bar{I}\bar{\omega} \tag{4.56}$$

where $\bar{I}$ is the aircraft's inertia tensor and $\bar{\omega}$ is the aircraft's angular rate. The inertia tensor for an aircraft is

$$\bar{I} = \begin{bmatrix} I_{xx} & -I_{xy} & -I_{xz} \\ -I_{xy} & I_{yy} & -I_{yz} \\ -I_{xz} & -I_{yz} & I_{zz} \end{bmatrix}_{\text{Body}} \tag{4.57}$$

Therefore,

$$H_B = \begin{bmatrix} I_{xx} & -I_{xy} & -I_{xz} \\ -I_{xy} & I_{yy} & -I_{yz} \\ -I_{xz} & -I_{yz} & I_{zz} \end{bmatrix} \begin{bmatrix} P \\ Q \\ R \end{bmatrix} \tag{4.58}$$

so,

$$\begin{aligned} H_x &= PI_{xx} - QI_{xy} - RI_{xz} \\ H_y &= QI_{yy} - RI_{yz} - PI_{xy} \\ H_z &= RI_{zz} - PI_{xz} - QI_{yz} \end{aligned} \tag{4.59}$$

Note this is the exact same result as Eq. (4.54) that resulted from applying Steps 1–5.

If the aircraft is assumed to have an *xz* plane of symmetry, the I_{xy} and I_{yz} products of inertia are zero. An aircraft has an *xz* plane of symmetry when the left side of the aircraft is a mirror image of the right side about the *xz* plane. The I_{xz} is not necessarily zero because the aircraft is not symmetrical from top to bottom about the *xy* plane and not symmetrical from front to rear about the *yz* plane. These concepts are illustrated in Fig. 4.10. Notice the reflection plane symmetry for I_{xy} and I_{yz} between quadrants I and IV, and II and III. This leads to a zero value for both these products of inertia. Also notice that we do not have reflection plane symmetry for the case of I_{xz}; therefore, it has a non-zero value.

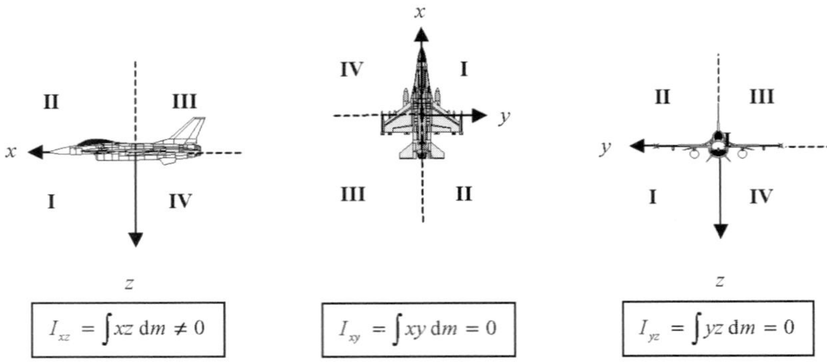

Fig. 4.10 Aircraft products of inertia.

The angular momentum components for the aircraft become

$$\begin{aligned} H_x &= PI_{xx} - RI_{xz} \\ H_y &= QI_{yy} \\ H_z &= RI_{zz} - PI_{xz} \end{aligned} \tag{4.60}$$

or

$$\bar{H} = (PI_{xx} - RI_{xz})\hat{i} + (QI_{yy})\hat{j} + (RI_{zz} - PI_{xz})\hat{k} \tag{4.61}$$

4.4.1.6 Step 6. After the angular momentum vector of the aircraft has been determined, the final step is to take the time rate of change of the angular momentum vector with respect to inertial space but represented in the aircraft body axis system. The same relationship used in developing the acceleration with respect to an inertial reference frame from the force equations can be used, namely,

$$\left[\frac{d\bar{H}}{dt}\right]_{\text{Inertial}} = \left[\frac{d\bar{H}}{dt}\right]_{\text{Body}} + \omega_{\text{Body}} \times \bar{H}_{\text{Body}} \tag{4.62}$$

$$\left[\frac{d\bar{H}}{dt}\right]_{\text{Body}} = \begin{bmatrix} \dot{P}I_{xx} - \dot{R}I_{xz} + P\dot{I}_{xx} - R\dot{I}_{xz} \\ \dot{Q}I_{yy} + Q\dot{I}_{yy} \\ \dot{R}I_{zz} - \dot{P}I_{xz} + R\dot{I}_{zz} - P\dot{I}_{xz} \end{bmatrix}_{\text{Body}} \tag{4.63}$$

Assuming that the mass distribution of the aircraft is constant, such as neglecting fuel slosh, the moments and products of inertia do not change with time, that is, $\dot{I}_{xx}$, $\dot{I}_{yy}$, $\dot{I}_{zz}$, and $\dot{I}_{xz}$ are all zero. Therefore,

$$\left[\frac{d\bar{H}}{dt}\right]_{\text{Body}} = \begin{bmatrix} \dot{P}I_{xx} - \dot{R}I_{xz} \\ \dot{Q}I_{yy} \\ \dot{R}I_{zz} - \dot{P}I_{xz} \end{bmatrix}_{\text{Body}} \tag{4.64}$$

Finally,

$$\omega \times \bar{H}_{\text{Body}} = \begin{vmatrix} i & j & k \\ P & Q & R \\ (PI_{xx} - RI_{xz}) & (QI_{yy}) & (RI_{zz} - PI_{xz}) \end{vmatrix}_{\text{Body}} \tag{4.65}$$

$$= \begin{bmatrix} Q(RI_{zz} - PI_{xz}) - RQI_{yy} \\ R(PI_{xx} - RI_{xz}) - P(RI_{zz} - PI_{xz}) \\ PQI_{yy} - Q(PI_{xx} - RI_{xz}) \end{bmatrix}_{\text{Body}} \tag{4.66}$$

Grouping terms yields

$$\left[\frac{d\bar{H}}{dt}\right]_{\text{Inertial}_{\text{Body}}} = \begin{bmatrix} \dot{P}I_{xx} + QR(I_{zz} - I_{yy}) - (\dot{R} + PQ)I_{xz} \\ \dot{Q}I_{yy} - PR(I_{zz} - I_{xx}) + (P^2 - R^2)I_{xz} \\ \dot{R}I_{zz} + PQ(I_{yy} - I_{xx}) + (QR - \dot{P})I_{xz} \end{bmatrix}_{\text{Body}} \tag{4.67}$$

Therefore, Eq. (4.68) yields the three moment equations of motion in the body axis system, where the left-hand side represents the response of the aircraft and the right-hand side consists of the applied moments.

$$\begin{aligned} &\dot{P}I_{xx} + QR(I_{zz} - I_{yy}) - (\dot{R} + PQ)I_{xz} = L \\ &\dot{Q}I_{yy} - PR(I_{zz} - I_{xx}) + (P^2 - R^2)I_{xz} = M \\ &\underbrace{\dot{R}I_{zz}}_{\substack{\text{angular}\\ \text{acceleration}\\ \text{terms}}} + \underbrace{PQ(I_{yy} - I_{xx})}_{\substack{\text{gyro}\\ \text{precession}\\ \text{terms}}} + \underbrace{(QR - \dot{P})I_{xz}}_{\text{coupling terms}} = N \end{aligned} \tag{4.68}$$

L, M, and N are the rolling moment, pitching moment, and yawing moment, respectively. Unfortunately, the letter L is used to also represent lift. This can be confusing and the reader is advised to carefully check the context of its use in any aeronautical engineering text. Recall that the assumptions made in developing the equations of motion were: the mass of the aircraft is constant, the aircraft is a rigid airframe, the Earth axis system is an inertial reference frame, the mass distribution of the aircraft is constant, and the aircraft has an xz plane of symmetry. It is, therefore, extremely important to realize that these equations are valid for flight conditions where these assumptions are reasonable.

The aircraft response side of Eq. (4.68) can be divided into three types of terms: angular acceleration, gyro precession, and coupling. The angular acceleration terms are the easiest to understand. For example, if we take the rolling moment EOM and assume the gyro precession and coupling terms are negligible, we have

$$\dot{P}I_{xx} = L$$

If a rolling moment (L) is applied to the aircraft (such as with an aileron deflection), this equation predicts that an angular acceleration in roll ($\dot{P}$) will result. If a positive rolling moment is applied, a positive roll angular acceleration will result. For a given applied rolling moment, the larger the moment of inertia, I_{xx}, the smaller the roll angular acceleration. The angular acceleration terms can be thought of as describing the motion that results from the application of torque (the applied moment) to a rotating body with a moment of inertia. For example, if we apply a torque to a flywheel, it experiences an angular acceleration as it spins up.

The gyroscopic precession terms describe precession of the aircraft because of the combination of angular momentum about an axis and an applied moment. For example, consider the rolling moment EOM for the case of a rolling pull-up. We will, for the moment, assume the angular acceleration and coupling terms are negligible along with I_{zz}. This leaves us with

$$-I_{yy}QR = L$$

We next identify the angular momentum term for an aircraft in a pull-up (with positive pitch rate Q) as $I_{yy}Q$. Notice that I_{yy} is the moment of inertia (always positive) about the y axis and Q is the angular velocity about the y axis (positive for a pull-up). Multiplied together, we have angular momentum. We will rewrite this as

$$-R\underbrace{(I_{yy}Q)}_{\text{angular momentum in pitch}} = L$$

If the pilot now applies a positive rolling moment to the aircraft by deflecting the ailerons, this equation predicts that a negative yaw rate (R) must result. The precession terms describe the same type of precession that is experienced by a gyroscope when a torque is applied. We will look at a second example of a gyro precession term for the case of tail dragger propeller aircraft. Consider the pitching moment EOM. We will assume the angular acceleration and coupling terms are negligible along with I_{zz}.

$$R\underbrace{(I_{xx}P)}_{\text{angular momentum in roll}} = M$$

A propeller aircraft, such as a P-51, has a significant amount of angular momentum in roll because of the rotation of the propeller. Because the propeller usually rotates clockwise as seen from the cockpit, the roll angular momentum is positive. With the aircraft on takeoff roll, the pilot will apply a small nose down stick input (negative M) to raise the tail. Our abbreviated equation predicts that a negative yaw rate (R) must result. Skillful pilots are ready for this precession and will apply right rudder to counteract the negative yaw rate.

The coupling terms describe inertial coupling tendency of the aircraft. They can be easily identified by the I_{xz} product of inertia. A nonzero value of I_{xz} indicates inertial nonsymmetry of the aircraft. Many modern high-performance aircraft have a negative I_{xz}, which indicates a larger concentration of mass in the two negative quadrants formed by xz plane of the aircraft. Consider the pitching moment EOM with the gyro precession and applied pitching moment terms assumed negligible and zero yaw rate (R).

$$P^2 I_{xz} = -I_{yy}\dot{Q}$$

We will assume the aircraft is doing a high-speed fly-by and has a negative I_{xz}. If the pilot does a snap roll to either direction, this equation predicts that a positive angular acceleration ($\dot{Q}$) will result. This will cause the aircraft to pitch up and could lead to serious effects if the pilot does not anticipate the pitch up. It can, of course, be counteracted with a nose down stick input. This particular case is referred to as roll coupling, which resulted in the crash of several aircraft in the 1940s and 1950s, before a full understanding of coupling.

4.4.2 Applied Moments

The applied moments consist of the aerodynamic rolling, pitching, and yawing moments, L_A, M_A, and N_A, respectively, and the rolling, pitching, and yawing moments because of thrust, L_T, M_T, and N_T, respectively. There are no moments because of gravity because the weight vector acts through the center of gravity and the moment arms are zero. Also, any moments because of rotating masses (such as jet engines) on or within the aircraft have been neglected. Therefore,

$$\begin{aligned}
\dot{P}I_{xx} + QR(I_{zz} - I_{yy}) - (\dot{R} + PQ)I_{xz} &= L_A + L_T \\
\dot{Q}I_{yy} - PR(I_{zz} - I_{xx}) + (P^2 - R^2)I_{xz} &= M_A + M_T \\
\dot{R}I_{zz} + PQ(I_{yy} - I_{xx}) + (QR - \dot{P})I_{xz} &= N_A + N_T
\end{aligned} \tag{4.69}$$

The makeup of each of these moments will be discussed in detail in subsequent chapters.

4.5 Longitudinal and Lateral-Directional Equations of Motion

The six aircraft equations of motion (EOM) can be decoupled into two sets of three equations. These are the three longitudinal EOM and the three lateral-directional EOM. This is convenient in that it requires only three equations to be solved simultaneously for many flight conditions. For example, an aircraft in wings-level flight with no sideslip and a pitching motion can be analyzed using only the longitudinal EOM because the aircraft does not have any lateral-directional motion.

4.5.1 Longitudinal Equations of Motion

The three longitudinal EOM consist of the x force, z force, and y moment equations, namely,

$$\begin{aligned}
&m(\dot{U} + QW - RV) = -mg\sin\Theta + (-D\cos\alpha + L\sin\alpha) + T\cos\Phi_T \\
&\dot{Q}I_{yy} - PR(I_{zz} - I_{xx}) + (P^2 - R^2)I_{xz} = M_A + M_T \\
&m(\dot{W} + PV - QU) = mg\cos\Phi\cos\Theta + (-D\sin\alpha - L\cos\alpha) - T\sin\Phi_T
\end{aligned} \tag{4.70}$$

One way of thinking of the longitudinal EOM is to picture an aircraft with its xz plane coincident with an xz plane fixed in space. Longitudinal motion consists of those movements where the aircraft would only move within that xz plane, that is, translation in the x direction, translation in the z direction, and rotation about the y axis. In each of these cases, the xz plane of the aircraft would be moving within a xz plane fixed in space. It should be noted that the L in Eq. (4.70) refers to lift and not rolling moment.

4.5.2 Lateral-Directional Equations of Motion

The lateral-directional EOM consist of the y force, x moment, and z moment equations, namely,

$$\begin{aligned}
&\dot{P}I_{xx} + QR(I_{zz} - I_{yy}) - (\dot{R} + PQ)I_{xz} = L_A + L_t \\
&m(\dot{V} + RU - PW) = mg\sin\Phi\cos\Theta + F_{A_y} + F_{T_y} \\
&\dot{R}I_{zz} + PQ(I_{yy} - I_{xx}) + (QR - \dot{P})I_{xz} = N_A + N_T
\end{aligned} \tag{4.71}$$

For any lateral-directional motion the xz plane would move out of some xz plane fixed in space. Translation in the y direction, roll about the x axis, and yaw about the z axis would all cause the xz plane of the aircraft to move out of that arbitrarily fixed xz plane in space.

4.6 Kinematic Equations

In addition to the six force and moment EOM, additional equations are required in order to completely solve the aircraft problem. These additional equations are necessary because there are more than six unknowns due to the presence of the Euler angles in the force equations. Three equations are obtained by relating the three body axis system rates, P, Q, and R to the three Euler rates, $\dot{\Psi}$, $\dot{\Theta}$, and $\dot{\Phi}$. Note that the Euler rates are just the time rate of change of the Euler angles.

To develop the relationship between the body rates and the Euler rates, the following equality must be satisfied because the magnitude of the three body rates must equal the magnitude of three Euler rates. Note that these are vector equations.

$$\bar{\omega}_{\text{Body}} = P\hat{i} + Q\hat{j} + R\hat{k} = \dot{\bar{\Psi}} + \dot{\bar{\Theta}} + \dot{\bar{\Phi}} \tag{4.72}$$

In other words,

$$\sqrt{P^2 + Q^2 + R^2} = \sqrt{\dot{\Psi}^2 + \dot{\Theta}^2 + \dot{\Phi}^2}$$

Each of the three Euler rates can be conveniently displayed in one of the axis systems used in transforming a vector from the Earth axis system to the body axis system. Because $\dot{\bar{\Psi}}$ represents an angular rate about the Z_E or Z' axis

$$\dot{\bar{\Psi}} = \dot{\Psi}\hat{k}_E = \dot{\Psi}\hat{k}' \tag{4.73}$$

This Earth axis heading angular rate is illustrated in Fig. 4.11.

The earth axis system is first rotated about the Z_E (Z') axis through the heading angle (Ψ) into the X'-Y'-Z' axis system as shown in Fig. 4.12. The new X' axis lies directly beneath the x-body axis but is offset by the pitch angle (Θ).

This first interim coordinate system is then rotated about the Y' axis through the pitch angle (Θ) into the X''-Y''-Z'' axis system as shown in Fig. 4.13. Note that because the Y' axis is the axis of rotation, the new Y'' axis is the same as

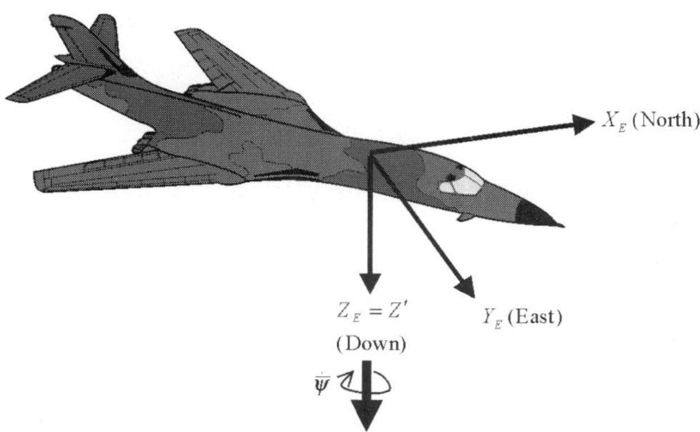

Fig. 4.11 Illustration of heading angular rate in Earth axis system.

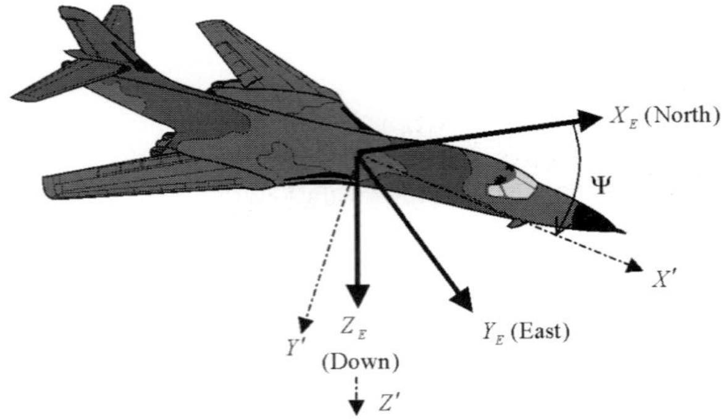

Fig. 4.12 X'-Y'-Z' interim coordinate system.

Y'. $\dot{\bar{\Theta}}$ then represents an angular rate about the Y' or Y'' axis. Mathematically, this can be summarized as

$$\dot{\bar{\Theta}} = \dot{\Theta}\hat{j}' = \dot{\Theta}\hat{j}'' \tag{4.74}$$

This second interim coordinate system aligns the X'' axis with the body x-axis as shown in Fig. 4.14.

This second interim coordinate system is then rotated about the X'' axis through the roll angle (Φ) into the body axis system as shown in Fig. 4.15. $\dot{\bar{\Phi}}$ then represents an angular rate about the X'' or X_B axis. Mathematically, this can be summarized as

$$\dot{\bar{\Phi}} = \dot{\Phi}\hat{i}'' = \dot{\Phi}\hat{i} \tag{4.75}$$

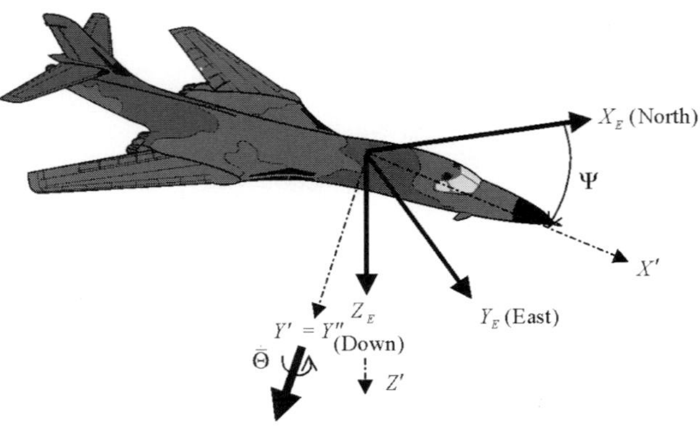

Fig. 4.13 Illustration of pitch attitude angular rate.

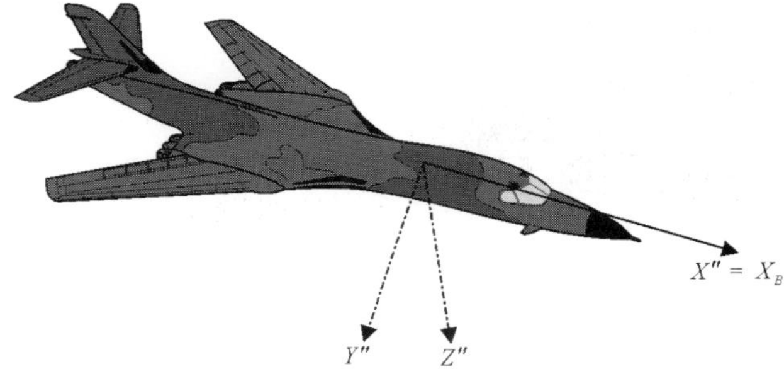

Fig. 4.14 X'-Y''-Z'' **interim coordinate system.**

By using the transformations discussed in Sec. 4.2.1, each of these angular rate vectors can be transformed into the body axis system. Therefore,

$$\bar{\omega}_{\text{Body}} = P\hat{i} + Q\hat{j} + R\hat{k} = R_1(\Phi)R_2(\Theta)\dot{\bar{\Psi}} + R_1(\Phi)\dot{\bar{\Theta}} + \dot{\bar{\Phi}} \tag{4.76}$$

The previous relationship is true for the following reasons. To transform $\dot{\bar{\Psi}}$ from the $\hat{k}'$ axis to the body axis system requires a positive rotation through Θ, followed by a positive rotation through Φ. To transform $\dot{\bar{\Theta}}$ from the $\hat{j}''$ axis to the body axis system requires a positive rotation through Φ only. Finally, $\dot{\bar{\Phi}}$ is already represented in the body axis system, so no transformation is necessary. Mathematically, Eq. (4.76) is carried out as shown in Eqs. (4.77–4.79).

$$\begin{bmatrix} P \\ Q \\ R \end{bmatrix} = \begin{bmatrix} 1 & 0 & 0 \\ 0 & \cos\Phi & \sin\Phi \\ 0 & -\sin\Phi & \cos\Phi \end{bmatrix} \begin{bmatrix} \cos\Theta & 0 & -\sin\Theta \\ 0 & 1 & 0 \\ \sin\Theta & 0 & \cos\Theta \end{bmatrix} \begin{bmatrix} 0 \\ 0 \\ \dot{\Psi} \end{bmatrix} + \begin{bmatrix} 1 & 0 & 0 \\ 0 & \cos\Phi & \sin\Phi \\ 0 & -\sin\Phi & \cos\Phi \end{bmatrix} \begin{bmatrix} 0 \\ \dot{\Theta} \\ 0 \end{bmatrix} + \begin{bmatrix} \dot{\Phi} \\ 0 \\ 0 \end{bmatrix} \tag{4.77}$$

$$\begin{bmatrix} P \\ Q \\ R \end{bmatrix} = \begin{bmatrix} 1 & 0 & 0 \\ 0 & \cos\Phi & \sin\Phi \\ 0 & -\sin\Phi & \cos\Phi \end{bmatrix} \begin{bmatrix} -\sin\Theta\dot{\Psi} \\ 0 \\ \cos\Theta\dot{\Psi} \end{bmatrix} + \begin{bmatrix} 0 \\ \cos\Phi\dot{\Theta} \\ -\sin\Phi\dot{\Theta} \end{bmatrix} + \begin{bmatrix} \dot{\Phi} \\ 0 \\ 0 \end{bmatrix} \tag{4.78}$$

$$\begin{bmatrix} P \\ Q \\ R \end{bmatrix} = \begin{bmatrix} -\sin\Theta\dot{\Psi} + \dot{\Phi} \\ \sin\Phi\cos\Theta\dot{\Psi} + \cos\Phi\dot{\Theta} \\ \cos\Phi\cos\Theta\dot{\Psi} - \sin\Phi\dot{\Theta} \end{bmatrix} \tag{4.79}$$

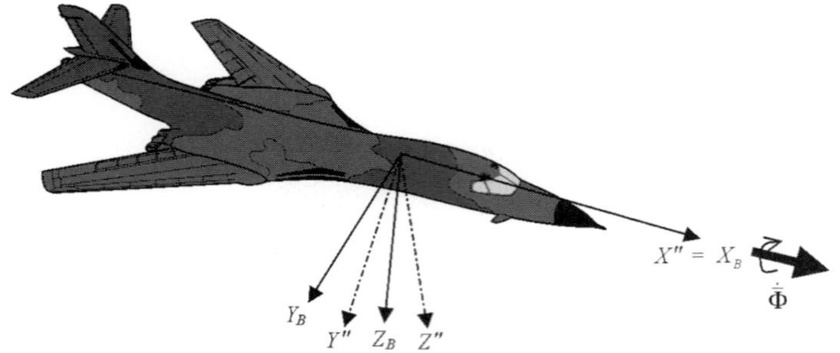

Fig. 4.15 Illustration of roll attitude angular rate.

Therefore, the three kinematic equations are

$$
\begin{aligned}
P &= -\sin\Theta\dot{\Psi} + \dot{\Phi} \\
Q &= \sin\Phi\cos\Theta\dot{\Psi} + \cos\Phi\dot{\Theta} \\
R &= \cos\Phi\cos\Theta\dot{\Psi} - \sin\Phi\dot{\Theta}
\end{aligned}
\tag{4.80}
$$

Examples are provided to clearly illustrate results from the kinematic equations.

Example 4.1

An aircraft has the following Euler angles and Euler rates:

$$
\begin{aligned}
\Psi &= 0 \text{ deg} \quad & \dot{\Psi} &= 10 \text{ deg/s} \\
\Theta &= 0 \text{ deg} \quad & \dot{\Theta} &= 0 \text{ deg/s} \\
\Phi &= 90 \text{ deg} \quad & \dot{\Phi} &= 0 \text{ deg/s}
\end{aligned}
$$

Applying Eq. (4.80) yields:

$$
\begin{aligned}
P &= 0 \text{ deg/s} \\
Q &= 10 \text{ deg/s} \\
R &= 0 \text{ deg/s}
\end{aligned}
$$

Note that for this flight condition, a 10 deg/s Euler yaw rate, $\dot{\Psi}$, is felt by the pilot as a 10 deg/s pitch rate, Q.

Example 4.2

An aircraft has the following Euler angles and Euler rates:

$$\Psi = 0 \text{ deg} \qquad \dot{\Psi} = 0 \text{ deg/s}$$
$$\Theta = 0 \text{ deg} \qquad \dot{\Theta} = 20 \text{ deg/s}$$
$$\Phi = 90 \text{ deg} \qquad \dot{\Phi} = 0 \text{ deg/s}$$

Applying Eq. (4.80) yields

$$P = 0 \text{ deg/s}$$
$$Q = 0 \text{ deg/s}$$
$$R = -20 \text{ deg/s}$$

Note that for this flight condition, a 20 deg/s Euler pitch rate, $\dot{\Theta}$, is felt by the pilot as a -20 deg/s yaw rate, R.

4.7 Historical Snapshot—Genesis 2000 Flight Simulator

In the late 1980s, Veda, Incorporated, of Lexington Park, Maryland, and the U.S. Air Force Academy Department of Aeronautics developed an engineering flight simulator specifically tailored to support educational requirements.[1] The simulator was named the Genesis 2000 and solved the full six degree-of-freedom EOM Eqs. (4.38) and (4.69) in nearly real time to continuously represent almost any mission. The delay time between a pilot input and representation of aircraft motion was less than 0.12 s based on available computer speed at that

Fig. 4.16 Genesis 2000 flight station.

time. The fixed base flight station included an outside visual display complete with heads-up display (HUD), a display for flight instruments, and standard cockpit controls including a sidestick. Figure 4.16 shows the flight station.

A typical expansion of the applied force and moment side of each EOM included the influences of angle of attack, sideslip, elevator deflection, aileron deflection, rudder deflection, differential tail, speedbrakes, flaps, and spoilers through the use of appropriate derivatives, as will be discussed in Chapter 5. A key aspect of the Genesis 2000 was easy and quick access to each individual derivative. This allowed flight evaluation of the derivative's contribution to the overall handling qualities (discussed in Chapter 7) of an aircraft. To do this, another key feature of the Genesis 2000 was incorporation of specific mission tasks such as approach and landing, air-to-ground tracking, and air-to-air tracking. The system was also used by the Air Force Test Pilot School, the U.S. Naval Academy, and the British Empire Test Pilot School throughout the 1990s. At the Air Force Academy, the system was upgraded in 2000 to incorporate current computer technology.

Reference

[1]Russell, J. H., Mouch, T. N., and Yechout, T. R., "Integration of Flight Simulation Into the Undergraduate Design Experience," AIAA TP 90-3263, *AIAA/AHS/ASEE Aircraft Design Systems and Operations Conference*, Dayton, OH, Sept. 1990.

Problems

4.1 Consider the T-37 at the following Euler angles:

$$\Psi = 90 \text{ deg} \qquad \Theta = +10 \text{ deg} \qquad \Phi = +10 \text{ deg}$$

Describe the aircraft attitude and transform the weight force through these angles to the body axis system. The gross weight is 6600 lb_f.

4.2 A T-37 is executing a loop at the following conditions:

$$\text{Euler angles: } \Psi = 0 \text{ deg}, \Theta = 30 \text{ deg}, \Phi = 0 \text{ deg}$$

The pilot observes a pure pitch rate at a constant velocity in the body axis system:

$$\bar{\omega}_B = \begin{Bmatrix} 0 \\ 0.1 \\ 0 \end{Bmatrix}_B \text{rad/s} \qquad \bar{V}_B = \begin{Bmatrix} 200 \\ 0 \\ 0 \end{Bmatrix}_B \text{ft/s}$$

What is the acceleration in the Earth-fixed reference system?

4.3 Given the following vectors, find the inertial acceleration in the body axis system:

$$\bar{a}_I = \dot{\bar{V}}_B + \bar{\omega}\, x\, \bar{V}_B$$

$$\dot{\bar{V}}_B = \begin{Bmatrix} 10 \\ 0 \\ 0 \end{Bmatrix} \frac{\text{ft}}{\text{s}^2} \qquad \bar{\omega} = \begin{Bmatrix} 0 \\ 0 \\ 0.3 \end{Bmatrix} \frac{\text{rad}}{\text{s}} \qquad \bar{V}_B = \begin{Bmatrix} 300 \\ 0 \\ 0 \end{Bmatrix} \frac{\text{ft}}{\text{s}}$$

4.4 Express all forces (namely, weight, aerodynamic, and thrust forces) for sea-level standard day at military thrust on the T-37 in its most convenient axis system. Assume the thrust lines are parallel to the longitudinal axis and in the plane of the CG. The aircraft weighs 5500 lb_f and each engine is delivering 700 lb_f of thrust.

4.5 How many degrees of freedom does an aircraft have? How many are translational and how many are rotational?

4.6 What forces and moments contribute to the pitching moment equation for a conventional aircraft? Which ones do we generally ignore?

5
Aircraft Static Stability

The static stability of an aircraft is generally the first type of stability that a designer evaluates in a new aircraft configuration. Static stability criteria for the three rotational modes of the aircraft (pitch, roll, and yaw) must be considered individually. Accepted design practice for most aircraft has been to achieve some degree of static stability for each of these rotational degrees of freedom. A tradeoff typically exists between static stability and maneuverability. A high degree of static stability would normally result in an aircraft that was easy to fly but that had low maneuverability (trainer aircraft, for example). Fighter aircraft designs would incorporate low levels of static stability so that enhanced maneuvering capabilities could be achieved. However, with the advent of high authority fly-by-wire control systems in aircraft such as the F-16, the Space Shuttle, and the X-29, acceptable levels of both static stability and maneuverability can be achieved. Some degree of static instability in the basic airframe design is intentionally incorporated in these aircraft to achieve enhanced maneuvering capabilities, while the fly-by-wire system provides automatic control inputs to achieve the appearance of static stability from the pilot perspective. Static stability concepts provide a first step in achieving a design with acceptable flying qualities.

5.1 Static Stability Overview

Static stability is generally defined as the initial tendency of an airplane, following a perturbation from a steady-state flight condition, to develop aerodynamic forces or moments that are in a direction to return the aircraft to the steady-state flight condition. This somewhat complex definition can be simply illustrated with an example. If each of the balls in Fig. 5.1 begin in steady-

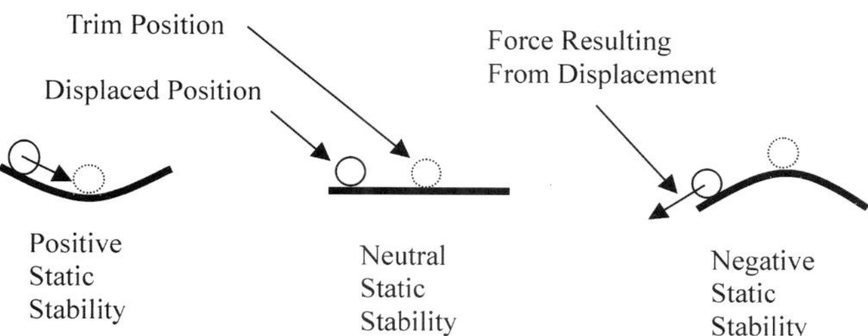

Fig. 5.1 Example of positive static stability, neutral static stability, and negative static stability.

state equilibrium or a trimmed condition (as illustrated by the dotted ball), then the direction of the resulting force that develops after the ball is perturbed from the trim condition determines the type of static stability the system has. In the first case, the ball is perturbed from the bottom of the bowl, and the resulting force tends to return it to the steady-state condition; thus, the situation demonstrates positive static stability or simply static stability. In the second case, when the ball is perturbed on a flat surface, no restoring force develops and the situation demonstrates neutral static stability. The third case illustrates negative static stability or static instability because when the ball is perturbed, the resulting force tends to make the ball further diverge from the trimmed position.

For an aircraft, static stability is generally evaluated relative to a steady-state trimmed flight condition. If the aircraft is perturbed or displaced from the trimmed flight condition with a gust, for example, then the initial aerodynamic moment that results will determine the type of static stability that the aircraft has. As illustrated in Fig. 5.2, if a gust produces a perturbation from the trimmed angle of attack, and an aerodynamic moment results that would rotate the aircraft back toward the trimmed condition, then we have positive static stability. If no aerodynamic moment results, we have neutral static stability. Finally, if an aerodynamic moment results that tends to increase the perturbation, we have negative static stability or static instability. Because pitching or longitudinal motion is depicted Fig. 5.2, we would refer to the aircraft as having longitudinal static stability, neutral longitudinal static stability, or negative longitudinal static stability.

An arrow in flight is another example of static stability. An arrow with fins on the back will always produce an aerodynamic restoring moment when perturbed from its trimmed angle of attack (which is normally near zero). That is why an arrow is very stable when flying through the air. However, removal of the fins will result in a statically unstable situation where the arrow tumbles.

5.2 Stability, Control Power, and Cross-Control Derivatives, and Control Deflection Sign Convention

The first step in evaluating the static stability of an aircraft involves developing an expression for the applied aerodynamic forces and moments that act

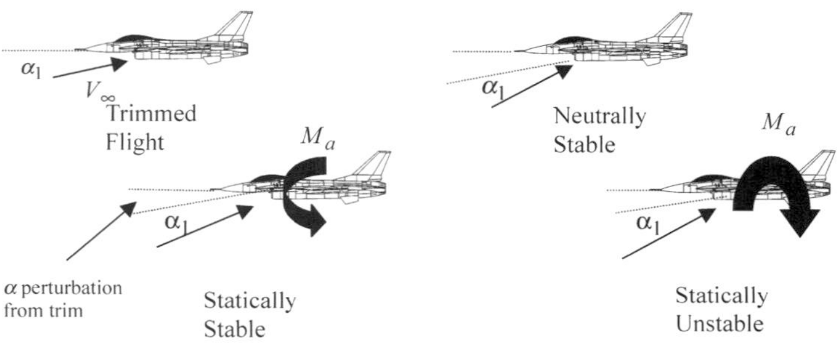

Fig. 5.2 Examples of an aircraft in perturbed flight.

on the aircraft. These were developed in general form in Secs. 4.3.2 and 4.4.2. To expand terms such as M_A (the aerodynamic pitching moment) for stability analysis, the stability axis is used along with a Taylor series expansion. We know that M_A can be expressed in terms of the pitching moment coefficient as

$$M_A = C_m(\bar{q}S\bar{c}) \tag{5.1}$$

and

$$L_A = C_l\bar{q}Sb \tag{5.2}$$

$$N_A = C_n\bar{q}Sb \tag{5.3}$$

C_m is then expressed with a first-order Taylor series as

$$C_m = C_{m_0} + C_{m_\alpha}\alpha + C_{m_{\delta_e}}\delta_e + C_{m_{i_h}}i_h \tag{5.4}$$

because C_m varies with α, δ_e, and i_h for steady-state trimmed analysis, α is of course, angle of attack, δ_e is the elevator deflection, and i_h is the incidence angle of the horizontal stabilizer (the angle between the horizontal stabilizer chord line and the fuselage reference line or x-body axis, defined as **positive** for leading edge up deflections). C_{m_0} is the value of C_m when α, δ_e, and i_h are equal to zero as illustrated in Fig. 5.3. Additional terms such as $C_{m_q}q$ will be added to this Taylor series when we consider dynamic stability but are not needed now for our analysis of static stability.

In a similar manner,

$$C_l = C_{l_0} + C_{l_\beta}\beta + C_{l_{\delta_a}}\delta_a + C_{l_{\delta_r}}\delta_r \tag{5.5}$$

and

$$C_n = C_{n_0} + C_{n_\beta}\beta + C_{n_{\delta_a}}\delta_a + C_{n_{\delta_r}}\delta_r \tag{5.6}$$

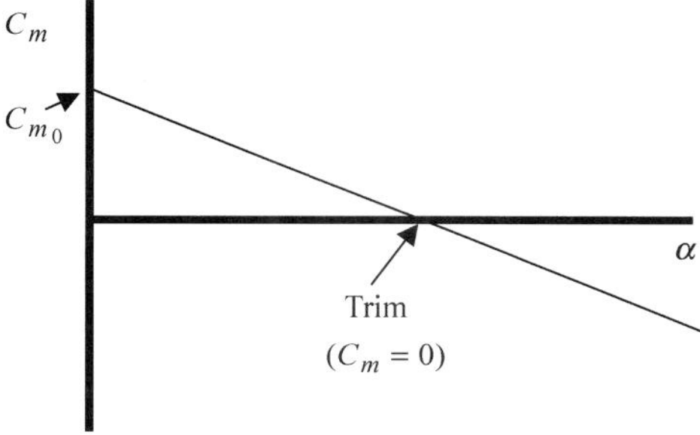

Fig. 5.3 Typical pitching moment coefficient vs angle of attack characteristics.

because C_l and C_n are a function of β, δ_a and δ_r. β is the sideslip angle (angle between the relative wind and the x-body axis in the xy plane—see Fig. 5.26), δ_a is the aileron deflection, and δ_r is rudder deflection. C_{l_0} and C_{n_0} are the values of C_l and C_n respectively when β, δ_a and δ_r are equal to zero. For symmetrical aircraft, C_{l_0} and C_{n_0} are usually zero.

C_{m_α}, C_{l_β}, and C_{n_β} are referred to as **stability derivatives**. They represent the change in the applicable moment coefficient with respect to the change in direction of the relative wind (α for longitudinal motion, β for lateral and directional motion). We will see that the sign of each of these derivatives will determine the longitudinal, lateral, and directional static stability, respectively, of an aircraft. For example, Fig. 5.3 presents a negative slope for C_m vs angle of attack characteristics. This slope is the change in C_m divided by the change in angle of attack, or simply the stability derivative C_{m_α}. With C_{m_α} negative, notice that a gust that causes a perturbation in trim angle of attack to a slightly lower value will result in a positive aerodynamic pitching moment, which gives the aircraft an initial tendency back toward the trim condition. Likewise, a gust that causes a perturbed increase in angle of attack will result in a negative pitching moment, which also gives the aircraft an initial tendency back toward the trim condition. Thus, a negative C_{m_α} can be seen as a requirement for longitudinal static stability.

$C_{m_{\delta_e}}$, $C_{m_{i_h}}$, $C_{l_{\delta_a}}$, and $C_{n_{\delta_r}}$ are referred to as **primary control derivatives** or simply **control powers**. For example, $C_{m_{\delta_e}}$ is the longitudinal control power or longitudinal control derivative that defines the change in C_m that results from a change in elevator deflection (δ_e). The higher the absolute value of a control derivative, the more moment is generated for a given control deflection. One could think of this as higher control sensitivity for a given moment of inertia.

Primary control derivatives are defined using the following **sign convention** to define a **positive** control surface deflection:

1) **Positive elevator deflection** (δ_e) is **trailing edge down** (TED).
2) **Positive incidence angle** of the horizontal stabilizer (i_h) is **leading edge up** (LEU).
3) **Positive aileron deflection** is **TED** on either aileron, and a composite aileron deflection (δ_a) is defined as the difference between the left aileron deflection and the right aileron deflection divided by two. A formula will help clarify this definition:

$$\delta_a = \frac{1}{2}(\delta_{a_{\text{left}}} - \delta_{a_{\text{right}}})$$

4) **Positive rudder deflection** (δ_r) is **trailing edge left (TEL)**.

For example, a positive elevator deflection (δ_e) will typically result in a negative (nose down) pitching moment, a positive i_h will typically result in a negative pitching moment, a positive aileron deflection (δ_a) will typically result in a positive (right wing down) rolling moment, and a positive rudder deflection (δ_r) will typically result in a negative (nose left) yawing moment. Deflections in the opposite direction are negative. Most flight mechanics texts and aerospace companies use this sign convention. A way to help remember this convention is to realize that it conforms to the right-hand rule if applied to the hinge line of each control surface. For example, if the right thumb is placed along the elevator hinge line pointing in the direction of the positive y axis, the fingers curl in the trailing edge down direction, which is defined as positive. A similar argument holds for the aileron deflection. For the rudder, the right thumb is pointed toward the direction of the positive z axis (out the bottom of the aircraft) and the fingers curl in the trailing edge left direction. Using this sign convention, the primary control derivatives have the following sign: $C_{m_{\delta_e}}$ is **negative**, $C_{l_{\delta_a}}$ is **positive**, and $C_{n_{\delta_r}}$ is **negative**. *Because the control surface sign convention for a specific application external to this text may be defined differently, the reader is cautioned to check the control deflection sign convention for consistency in any application.*

$C_{l_{\delta_r}}$ and $C_{n_{\delta_a}}$ are called **cross-control derivatives** because they define the change in moment that results from the change in a control surface that is not the primary control surface for that axis. For example, $C_{l_{\delta_r}}$ defines the change in rolling moment that results from a change in rudder deflection. With the aileron being the primary control surface for the roll axis, $C_{l_{\delta_r}}$ defines the rolling moment that can be generated with a nonprimary control surface, the rudder.

5.3 Longitudinal Applied Forces and Moments

In Sec. 4.5, we separated aircraft motion into two independent cases, longitudinal (pitch rotation, x- and z-axis translation), and lateral-directional (roll and yaw rotation, y-axis translation). As discussed previously, conventional aircraft configurations generally exhibit longitudinal motion, which is relatively independent of lateral directional motion. This is not a perfect assumption, especially for modern highly coupled aircraft, but it does allow us to analyze each type of motion with three rather than six equations of motion (EOM). Beginning with Eqs. (4.37) and (4.70) from Secs. 4.32 and 4.5, we now expand the applied aero force and moment terms with conventional aerodynamic coefficients for the case of longitudinal motion. In the stability axis,

$$F_{A_{x_{1_S}}} = -D \tag{5.7}$$

$$F_{A_{z_{1_S}}} = -L \tag{5.8}$$

$$M_{A_{1_S}} = M_A \tag{5.9}$$

The added subscripting may seem confusing at first, but it simply indicates the source (A for aero, T for thrust), the applicable axis (x, y, or z) for force directions, the "1" indicates a steady-state condition (as opposed to a condition that varies with time), and the final subscript indicates the axis system for reference (s for stability axis, b for body axis). For example, $F_{A_{x_{1_s}}}$ indicates an aero force along the x stability axis, which is steady state. Moments eliminate the middle subscript because the moment symbol designates the axis that the moment is about. For example, $M_{A_{1_s}}$ indicates an aero pitching moment that is steady state and is defined about the y stability axis. $L_{A_{1_b}}$ would indicate an aero rolling moment that is steady state and defined about the x body axis. Fortunately, this detail is only intended to reinforce the definition of each force or moment contribution and will be quickly dropped to maintain simplicity.

The lift and drag forces could be translated to the body axis through a transformation matrix to maintain compatibility with Eqs. (4.37) and (4.70) that were developed in the body axis. If a small angle of attack is assumed, the difference between body axis forces and stability axis forces in terms of lift and drag would be small and the following approximations could be used:

$$F_{A_{x_{1_b}}} \cong -D \tag{5.10}$$

$$F_{A_{z_{1_b}}} \cong -L \tag{5.11}$$

However, from this point on we will develop the EOM in the stability axis, primarily to keep the representation of lift and drag forces simplified. This approach provides the advantage of simplifying the applied force side of the EOM. Lift can be represented entirely along the z stability axis, and drag can be represented entirely along the x stability axis. In essence, we will use a body-fixed stability axis system for the steady-state, static-stability situations evaluated. The body-fixed stability axis is best thought of as the stability axis permanently fixed to the aircraft at a given trim or steady-state flight condition. Additional discussion of this may be found in Sec. 6.2. To be perfectly correct, the translational velocities U, V, and W, the angular velocities P, Q, and R, and the appropriate moments and products of inertia, all found on the aircraft response side of the six EOM, should now be defined relative to the body-fixed stability axis system. This would keep both sides of the equations of motion in the same axis system. However, if we assume the angle of attack is small, little error is introduced by keeping the aircraft response side of the EOM defined relative to the body axis. This provides the advantage of not having to redefine the aircraft motion parameters (U, V, W, P, Q, and R) and the moments and products of inertia every time a new trim condition (new trim angle of attack) is analyzed. Use of the body-fixed stability axis system also simplifies linearized representation of the applied aerodynamic forces and moments, as we will discuss in Chapter 6.

5.3.1 Aircraft Drag

As seen in Sec. 1.3.4.2, aircraft drag is expressed using the drag coefficient as

$$D = C_D \bar{q} S \tag{5.12}$$

C_D represents the total aircraft drag coefficient and is assumed to be a function of the same parameters which affected C_m in Eq. (5.4), alpha, i_h, and δ_e, so that a similar first-order Taylor series may be used to define C_D.

$$C_D = C_{D_0}{}^* + C_{D_\alpha}\alpha + C_{D_{i_h}} i_h + C_{D_{\delta_e}}\delta_e \tag{5.13}$$

$C_{D_0}{}^*$ is defined as the value of C_D when alpha, i_h, and δ_e are all zero. C_{D_α} is the change in C_D because of a change in alpha; $C_{D_{i_h}}$ is the change in C_D because of a change in stabilizer incidence angle (defined at alpha and δ_e equal to zero); and $C_{D_{\delta_e}}$ is the change in C_D because of a change in δ_e (defined at alpha and i_h equal to zero).

The drag coefficient is also a function of dynamic pressure (to account for aeroelastic effects if the assumption of a rigid body is not valid), Mach number, and Reynolds number. The previous expression for C_D requires that $C_{D_0}{}^*$, C_{D_α}, $C_{D_{i_h}}$, and $C_{D_{\delta_e}}$ are defined at appropriate Mach number and Reynolds number conditions, and that aeroelastic effects are negligible. It also assumes that $C_{D_0}{}^*$ includes appropriate skin friction and pressure drag contributions. The Taylor series of Eq. (5.13) provides a straightforward representation of C_D suitable for 1) stability analysis (because alpha is our perturbed longitudinal motion parameter); and 2) trimmed flight analysis (because i_h and δ_e are the longitudinal control surfaces used to trim the aircraft). This representation of C_D is also compatible with the lift and pitching moment expressions we will develop in the next sections. For most cases, $C_{D_{i_h}}$ and $C_{D_{\delta_e}}$ may be neglected (assumed to have a value approximately equal to zero). Exceptions to this simplification include minimum control speed problems and problems where trim drag is important. With this assumption, Eq. (5.13) becomes

$$C_D = C_{D_0}{}^* + C_{D_\alpha}\alpha \tag{5.14}$$

It is important to realize the differences in this representation of C_D and the parabolic form of the drag polar discussed in Sec. 1.5.2. Recalling the parabolic drag polar

$$C_D = C_{D_0} + \frac{C_L^2}{\pi e AR} \tag{5.15}$$

and referring to Fig. 5.4 (α and C_L can be thought of as interchangeable on the vertical axis), we see that Eq. (5.14) represents a linearized approximation of the parabolic drag polar at a tangent point defined by the steady-state angle of attack and C_D conditions being analyzed for stability. This is a good approximation if the perturbations from the steady-state trim condition are small, as is typically the case in stability analysis.

As can be seen from Fig. 5.4, the value of $C_{D_0}{}^*$ and C_{D_α} change with the steady-state trim point being analyzed. It is also important to realize that C_{D_0}, the parasite drag coefficient in the drag polar equation, and $C_{D_0}{}^*$, the value of C_D with alpha, i_h, and δ_e equal to zero in the linearized drag approximation, are not the same. The use of the asterisk (*) is intended to highlight this.

To estimate the two terms in Eq. (5.14), the drag polar Eq. (5.15) is needed. This may be defined from analytical estimates, wind tunnel testing, or flight test. $C_{D_0}{}^*$ can then be defined from a plot such as Fig. 5.4, and C_{D_α} from

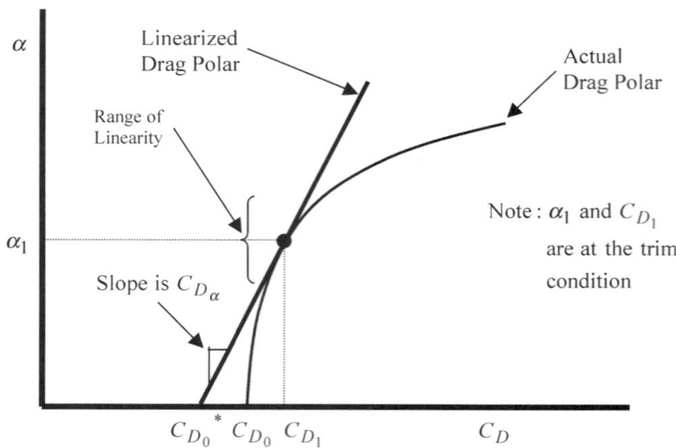

Fig. 5.4 Comparison of the drag polar and the linearized drag approximation.

taking the derivative of the drag polar equation with respect to alpha (using the assumption that $C_L = C_{L_\alpha}\alpha$).

$$C_{D_\alpha} = 2C_{L_1}\frac{C_{L_\alpha}}{\pi eAR} \tag{5.16}$$

5.3.2 Aircraft Lift

From Sec. 2.4.1, lift is expressed in terms of the lift coefficient as

$$L = C_L\bar{q}S \tag{5.17}$$

C_L represents the total aircraft lift coefficient and will be assumed to be a function of the same parameters as C_D and C_M, namely alpha, i_h, and δ_e. The first-order Taylor series expansion for C_L becomes

$$C_L = C_{L_0} + C_{L_\alpha}\alpha + C_{L_{i_h}}i_h + C_{L_{\delta_e}}\delta_e \tag{5.18}$$

C_{L_0} is defined as the value of C_L when alpha, i_h, and δ_e are all zero; C_{L_α} is the change in C_L because of a change in alpha; $C_{L_{i_h}}$ is the change in C_L because of a change in stabilizer incidence angle (defined at alpha and δ_e equal to zero); and $C_{L_{\delta_e}}$ is the change in C_L because of a change in δ_e (defined at alpha and i_h equal to zero).

The lift coefficient is also a function of dynamic pressure (to account for aeroelastic effects if the assumption of a rigid body is not valid), Mach number, and Reynolds number. As with Eq. (5.13), Eq. (5.18) requires that C_{L_0}, C_{L_α}, $C_{L_{i_h}}$, and $C_{L_{\delta_e}}$ are defined at appropriate Mach number and Reynolds number conditions, and that aeroelastic effects are negligible.

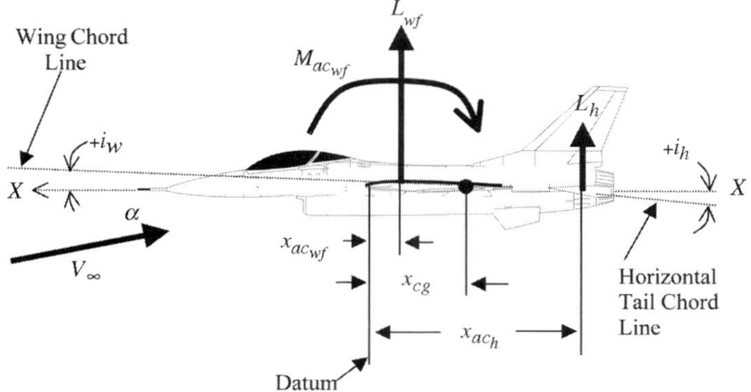

Fig. 5.5 Geometry for calculating lift and pitching moment derivatives.

To estimate C_{L_0} and the derivative terms in Eq. (5.18) in terms of geometric characteristics of an aircraft, we will use a conventional airplane (wing forward, tail aft) as the example and lift predictions we have already developed. To begin, we will consider the aircraft as being made up of two components: the wing/fuselage and the horizontal tail.

The angle of attack acting on the lifting devices (wing and horizontal tail) of each of these components must be defined. For many airplanes, the overall aircraft angle of attack is defined as the angle between the relative wind and the fuselage reference line (sometimes called the fuselage water line) or the x-body axis. The angle of attack experienced by the wing is, in simple terms, the aircraft angle of attack plus the wing incidence angle (as illustrated in Fig. 5.5)

$$\alpha_w = \alpha + i_w \tag{5.19}$$

To calculate the angle of attack at the horizontal tail (α_h), the downwash resulting from wingtip vortices must be considered. Downwash decreases the effective incidence angle of the relative wind at the horizontal tail from that experienced at the nose of the aircraft (see Fig. 5.6).

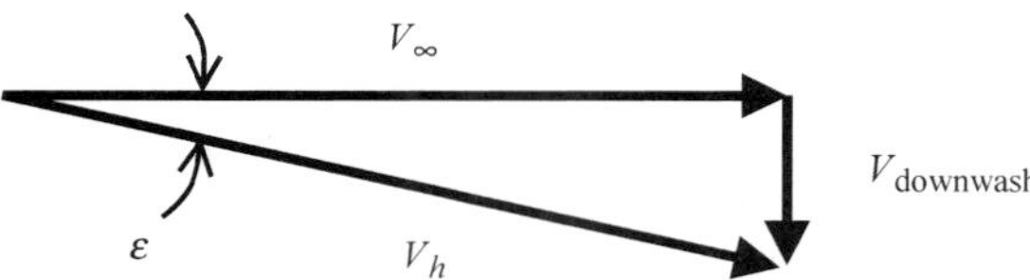

Fig. 5.6 Velocity components at the horizontal tail.

The downwash effect is described with the angle epsilon, which is the average downwash angle induced by the wing on the tail. It is normally expressed as

$$\varepsilon = \varepsilon_0 + \frac{\partial \varepsilon}{\partial \alpha}\alpha \tag{5.20}$$

Where ε_0 is the downwash angle at zero aircraft angle of attack. ε_0 and $\frac{\partial \varepsilon}{\partial \alpha}$ are normally estimated using an empirical formula or wind tunnel results.

To compute the angle of attack experienced by the horizontal tail (α_h), we must also take into account the horizontal tail incidence angle (i_h) relative to the fuselage reference line as shown in Fig. 5.5 and the effect of elevator deflection (δ_e). i_h is defined as positive with a leading edge up deflection. α_h then becomes

$$\alpha_h = \alpha + i_h + \tau_e \delta_e - \varepsilon \tag{5.21}$$

τ_e is called the elevator effectiveness. It is a ratio that relates a change in angle of attack to a change in δ_e. A physical understanding of τ_e is presented in Fig. 5.7. With the orientation of the relative wind and the front portion of an airfoil staying fixed, an elevator deflection, δ_e, produces a change in the orientation of the chord line which, in turn, changes the angle of attack of the airfoil. The larger the chord of the elevator is relative to the overall chord of the airfoil, the larger τ_e will be.

The total lift acting on the aircraft in terms of the two components, wing fuselage, and horizontal tail can then be expressed as

$$L = L_{wf} + L_h \cos \varepsilon \cong L_{wf} + L_h \tag{5.22}$$

In coefficient form, this becomes

$$C_L \bar{q} S = C_{L_{wf}} \bar{q} S + C_{L_h} \bar{q}_h S_h \tag{5.23}$$

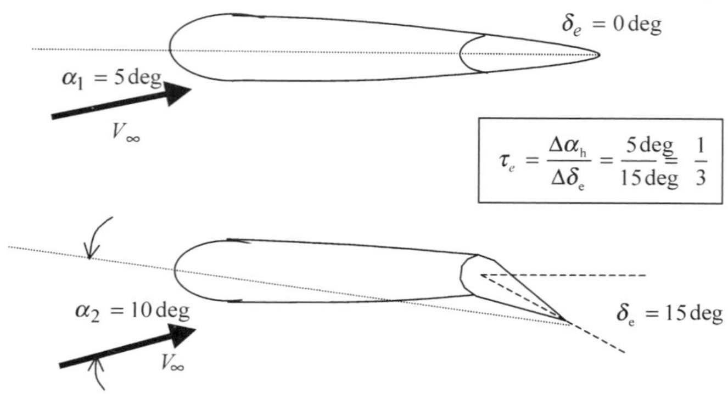

Fig. 5.7 Geometric description of elevator effectiveness.

The dynamic pressure at the horizontal tail ($\bar{q}_h = \frac{1}{2}\rho V_h^2$) may be different than that acting on the wing-fuselage because of the boundary layer and engine effects. The dynamic pressure ratio (η_h) is used to describe this difference.

$$\eta_h = \frac{\bar{q}_h}{\bar{q}} \tag{5.24}$$

Combining Eqs. (5.23) and (5.24) to solve for C_L:

$$C_L = C_{L_{wf}} + C_{L_h}\eta_h\frac{S_h}{S} \tag{5.25}$$

and expanding $C_{L_{wf}}$ and C_{L_h} using a familiar Taylor series (and assuming $i_w = 0$), we have

$$C_{L_{wf}} = C_{L_{0_{wf}}} + C_{L_{\alpha_{wf}}}\alpha \tag{5.26}$$

$$C_{L_h} = C_{L_{0_h}} + C_{L_{\alpha_h}}\alpha_h \tag{5.27}$$

Substituting Eqs. (5.26) and (5.27) into Eq. (5.25), we have

$$\begin{aligned} C_L = {} & C_{L_{0_{wf}}} + C_{L_{\alpha_{wf}}}\alpha \\ & + C_{L_{\alpha_h}}\eta_h\frac{S_h}{S}\left[\alpha - \left(\varepsilon_0 + \frac{d\varepsilon}{d\alpha}\alpha\right) + i_h + \tau_e\delta_e\right] + \eta_h\frac{S_h}{S}C_{L_{0_h}} \end{aligned} \tag{5.28}$$

This is a detailed expression for the lift coefficient for the overall aircraft and is very useful because most of the parameters in it can be calculated or estimated given the geometry of an aircraft configuration. We can now use Eq. (5.28) to calculate the key parameters in Eq. (5.18). For C_{L_0}, we set alpha, i_h and δ_e equal to zero.

$$C_{L_0} = C_{L_{0_{wf}}} - C_{L_{\alpha_h}}\eta_h\frac{S_h}{S}\varepsilon_0 + \eta_h\frac{S_h}{S}C_{L_{0_h}} \approx C_{L_{0_{wf}}} \tag{5.29}$$

The approximation in Eq. (5.29) can be made if ε_0 and $C_{L_{0_h}}$ are small. In the case of $C_{L_{0_h}}$, most aircraft use a symmetrical airfoil for the horizontal tail and, for this case, $C_{L_{0_h}}$ is zero.

To obtain C_{L_α} (the total aircraft lift-curve slope), we simply take the partial derivative of Eq. (5.28) with respect to alpha.

$$C_{L_\alpha} = C_{L_{\alpha_{wf}}} + C_{L_{\alpha_h}}\eta_h\frac{S_h}{S}\left(1 - \frac{d\varepsilon}{d\alpha}\right) \tag{5.30}$$

$C_{L_{i_h}}$ and $C_{L_{\delta_e}}$ are obtained through partial differentiation in a similar manner.

$$C_{L_{i_h}} = C_{L_{\alpha_h}} \eta_h \frac{S_h}{S} \tag{5.31}$$

$$C_{L_{\delta_e}} = C_{L_{\alpha_h}} \eta_h \frac{S_h}{S} \tau_e \tag{5.32}$$

In summary, we have developed a detailed method to approximate lift using a first order Taylor series expansion. To once again put things in perspective, remember that

$$F_{A_{z_{1_s}}} = -L = -C_L \bar{q} S = -(C_{L_0} + C_{L_\alpha} \alpha + C_{L_{i_h}} i_h + C_{L_{\delta_\varepsilon}} \delta_e) \bar{q} S \tag{5.33}$$

Figure 5.8 presents wind tunnel data for the F-15B for four tail deflections.[1] An angle of attack range of zero to 90 deg was evaluated. Notice that the maximum lift coefficient for each configuration occurs between 25 and 40 deg angle of attack. Notice also that higher values of i_h result in higher values of C_L.

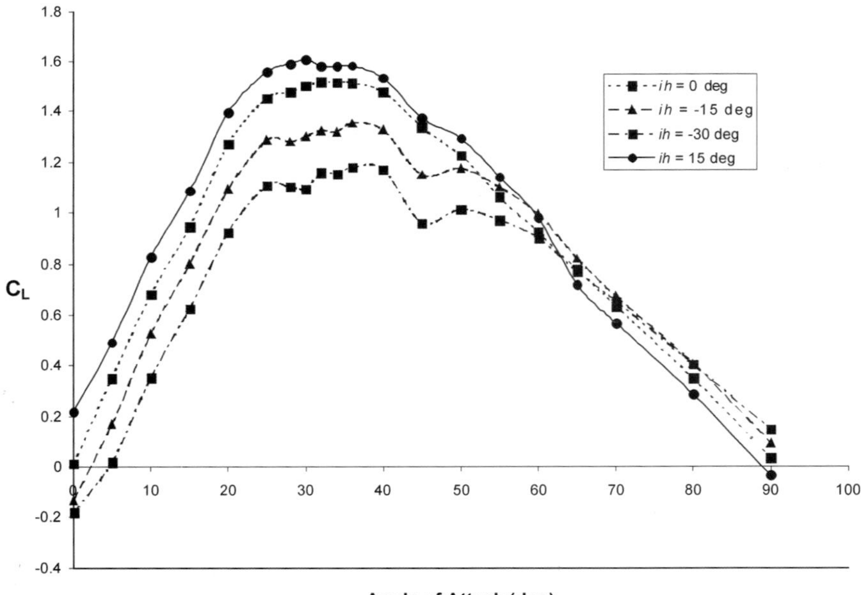

Fig. 5.8 F-15B lift coefficient wind tunnel data.

5.3.3 *Aircraft Aerodynamic Pitching Moment*

As seen in Sec. 1.3.4.2, aircraft pitching moment may be expressed using the pitching moment coefficient as

$$M_A = C_m \bar{q} S \bar{c} \tag{5.34}$$

C_m represents the total aircraft pitching moment coefficient, and we will again express it using a first-order Taylor series expansion with α, i_h, and δ_e.

$$C_m = C_{m_0} + C_{m_\alpha}\alpha + C_{m_{i_h}} i_h + C_{m_{\delta_e}}\delta_e \tag{5.35}$$

The moment coefficient is also a function of dynamic pressure (to account for aeroelastic effects), Mach number, Reynolds number, and the center of gravity location (or moment reference center) of the aircraft. C_{m_0}, C_{m_α}, $C_{m_{i_h}}$, and $C_{m_{\delta_e}}$ must be defined for the appropriate Mach number, Reynolds number, *cg* location, and dynamic pressure (if aeroelastic effects are considered).

5.3.3.1 C_{m_0} and C_{m_α}. To estimate C_{m_0} and the derivative terms in Eq. (5.35), we will again consider the aircraft as two components: the wing-fuselage and the horizontal tail. Figure 5.5 should be referred to using the *cg* as the moment reference center. Neglecting the effect on pitching moment of wing-fuselage and horizontal tail drag, it can be seen that the aerodynamic pitching moment about the *cg* is

$$\begin{aligned} M_A = M_{AC_{wf}} &+ L_{wf}(x_{cg} - x_{AC_{wf}})\cos\alpha \\ &- L_h(x_{AC_h} - x_{cg})\cos(\alpha - \varepsilon) \end{aligned} \tag{5.36}$$

For most cases, the cosine terms in Eq. (5.36) are very close to one because the angles are small. With this assumption, we recast Eq. (5.36) in coefficient form by nondimensionalizing with respect to $\bar{q}S\bar{c}$.

$$C_m = C_{m_{AC_{wf}}} + C_{L_{wf}}\frac{(x_{cg} - x_{AC_{wf}})}{\bar{c}} - C_{L_h}\eta_h\frac{S_h}{S}\frac{(x_{AC_h} - x_{cg})}{\bar{c}} \tag{5.37}$$

The distances x_{cg}, $x_{AC_{wf}}$, and x_{AC_h} are now conveniently expressed in nondimensional form as

$$\bar{x}_{cg} = \frac{x_{cg}}{\bar{c}} \tag{5.38}$$

$$\bar{x}_{AC_{wf}} = \frac{x_{AC_{wf}}}{\bar{c}} \tag{5.39}$$

$$\bar{x}_{AC_h} = \frac{x_{AC_h}}{\bar{c}} \tag{5.40}$$

remembering that $\bar{c}$ is the wing mean chord.

We now assume $C_{L_{0_h}}$ equals zero and substitute Eqs. (5.20), (5.21), (5.26), (5.27), and (5.38–5.40) into Eq. (5.37) to obtain

$$\begin{aligned} C_m = {} & C_{m_{AC_{wf}}} + (C_{L_{0_{wf}}} + C_{L_{\alpha_{wf}}}\alpha)(\bar{x}_{cg} - \bar{x}_{AC_{wf}}) \\ & - C_{L_{\alpha_h}}\eta_h \frac{S_h}{S}(\bar{x}_{AC_h} - \bar{x}_{cg})\left[\alpha - \left(\varepsilon_0 + \frac{d\varepsilon}{d\alpha}\alpha\right) + i_h + \tau_e\delta_e\right] \end{aligned} \tag{5.41}$$

We are now in a position to define C_{m_0} and the derivative terms in Eq. (5.35). Setting α, i_h, and δ_e equal to zero in Eq. (5.41), we have

$$\begin{aligned} C_{m_0} & = C_{m_{AC_{wf}}} + C_{L_{0_{wf}}}(\bar{x}_{cg} - \bar{x}_{AC_{wf}}) + C_{L_{\alpha_h}}\eta_h \frac{S_h}{S}(\bar{x}_{AC_h} - \bar{x}_{cg})\varepsilon_0 \\ & \approx C_{m_{AC_{wf}}} + C_{L_{0_{wf}}}(\bar{x}_{cg} - \bar{x}_{AC_{wf}}) \text{ if } \varepsilon_0 \text{ is negligible} \end{aligned} \tag{5.42}$$

Taking the partial derivative of Eq. (5.41) with respect to α, we have

$$C_{m_\alpha} = C_{L_{\alpha_{wf}}}(\bar{x}_{cg} - \bar{x}_{AC_{wf}}) - C_{L_{\alpha_h}}\eta_h \frac{S_h}{S}(\bar{x}_{AC_h} - \bar{x}_{cg})\left(1 - \frac{d\varepsilon}{d\alpha}\right) \tag{5.43}$$

5.3.3.2 $C_{m_{i_h}}$ and $C_{m_{\delta_e}}$. $C_{m_{i_h}}$ and $C_{m_{\delta_e}}$ are obtained through partial differentiation in a similar manner. The tail volume ratio, $\bar{V}$, is introduced to simplify these expressions. It is also useful early in the aircraft design process to do preliminary sizing of the horizontal tail.

$$\bar{V}_h = \left(\frac{S_h}{S}\right)(\bar{x}_{AC_h} - \bar{x}_{cg}) \tag{5.44}$$

$$C_{m_{i_h}} = -C_{L_{\alpha_h}}\eta_h \frac{S_h}{S}(\bar{x}_{AC_h} - \bar{x}_{cg}) = -C_{L_{\alpha_h}}\eta_h \bar{V}_h \tag{5.45}$$

$$C_{m_{\delta_e}} = -C_{L_{\alpha_h}}\eta_h \bar{V}_h \tau_e \tag{5.46}$$

As discussed in Sec. 5.2, C_{m_α} is the static longitudinal stability derivative. $C_{m_{i_h}}$ and $C_{m_{\delta_e}}$ are longitudinal control power derivatives.

In summary, we have developed a detailed method to approximate the aerodynamic pitching moment using a first-order Taylor series expansion. To once again put things in perspective, remember that

$$M_{a_{1_s}} = M_a = C_m\bar{q}S\bar{c} = (C_{m_0} + C_{m_\alpha}\alpha + C_{m_{i_h}}i_h + C_{m_{\delta_\varepsilon}}\delta_e)\bar{q}S\bar{c} \tag{5.47}$$

Notice that M_a is the same for either the body or stability axis because the transformation rotates about the number two (or y) axis. The y axis is the same for either system and is also the axis that pitching moment is defined about.

Figure 5.9 illustrates the effect of changing i_h and δ_e on a C_m vs alpha graph. As shown, the magnitude of the curve shift is directly proportional to

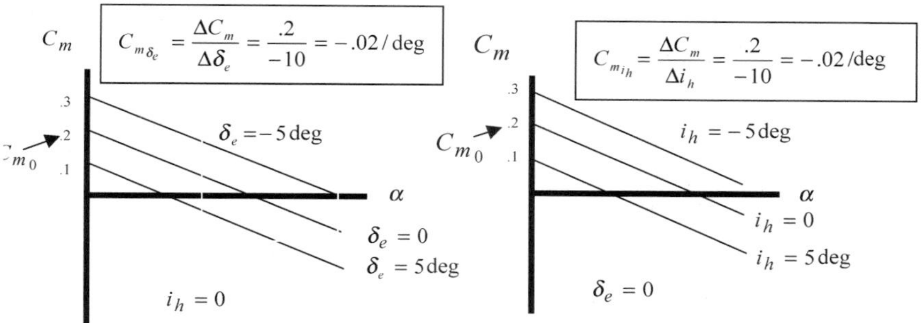

Fig. 5.9 Effect of i_h and δ_e on C_m vs alpha characteristics.

the magnitude of $C_{m_{i_h}}$ and $C_{m_{\delta_e}}$. A positive i_h deflection is defined as leading edge up so that a positive i_h deflection produces a negative increment on pitching moment. Remember that a positive δ_e deflection is defined as trailing edge down that produces a negative increment on pitching moment.

The relative magnitude of $C_{m_{i_h}}$ and $C_{m_{\delta_e}}$ is dependent on the ratio of the horizontal tail chord percentage to the elevator chord percentage. For example, if the elevator is 33% chord, the horizontal tail will be 67% chord and the ratio of $C_{m_{i_h}}$ to $C_{m_{\delta_e}}$ will be approximately two to one.

Figure 5.10 presents pitching moment vs angle of attack data for the F-15B aircraft.[1] Notice the negative slope of each curve indicating static longitudinal stability, and also notice that the curves shift upward as i_h becomes more negative.

Some aircraft, such as the F-18 with canted twin vertical tails, use inboard or outboard rudder deflection to augment pitching moment control power.[2] If this is the case, an additional term is included in the pitching moment Taylor series Eq. (5.47) to account for the pitching moment because of asymmetrical rudder deflection. For such aircraft, longitudinal and lateral-directional motion is referred to as coupled.

5.3.3.3 Aircraft aerodynamic center. The aerodynamic pitching moment characteristics of the wing-fuselage and horizontal tail may together be represented at a very convient point for the entire aircraft. This point is called the aircraft aerodynamic center (AC). Similar to the AC for an airfoil, discussed in Sec. 1.3.4.3, the aircraft AC is that point on the aircraft where the variation of aircraft pitching moment coefficient with angle of attack is zero. The location of the aircraft AC is normally defined with respect to the mean chord of the wing using the bar notation. For example,

$$\bar{x}_{AC} \approx \frac{x_{AC}}{\bar{c}} \tag{5.48}$$

To develop an expression for the aircraft AC, we use: 1) the definition which, in other words, says that C_{m_α} must be equal to zero at the aircraft AC if the

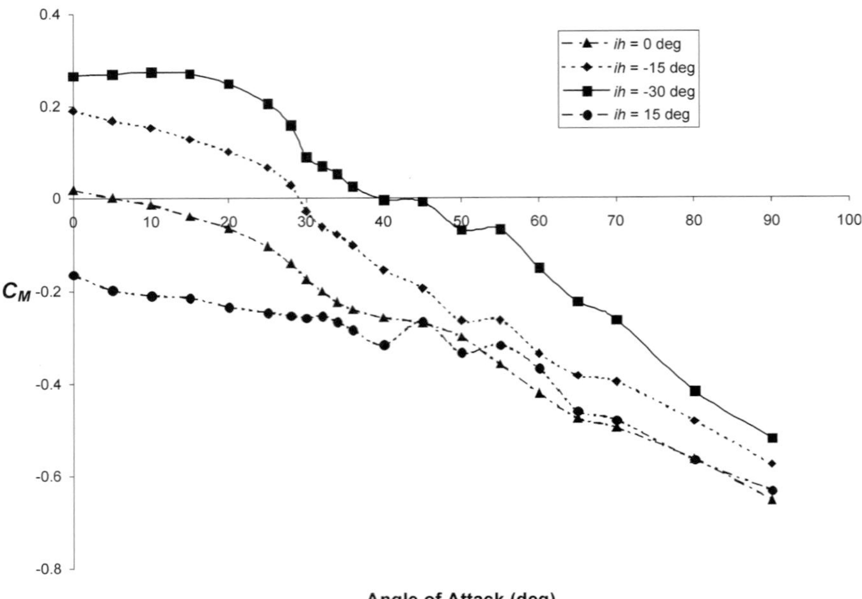

Fig. 5.10 Pitching moment wind tunnel data for the F-15B aircraft.

aircraft is rotated about the AC; and 2) the fact that the c.g. of the aircraft (the moment reference center) must be at the AC if the aircraft is rotating about the AC. Thus, we have

$$\text{If} \quad C_{m_\alpha} = 0, \tag{5.49}$$

$$\text{then} \quad \bar{x}_{cg} = \bar{x}_{AC} \tag{5.50}$$

Applying these constraints to Eq. (5.43) and solving for the aircraft AC, we have

$$\bar{x}_{AC} = \frac{\bar{x}_{AC_{wf}} + \dfrac{C_{L_{\alpha_h}}}{C_{L_{\alpha_{wf}}}} \eta_h \dfrac{S_h}{S} \bar{x}_{AC_h} \left(1 - \dfrac{d\varepsilon}{d\alpha}\right)}{1 + \dfrac{C_{L_{\alpha_h}}}{C_{L_{\alpha_{wf}}}} \eta_h \dfrac{S_h}{S} \left(1 - \dfrac{d\varepsilon}{d\alpha}\right)} \tag{5.51}$$

Notice that the aircraft AC may be moved aft on the aircraft by increasing $\bar{x}_{AC_h}$, S_h, and/or $C_{L_{\alpha_h}}$. This approach can be very useful in designing an aircraft for static stability. As we will see, moving the aircraft AC aft will increase static longitudinal stability, and moving it forward decreases static stability for a given c.g. location.

Example 5.1

Estimate the location of the aircraft aerodynamic center for the A-10 given the following:

$$S = 506\ \text{ft}^2 \quad S_h = 120\ \text{ft}^2 \quad \eta_h \approx 0.9$$
$$AR = 6.54 \quad AR_h = 3.0 \quad \frac{d\varepsilon}{d\alpha} \approx 0.1$$
$$\bar{c} = 8.8\ \text{ft} \quad x_{AC_h} = 26.25\ \text{ft}$$

Neglect any fuselage effects and assume a span efficiency factor of 0.9 for the wing and tail.

We first look at Eq. (5.51) and see that we must determine $C_{L_{\alpha_{wf}}}$ and $C_{L_{\alpha_h}}$. Using Eq. (1.30),

$$C_{L_{\alpha_{wf}}} = \frac{C_{l_\alpha}}{1 + \dfrac{57.3C_{l_\alpha}}{\pi eAR}} = \frac{0.11}{1 + \dfrac{57.3(0.11)}{(3.14)(0.9)(6.54)}} = 0.082/\text{deg}$$
$$C_{L_{\alpha_h}} = \frac{0.11}{1 + \dfrac{57.3(0.11)}{(3.14)(0.9)(3.0)}} = 0.0631/\text{deg}$$

For a subsonic aircraft such as the A-10 and neglecting fuselage effects, we know that

$$\bar{x}_{AC_{wf}} \approx 0.25$$

Also,

$$\bar{x}_{AC_h} = \frac{x_{AC_h}}{\bar{c}} = \frac{26.25}{8.8} = 2.98$$

We are now ready to use Eq. (5.51).

$$\bar{x}_{AC} = \frac{\bar{x}_{AC_{wf}} + \dfrac{C_{L_{\alpha_h}}}{C_{L_{\alpha_{wf}}}}\eta_h \dfrac{S_h}{S}\bar{x}_{AC_h}\left(1 - \dfrac{d\varepsilon}{d\alpha}\right)}{1 + \dfrac{C_{L_{\alpha_h}}}{C_{L_{\alpha_{wf}}}}\eta_h \dfrac{S_h}{S}\left(1 - \dfrac{d\varepsilon}{d\alpha}\right)}$$
$$= \frac{0.25 + \dfrac{0.0631}{0.082}(0.9)\dfrac{120}{506}(2.98)(1 - 0.1)}{1 + \dfrac{0.0631}{0.082}(0.9)\dfrac{120}{506}(1 - 0.1)}$$
$$\bar{x}_{AC} = 0.602$$

Notice that the AC of the aircraft is aft of the wing AC. This is to be expected with aft tail aircraft.

5.3.4 *Thrust Forces and Moments*

Longitudinal forces and moments resulting from engine thrust must also be defined to complete the applied forces and moments side of the aircraft equations of motion. We will only consider direct thrust effects on the aircraft. Indirect thrust effects, such as jet exhaust impinging on lifting surfaces, will be ignored. In addition, the orientation of the thrust vector produced by the engine or engines will be assumed to be in the *xz* body axis plane (no side force components). These assumptions lead to a simple representation of the thrust forces and moments in the body and stability axis using Fig. 5.11.

In the stability axis, the thrust forces and moments are

$$F_{T_{x_{1_s}}} = T\cos(\phi_T + \alpha_1) \tag{5.52}$$

$$F_{T_{z_{1_s}}} = -T\sin(\phi_T + \alpha_1) \tag{5.53}$$

$$M_{T_{1_s}} = M_{T_1} = -T\mathrm{d}_T \tag{5.54}$$

In the body axis, they simplify to

$$F_{T_{x_{1_b}}} = T\cos\phi_T \tag{5.55}$$

$$F_{T_{z_{1_b}}} = -T\sin\phi_T \tag{5.56}$$

$$M_{T_{1_b}} = M_{T_{1_s}} = -T\mathrm{d}_T \tag{5.57}$$

Note: M_T is shown in the positive direction, however, for this example, M_T would be negative.

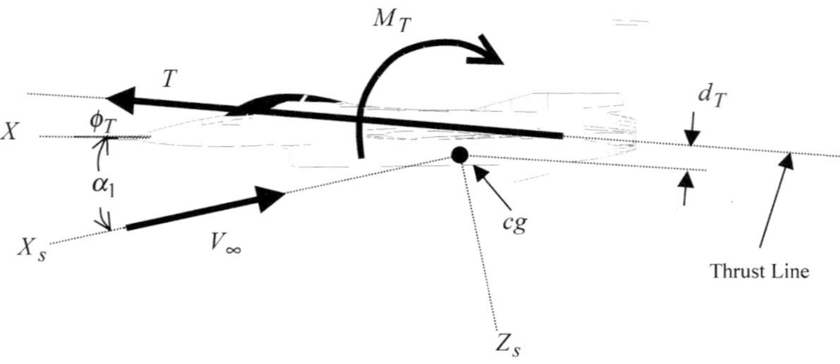

Fig. 5.11 Thrust forces and moments.

5.4 Longitudinal Static Stability

As discussed in Sec. 5.1, static stability refers to the initial tendency of an airplane, following a disturbance from steady-state flight, to develop aerodynamic forces and moments that are in a direction to return the aircraft to the steady-state flight condition. For purposes of this text, longitudinal static stability will primarily refer to aircraft pitching moment characteristics and will be analyzed for the **stick fixed** condition (see Sec. 10.5). As discussed in Sec. 5.2, the sign of the stability derivative, C_{m_α}, is key in determining the static longitudinal stability of an aircraft. The requirement to trim the aircraft at usable angles of attack is also discussed with the longitudinal stability requirement because both are generally necessary to achieve acceptable flight characteristics.

5.4.1 Trim Conditions

The general requirement for longitudinal trim is that the overall pitching moment acting on the aircraft be equal to zero, or, in coefficient form

$$C_m = 0 \tag{5.58}$$

The total pitching moment includes both aerodynamic and thrust contributions. Referring to Fig. 5.12 for a tail-aft aircraft configuration, we can do a simple moment balance to satisfy the trim condition.

$$L_{wf}(x_{cg} - x_{AC_{wf}}) + M_{AC_{wf}} - L_h(x_{AC_h} - x_{cg}) - T\mathrm{d}_T = 0 \tag{5.59}$$

This relationship allows us to solve for L_h, the lift on the horizontal tail required to trim.

$$L_h = \frac{L_{wf}(x_{cg} - x_{AC_{wf}}) + M_{AC_{wf}} - T\mathrm{d}_T}{(x_{AC_h} - x_{cg})} \tag{5.60}$$

For a tail-aft airplane, $(x_{AC_h} - x_{cg})$ is positive, $M_{AC_{wf}}$ is normally negative because of positive wing camber, L_{wf} is positive, and $T\mathrm{d}_T$ will be assumed

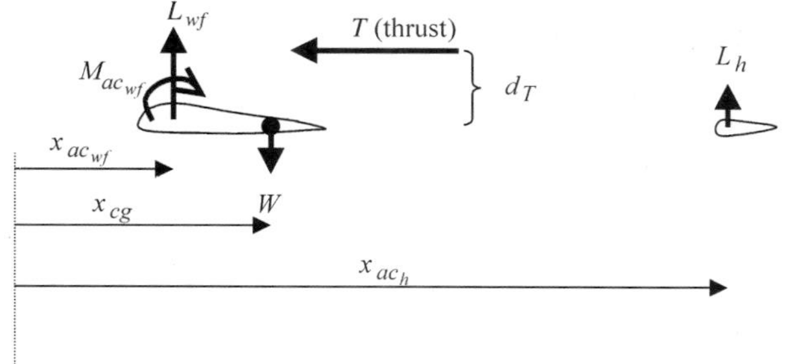

Fig. 5.12 Trim moments for a conventional aircraft.

small. With x_{cg} less than or equal $x_{AC_{wf}}$, it can be seen that a down load ($L_h < 0$) is needed to trim. If x_{cg} is greater than $x_{AC_{wf}}$, (an unstable wing-fuselage combination), the load on the horizontal tail may be either up or down depending on the magnitude of $M_{AC_{wf}}$.

We can also think of the trim condition in terms of Eq. (5.4). For a trim condition,

$$C_{m_0} + C_{m_\alpha}\alpha + C_{m_{\delta_e}}\delta_e + C_{m_{i_h}}i_h = 0 \tag{5.61}$$

Referring to Fig. 5.3, it can be seen that another requirement for trim is the ability to trim the aircraft at a positive angle of attack. This is generally accomplished by keeping

$$C_{m_0} + C_{m_{i_h}}i_h > 0 \tag{5.62}$$

in Eq. (5.4). Because the primary requirement for longitudinal static stability is C_{m_α} negative (the slope of the graph in Fig. 5.3), it can be seen that the Eq. (5.61) requirement keeps the vertical intercept positive so that the aircraft can be trimmed at a positive angle of attack. The $C_{m_{\delta_e}}\delta_e$ term in Eq. (5.4) allows adjustment of the trim angle of attack by varying the elevator deflection (δ_e) as illustrated in Fig. 5.13.

Equation (5.61) can be used to develop an expression for the elevator required to trim the aircraft. By simply solving for δ_e, we have

$$\delta_{e_{\text{trim}}} = -\frac{C_{m_0} + C_{m_\alpha}\alpha + C_{m_{i_h}}i_h}{C_{m_{\delta_e}}} \tag{5.63}$$

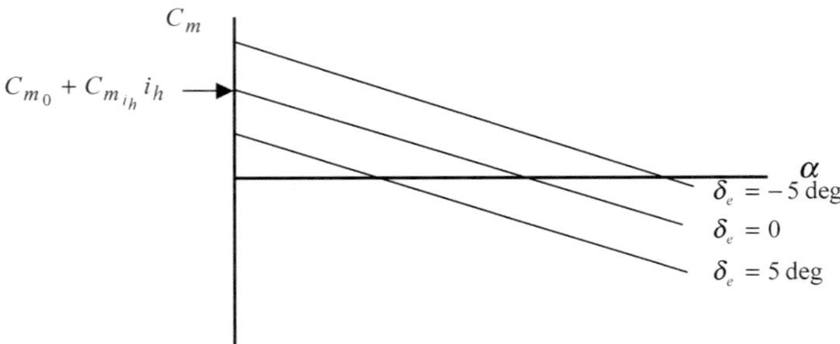

Fig. 5.13 Pitching moment vs angle of attack with various elevator deflections.

Because C_L and alpha have approximately a linear relationship for angles of attack below the stall, the contribution of the term $C_{M_\alpha}\alpha$ can also be represented by $C_{M_{C_L}} C_L$, or

$$\delta_{e_{\text{trim}}} = -\frac{C_{m_0} + C_{m_{C_L}} C_L + C_{m_{i_h}} i_h}{C_{m_{\delta_e}}} \tag{5.64}$$

Figure 5.14 presents a typical graph of $\delta_{e_{\text{trim}}}$ as a function of C_L.

Typically, at lower values of lift coefficient (higher speeds), positive values of δ_e (trailing edge down) are required. This equates to coming forward with the stick. At higher values of lift coefficient (lower speeds), negative values of δ_e (trailing edge up) are required, which equates to aft stick.

If we combine C_{m_0} and $C_{m_{i_h}} i_h$ into one term that meets the criteria of Eq. (5.62), Eq. (5.64) becomes

$$\delta_{e_{\text{trim}}} = -\frac{C_{m_0} + C_{m_{i_h}} i_h}{C_{m_{\delta_e}}} - \frac{C_{m_{C_L}} C_{L_{\text{trim}}}}{C_{m_{\delta_e}}}$$

or, in simpler terms

$$\delta_{e_{\text{trim}}} = \delta_{e_0} - \frac{C_{m_{C_L}} C_{L_{\text{trim}}}}{C_{m_{\delta_e}}} \tag{5.65}$$

where

$$\delta_{e_0} = -\frac{C_{m_0} + C_{m_{i_h}} i_h}{C_{m_{\delta_e}}}$$

δ_{e_0} can be thought of as the elevator required to trim the aircraft at a lift coefficient of zero.

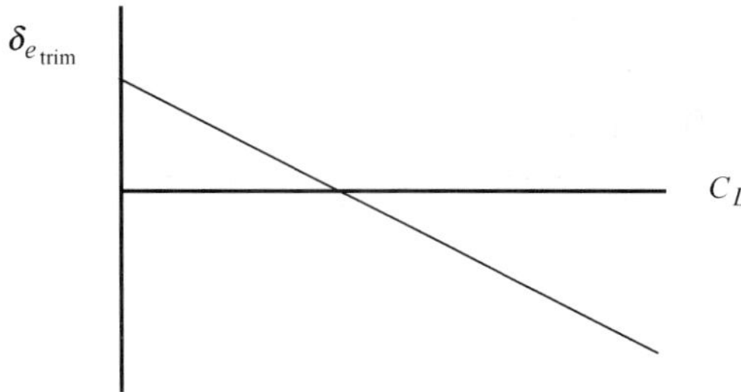

Fig. 5.14 Elevator deflection required to trim as a function of lift coefficient.

5.4.2 Stability Requirements

The primary requirement for longitudinal static stability is

$$C_{m_\alpha} < 0 \tag{5.66}$$

As discussed in Sec. 5.2, this requirement provides an aerodynamic restoring moment for angle of attack perturbations to steady-state flight. As seen in Fig. 5.3, a negative C_{m_α} results in a negative slope on C_m vs alpha plots.

For a simple vehicle in flight, like a projectile, a negative C_{m_α} is achieved by simply **locating the c.g. in front of the AC**. From Fig. 5.15, we can see that an airfoil with negative camber (positive $C_{m_{AC}}$) will be able to achieve trim ($C_m = 0$) and will also be statically stable if the c.g. is located in front of the AC.

Because the airfoil will rotate about the c.g. in flight, we can analyze a perturbed increase in angle of attack from trimmed flight (Fig. 5.15b) and a perturbed decrease in angle of attack from trimmed flight (Fig. 5.15c). Notice in Fig. 5.15b that a perturbed increase in angle of attack results in an increase in the lift force. Because the lift vector is represented at the AC, and is located behind the c.g., a negative (nose down) pitching moment results in response to the perturbation. This design is statically stable because this moment provides an initial tendency for the airfoil to return to trim (equilibrium). The same analysis applies to Fig. 5.15c for a perturbed decrease in angle of attack. Figure 5.16 presents each case on a C_m vs alpha plot. Note again that a negative C_{m_α} is required for static stability.

For a conventional tail-aft aircraft, Eq. (5.43) may be used to compute C_{m_α} so that the static stability requirement can be checked. However, use of the overall aircraft AC Eq. (5.51) is generally more useful from an analysis

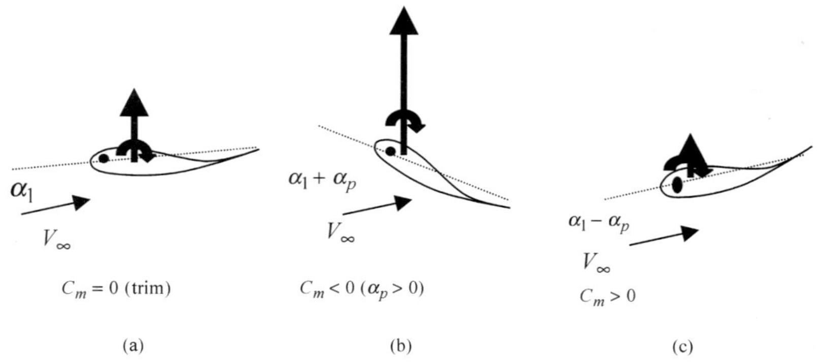

Fig. 5.15 Trimmed and perturbed flight conditions for a statically stable airfoil.

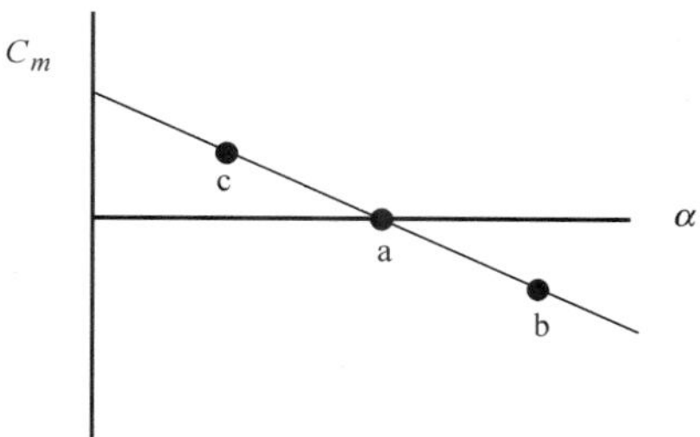

Fig. 5.16 C_m vs alpha characteristics for a statically stable airfoil.

standpoint. Because $C_{m_{AC}}$ remains constant with changes in angle of attack (per the definition of aerodynamic center), C_{m_α} becomes

$$M_{cg} = L(x_{cg} - x_{AC}) + M_{AC}$$
$$C_{m_{cg}} = C_L(\bar{x}_{cg} - \bar{x}_{AC}) + C_{m_{AC}}$$
$$C_{m_\alpha} = C_{L_\alpha}(\bar{x}_{cg} - \bar{x}_{AC}) \tag{5.67}$$

Because C_{L_α} is positive for lift coefficients below $C_{L_{\max}}$ (the normal C_L range for flight), it can be seen that $\bar{x}_{AC}$ (for the aircraft) must be larger than $\bar{x}_{cg}$ for C_{m_α} to be negative. In simple terms, the c.g. must be in front of the aircraft a.c. for longitudinal static stability (C_{m_α} negative).

A conventional tail-aft balsa wood glider illustrates this principle nicely. The c.g. is moved forward with a weight such as a metal clip or ball of clay so that it is front of the AC. This allows it to achieve stable flight. If the weight is removed, the glider will pitch up or down (depending on the initial angle of attack) when launched because the c.g. is now behind the a.c. and the aircraft is statically unstable.

5.4.3 Neutral Point and Static Margin

For neutral static stability, C_{m_α} will be equal to zero. This equates to a horizontal line on a C_m vs alpha graph as shown in Fig. 5.17.

As shown, the aircraft will pitch up for any angle of attack condition. This may not seem consistent with the definition of neutral static stability discussed in Sec. 5.1. However, because of the Eq. (5.62) requirement for $C_{m_0} + C_{m_{i_h}} i_h$ to be greater than zero, notice that a C_{m_α} equal to zero (a zero slope) results in C_m being equal to the constant value of $C_{m_0} + C_{m_{i_h}} i_h$. If $C_{m_0} + C_{m_{i_h}} i_h$ were set equal to zero and we had C_{m_α} equal to zero, then the C_m characteristics would remain zero for any angle of attack (a horizontal line coinciding with the horizontal axis) and theoretically the airplane could be trimmed at any angle of attack and would not return or diverge following a disturbance.

The condition for neutral static stability is important because it represents the boundary between static stability and instability. As seen from Eq. (5.67),

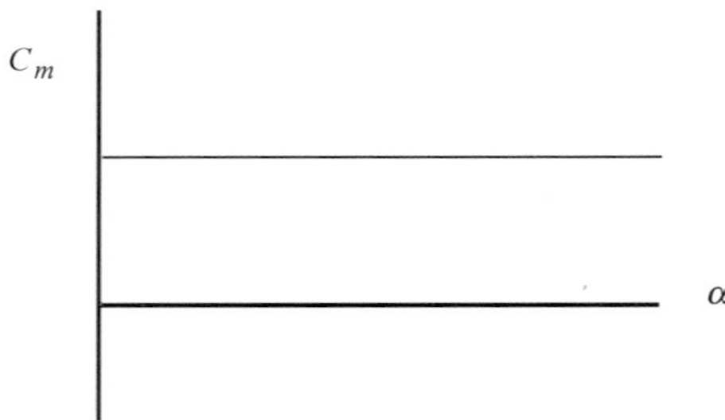

Fig. 5.17 C_m vs alpha characteristics for a neutral longitudinal stability.

C_{m_α} equal to zero is achieved if $\bar{x}_{cg}$ equals $\bar{x}_{AC}$. In physical terms, locating the c.g. at the A.C. of the aircraft will result in an aircraft with neutral static stability. As a result, the aircraft a.c. location is also called the stick fixed neutral point (see Sec. 10.5), or simply the Neutral Point of the aircraft.

$$\bar{x}_{AC} = \bar{x}_{NP} \tag{5.68}$$

If the c.g. is located aft of the neutral point, the aircraft will be statically unstable (longitudinally) and C_{m_α} will be positive. Equation (5.51) allows prediction of the neutral point or aerodynamic center for an aircraft.

Another useful concept to longitudinal static stability is that of static margin. Starting with Eq. (5.67), we define static margin (SM) as

$$SM = \bar{x}_{AC} - \bar{x}_{cg} = -(\bar{x}_{cg} - \bar{x}_{AC}) = \bar{x}_{NP} - \bar{x}_{cg} \tag{5.69}$$

Notice that the static margin is simply the distance that the c.g. is in front of the aircraft a.c. normalized (divided) with respect to the mean wing chord. Because C_{L_α} will be positive for lift coefficients below stall, it can be seen that a positive static margin will result in a negative C_{m_α}. Thus, a positive static margin results in positive longitudinal static stability. A negative static margin results in negative static stability and neutral longitudinal static stability is associated with a static margin equal to zero.

Referring to Eqs. (5.67) and (5.69), we have

$$C_{m_\alpha} = -C_{L_\alpha}(SM) \tag{5.70}$$

and

$$SM = -\frac{C_{m_\alpha}}{C_{L_\alpha}} = -C_{m_{C_L}} \tag{5.71}$$

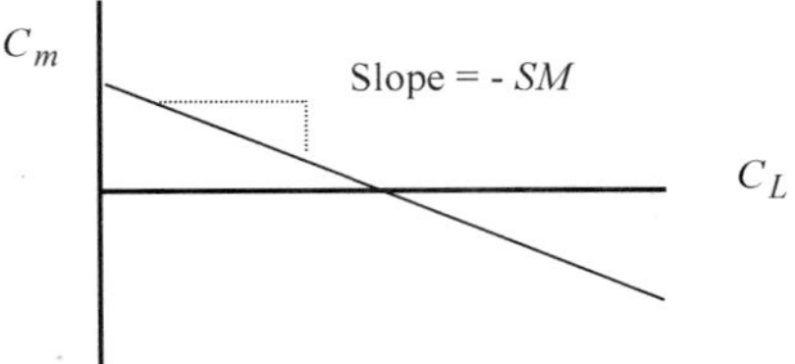

Fig. 5.18 C_m vs C_L characteristics for an aircraft with a positive static margin.

Thus, the derivative $C_{m_{C_L}}$, the slope of a plot of C_m vs C_L (which is similar to a C_m vs alpha plot), is directly proportional to the static margin. The negative of the slope of a C_m vs C_L graph will be the static margin as indicated in Fig. 5.18.

If we take the derivative of Eq. (5.64) with respect to lift coefficient, we have

$$\frac{\partial \delta_e}{\partial C_L} = -\frac{C_{m_{C_L}}}{C_{m_{\delta_e}}} \tag{5.72}$$

This relationship becomes useful from a flight test standpoint because δ_e can be instrumented and measured on an aircraft, and C_L can be computed for trimmed flight if weight, airspeed, and altitude are recorded. As the c.g. of the aircraft is moved aft toward the tail, $C_{m_{C_L}}$ becomes less negative and eventually equals zero when the c.g. is located at the neutral point. Thus, the neutral point is achieved when the change in δ_e with respect to C_L is equal to zero. This relationship is illustrated in Fig. 5.19.

To determine the neutral point location in flight test, a series of trim points are flown with several different c.g. locations to generate curves similar to those presented in Fig. 5.19. Of course, the aircraft is not flown with the c.g. at the neutral point because neutral stability would result. The location of the

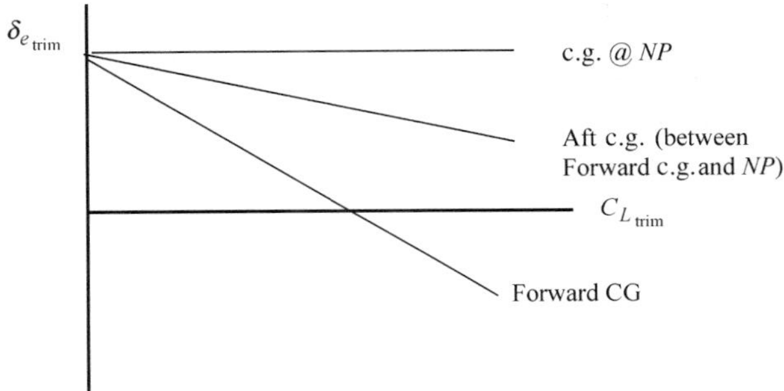

Fig. 5.19 Trim δ_e vs C_L.

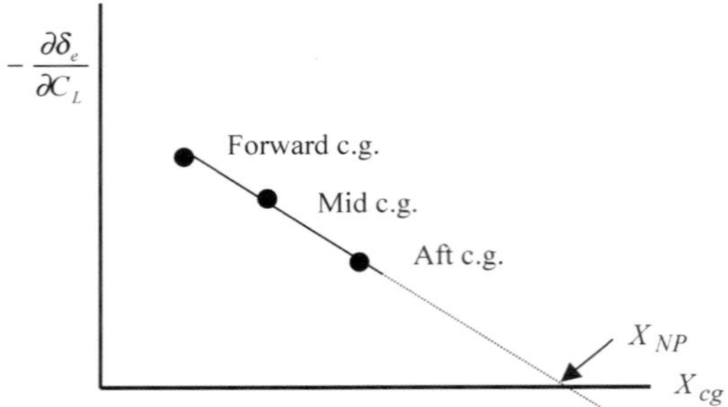

Fig. 5.20 Flight test neutral point determination.

neutral point is extrapolated by plotting the slope of each δ_e vs C_L line (for each c.g. location) with respect to c.g. location as illustrated in Fig. 5.20.

Example 5.2

For the A-10 of Example 5.1, determine the static margin, C_{m_α}, and $C_{m_{C_L}}$ if $\bar{x}_{cg}$ is 0.5. Is the aircraft statically stable in pitch?

From Eq. (5.69), the static margin is

$$SM = \bar{x}_{AC} - \bar{x}_{cg} = 0.602 - 0.5 = 0.102$$

Because the static margin is positive, the aircraft is statically stable in pitch. We can also see this because the c.g. is in front of the aircraft AC. To determine C_{m_α} we will use Eq. (5.70) but first we must determine C_{L_α} for the aircraft. From Eq. (5.30) we have,

$$C_{L_\alpha} = C_{L_{\alpha_{wf}}} + C_{L_{\alpha_h}} \eta_h \frac{S_h}{S}\left(1 - \frac{d\varepsilon}{d\alpha}\right) = 0.082 + 0.0631(0.9)\frac{120}{506}(1 - 0.1)$$

$$C_{L_{\alpha_{\text{aircraft}}}} = 0.0941/\text{deg}$$

Using Eq. (5.70), we have

$$C_{m_\alpha} = -C_{L_\alpha}(SM) = -(0.0941)(0.102) = -0.0096/\text{deg}$$

Eq. (5.71) is used to determine $C_{m_{C_L}}$.

$$C_{m_{C_L}} = -SM = -0.102$$

5.4.4 Maneuvering Flight and Maneuver Point

Longitudinal maneuvering flight includes maneuvers where the load factor is not equal to one and the aircraft has a pitch rate. Typical maneuvering flight involves load factors above one where the aircraft is pulling g. To analyze this condition from a pitching moment standpoint, Eqs. (5.4) and (5.61) must be modified with an additional term called the pitch damping derivative (C_{m_q}). Equation (5.4) becomes

$$C_m = C_{m_0} + C_{m_\alpha}\alpha + C_{m_{\delta_e}}\delta_e + C_{m_{i_h}} i_h + C_{m_q} Q \tag{5.73}$$

For trim, Eq. (5.61) becomes

$$0 = C_{m_0} + C_{m_\alpha}\alpha + C_{m_{\delta_e}}\delta_e + C_{m_{i_h}} i_h + C_{m_q} Q \tag{5.74}$$

C_{m_q} provides a damping moment that opposes the pitch rate of the aircraft. It results from the incremental change in horizontal tail angle of attack because the aircraft has a pitch rate. This damping effect is similar to the effect a shock absorber would have if it were attached to the tail of a suspended aircraft that was allowed to rotate in pitch about the c.g. Figure 5.21 illustrates the change in horizontal tail angle of attack that provides the damping effect on a tail aft aircraft configuration.

A positive (nose up) pitch rate on the aircraft (Q_1) produces a downward velocity at the horizontal tail, $Q_1 X_h$. This downward velocity produces an increase in the angle of attack at the horizontal tail, $\Delta\alpha_h$.

$$\Delta\alpha_h \cong \frac{Q_1 X_h}{U_1} \tag{5.75}$$

using the small angle assumption and keeping $\Delta\alpha_h$ in radians.

The increased angle of attack, in turn, produces an increase in lift, which results in a nose down moment ($\Delta L(X_h)$) opposing the direction of the pitch rate, Q_1. This is where the pitch damping effect comes from. Pitch damping provides a stabilizing effect on the dynamic stability of the aircraft because it tends to damp out oscillations. It also changes the elevator required for trim in maneuvering flight over that predicted with Eq. (5.65) for level 1-g flight.

We previously defined in Eq. (5.21) the angle of attack at the horizontal tail as

$$\alpha_h = \alpha + i_h + \tau_e\delta_e - \varepsilon$$

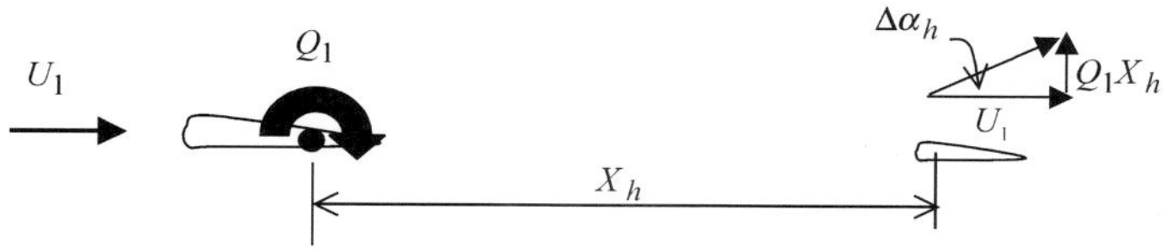

Fig. 5.21 Change in horizontal tail angle of attack because of pitch rate.

where

$$\tau_e = \frac{\partial \alpha_h}{\partial \delta_e}$$

For maneuvering flight, $\Delta\alpha_h$ must be added to the expression for α_h, or

$$\alpha_h = \alpha + i_h + \tau_e \delta_e - \varepsilon + \Delta\alpha_h \tag{5.76}$$

To determine the change in elevator deflection required to maintain a maneuvering trim lift coefficient vs a level flight trim lift coefficient (or the elevator deflection required to offset $\Delta\alpha_h$), we first realize that τ_e, i_h, and ε are constant for a given trim lift coefficient. The change in δ_e ($\Delta\delta_e$) required to maintain a maneuvering flight trim lift coefficient can be found by realizing that an additional aerodynamic force must be generated on the horizontal tail in the opposite direction to the increased lift produced by $\Delta\alpha_h$. This offsetting force is generated with a change in elevator deflection where

$$\tau_e(\Delta\delta_e) + \Delta\alpha_h = 0$$

or

$$\Delta\delta_e = -\frac{\Delta\alpha_h}{\tau_e}$$

From Eq. (5.75),

$$\Delta\delta_e = -\frac{Q_1 X_h}{U_1 \tau_e} \tag{5.77}$$

Equation (5.77) will have a negative sign (because τ_e is positive), which indicates that increased trailing edge up elevator (aft stick) is required to hold an aircraft in a positive (nose up) pitch rate maneuver. This increased elevator deflection offsets the pitch damping moment produced by $\Delta\alpha_h$ so that a given trimmed lift coefficient can be maintained in steady-state maneuvering flight.

Equation (5.77) can be combined with Eq. (5.65) to develop the expression for the elevator deflection required to trim in maneuvering flight.

$$\delta_{e_{\text{maneuvering flight}}} = \delta_{e_0} - \frac{C_{m_{C_L}} C_{L_{\text{trim}}}}{C_{m_{\delta_e}}} + \Delta\delta_e$$

or

$$\delta_{e_{\text{maneuvering flight}}} = \delta_{e_0} - \frac{C_{m_{C_L}} C_{L_{\text{trim}}}}{C_{m_{\delta_e}}} - \frac{Q_1 X_h}{U_1 \tau_e} \tag{5.78}$$

For a pull-up, recall from Sec. 3.9.3.

$$Q_1 = \frac{g(n-1)}{U_1}$$

Equation (5.78) then becomes

$$\delta_{e_{\text{maneuvering flight}}} = \delta_{e_0} - \frac{C_{m_{C_L}} C_{L_{\text{trim}}}}{C_{m_{\delta_e}}} - \frac{g(n-1)X_h}{U_1^2 \tau_e} \tag{5.79}$$

for a pull-up.

In addition, we know that $C_{L_{\text{trim}}}$ for maneuvering flight is

$$C_{L_{\text{trim}}} = \frac{nW}{\bar{q}S}$$

Combining, we have

$$\delta_{e_{\text{maneuvering flight}}} = \delta_{e_0} - \frac{C_{m_{C_L}} nW}{C_{m_{\delta_e}} \bar{q}S} - \frac{g(n-1)X_h}{U_1^2 \tau_e} \tag{5.80}$$

Equation (5.80) is interesting to evaluate. As load factor increases, the elevator deflection required for trim increases in a negative sense (more trailing edge up). Remember, τ_e is positive. Also, as the elevator effectiveness (τ_e) increases, the elevator required for trim decreases based on the last term decreasing. As the c.g. is moved aft and $C_{m_{C_L}}$ becomes less negative, less trailing edge up δ_e is required for trim. As the velocity (U_1) increases, $\delta_{e_{\text{maneuvering flight}}}$ increases, and as the wing loading (W/S) increases, $\delta_{e_{\text{maneuvering flight}}}$ becomes more negative.

The load factor sensitivity of an aircraft is a characteristic that pilots relate to. This can be expressed as the derivative of Eq. (5.80) with respect to load factor.

$$\frac{\partial \delta_e}{\partial n} = -\frac{C_{m_{C_L}} W}{C_{m_{\delta_e}} \bar{q}S} - \frac{gX_h}{U_1^2 \tau_e} \tag{5.81}$$

The second term in Eq. (5.81) is the maneuvering stability contribution. Analogous to the one g case presented in Eq. (5.72), when $\frac{\partial \delta_e}{\partial n}$ is equal to zero, we have neutral static stability for the maneuvering aircraft case. The c.g. location where this occurs is called the **maneuver point**.

$$\frac{\partial \delta_e}{\partial n} = 0 \text{ @ the maneuver point} \tag{5.82}$$

It is logical to ask at this point how the maneuver point compares to the neutral point. The maneuver point is located aft of the neutral point because the added damping provided by C_{m_q} provides more stability for the maneuvering flight case. C_{m_q} essentially makes the aircraft appear to have a more negative C_{m_α} (or $C_{m_{C_L}}$). To define the maneuver point in flight test, a range of c.g. are flown with the pilot gradually increasing load factor for each c.g. configuration so that the derivative $\partial\delta_e/\partial n$ can be evaluated. This is accomplished by recording load factor and elevator position during the maneuver. The data is then plotted vs c.g. position and extrapolated to determine the maneuver point as shown in Fig. 5.22.

In summary, we used the relationship $\partial\delta_e/\partial C_L$ equal to zero to find the neutralpoint, and the relationship $\partial\delta_e/\partial n$ equal to zero to find the maneuver point. All of these discussions are for the stick-fixed case where the elevator is not allowed to float when the aircraft is in one g or maneuvering flight. This is the case for aircraft with irreversible flight control systems (that is, hydraulic powered) and for aircraft with reversible control systems where the pilot maintains the stick in a fixed position (see Sec. 10.5 for a discussion of reversible and irreversible flight control systems). A discussion of the stick-free case may be found in Ref. 1. As might be expected, the stick-free case occurs with reversible flight control systems where the stick is not fixed and the elevator is allowed to float. For this situation, the static stability of the aircraft is degraded and different relationships apply for determination of the stick-free neutral and maneuver points.

5.5 Lateral-Directional Applied Forces and Moments

Because we have assumed that longitudinal and lateral-directional motion are independent of each other, lateral-directional motion is assumed to consist of roll and yaw rotation and y-axis translation. These two rotations and the translation are typically coupled (that is, they occur together). We will now expand the applied lateral-directional aero force and moment terms with conventional aerodynamic coefficients. In the stability axis, we have

$$F_{A_{y_{1_s}}} = F_{A_y} \quad \text{(side force)} \tag{5.83}$$

$$L_{A_{1_s}} = L_A \quad \text{(rolling moment)} \tag{5.84}$$

$$N_{A_{1_s}} = N_A \quad \text{(yawing moment)} \tag{5.85}$$

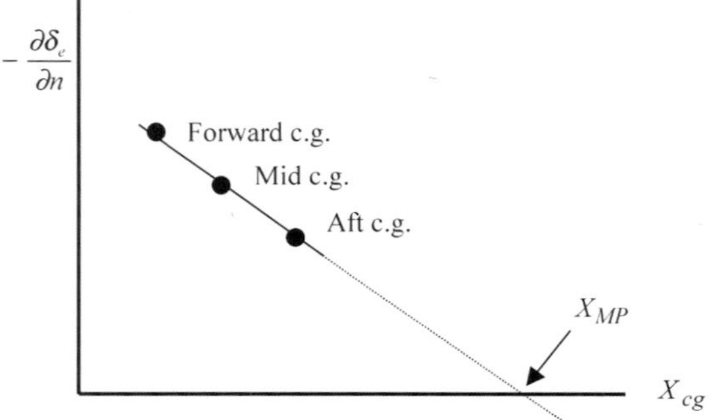

Fig. 5.22 Flight test determination of maneuver point.

Because the body-fixed stability axis will be used consistently for this steady-state analysis, the subscript 1_s will not be carried along in the analysis, but implied as previously indicated.

5.5.1 Aircraft Side Force

Aerodynamic side-force acts along the number two stability axis (positive out the right wing) and may be expressed using the side-force coefficient (C_y) as

$$F_{A_y} = C_y \bar{q} S \tag{5.86}$$

Side force is a function of the angle of sideslip (beta or β), aileron deflection (δ_a), rudder deflection (δ_r), angle of attack (α), Mach number, and Reynolds number. A positive sideslip angle (β) is defined in Fig. 5.26. It can be easily remembered as positive β is 'wind in the right ear' for the pilot. Our Taylor series expansion of the side-force coefficient will include the first three terms.

$$C_y = C_{y_0} + C_{y_\beta}\beta + C_{y_{\delta_a}}\delta_a + C_{y_{\delta_r}}\delta_r \tag{5.87}$$

C_{y_0} is the value of C_y for β, δ_a, and δ_r all equal to zero. It is typically equal to zero for symmetrical aircraft; however, at high angles of attack, aircraft with long slender nose configurations may have a nonzero C_{y_0} because of the shedding of asymmetrical vortices. The derivatives C_{y_β}, $C_{y_{\delta_a}}$, $C_{y_{\delta_r}}$ and C_{y_0} must be defined at appropriate Mach number, Reynolds number, and angle of attack conditions.

The derivative C_{y_β} is the change in side force coefficient because of a change in sideslip angle (at constant angle of attack). It has an important influence on dutch roll dynamics. The vertical tail is the primary aircraft component that influences this derivative. C_{y_β} is normally negative because positive sideslip will typically result in a side force along the negative y axis (out the left wing), and a negative sideslip angle will result in a positive side force. The vertical tail can be thought of as a vertical wing with sideslip playing an analogous role to angle of attack. A method to estimate C_{y_β} based on aircraft configuration begins with the definition of the aero sideforce acting on the vertical tail using the sideforce coefficient.

$$F_{A_{y_{\text{vertical tail}}}} = C_y \bar{q}_1 S$$

and

$$\frac{\partial F_{A_{y_{\text{vertical tail}}}}}{\partial \beta} = \frac{\partial C_{y_{\text{vertical tail}}}}{\partial \beta} \bar{q}_1 S = C_{y_{\beta_{\text{vertical tail}}}} \bar{q}_1 S$$

The contribution of the vertical tail to C_{y_β} can be estimated using an approach similar to that used to develop Eq. (5.30). This results in the following equation

$$C_{y_{\beta_{\text{vertical tail}}}} = -C_{L_{\alpha_{\text{vertical tail}}}} \left(1 - \frac{\partial\sigma}{\partial\beta}\right) \eta_{\text{vertical tail}} \frac{S_{\text{vertical tail}}}{S} \tag{5.88}$$

where σ is the sidewash angle (analogous to the downwash angle) and $\eta_{\text{vertical tail}}$ is $\bar{q}_{\text{vertical tail}}/\bar{q}_\infty$. Note the similarity of this equation to Eq. (5.30) for C_{L_α}. The sidewash derivative, $\partial\sigma/\partial\beta$, is normally small. A complete estimate of C_{y_β} should also include the contribution of the wing and fuselage. When computing, $C_{L_{\alpha_{\text{vertical tail}}}}$, Eq. (1.30) may be used with the following approximation for the aspect ratio of the vertical tail:

$$AR_{\text{vertical tail}} \approx \frac{2(h_{\text{vertical tail}})^2}{S_{\text{vertical tail}}} \tag{5.89}$$

Equation (5.89) provides a rough estimate for the effective aspect ratio of the vertical tail based on experimental data for simple aircraft configurations. It attempts to account for the end-plate effect caused by the horizontal stabilizer and fuselage. In Eq. (5.89) $h_{\text{vertical tail}}$ is the height of the vertical tail and $S_{\text{vertical tail}}$ is the planform area of the vertical tail.

The derivative $C_{y_{\delta_a}}$ is generally insignificant or negligible. However, it can be significant if rolling moment is generated with control surfaces that are in close proximity to a vertical surface. For example, on the F-111 aircraft differential deflection of the left and right stabilators was used to produce rolling moment when the wings were in the swept back position. This produced a differential pressure on the vertical tail, which resulted in a side force as illustrated in Fig. 5.23 with the B-1 aircraft.

The derivative $C_{y_{\delta_r}}$ is positive because a positive rudder deflection (trailing edge left) will generate a side force along the positive y axis. The size of the vertical tail in relation to the wing, along with the aspect ratio and sweepback of the vertical tail, determine the magnitude of this derivative.

5.5.2 Aircraft Rolling Moment

Aircraft rolling moment acts about the x axis and may be expressed using the rolling moment coefficient as

$$L_a = C_l \bar{q} S b \tag{5.90}$$

The rolling moment coefficient is a function of the same parameters we considered for side force; namely, sideslip angle, aileron deflection, rudder deflection, angle of attack, Mach number, and Reynolds number. In addition, dynamic pressure and center of gravity also influence the rolling moment coefficient.

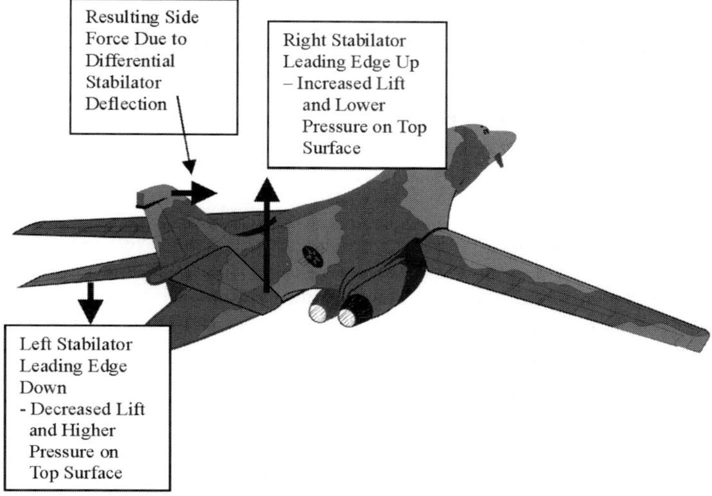

Fig. 5.23 Side force resulting from differential stabilator deflection.

We will again use sideslip angle, aileron deflection, and rudder deflection in our first-order Taylor series expansion

$$C_l = C_{l_0} + C_{l_\beta}\beta + C_{l_{\delta_a}}\delta_a + C_{l_{\delta_r}}\delta_r \tag{5.91}$$

C_{l_0} is the value of C_l for β, δ_a and δ_r all equal to zero. It is typically equal to zero for symmetrical aircraft; however, it may not be typical for aircraft with long, slender fore-bodies where asymmetric vortex shedding is present. The derivatives C_{l_β}, $C_{l_{\delta_a}}$, $C_{l_{\delta_r}}$, and C_{l_0} must be defined at appropriate angle of attack, Mach number, Reynolds number, dynamic pressure, and c.g. conditions.

5.5.2.1 C_{l_β}. C_{l_β} is the lateral (roll) static stability derivative. It is also sometimes called the dihedral effect. As will be seen in Sec. 5.6.2.1, the sign of C_{l_β} must be **negative** if an aircraft has roll static stability. A negative C_{l_β} simply implies that the aircraft generates a rolling moment that rolls the aircraft away from the direction of sideslip.

Four aspects of an aircraft design primarily influence C_{l_β}: geometric dihedral, wing position, wing sweep, and the contribution of the vertical tail. In other words,

$$C_{l_\beta} = C_{l_{\beta_{\text{dihedral}}}} + C_{l_{\beta_{\text{wing position}}}} + C_{l_{\beta_{\text{wing sweep}}}} + C_{l_{\beta_v}}$$

First, the wing **geometric dihedral** angle, Γ (capital gamma), as illustrated in Fig. 5.24, provides a significant negative contribution to C_{l_β}. The larger the dihedral angle, the more negative rolling moment will result from a positive sideslip angle and the more positive rolling moment will result from a negative sideslip angle. This occurs because the wing toward the relative wind (right wing for positive sideslip and left wing for negative sideslip) experiences a higher angle of attack than that experienced by the opposite wing. The dihedral contribution to C_{l_β} may be estimated with the following approximation (reference USAF Academy Glider Design Program) where λ is the wing taper ratio as defined in Eq. (1.25) and Γ must be in radians.

$$C_{l_{\beta_{\text{dihedral}}}} \approx -\frac{1}{6} C_{L_{\alpha_{\text{wing}}}} \Gamma \frac{1+2\lambda}{1+\lambda} \tag{5.92}$$

Several aircraft such as the F-104, C-5, and C-141, are designed with a negative dihedral angle or anhedral (the wings are bent down). This design feature provides a positive contribution to C_{l_β} and is normally used to reduce the magnitude of roll stability because too much roll stability (a large negative value of C_{l_β}) will have an adverse influence on dutch roll, engine out, and crosswind landing characteristics.

Wing position on the fuselage is the second design aspect that influences C_{l_β}. A high wing position will provide a negative contribution to C_{l_β}, a low wing position will provide a positive contribution, and a mid-wing position will provide a fairly neutral contribution. Because sideslip induces a crossflow over the fuselage as shown in Fig. 5.25, a high wing configuration will experience a higher wing angle of attack near the wing-fuselage intersection on the wing toward the relative wind. Likewise, a lower angle of attack will be experienced on the opposite wing near the wing-fuselage juncture, and the asymmetric lift that results will roll the aircraft away from the direction of sideslip. The opposite effect will occur with a low wing configuration because the wing toward the relative wind will experience a reduction in angle of attack and the wing away from the relative wind will experience an increase in angle of attack. Thus, a low wing position will tend to roll the aircraft toward the direction of sideslip. A mid-wing experiences little change in angle of attack near the wing-fuselage juncture and consequently has little effect on C_{l_β}.

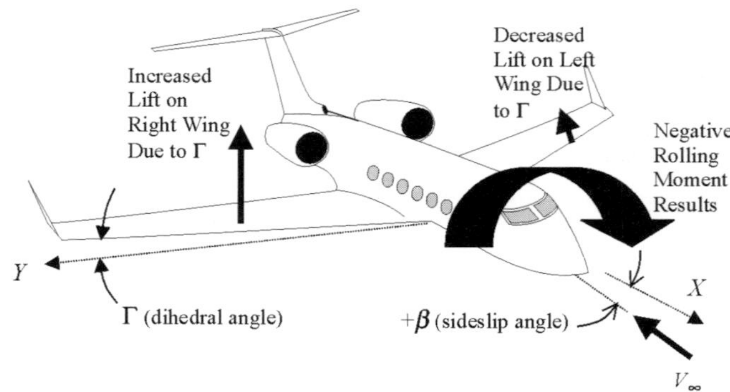

Fig. 5.24 Wing geometric dihedral effects.

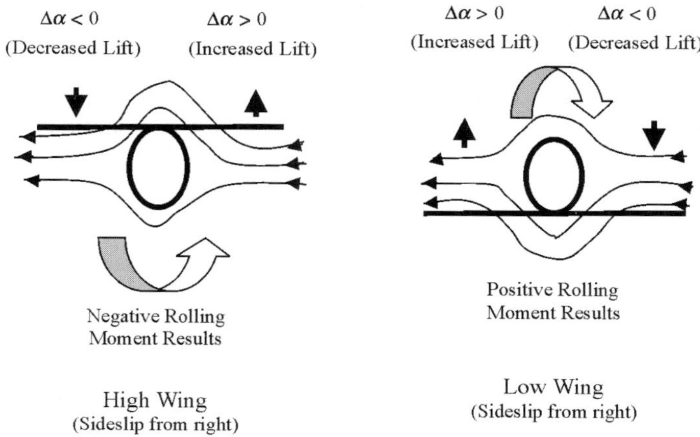

Fig. 5.25 Rolling moment because of sideslip with high and low wing positions.

A fairly complex estimate for $C_{l_{\beta_{\text{wing position}}}}$ in terms of aircraft geometry parameters can be found in Ref. 3 Eq. (10.34). Horizontal tail position may also contribute to C_{l_β} in a similar manner, and the referenced equation may be used to estimate this effect.

Wing sweep angle is the third design aspect that influences C_{l_β}. As seen in Fig. 5.26, a sideslip angle results in a side velocity that can be broken into vector components normal and parallel to the leading edge of each wing. With aft sweep, the wing toward the velocity vector (the leading wing) has a larger

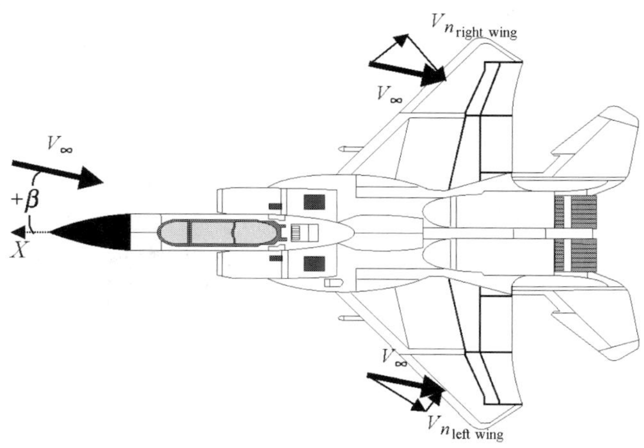

Fig. 5.26 Rolling moment because of sideslip with an aft-swept wing.

normal velocity component than the wing opposite the velocity vector (the trailing wing). As a result, the upstream wing will produce more lift than the downstream wing (resulting in a rolling moment away from the sideslip direction), and a negative contribution will result for C_{l_β}. A simplified estimate of the wing sweep contribution to C_{l_β} is presented in Eq. (5.93) for aft sweep (Reference USAF Academy Glider Design Program).

$$C_{l_{\beta_{\text{wing sweep}}}} = -2C_{L_{\text{wing}}}\left(\frac{y_{AC_{\text{wing}}}}{b}\right)\sin(2\Lambda_{\text{leading edge}}) \tag{5.93}$$

In using this equation, $y_{AC_{\text{wing}}}$ is the positive y-body axis distance to the wing a.c., $\Lambda_{\text{leading edge}}$ is the sweep of the wing leading edge in radians, and b is the wing span. Notice that $C_{l_{\beta_{\text{wing sweep}}}}$ is directly dependent on the wing lift coefficient. The analysis is reversed for forward-swept wings. Forward wing sweep provides a positive contribution to C_{l_β}, and Eq. (5.93) may be used as an estimate of this contribution by simply dropping the negative sign.

The last design aspect we will consider is the **vertical tail**. A positive sideslip angle will result in an aerodynamic force on the vertical tail in the negative y-axis direction. Because the vertical tail is normally above the x (or rolling) axis of the aircraft, this aerodynamic force produces a negative rolling moment that results in a negative contribution to C_{l_β}. A similar analysis holds for negative sideslip angles. The larger and higher the vertical tail, the more negative the contribution to C_{l_β}. An estimate of the vertical tail contribution to C_{l_β} begins with Eq. (5.88) which defines $C_{y_{\beta_{\text{vertical tail}}}}$.

$$C_{l_{\beta_{\text{vertical tail}}}} = \frac{z_v}{b}C_{y_{\beta_{\text{vertical tail}}}} \tag{5.94}$$

z_v is the vertical distance (z stability axis direction with a change in sign convention: "up" is positive in the definition of z_v) that the aerodynamic center of the vertical tail is above the aircraft c.g. z_v is normally positive; however, at high angles of attack, some aircraft may have a negative value for z_v (the a.c. of the tail below the c.g., see Fig. 5.27).

z_v may be estimated with the following equation for the low angle of attack case:

$$z_v = \frac{1}{3}(h_{\text{vertical tail}})\frac{[1 + 2\lambda_{\text{vertical tail}}]}{[1 + \lambda_{\text{vertical tail}}]}$$

The span, b, is included in Eq. (5.93) to normalize z_v. Substituting Eq. (5.88) into Eq. (5.94), we have:

$$C_{l_{\beta_{\text{vertical tail}}}} = -\frac{z_v}{b}C_{L_{\alpha_{\text{vertical tail}}}}\left(1 - \frac{\partial\sigma}{\partial\beta}\right)\eta_{\text{vertical tail}}\frac{S_{\text{vertical tail}}}{S} \tag{5.95}$$

A T-tail, such as on the F-104, C-141, and C-5, will increase $C_{L_{\alpha_{\text{vertical tail}}}}$ because the T-tail serves as an end plate to effectively increase the aspect ratio of the

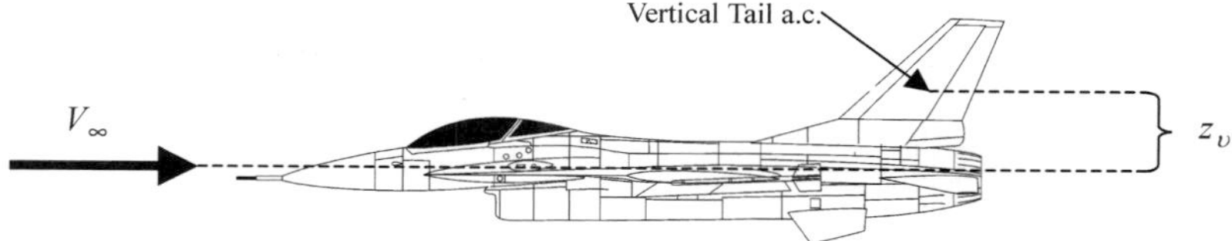

(a) Illustration of a Positive z_v at a Low Angle of Attack

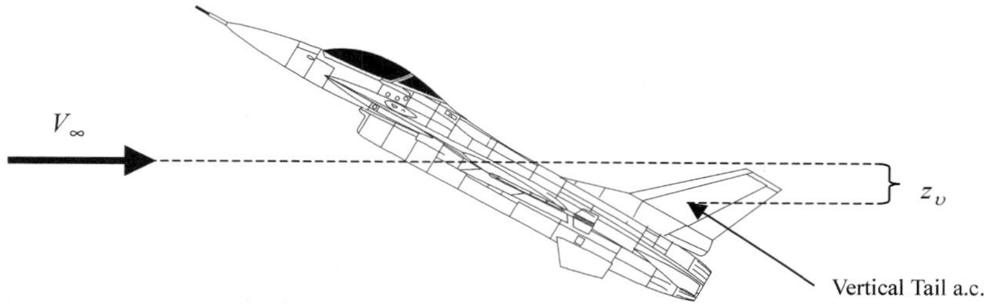

(b) Illustration of a Negative z_v at a High Angle of Attack

Fig. 5.27 Illustration of vertical tail moment arm (z_v).

vertical tail by minimizing the tip vortex. This effect results in an increase in the magnitude of $C_{l_{\beta_{\text{vertical tail}}}}$.

To help put these design features in perspective, consider an aircraft such as the A-7 (Fig. 5.28). It has a high wing, aft sweep, and a large vertical tail. These features result in a large negative value for C_{l_β}. To reduce the absolute value of C_{l_β} (make it less negative) so that dutch roll characteristics and cross-wind landings will not be a problem, anhedral was included in the design.

Figure 5.29 presents rolling moment wind tunnel data for the F-16 VISTA aircraft at zero and 35 deg angle of attack.[4] Notice the decrease in slope (C_{l_β}) as angle of attack is increased. Angle of attack usually has a very pronounced effect on C_{l_β}.

5.5.2.2 $C_{l_{\delta_a}}$. Ailerons are typically the primary control surface for producing rolling moment in response to a pilot command. A positive aileron deflection results in a positive rolling moment about the x axis (see Sec. 5.2). Ailerons are generally not deflected symmetrically so that adverse yaw effects can be minimized. For example, in response to a right stick input, the right aileron may have a larger trailing edge up deflection that the left aileron has a trailing edge down deflection. As discussed in Sec. 5.2, we define the magnitude of aileron deflection (using the convention that trailing edge down is positive) as

$$\delta_a = \frac{1}{2}[\delta_{a_{\text{left}}} - \delta_{a_{\text{right}}}] \tag{5.96}$$

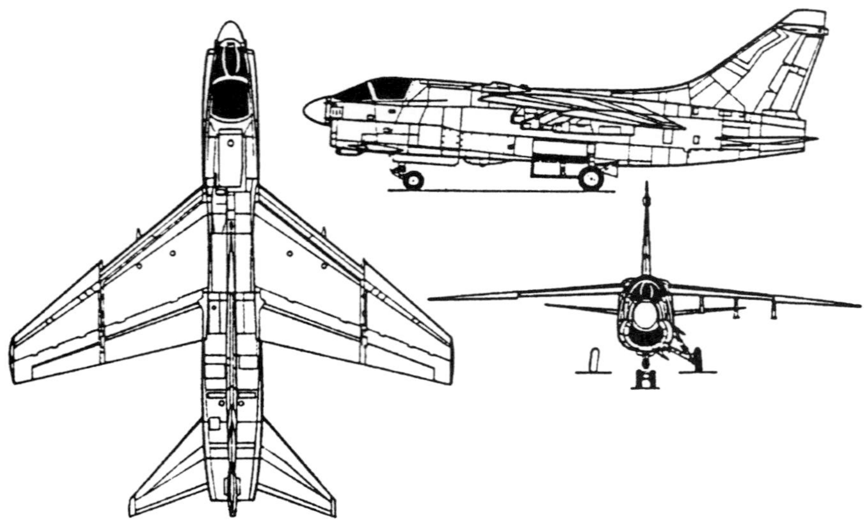

Fig. 5.28 Three-view drawing of A-7 corsair.

The derivative $C_{l_{\delta_a}}$ defines the change in rolling moment that results from aileron deflection. It is also called the **aileron control power**. $C_{l_{\delta_a}}$ is **positive** based on the definition of a positive aileron deflection. The magnitude of $C_{l_{\delta_a}}$ depends on several factors. The aileron chord to wing chord ratio is a measure of the relative size of the aileron in terms of wing chord. The larger the ratio, the larger $C_{l_{\delta_a}}$ becomes. The aileron span location on the wing determines the moment arm and length of the ailerons. The larger the moment arm (the further outboard) and the longer the length, the larger $C_{l_{\delta_a}}$ becomes. The magnitude of aileron deflection is also a factor in defining the magnitude of $C_{l_{\delta_a}}$. Aileron deflections greater than about 20 deg may result in flow separation over the top surface, which will decrease $C_{l_{\delta_a}}$. The wing angle of attack must also be considered. For conditions near wing stall, a small downward aileron deflection can result in flow separation and a decrease in $C_{l_{\delta_a}}$. Another factor to be considered is the wing sweep angle. Sweep angles greater than approximately 55 deg generally result in outward spanwise flow, which is nearly parallel to the aileron hinge line. This will reduce the magnitude of $C_{l_{\delta_a}}$. $C_{l_{\delta_a}}$ also depends on dynamic pressure because of aeroelastic effects when our assumption of a rigid body does not hold. Many high-performance aircraft experience aileron reversal above a specific dynamic pressure because the twisting loads on the wing resulting from an aileron deflection (combined with the flexibility of the wing structure) are sufficient to make $C_{l_{\delta_a}}$ negative. Mach number is another factor that influences $C_{l_{\delta_a}}$. For example, in the tran-sonic region shock-wave formation can reduce the magnitude of $C_{l_{\delta_a}}$.

Some airplane designs also use spoiler and/or differential stabilizer control to generate rolling moment. For these designs, appropriate additional terms

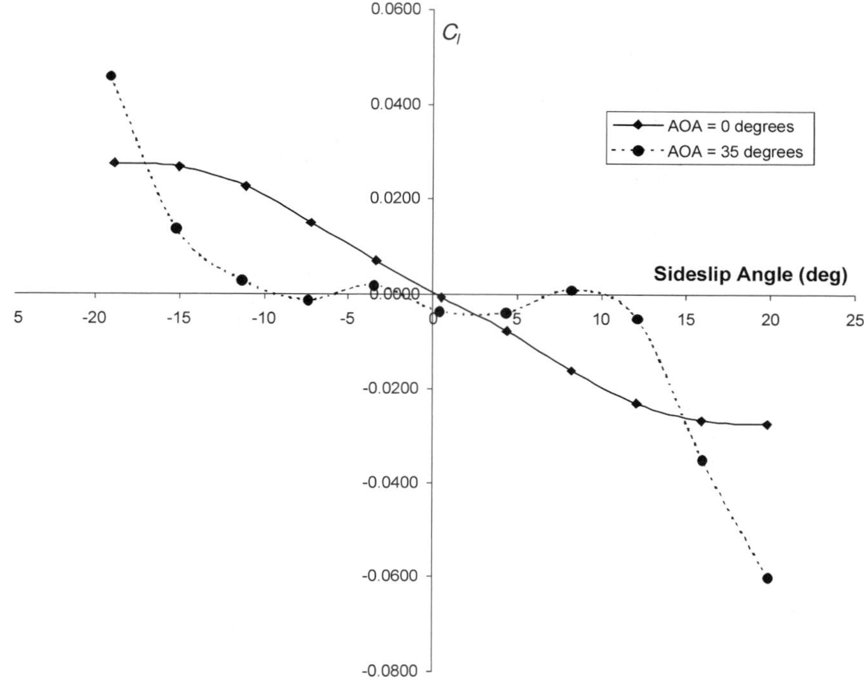

Fig. 5.29 Rolling moment wind tunnel data for F-16 VISTA aircraft.

must be included in the Taylor series expansion for rolling moment [Eq. (5.91)]. Most of the factors discussed previously also influence the effectiveness of spoilers and/or differential stabilizer control.

Roll control power is generally an important requirement in high-performance aircraft. As an example, the roll performance of the YF-17 was found to be unacceptable during initial Air Force flight evaluations because of the aeroelastic aileron reversal effect. The flight control system was modified during the flight program to use additional differential horizontal tail to increase the overall roll control power. This modification improved the roll control power, however additional improvements were also required.[2] Almost all modern fighter aircraft use differential horizontal tail for roll control in some portion of the flight envelope. In fact, the supersonic RA-5C Vigilante abandoned ailerons altogether. It used multiple sets of spoilers, deflectors, and differential horizontal tail for roll control. Another approach is that of the F-16, which has a single trailing edge control device, a flaperon, that doubles as both aileron and trailing edge flap.

5.5.2.3 $C_{l_{\delta_r}}$. A rudder deflection can also generate a rolling moment. As discussed in Sec. 5.5.1, a positive rudder deflection (trailing edge left) will generate a side force in the direction of the positive y axis. Because the rudder and vertical tail are normally above the rolling axis of the aircraft, this side force also results in a positive rolling moment. Thus, $C_{l_{\delta_r}}$ will be positive for most aircraft. Using the same rationale, a trailing edge right (negative) rudder deflection will result in a negative rolling moment. $C_{l_{\delta_r}}$ can reverse sign from positive to negative for high angle of attack conditions. At high angles of attack, the vertical tail can be positioned below the x stability axis, causing the moment arm to be below the rolling axis of the aircraft.

5.5.3 Aircraft Yawing Moment

Aircraft yawing moment acts about the z axis and may be expressed using the yawing moment coefficient as

$$N_a = C_n \bar{q} S b \tag{5.97}$$

The yawing moment coefficient is a function of the same parameters we considered for rolling moment coefficient, with the exception of dynamic pressure that usually has an insignificant effect. These parameters are again sideslip angle, aileron deflection, rudder deflection, angle of attack, Mach number, Reynolds number, and center of gravity location. Sideslip angle, aileron deflection, and rudder deflection will again be used in our first-order Taylor series expansion.

$$C_n = C_{n_0} + C_{n_\beta}\beta + C_{n_{\delta_a}}\delta_a + C_{n_{\delta_r}}\delta_r \tag{5.98}$$

C_{n_0} is the value of C_n for β, δ_a, and δ_r all equal to zero. Like C_{l_0}, it is typically equal to zero for symmetrical aircraft; however, it may not be for aircraft with long, slender forebodies where asymmetric vortex shedding is present. The derivatives C_{n_β}, $C_{n_{\delta_a}}$, $C_{n_{\delta_r}}$, and C_{n_0} must be defined at appropriate angle of attack, Mach number, Reynolds number, and c.g. conditions.

5.5.3.1 C_{n_β}. C_{n_β} is the directional (yaw) static stability derivative. It is sometimes called the weathercock stability derivative. As will be seen in Sec. 5.6.2.2, the sign of C_{n_β} must be positive if the aircraft has yaw static stability. A positive C_{n_β} implies that in response to a sideslip angle, the aircraft will generate an aerodynamic yawing moment, which tends to reduce or zero-out the sideslip. For example, a positive C_{n_β} will result in a positive yawing moment being generated in response to a positive sideslip angle. This yawing moment will tend to yaw the aircraft toward the relative wind and reduce the sideslip angle. We can also think of this as the weathervane effect.

The **vertical tail** is the primary aircraft component that drives the magnitude of C_{n_β}. The larger the vertical tail, the more positive C_{n_β} will be. The x-axis distance between the c.g. and the a.c. of the tail is another design feature that influences C_{n_β}. The larger this distance, the more positive C_{n_β} will be. $C_{n_{\beta_{\text{vertical tail}}}}$ may be estimated by again starting with Eq. (5.88).

$$C_{n_{\beta_{\text{vertical tail}}}} = -\frac{x_v}{b} C_{y_{\beta_{\text{vertical tail}}}} \tag{5.99}$$

where x_v is the distance from the aerodynamic center of the vertical tail to the aircraft c.g. (along the x stability axis). Substituting Eq. (5.88) into Eq. (5.99), we have

$$C_{n_{\beta_{\text{vertical tail}}}} = C_{L_{\alpha_{\text{vertical tail}}}} \left(1 - \frac{\partial \sigma}{\partial \beta}\right) \eta_{\text{vertical tail}} \frac{S_{\text{vertical tail}} x_v}{Sb} \tag{5.100}$$

A complete estimate of C_{n_β} should also include the contribution of the wing and fuselage.

Aft wing sweep also provides a positive contribution to C_{n_β}. As seen in Fig. 5.26, higher lift is generated on the upstream wing (wing toward the sideslip), resulting in more induced drag than on the downstream wing. This results in a stabilizing yawing moment that will yaw the aircraft toward the relative wind, thus reducing sideslip. Aft wing sweep is the major contribution to C_{n_β} for a flying wing aircraft such as the B-2.

Maintaining a sufficiently positive C_{n_β} at high angles of attack presents a challenge to aircraft designers. Flow separation from the wing may reduce the dynamic pressure experienced by a significant portion of the vertical tail. If this happens, C_{n_β} is reduced in magnitude. A large vertical stabilizer may be designed for the aircraft to maintain sufficient directional stability under these conditions. A better option might be twin vertical tails as used on the F-18. For example, the F-16, with its single vertical tail, loses directional stability at high angles of attack and, as a result, incorporated an angle of attack limiter in the flight control system. Its competitor, the YF-17 (F-18 predecessor) had twin vertical tails canted outboard, which retained the directional stability of the aircraft at high angles of attack better than the F-16 because the lateral location and cant of the tails placed them out of the separated flow from the fuselage.[2] This is one of the reasons that the F-22 and some Russian aircraft, such as the MIG-29, use similar tail designs.

Figure 5.30 presents yawing moment wind tunnel data for the F-16 VISTA aircraft at zero and 35 deg angle of attack.[4] Notice the positive slope indicating a positive C_{n_β} for an angle of attack of 0 deg. Also notice the negative C_{n_β} for 35 deg angle of attack (indicating an unstable situation). Angle of attack usually has a very pronounced effect on C_{n_β} because of the blanking of the vertical tail.

5.5.3.2 $C_{n_{\delta_a}}$. The derivative $C_{n_{\delta_a}}$ defines how yawing moment changes with aileron deflection. For aircraft equipped with conventional ailerons, $C_{n_{\delta_a}}$ is typically **negative**, indicating that **adverse yaw** is generated as a result of the control input. This means that a positive aileron input (right wing down) will have

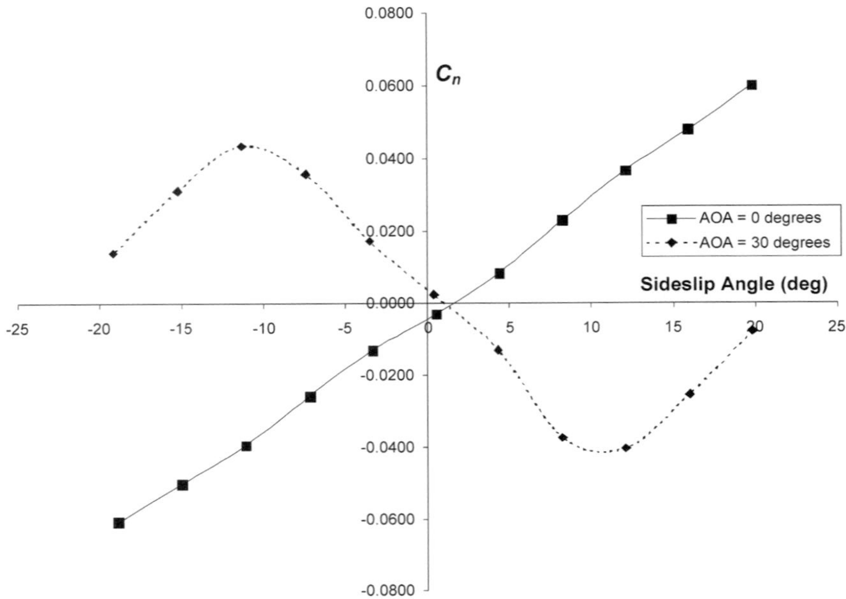

Fig. 5.30 Yawing moment wind tunnel data for F-16 VISTA aircraft.

a nose left yawing moment result. This yawing moment away from the direction of the turn results from the differential induced drag. A TEU aileron deflection reduces the lift on the wing being rolled into, while a TED aileron deflection increases the lift on the wing coming up. With the induced drag higher on the wing coming up, a yawing moment away from the direction of the turn is generated. This is illustrated in Fig. 5.31. Many aircraft incorporate an aileron to rudder interconnect (ARI), which automatically inputs a rudder deflection in the direction of the turn to compensate for adverse yaw.

$C_{n_{\delta_a}}$ may also be **positive**. This is called a **proverse yaw** condition and results when roll control surfaces such as spoilers are used. For example, many sailplanes use differential spoilers to generate a rolling moment. Lift is decreased using spoiler deployment on the wing being rolled into. The spoiler deployment increases drag on the wing at the same time it is decreasing lift. This increased drag generates a yawing moment in the direction of the turn.

The F-4 Phantom incorporated a combination of these ideas to minimize adverse yaw. The lateral control system incorporated both ailerons, spoilers, and an aileron to rudder interconnect. The ailerons were designed to only deflect trailing edge down, while the spoilers would only deploy on the wing being rolled into.

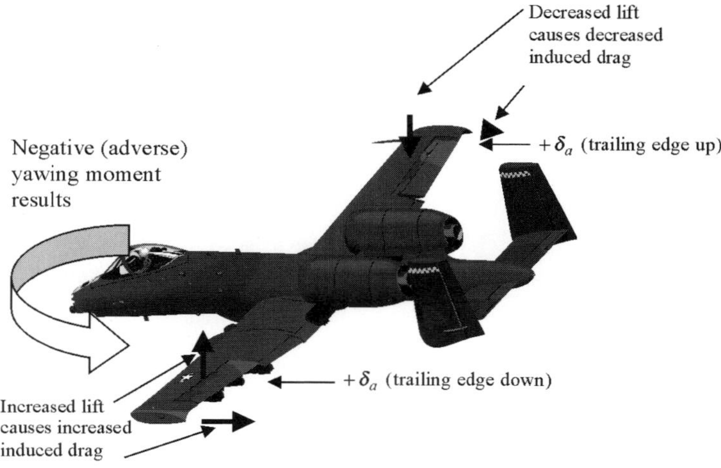

Fig. 5.31 Adverse yawing moment because of aileron deflection.

5.5.3.3 $C_{n_{\delta_r}}$. The rudder is typically the primary control surface for producing a yawing moment in response to a pilot command. As discussed in Sec. 5.2, a positive rudder deflection is defined as trailing edge left. The derivative $C_{n_{\delta_r}}$ defines the change in yawing moment that results from rudder deflection. It is also called the **rudder control power**. $C_{n_{\delta_r}}$ is **negative** because a positive rudder deflection results in a negative yawing moment. The magnitude of $C_{n_{\delta_r}}$ depends on several factors. The rudder chord to vertical tail chord ratio is a measure of the relative size of the rudder in terms of the vertical tail chord. The larger the ratio, the larger the magnitude of $C_{n_{\delta_r}}$. The relative size of the vertical tail/rudder in relation to the wing size is also a directly proportional factor in increasing $C_{n_{\delta_r}}$. In addition, the aspect ratio and sweep of the vertical tail will influence the lift–curve slope. The distance between the c.g. and the a.c. of the vertical tail represents the moment arm that the rudder acts through to produce a yawing moment. The larger this distance, the larger the magnitude of $C_{n_{\delta_r}}$.

Aircraft without a vertical tail such as the B-2 bomber (Fig. 5.32) may use differential split ailerons (sometimes called drag rudders) to generate a yawing moment in response to a pilot command. For advanced configurations such as this, a yawing moment control power derivative is still present with the control input being differential aileron deflection rather than rudder deflection.

5.6 Lateral-Directional Static Stability

As discussed in Sec. 5.1, static stability refers to the initial tendency of an airplane, following a disturbance from steady-state flight, to develop aerodynamic forces and moments that are in a direction to return the aircraft to the

Fig. 5.32 B-2 Bomber.

steady-state flight condition. For purposes of this text, lateral-directional static stability will primarily refer to aircraft rolling moment and yawing moment characteristics. Lateral and directional stability will be discussed separately, but it should be realized that rolling and yawing motions are inherently coupled. This highly coupled behavior necessitates consideration of these motions together, especially when analyzing and designing lateral-directional handling qualities.

5.6.1 Trim Conditions

Lateral-directional trim requirements can be simply stated as achieving a total aircraft rolling moment and yawing moment of zero. In coefficient terms, trim equates to

$$C_l \text{ and } C_n = 0 \quad \text{for trim} \tag{5.101}$$

Implied in the idea of lateral-directional trim is typically the condition of a zero sideslip angle. This condition is more correctly referred to as coordinated flight (beta equal to zero). Most trim conditions are achieved in coordinated flight, but it is certainly possible to achieve the trim requirement of Eq. (5.101) with nonzero sideslip (uncoordinated flight).

The rolling moment coefficient and yawing moment coefficient Taylor series expansions Eqs. (5.91) and (5.98) provide the first step in satisfying the lateral-directional trim requirement, Eq. (5.101). With the assumption of a symmetrical aircraft (C_{l_0} and C_{n_0} equal to zero) and coordinated flight (β equal to zero), zero roll and yaw coefficients are achieved simply with δ_a and δ_r equal to zero. However, for a nonsymmetrical aircraft or a flight condition where sideslip is not equal to zero, Eqs. (5.91) and (5.98) must be worked together to determine appropriate values of aileron and rudder deflection for trim.

Example 5.3

Determine the aileron and rudder deflections required for an F-15 to maintain a +1 degree "wings level" sideslip at 0.9 Mach and 20,000 ft. Determine the value of the sideforce coefficient under these conditions. Applicable derivatives follow:

$$C_{y_0} = 0 \quad C_{y_\beta} = -0.9056/\text{rad} \quad C_{y_{\delta_a}} = -0.0047/\text{rad} \quad C_{y_{\delta_r}} = 0.1492/\text{rad}$$
$$C_{l_0} = 0 \quad C_{l_\beta} = -0.0732/\text{rad} \quad C_{l_{\delta_a}} = 0.0226/\text{rad} \quad C_{l_{\delta_r}} = 0.0029/\text{rad}$$
$$C_{n_0} = 0 \quad C_{n_\beta} = 0.1638/\text{rad} \quad C_{n_{\delta_a}} = 0.0026/\text{rad} \quad C_{n_{\delta_r}} = -0.0712/\text{rad}$$

We start with Eqs. (5.91) and (5.98).

$$C_l = C_{l_0} + C_{l_\beta}\beta + C_{l_{\delta_a}}\delta_a + C_{l_{\delta_r}}\delta_r$$
$$C_n = C_{n_0} + C_{n_\beta}\beta + C_{n_{\delta_a}}\delta_a + C_{n_{\delta_r}}\delta_r$$

For trim, $C_l = C_n = 0$, and the two equations become

$$0 = -0.0732\left(\frac{1 \text{ deg}}{57.3}\right) + 0.0226\delta_a + 0.0029\delta_r$$
$$0 = 0.1638\left(\frac{1 \text{ deg}}{57.3}\right) + 0.0026\delta_a + -0.0712\delta_r$$

We have two equations with two unknowns, δ_a and δ_r. Using conventional or matrix methods such as Cramer's rule to solve, we have,

$$\delta_a = 0.051 \text{ rad} = 2.94 \text{ deg}$$
$$\delta_r = 0.042 \text{ rad} = 2.4 \text{ deg}$$

The result makes sense from a sign standpoint. Left rudder (positive) is needed to generate the sideslip and opposite aileron (positive) is needed to offset the rolling moment generated by the aircraft's lateral stability. To calculate the sideforce coefficient, we use Eq. (5.87).

$$C_y = C_{y_0} + C_{y_\beta}\beta + C_{y_{\delta_a}}\delta_a + C_{y_{\delta_r}}\delta_r$$
$$C_y = -0.9056\left(\frac{1 \text{ deg}}{57.3}\right) + -0.0047(0.051) + 0.1492(0.042)$$
$$C_y = -0.00978$$

Here again, the result makes sense from a direction standpoint. The negative sign indicates a side force in the negative y direction, which is what will result from a positive sideslip angle.

5.6.2 Stability Requirements

As discussed in Secs. 5.5.2.1 and 5.5.3.1, the sign of the stability derivatives C_{l_β} and C_{n_β} is key in determining the lateral and directional static stability of the aircraft.

5.6.2.1 Lateral motion.

The requirement for lateral (roll) static stability is

$$C_{l_\beta} < 0 \tag{5.102}$$

and is illustrated in Fig. 5.33.

This requirement results in an aircraft that generates a rolling moment that rolls the aircraft away from the direction of sideslip. This may seem contrary to common intuition for coordinated turning flight, but we must think about the aircraft without pilot inputs and with fixed control surfaces. If an aircraft is flying straight and level in trimmed flight and encounters air turbulence, which induces a nonzero bank angle, we would like the aircraft to return to wings-level flight through its inherent lateral static stability with no pilot input. If the aircraft is banked to the right, for example, a component of the weight vector produces a sideforce to the right, which results in a velocity along the positive y axis (W). Figure 5.34 illustrates this sideforce with the aircraft experiencing a perturbation in bank angle. The side velocity, when combined with U, the forward velocity, results in a positive sideslip condition. If C_{l_β} is negative, the aircraft will roll away from the sideslip and return to wings level, where it should regain trimmed flight.

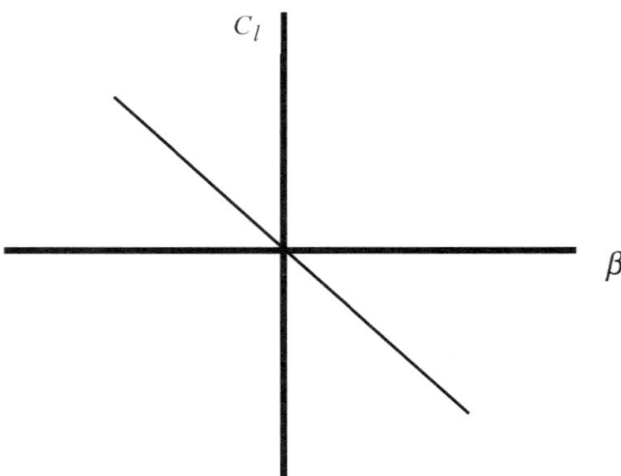

Fig. 5.33 Rolling moment coefficient vs sideslip characteristics for an aircraft with static lateral stability.

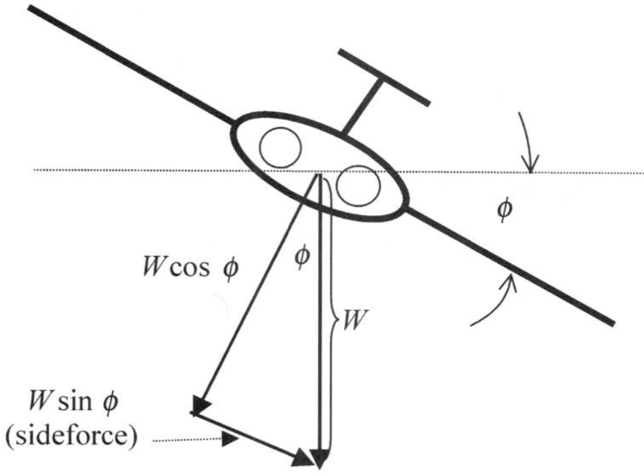

Fig. 5.34 Sideforce vector for an aircraft in a bank.

The lateral stability design features discussed in Sec. 5.5.2.1, such as geometric dihedral, wing position, wing sweep, and vertical tail size, must be balanced in an overall aircraft design to achieve an acceptable degree of lateral stability. Too much lateral stability typically results in unacceptable dutch roll and crosswind landing characteristics. Figure 5.35 compares high-wing and low-wing general aviation aircraft, namely the Cessna 172 and Piper Cherokee (both with zero leading-edge sweep and similar vertical tail size). Notice that the high wing Cessna 172 has no dihedral, whereas the low-wing Cherokee has significant dihedral. On the Cessna, the high wing position provides sufficient lateral stability without dihedral, whereas the Cherokee needs dihedral to overcome the destabilizing influence of the low wing position. Both designs result in acceptable lateral stability.

5.6.2.2 Directional motion. The requirement for directional (yaw) static stability is

$$C_{n_\beta} > 0 \tag{5.103}$$

and is illustrated in Fig. 5.36.

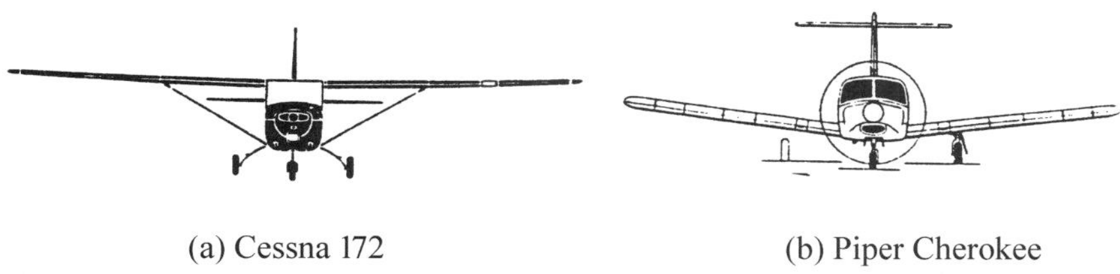

(a) Cessna 172 (b) Piper Cherokee

Fig. 5.35 Two general aviation aircraft.

This requirement results in an aircraft that generates a yawing moment that yaws the nose of the aircraft toward the direction of sideslip. An aircraft with a positive C_{n_β} will generate a positive aerodynamic yawing moment in response to a positive sideslip angle. Directional stability attempts to keep the aircraft in a coordinated flight condition where the sideslip angle is equal to zero.

As discussed in Sec. 5.5.3.1, many aircraft experience a decrease in directional stability (C_{n_β} becoming less positive) at high angles of attack because of separated flow from the wing reducing the dynamic pressure on the vertical tail. The A-7 Corsair (Fig. 5.28) lost directional stability ($C_{n_\beta} < 0$) at sufficiently high angles of attack and would experience a "nose slice" in which the aircraft would yaw away from the direction of sideslip and depart controlled flight.

5.6.3 Engine-Out Analysis

The lateral-directional force and moment equations can be used to analyze the case of an engine failure in flight that results in a yawing moment. Consider a twin-engine aircraft that has experienced a right engine failure (Fig. 5.37).

The left engine is still operating while the right engine is producing windmilling drag. The yawing moment that results is

$$N(\text{engine out}) = N_T + N_{\Delta D} = F_T y_e + \Delta D y_e \tag{5.104}$$

Notice that for the engine out case presented, the asymmetrical thrust and ram drag moments are both in the positive direction. For trimmed flight and assuming C_{n_0} is zero, these terms are included in the directional Taylor series.

$$C_{n_\beta}\beta + C_{n_{\delta_a}}\delta_a + C_{n_{\delta_r}}\delta_r + \frac{F_T y_e}{\bar{q}Sb} + \frac{\Delta D y_e}{\bar{q}Sb} = 0 \tag{5.105}$$

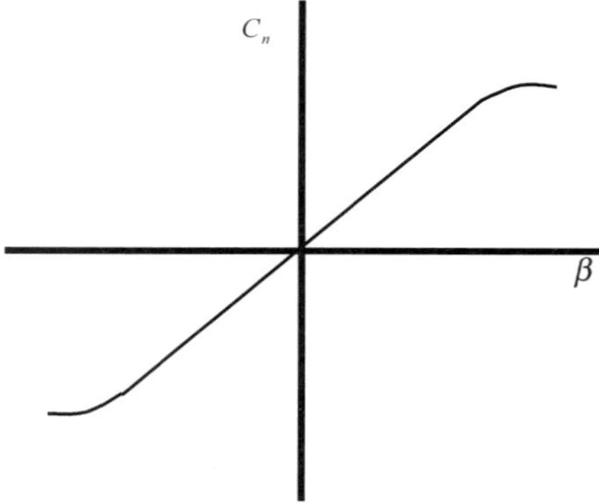

Fig. 5.36 Yawing moment vs sideslip characteristics for an aircraft with static directional stability.

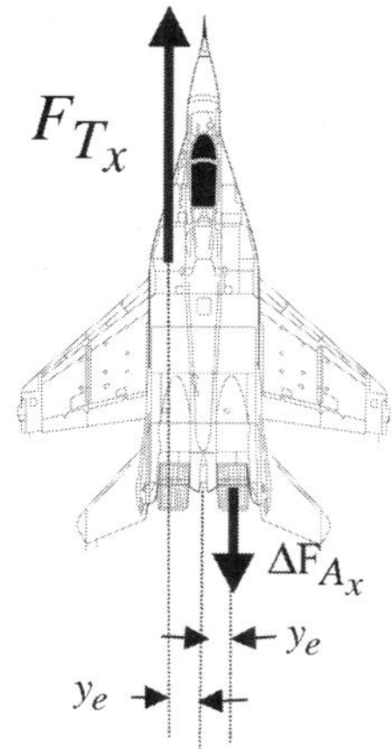

Fig. 5.37 Twin-engine aircraft with the right engine out.

For a jet aircraft, ΔD for a windmilling engine can be estimated as 10 to 15% of the thrust normally produced by the engine. For propeller-powered aircraft, the windmilling drag is significantly higher than this.

The rolling moment resulting from an asymmetric thrust configuration should also be considered. For a jet engine configuration with the engines mounted forward and below the wing, a right engine out configuration will probably result in a negative rolling moment because at the lower pressure generated below the left wing by the high velocity exhaust from the operating engine. Likewise, if the engines are mounted above the wings, like on the A-10, a right engine failure will probably result in a positive rolling moment. An aircraft such as the C-17 that incorporates thrust augmented lift will have a rolling moment toward the inoperative engine because of the loss of lift. The symbol L_T will be used to quantify the rolling moment that results from an asymmetric thrust configuration. The following lateral Taylor series (assuming C_{l_0} is zero) is then appropriate for trimmed flight.

$$C_{l_\beta}\beta + C_{l_{\delta_a}}\delta_a + C_{l_{\delta_r}}\delta_r + \frac{L_T}{\bar{q}Sb} = 0 \tag{5.106}$$

We will also present the sideforce Taylor series for trimmed flight, including the gravity term.

$$C_{y_\beta}\beta + C_{y_{\delta_a}}\delta_a + C_{y_{\delta_r}}\delta_r + \frac{mg\sin\Phi\cos\gamma}{\bar{q}S} = 0 \tag{5.107}$$

Two options for engine-out flight will now be considered.

Option 1: The aileron and rudder control deflections remain at zero after engine failure, and the aircraft is allowed to attain a steady-state trim condition with asymmetric thrust.

A 1-degree-of-freedom (DOF) estimate of the resulting steady-state sideslip angle can be obtained using only Eq. (5.105) with δ_a and δ_r equal to zero.

$$\beta \cong -\frac{F_T y_e + \Delta D y_e}{C_{n_\beta}(\bar{q}Sb)} \tag{5.108}$$

For the right engine-out case, the resulting steady-state sideslip angle is negative. Two simplifying assumptions were made in this prediction. First, the steady-state sideslip angle predicted by Eq. (5.108) will also induce a rolling moment on the aircraft because of the C_{l_β} term in Eq. (5.106). Because β is negative and C_{l_β} is negative, a positive rolling moment will result. It is assumed that the rolling moment that results from asymmetric thrust (L_T) will be of equal magnitude and in the opposite direction. If this is not the case, then Eqs. (5.105–5.107) must be worked together to obtain a 3-DOF solution. Solution parameters would be sideslip angle, δ_a, and bank angle (ϕ). Second, it is assumed that the sideslip angle predicted by Eq. (5.108) is within the range where C_{n_β} is linear. Large sideslip angles may stall the vertical tail, resulting in a lower or even negative value of C_{n_β}. If this is the case, the aircraft may not achieve a steady-state sideslip angle but may diverge directionally.

Option 2: The rudder is deflected to zero out the sideslip. For the right engine-out case, this would require left rudder. A 1-DOF estimate of the rudder required can be obtained using Eq. (5.105) with β and δ_a equal to zero.

$$\delta_r = -\frac{F_T y_e + \Delta D y_e}{C_{n_{\delta_r}}(\bar{q}Sb)} \tag{5.109}$$

Of course, an exact 3-DOF solution would include Eqs. (5.105–5.107) with solution parameters of δ_a, δ_r, and bank angle. Because a trailing-edge left rudder deflection also produces a sideforce to the right, a bank angle to the left should be expected from this solution to offset the sideforce because of rudder. Pilots refer to this condition as "banking into the good engine" to achieve trimmed flight.

Example 5.4

A C-21 (Learjet) flying with a dynamic pressure ($\bar{q}$) of 86.6 psf has the following characteristics:

$W = 8750$ lb	$S = 253$ ft^2	$b = 38$ ft	
$C_{y_0} = 0$	$C_{y_\beta} = -0.0105/\text{deg}$	$C_{y_{\delta_a}} = 0$	$C_{y_{\delta_r}} = 0.0021/\text{deg}$
$C_{l_0} = 0$	$C_{l_\beta} = -0.00293/\text{deg}$	$C_{l_{\delta_a}} = 0.0024/\text{deg}$	$C_{l_{\delta_r}} = 0.00015/\text{deg}$
$C_{n_0} = 0$	$C_{n_\beta} = 0.0018/\text{deg}$	$C_{n_{\delta_a}} = 0.0005/\text{deg}$	$C_{n_{\delta_r}} = -0.00096/\text{deg}$

For a right engine failure, $F_T y_e = 11{,}940$ ft·lb and $L_T = 1682$ ft·lb. Determine the sideslip angle that results after the engine failure and the rudder deflection necessary to return the aircraft to zero sideslip using 1-DOF approximations. The drag on the failed engine can be neglected ($\Delta D = 0$).

Equation (5.108) will be used to determine a 1-DOF approximation of the sideslip that results after engine failure with $\delta_r = 0$.

$$\beta \cong -\frac{F_T y_e + \Delta D y_e}{C_{n_\beta}(\bar{q}Sb)} = -\frac{11940}{(0.0018)[(86.6)(253)(38)]} = -7.97 \text{ deg}$$

Notice that the sideslip is negative for a right engine failure. Equation (5.109) will be used to determine a 1-DOF approximation of how much rudder is needed to zero out the sideslip with the engine failure.

$$\delta_r = -\frac{F_T y_e + \Delta D y_e}{C_{n_{\delta_r}}(\bar{q}Sb)} = -\frac{11940}{(-0.00096)[(86.6)(253)(38)]} = 14.94 \text{ deg}$$

The positive sign indicates that left rudder is required for a right engine failure.

Example 5.5

For the engine-out conditions of Example 5.4, determine the sideslip angle, rudder deflection, and aileron deflection necessary to keep the wings level using a 3-DOF solution.

Equations (5.105–5.107) must be solved simultaneously for the three unknowns: β, δ_r, and δ_a. The yawing moment equation becomes

$$C_{n_\beta}\beta + C_{n_{\delta_a}}\delta_a + C_{n_{\delta_r}}\delta_r + \frac{F_T y_e}{\bar{q}Sb} + \frac{\Delta D y_e}{\bar{q}Sb} = 0$$

$$0.0018\beta + 0.0005\delta_a - 0.00096\delta_r + \frac{11940}{(86.6)(253)(38)} = 0$$

The rolling moment equation becomes

$$C_{l_\beta}\beta + C_{l_{\delta_a}}\delta_a + C_{l_{\delta_r}}\delta_r + \frac{L_T}{\bar{q}Sb} = 0$$

$$-0.00293\beta + 0.0024\delta_a + (0.00015)\delta_r + \frac{1682}{(86.6)(253)(38)} = 0$$

and the sideforce equation becomes

$$C_{y_\beta}\beta + C_{y_{\delta_a}}\delta_a + C_{y_{\delta_r}}\delta_r + \frac{mg\sin\Phi\cos\gamma}{\bar{q}S} = 0$$

$$-0.0105\beta + (0.0021)\delta_r = 0$$

for wings level ($\phi = 0$). Conventional techniques (such as Cramer's rule) are used to solve this three equation/three unknown problem. The result is

$$\beta = 5.45 \text{ deg} \qquad \delta_r = 27.26 \text{ deg} \qquad \delta_a = 4.055 \text{ deg}$$

Notice that with the 3-DOF solution and the constraint of wings level, that a large rudder deflection is required. The positive sideslip is a result of the positive sideforce being generated by the rudder. The positive aileron deflection is required to counteract the negative rolling moment produced by the lateral stability (C_{l_β}). A concern is the large rudder deflection required for this case. It may be near or exceed the effective rudder travel of the aircraft.

Example 5.6

For the engine-out conditions of Example 5.4, determine the sideslip angle, rudder deflection, and aileron deflection required to maintain a bank angle of $\phi = -5$ deg using a 3-DOF solution.

We will take the same approach as in Example 5.5 but with a bank angle of -5 deg. The yawing moment and rolling moment equations remain

$$C_{n_\beta}\beta + C_{n_{\delta_a}}\delta_a + C_{n_{\delta_r}}\delta_r + \frac{F_T y_e}{\bar{q}Sb} + \frac{\Delta D y_e}{\bar{q}Sb} = 0$$

$$0.0018\beta + 0.0005\delta_a - 0.00096\delta_r + \frac{11940}{(86.6)(253)(38)} = 0$$

and

$$C_{l_\beta}\beta + C_{l_{\delta_a}}\delta_a + C_{l_{\delta_r}}\delta_r + \frac{L_T}{\bar{q}Sb} = 0$$

$$-0.00293\beta + 0.0024\delta_a + (0.00015)\delta_r + \frac{1682}{(86.6)(253)(38)} = 0$$

Assuming level flight ($\gamma = 0$), the sideforce equation becomes

$$C_{y_\beta}\beta + C_{y_{\delta_a}}\delta_a + C_{y_{\delta_r}}\delta_r + \frac{mg\sin\Phi\cos\gamma}{\bar{q}S} = 0$$

$$-0.0105\beta + (0.0021)\delta_r + \frac{(8750)\sin(-5\text{ deg})}{(86.6)(253)} = 0$$

A three-equation/three-unknown solution yields

$$\beta = -0.98\text{ deg} \qquad \delta_r = 11.67\text{ deg} \qquad \delta_a = -2.74\text{ deg}$$

In comparing these results with those of Example 5.5, notice that a moderate bank angle of 5 deg into the good engine has reduced rudder required by nearly 16 deg, and that the sideslip angle has become negative. It should also be realized that a sideslip angle of zero could be achieved with slightly less bank angle.

Another consideration associated with engine-out flight is the **minimum directional control speed** (V_{mc}). This must be considered because all control surface deflections have a limited effective range. Simply stated, the minimum directional control speed is the minimum speed at which an aircraft can maintain straight flight for an asymmetrical engine-out condition. The minimum directional control speed normally involves max rudder deflection and a bank angle not exceeding 5 deg (with the aircraft banked into the good engine). A 1-DOF estimate of V_{mc} begins with Eq. (5.105) and the assumption of zero sideslip, zero bank angle, and that $C_{n_{\delta_a}}$ is negligible.

$$C_{n_\beta}\beta + C_{n_{\delta_r}}\delta_r + \frac{F_T y_e}{\bar{q}Sb} + \frac{\Delta D y_e}{\bar{q}Sb} = 0$$

The amount of rudder deflection required to maintain zero sideslip ($\beta = 0$) is then

$$\delta_r = -\frac{\dfrac{F_T y_e}{\bar{q}Sb} + \dfrac{\Delta D y_e}{\bar{q}Sb}}{C_{n_{\delta_r}}} = -\frac{F_T y_e + \Delta D y_e}{C_{n_{\delta_r}}\bar{q}Sb} \tag{5.110}$$

It can be seen from Eq. (5.110) that the magnitude of rudder deflection required to keep sideslip zero increases as aircraft speed decreases (decreasing $\bar{q}$). Because maximum rudder travel is generally limited to approximately ± 25 deg to prevent stalling of the vertical tail, there is a minimum speed at which zero sideslip can be maintained. If the maximum rudder deflection is designated $\delta_{r_{\max}}$, Eq. (5.110) can be solved for V_{mc}.

$$V_{mc} = \sqrt{\frac{2(F_T y_e + \Delta D y_e)}{-\rho C_{n_{\delta_r}}\delta_{r_{\max}} Sb}} \quad \text{(for wings level flight)} \tag{5.111}$$

Generally, military requirements dictate that V_{mc} be less than or equal to $1.1 V_{\text{stall}}$ or not more than 10 kn above V_{stall}. Equation (5.111) also points out that V_{mc} is inversely proportional to $C_{n_{\delta_r}}$, the rudder control power. The requirement to keep V_{mc} low may be a driving factor in the design of the vertical tail and rudder area because $C_{n_{\delta_r}}$ must be large enough to meet V_{mc} requirements. The effect of bank angle, when determining V_{mc}, is very influential, and a full 3-DOF analysis, similar to Example 5.6, should be conducted for an accurate determination. A significant reduction in V_{mc} will be found for the wing low situation versus that found with a 1-DOF analysis that assumes zero sideslip and which leads to Eq. (5.111).

5.6.4 Crosswind Landings

Landing approaches with a component of the wind across the runway can generally be handled in two ways by a pilot. The first approach is to "crab into the wind" as illustrated in Fig. 5.38 while maintaining the aircraft in coordinated (zero sideslip angle), wings level flight. The degree of crab is adjusted until the aircraft ground track aligns with the direction of the runway. This

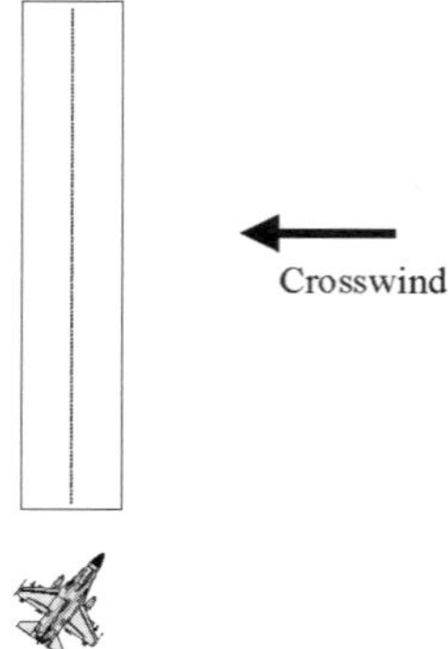

Fig. 5.38 Illustration of crabbing into the wind for a crosswind landing.

approach works well until the aircraft is at the point of touchdown on the runway. Then the aircraft must generally align the x-body axis with the runway direction so that the landing gear wheels are aligned with the direction of touchdown.

We will analyze the second approach to dealing with a crosswind landing. In this approach, the aircraft is trimmed with sideslip while the x axis is kept aligned with the runway direction. As a result, the landing gear is lined up with the runway direction and the pilot normally only needs to level the wings of the aircraft before landing. With the aircraft trimmed in a sideslip condition, a steady-state side velocity is generated that can be adjusted to be equal and in the opposite direction to the crosswind. Figure 5.39 illustrates the desired side velocity for a crosswind situation.

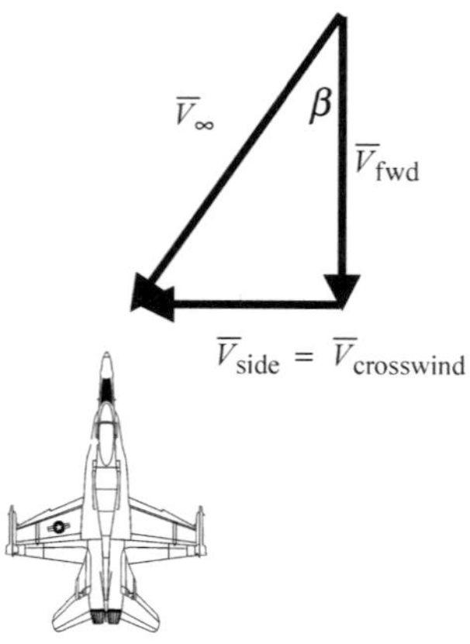

Fig. 5.39 Side velocity generated for a crosswind situation.

The needed sideslip angle can be computed from the geometry presented in Fig. 5.39.

$$\beta = \tan^{-1}\left(\frac{V_{\text{crosswind}}}{V_{\text{forward}}}\right) = \sin^{-1}\left(\frac{V_{\text{crosswind}}}{V_\infty}\right) \tag{5.112}$$

The control deflections necessary to achieve this trimmed sideslip condition can be found using the lateral and directional Taylor series and assuming C_{l_0} and C_{n_0} are zero.

$$C_{l_\beta}\beta + C_{l_{\delta_a}}\delta_a + C_{l_{\delta_r}}\delta_r = 0 \tag{5.113}$$

$$C_{n_\beta}\beta + C_{n_{\delta_a}}\delta_a + C_{n_{\delta_r}}\delta_r = 0 \tag{5.114}$$

These two equations lead to a fairly accurate 2-DOF solution of the crosswind problem. A more accurate 3-DOF solution would include the sideforce equation, and bank angle would become one of the parameters. Because we are using two equations, we can solve for two unknowns. The equations are typically used in two ways. First, given a crosswind and landing speed (that is, β can be computed from this information and thus is known), the control deflections (δ_a and δ_r) necessary to hold the needed sideslip angle can be found.

Second, the crosswind limit of the aircraft can be found based on determining the limiting control deflection (generally the rudder, but not always). This is a two-step process. Assume that the rudder is at maximum deflection, then solve for β and δ_a. Next, assume that the aileron is at maximum deflection, then solve for β and δ_r. The lower value of β defines the max crosswind condition and the maximum control deflection associated with it is the limiting control deflection. For example, if an aircraft has a maximum aileron and rudder deflection of 25 deg, Eqs. (5.113) and (5.114) should be solved simultaneously twice. The first solution fixes the rudder deflection at maximum, and the second solution fixes the aileron deflection at maximum. Results for the two cases might look like the following:

β (deg)	δ_r	δ_a
5	25	17
15	50	25

A δ_a of 25 deg requires that δ_r be 50 deg to hold 15 deg of sideslip. Because 50 deg of rudder is not available, the first case with 25 deg of rudder and 17 deg of aileron (both within the capabilities of the aircraft) defines the maximum crosswind condition. Therefore, with a maximum sideslip angle of 5 deg, the aircraft crosswind limit is

$$V_{\text{crosswind}} = V_\infty \sin\beta \tag{5.115}$$

The rudder is the limiting control deflection.

Example 5.7

Using a 2-DOF solution, determine the crosswind limit of the T-37 assuming a final approach equivalent airspeed of 100 kn and the following characteristics:

$$C_{l_0} = 0 \qquad C_{l_\beta} = -0.11/\text{deg} \quad C_{l_{\delta_a}} = 0.178/\text{deg} \qquad C_{l_{\delta_r}} = 0.172/\text{deg}$$
$$C_{n_0} = 0 \qquad C_{n_\beta} = 0.127/\text{deg} \quad C_{n_{\delta_a}} = -0.0172/\text{deg} \quad C_{n_{\delta_r}} = -0.0747/\text{deg}$$
$$\delta_{r_{\max}} = \pm 20 \text{ deg} \quad \delta_{a_{\max}} = \pm 15 \text{ deg}$$

We first assume that the rudder is the limiting deflection and substitute $\delta_{r_{\max}} = 20$ deg into Eqs. (5.113) and (5.114).

$$C_{l_\beta}\beta + C_{l_{\delta_a}}\delta_a + C_{l_{\delta_r}}\delta_r = 0$$
$$-0.11\beta + 0.178\delta_a + (0.172)(20) = 0$$

$$C_{n_\beta}\beta + C_{n_{\delta_a}}\delta_a + C_{n_{\delta_r}}\delta_r = 0$$
$$0.127\beta + (-0.0172)\delta_a - 0.0747(20) = 0$$

Solving this two-equation/two-unknown problem yields

$$\beta = 12.55 \text{ deg} \qquad \delta_a = 5.82 \text{ deg}$$

Because the aileron deflection is within the maximum limits, a rudder deflection of 20 deg represents the limiting control for crosswinds, and a steady-state sideslip angle of 12.55 deg is the maximum which the aircraft can hold. To further illustrate this, we will assume that the aircraft is aileron limited and rework the problem with $\delta_{a_{\max}} = 15$ deg. Using the same approach, we have

$$C_{l_\beta}\beta + C_{l_{\delta_a}}\delta_a + C_{l_{\delta_r}}\delta_r = 0$$
$$-0.11\beta + 0.178(15) + (0.172)\delta_r = 0$$

$$C_{n_\beta}\beta + C_{n_{\delta_a}}\delta_a + C_{n_{\delta_r}}\delta_r = 0$$
$$0.127\beta + (-0.0172)(15) - 0.0747\delta_r = 0$$

and the solution is

$$\beta = 32.33 \text{ deg} \qquad \delta_r = 51.5 \text{ deg}$$

A δ_r of 51.5 deg is not possible, of course, because of $\delta_{r_{\max}}$ being 20 deg. Thus, our initial conclusion is correct—a rudder deflection of 20 deg represents the limiting control for crosswinds, and a steady-state sideslip angle of 12.55 deg is the maximum that the aircraft can hold. We next use Eq. (5.115) to determine the maximum crosswind under which the aircraft can maintain a ground track parallel to the runway.

$$V_{\text{crosswind}} = V_\infty \sin\beta = 100 \sin 12.55 \text{ deg} = 21.73 \text{ kn (equivalent airspeed)}$$

Of course, the maximum crosswind limit published in the flight manual will be a little less than 21.73 kn to allow a margin of safety.

5.7 Summary of Steady-State Force and Moment Derivatives

Table 5.1 summarizes the longitudinal and lateral-directional, steady-state, force and moment derivatives discussed in this section. The signs for each derivative are presented, but they should be used as a reference only and not memorized. The reader should be able to define each derivative sign by applying the appropriate (positive) angle or control surface displacement, and then determining the sign of the force or moment that results. A model airplane can be useful in this exercise.

Table 5.1 Steady-state force and moment derivatives

Derivative	Sign
Longitudinal motion:	
C_{L_0}	> 0 (normally); can be < 0
C_{L_α} (aircraft lift curve slope)	> 0
$C_{L_{\delta_e}}$	> 0
$C_{L_{i_h}}$	> 0
$C^*_{D_0}$	> 0
C_{D_α}	> 0
$C_{D_{\delta_e}}$	Depends on i_h
$C_{D_{i_h}}$	> 0 for $i_h > 0$; < 0 for $i_h < 0$
C_{m_0}	Can be either > 0 or < 0
C_{m_α} (longitudinal static stability)	< 0 (for static stability)
$C_{m_{\delta_e}}$ (longitudinal/elevator control power)	< 0
$C_{m_{i_h}}$	< 0
Lateral-directional motion:	
C_{y_β}	< 0
$C_{y_{\delta_a}}$	≈ 0
$C_{y_{\delta_r}}$	> 0
C_{l_β} (lateral static stability)	< 0 (for static stability)
$C_{l_{\delta_a}}$ (lateral or aileron control power)	> 0
$C_{l_{\delta_r}}$ (cross-control derivative)	> 0
C_{n_β} (directional static stability)	> 0 (for static stability)
$C_{n_{\delta_a}}$ (cross-control derivative)	< 0 (adverse yaw); > 0 (proverse yaw)
$C_{n_{\delta_r}}$ (directional or rudder control power)	< 0

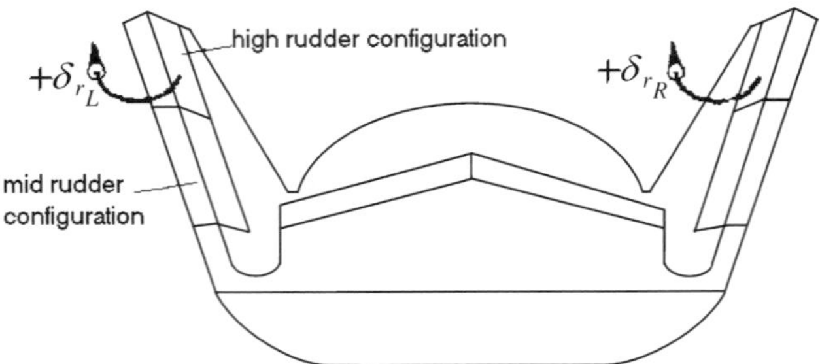

Fig. 5.40 X-38 mid-rudder and high-rudder geometry.

5.8 Historical Snapshot—The X-38 Mid-Rudder Investigation

The U.S. Air Force Academy Department of Aeronautics, under sponsorship from the NASA Johnson Space Center (JSC) Aerosciences and Flight Mechanics Division, conducted wind tunnel research on the stability and control of the X-38 Crew Return Vehicle (CRV) during the late 1990s and early 2000s.[5] In particular, one effort focused on an alternate rudder design for the X-38 that placed the rudders lower on the vertical stabilizer in a mid-rudder configuration and increased the rudder area by 32%. The alternate rudder design was investigated as an attempt to reduce the significant roll coupling effect produced by the X-38 high-rudder design, which had the rudders located at the top of the vertical stabilizers (see Fig. 5.40).

All tests were conducted in the Air Force Academy Subsonic Wind Tunnel (see Fig. 1.59) with a 4.62% scale model of the X-38 as shown in Fig. 5.41.

Of particular interest were roll coupling ($C_{l_{\delta_r}}$) and rudder yaw control ($C_{n_{\delta_r}}$) effects as discussed in Secs. 5.5.2.3 and 5.5.3.3. Figure 5.42 presents a graph of roll coupling as a function of angle of attack and Mach number, which shows that roll coupling decreased with increasing angles of attack.

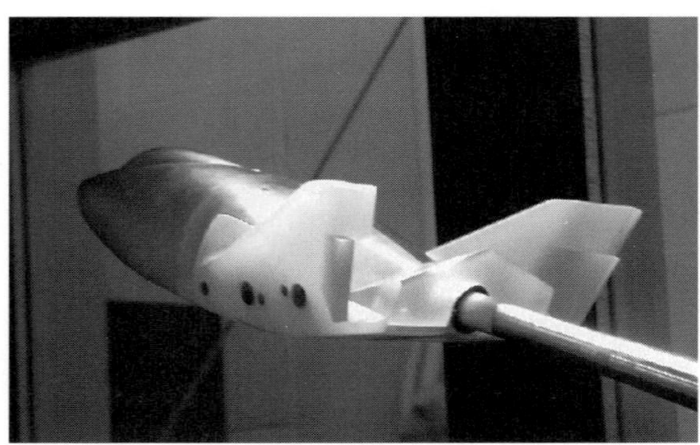

Fig. 5.41 X-38 wind tunnel model.

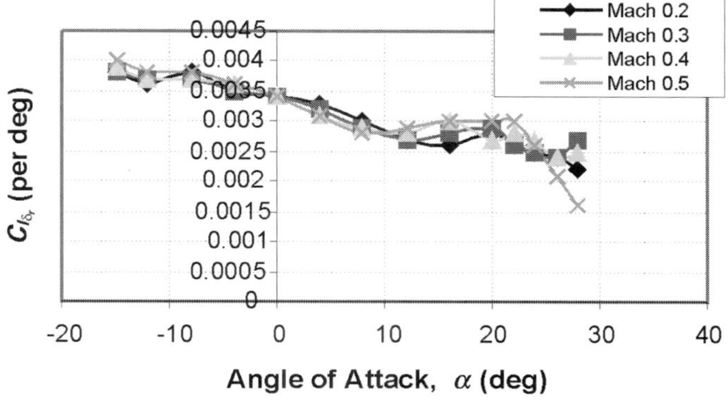

Fig. 5.42 X-38 roll coupling vs angle of attack at various Mach numbers.

When the roll coupling data were compared with that of the high-rudder configuration, it was found that the mid-rudder configuration had approximately an equivalent roll coupling, $C_{l_{\delta_r}}$, compared to the high rudder. Figure 5.43 illustrates the comparison.

The reason for the similarity in the roll coupling between the high-rudder and mid-rudder configurations was a result of several influences. Because the increase in rudder area with the mid-rudder configuration exposed a larger surface area to the freestream, there was an increase in side force. The lower placement, on the other hand, produced a decrease in moment arm. These two counteracting effects were a significant factor in keeping the roll coupling nearly the same. In addition, flow separation effects near the mid-rudder also contributed.

Figure 5.44 presents $C_{n_{\delta_r}}$ as a function of angle of attack at 0.2 Mach. Similar to $C_{l_{\delta_r}}$, $C_{n_{\delta_r}}$, decreased as angle of attack increased. In comparing the

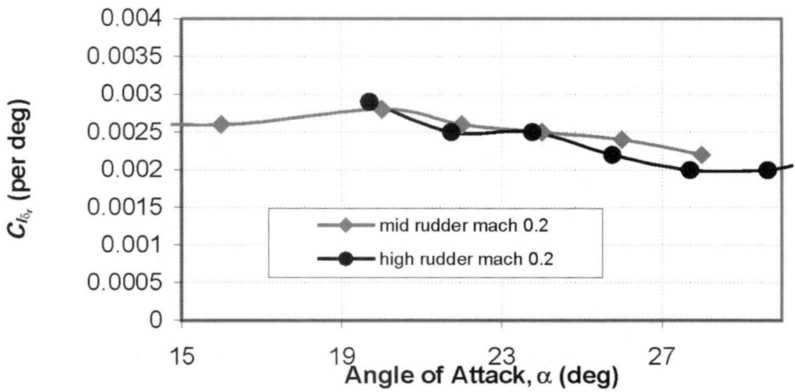

Fig. 5.43 Roll coupling comparison for mid- and high-rudder configurations.

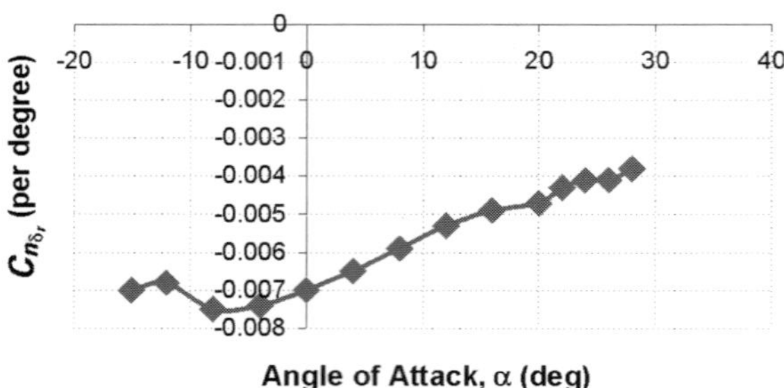

Fig. 5.44 Yaw control power as a function of angle of attack at $M = 0.2$.

directional control power of the mid-rudder configuration with high-rudder configuration, it was found that the values of the mid-rudder data points were more negative than those of the high-rudder configuration, signifying slightly higher yaw control power. Figure 5.45 demonstrates this result.

The increase in overall yaw control power with the mid-rudder configuration was primarily because of the increase in rudder area. However, the yaw control power for the mid-rudder configuration increased by only 12% from the high rudder, compared to a 32% increase in rudder area. This indicated a decrease in effectiveness per unit area for the mid-rudder configuration. It was determined that the decreased effectiveness per unit area resulted from a flow separation region near the mid-rudder.

After analyzing the differences between the mid-rudder and high-rudder configuration, it was recommended that a smaller area mid-rudder design be investigated to decrease roll coupling. Yaw control power was expected to also decrease which would provide a value closer to the high-rudder yaw control power. The program provided valuable inputs to the X-38 design team regarding control design options.

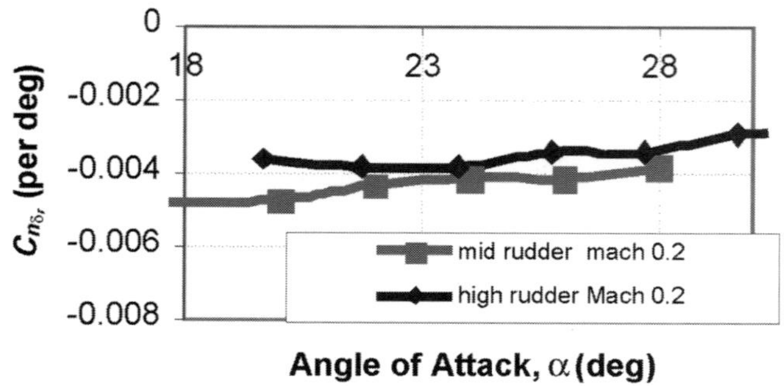

Fig. 5.45 Yaw control power comparison for mid- and high-rudder configurations.

References

[1]Murri, D. G., Grafton, S. B., and Hoffler, K. D., *Wind Tunnel Investigation and Free-Flight Evaluation of the F-15 STOL and Maneuver Technology Demonstrator*, NASA TP 3003, Aug. 1990.

[2]Olson, W. M., Wood, R. A., and Clarke, M. J., "YF-17 Performance and Flying Qualities Evaluation," AFFTC TR 75-18, June 1975.

[3]Roskam, J., *Airplane Design, Part VI*, Design, Analysis, and Research Corporation, Lawrence, KS, 1990.

[4]Simon, J. M., LeMay, S., and Brandon, J. M., *Results of Exploratory Wind Tunnel Tests of F-16/VISTA Forebody Vortex Control Devices*, WL-TR-93-3013, Jan. 1993.

[5]Johnston, C. N., Nettleblad, T. A., and Yechout, T. R., "X-38 Mid-Rudder Feasibility and Parafoil Cavity Investigations," U.S. Air Force Academy Dept. of Aeronautics TR 01-02, Sept. 2001.

Problems

5.1 What are the two requirements for longitudinal static stability?

5.2 Consider the following aircraft flying in straight and level unaccelerated trimmed flight at $M = 0.2$ at sea level.

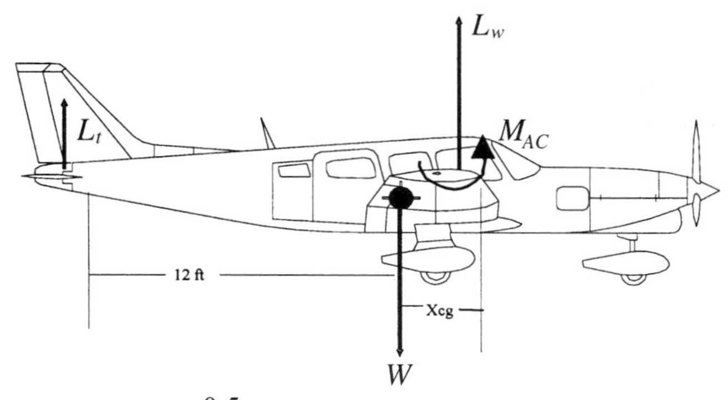

$$x_{cg} = 0.5c_{\text{wing}}$$

$$M_{ac_{\text{tail}}} = 0 \qquad M_{ac_{\text{wing}}} = -5000 \text{ ft} \cdot \text{lb}$$

$$T = 3000 \text{ lb} \qquad W = 12{,}000 \text{ lb}$$

$$c_{\text{wing}} = 8 \text{ ft} \qquad c_{\text{tail}} = 4 \text{ ft} \qquad b_{\text{wing}} = 30 \text{ ft}$$

(a) Find the lift of the wing and the lift of the tail. Assume L_{wing} and L_{tail} act at the quarter chord points.
(b) Find the aircraft drag.
(c) Assuming a rectangular wing, what is the total aircraft C_L?
(d) What is the pitching moment coefficient for the entire aircraft?

5.3 For the aircraft in Problem 5.2, with the additional information

$$AR_w = 7.65 \qquad \alpha_{OL_W} = -2 \text{ deg} \qquad e = 1 \qquad \alpha_{OL_t} = 0 \text{ deg}$$
$$AR_t = 5 \qquad S_w = 240 \text{ ft}^2 \qquad S_t = 40 \text{ ft}^2$$

(a) What is the trimmed angle of attack of the tail?
(b) What is the trimmed angle of attack of the wing?

5.4 Given the following information, find the lift of the wing and lift of the tail. Also, determine the trimmed angle of attack of the wing and horizontal tail. Assume that the lift of the wing and tail act at their respective quarter chords.

$$AR_{\text{wing}} = 7.65 \qquad \alpha_{OL_{\text{wing}}} = -2 \text{ deg} \qquad e = 1 \qquad x_{cg} = 0.5\text{c}$$
$$AR_{\text{tail}} = 5.08 \qquad \alpha_{OL_{\text{tail}}} = 0 \qquad S_w = 240 \text{ ft}^2 \qquad W = 12{,}000 \text{ lb}$$
$$S_{\text{tail}} = 20.32 \text{ ft}^2 \qquad \text{Mach} = 0.2 \text{ (@ S.L.)} \qquad M_{AC_{\text{wing}}} = -5000 \text{ ft} \cdot \text{lbf}$$
$$l = 17 \text{ ft} \qquad c = 8 \text{ ft}$$
$$(L_w = 11{,}000 \text{ lb},\ L_t = 1000 \text{ lb},\ \alpha_t = 10.56 \text{ deg},\ \alpha_w = 6.90 \text{ deg})$$

5.5 Derive an expression for $C_{m_{C_L}}$ in terms of static margin, noting

$$C_{m_{C_L}} = \frac{\partial C_m}{\partial C_L}$$

5.6 The following questions all pertain to a single aircraft:
(a) When the c.g. is at 29% chord, and $C_{m_{C_L}} = -0.10$, what is the c.g. position for $C_{m_{C_L}} = 0$?
(b) If the c.g. is shifted to 26% chord, is the aircraft more or less stable? What is the new value of $C_{m_{C_L}}$?
(c) With the c.g. at 26% chord, the C_m was found to be equal to 0.04 when $C_L = 0$ and $\delta_e = 0$. What is the value of C_L for trimmed flight with $\delta_e = 0$? [Hint: Use the value of $C_{m_{C_L}}$ from Part (b).]
(d) If $\bar{X}_{cg} = 0.26$, $C_{L_{\max}} = 1.0$, $C_{m_{\delta_e}} = -0.01/\text{deg}$, and $\delta_{e_{\max}} = \pm 10$ deg, is it possible to attain $C_{L_{\max}}$ in this aircraft? If so, solve for $\delta_{e_{\text{trim}}}$ at $C_{L_{\max}}$. [Hint: Use C_{m_0} from Part (c).]

5.7 Find the maneuver point. Use the following plot to determine $d\delta_e/dn$ for each c.g. location and the blank graph to estimate where $d\delta_e/dn = 0$.

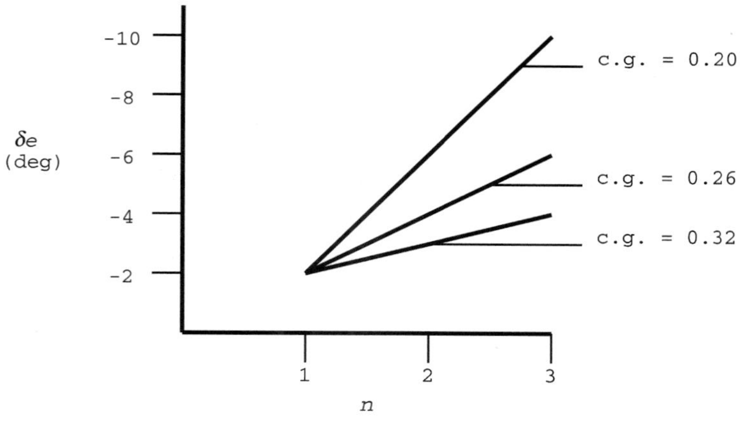

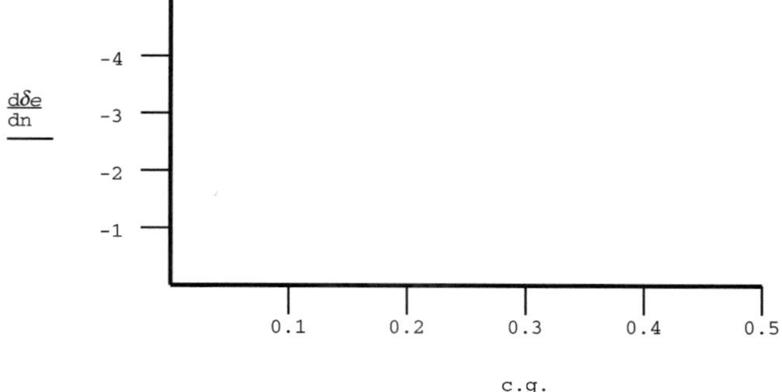

5.8 (a) Define maneuver point in your own words.
(b) Where should the maneuver point be located relative to the neutral point?
(c) What should the sign of C_{m_q} be to have a stabilizing effect? Why?

5.9 Label the relative positions of the neutral and maneuver points:

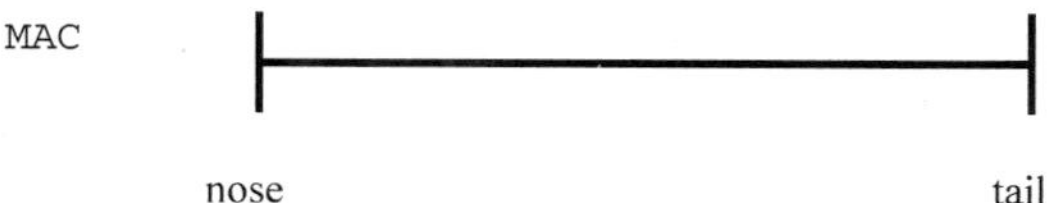

5.10 Which of the following design features increases roll stability?

High Wing (T-41)	Geometric Dihedral
Ventral Fin	Dorsal Fin
Sweep Back	Fuselage

5.11 The right engine on an aircraft with two 10,000 lb_f thrust engines fails. The aircraft is at sea level.

$$C_{n_\beta} = 0.002/\text{deg} \qquad S = 300\ \text{ft}^2 \qquad b = 50\ \text{ft}$$
$$C_{n_{\delta_r}} = -0.0033/\text{deg} \qquad q = 100\ \text{lb/ft}^2 \qquad Y_e = 5\ \text{ft}$$

(a) If the pilot takes no corrective action, what will the sideslip angle (β) be?
(b) How many degrees and which direction should the pilot deflect the rudder to realign the nose with the relative wind?
(c) If the max rudder deflection is 15 deg, at what airspeed would the pilot no longer be able to maintain $\beta = 0$ deg?

5.12 An aircraft has the following directional stability and control characteristics:

$$C_{n_\beta} = +0.0035/\text{deg} \qquad C_{n_{\delta_r}} = -0.003/\text{deg} \qquad \delta_{r_{\max}} = 30\ \text{deg}$$

(a) Determine the rudder deflection required to maintain a sideslip angle of $\beta = 4$ deg. Which rudder pedal would you push to get $\beta = 4$ deg?
(b) Given the following conditions: $\left.\dfrac{W}{S}\right|_{\text{landing}} = 60\dfrac{\text{lb}_f}{\text{ft}^2},\ C_{L_{\max}} = 1.0$

Assuming $V_{\text{land}} = 1.2\,V_{\text{stall}}$, determine the maximum crosswind component that can be handled by the rudder at sea level. We want to land the aircraft with the longitudinal axis aligned with the runway.

5.13 The aircraft in Problem 5.12 has the following additional characteristics:

$$C_{n_\beta} = +0.0035/\text{deg} \qquad C_{n_{\delta_r}} = -0.003/\text{deg} \qquad \delta_{r_{\max}} = 30\ \text{deg}$$
$$C_{l_\beta} = -0.0024/\text{deg} \qquad C_{l_{\delta_r}} = 0 \qquad C_{l_{\delta_a}} = 0.0008/\text{deg}$$
$$\delta_{a_{\max}} = \pm\frac{\delta_{a_{\text{left}}} + \delta_{a_{\text{right}}}}{2} = \pm 30\ \text{deg}$$

(a) If the landing speed is $1.2\,V_{\text{stall}}$ and we keep the upwind wing low (fuselage aligned with the runway), determine the max crosswind component that can be handled by the ailerons.
(b) Determine the aileron deflection to maintain a sideslip of $\beta = 4$ deg.
(c) Which control system, ailerons or rudder, will limit the crosswind component for the aircraft?

5.14 A T-37 has the following stability derivative values during a no-flap final approach at 100 kn equivalent airspeed:

$$C_{l_\beta} = -0.11/\text{deg} \qquad C_{l_{\delta_r}} = +0.0172/\text{deg}$$

$$C_{n_\beta} = +0.127/\text{deg} \qquad C_{l_{\delta_a}} = +0.178/\text{deg}$$

$$C_{n_{\delta_r}} = -0.0747/\text{deg} \qquad C_{n_{\delta_a}} = -0.0172/\text{deg}$$

Assume that the pilot lands with the fuselage aligned with the runway. With $\delta_{r_{\max}} = \pm 20$ deg and $\delta_{a_{\max}} = +15$ deg what is the maximum cross-wind component allowable and which is the limiting control (aileron or rudder)?

6
Linearizing the Equations of Motion

The six aircraft equations of motion developed in Chapter 4 [see Eqs. (4.70) and (4.71)] are nonlinear differential equations. They can be solved with a variety of numerical integration techniques to obtain time histories of motion variables, but it is nearly impossible to obtain closed solutions (equations for each variable). Because valuable insight can be obtained from closed solutions regarding the dynamic response of the aircraft, this chapter will use the small perturbation approach to linearize the equations of motion and facilitate the definition of closed solutions. In addition, the dynamic derivatives associated with definition of applied forces and moments on the aircraft will be discussed.

6.1 Small Perturbation Approach

Linearization of the aircraft equations of motion begins with consideration of **perturbed flight**. Perturbed flight is defined relative to a steady-state (trimmed) flight condition using a combination of steady-state and perturbed variables for aircraft motion parameters and for forces and moments. Simply stated, each motion variable, Euler angle, force, and moment in the equations of motion (EOM) are redefined as the summation of a steady-state value (designated with the subscript "1") and a perturbed value (designated with lower case symbols) as summarized in Eq. (6.1).

$$\begin{array}{lll} U = U_1 + u & V = V_1 + v & W = W_1 + w \\ P = P_1 + p & Q = Q_1 + q & R = R_1 + r \\ \Psi = \Psi_1 + \psi & \Theta = \Theta_1 + \theta & \Phi = \Phi_1 + \phi \\ F_A = F_{A_1} + f_A & F_T = F_{T_1} + f_T & \\ L_A = L_{A_1} + l_A & M_A = M_{A_1} + m_A & N_A = N_{A_1} + n_A \\ L_T = L_{T_1} + l_T & M_T = M_{T_1} + m_T & N_T = N_{T_1} + n_T \end{array} \tag{6.1}$$

For example, if an aircraft has a steady-state trimmed value for U of 400 ft/s and then encounters turbulence which increases U to 402 ft/s, U at that instant would be

$$U = U_1 + u = 400 + 2$$

The "perturbed" x-axis velocity, u, would be 2 ft/s in this case. The assumption of **small perturbations** (small values for u, v, w, p, etc.), allows linearization of the aircraft EOM. The following four-step approach summarizes the linearization technique:

Step 1: Recast each variable in terms of a steady-state value and a perturbed value ($A = A_1 + a$). Assume small perturbations (a is small). Multiply out and use trig identities.

Step 2: Apply the small-angle assumption to trig functions of perturbed angles [$\cos\theta \approx 1$; $\sin\theta \approx \theta$ (in radians)]

Step 3: Assume products of small perturbations are negligible ($ab \approx 0$).

Step 4: Remove the steady-state equation from the perturbed equation. The remaining perturbed equation is a linearized differential equation with the perturbed variables as the unknowns.

This approach is illustrated in Example 6.1.

Example 6.1

Linearize the following nonlinear differential equation.

$$\dot{C} = AB\sin\Omega$$

Step 1: $A = A_1 + a$, $B = B_1 + b$, $C = C_1 + c$, $\Omega = \Omega_1 + \omega$; substitute these variables into the original equation and multiply out.

$$\dot{C}_1 + \dot{c} = (A_1 + a)(B_1 + b)\sin(\Omega_1 + \omega)$$
$$\dot{C}_1 + \dot{c} = [A_1B_1 + A_1b + aB_1 + ab]\sin(\Omega_1 + \omega)$$
$$\dot{C}_1 + \dot{c} = [A_1B_1 + A_1b + aB_1 + ab][\sin\Omega_1\cos\omega + \cos\Omega_1\sin\omega]$$

Step 2: $\cos\omega \approx 1$; $\sin\omega \approx \omega$ (in radians)

$$\dot{C}_1 + \dot{c} = [A_1B_1 + A_1b + aB_1 + ab][\sin\Omega_1 + \omega\cos\Omega_1]$$

Step 3: $ab \approx 0$

$$\dot{C}_1 + \dot{c} = [A_1B_1 + A_1b + aB_1][\sin\Omega_1 + \omega\cos\Omega_1]$$

and $a\omega$ and $b\omega \approx 0$,

$$\dot{C}_1 + \dot{c} = A_1B_1\sin\Omega_1 + A_1b\sin\Omega_1 + aB_1\sin\Omega_1 + A_1B_1\omega\cos\Omega_1$$

Step 4: The steady-state equation can be easily identified by going back to the original differential equation and inserting subscripts of "1" on each variable.

$$\dot{C}_1 = A_1B_1\sin\Omega_1$$

Subtracting the previous equation from the differential equation at the end of Step 3, we have:

$$\dot{c} = A_1b\sin\Omega_1 + aB_1\sin\Omega_1 + A_1B_1\omega\cos\Omega_1$$

which is the linearized form of the differential equation in terms of the perturbation variables a, b, c, and ω.

6.2 Developing the Linearized Aircraft Equations of Motion

The nonlinear longitudinal EOM presented in Eq. (4.70) can be rewritten as

$$\begin{aligned}
m(\dot{U} + QW - RV) &= -mg\sin\Theta + F_{A_x} + F_{T_x} \\
\dot{Q}I_{yy} - PR(I_{zz} - I_{xx}) + (P^2 - R^2)I_{xz} &= M_A + M_T \\
m(\dot{W} + PV - QU) &= mg\cos\Phi\cos\Theta + F_{A_z} + F_{T_z}
\end{aligned} \tag{6.2}$$

where F_{A_x} and F_{A_z} are the aero forces acting along the respective x and z axis directions. Likewise, F_{T_x} and F_{T_z} represent the thrust forces acting along their respective directions. Notice also that the Euler angles have been designated with capital letters to conform with the approach defined by Eq. (6.1).

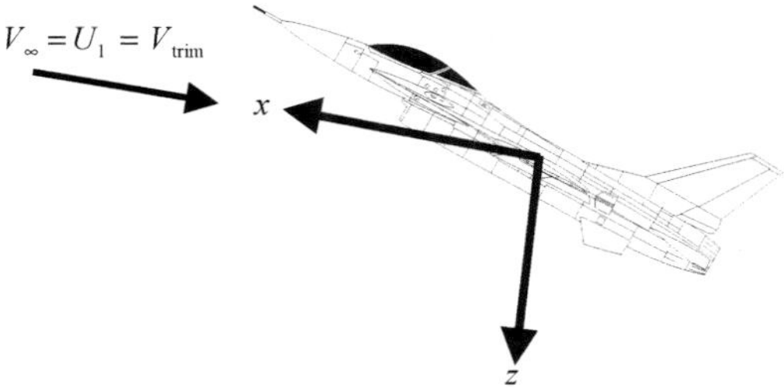

a) Body-Fixed Stability Axis During Trimmed Flight

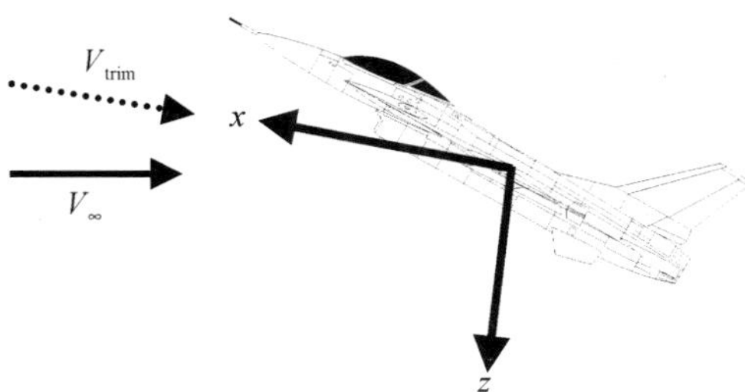

b) Body-Fixed Stability Axis During Perturbed Flight

Fig. 6.1 Illustration of body-fixed stability axis.

The nonlinear lateral-directional equations of motion presented in Eq. (4.71) are presented again here for convenient reference.

$$
\begin{aligned}
&\dot{P}I_{xx} + QR(I_{zz} - I_{yy}) - (\dot{R} + PQ)I_{xz} = L_A + L_T \\
&m(\dot{V} + RU - PW) = mg\sin\Phi\cos\Theta + F_{A_y} + F_{T_y} \\
&\dot{R}I_{zz} + PQ(I_{yy} - I_{xx}) + (QR - \dot{P})I_{xz} = N_A + N_T
\end{aligned}
\tag{6.3}
$$

As discussed in Sec. 5.3, we will use a body-fixed stability axis system where the x-stability axis is oriented directly with the steady-state relative wind. Thus, U_1 will be equal to V_{trim} while V_1 and W_1 are equal to zero. This approach simplifies representation of the aero forces and moments. However, the reader is reminded that, to be perfectly correct, the moments and products of inertia in Eqs. (6.2) and (6.3) must be defined relative to this body-fixed stability axis. Figure 6.1 illustrates the body-fixed stability axis.

6.2.1 Longitudinal Linearized EOMs

To linearize the nonlinear longitudinal EOM of Eq. (6.2), the small perturbation approach discussed in Sec. 6.1 will be used. The substitutions of Eq. (6.1) are made in Eq. (6.2) to yield

$$
\begin{aligned}
&m[\dot{U}_1 + \dot{u} + (Q_1 + q)(W_1 + w) - (R_1 + r)(V_1 + v)] \\
&\qquad = -mg\sin(\Theta_1 + \theta) + F_{A_{x_1}} + f_{A_x} + F_{T_{x_1}} + f_{T_x} \\
&I_{yy}(\dot{Q}_1 + \dot{q}) - (P_1 + p)(R_1 + r)(I_{zz} - I_{xx}) + [(P_1 + p)^2 - (R_1 + r)^2]I_{xz} \\
&\qquad = M_{A_1} + m_A + M_{T_1} + m_T \\
&m[\dot{W}_1 + \dot{w} + (P_1 + p)(V_1 + v) - (Q_1 + q)(U_1 + u)] \\
&\qquad = mg\cos(\Phi_1 + \phi)\cos(\Theta_1 + \theta) + F_{A_{z_1}} + f_{A_z} + F_{T_{z_1}} + f_{T_z}
\end{aligned}
$$

Applying Steps 2–4 of Sec. 6.1, these reduce to

$$
\begin{aligned}
&m(\dot{u} - V_1 r - R_1 v + W_1 q + Q_1 w) = -mg\theta\cos\Theta_1 + f_{A_x} + f_{T_x} \\
&I_{yy}\dot{q} + (I_{xx} - I_{zz})(P_1 r + R_1 p) + I_{xz}(2P_1 p - 2R_1 r) = m_A + m_T \\
&m(\dot{w} - U_1 q - Q_1 u + V_1 p + P_1 v) = -mg\theta\cos\Phi_1\sin\Theta_1 \\
&\qquad\qquad - mg\phi\sin\Phi_1\cos\Theta_1 + f_{A_z} + f_{T_z}
\end{aligned}
\tag{6.4}
$$

Equation (6.4) represents the linearized longitudinal EOM as a function of the variables u, v, w, p, q, r, θ and ϕ.

6.2.2 Lateral-Directional Linearized EOMs

Linearization of the nonlinear lateral-directional EOM presented in Eq. (6.3) also use the small perturbation approach of Sec. 6.1. The small perturbation substitutions result in:

$$\begin{aligned}
&(\dot{P}_1 + \dot{p})I_{xx} + (Q_1 + q)(R_1 + r)(I_{zz} - I_{yy}) - [\dot{R}_1 + \dot{r} + (P_1 + p)(Q_1 + q)]I_{xz} \\
&\qquad = L_A + l_A + L_T + l_T \\
&m[\dot{V}_1 + v + (R_1 + r)(U_1 + u) - (P_1 + p)(W_1 + w)] \\
&\qquad = mg\sin(\Phi_1 + \phi)\cos(\Theta_1 + \theta) + F_{A_{y_1}} + f_{A_y} + F_{T_{y_1}} + f_{T_y} \\
&(\dot{R}_1 + r)I_{zz} + (P_1 + p)(Q_1 + q)(I_{yy} - I_{xx}) + [(Q_1 + q)(R_1 + r) - (\dot{P}_1 + \dot{p})]I_{xz} \\
&\qquad = N_{A_1} + n_A + N_{T_1} + n_T
\end{aligned}$$

Applying Steps 2–4 of Sec. 6.1, these reduce to

$$\begin{aligned}
&I_{xx}\dot{p} - I_{xz}\dot{r} - I_{xz}(P_1 q + Q_1 p) + (I_{zz} - I_{yy})(R_1 q + Q_1 r) = l_A + l_T \\
&m(\dot{v} + U_1 r + R_1 u - W_1 p - P_1 w) = -mg\theta\sin\Phi_1\sin\Theta_1 \\
&\qquad + mg\phi\cos\Phi_1\cos\Theta_1 + f_{A_y} + f_{T_y} \\
&I_{zz}\dot{r} - I_{xz}\dot{p} + (I_{yy} - I_{xx})(P_1 q + Q_1 p) + I_{xz}(Q_1 r + R_1 q) = n_A + n_T
\end{aligned} \tag{6.5}$$

Equation (6.5) represents the linearized lateral-directional EOM as a function of the variables u, v, w, p, q, r, θ and ϕ, the same variables in the linearized longitudinal EOM. We thus have a situation with six equations and eight unknowns. The kinematic equations [Eq. (4.80)] may be linearized using a similar approach[1] to yield three additional equations and one additional unknown (ψ)—a solvable nine-equation/nine-unknown problem for the perturbed aircraft EOM.

6.2.3 Simplifying the Linearized EOMs for Wings Level, Straight Flight

Equations (6.4) and (6.5) may be simplified with the assumption of wings level, straight line flight for the initial trim, or steady state, condition. With this constraint,

$$\Phi_1 = P_1 = Q_1 = R_1 = \beta = 0$$

In addition, our choice of the body-fixed stability axis, as discussed in Sec. 6.2, leads to the additional simplification of:

$$V_1 = W_1 = 0$$

With these assumptions, the linearized longitudinal EOM of Eq. (6.4) become

$$\begin{aligned} m\dot{u} &= -mg\theta\cos\Theta_1 + f_{A_x} + f_{T_x} \\ I_{yy}\dot{q} &= m_A + m_T \\ m(\dot{w} - U_1 q) &= -mg\theta\sin\Theta_1 + f_{A_z} + f_{T_z} \end{aligned} \tag{6.6}$$

and the linearized lateral-directional EOM of Eq. (6.5) become

$$\begin{aligned} I_{xx}\dot{p} - I_{xz}\dot{r} &= l_A + l_T \\ m(\dot{v} + U_1 r) &= mg\phi\cos\Theta_1 + f_{A_y} + f_{T_y} \\ I_{zz}\dot{r} - I_{xz}\dot{p} &= n_A + n_T \end{aligned} \tag{6.7}$$

At this point, it is reasonable to ask if the restrictive assumptions of wings-level, straight flight will restrict our study of aircraft dynamic stability characteristics. Fortunately, the answer is no, because the fundamental dynamic modes of the aircraft are still present with these assumptions and dynamic stability characteristics observed about a wings-level, straight-line, trimmed flight condition are representative of those experienced during maneuvering flight. In short, these assumptions and choice of the body-fixed stability axis system have allowed simplification of the EOM to a manageable form so that understanding of dynamic stability concepts can be maximized.

6.3 First-Order Approximation of Applied Aero Forces and Moments

We will now focus on the applied aerodynamic force and moment terms (include f_{A_x} and m_A) in Eqs. (6.6) and (6.7). These represent the perturbed change in an aerodynamic force or moment that results from a nonzero value of a perturbed motion variable like u. We begin with the observation that the longitudinal perturbed forces and moment are a function primarily of five parameters:

$$f_{A_x}, m_A, f_{A_z} = f(u, \hat{\alpha}, \dot{\alpha}, q, \hat{\delta}_e) \tag{6.8}$$

As a reminder, the parameters u, $\hat{\alpha}$, $\dot{\alpha}$, q, and $\hat{\delta}_e$ are perturbation variables. $\hat{\alpha}$, $\dot{\alpha}$, and $\hat{\delta}_e$ deserve explanation because we have not discussed them before. The angle of attack is defined using the perturbation variable $\hat{\alpha}$ with the same approach as in Eq. (6.1).

$$\alpha = \alpha_1 + \hat{\alpha}$$

$\dot{\alpha}$ could be thought about in the same manner:

$$\dot{\alpha} = \dot{\alpha}_1 + \dot{\hat{\alpha}}$$

Because $\dot{\alpha}_1$ is equal to zero (the derivative of a constant), we have $\dot{\alpha} = \dot{\hat{\alpha}}$, and we will dispense with the hat for simplicity. The perturbed elevator deflection, $\hat{\delta}_e$, is defined in a similar manner as used with Eq. (6.1):

$$\delta_e = \delta_{e_1} + \hat{\delta}_e$$

Recall that the subscript "1" indicates the steady-state or trimmed value of the parameter.

We will use a first-order Taylor series to approximate how the longitudinal perturbed forces and moment vary as a function of the five perturbed variables presented in Eq. (6.8).

$$\begin{aligned}
f_{A_x} &= \frac{\partial F_{A_x}}{\partial u} u + \frac{\partial F_{A_x}}{\partial \hat{\alpha}} \hat{\alpha} + \frac{\partial F_{A_x}}{\partial \dot{\alpha}} \dot{\alpha} + \frac{\partial F_{A_x}}{\partial q} q + \frac{\partial F_{A_x}}{\partial \hat{\delta}_e} \hat{\delta}_e \\
m_A &= \frac{\partial M_A}{\partial u} u + \frac{\partial M_A}{\partial \hat{\alpha}} \hat{\alpha} + \frac{\partial M_A}{\partial \dot{\alpha}} \dot{\alpha} + \frac{\partial M_A}{\partial q} q + \frac{\partial M_A}{\partial \hat{\delta}_e} \hat{\delta}_e \\
f_{A_z} &= \frac{\partial F_{A_z}}{\partial u} u + \frac{\partial F_{A_z}}{\partial \hat{\alpha}} \hat{\alpha} + \frac{\partial F_{A_z}}{\partial \dot{\alpha}} \dot{\alpha} + \frac{\partial F_{A_z}}{\partial q} q + \frac{\partial F_{A_z}}{\partial \hat{\delta}_e} \hat{\delta}_e
\end{aligned} \tag{6.9}$$

Next, we will use the same approach to define the perturbed lateral-directional force and moments. We begin with the observation that six parameters primarily influence the lateral-directional perturbed force and moments:

$$l_A, f_{A_y}, n_A = f(\hat{\beta}, \dot{\beta}, p, r, \hat{\delta}_a, \hat{\delta}_r) \tag{6.10}$$

Again, $\hat{\beta}$, $\dot{\beta}$, p, r, $\hat{\delta}_a$, and $\hat{\delta}_r$ are perturbation variables. The perturbed sideslip, $\hat{\beta}$, is defined using the same approach as in Eq. (6.1).

$$\beta = \beta_1 + \hat{\beta}$$

For wings-level, trimmed, straight-line flight, β_1 equals zero and we have $\beta = \hat{\beta}$. We will dispense, at this point, with the hat for simplicity. $\dot{\beta}$ should be considered in the same manner.

$$\dot{\beta} = \dot{\beta}_1 + \dot{\hat{\beta}}$$

Because $\dot{\beta}_1$ is equal to zero (the derivative of a constant), we have $\dot{\beta} = \dot{\hat{\beta}}$, and we will again dispense with the hat. The perturbed aileron and rudder deflections, $\hat{\delta}_a$ and $\hat{\delta}_r$, are also defined as perturbations to the steady-state value:

$$\delta_a = \delta_{a_1} + \hat{\delta}_a$$
$$\delta_r = \delta_{r_1} + \hat{\delta}_r$$

We will again use a first-order Taylor series to approximate how the lateral-directional perturbed force and moments vary as a function of the six perturbed variables presented in Eq. (6.10).

$$\begin{aligned}
l_A &= \frac{\partial L_A}{\partial \beta}\beta + \frac{\partial L_A}{\partial \dot{\beta}}\dot{\beta} + \frac{\partial L_A}{\partial p}p + \frac{\partial L_A}{\partial r}r + \frac{\partial L_A}{\partial \hat{\delta}_a}\hat{\delta}_a + \frac{\partial L_A}{\partial \hat{\delta}_r}\hat{\delta}_r \\
f_{A_y} &= \frac{\partial F_{A_y}}{\partial \beta}\beta + \frac{\partial F_{A_y}}{\partial \dot{\beta}}\dot{\beta} + \frac{\partial F_{A_y}}{\partial p}p + \frac{\partial F_{A_y}}{\partial r}r + \frac{\partial F_{A_y}}{\partial \hat{\delta}_a}\hat{\delta}_a + \frac{\partial F_{A_y}}{\partial \hat{\delta}_r}\hat{\delta}_r \\
n_A &= \frac{\partial N_A}{\partial \beta}\beta + \frac{\partial N_A}{\partial \dot{\beta}}\dot{\beta} + \frac{\partial N_A}{\partial p}p + \frac{\partial N_A}{\partial r}r + \frac{\partial N_A}{\partial \hat{\delta}_a}\hat{\delta}_a + \frac{\partial N_A}{\partial \hat{\delta}_r}\hat{\delta}_r
\end{aligned} \tag{6.11}$$

The Taylor series representation of Eqs. (6.9) and (6.11) makes the quasi-steady assumption that the perturbed forces and moment are only a function of the instantaneous values of the perturbed motion variables. For the majority of rigid airplane dynamic stability analysis (at frequencies below 10 rad/s), this assumption provides accurate results.

6.3.1 Nondimensionalizing the First-Order Approximations

At this point, it is customary to nondimensionalize the perturbed variables in Eqs. (6.9) and (6.11). This is typically done to facilitate comparisons. It is accomplished in a straightforward manner. First, all angles (α, $\hat{\delta}_e$, β, $\hat{\delta}_a$, and $\hat{\delta}_r$) are represented in radians, which are dimensionless. Second, the perturbed x-axis velocity u is divided by the steady-state velocity U_1 to yield u/U_1, which is a dimensionless ratio. Third, the perturbed longitudinal angular rates ($\dot{\alpha}$ and q) are multiplied by $\bar{c}/2U_1$ to yield $\dot{\alpha}\bar{c}/2U_1$ and $q\bar{c}/2U_1$, both nondimensional ratios when $\dot{\alpha}$ and q have units of radians/second. Fourth, the perturbed lateral-directional angular rates ($\dot{\beta}$, p, and r) are multiplied by $b/2U_1$ to form $\dot{\beta}b/2U_1$, $pb/2U_1$, and $rb/2U_1$. Again, $\dot{\beta}$, p, and r must have units of radians/second to make the ratios dimensionless. This is the conventional approach to nondimensionalizing the perturbation variables. The "2" in the denominator of the nondimensional angular rate terms comes from the roll helix angle (the angle that a wing tip makes with the forward velocity vector during a rolling maneuver). It is maintained in the other angular rate terms for consistency. With the nondimensional perturbation variables incorporated, Eqs. (6.9) and (6.11) become

$$
\begin{aligned}
f_{A_x} &= \frac{\partial F_{A_x}}{\partial \frac{u}{U_1}}\left(\frac{u}{U_1}\right) + \frac{\partial F_{A_x}}{\partial \hat{\alpha}}\hat{\alpha} + \frac{\partial F_{A_x}}{\partial \frac{\dot{\alpha}\bar{c}}{2U_1}}\left(\frac{\dot{\alpha}\bar{c}}{2U_1}\right) + \frac{\partial F_{A_x}}{\partial \frac{q\bar{c}}{2U_1}}\left(\frac{q\bar{c}}{2U_1}\right) + \frac{\partial F_{A_x}}{\partial \hat{\delta}_e}\hat{\delta}_e \\
m_A &= \frac{\partial M_A}{\partial \frac{u}{U_1}}\left(\frac{u}{U_1}\right) + \frac{\partial M_A}{\partial \hat{\alpha}}\hat{\alpha} + \frac{\partial M_A}{\partial \frac{\dot{\alpha}\bar{c}}{2U_1}}\left(\frac{\dot{\alpha}\bar{c}}{2U_1}\right) + \frac{\partial M_A}{\partial \frac{q\bar{c}}{2U_1}}\left(\frac{q\bar{c}}{2U_1}\right) + \frac{\partial M_A}{\partial \hat{\delta}_e}\hat{\delta}_e \qquad (6.12) \\
f_{A_z} &= \frac{\partial F_{A_z}}{\partial \frac{u}{U_1}}\left(\frac{u}{U_1}\right) + \frac{\partial F_{A_z}}{\partial \hat{\alpha}}\hat{\alpha} + \frac{\partial F_{A_z}}{\partial \frac{\dot{\alpha}\bar{c}}{2U_1}}\left(\frac{\dot{\alpha}\bar{c}}{2U_1}\right) + \frac{\partial F_{A_z}}{\partial \frac{q\bar{c}}{2U_1}}\left(\frac{q\bar{c}}{2U_1}\right) + \frac{\partial F_{A_z}}{\partial \hat{\delta}_e}\hat{\delta}_e
\end{aligned}
$$

$$
\begin{aligned}
l_A &= \frac{\partial L_A}{\partial \beta}\beta + \frac{\partial L_A}{\partial \frac{\dot{\beta}b}{2U_1}}\left(\frac{\dot{\beta}b}{2U_1}\right) + \frac{\partial L_A}{\partial \frac{pb}{2U_1}}\left(\frac{pb}{2U_1}\right) + \frac{\partial L_A}{\partial \frac{rb}{2U_1}}\left(\frac{rb}{2U_1}\right) \\
&\quad + \frac{\partial L_A}{\partial \delta_a}\hat{\delta}_a + \frac{\partial L_A}{\partial \delta_r}\hat{\delta}_r \\
f_{A_y} &= \frac{\partial F_{A_y}}{\partial \beta}\beta + \frac{\partial F_{A_y}}{\partial \frac{\dot{\beta}b}{2U_1}}\left(\frac{\dot{\beta}b}{2U_1}\right) + \frac{\partial F_{A_y}}{\partial \frac{pb}{2U_1}}\left(\frac{pb}{2U_1}\right) + \frac{\partial F_{A_y}}{\partial \frac{rb}{2U_1}}\left(\frac{rb}{2U_1}\right) \\
&\quad + \frac{\partial F_{A_y}}{\partial \delta_a}\hat{\delta}_a + \frac{\partial F_{A_y}}{\partial \delta_r}\hat{\delta}_r \qquad (6.13) \\
n_A &= \frac{\partial N_A}{\partial \beta}\beta + \frac{\partial N_A}{\partial \frac{\dot{\beta}b}{2U_1}}\left(\frac{\dot{\beta}b}{2U_1}\right) + \frac{\partial N_A}{\partial \frac{pb}{2U_1}}\left(\frac{pb}{2U_1}\right) + \frac{\partial N_A}{\partial \frac{rb}{2U_1}}\left(\frac{rb}{2U_1}\right) \\
&\quad + \frac{\partial N_A}{\partial \delta_a}\hat{\delta}_a + \frac{\partial N_A}{\partial \delta_r}\hat{\delta}_r
\end{aligned}
$$

Equation (6.12) is the nondimensionalized, longitudinal equation for perturbed forces and moment, and Eq. (6.13) is the nondimensionalized, lateral-directional, perturbed force and moments equation.

6.3.2 Longitudinal Perturbed Force and Moment Derivatives

We will next analyze each of the partial derivative terms in Eq. (6.12) so that they may be expressed with common longitudinal aerodynamic coefficients such as C_L, C_D, and C_m. To do this, we will analyze a perturbation in angle of attack ($\hat{\alpha}$) about the body-fixed stability axis. Figure 6.2 presents the axis systems associated with an exaggerated $\hat{\alpha}$ perturbation. Notice that the instantaneous velocity, V_∞, defines the direction of the lift and drag coefficients (C_L

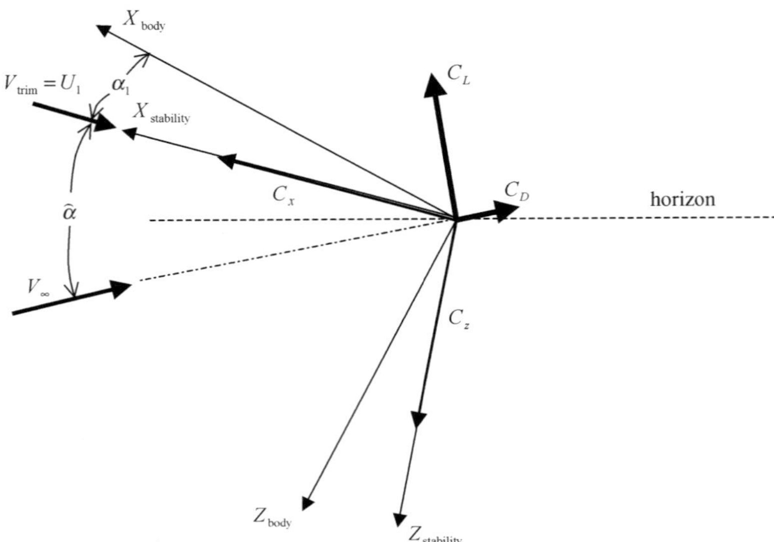

Fig. 6.2 Axis systems associated with an angle of attack perturbation.

and C_D). The coefficients C_x and C_z are defined relative to the body-fixed stability axis, which has its x axis aligned with V_{trim}.

6.3.2.1 u/U_1 derivatives. The u/U_1 derivatives consist of $\partial F_{A_x}/\partial(u/U_1)$, $\partial M_A/\partial(u/U_1)$, and $\partial F_{A_z}/\partial(u/U_1)$ in Eq. (6.12). We will begin with $\partial F_{A_x}/\partial(u/U_1)$. F_{A_x} is defined along $X_{\text{stability}}$, the body-fixed stability x axis. Figure 6.3 presents the vectors associated with the analysis. Notice that the vectors U_1 and V_∞ are now shown relative to fixed space.

F_{A_x} may be defined in terms of the coefficient C_x, which is shorthand for $C_{F_{A_x}}$, as

$$F_{A_x} = C_x \bar{q} S$$

We then have

$$\frac{\partial F_{A_x}}{\partial \dfrac{u}{U_1}} = \left. \frac{\partial (C_x \bar{q} S)}{\partial \dfrac{u}{U_1}} \right|_1$$

where the $|_1$ indicates that the partial derivative must be evaluated at the steady-state condition where the perturbation variables such as $\hat{\alpha}$ and u are zero. We must do this because the partial derivatives in the Taylor series expansions of Eqs. (6.12) and (6.13) are simply slopes used for a linear projec-

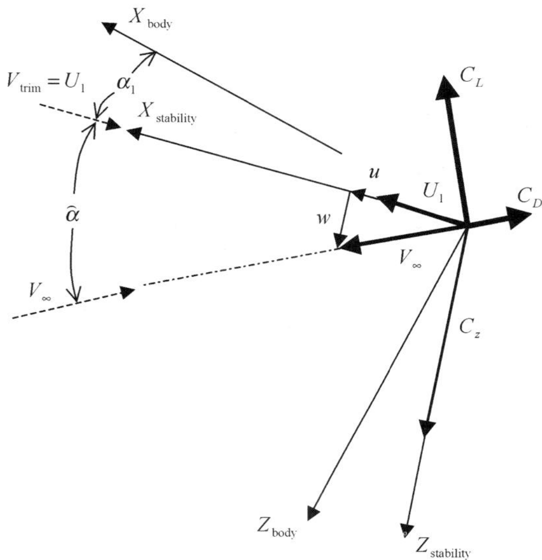

Fig. 6.3 Illustration of a *u* perturbation.

tion of perturbation values about the steady-state trimmed condition. Because both C_x and $\bar{q}$ may vary with u, we take the derivative in two parts:

$$\frac{\partial F_{A_x}}{\partial \dfrac{u}{U_1}} = \left.\frac{\partial (C_x \bar{q} S)}{\partial \dfrac{u}{U_1}}\right|_1 = \underbrace{\left.\frac{\partial C_x}{\partial \dfrac{u}{U_1}} \bar{q} S\right|_1}_{A} + \underbrace{\left. C_x S \frac{\partial \bar{q}}{\partial \dfrac{u}{U_1}}\right|_1}_{B} \tag{6.14}$$

We will address part A of Eq. (6.14) first.

From Fig. 6.4, we have

$$C_x = -C_D \cos\hat{\alpha} + C_L \sin\hat{\alpha}$$

and with the small perturbation assumption, this becomes

$$C_x \approx -C_D + C_L \hat{\alpha} \tag{6.15}$$

Using Eq. (6.15),

$$\frac{\partial C_x}{\partial \dfrac{u}{U_1}} \approx -\left.\frac{\partial C_D}{\partial \dfrac{u}{U_1}}\right|_1 + \left.\frac{\partial C_L}{\partial \dfrac{u}{U_1}} \hat{\alpha}\right|_1 \tag{6.16}$$

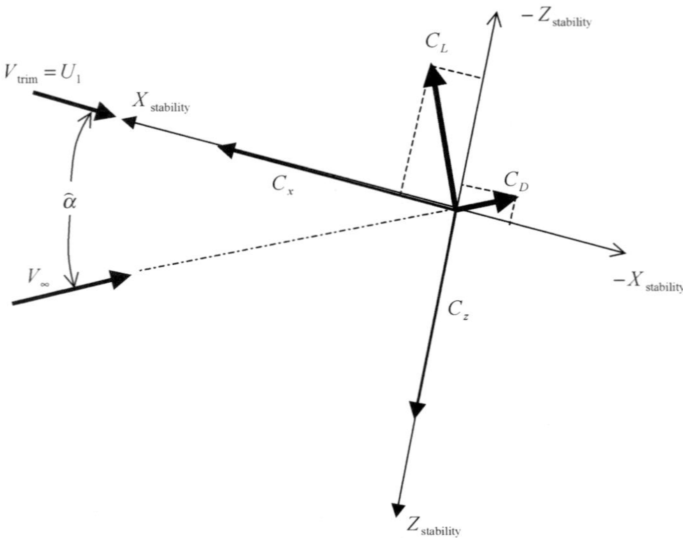

Fig. 6.4 Illustration of resolution of lift and drag coefficient.

Evaluating Eq. (6.16) at the steady-state condition, $\hat{\alpha} = 0$, and, for part A we have

$$\frac{\partial C_x}{\partial \dfrac{u}{U_1}} \approx -\frac{\partial C_D}{\partial \dfrac{u}{U_1}} = -C_{D_u} \tag{6.17}$$

C_{D_u} is called the **speed damping derivative**. It represents the change in drag coefficient with respect to u/U_1 and its value is dependent on Mach number. The value of C_{D_u} is generally zero or very small for pretransonic, subsonic Mach numbers. In the subsonic, transonic regime approaching Mach 1, C_{D_u} is generally positive indicating the significant drag rise as sonic flight is approached. Above Mach 1, C_{D_u} is generally negative.

Returning to Eq. (6.14), we now look at part B.

$$\frac{\partial \bar{q}}{\partial \dfrac{u}{U_1}} = U_1 \frac{\partial \bar{q}}{\partial u} = U_1 \left. \frac{\partial [\frac{1}{2} \rho \{(U_1 + u)^2 + v^2 + w^2\}]}{\partial u} \right|_1 = U_1 \left(\frac{1}{2}\right) \rho (2\{U_1 + u\})|_1$$

Evaluation at the steady state condition were $u = 0$ yields

$$\frac{\partial \bar{q}}{\partial \dfrac{u}{U_1}} = \rho U_1^2 = 2\bar{q}_1 \tag{6.18}$$

where $\bar{q}_1$ is the steady-state dynamic pressure. We now are able to incorporate Eqs. (6.17) and (6.18) into Eq. (6.14).

$$\frac{\partial F_{A_x}}{\partial \dfrac{u}{U_1}} = -C_{D_u}\bar{q}S|_1 + C_x S(2\bar{q}_1)|_1$$

When evaluated at the steady-state condition, $C_x = -C_{D_1}$ and $\bar{q} = \bar{q}_1$. These substitutions result in

$$\frac{\partial F_{A_x}}{\partial \dfrac{u}{U_1}} = -(C_{D_u} + 2C_{D_1})\bar{q}_1 S \tag{6.19}$$

The beauty of Eq. (6.19) is that the derivative is now expressed in terms of aerodynamic characteristics such as C_{D_u} and C_{D_1}, which can be estimated analytically, determined from wind tunnel testing, or computed with computational fluid dynamics (CFD) techniques.

The next u/U_1 derivative to be considered is $\partial M_A/\partial(u/U_1)$. Because $M_A = C_m\bar{q}S\bar{c}$ and both C_m and $\bar{q}$ may vary with u, we have

$$\frac{\partial M_A}{\partial \dfrac{u}{U_1}} = \frac{\partial C_m}{\partial\left(\dfrac{u}{U_1}\right)}\bar{q}S\bar{c}\,\Bigg|_1 + C_m S\bar{c}\frac{\partial \bar{q}}{\partial\left(\dfrac{u}{U_1}\right)}\Bigg|_1 \tag{6.20}$$

We will define

$$\frac{\partial C_m}{\partial\left(\dfrac{u}{U_1}\right)} = C_{m_u} \tag{6.21}$$

Note again that the subscript u in C_{m_u} really implies partial differentiation of C_m with respect to "u/U_1". With the substitutions of Eqs. (6.18) and (6.21), along with evaluation at the steady-state condition, Eq. (6.20) becomes

$$\frac{\partial M_A}{\partial \dfrac{u}{U_1}} = C_{m_u}\bar{q}_1 S\bar{c} + C_{m_1}S\bar{c}(2\bar{q}_1)$$

and, in combined form

$$\frac{\partial M_A}{\partial \dfrac{u}{U_1}} = (C_{m_u} + 2C_{m_1})\bar{q}_1 S\bar{c} \tag{6.22}$$

The derivative C_{m_u} results from changes in C_{m_0} and the aerodynamic center location with changes with forward speed. Probably the most notable effect on

C_{m_u} is the aft shift in aerodynamic center that occurs in the subsonic, transonic speed range. As the aerodynamic center moves aft, a negative pitching moment results that typically results in a negative C_{m_u}. Thus, in this range, the aircraft tends to experience a nose-down pitching moment with increasing speed. This phenomena is commonly referred to as "Mach Tuck," a characteristic that caused the crash of several high-speed subsonic aircraft that were not able to pull out of steep dives as the speed increased and C_{m_u} became more negative. As a result, C_{m_u} is commonly called the **Mach tuck derivative**. C_{m_1} is the steady-state aerodynamic pitching moment. It will be nonzero for cases where a thrust pitching moment must be counteracted by an aerodynamic pitching moment to trim the aircraft to a total pitch moment of zero. For all other cases such as gliders and power-off flight, C_{m_u} will be equal to zero.

The last u/U_1 derivative is $\partial F_{A_z}/\partial(u/U_1)$. We begin by referring to Fig. 6.2 and defining

$$F_{A_z} = C_z\bar{q}S$$

Using the same approach as with Eq. (6.14), we have

$$\frac{\partial F_{A_z}}{\partial \dfrac{u}{U_1}} = \left.\frac{\partial(C_z\bar{q}S)}{\partial \dfrac{u}{U_1}}\right|_1 = \left.\frac{\partial C_z}{\partial \dfrac{u}{U_1}}\bar{q}S\right|_1 + \left.C_zS\frac{\partial \bar{q}}{\partial \dfrac{u}{U_1}}\right|_1$$

where (referring to Fig. 6.4)

$$C_z = -C_L\cos\hat{\alpha} - C_D\sin\hat{\alpha}$$

With the assumption of small perturbations, this becomes

$$C_z \approx -C_L - C_D\hat{\alpha} \tag{6.23}$$

and

$$\frac{\partial C_z}{\partial \dfrac{u}{U_1}} \approx -\left.\frac{\partial C_L}{\partial \dfrac{u}{U_1}}\right|_1 - \left.\frac{\partial C_L}{\partial \dfrac{u}{U_1}}\hat{\alpha}\right|_1 \tag{6.24}$$

Evaluating Eq. (6.24) at the steady-state condition, $\hat{\alpha} = 0$, we have

$$\frac{\partial C_z}{\partial \dfrac{u}{U_1}} \approx -\frac{\partial C_L}{\partial \dfrac{u}{U_1}} = -C_{L_u} \tag{6.25}$$

C_{L_u} represents the change in lift coefficient with respect to velocity. The $\partial\bar{q}/\partial(u/U_1)$ derivative in Eq. (6.23) follows the same development as found with Eq. (6.18), $\partial\bar{q}/\partial(u/U_1) = 2\bar{q}_1$. Incorporating Eqs. (6.18) and (6.25) into the original expression for $\partial F_{A_z}/\partial(u/U_1)$ and realizing that $C_z = C_{L_1}$ at steady state, we have

$$\frac{\partial F_{A_z}}{\partial \frac{u}{U_1}} = -(C_{L_u} + 2C_{L_1})\bar{q}_1 S \tag{6.26}$$

Example 6.2

Find the u/U_1 derivative $\partial F_{A_x}/\partial(u/U_1)$ for the F-4C aircraft at 35,000 ft and Mach 0.9 ($U_1 = 876$ ft/s, $\bar{q} = 283.2$ lb/ft^2, $S = 530$ ft^2) if $C_{D_1} = 0.03$ and $C_{D_u} = 0.027$. If U is perturbed to 880 ft/s, find the perturbed applied aero force along the x stability axis (f_{A_x}).

Starting with Eq. (6.19),

$$\frac{\partial F_{A_x}}{\partial \frac{u}{U_1}} = -(C_{D_u} + 2C_{D_1})\bar{q}_1 S = -(0.027 + 2[0.03])283.2(530)$$

$$\frac{\partial F_{A_x}}{\partial \frac{u}{U_1}} = -13{,}058.4 \text{ lb}$$

To find f_{A_x}, we first find the perturbed velocity u.

$$u = U - U_1 = 880 - 876 = 4 \text{ ft/s}$$

and, from Eq. (6.12) for only a u perturbation

$$f_{A_x} = \frac{\partial F_{A_x}}{\partial \frac{u}{U_1}}\left(\frac{u}{U_1}\right) = -13{,}058.4\left(\frac{4}{876}\right) = -59.6 \text{ lb}$$

Because f_{A_x} is positive along the positive x stability axis, we have predicted a 59.6-lb increase in the drag of the aircraft if the velocity perturbs by 4 ft/s.

6.3.2.2 $\hat{\alpha}$ derivatives. The $\hat{\alpha}$, or perturbed angle of attack derivatives, consist of $\partial F_{A_x}/\partial\hat{\alpha}$, $\partial M_A/\partial\hat{\alpha}$, and $\partial F_{A_z}/\partial\hat{\alpha}$ in Eq. (6.12). We begin with $\partial F_{A_x}/\partial\hat{\alpha}$ and Eq. (6.15) for C_x.

$$\frac{\partial F_{A_x}}{\partial\hat{\alpha}} = \frac{\partial C_x}{\partial\hat{\alpha}}\bar{q}S$$

and

$$\frac{\partial C_x}{\partial \hat{\alpha}} = -\frac{\partial C_D}{\partial \hat{\alpha}} + \frac{\partial C_L}{\partial \hat{\alpha}}\hat{\alpha} + C_L \tag{6.27}$$

Evaluating Eq. (6.27) at the steady-state flight condition ($\hat{\alpha} = 0$),

$$\frac{\partial C_x}{\partial \hat{\alpha}} = -C_{D_{\hat{\alpha}}} + C_{L_1}$$

and

$$\frac{\partial F_{A_x}}{\partial \hat{\alpha}} = (-C_{D_{\hat{\alpha}}} + C_{L_1})\bar{q}_1 S \tag{6.28}$$

$C_{D_{\hat{\alpha}}}$ is the same as the C_{D_α} discussed in Sec. 5.3.1 and Eq. (5.16). It is basically the slope of the C_D vs α plot at the trim condition, α_1 (see Fig. 5.4).

The next $\hat{\alpha}$ derivative to be considered is $\partial M_A/\partial\hat{\alpha}$. This becomes

$$\frac{\partial M_A}{\partial \hat{\alpha}} = \frac{\partial C_m}{\partial \hat{\alpha}}\bar{q}S\bar{c}\bigg|_1 = C_{m_{\hat{\alpha}}}\bar{q}_1 S\bar{c} \tag{6.29}$$

$C_{m_{\hat{\alpha}}}$ is the same as the C_{m_α} discussed in Sec. 5.3.3.1 and Eq. (5.43). It is the **longitudinal static stability** derivative, which must be negative in value for longitudinal static stability.

The last $\hat{\alpha}$ derivative is $\partial F_{A_z}/\partial\hat{\alpha}$. We begin by referring to Fig. 6.4 and Eq. (6.23).

$$\frac{\partial F_{A_z}}{\partial \hat{\alpha}} = \frac{\partial C_z}{\partial \hat{\alpha}}\bar{q}S = \left(-\frac{\partial C_L}{\partial \hat{\alpha}} - \frac{\partial C_D}{\partial \hat{\alpha}}\hat{\alpha} - C_D\right)\bar{q}S\bigg|_1 \tag{6.30}$$

Evaluating Eq. (6.30) at the steady state flight condition, $\hat{\alpha} = 0$, we have

$$\frac{\partial F_{A_z}}{\partial \hat{\alpha}} = -(C_{L_{\hat{\alpha}}} + C_{D_1})\bar{q}_1 S \tag{6.31}$$

$C_{L_{\hat{\alpha}}}$ is the same as the C_{L_α} discussed in Sec. 5.3.2 and Eq. (5.30). It is commonly referred to as the **lift curve slope**.

Example 6.3

Find the $\hat{\alpha}$ derivative $\partial F_{A_z}/\partial\hat{\alpha}$ for the F-4C aircraft at the same flight conditions as those of Example 6.2. C_{L_α} is equal to 3.75/rad. If the F-4C is trimmed at an angle of attack of 2.6 deg and then is perturbed to 3.1 deg, find the perturbed aero force along the z stability axis (f_{A_z}).

Starting with Eq. (6.31),

$$\frac{\partial F_{A_z}}{\partial \hat{\alpha}} = -(C_{L_{\hat{\alpha}}} + C_{D_1})\bar{q}_1 S = -(3.75 + 0.03)(283.2)(530)$$

$$\frac{\partial F_{A_z}}{\partial \hat{\alpha}} = -567{,}363 \text{ lb/rad}$$

To find f_{A_z}, we first find the perturbed angle of attack $\hat{\alpha}$

$$\hat{\alpha} = \alpha - \alpha_1 = 3.1 \text{ deg} - 2.6 \text{ deg} = 0.5 \text{ deg} = 0.00873 \text{ rad}$$

and from Eq. (6.12) for only an $\hat{\alpha}$ perturbation

$$f_{A_z} = \frac{\partial F_{A_z}}{\partial \hat{\alpha}}\hat{\alpha} = -567{,}363(0.00873) = -4951 \text{ lb}$$

Because f_{A_z} is positive along the positive z stability axis, we have predicted a 4951 lb increase in the lift of the aircraft if the angle of attack perturbs by one-half degree. This may seem very large, but remember the aircraft is at Mach 0.9.

6.3.2.3 Quasi-steady $\dot{\alpha}\bar{c}/2U_1$ derivatives. If a rate of change in angle of attack ($\dot{\alpha}$) is present, a lag in the development of downwash at the horizontal tail occurs. Because the α derivatives assume that the downwash is fully developed, the "$\dot{\alpha}\bar{c}/2U_1$" derivatives provide a correction to the α derivatives when the aircraft is undergoing a rate of change in angle of attack.

The first $\dot{\alpha}\bar{c}/2U_1$ derivative in Eq. (6.12) to be considered is $\partial F_{A_x}/\partial(\dot{\alpha}\bar{c}/2U_1)$. Using a similar approach to the u/U_1 and α derivatives, we have

$$\frac{\partial F_{A_x}}{\partial \dfrac{\dot{\alpha}\bar{c}}{2U_1}} = \frac{\partial(C_x \bar{q} S)}{\partial \dfrac{\dot{\alpha}\bar{c}}{2U_1}} = C_{x_{\dot{\alpha}}}\bar{q}S = -C_{D_{\dot{\alpha}}}\bar{q}_1 S \tag{6.32}$$

The derivative $C_{D_{\dot{\alpha}}}$ represents the change in drag coefficient with respect to nondimensional $\dot{\alpha}$. For most applications, the lag in downwash because of $\dot{\alpha}$ has little effect on the drag coefficient, therefore, it is typically assumed that $C_{D_{\dot{\alpha}}} = 0$.

The next $\dot{\alpha}\bar{c}/2U_1$ derivative that we will consider is $\partial F_{A_z}/\partial(\dot{\alpha}\bar{c}/2U_1)$. Again, we have

$$\frac{\partial F_{A_z}}{\partial \dfrac{\dot{\alpha}\bar{c}}{2U_1}} = \frac{\partial(C_z \bar{q} S)}{\partial \dfrac{\dot{\alpha}\bar{c}}{2U_1}} = C_{z_{\dot{\alpha}}}\bar{q}S = -C_{L_{\dot{\alpha}}}\bar{q}_1 S \tag{6.33}$$

The derivative $C_{L_{\dot{\alpha}}}$ is significant, and we will develop an approach to estimate it. Remember that $C_{L_{\dot{\alpha}}}$ should be considered a correction to C_{L_α} for nonsteady-state conditions. Figure 6.5 presents an aircraft experiencing an $\dot{\alpha}$ as it transitions from α_{initial} to α_{final}. If this change in α takes place in Δt seconds, we have

$$\dot{\alpha} = \frac{\alpha_{\text{final}} - \alpha_{\text{initial}}}{\Delta t}$$

This figure also shows the change in downwash angle that occurs at the horizontal tail. The change in downwash angle will be defined as

$$\Delta\varepsilon = \varepsilon_{\text{initial}} - \varepsilon_{\text{final}}$$

$\Delta\varepsilon$ can also be viewed as the correction needed to the steady-state downwash angle to compensate for the downwash lag. It can also be estimated with

$$\Delta\varepsilon = -\frac{\mathrm{d}\varepsilon}{\mathrm{d}\alpha}\frac{\mathrm{d}\alpha}{\mathrm{d}t}\Delta t = -\frac{\mathrm{d}\varepsilon}{\mathrm{d}\alpha}\dot{\alpha}\Delta t \tag{6.34}$$

$\mathrm{d}\varepsilon/\mathrm{d}\alpha$ is the rate of change of downwash angle with angle of attack as discussed in Sec. 5.3.2. Δt is the time it takes for the final downwash to travel back to the horizontal tail. Referring to Fig. 6.5, Δt is normally estimated as

$$\Delta t = \frac{X_h}{U_1} = \frac{x_{AC_h} - x_{cg}}{U_1} \tag{6.35}$$

Thus, Eq. (6.34) becomes

$$\Delta\varepsilon = -\frac{\mathrm{d}\varepsilon}{\mathrm{d}\alpha}\dot{\alpha}\frac{X_h}{U_1} \tag{6.36}$$

Because the estimate of the lift coefficient at the horizontal tail is based on the steady-state angle of attack (by using C_{L_α}), we next estimate the correction (ΔC_{L_h}) needed to the lift coefficient because of the lag in downwash (resulting from $\dot{\alpha}$). This becomes

$$\Delta C_{L_h} = -C_{L_{\alpha_h}}(\Delta\varepsilon) = (C_{L_{\alpha_h}})\frac{\mathrm{d}\varepsilon}{\mathrm{d}\alpha}\dot{\alpha}\frac{X_h}{U_1} \tag{6.37}$$

The correction to lift coefficient is positive because the steady-state lift coefficient assumes a fully developed downwash angle that reduces the lift. Downwash lag results in the original downwash angle being maintained, resulting in additional lift over that predicted for the steady-state angle of attack during this interim period. We next use the techniques of Sec. 5.3.2 and Eq. (6.37) to predict the increase in lift coefficient for the entire aircraft.

$$\Delta C_L = \Delta C_{L_h}\eta_h\frac{S_h}{S} = (C_{L_{\alpha_h}})\frac{\mathrm{d}\varepsilon}{\mathrm{d}\alpha}\dot{\alpha}\frac{X_h}{U_1}\eta_h\frac{S_h}{S} \tag{6.38}$$

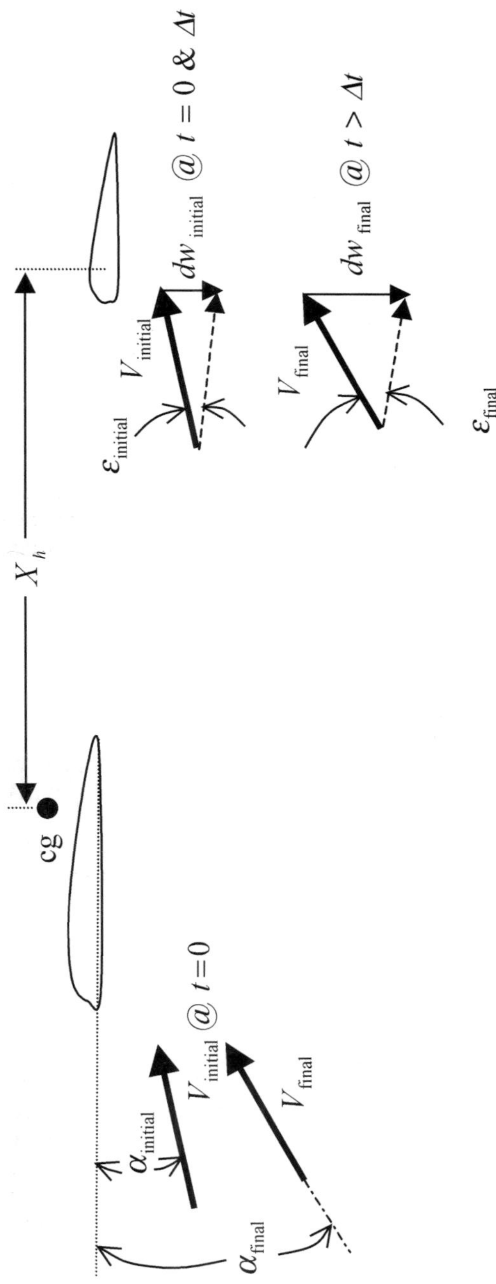

Fig. 6.5 Illustration of downwash lag.

Recalling from Eq. (6.33) that we wanted to develop an expression for $C_{L_{\dot{\alpha}}}$, we have

$$C_{L_{\dot{\alpha}}} = \frac{\partial C_L}{\partial \dfrac{\dot{\alpha}\bar{c}}{2U_1}} = \frac{2U_1}{\bar{c}} \frac{\partial C_L}{\partial \dot{\alpha}} \tag{6.39}$$

Finally, we can take the partial derivative of Eq. (6.38) with respect to $\dot{\alpha}$ and substitute the result into Eq. (6.39)

$$C_{L_{\dot{\alpha}}} = \frac{2U_1}{\bar{c}}(C_{L_{\alpha_h}}) \frac{\mathrm{d}\varepsilon}{\mathrm{d}\alpha} \frac{X_h}{U_1} \eta_h \frac{S_h}{S} \tag{6.40}$$

Recalling from Eq. (5.44) that the tail volume ratio, $\bar{V}_h$, is equal to $\dfrac{X_h}{\bar{c}}\dfrac{S_h}{S}$, Eq. (6.40) becomes

$$C_{L_{\dot{\alpha}}} = 2(C_{L_{\alpha_h}}) \frac{\mathrm{d}\varepsilon}{\mathrm{d}\alpha} \eta_h \bar{V}_h \tag{6.41}$$

The last $\dot{\alpha}\bar{c}/2U_1$ derivative to be considered is $\partial M_A/\partial(\dot{\alpha}\bar{c}/2U_1)$. Again, we have

$$\frac{\partial M_A}{\partial \dfrac{\dot{\alpha}\bar{c}}{2U_1}} = \frac{\partial C_m \bar{q} S \bar{c}}{\partial \dfrac{\dot{\alpha}\bar{c}}{2U_1}} = C_{m_{\dot{\alpha}}} \bar{q}_1 S \bar{c} \tag{6.42}$$

$C_{m_{\dot{\alpha}}}$ results from the same downwash lag phenomena that was analyzed to obtain the estimate of $C_{L_{\dot{\alpha}}}$. Thus, Eq. (6.41) is multiplied by the nondimensional moment arm $X_h/\bar{c}$ along with a negative sign indicating that positive lift on the horizontal tail produces a nose-down (negative) pitching moment to obtain $C_{m_{\dot{\alpha}}}$.

$$C_{m_{\dot{\alpha}}} = -2(C_{L_{\alpha_h}}) \frac{\mathrm{d}\varepsilon}{\mathrm{d}\alpha} \eta_h \bar{V}_h \frac{X_h}{\bar{c}} \tag{6.43}$$

As a rule of thumb, for many airplanes $C_{m_{\dot{\alpha}}}$ is approximately equal to one third the value of C_{m_q} [Eq. (6.49)].

Example 6.4

Find the $\dot{\alpha}\bar{c}/2U_1$ derivative $\partial M_A/\partial(\dot{\alpha}\bar{c}/2U_1)$ for the F-4C at the same conditions as presented in Example 6.2. $\bar{c}$ for the F-4C is 16 ft and $C_{m_{\dot{\alpha}}}$ is -1.3 per rad. If $\dot{\alpha}$ is 0.5 deg/s, find the perturbed pitching moment m_A.

Starting with Eq. (6.42),

$$\frac{\partial M_A}{\partial \dfrac{\dot{\alpha}\bar{c}}{2U_1}} = C_{m_{\dot{\alpha}}} \bar{q}_1 S \bar{c} = -1.3(283.2)530(16) = -3{,}121{,}997 \text{ ft} \cdot \text{lb/rad}$$

To find m_A, we use Eq. (6.12) for only an $\dot{\alpha}$ perturbation

$$m_A = \frac{\partial M_A}{\partial \frac{\dot{\alpha}\bar{c}}{2U_1}}\left(\frac{\dot{\alpha}\bar{c}}{2U_1}\right) = -3{,}121{,}997\left(\frac{(0.5/57.3)(16)}{2(876)}\right) = -248.8 \text{ ft} \cdot \text{lb}$$

Notice that degrees/second are converted to radians/second to maintain consistent units. We have predicted a nose-down pitching moment of 248.8 ft/lb resulting from a positive 0.5 deg/s $\dot{\alpha}$.

6.3.2.4 Pitch rate $q\bar{c}/2U_1$ derivatives. The $q\bar{c}/2U_1$ derivatives consist of $\partial F_{A_x}/\partial(q\bar{c}/2U_1)$, $\partial M_A/\partial(q\bar{c}/2U_1)$ and $\partial F_{A_z}/\partial(q\bar{c}/2U_1)$ in Eq. (6.12). We begin with $\partial F_{A_x}/\partial(q\bar{c}/2U_1)$.

$$\frac{\partial F_{A_x}}{\partial \frac{q\bar{c}}{2U_1}} = \frac{\partial(C_x \bar{q} S)}{\partial \frac{q\bar{c}}{2U_1}} = C_{x_q}\bar{q}S = -C_{D_q}\bar{q}_1 S \tag{6.44}$$

The derivative C_{D_q} represents the change in drag coefficient with respect to nondimensional pitch rate. For most applications, this derivative is very small and assumed to be equal to zero ($C_{D_q} \approx 0$).

The next "$q\bar{c}/2U_1$" derivative to be considered is $\partial F_{A_z}/\partial(q\bar{c}/2U_1)$.

$$\frac{\partial F_{A_z}}{\partial \frac{q\bar{c}}{2U_1}} = \frac{\partial(C_z \bar{q} S)}{\partial \frac{q\bar{c}}{2U_1}} = C_{z_q}\bar{q}S = -C_{L_q}\bar{q}_1 S \tag{6.45}$$

C_{L_q} should be thought of as the change in lift coefficient because of pitch rate. Referring to Fig. 6.6, it can be seen that a positive pitch rate, q, results in a downward velocity, qX_h, at the horizontal tail.

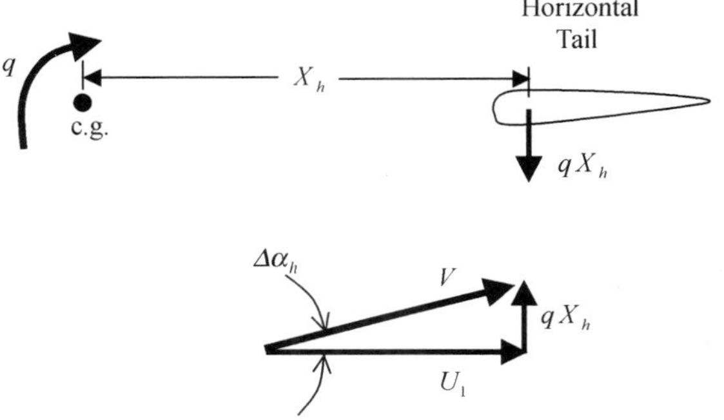

Fig. 6.6 Illustration of change in angle of attack at the horizontal tail because of pitch rate.

This downward velocity induces an increase in the angle of attack for the horizontal tail, $\Delta\alpha_h$, which can be defined as

$$\Delta\alpha_h = \tan^{-1}\frac{qX_h}{U_1} \approx \frac{qX_h}{U_1}$$

This increase in angle of attack results in an increase in horizontal tail lift, which can be quantified in terms of an increase in lift coefficient because of pitch rate

$$\Delta C_{L_h} = C_{L_{\alpha_h}}(\Delta\alpha_h) = C_{L_{\alpha_h}}\left(\frac{qX_h}{U_1}\right)$$

The change in lift coefficient for the entire aircraft then becomes (using the techniques of Sec. 5.3.2)

$$\Delta C_L = C_{L_{\alpha_h}}\left(\frac{qX_h}{U_1}\right)\eta_h\frac{S_h}{S} \tag{6.46}$$

C_{L_q} may be determined using Eq. (6.47)

$$C_{L_q} = \frac{\partial C_L}{\partial\dfrac{q\bar{c}}{2U_1}} = \frac{\Delta C_L}{\Delta\dfrac{q\bar{c}}{2U_1}} = 2C_{L_{\alpha_h}}\frac{X_h}{\bar{c}}\eta_h\frac{S_h}{S} = 2C_{L_{\alpha_h}}\eta_h\bar{V}_h \tag{6.47}$$

The derivative C_{L_q} will typically have a positive value and vary with Mach number.

The final $q\bar{c}/2U_1$ derivative to be considered is $\partial M_A/\partial(q\bar{c}/2U_1)$. Again, we have

$$\frac{\partial M_A}{\partial\dfrac{q\bar{c}}{2U_1}} = \frac{\partial C_m\bar{q}S\bar{c}}{\partial\dfrac{q\bar{c}}{2U_1}} = C_{m_q}\bar{q}_1S\bar{c} \tag{6.48}$$

C_{m_q} results from the same increase in horizontal tail lift because of pitch rate as was discussed for C_{L_q}. Thus, Eq. (6.47) is simply multiplied by the nondimensional moment arm $X_h/\bar{c}$ and a negative sign is added to indicate that a positive pitch rate results in a nose-down (negative) pitching moment.

$$C_{m_q} = \frac{\partial C_m}{\partial\dfrac{q\bar{c}}{2U_1}} = -2C_{L_{\alpha_h}}\eta_h\bar{V}_h\frac{X_h}{\bar{c}} \tag{6.49}$$

C_{m_q} is called the **pitch damping derivative**. It is a very important factor for longitudinal dynamic stability characteristics. It is negative (providing a moment that opposes the direction of the pitch rate) and will be the primary factor (along with $C_{m_{\dot{\alpha}}}$) for damping out pitch oscillations. For most aircraft, the wing and fuselage also contribute to pitch damping; to account for this effect, the value of C_{m_q} predicted by Eq. (6.49) is typically increased by approximately 10%. Of course, for tailless aircraft the wing body contribution becomes the primary contributor to C_{m_q}. An analysis of Eq. (6.49) reveals that C_{m_q} is proportional to the square of X_h ($\bar{V}_h$ contains an X_h term). Thus, X_h becomes an important design parameter when considering longitudinal dynamic stability.

Example 6.5

Find the $q\bar{c}/2U_1$ derivative $\partial F_{A_z}/\partial(q\bar{c}/2U_1)$ for the F-4C aircraft at the same flight conditions as those of Example 6.2. C_{L_q} is equal to 1.80. If q is 2.5 deg/s, find the perturbed aero force along the z stability axis (f_{A_z}).

Starting with Eq. (6.45),

$$\frac{\partial F_{A_z}}{\partial \dfrac{q\bar{c}}{2U_1}} = -C_{L_q}\bar{q}_1 S = -1.8(283.2)(530)$$

$$\frac{\partial F_{A_z}}{\partial \dfrac{q\bar{c}}{2U_1}} = -270{,}172.8 \text{ lb/rad}$$

To find f_{A_z}, we use Eq. (6.12) for only a q perturbation

$$f_{A_z} = \frac{\partial F_{A_z}}{\partial \dfrac{q\bar{c}}{2U_1}}\left(\frac{q\bar{c}}{2U_1}\right) = -270{,}172.8\left(\frac{(2.5/57.3)(16)}{2(876)}\right) = -107.7 \text{ lb}$$

Notice that degrees/second are converted to radians/second to maintain consistent units. Because f_{A_z} is positive along the positive z stability axis, we have predicted 107.7 lb increase in lift if pitch rate perturbs by 2.5 deg/s.

Example 6.6

Estimate the pitch damping derivative, C_{m_q}, for an aircraft with the following characteristics: $C_{L_{\alpha_h}} = 0.075/\text{deg}$, $\eta_h = 0.98$, $\bar{V}_h = 0.375$, $(X_h/\bar{c}) = 3.0$.

Starting with Eq. (6.49),

$$C_{m_q} = -2C_{L_{\alpha_h}}\eta_h\bar{V}_h\frac{X_h}{\bar{c}} = -2(0.075)(57.3)0.98(0.375)3.0$$

$$C_{m_q} = -9.48/\text{rad}$$

Notice that $C_{L_{\alpha_h}}$ has been converted to per radian.

6.3.2.5 $\hat{\delta}_e$ derivatives. The $\hat{\delta}_e$ derivatives consist of $\partial F_{A_x}/\partial\hat{\delta}_e$, $\partial M_A/\partial\delta_e$, and $\partial F_{A_z}/\partial\hat{\delta}_e$ in Eq. (6.12). Using a similar development to that for the previous derivatives, we have

$$\frac{\partial F_{A_x}}{\partial\hat{\delta}_e} = \frac{\partial(C_x\bar{q}S)}{\partial\hat{\delta}_e} = C_{x_{\hat{\delta}_e}}\bar{q}S = -C_{D_{\hat{\delta}_e}}\bar{q}_1 S \tag{6.50}$$

$C_{D_{\hat{\delta}_e}}$ is the change in drag coefficient because of elevator deflection. As discussed in Sec. 5.3.1, it is typically very small and usually assumed to be equal to zero.

$\partial M_A/\partial\hat{\delta}_e$ becomes

$$\frac{\partial M_A}{\partial\hat{\delta}_e} = \frac{\partial C_m\bar{q}S\bar{c}}{\partial\delta_e} = C_{m_{\hat{\delta}_e}}\bar{q}_1 S\bar{c} \tag{6.51}$$

$C_{m_{\hat{\delta}_e}}$ is the change in pitching moment coefficient because of elevator deflection and is a **primary control derivative**, as discussed in Sec. 5.2. It is also referred to as the **elevator control power** derivative. It was previously defined with Eq. (5.46).

Finally, $\partial F_{A_z}/\partial\hat{\delta}_e$ becomes

$$\frac{\partial F_{A_z}}{\partial\hat{\delta}_e} = \frac{\partial(C_z\bar{q}S)}{\partial\hat{\delta}_e} = C_{z_{\hat{\delta}_e}}\bar{q}S = -C_{L_{\hat{\delta}_e}}\bar{q}_1 S \tag{6.52}$$

$C_{L_{\hat{\delta}_e}}$ is the change in lift coefficient because of elevator deflection. This derivative was discussed in Sec. 5.3.2 and was defined with Eq. (5.32).

The $\hat{\delta}_e$ derivatives were developed assuming a conventional (tail aft) aircraft with a horizontal tail and elevator configuration. Of course, a variety of other longitudinal control configurations may be used with modern aircraft. These include stabilators, canards, and flaps. For these cases, appropriate control derivatives must be developed using the same approach presented here for the $\hat{\delta}_e$ derivatives.

Example 6.7

Find the $\hat{\delta}_e$ derivative $\partial M_A/\partial\hat{\delta}_e$ for the F-4C aircraft at the same flight conditions as those of Example 6.2. $C_{m_{\hat{\delta}_e}}$ is equal to -0.058/rad. If $\hat{\delta}_e$ is 1 deg, find the perturbed pitching moment, m_A.

Starting with Eq. (6.51),

$$\frac{\partial M_A}{\partial\hat{\delta}_e} = C_{m_{\hat{\delta}_e}}\bar{q}_1 S\bar{c} = -0.058(283.2)530(16) = -139{,}289 \text{ ft}\cdot\text{lb/rad}$$

To find m_A, we use Eq. (6.12) for only a $\hat{\delta}_e$ perturbation

$$m_A = \frac{\partial M_A}{\partial \delta_e}\hat{\delta}_e = (-139{,}289)(1/57.3) = -2430.9 \text{ ft} \cdot \text{lb}$$

Notice that degrees are converted to radians to maintain consistent units.

6.3.2.6 Summary. We now are able to define the longitudinal perturbed forces and moments from Eqs. (6.12) using the derivatives developed in Secs. 6.3.2.1–6.3.2.5. This recasting of the equations will use matrix format and Eqs. (6.19), (6.22), (6.26), (6.28), (6.29), (6.31–6.33), (6.42), (6.44), (6.45), (6.48), and (6.50–6.52), resulting in

$$\begin{bmatrix} \dfrac{f_{A_x}}{\bar{q}_1 S} \\ \dfrac{m_A}{\bar{q}_1 S\bar{c}} \\ \dfrac{f_{A_z}}{\bar{q}_1 S} \end{bmatrix} = \begin{bmatrix} -(C_{D_u} + 2C_{D_1}) & (-C_{D_{\hat{\alpha}}} + C_{L_1}) & -C_{D_{\dot{\alpha}}} & -C_{D_q} & -C_{D_{\hat{\delta}_e}} \\ (C_{m_u} + 2C_{m_1}) & C_{m_{\hat{\alpha}}} & C_{m_{\dot{\alpha}}} & C_{m_q} & C_{m_{\hat{\delta}_e}} \\ -(C_{L_u} + 2C_{L_1}) & -(C_{L_{\hat{\alpha}}} + C_{D_1}) & -C_{L_{\dot{\alpha}}} & -C_{L_q} & -C_{L_{\hat{\delta}_e}} \end{bmatrix} \begin{bmatrix} \dfrac{u}{U_1} \\ \hat{\alpha} \\ \dfrac{\dot{\alpha}\bar{c}}{2U_1} \\ \dfrac{q\bar{c}}{2U_1} \\ \hat{\delta}_e \end{bmatrix} \tag{6.53}$$

The advantage of Eq. (6.53) over Eq. (6.12) is that the longitudinal perturbed forces and moments are now expressed in terms of common aero derivatives such as C_{D_u}, $C_{m_{\hat{\alpha}}}$, and C_{L_q}. The value of these derivatives can be estimated with analytical or experimental techniques. Remember that each derivative in Eq. (6.53) is dimensionless—for example, C_{m_u} is the abbreviated form of $\partial C_m/\partial(u/U_1)$, as discussed in Sec. 6.3.1. Table 6.1 summarizes the coefficients and derivatives discussed for the perturbed longitudinal forces and moment estimates.

Finally, the derivatives associated with the perturbed quantities $\hat{\alpha}$, q, and $\hat{\delta}_e$ will, in most cases, be equal to the same derivatives with respect to the absolute quantities α, Q, and δ_e. For example, $C_{m_{\hat{\alpha}}} \approx C_{m_\alpha}$, $C_{m_q} \approx C_{m_Q}$, and $C_{m_{\hat{\delta}_e}} \approx C_{m_{\delta_e}}$, because each derivative represents a change in the value of the pitching moment coefficient with respect to a change in the value of angle of attack, pitch rate, and elevator deflection, respectively.

6.3.3 Lateral-Directional Perturbed Force and Moment Derivatives

We will next analyze each of the partial derivative terms in Eq. (6.13) so that they may also be expressed with common lateral-directional aerodynamic coefficients such as C_y, C_1 and C_n. The process is similar to that presented in Sec. 6.3.2 for the longitudinal derivatives. It is helpful to recall from Fig. 5.24 that a positive sideslip angle (β) is defined as the relative wind oriented to the

Table 6.1 Summary of longitudinal derivatives

Derivative	Name	Normal sign
C_{D_u}	Speed damping derivative	+ or −
C_{m_u}	Mach tuck derivative	+ or −
C_{L_u}	None	+ or −
$C_{D_{\hat{\alpha}}}$	None	+
$C_{m_{\hat{\alpha}}}$	Longitudinal static stability derivative	−
$C_{L_{\hat{\alpha}}}$	Lift curve slope	+
$C_{D_{\dot{\alpha}}}$	Quasi-steady derivative	≈ 0
$C_{m_{\dot{\alpha}}}$	Quasi-steady derivative	−
$C_{L_{\dot{\alpha}}}$	Quasi-steady derivative	+
C_{D_q}	None	≈ 0
C_{m_q}	Pitch damping derivative	−
C_{L_q}	None	+
$C_{D_{\hat{\delta}_e}}$	None	≈ 0
$C_{m_{\hat{\delta}_e}}$	Elevator control power	−
$C_{L_{\hat{\delta}_e}}$	None	+

right of the aircraft nose or, from the pilot's perspective, "wind in the right ear." A review of the lateral-directional control deflection sign convention from Sec. 5.2 is also helpful. Trailing edge up right aileron deflection and trailing edge down left aileron deflection is defined as a positive δ_a. Trailing edge left rudder deflection is defined as a positive δ_r.

6.3.3.1 Sideslip "β" derivatives. The β derivatives consist of $\partial L_A/\partial\beta$, $\partial F_{A_y}/\partial\beta$, and $\partial N_A/\partial\beta$ in Eq. (6.13). We will begin with $\partial L_A/\partial\beta$. L_A is the aerodynamic rolling moment and may be defined in terms of the rolling moment coefficient C_1 with Eq. (5.87).

$$L_A = C_l\bar{q}_1 Sb$$

We then have

$$\frac{\partial L_A}{\partial\beta} = \frac{\partial C_1}{\partial\beta}\bar{q}_1 Sb = C_{l_\beta}\bar{q}_1 Sb \tag{6.54}$$

C_{l_β} is the **lateral** (roll) **static stability** derivative as discussed in Sec. 5.5.2.1. It must be negative if an aircraft has roll static stability. An estimate of C_{l_β} can be obtained through analysis of the four aircraft design aspects that have the greatest influence on C_{l_β}, namely, geometric dihedral, wing position, wing sweep angle, and the vertical tail. Reference 1 presents such an approach. However, because of the complex interaction of each design feature, wind tunnel and/or computational fluid dynamic analysis is also suggested.

In a similar manner, development of $\partial F_{A_y}/\partial\beta$ begins with a restatement of Eq. (5.85):

$$F_{A_y} = C_y \bar{q}_1 S$$

and

$$\frac{\partial F_{A_y}}{\partial\beta} = \frac{\partial C_y}{\partial\beta}\bar{q}_1 S = C_{y_\beta}\bar{q}_1 S \tag{6.55}$$

The derivative C_{y_β} is normally negative and was discussed in Sec. 5.5.1.

Finally, $\partial N_A/\partial\beta$ is developed in the same manner starting with Eq. (5.94).

$$N_A = C_n \bar{q}_1 Sb$$

We then have

$$\frac{\partial N_A}{\partial\beta} = \frac{\partial C_n}{\partial\beta}\bar{q}_1 Sb = C_{n_\beta}\bar{q}_1 Sb \tag{6.56}$$

The derivative C_{n_β} is the **directional** (yaw) **static stability** as discussed in Sec. 5.5.3.1. It must be positive for the aircraft to have directional static stability.

Example 6.8

Find the β derivative $\partial L_A/\partial\beta$ for the F-4C aircraft at 35,000 ft and Mach 0.9 ($U_1 = 876$ ft/s, $\bar{q} = 283.2$ lb/ft^2, $S = 530$ ft^2, $b = 38.7$ ft) if $C_{l_\beta} = -0.08$. If β is perturbed to 1 deg, find the perturbed rolling moment l_A.

Starting with Eq. (6.54),

$$\frac{\partial L_A}{\partial\beta} = C_{l_\beta}\bar{q}_1 Sb = -0.08(283.2)530(38.7) = -464{,}697 \text{ ft}\cdot\text{lb/rad}$$

To find l_A, we use Eq. (6.13) for only a β perturbation

$$l_A = \frac{\partial L_A}{\partial\beta}\beta = -464{,}697(1/57.3) = -8{,}109.9 \text{ ft/lb}$$

Notice that degrees are converted to radians to maintain consistent units.

6.3.3.2 Quasi-steady $\dot{\beta}b/2U_1$ derivatives. Similar to the discussion of Sec. 6.3.2.3, if a rate of change in sideslip ($\dot{\beta}$) is present, a lag in the development of sidewash at the vertical tail occurs. Because the β derivatives assume that the sidewash is fully developed, the $\dot{\beta}b/2U_1$ derivatives provide a correction to the β derivatives when the aircraft is undergoing a rate of change in sideslip. These derivatives are generally considered negligible because of the relatively unobstructed flow that is present at most vertical tails. However, they may be significant in the high subsonic speed region.

The first $\dot{\beta}b/2U_1$ derivative in Eq. (6.13) to be considered is $\partial L_A/\partial(\dot{\beta}b/2U_1)$. Using a similar approach as with the β derivatives, we have:

$$\frac{\partial L_A}{\partial \dfrac{\dot{\beta}b}{2U_1}} = \frac{\partial C_l \bar{q}_1 Sb}{\partial \dfrac{\dot{\beta}b}{2U_1}} = C_{l_{\dot{\beta}}} \bar{q}_1 Sb \tag{6.57}$$

The next two $\dot{\beta}b/2U_1$ derivatives follow the same pattern.

$$\frac{\partial F_{A_y}}{\partial \dfrac{\dot{\beta}b}{2U_1}} = \frac{\partial C_y \bar{q}_1 S}{\partial \dfrac{\dot{\beta}b}{2U_1}} = C_{y_{\dot{\beta}}} \bar{q}_1 S \tag{6.58}$$

$$\frac{\partial N_A}{\partial \dfrac{\dot{\beta}b}{2U_1}} = \frac{\partial C_n \bar{q}_1 Sb}{\partial \dfrac{\dot{\beta}b}{2U_1}} = C_{n_{\dot{\beta}}} \bar{q}_1 Sb \tag{6.59}$$

Methods for estimating the $\dot{\beta}b/2U_1$ derivatives are presented in several flight mechanics and aircraft design texts.

6.3.3.3 Roll rate $pb/2U_1$ derivatives. The $pb/2U_1$ derivatives consist of $\partial L_A/\partial(pb/2U_1)$, $\partial F_{A_y}/\partial(pb/2U_1)$, and $\partial N_A/\partial(pb/2U_1)$ in Eq. (6.13). We begin with $\partial L_A/\partial(pb/2U_1)$.

$$\frac{\partial L_A}{\partial \dfrac{pb}{2U_1}} = \frac{\partial C_l}{\partial \dfrac{pb}{2U_1}} \bar{q}_1 Sb = C_{l_p} \bar{q}_1 Sb \tag{6.60}$$

The derivative C_{l_p} is called the **roll damping derivative**. It represents the change in rolling moment coefficient with respect to nondimensional roll rate and is usually negative (providing a moment that opposes the direction of the roll rate). C_{l_p} is a very important factor for lateral-directional dynamic stability characteristics. Three aircraft components have a primary influence on the value of C_{l_p}: the wing, the horizontal tail, and the vertical tail.

As illustrated in Fig. 6.7, roll rate induces a vertical velocity contribution on the wing and horizontal tail. At the wing tips, this vertical velocity because of roll rate has a magnitude of $pb/2$. Of course, the vertical velocity because of roll rate decreases as the distance from the fuselage decreases.

Figure 6.8 illustrates how this vertical velocity because of roll rate changes the angle of attack at the left and right wing tip.

Thus, with this illustration for a positive roll rate, an increase in angle of attack is experienced on the right wing and a decrease in angle of attack is experienced on the left wing. This change in angle of attack on the wings and horizontal tail because of roll rate results in an increase in lift on the right

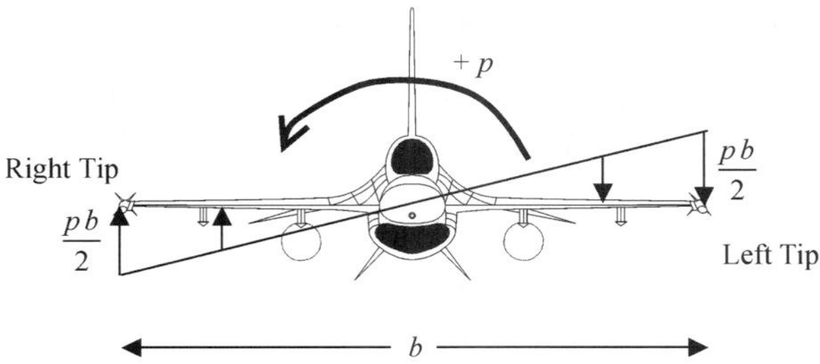

Fig. 6.7 Wing velocity distribution because of roll rate.

wing and a decrease in lift on the left wing. Thus, a roll damping moment is created that is in the opposite direction of the roll rate and C_{l_p} is negative. Of course, this analysis assumes that the angle of attack remains below the stall angle of attack on both wings. Methods for estimating the wing and horizontal tail contribution to C_{l_p} are presented in several flight mechanics and aircraft design texts. In general terms, increases in the span and/or area of the wing and horizontal tail will result in an increase in C_{l_p}.

The absolute value of this change in angle of attack at the wing tips due to roll rate $(pb/2U_1)$ is called the **roll helix angle**. It provides the basis for the form of the nondimensionalization approach used for angular rates in Sec. 6.3.1. The roll helix angle has physical meaning as well. It can be thought of generally as the angle that the wing tip light would make with the horizon for an aircraft undergoing a roll rate (if observed as the aircraft passes through wings level).

The contribution of the vertical tail to C_{l_p} may be estimated by first finding the force on the vertical tail because of roll rate. Looking at the aircraft from the rear using Fig. 6.9, we see that a positive roll rate creates a force on the tail in the negative y direction. We call this sideforce F_s, where $F_{A_{y_{\substack{\text{vertical}\\\text{tail}}}}} = -F_s$.

Figure 6.10 analyzes the velocity components at the center of pressure for the aircraft shown in Fig. 6.9. There is a velocity component in the x direction because of the forward velocity and a velocity component in the y direction because of the roll rate, p.

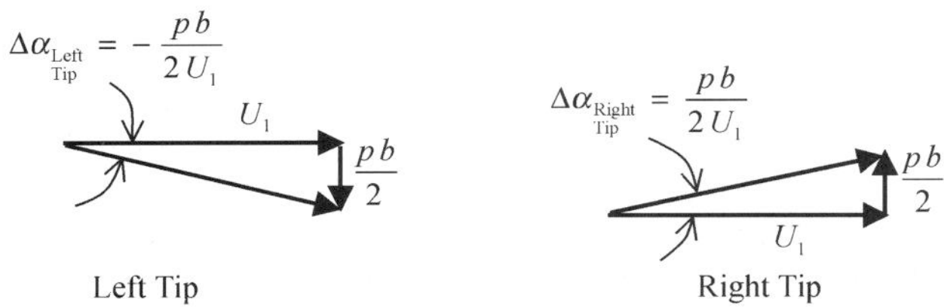

Fig. 6.8 Wing tip angle of attack change because of roll rate.

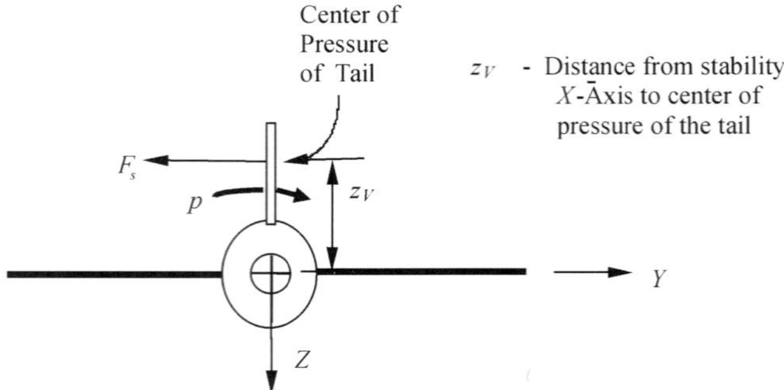

Fig. 6.9 Rear view of aircraft undergoing positive roll rate.

The angle $\Delta\alpha_v$ is the same as an effective sideslip on the vertical tail and may be approximated with the following equation:

$$\Delta\alpha_v \approx \frac{pz_v}{U_1} \tag{6.61}$$

$\Delta\alpha_v$ results in generation of the sideforce, F_s, based on the lift curve slope of the vertical tail, $C_{L_{\alpha_v}}$. F_s therefore becomes

$$F_s = C_{L_{\alpha_v}} \Delta\alpha_v \bar{q}_v S_v = C_{L_{\alpha_v}} \left(\frac{pz_v}{U_1}\right) \bar{q}_v S_v \tag{6.62}$$

This sideforce on the vertical tail because of roll rate also produces a negative rolling moment (L_{A_v}) about the center of gravity because it acts at z_v above the c.g. Recalling that, $F_{A_{y_{\text{vertical tail}}}} = -F_s$,

$$L_{A_v} = F_{A_{y_v}} z_v = -F_s z_v = C_{l_v} \bar{q} S b \tag{6.63}$$

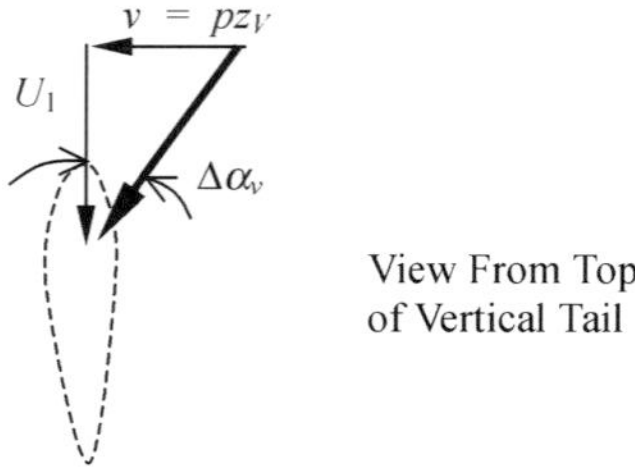

Fig. 6.10 Velocity components at the vertical tail resulting from positive roll rate.

Combining Eqs. (6.62) and (6.63), and solving for C_{l_v}, we have

$$C_{l_v} = -C_{L_{\alpha_v}} \left(\frac{pz_v}{U_1}\right) \eta_v \frac{S_v}{S} \left(\frac{z_v}{b}\right) \tag{6.64}$$

Taking the partial derivative of Eq. (6.64) with respect to p,

$$\frac{\partial C_{l_v}}{\partial p} = -C_{L_{\alpha_v}} \left(\frac{z_v}{U_1}\right) \eta_v \frac{S_v}{S} \left(\frac{z_v}{b}\right)$$

Finally, we obtain an estimate for the vertical tail contribution to the roll damping derivative in nondimensional form

$$C_{l_{p_v}} = \frac{\partial C_1}{\partial\left(\dfrac{pb}{2U_1}\right)} = \frac{2U_1}{b} \frac{\partial C_{l_v}}{\partial p} = -2C_{L_{\alpha_v}} \left(\frac{z_v}{b}\right)^2 \eta_v \frac{S_v}{S} \tag{6.65}$$

Because $C_{L_{\alpha_v}}$, $(z_v/b)^2$, η_v, and S_v/S are all positive, $C_{l_{p_v}}$ must be negative.

The next $pb/2U_1$ derivative to be considered is $\partial F_{A_y}/\partial(pb/2U_1)$. We begin with

$$\frac{\partial F_{A_y}}{\partial \dfrac{pb}{2U_1}} = \frac{\partial C_y \bar{q} S}{\partial \dfrac{pb}{2U_1}} = C_{y_p} \bar{q}_1 S \tag{6.66}$$

The vertical tail is the major contributor to C_{y_p} and the preceding analysis to estimate C_{l_p} is appropriate. Recalling that $F_{A_{y_{\text{vertical tail}}}} = -F_s$, we express F_s (the side force because of roll rate) in terms of the side force coefficient of the entire aircraft:

$$F_s = -F_{A_{y_v}} = -C_{y_v} \bar{q}_1 S \tag{6.67}$$

Equation (6.62) is then substituted into Eq. (6.67) and solved for C_{y_v}

$$C_{y_v} = -C_{L_{\alpha_v}} \left(\frac{pz_v}{U_1}\right) \frac{\bar{q}_v}{\bar{q}_1} \frac{S_v}{S}$$

Taking the partial derivative with respect to p, we have

$$\frac{\partial C_{y_v}}{\partial p} = -C_{L_{\alpha_v}} \left(\frac{z_v}{U_1}\right) \eta_v \frac{S_v}{S} \tag{6.68}$$

Finally, we obtain an estimate for the vertical tail contribution to the side force with respect to roll rate derivative in nondimensional form

$$\frac{\partial C_{y_v}}{\partial\left(\dfrac{pb}{2U_1}\right)} = \frac{2U_1}{b}\frac{\partial C_{y_v}}{\partial p} = C_{y_{p_v}} \approx C_{y_p} \tag{6.69}$$

Combining Eqs. (6.68) and (6.69), we have

$$C_{y_p} \approx \frac{2U_1}{b}(-C_{L_{\alpha_v}})\left(\frac{z_v}{U_1}\right)\eta_v\frac{S_v}{S}$$

Simplifying and rearranging, we have

$$C_{y_p} \approx -2C_{L_{\alpha_v}}\left(\frac{z_v}{b}\right)\eta_v\frac{S_v}{S} \tag{6.70}$$

C_{y_p} is generally negative; however, at high angles of attack, z_v may become negative as the center of pressure of the vertical tail drops below the x stability axis. For that case, C_{y_p} will be positive.

The final $pb/2U_1$ derivative to be considered is $\partial N_A/\partial(pb/2U_1)$. In a similar manner, we have

$$\frac{\partial N_A}{\partial\dfrac{pb}{2U_1}} = \frac{\partial C_n\bar{q}Sb}{\partial\dfrac{pb}{2U_1}} = C_{n_p}\bar{q}_1Sb \tag{6.71}$$

C_{n_p} is called a **cross derivative** because it represents the change in yawing moment coefficient (a moment about the z axis) because of a nondimensional roll rate (an angular rate about the x axis). The wing and vertical tail are the primary components that contribute to C_{n_p}. The contribution of the horizontal tail is typically small compared to the wing because of its smaller area.

The wing contributes to C_{n_p} in three ways that will be addressed qualitatively. The first contribution comes from the 1) increase in drag that results from the increase in angle of attack on the wing being rolled into, and 2) decrease on drag that results from the decrease in angle of attack on the wing being rolled away from. For example, a positive right wing down roll rate will increase the angle of attack on the right wing and decrease the angle of attack on the left wing. The increased drag on that results on the right wing and decreased drag that results on the left wing will provide a positive yawing moment to the aircraft, resulting in a positive contribution to C_{n_p}. The second contribution to C_{n_p} results from tilting of the lift vector on each wing because of the change in angle of attack. Recall that lift is defined in a direction perpendicular to the relative wind. For our example, the increase in angle of attack on the right wing results in tilting of the lift vector forward, while the decrease in angle of attack on the left wing provides an aft tilting of the lift vector. The net result is a negative contribution to yawing moment; thus, a

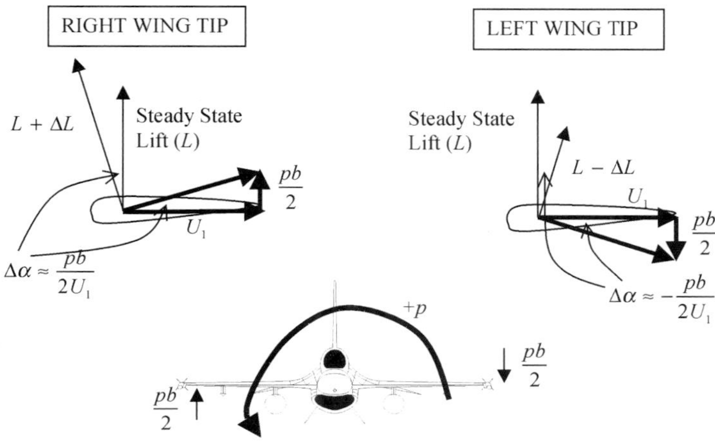

Fig. 6.11 Illustration of lift vector tilting because of roll rate.

negative contribution to C_{n_p}. The lift vector tilting effect is illustrated in Fig. 6.11.

The final contribution may result from an asymmetrical sideforce generated at each wing tip. Again returning to our example with positive roll rate, the right wing experiences higher lift (and lower upper surface pressure) because of the increased angle of attack. Conversely, the left wing experiences a decrease in lift (and higher upper surface pressure) because of the decreased angle of attack. As a result, there is a greater tendency for a positive sideforce to develop at the right wing tip as the flow migrates from the lower surface to the low-pressure upper surface. At the left wing tip, a lower magnitude negative sideforce develops because the pressure on the left wing upper surface is higher (a smaller differential pressure than on the right wing). The net result should be a positive sideforce acting through the right wing tip. If this sideforce acts behind the c.g., a negative yawing moment results and the wing tip sideforce effect makes a negative contribution to C_{n_p} as illustrated in Fig. 6.12. If this sideforce acts in front of the c.g. (unusual), a positive yawing moment results and the wing tip sideforce effect makes a positive contribution to C_{n_p}. The wing tip sideforce effect is most pronounced on low aspect ratio wings (strong wing tip vorticies) with relatively thick wing tips.

The contribution of the vertical tail to C_{n_p} results from the sideforce because of roll rate (F_s) illustrated in Fig. 6.9 and defined by Eq. (6.62). As discussed for a positive roll rate, a negative sideforce at the vertical tail results. This negative sideforce produces a positive yawing moment provided the distance z_v is positive (the case for low to moderate angles of attack). Thus, the contribution of the vertical tail to C_{n_p} is generally positive but may be negative at high angles of attack. If we define x_v as the distance from the c.g. to the aerody-

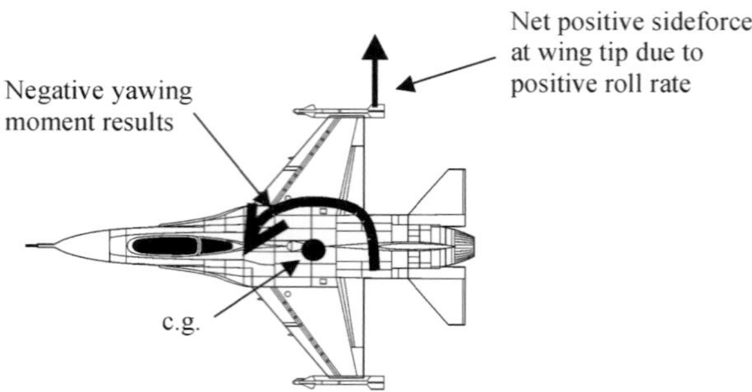

Fig. 6.12 Illustration of wing tip sideforce effect because of positive roll rate.

namic center of the vertical tail, then a similar analysis to that which led to Eq. (6.65) yields an estimate for the vertical contribution to C_{n_p}.

$$C_{n_{p_v}} = 2C_{L_{\alpha_v}}\left(\frac{z_v}{b}\right)\left(\frac{x_v}{b}\right)\eta_v\frac{S_v}{S} \tag{6.72}$$

An estimate for the wing contribution to C_{n_p}, $C_{n_{p_w}}$, developed in some texts[1] using wing strip theory, follows:

$$C_{n_{p_w}} = -\frac{C_L}{8}$$

Two observations may be made regarding this estimate. First, $C_{n_{p_w}}$ is directly proportional to the overall aircraft lift coefficient; second, the estimate assumes that lift vector tilting is the dominant contribution based on the negative sign.

C_{n_p} is one of the more difficult derivatives to estimate because some aircraft components make positive contributions, others make negative contributions, and in some cases the sign of the contribution depends on angle of attack. Fortunately, in most applications C_{n_p} has a relatively small influence on dynamic stability characteristics.

Example 6.9

Find the $pb/2U_1$ derivative $\partial N_A/\partial(pb/2U_1)$ for the F-4C at the same conditions as presented in Example 6.7. C_{n_p} for the F-4C is -0.036. If p is 5 deg/s, find the perturbed yawing moment, n_A.

Starting with Eq. (6.71),

$$\frac{\partial N_A}{\partial \dfrac{pb}{2U_1}} = C_{n_p}\bar{q}_1 Sb = -0.036(283.2)530(38.7) = -209{,}114 \text{ ft} \cdot \text{lb/rad}$$

To find n_A, we use Eq. (6.13) for only a roll rate perturbation:

$$n_A = \frac{\partial N_A}{\partial \dfrac{pb}{2U_1}}\left(\frac{pb}{2U_1}\right) = -209{,}114\left(\frac{(5/57.3)(38.7)}{2(876)}\right) = -403.1 \text{ ft} \cdot \text{lb}$$

Notice that degrees/second are converted to radians/second to maintain consistent units.

6.3.3.4 Yaw rate $rb/2U_1$ derivatives. The $rb/2U_1$ derivatives consist of $\partial L_A/\partial(rb/2U_1)$, $\partial F_{A_y}/\partial(rb/2U_1)$, and $\partial N_A/\partial(rb/2U_1)$ in Eq. (6.13). We begin with $\partial L_A/\partial(rb/2U_1)$.

$$\frac{\partial L_A}{\partial \dfrac{rb}{2U_1}} = \frac{\partial C_l}{\partial \dfrac{rb}{2U_1}}\bar{q}_1 Sb = C_{l_r}\bar{q}_1 Sb \tag{6.73}$$

The derivative C_{l_r} is called a **cross derivative**. It represents the change in rolling moment coefficient (a moment about the x axis) due to nondimensional yaw rate (an angular rate about the z axis). The wing and vertical tail are the primary aircraft components that contribute to C_{l_r}.

The wing contribution to C_{l_r} results from the yaw rate increasing the effective velocity on one wing and decreasing the effective velocity on the opposite wing. For example, a positive nose right yaw rate will provide an angular rate that increases the effective velocity on the left wing and that decreases the effective velocity on the right wing. The increase in velocity results in increased lift on the left wing, and the decrease in velocity results in decreased lift on the right wing. The net result is a positive rolling moment (right wing down). Thus, the wings make a positive contribution to C_{l_r}.

The vertical tail contribution to C_{l_r} results from the change in angle of attack (actually a sideslip angle) experienced by the vertical tail because of yaw rate. For example, for a positive yaw rate, the vertical tail will experience an increase in angle of attack—actually sideslip—($\Delta\alpha_v$) on the left side of the vertical tail, which produces a side force (F_s) in the positive y direction. This is illustrated in Fig. 6.13.

The angle $\Delta\alpha_v$ is the effective sideslip on the vertical tail and may be approximated with the following equation:

$$\Delta\alpha_v \approx \frac{rx_v}{U_1} \tag{6.74}$$

where x_v is the distance from the c.g. to the a.c. of the vertical tail. F_s therefore becomes

$$F_s = C_{L_{\alpha_v}}\Delta\alpha_v\bar{q}_v S_v = C_{L_{\alpha_v}}\left(\frac{rx_v}{U_1}\right)\bar{q}_v S_v \tag{6.75}$$

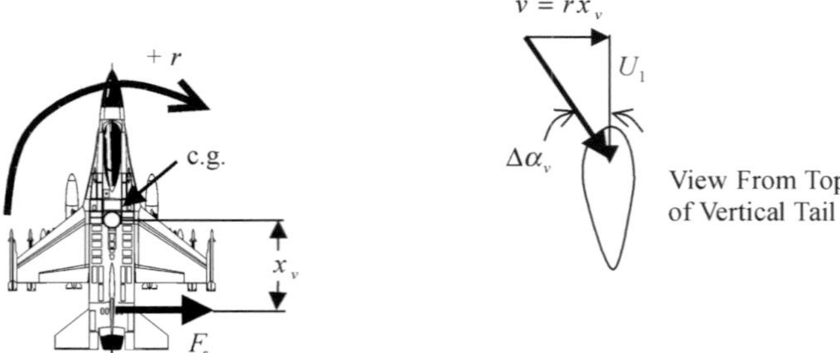

Fig. 6.13 Illustration of sideforce and change in angle of attack at the vertical tail resulting from positive yaw rate.

This sideforce on the vertical tail because of yaw rate also produces a positive rolling moment (L_{A_v}) about the center of gravity because it acts at z_v (see Fig. 6.9) above the c.g. Thus,

$$L_{A_v} = F_s z_v = C_{l_v}\bar{q}Sb \tag{6.76}$$

Combining Eqs. (6.75) and (6.76), and solving for C_{l_v}, we have

$$C_{l_v} = C_{L_{\alpha_v}}\left(\frac{rx_v}{U_1}\right)\eta_v\frac{S_v}{S}\left(\frac{z_v}{b}\right) \tag{6.77}$$

Taking the partial derivative of Eq. (6.77) with respect to r,

$$\frac{\partial C_{l_y}}{\partial r} = C_{L_{\alpha_v}}\left(\frac{x_v}{U_1}\right)\eta_v\frac{S_v}{S}\left(\frac{z_v}{b}\right)$$

Finally, we obtain an estimate for the vertical tail contribution to C_{l_r}. In non-dimensional form

$$C_{l_{r_v}} = \frac{\partial C_1}{\partial\left(\dfrac{rb}{2U_1}\right)} = \frac{2U_1}{b}\frac{\partial C_{l_v}}{\partial r} = 2C_{L_{\alpha_v}}\left(\frac{x_v z_v}{b^2}\right)\eta_v\frac{S_v}{S} \tag{6.78}$$

As seen from Eq. (6.78), the vertical tail makes a positive contribution to C_{l_r} at low to moderate values of angle of attack where z_v is positive. However, at high angles of attack, z_v may be negative and then the vertical tail contribution to C_{l_r} will be negative. Because the wing contribution normally outweighs the vertical tail contribution to C_{l_r}, C_{l_r} is usually positive for most flight conditions.

The next $rb/2U_1$ derivative to be considered is $\partial F_{A_y}/\partial(rb/2U_1)$. We begin with

$$\frac{\partial F_{A_y}}{\partial \dfrac{rb}{2U_1}} = \frac{\partial C_y \bar{q} S}{\partial \dfrac{rb}{2U_1}} = C_{y_r} \bar{q}_1 S \tag{6.79}$$

The vertical tail is the major contributor to C_{y_r} and the preceding analysis to estimate C_{l_r} is appropriate. We begin with the estimate of sideforce on the vertical tail because of roll rate (F_s) defined by Eq. (6.75).

$$F_s = C_{L_{\alpha_v}} \left(\frac{r x_v}{U_1}\right) \bar{q}_v S_v = C_{y_v} \bar{q}_1 S \tag{6.80}$$

Solving for C_{y_v},

$$C_{y_v} = C_{L_{\alpha_v}} \left(\frac{r x_v}{U_1}\right) \eta_v \frac{S_v}{S}$$

Taking the partial derivative with respect to r, we have

$$\frac{\partial C_{y_v}}{\partial r} = C_{L_{\alpha_v}} \left(\frac{x_v}{U_1}\right) \eta_v \frac{S_v}{S} \tag{6.81}$$

Finally, C_{y_r} may be obtained in nondimensional form

$$\frac{\partial C_{y_v}}{\partial \left(\dfrac{rb}{2U_1}\right)} = \frac{2U_1}{b} \frac{\partial C_{y_v}}{\partial r} = C_{y_{r_v}} \approx C_{y_r} \tag{6.82}$$

Combining Eqs. (6.81) and (6.82), we have

$$C_{y_r} \approx 2 C_{L_{\alpha_v}} \left(\frac{x_v}{b}\right) \eta_v \frac{S_v}{S} \tag{6.83}$$

C_{y_r} is a positive derivative because a positive yaw rate results in a positive sideforce on the vertical tail.

The final $rb/2U_1$ derivative to be considered is $\partial N_A/\partial(rb/2U_1)$. In a similar manner, we have

$$\frac{\partial N_A}{\partial \dfrac{rb}{2U_1}} = \frac{\partial C_n \bar{q} S b}{\partial \dfrac{rb}{2U_1}} = C_{n_r} \bar{q}_1 S b \tag{6.84}$$

The derivative C_{n_r} is called the **yaw damping derivative**. It represents the change in yawing moment coefficient with respect to nondimensional yaw rate and will always be **negative** (providing a moment which opposes the direction of the yaw rate). C_{n_r} is also an important factor in lateral-directional stability characteristics. The wing and vertical tail are the primary components that contribute to C_{n_r}.

The wing contribution to C_{n_r} results from the yaw rate, increasing the effective velocity on one wing and decreasing the effective velocity on the opposite wing as discussed previously. A positive "nose right" yaw rate will provide an angular rate that increases the effective velocity on the left wing and decreases the effective velocity on the right wing. The increase in velocity results in increased lift and induced drag on the left wing, and the decrease in velocity results in decreased lift and decreased induced drag on the right wing. The net result is a negative yawing moment (nose left). Thus, the wings make a negative contribution to C_{n_r}.

The vertical tail contribution to C_{n_r} results from the sideforce (F_s) on the vertical tail resulting from yaw rate as presented in Eq. (6.75). Referring also to Fig. 6.13, the yawing moment resulting from a positive yaw rate on the aircraft is

$$N_{A_v} = -F_s x_v = C_{n_v} \bar{q} S b \tag{6.85}$$

Combining Eqs. (6.75) and (6.85) and solving for C_{n_v}, we have

$$C_{n_v} = -C_{L_{\alpha_v}} \left(\frac{r x_v}{U_1} \right) \eta_v \frac{S_v}{S} \left(\frac{x_v}{b} \right) \tag{6.86}$$

Taking the partial derivative of Eq. (6.86) with respect to r,

$$\frac{\partial C_{n_v}}{\partial r} = -C_{L_{\alpha_v}} \left(\frac{x_v}{U_1} \right) \eta_v \frac{S_v}{S} \left(\frac{x_v}{b} \right) \tag{6.87}$$

Finally, C_{n_r} may be obtained in nondimensional form

$$\frac{\partial C_{n_v}}{\partial \left(\dfrac{rb}{2U_1} \right)} = \frac{2U_1}{b} \frac{\partial C_{n_v}}{\partial r} = C_{n_{r_v}} \tag{6.88}$$

Combining Eqs. (6.87) and (6.88), we have

$$C_{n_{r_v}} = -C_{L_{\alpha_v}} \left(\frac{2 x_v^2}{b^2} \right) \eta_v \frac{S_v}{S} \tag{6.89}$$

Equation (6.89) provides an estimate of the vertical tail contribution to C_{n_r}, which can also be seen to be negative. For a complete estimate of C_{n_r}, the wing contribution must be added to this estimate.

Example 6.10

Estimate the yaw damping derivative, C_{n_r}, for an aircraft based on the contribution of the vertical tail. The aircraft has the following characteristics

$$C_{L_{\alpha_v}} = 0.08/\text{deg}, \quad \frac{x_v}{b} = 0.6, \quad \eta_v = 0.95, \quad \frac{S_v}{S} = 0.125$$

Starting with Eq. (6.89),

$$C_{n_{r_v}} = -C_{L_{\alpha_v}}\left(\frac{2x_v^2}{b^2}\right)\eta_v\frac{S_v}{S} = (-0.08)(57.3)2(0.6)^2(0.95)(0.125)$$

$$C_{n_{r_v}} = -0.392$$

Notice that $C_{L_{\alpha_v}}$ has been converted to "per radian" to keep consistent units.

6.3.3.5 δ_a derivatives. The δ_a derivatives consist of $\partial L_A/$, $\partial F_{A_y}/\partial\delta_a$, and $\partial N_A/\partial\delta_a$ in Eq. (6.13). Using a similar development to that for the previous derivatives, we have

$$\frac{\partial L_A}{\partial\delta_a} = \frac{\partial C_l}{\partial\delta_a}\bar{q}_1 Sb = C_{l_{\delta_a}}\bar{q}_1 Sb \tag{6.90}$$

$C_{l_{\delta_a}}$ is a **primary control derivative** and is also called the **aileron control power**. It was discussed in Sec. 5.5.2.2. The sign of $C_{l_{\delta_a}}$ is **positive** with our sign convention. Thus, a positive aileron deflection (normally right aileron trailing edge up/left aileron trailing edge down) will produce a positive rolling moment.

$\partial F_{A_y}/\partial\delta_a$ is developed in a similar manner.

$$\frac{\partial F_{A_y}}{\partial\delta_a} = \frac{\partial C_y}{\partial\delta_a}\bar{q}_1 S = C_{y_{\delta_a}}\bar{q}_1 S \tag{6.91}$$

$C_{y_{\delta_a}}$ is the change in sideforce coefficient resulting from an aileron deflection. It generally has a negligible value. It may have a negative value for situations where differential horizontal tail is used to generate rolling moment as discussed in Sec. 5.5.1.

Finally, the development of $\partial N_A/\partial\delta_a$ follows the same approach.

$$\frac{\partial N_A}{\partial\delta_a} = \frac{\partial C_n}{\partial\delta_a}\bar{q}_1 Sb = C_{n_{\delta_a}}\bar{q}_1 Sb \tag{6.92}$$

$C_{n_{\delta_a}}$ is a **cross-control derivative** that was discussed in Sec. 5.5.3.2. If it is positive, the aircraft exhibits proverse yaw. More typically, it is negative, indicating that adverse yaw is generated as a result of an aileron input.

6.3.3.6 δ_r derivatives. The δ_r derivatives are $\partial L_A/\partial\delta_r$, $\partial F_{A_y}/\partial\delta_r$, and $\partial N_A/\partial\delta_r$ in Eq. (6.13). Using the same approach as with the δ_a derivatives,

$$\frac{\partial L_A}{\partial\delta_r} = \frac{\partial C_l}{\partial\delta_r}\bar{q}_1 Sb = C_{l_{\delta_r}}\bar{q}_1 Sb \tag{6.93}$$

$C_{l_{\delta_r}}$ is a **cross-control derivative** that was discussed in Sec. 5.5.2.3. It is usually positive because the rudder is normally above the x-body axis.

$\partial F_{A_y}/\partial\delta_r$ is developed in a similar manner.

$$\frac{\partial F_{A_y}}{\partial\delta_r} = \frac{\partial C_y}{\partial\delta_r}\bar{q}_1 S = C_{y_{\delta_r}}\bar{q}_1 S \tag{6.94}$$

$C_{y_{\delta_r}}$ is the change in sideforce coefficient resulting from a rudder deflection. It was discussed in Sec. 5.5.1. and generally has a positive value because a positive rudder deflection will generate a sideforce along the positive y axis.

Finally, the development of $\partial N_A/\partial\delta_r$ follows the same approach.

$$\frac{\partial N_A}{\partial\delta_r} = \frac{\partial C_n}{\partial\delta_r}\bar{q}_1 Sb = C_{n_{\delta_r}}\bar{q}_1 Sb \tag{6.95}$$

$C_{n_{\delta_r}}$ is a **primary control derivative** and is also called the **rudder control power**. It was discussed in Sec. 5.5.3.3. The sign of $C_{n_{\delta_r}}$ is **negative** with our sign convention. Thus, a positive rudder deflection (trailing edge left) will produce a negative yawing moment.

6.3.3.7 Summary. We now define the lateral-directional forces and moments from Eq. (6.13) using the derivatives developed in Secs. 6.3.3.1–6.3.3.6. This recasting of the equations will use matrix format and Eqs. (6.54–6.60), (6.66), (6.71), (6.73), (6.79), (6.84), and (6.90–6.95), resulting in

$$\begin{bmatrix} \dfrac{l_A}{\bar{q}_1 Sb} \\ \dfrac{f_{A_y}}{\bar{q}_1 S} \\ \dfrac{n_A}{\bar{q}_1 Sb} \end{bmatrix} = \begin{bmatrix} C_{l_\beta} & C_{l_{\dot\beta}} & C_{l_p} & C_{l_r} & C_{l_{\delta_a}} & C_{l_{\delta_r}} \\ C_{y_\beta} & C_{y_{\dot\beta}} & C_{y_p} & C_{y_r} & C_{y_{\delta_a}} & C_{y_{\delta_r}} \\ C_{n_\beta} & C_{n_{\dot\beta}} & C_{n_p} & C_{n_r} & C_{n_{\delta_a}} & C_{n_{\delta_r}} \end{bmatrix} \begin{bmatrix} \beta \\ \dfrac{\dot\beta b}{2U_1} \\ \dfrac{pb}{2U_1} \\ \dfrac{rb}{2U_1} \\ \hat\delta_a \\ \hat\delta_r \end{bmatrix} \tag{6.96}$$

Table 6.2 Summary of lateral-directional derivatives

Derivative	Name	Normal sign
C_{l_β}	Lateral static stability derivative	$-$
C_{y_β}	None	$-$
C_{n_β}	Directional static stability derivative	$+$
$C_{l_{\dot\beta}}$	Quasi-steady derivative	≈ 0
$C_{y_{\dot\beta}}$	Quasi-steady derivative	≈ 0
$C_{n_{\dot\beta}}$	Quasi-steady derivative	≈ 0
C_{l_p}	Roll damping derivative	$-$
C_{y_p}	None	$-$
C_{n_p}	Cross derivative	$+$ or $-$
C_{l_r}	Cross derivative	$+$
C_{y_r}	None	$+$
C_{n_r}	Yaw damping derivative	$-$
$C_{l_{\delta_a}}$	Aileron control power	$+$
$C_{y_{\delta_a}}$	None	≈ 0
$C_{n_{\delta_a}}$	Cross control derivative (proverse or adverse yaw)	$+$ or $-$
$C_{l_{\delta_r}}$	Cross control derivative	$+$
$C_{y_{\delta_r}}$	None	$+$
$C_{n_{\delta_r}}$	Rudder control power	$-$

The advantage of Eq. (6.96) over Eq. (6.13) is that the lateral-directional perturbed forces and moments are now expressed in terms of common "aero" derivatives such as C_{l_β}, $C_{y_{\delta_r}}$, and C_{n_r}. The value of these derivatives can be estimated with analytical or experimental techniques. Remember that each derivative in Eq. (6.96) is dimensionless—for example, C_{n_r} is the abbreviated form of $\partial C_n/\partial(rb/2U_1)$. Table 6.2 summarizes the derivatives discussed for the perturbed lateral-directional force and moment estimates.

6.4 First-Order Approximation of Perturbed Thrust Forces and Moments

We will now focus on the perturbed thrust force and moment terms (such as f_{T_x} and m_T) in Eqs. (6.6) and (6.7). These represent the perturbed change in a thrust force or moment that results from a nonzero value of a perturbation variable like u. We begin with the assumption that these perturbed thrust terms are only a function of u, $\hat{\alpha}$, and β, which is generally the case but does neglect effects from p, q, r, $\dot{\alpha}$, $\dot{\beta}$, and the control deflections. The longitudinal perturbed thrust forces and moment can thus be represented using a Taylor series and the nondimensional perturbed longitudinal variables (defined along the x, y, and z stability axis) as

$$f_{T_x} = \frac{\partial F_{T_x}}{\partial\left(\frac{u}{U_1}\right)}\left(\frac{u}{U_1}\right) + \frac{\partial F_{T_x}}{\partial\hat{\alpha}}\hat{\alpha} \tag{6.97}$$

$$m_T = \frac{\partial M_T}{\partial\left(\frac{u}{U_1}\right)}\left(\frac{u}{U_1}\right) + \frac{\partial M_T}{\partial\hat{\alpha}}\hat{\alpha} \tag{6.98}$$

$$f_{T_z} = \frac{\partial F_{T_z}}{\partial\left(\frac{u}{U_1}\right)}\left(\frac{u}{U_1}\right) + \frac{\partial F_{T_z}}{\partial\hat{\alpha}}\hat{\alpha} \tag{6.99}$$

The longitudinal thrust forces and moment can be expressed in coefficient form as

$$F_{T_x} = C_{T_x}\bar{q}_1 S \tag{6.100}$$

$$M_T = C_{m_T}\bar{q}_1 S\bar{c} \tag{6.101}$$

$$F_{T_z} = C_{T_z}\bar{q}_1 S \tag{6.102}$$

The lateral-directional perturbed thrust force and moments can be represented in a similar manner as a function of β

$$l_T = \frac{\partial L_T}{\partial\beta}\beta \tag{6.103}$$

$$f_{T_y} = \frac{\partial F_{T_y}}{\partial\beta}\beta \tag{6.104}$$

$$n_T = \frac{\partial N_T}{\partial\beta}\beta \tag{6.105}$$

The lateral-directional thrust force and moments can be expressed in coefficient form as

$$L_T = C_{l_T}\bar{q}_1 Sb \tag{6.106}$$

$$F_{T_y} = C_{T_y}\bar{q}_1 S \tag{6.107}$$

$$N_T = C_{n_T}\bar{q}_1 Sb \tag{6.108}$$

6.4.1 Longitudinal Perturbed Thrust Force and Moment Derivatives

We will next analyze each of the partial derivative terms in Eqs. (6.97–6.99) so that they may be expressed with common thrust coefficient derivatives.

6.4.1.1 u/U_1 derivatives. The u/U_1 derivatives consist of $\partial F_{T_x}/\partial(u/U_1)$, $\partial M_T/\partial(u/U_1)$, and $\partial F_{T_z}/\partial(u/U_1)$ in Eqs. (6.97–6.99). We will begin with $\partial F_{T_x}/\partial(u/U_1)$. F_{T_x} is defined along the body-fixed stability x axis as discussed in Secs. 6.3.2. and 6.3.2.1. Using a similar approach as that for $\partial F_{A_x}/\partial(u/U_1)$ in Sec. 6.3.2.1, we have

$$\frac{\partial F_{T_x}}{\partial \dfrac{u}{U_1}} = \frac{\partial C_{T_x}}{\partial \dfrac{u}{U_1}} \bar{q} S + C_{T_x} S \left(\frac{\partial \bar{q}}{\partial \dfrac{u}{U_1}} \right) = C_{T_{x_u}} \bar{q}_1 S + 2 C_{T_{x_1}} \bar{q}_1 S \tag{6.109}$$

In comparing Eq. (6.109) to Eq. (6.19), notice the similarities with the exception of the negative sign in Eq. (6.19). This results because drag is defined as positive in the negative x direction while thrust is defined as positive in the positive x direction. For gliders or power-off flight, $C_{T_{x_u}}$ and $C_{T_{x_1}}$ are equal to zero and thus $\partial F_{T_x}/\partial(u/U_1)$ becomes zero. Estimates of $C_{T_{x_u}}$ and $C_{T_{x_1}}$ for powered cases are dependent of the type of propulsion system used in the aircraft.

The next u/U_1 derivative to be considered is $\partial M_T/\partial(u/U_1)$. Using a similar approach as that for $\partial M_A/\partial(u/U_1)$ in Sec. 6.3.2.1, we have

$$\frac{\partial M_T}{\partial \dfrac{u}{U_1}} = C_{m_{T_u}} \bar{q}_1 S \bar{c} + C_{m_{T_1}} S \bar{c} (2 \bar{q}_1) \tag{6.110}$$

and, in combined form

$$\frac{\partial M_T}{\partial \dfrac{u}{U_1}} = (C_{m_{T_u}} + 2 C_{m_{T_1}}) \bar{q}_1 S \bar{c} \tag{6.111}$$

For steady-state trimmed flight, the total pitching moment acting on the aircraft should be zero. Thus, the sum of the steady-state thrust pitching moment and steady-state aerodynamic pitching moment [referring to Eq. (6.22)] should be

$$C_{m_{T_1}} + C_{m_1} = 0 \qquad @ \text{ trim} \tag{6.112}$$

Combining Eqs. (6.22), (6.111), and (6.112), we have

$$\frac{\partial (M_A + M_T)}{\partial \dfrac{u}{U_1}} = (C_{m_u} + C_{m_{T_u}}) \bar{q}_1 S \bar{c} \tag{6.113}$$

The derivative $C_{m_{T_u}}$ has a negligible value for situations where the thrust vector passes through the center of gravity.

The last u/U_1 derivative is $\partial F_{T_z}/\partial(u/U_1)$. Again, using a similar approach as that for $\partial F_{A_z}/\partial(u/U_1)$ in Sec. 6.3.2.1, we have

$$\frac{\partial F_{T_z}}{\partial\left(\dfrac{u}{U_1}\right)} = (C_{T_{z_u}} + 2C_{T_{z_1}})\bar{q}_1 S \tag{6.114}$$

In comparing Eq. (6.114) to Eq. (6.26), notice the similarities with the exception of the negative sign in Eq. (6.26). This results because lift is defined as positive in the negative z direction while the z component of thrust is defined as positive in the positive z direction.

6.4.1.2 $\hat{\alpha}$ derivatives. The $\hat{\alpha}$, or perturbed angle of attack derivatives, consist of $\partial F_{T_x}/\partial\hat{\alpha}$, $\partial M_T/\partial\hat{\alpha}$, and $\partial F_{T_z}/\partial\hat{\alpha}$ in Eqs. (6.97–6.99). We begin with $\partial F_{T_x}/\partial\hat{\alpha}$, which is simply

$$\frac{\partial F_{T_x}}{\partial\hat{\alpha}} = C_{T_{x_{\hat{\alpha}}}}\bar{q}_1 S \tag{6.115}$$

$C_{T_{x_{\hat{\alpha}}}}$ is typically negligible for normal angle of attack ranges.

$\partial M_T/\partial\hat{\alpha}$ follows a similar approach.

$$\frac{\partial M_T}{\partial\hat{\alpha}} = C_{m_{T_{\hat{\alpha}}}}\bar{q}_1 S\bar{c} \tag{6.116}$$

$C_{m_{T_{\hat{\alpha}}}}$ is a significant derivative because of momentum forces in the z direction that result from turning the flow through the engine. For example, if $\hat{\alpha}$ is positive an increase in the turning angle of the flow through the engine results, along with an increase in the force associated with the change in momentum direction (which is in the negative z direction). Because this force generally acts at the nacelle of the engine, a pitching moment results about the aircraft c.g. because of the moment arm x_d. This effect is illustrated in Fig. 6.14.

For aircraft with the engine nacelle located in front of c.g., a positive $C_{m_{T_{\hat{\alpha}}}}$ results that provides a destabilizing contribution to the overall C_{m_α} (longitudinal

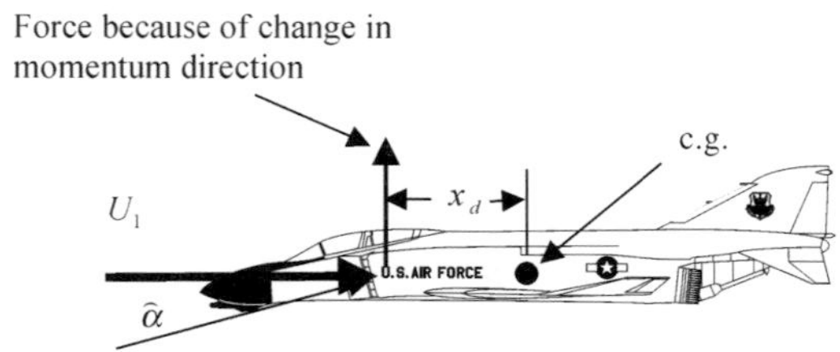

Fig. 6.14 Illustration of change in C_{m_T} because of $\hat{\alpha}$.

stability) of the aircraft. For aircraft with the engine nacelles located aft of the c.g. (such as the A-10), a negative $C_{m_{T_{\hat{\alpha}}}}$ results that provides a stable contribution to overall longitudinal stability. Finally, the magnitude of $C_{m_{T_{\hat{\alpha}}}}$ is also dependent on the thrust (and airflow) being produced by the engine at the time of evaluation because the force due to the change in momentum direction is dependent on the magnitude of the momentum.

The last $\hat{\alpha}$ derivative to be considered is $\partial F_{T_z}/\partial\hat{\alpha}$, which is simply

$$\frac{\partial F_{T_z}}{\partial \hat{\alpha}} = C_{T_{z_{\hat{\alpha}}}} \bar{q}_1 S \tag{6.117}$$

$C_{T_{z_{\hat{\alpha}}}}$ results from the same z-direction force accompanying the change in momentum direction discussed for $C_{m_{T_{\hat{\alpha}}}}$. Again, this force is illustrated in Fig. 6.14.

Actual determination of the thrust derivatives depends on the specific engine and inlet characteristics of a particular aircraft. These must be available before accurate estimates of the thrust derivatives can be made.

6.4.2 Lateral-Directional Perturbed Thrust Force and Moment Derivatives

The β derivatives consist of $\partial L_T/\partial\beta$, $\partial F_{T_y}/\partial\beta$, and $\partial N_T/\partial\beta$ in Eqs. (6.103–6.105). We begin with $\partial L_T/\partial\beta$, which is simply

$$\frac{\partial L_T}{\partial \beta} = C_{l_{T_\beta}} \bar{q}_1 S b \tag{6.118}$$

$C_{l_{T_\beta}}$ is typically negligible for normal angle of attack and sideslip ranges. $\partial F_{T_y}/\partial\beta$ follows a similar approach

$$\frac{\partial F_{T_y}}{\partial \beta} = C_{T_{y_\beta}} \bar{q}_1 S \tag{6.119}$$

$C_{T_{y_\beta}}$ results from the sideforce associated with the change in momentum direction when the flow is turned through the engine. For example, if β is positive the turning angle of the flow through the engine results in a sideforce in the negative y direction, as illustrated in Fig. 6.15. For most flight conditions, $C_{T_{y_\beta}}$ is small and can be considered negligible.

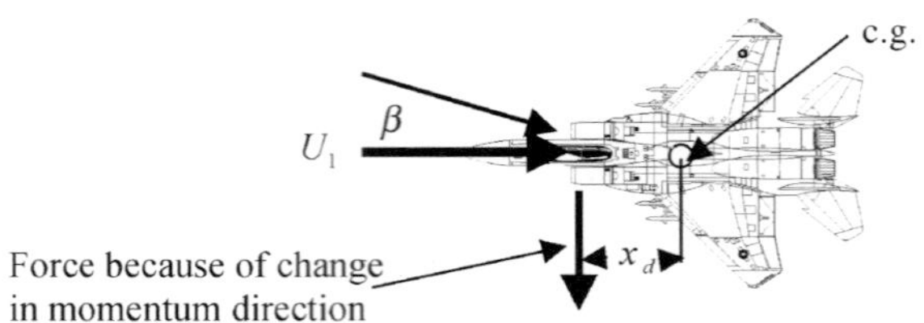

Fig. 6.15 Illustration of C_{T_y} because of β.

The last β derivative to be considered is $\partial N_T/\partial\beta$, which is simply

$$\frac{\partial N_T}{\partial \beta} = C_{n_{T_\beta}} \bar{q}_1 Sb \tag{6.120}$$

$C_{n_{T_\beta}}$ also results from the sideforce associated with the change in momentum direction when the flow is turned through the engine. By referring to Fig. 6.15, a positive β results in a sideforce in the negative y direction applied at the engine nacelle. For the case illustrated, this sideforce is located in front of the c.g. As a result, a negative yawing moment results ($C_{n_{T_\beta}}$ negative) and a negative contribution is made to the overall directional stability of the aircraft. For aircraft with the engine nacelles located aft of the c.g., a positive $C_{n_{T_\beta}}$ results, which provides a stable contribution to overall directional stability. Of course, the magnitude of $C_{n_{T_\beta}}$ is also dependent on the thrust (and airflow) being produced by the engine at the time of evaluation.

6.4.3 Summary

We are now able to define first-order approximations of the perturbed thrust forces and moments in Eqs. (6.6) and (6.7) using the derivatives developed in Sec. 6.4.1. This recasting of the equations will use matrix format and Eqs. (6.109), (6.111), and (6.114–6.120). For longitudinal motion, we have

$$\begin{bmatrix} \dfrac{f_{T_x}}{\bar{q}_1 S} \\ \dfrac{m_T}{\bar{q}_1 S\bar{c}} \\ \dfrac{f_{T_z}}{\bar{q}_1 S} \end{bmatrix} = \begin{bmatrix} (C_{T_{x_u}} + 2C_{T_{x_1}}) & C_{T_{x_{\hat{\alpha}}}} \\ (C_{m_{T_u}} + 2C_{m_{T_1}}) & C_{m_{T_{\hat{\alpha}}}} \\ (C_{T_{z_u}} + 2C_{T_{z_1}}) & C_{T_{z_{\hat{\alpha}}}} \end{bmatrix} \begin{bmatrix} \dfrac{u}{U_1} \\ \hat{\alpha} \end{bmatrix} \tag{6.121}$$

For lateral-directional motion, we have

$$\begin{bmatrix} \dfrac{l_T}{\bar{q}_1 Sb} \\ \dfrac{f_{T_y}}{\bar{q}_1 S} \\ \dfrac{n_T}{\bar{q}_1 Sb} \end{bmatrix} = \begin{bmatrix} C_{l_{T_\beta}} \\ C_{T_{y_\beta}} \\ C_{n_{T_\beta}} \end{bmatrix} \beta \tag{6.122}$$

Equations (6.121) and (6.122) are typically simplified based on the negligible value of several derivatives to the following:

$$\begin{bmatrix} \dfrac{f_{T_x}}{\bar{q}_1 S} \\ \dfrac{m_T}{\bar{q}_1 S\bar{c}} \\ \dfrac{f_{T_z}}{\bar{q}_1 S} \end{bmatrix} = \begin{bmatrix} (C_{T_{x_u}} + 2C_{T_{x_1}}) & 0 \\ (C_{m_{T_u}} + 2C_{m_{T_1}}) & C_{m_{T_{\hat{\alpha}}}} \\ 0 & 0 \end{bmatrix} \begin{bmatrix} \dfrac{u}{U_1} \\ \hat{\alpha} \end{bmatrix} \tag{6.123}$$

$$\begin{bmatrix} \dfrac{l_T}{\bar{q}_1 Sb} \\ \dfrac{f_{T_y}}{\bar{q}_1 S} \\ \dfrac{n_T}{\bar{q}_1 Sb} \end{bmatrix} = \begin{bmatrix} 0 \\ 0 \\ C_{n_{T_\beta}} \end{bmatrix} \beta \tag{6.124}$$

6.5 Recasting the Equations of Motion in Acceleration Format

It is now time to combine our development of the linearized aircraft EOM simplified for wings level, straight flight Eqs. (6.6) and (6.7) with our development of first-order approximations for the perturbed aero and thrust forces/moments (see Secs. 6.3 and 6.4).

6.5.1 Longitudinal EOM

For the longitudinal EOM, we begin with Eq. (6.6) and substitute in Eq. (6.53) for the perturbed aero forces and moments, and Eq. (6.123) for the perturbed thrust forces and moments. In addition, we neglect derivatives that typically have negligible values. This results in:

$$
\begin{aligned}
m\dot{u} &= -mg\theta\cos\Theta_1 \\
&\quad + \bar{q}_1 S\begin{bmatrix} -(C_{D_u} + 2C_{D_1})\dfrac{u}{U_1} + (C_{T_{x_u}} + 2C_{T_{x_1}})\dfrac{u}{U_1} \\ +(-C_{D_{\hat{\alpha}}} + C_{L_1})\hat{\alpha} - C_{D_{\hat{\delta}_e}}\hat{\delta}_e \end{bmatrix} \\
I_{yy}\dot{q} &= \bar{q}_1 S\bar{c}\begin{bmatrix} (C_{m_u} + 2C_{m_1})\dfrac{u}{U_1} + (C_{m_{T_u}} + 2C_{m_{T_1}})\dfrac{u}{U_1} \\ +C_{m_{\hat{\alpha}}}\hat{\alpha} + C_{m_{T_{\hat{\alpha}}}}\hat{\alpha} + C_{m_{\dot{\alpha}}}\dfrac{\dot{\alpha}\bar{c}}{2U_1} + C_{m_q}\dfrac{q\bar{c}}{2U_1} + C_{m_{\hat{\delta}_e}}\hat{\delta}_e \end{bmatrix} \\
m(\dot{w} - U_1 q) &= -mg\theta\sin\Theta_1 \\
&\quad + \bar{q}_1 S\begin{bmatrix} -(C_{L_u} + 2C_{L_1})\dfrac{u}{U_1} - (C_{L_{\hat{\alpha}}} + C_{D_1})\hat{\alpha} \\ -C_{L_{\dot{\alpha}}}\dfrac{\dot{\alpha}\bar{c}}{2U_1} - C_{L_q}\dfrac{q\bar{c}}{2U_1} - C_{L_{\hat{\delta}_e}}\hat{\delta}_e \end{bmatrix}
\end{aligned}
\tag{6.125}
$$

Next, the x and z force equations of Eq. (6.125) are divided by the mass (m) and grouped first and second, while the pitching moment equation of Eq. (6.125) is divided by the moment of inertia about the y axis (I_{yy}) and presented last. This results in

$$
\begin{aligned}
\dot{u} &= -g\theta\cos\Theta_1 \\
&\quad + \frac{\bar{q}_1 S}{m}\begin{bmatrix} -\left(C_{D_u} + 2C_{D_1}\right)\dfrac{u}{U_1} + \left(C_{T_{x_u}} + 2C_{T_{x_1}}\right)\dfrac{u}{U_1} \\ +\left(-C_{D_{\hat{\alpha}}} + C_{L_1}\right)\hat{\alpha} - C_{D_{\hat{\delta}_e}}\hat{\delta}_e \end{bmatrix} \\
(\dot{w} - U_1 q) &= -g\theta\sin\Theta_1 \\
&\quad + \frac{\bar{q}_1 S}{m}\begin{bmatrix} -\left(C_{L_u} + 2C_{L_1}\right)\dfrac{u}{U_1} - \left(C_{L_{\hat{\alpha}}} + C_{D_1}\right)\hat{\alpha} \\ -C_{L_{\dot{\alpha}}}\dfrac{\dot{\alpha}\bar{c}}{2U_1} - C_{L_q}\dfrac{q\bar{c}}{2U_1} - C_{L_{\hat{\delta}_e}}\hat{\delta}_e \end{bmatrix} \\
\dot{q} &= \frac{\bar{q}_1 S\bar{c}}{I_{yy}}\begin{bmatrix} \left(C_{m_u} + 2C_{m_1}\right)\dfrac{u}{U_1} + \left(C_{m_{T_u}} + 2C_{m_{T_1}}\right)\dfrac{u}{U_1} \\ +C_{m_{\hat{\alpha}}}\hat{\alpha} + C_{m_{T_{\hat{\alpha}}}}\hat{\alpha} + C_{m_{\dot{\alpha}}}\dfrac{\dot{\alpha}\bar{c}}{2U_1} + C_{m_q}\dfrac{q\bar{c}}{2U_1} + C_{m_{\hat{\delta}_e}}\hat{\delta}_e \end{bmatrix}
\end{aligned}
\tag{6.126}
$$

This results in each term on the right-hand side of Eq. (6.126) having units of **linear** or **angular acceleration**.

6.5.2 Longitudinal Stability Parameters

To simplify Eq. (6.126) and to gain understanding of the relative contributions of each term in these equations, we next introduce the concept of **stability parameters** or **dimensional stability derivatives**. Stability parameters represent the linear or angular acceleration per motion or control variable (u, $\hat{\alpha}$, $\dot{\alpha}$, q, and $\hat{\delta}_e$) for each term in the right-hand side of Eq. (6.126). For example, we will rewrite the x force equation of Eq. (6.126) so that the appropriate stability parameters can be identified

$$\dot{u} = -g\theta\cos\Theta_1 - \underbrace{\frac{\bar{q}_1 S}{mU_1}(C_{D_u} + 2C_{D_1})}_{X_u} u + \underbrace{\frac{\bar{q}_1 S}{mU_1}(C_{T_{x_u}} + 2C_{T_{x_1}})}_{X_{T_u}} u$$

$$+ \underbrace{\frac{\bar{q}_1 S}{m}(-C_{D_{\hat{\alpha}}} + C_{L_1})}_{X_\alpha} \hat{\alpha} - \underbrace{\frac{\bar{q}_1 S}{m} C_{D_{\hat{\delta}_e}}}_{X_{\delta_e}} \hat{\delta}_e$$

The stability parameters X_u, X_{T_u}, X_α, and X_{δ_e} can be seen to be simply a grouping of the terms multiplying each motion/control parameter when the EOM is in acceleration format. The relative importance of each motion/control parameter to the overall response of the aircraft can be assessed from the relative magnitude of its respective stability parameter. The units of a stability parameter are acceleration (linear or angular) per the appropriate units of the motion/control parameter. For example, X_u represents the linear acceleration along the x axis per unit of the perturbed x axis velocity u. The appropriate units for X_u are ft/s^2/ft/s or s^{-1}. Finally, notice that we will be dropping the hat on the perturbed variables $\hat{\alpha}$ and $\hat{\delta}_e$ from this point on for simplicity. The reader should keep in mind that whenever we are dealing with the linearized EOM, the motion/control parameters always represent perturbed quantities.

A similar approach will be taken with the z force equation of Eq. (6.126).

$$(\dot{w} - U_1 q) = -g\theta\sin\Theta_1 - \underbrace{\frac{\bar{q}_1 S}{mU_1}(C_{L_u} + 2C_{L_1})}_{Z_u} u - \underbrace{\frac{\bar{q}_1 S}{m}(C_{L_{\hat{\alpha}}} + C_{D_1})}_{Z_\alpha} \hat{\alpha}$$

$$- \underbrace{\frac{\bar{q}_1 S\bar{c}}{2mU_1} C_{L_{\dot{\alpha}}}}_{Z_{\dot{\alpha}}} \dot{\alpha} - \underbrace{\frac{\bar{q}_1 S\bar{c}}{2mU_1} C_{L_q}}_{Z_q} q - \underbrace{\frac{\bar{q}_1 S}{m} C_{L_{\hat{\delta}_e}}}_{Z_{\delta_e}} \hat{\delta}_e$$

Notice that the $\bar{c}/2U_1$ term used to nondimensionalize $\dot{\alpha}$ and q has been combined into the stability parameters $Z_{\dot{\alpha}}$ and Z_q. As an example, Z_q represents the linear acceleration along the z axis per unit of the perturbed variable q. It has units of ft/s^2/rad/s or simply ft/s.

Finally, we will incorporate stability parameters into the pitching moment equation of Eq. (6.126).

$$\dot{q} = \underbrace{\frac{\bar{q}_1 S\bar{c}}{I_{yy}U_1}(C_{m_u} + 2C_{m_1})u}_{M_u} + \underbrace{\frac{\bar{q}_1 S\bar{c}}{I_{yy}U_1}(C_{m_{T_u}} + 2C_{m_{T_1}})u}_{M_{T_u}}$$
$$+ \underbrace{\frac{\bar{q}_1 S\bar{c}}{I_{yy}}C_{m_{\hat{\alpha}}}\hat{\alpha}}_{M_\alpha} + \underbrace{\frac{\bar{q}_1 S\bar{c}}{I_{yy}}C_{m_{T_{\hat{\alpha}}}}\hat{\alpha}}_{M_{T_\alpha}} + \underbrace{\frac{\bar{q}_1 S\bar{c}^2}{2I_{yy}U_1}C_{m_{\dot{\alpha}}}\dot{\alpha}}_{M_{\dot{\alpha}}}$$
$$+ \underbrace{\frac{\bar{q}_1 S\bar{c}^2}{2I_{yy}U_1}C_{m_q}q}_{M_q} + \underbrace{\frac{\bar{q}_1 S\bar{c}}{I_{yy}}C_{m_{\hat{\delta}_e}}\hat{\delta}_e}_{M_{\delta_e}}$$

Using the example of the M_α stability parameter, it represents the pitch angular acceleration per unit of the perturbed variable $\hat{\alpha}$. It has units of rad/s^2/rad or simply s^{-2}. Equation (6.126) may now be recast using stability parameters:

$$\begin{aligned}
\dot{u} &= -g\theta\cos\Theta_1 + X_u u + X_{T_u}u + X_\alpha\alpha + X_{\delta_e}\hat{\delta}_e \\
\dot{w} - U_1 q &= -g\theta\sin\Theta_1 + Z_u u + Z_\alpha\alpha + Z_{\dot{\alpha}}\dot{\alpha} + Z_q q + Z_{\delta_e}\hat{\delta}_e \\
\dot{q} &= M_u u + M_{T_u}u + M_\alpha\alpha + M_{T_\alpha}\alpha + M_{\dot{\alpha}}\dot{\alpha} + M_q q + M_{\delta_e}\hat{\delta}_e
\end{aligned} \tag{6.127}$$

Table 6.3 summaries the longitudinal stability parameters associated with Eq. (6.127).

6.5.3 Lateral-Directional EOM

For the lateral-directional EOM, we begin with Eq. (6.7) and substitute in Eq. (6.96) for the perturbed aero forces and moments, and Eq. (6.124) for the perturbed thrust forces and moments. In addition, we neglect derivatives that typically have negligible values. This results in

$$\begin{aligned}
I_{xx}\dot{p} - I_{xz}\dot{r} &= \bar{q}_1 Sb\left[C_{l_\beta}\beta + C_{l_p}\frac{pb}{2U_1} + C_{l_r}\frac{rb}{2U_1} + C_{l_{\delta_a}}\hat{\delta}_a + C_{l_{\delta_r}}\hat{\delta}_r\right] \\
m(\dot{v} + U_1 r) &= mg\phi\cos\Theta_1 + \bar{q}_1 S\left[C_{y_\beta}\beta + C_{y_p}\frac{pb}{2U_1} + C_{y_r}\frac{rb}{2U_1} + C_{y_{\delta_a}}\hat{\delta}_a + C_{y_{\delta_r}}\hat{\delta}_r\right] \\
I_{zz}\dot{r} - I_{xz}\dot{p} &= \bar{q}_1 Sb\left[C_{n_\beta}\beta + C_{n_{T_\beta}}\beta + C_{n_p}\frac{pb}{2U_1} + C_{n_r}\frac{rb}{2U_1} + C_{n_{\delta_a}}\hat{\delta}_a + C_{n_{\delta_r}}\hat{\delta}_r\right]
\end{aligned} \tag{6.128}$$

Next, the y force equation of Eq. (6.128) is divided by the mass (m) and grouped first, while the rolling moment and yawing moment equations of Eq.

Table 6.3 Longitudinal stability parameters

Stability parameter	Definition	Units
X_u	$-\frac{\bar{q}_1 S}{mU_1}(C_{D_u}+2C_{D_l})$	$\frac{\text{ft/s}^2}{\text{ft/s}}=\text{s}^{-1}$
X_{T_u}	$\frac{\bar{q}_1 S}{mU_1}(C_{T_{x_u}}+2C_{T_{x_l}})$	$\frac{\text{ft/s}^2}{\text{ft/s}}=\text{s}^{-1}$
X_α	$+\frac{\bar{q}_l S}{m}(-C_{D_{\hat{\alpha}}}+C_{L_l})$	$\frac{\text{ft/s}^2}{\text{rad}}=\text{ft/s}^2$
X_{δ_e}	$-\frac{\bar{q}_1 S}{m}C_{D_{\hat{\delta}_e}}$	$\frac{\text{ft/s}^2}{\text{rad}}=\text{ft/s}^2$
Z_u	$-\frac{\bar{q}_1 S}{mU_1}(C_{L_u}+2C_{L_l})$	$\frac{\text{ft/s}^2}{\text{ft/s}}=\text{s}^{-1}$
Z_α	$-\frac{\bar{q}_1 S}{m}(C_{L_{\hat{\alpha}}}+2C_{D_l})$	$\frac{\text{ft/s}^2}{\text{rad}}=\text{ft/s}^2$
$Z_{\dot{\alpha}}$	$-\frac{\bar{q}_1 S\bar{c}}{2mU_1}C_{L_{\dot{\alpha}}}$	$\frac{\text{ft/s}^2}{\text{rad/s}}=\text{ft/s}$
Z_q	$-\frac{\bar{q}_1 S\bar{c}}{2mU_1}C_{L_q}$	$\frac{\text{ft/s}^2}{\text{rad/s}}=\text{ft/s}$
Z_{δ_e}	$-\frac{\bar{q}_1 S}{m}C_{L_{\hat{\delta}_e}}$	$\frac{\text{ft/s}^2}{\text{rad}}=\text{ft/s}^2$
M_u	$\frac{\bar{q}_1 S\bar{c}}{I_{yy}U_1}(C_{m_u}+2C_{m_1})$	$\frac{\text{rad/s}^2}{\text{ft/s}}=1/(\text{ft}\cdot\text{s})$
M_{T_u}	$\frac{\bar{q}_1 S\bar{c}}{I_{yy}U_1}(C_{m_{T_u}}+2C_{m_{T_1}})$	$\frac{\text{rad/s}^2}{\text{ft/s}}=1/(\text{ft}\cdot\text{s})$
M_α	$\frac{\bar{q}_1 S\bar{c}}{I_{yy}}C_{m_{\hat{\alpha}}}$	$\frac{\text{rad/s}^2}{\text{rad}}=\text{s}^{-2}$
M_{T_α}	$\frac{\bar{q}_1 S\bar{c}}{I_{yy}}C_{m_{T_{\hat{\alpha}}}}$	$\frac{\text{rad/s}^2}{\text{rad/s}}=\text{s}^{-1}$
$M_{\dot{\alpha}}$	$\frac{\bar{q}_1 S\bar{c}^2}{2I_{yy}U_1}C_{m_{\dot{\alpha}}}$	$\frac{\text{rad/s}^2}{\text{rad/s}}=\text{s}^{-1}$
M_q	$\frac{\bar{q}_1 S\bar{c}^2}{2I_{yy}U_1}C_{m_q}$	$\frac{\text{rad/s}^2}{\text{rad/s}}=\text{s}^{-1}$
M_{δ_e}	$\frac{\bar{q}_1 S\bar{c}}{I_{yy}}C_{m_{\hat{\delta}_e}}$	$\frac{\text{rad/s}^2}{\text{rad}}=\text{s}^{-2}$

(6.128) are divided by I_{xx} and I_{zz}, respectively, and presented second and third. This results in

$$(\dot{v} + U_1 r) = g\phi \cos\Theta_1 + \frac{\bar{q}_1 S}{m}\left[C_{y_\beta}\beta + C_{y_p}\frac{pb}{2U_1} + C_{y_r}\frac{rb}{2U_1} + C_{y_{\delta_a}}\hat{\delta}_a + C_{y_{\delta_r}}\hat{\delta}_r\right]$$

$$\dot{p} - \frac{I_{xz}}{I_{xx}}\dot{r} = \frac{\bar{q}_1 Sb}{I_{xx}}\left[C_{l_\beta}\beta + C_{l_p}\frac{pb}{2U_1} + C_{l_r}\frac{rb}{2U_1} + C_{l_{\delta_a}}\hat{\delta}_a + C_{l_{\delta_r}}\hat{\delta}_r\right] \tag{6.129}$$

$$\dot{r} - \frac{I_{xz}}{I_{zz}}\dot{p} = \frac{\bar{q}_1 Sb}{I_{zz}}\left[C_{n_\beta}\beta + C_{n_{T_\beta}}\beta + C_{n_p}\frac{pb}{2U_1} + C_{n_r}\frac{rb}{2U_1} + C_{n_{\delta_a}}\hat{\delta}_a + C_{n_{\delta_r}}\hat{\delta}_r\right]$$

This results in each term on the right-hand side of Eq. (6.129) having units of **linear** or **angular acceleration**.

Because the EOM have been developed in the body-fixed stability axis system (see Sec. 6.2), the two moments of inertia, I_{xx} and I_{zz}, and the one product of inertia, I_{xz}, must be calculated in that system. In most cases, these inertias are calculated in the body axis system so a transformation through the steady-state angle of attack, α_1, is required. This transformation is presented in Eq. (6.130) and is from Ref. 1.

$$\begin{bmatrix} I_{xx} \\ I_{zz} \\ I_{xz} \end{bmatrix}_{\text{Stability}} = \begin{bmatrix} \cos^2\alpha_1 & \sin^2\alpha_1 & -\sin 2\alpha_1 \\ \sin^2\alpha_1 & \cos^2\alpha_1 & \sin 2\alpha_1 \\ 0.5\sin 2\alpha_1 & -0.5\sin 2\alpha_1 & \cos^2\alpha_1 \end{bmatrix}\begin{bmatrix} I_{xx} \\ I_{zz} \\ I_{xz} \end{bmatrix}_{\text{Body}} \tag{6.130}$$

A transformation is not needed for the longitudinal case involving I_{yy} because $I_{yy_{\text{Stability}}} = I_{yy_{\text{Body}}}$.

6.5.4 Lateral-Directional Stability Parameters

To simplify Eq. (6.129) and to gain understanding of the relative contributions of each term in these equations, we again use the concept of stability parameters or dimensional stability derivatives. Again, stability parameters represent the linear or angular acceleration per motion or control variable (β, p, r, δ_a, and δ_r) for each term in the right-hand side of Eq. (6.129). We first rewrite the y force equation of Eq. (6.129):

$$(\dot{v} + U_1 r) = g\phi\cos\Theta_1 + \underbrace{\frac{\bar{q}_1 S}{m}C_{y_\beta}}_{Y_\beta}\beta + \underbrace{\frac{\bar{q}_1 Sb}{2mU_1}C_{y_p}}_{Y_p}p + \underbrace{\frac{\bar{q}_1 Sb}{2mU_1}C_{y_r}}_{Y_r}r + \underbrace{\frac{\bar{q}_1 S}{m}C_{y_{\delta_a}}}_{Y_{\delta_a}}\hat{\delta}_a + \underbrace{\frac{\bar{q}_1 S}{m}C_{y_{\delta_r}}}_{Y_{\delta_r}}\hat{\delta}_r$$

The stability parameters Y_β, Y_p, Y_r, Y_{δ_a}, and Y_{δ_r} are again simply a grouping of the terms multiplying each motion/control parameter when the EOM is in acceleration format. Notice that the $b/2U_1$ term used to nondimensionalize p

and r has been combined into the stability parameters Y_p and Y_r. A similar approach will be taken with the rolling moment equation of Eq. (6.129).

$$\dot{p} - \frac{I_{xz}}{I_{xx}}\dot{r} = \underbrace{\frac{\bar{q}_1 Sb}{I_{xx}} C_{l_\beta}}_{L_\beta} \beta + \underbrace{\frac{\bar{q}_1 Sb^2}{2I_{xx}U_1} C_{l_p}}_{L_p} p + \underbrace{\frac{\bar{q}_1 Sb^2}{2I_{xx}U_1} C_{l_r}}_{L_r} r + \underbrace{\frac{\bar{q}_1 Sb}{I_{xx}} C_{l_{\delta_a}}}_{L_{\delta_a}} \hat{\delta}_a + \underbrace{\frac{\bar{q}_1 Sb}{I_{xx}} C_{l_{\delta_r}}}_{L_{\delta_r}} \hat{\delta}_r$$

Finally, we will incorporate stability parameters into the yawing moment equation of Eq. (6.129).

$$\dot{r} - \frac{I_{xz}}{I_{xx}}\dot{p} = \underbrace{\frac{\bar{q}_1 Sb}{I_{zz}} C_{n_\beta}}_{N_\beta} \beta + \underbrace{\frac{\bar{q}_1 Sb}{I_{zz}} C_{n_{T_\beta}}}_{N_{T_\beta}} \beta + \underbrace{\frac{\bar{q}_1 Sb^2}{2I_{zz}U_1} C_{n_p}}_{N_p} p + \underbrace{\frac{\bar{q}_1 Sb^2}{2I_{zz}U_1} C_{n_r}}_{N_r} r + \underbrace{\frac{\bar{q}_1 Sb}{I_{zz}} C_{n_{\delta_a}}}_{N_{\delta_a}} \hat{\delta}_a + \underbrace{\frac{\bar{q}_1 Sb}{I_{zz}} C_{n_{\delta_r}}}_{N_{\delta_r}} \hat{\delta}_r$$

Equation (6.129) may now be recast using stability parameters:

$$\begin{aligned} \dot{v} + U_1 r &= g\phi\cos\Theta_1 + Y_\beta \beta + Y_p p + Y_r r + Y_{\delta_a}\hat{\delta}_a + Y_{\delta_r}\hat{\delta}_r \\ \dot{p} - \frac{I_{xz}}{I_{xx}}\dot{r} &= L_\beta \beta + L_p p + L_r r + L_{\delta_a}\hat{\delta}_a + L_{\delta_r}\hat{\delta}_r \\ \dot{r} - \frac{I_{xz}}{I_{zz}}\dot{p} &= N_\beta \beta + N_{T_\beta}\beta + N_r r + N_{\delta_a}\hat{\delta}_a + N_{\delta_r}\hat{\delta}_r \end{aligned} \qquad (6.131)$$

Table 6.4 summarizes the lateral-directional stability parameters associated with Eq. (6.131).

6.6 Historical Snapshot—The X-38 Parafoil Cavity Investigation

Another wind tunnel evaluation conducted at the U.S. Air Force Academy supported development of the X-38 and investigated the stability characteristics associated with a variety of parafoil cavity configurations.[2] The parafoil cavity study compared the stability characteristics of the clean "hatch on" configuration to that of four different parafoil cavity shapes representing the "hatch off" configuration. The parafoil cavity was located on the upper body of the X-38. Figure 6.16 illustrates the geometry of the four cavity configurations. Each cavity consisted of different geometry. The baseline hatch on configuration was denoted C1, and the four cavity configurations were denoted C2 through C5. C2 was a shallow cavity aft of the docking ring with deeper indentations closer to the fin region. C3 was similar to C2 but did not extend as close to the fin area. C4 was the smallest cavity and was concentrated toward the aft body, further away from the docking ring, and away from the fins. C5 was the deepest cavity with indentations close to the fins.

Table 6.4 Lateral-directional stability parameters

Stability parameter	Definition	Units
Y_β	$\dfrac{\bar{q}_1 S}{m} C_{y_\beta}$	$\dfrac{\text{ft/s}^2}{\text{rad}} = \text{ft/s}^2$
Y_p	$\dfrac{\bar{q}_1 Sb}{2mU_1} C_{y_p}$	$\dfrac{\text{ft/s}^2}{\text{rad/s}} = \text{ft/s}$
Y_r	$\dfrac{\bar{q}_1 Sb}{2mU_1} C_{y_r}$	$\dfrac{\text{ft/s}^2}{\text{rad/s}} = \text{ft/s}$
Y_{δ_a}	$\dfrac{\bar{q}_1 S}{m} C_{y_{\delta_a}}$	$\dfrac{\text{ft/s}^2}{\text{rad}} = \text{ft/s}^2$
Y_{δ_r}	$\dfrac{\bar{q}_1 S}{m} C_{y_{\delta_r}}$	$\dfrac{\text{ft/s}^2}{\text{rad}} = \text{ft/s}^2$
L_β	$\dfrac{\bar{q}_1 Sb}{I_{xx}} C_{l_\beta}$	$\dfrac{\text{rad/s}^2}{\text{rad}} = \text{s}^{-2}$
L_p	$\dfrac{\bar{q}_1 Sb^2}{2I_{xx}U_1} C_{l_p}$	$\dfrac{\text{rad/s}^2}{\text{rad/s}} = \text{s}^{-1}$
L_r	$\dfrac{\bar{q}_1 Sb^2}{2I_{xx}U_1} C_{l_r}$	$\dfrac{\text{rad/s}^2}{\text{rad/s}} = \text{s}^{-1}$
L_{δ_a}	$\dfrac{\bar{q}_1 Sb}{I_{xx}} C_{l_{\delta_a}}$	$\dfrac{\text{rad/s}^2}{\text{rad}} = \text{s}^{-2}$
L_{δ_r}	$\dfrac{\bar{q}_1 Sb}{I_{xx}} C_{l_{\delta_r}}$	$\dfrac{\text{rad/s}^2}{\text{rad}} = \text{s}^{-2}$
N_β	$\dfrac{\bar{q}_1 Sb}{I_{zz}} C_{n_\beta}$	$\dfrac{\text{rad/s}^2}{\text{rad}} = \text{s}^{-2}$
N_{T_β}	$\dfrac{\bar{q}_1 Sb}{I_{zz}} C_{n_{T_\beta}}$	$\dfrac{\text{rad/s}^2}{\text{rad}} = \text{s}^{-2}$
N_p	$\dfrac{\bar{q}_1 Sb^2}{2I_{zz}U_1} C_{n_p}$	$\dfrac{\text{rad/s}^2}{\text{rad/s}} = \text{s}^{-1}$
N_r	$\dfrac{\bar{q}_1 Sb^2}{2I_{zz}U_1} C_{n_r}$	$\dfrac{\text{rad/s}^2}{\text{rad/s}} = \text{s}^{-1}$
N_{δ_a}	$\dfrac{\bar{q}_1 Sb}{I_{zz}} C_{n_{\delta_a}}$	$\dfrac{\text{rad/s}^2}{\text{rad}} = \text{s}^{-2}$
N_{δ_r}	$\dfrac{\bar{q}_1 Sb}{I_{zz}} C_{n_{\delta_r}}$	$\dfrac{\text{rad/s}^2}{\text{rad}} = \text{s}^{-2}$

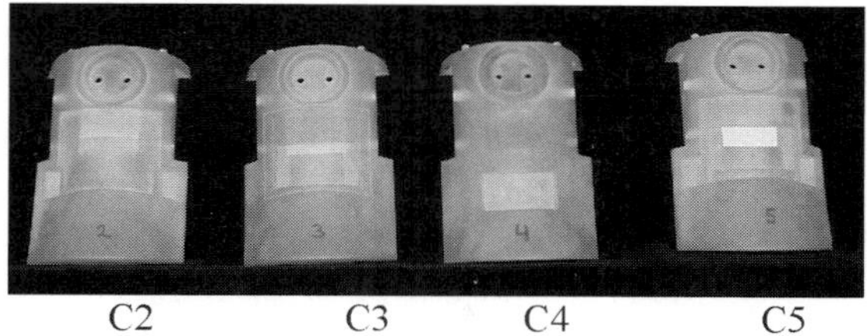

Fig. 6.16 Parafoil cavity shapes for the 4.5% scale X-38 wind tunnel model.

Longitudinal stability characteristics (C_{m_α}) of the various cavity configurations were investigated by plotting pitching moment coefficient vs angle of attack. Figure 6.17 illustrates this for all configurations at Mach 0.55. In comparing the five configurations, it can be observed that the slope of the line, and thus the value of C_{m_α}, is relatively constant. The primary difference found was in the trim angle of attack, which decreased in order from approximately 13 deg for the baseline hatch on configuration (C1) to 12 deg for C5. This was probably a result of flow disruption from the cavity reducing the effectiveness of the aft body ramp. The aft body ramp on top of the X-38 body is intended to provide a positive contribution to pitching moment, an effect that appears to be reduced by flow interference from the cavity. The data also indicated that the vehicles will experience a drop in angle of attack as the hatch cover is deployed.

The lateral stability characteristics of the various cavity configurations were investigated and compared to the hatch on configuration by graphing rolling

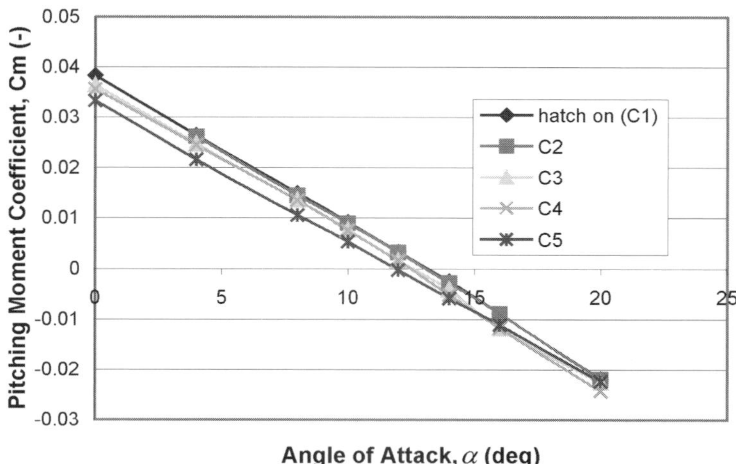

Fig. 6.17 Pitching moment coefficient vs alpha for all configurations, at 0 deg sideslip angle.

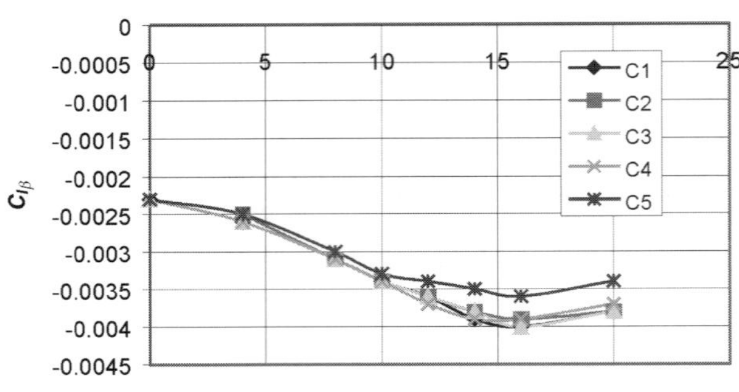

Fig. 6.18 C_{l_β} vs angle of attack for all configurations.

moment coefficient as a function of sideslip angle and finding the slope of the line (C_{l_β}). The stability derivative C_{l_β} was determined over a range of angles of attack.

Figure 6.18 presents a graph of C_{l_β} as a function of angle of attack for all configurations, and shows the increased lateral stability with increasing angle of attack. However, after 16 deg angle of attack, the lateral stability began to decrease.

In comparing C_{l_β} for the four configurations and the baseline hatch on configuration, it can be observed that C5 was less stable than the other four in the 10 to 16 deg angle of attack region. The increase in lateral stability with angles of attack below 16 deg was theorized to be a result of accelerated flow being channeled into the valley between the X-38 body and vertical fin. Typical values of C_{l_β} for various aircraft are shown in Table 6.5. Compared to an average value of C_{l_β} of -0.0037/deg for the X-38, the first three aircraft appear to be slightly less laterally stable than the X-38.

Another important parameter in analyzing X-38 lateral-directional characteristics was the directional stability derivative, C_{n_β}. To determine C_{n_β}, C_n was graphed as a function of sideslip angle at various angles of attack. Figure 6.19 presents the summary plot of C_{n_β} vs angle of attack for the five configurations. C_{n_β} appeared to be the same for all configurations at 0 deg angle of attack, then decrease as angle of attack increased to 4 deg. The behavior of C_{n_β} in the

Table 6.5 Typical C_{l_β} values for various aircraft (Ref. 1)

Aircraft	Cruise, C_{l_β}	Approach, C_{l_β}
Small Cessna	−0.00161	−0.00169
T-37	−0.00165	−0.00143
F-4	−0.0014	−0.0027
B-747	−0.0166	−0.0049

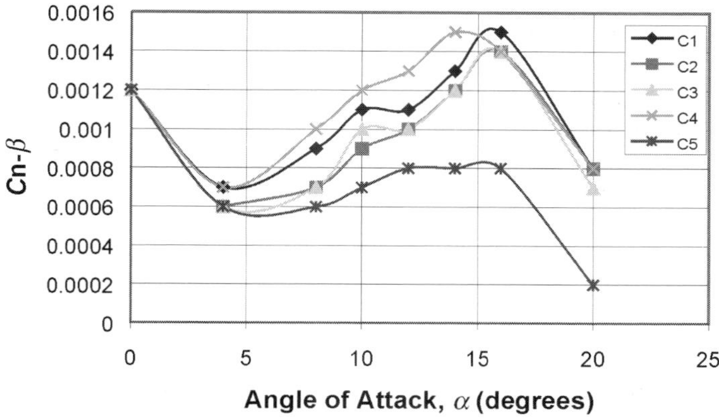

Fig. 6.19 C_{n_β} as a function of angle of attack at Mach = 0.55.

low angle of attack region was considered to be of more academic interest than practical interest because the X-38 flight angle of attack range was generally above 12 deg. C_{n_β} increased from 4 to 16 deg angle of attack, but beyond that point the derivative decreased. The drop in C_{n_β} reaching 20 deg may be because of a flow separation region masking a portion of the vertical fin at high angles of attack.

Comparing the five parafoil hatch configurations in Fig. 6.19, it can be observed that C5 seems to be less directionally stable in the 4 to 16 deg angle of attack range than the other four configurations. C4 provided the highest level of directional stability up to 14 deg angle of attack, even above that of the baseline. However, C2 and C3 tended to maintain their directional stability longer, up to approximately 16 deg angle of attack. C4 had a deeper cavity that was farther back on the body than any of the other cavity configurations. This was theorized to create more drag, which acts aft of the moment reference center and increases directional stability. C5 had a shallower cavity that extended farther out to the side. This may have disrupted the flow over a larger portion of the aft body ramp and possibly even the lower portion of the vertical fins, thus reducing the directional stability. C1, C2, and C3 fell in between C4

Table 6.6 Typical C_{n_β} values for various aircraft

Aircraft	Cruise, C_{n_β}	Approach, C_{n_β}
Small Cessna	0.001025	0.00122
T-37	0.00193	0.0019
F-4	0.00218	0.0035
B-747	0.00367	0.00321

and C5. Typical values of C_{n_β} for various aircraft are shown in Table 6.6. Compared to an average value of C_{n_β} of 0.001/deg for the X-38, these aircraft appear to be slightly more directionally stable than the X-38. In general, the aircraft comparison presented in Tables 6.5 and 6.6 indicate that the X-38 had higher roll stability and lower directional stability than the aircraft chosen for comparison. This may indicate degraded dutch roll characteristics (discussed in Chapter 7) for the X-38 when using these aircraft as a baseline.

Overall, cavity configuration C4 was considered the best of the four configurations evaluated because it had slightly higher directional stability. This shape had a smaller cavity size, and was located more toward the rear of the body, with the majority of the cavity away from the vertical fins. This was also the cavity configuration incorporated into the X-38 design.

References

[1]Schmidt, L. V., *Introduction to Aircraft Flight Dynamics*, AIAA Education Series, AIAA, Reston, Virginia, 1998.

[2]Johnston, C. N., Nettleblad, T. A., and Yechout, T. R., "X-38 Mid-Rudder Feasibility and Parafoil Cavity Investigations," U.S. Air Force Academy Dept. of Aeronautics TR 01-02, Sept. 2001.

Problems

6.1. Using vector mathematics, derive the expression for inertial acceleration as measured in the rotating body axis system starting with

$$\bar{V}_B = \begin{bmatrix} U \\ V \\ W \end{bmatrix}_B \quad \text{and} \quad \bar{\omega} = \begin{bmatrix} P \\ Q \\ R \end{bmatrix}_B$$

6.2 A T-37 is in a level turn. Sensors on the aircraft measure the following accelerations and rates:

$$\begin{array}{lll} U = 200 \text{ ft/s} & \dot{U} = 5 \text{ ft/s}^2 & P = 0 \text{ rad/s} \\ V = 0 \text{ ft/s} & \dot{V} = 0 \text{ ft/s}^2 & Q = 0.1 \text{ rad/s} \\ W = 10 \text{ ft/s} & \dot{W} = 0 \text{ ft/s}^2 & R = 0.1 \text{ rad/s} \end{array}$$

Find the T-37's inertial acceleration vector in the body axis.

6.3 An A-37 is loaded with two 500-lb bombs as shown. The distances of the bombs from the aircraft center of gravity are shown.

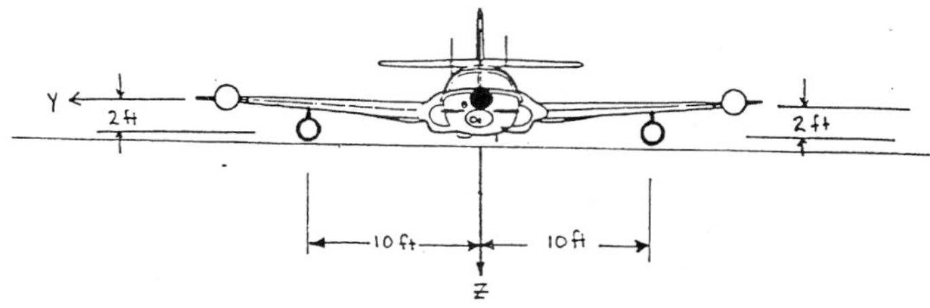

What happens to I_{xx} if *both* bombs are dropped? What is the value of the change in I_{xx}?

6.4 In Problem 6.3, what is the value of I_{yz} of the aircraft after the bomb on the right wing is dropped?

6.5 Given

$$\begin{bmatrix} U \\ V \\ W \end{bmatrix} = \begin{bmatrix} 200 \\ 0 \\ 10 \end{bmatrix} \text{ft/s}$$

in the body axis system with

$$\Phi = 25 \text{ deg}$$
$$\Theta = 10 \text{ deg}$$
$$\Psi = 0 \text{ deg}$$

(a) Find velocity components in earth axis system
(b) Show that this can be transformed back into the body axis system

6.6 Given

$$\begin{aligned} \dot{\Phi} &= 100 \text{ deg/s} & \Phi &= 45 \text{ deg} \\ \dot{\Psi} &= 10 \text{ deg/s} & \Psi &= 360 \text{ deg} \\ \dot{\Theta} &= 10 \text{ deg/s} & \Theta &= 5 \text{ deg} \end{aligned}$$

Find the body axis roll, pitch, and yaw rates using the kinematic equations.

6.7 Given the following nonlinear differential that represents some fictitious EOM for a vehicle

$$\dot{A} + BU = X$$

Linearize the equation using the perturbed form of the EOM and the same steps and assumptions used to develop the real aircraft EOM.

6.8 You would expect the term η_h to be
(a) Equal to one.
(b) Greater than one.
(c) Less than one.
(d) Approximately zero—at least a very small number.

6.9 Which of the following terms is not part of the horizontal tail volume coefficient?
(a) wing area
(b) tail area
(c) tail MAC
(d) tail moment arm
(e) dynamic pressure ratio
(f) tail lift curve slope
(g) wing lift curve slope

6.10 The tail sees the same AOA as the wing-body except for
(a) Nothing—they see the same AOA.
(b) Upwash and the tail incidence angle.
(c) Tail incidence angle.
(d) Tail incidence angle and downwash.

6.11 The elevator effectiveness term describes
(a) The effective change in aircraft AOA for a change in elevator deflection.
(b) The effective change in wing-body AOA for a change in elevator deflection.
(c) The effective change in horizontal tail AOA for a change in elevator deflection.
(d) None of the above.

6.12 If you saw the following expression for C_{m_0}

$$C_{m_0} = C_{m_{AC_{wf}}} + C_{L_{\alpha_h}} \eta_h \bar{V}_h [\varepsilon_0 + i_h]$$

You could assume that the aircraft
(a) Has a fixed tail incidence such as found on the T-41.
(b) Has a variable tail incidence like the B-1.

(c) Has a canard.
(d) None of the above.

6.13 The following is required for an aircraft to be stable and trimmed in 1-g flight:
(a) negative $(C_{m_0} + C_{m_{i_h}} i_h)$ and positive C_{m_α}
(b) positive $(C_{m_0} + C_{m_{i_h}} i_h)$ and positive C_{m_α}
(c) positive $(C_{m_0} + C_{m_{i_h}} i_h)$ and negative C_{m_α}
(d) negative $(C_{m_0} + C_{m_{i_h}} i_h)$ and negative C_{m_α}

6.14 How would a flying wing achieve the correct answer to Problem 6.13?

6.15 Given the following graph for a conventional aircraft, which of the following describes the relationship between line 1 and line 2?

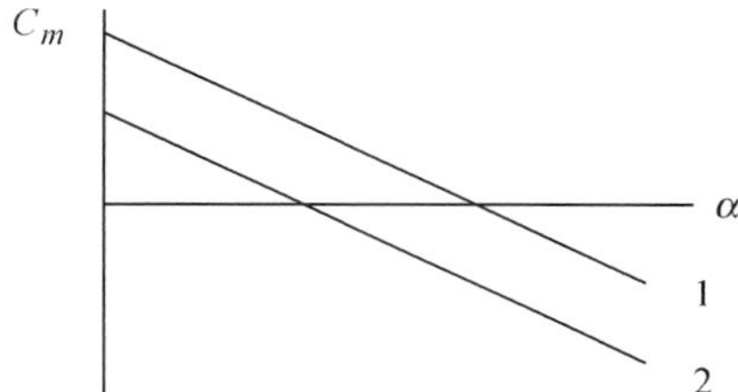

(a) Line 1 is for a more positive elevator deflection than line 2.
(b) Line 2 is for a more positive elevator deflection than line 1.
(c) The center of gravity is farther aft for line 2 than line 1.
(d) The center of gravity is farther aft for line 1 than line 2.

6.16 The neutral point location is almost always aft of the wing body a.c. (that is, the aircraft is unstable in pitch without the horizontal tail).
(a) True
(b) False

6.17 Increasing the tail volume would move the neutral point:
(a) farther forward.
(b) farther aft.
(c) need more information to tell.
(d) wouldn't change location.

6.18 Increasing elevator control power
(a) decreases the amount of elevator required to trim.
(b) increases the amount of elevator required to trim.
(c) has no effect on elevator required to trim.
(d) would only affect the stability level of the aircraft.

6.19 An aircraft has a lift curve slope of 0.10/deg and a $C_{m_\alpha} = -0.04$/deg. What is the aircraft's static margin?

6.20 An aircraft has longitudinal static stability at a given center of gravity location. Which of the following is true?
(a) the static margin is negative
(b) $\partial\delta_e/\partial C_L = 0$
(c) $C_{m_\alpha} < 0$
(d) all of the above

6.21 Assume a Lear Jet is cruising (level, unaccelerated flight) at 40,000 ft with $U_1 = 677$ ft/s, $S = 230$ ft^2, Weight = 13,000 lb, and $C_{T_{x_1}} = 0.0335$. Find C_{L_1} and C_{D_1}.

6.22 Compute the thrust being produced by the Lear Jet in Problem 6.21.

6.23 Given $f(x) = f(a) + f'(a)(x - a)$ and $f(z) = -0.1x^4 - 0.15x^3 - 0.5x^2 - 0.25x + 1.2$

Use the first-order Taylor series approximation to evaluate $f(x)$ given $f(a)$ where:

$$a = 1$$
$$x = 2$$

6.24 Compare the answer from Problem 6.23 with what you would obtain just evaluating $f(a)$ at $a = 2$.

6.25 To obtain the same answer for Problems 6.23 and 6.24, what order Taylor series would you use? Try it.

6.26 For each of the following terms, indicate if the term is important, approximately zero, or too small to worry about:

$$C_{l_\beta},\ C_{l_{\delta_a}},\ C_{y_p},\ C_{y_\beta},\ C_{n_\beta}\ C_{D_q},\ C_{D_\alpha},\ C_{D_u},\ C_{m_\alpha},\ C_{y_r},\ C_{y_{\delta_a}}$$

6.27 Which of the following is called a cross derivative?
(a) C_{m_α} (b) C_{l_p} (c) C_{n_p} (d) C_{n_r}

6.28 Write down the normal sign one would expect to see for each of the following:
(a) C_{l_p}
(b) C_{m_q}
(c) C_{l_r}
(d) C_{n_r}

6.29 Which of the following plays the most significant role in lateral-directional dynamics?
(a) C_{n_p}
(b) C_{l_r}
(c) C_{n_r}
(d) C_{m_α}

6.30 Most of the contribution (80 to 90%) to C_{n_r} comes from which component of the aircraft?

6.31 Normally, you would expect a fighter to have a higher or lower C_{l_p} than a glider?

6.32 Which of the following provides the largest contribution to C_{l_p}?
(a) wing
(b) horizontal tail
(c) vertical tail
(d) fuselage

6.33 When nondimensionalizing the lateral-directional stability derivatives, the characteristic length used is
(a) the MAC
(b) the wingspan
(c) Z_{vs}
(d) None of the above

6.34 Which of the following represent a nondimensional stability derivative?
(a) $C_{m_q} = \dfrac{\partial C_m}{\partial q}$
(b) $C_{m_q} = \dfrac{\partial C_m}{\partial\left(\dfrac{q\bar{c}}{2U_1}\right)}$
(c) $C_{m_{\hat{q}}} = \dfrac{\partial C_m}{\partial\left(\dfrac{q\bar{c}}{2U_1}\right)}$
(d) All of the options are dimensional stability derivatives.

6.35 An F-4 is cruising at 35,000 ft at a velocity of 876 ft/s (density is 0.000739 slug/ft^3). You are given

$$\begin{aligned} C_{L_1} &= 0.26 & S &= 503 \text{ ft}^2 \\ C_{D_1} &= 0.03 & W &= 39,000 \text{ lb} \\ C_{D_u} &= 0.027 \end{aligned}$$

Calculate:

(a) $$\frac{\partial F_{A_x}}{\partial\left(\frac{u}{U_1}\right)}$$

(b) The stability parameter X_u

(c) The contribution to $\dot{u}$ because of X_u, if $u = 2\,\text{ft/s}$.

6.36 Derive $\partial F_{A_x}/\partial\alpha = (-C_{D_\alpha} + C_{L_1})\bar{q}_1 S$.

6.37 Using Eq. (6.127) for the following aircraft, find $\dot{u}$ for a positive 1-deg step elevator input at $t = 0$.

$$X_{\delta_e} = 12.3976\ \text{ft/s}^2;\ X_{T_u} = -0.0123\ 1/\text{s};\ X_u = 0.0085\ 1/\text{s};$$
$$X_\alpha = -4.9591\ \text{ft/s}^2;\ U_1 = 876\ \text{ft/s}$$

(Hint: $\theta = u = \alpha = 0$ at $t = 0$)

At $t = \infty$, estimate U (the new stabilized velocity) if $\alpha = -0.1$ deg and $\theta = -0.15$ deg.

7
Aircraft Dynamic Stability

Aircraft dynamic stability focuses on the time history of aircraft motion after the aircraft is disturbed from an equilibrium or trim condition. This motion may be first order (exponential response) or second order (oscillatory response), and will have either positive dynamic stability (aircraft returns to the trim condition as time goes to infinity), neutral dynamic stability (aircraft neither returns to trim nor diverges further from the disturbed condition), or dynamic instability (aircraft diverges from the trim condition and the disturbed condition as time goes to infinity). The study of dynamic stability is important to understanding aircraft handling qualities and the design features that make an airplane fly well or not as well while performing specific mission tasks. The differential equations that define the aircraft equations of motion (EOM) form the starting point for the study of dynamic stability.

7.1 Mass–Spring–Damper System and Classical Solutions of Ordinary Differential Equations

The mass–spring–damper system illustrated in Fig. 7.1 provides a starting point for analysis of system dynamics and aircraft dynamic stability. This is an excellent model to begin the understanding of dynamic response.

We will first develop an expression for the sum of forces in the vertical direction. Notice that $x(t)$ is defined as positive for an upward displacement and that the zero position is chosen as the point where the system is initially at

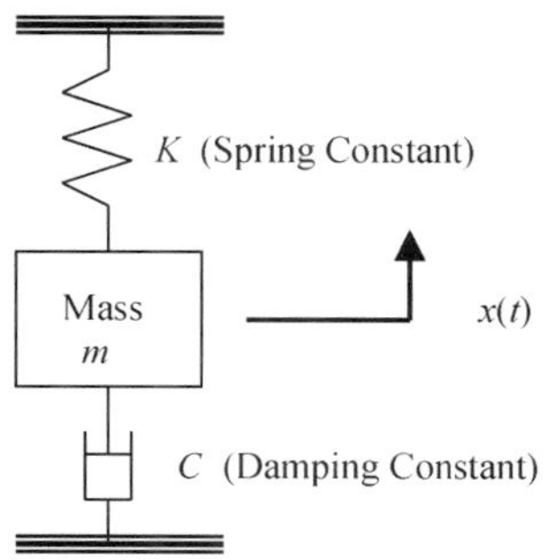

Fig. 7.1 Mass–spring–damper system.

rest or at equilibrium. We know that

$$\sum F_x = m\frac{\mathrm{d}^2x}{\mathrm{d}t^2} \tag{7.1}$$

There are two forces acting on the mass, the damping force, and the spring force. For the damping or frictional force (F_f), this can be approximated by a linear relationship of damping force as a function of velocity or $\mathrm{d}x/\mathrm{d}t$ (see Fig. 7.2).

A damper can be thought of as a "shock absorber" with a piston moving up and down inside a cylinder. The piston is immersed in a fluid and the fluid is displaced through a small orifice to provide a resistance force directly proportional to the velocity of the piston. This resistance force (F_f) can be expressed as

$$F_f = CV$$

where C is the slope in Fig. 7.2. The spring force (F_s) is directly proportional to the displacement (x) of the mass and can be represented as

$$F_s = Kx$$

where K is the spring constant. If the mass is displaced in the positive x direction, both the damping and spring forces act in a direction opposite to this displacement and can be represented by

$$F_f + F_s = -CV - Kx \tag{7.2}$$

Because

$$V = \frac{\mathrm{d}x}{\mathrm{d}t} \tag{7.3}$$

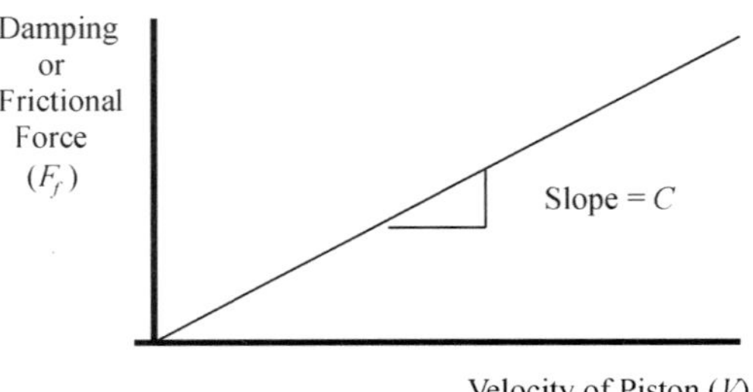

Fig. 7.2 Damper relationship.

we can combine Eqs. (7.1–7.3) to obtain

$$m\left(\frac{d^2x}{dt^2}\right) = -C\left(\frac{dx}{dt}\right) - Kx$$

or

$$m\left(\frac{d^2x}{dt^2}\right) + C\left(\frac{dx}{dt}\right) + Kx = 0 \tag{7.4}$$

which is the differential equation for the mass–spring–damper system with zero initial displacement ($x = 0$).

If we initially stretch the spring from its original position by a distance y as shown in Fig. 7.3, we build in a forcing function that must be added to Eq. (7.4).

Because the upper "tie down" point is moved up by a distance y to achieve this stretch or preload, the preload has a positive sign and a magnitude of Ky. It can be conveniently added to the right side of Eq. (7.4) to obtain

$$m\left(\frac{d^2x}{dt^2}\right) + C\left(\frac{dx}{dt}\right) + Kx = Ky \tag{7.5}$$

This is the differential equation for the spring–mass–damper system with a preload as shown. At this point, we should observe that if the mass is free to move, it will obtain a steady-state condition (a new equilibrium location) when d^2x/dt^2 and dx/dt equal zero; and the new equilibrium position will be $x = y$.

Now that the differential equation for the spring–mass–damper system has been defined, we will review classical approaches to solving ordinary differential equations of this type. Keep in mind that Eq. (7.5) is also representative of aircraft motion and that is why we are investigating it in depth.

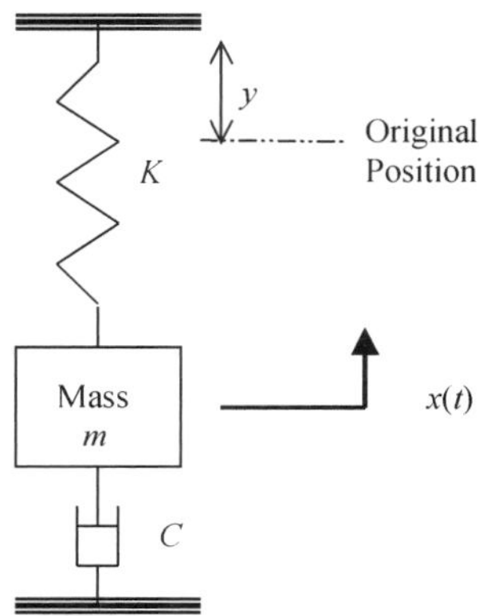

Fig. 7.3 Adding a forcing function to the spring–mass–damper system.

7.1.1 First-Order Systems

A special case of Eq. (7.5), which we will consider first, addresses a spring–mass–damper system where the mass is very small or negligible compared to the size of the spring and damper. We will call such a system a massless or first-order (referring to the order of the highest derivative) system. The following differential equation results when the mass is set equal to zero.

$$C\left(\frac{\mathrm{d}x}{\mathrm{d}t}\right) + Kx = Ky \tag{7.6}$$

To solve this differential equation, we will first describe the method of differential operators where P is defined as the differential operator, $\mathrm{d}/\mathrm{d}t$, so that

$$Px = \frac{\mathrm{d}x}{\mathrm{d}t} \qquad P^2x = \frac{\mathrm{d}^2x}{\mathrm{d}t^2} \qquad \frac{x}{P} = \int x\,\mathrm{d}t$$

We will first attack the homogeneous form (forcing function equal to zero) of Eq. (7.6),

$$C\left(\frac{\mathrm{d}x}{\mathrm{d}t}\right) + Kx = 0 \tag{7.7}$$

Substituting in the differential operator, P, Eq. (7.7) becomes

$$CPx + Kx = 0$$

We then solve for P, which now becomes a **root of the equation**,

$$(CP + K)x = 0$$
$$P = -(K/C)$$

The homogeneous solution is then of the form

$$x(t) = C_1e^{Pt} = C_1e^{(-K/C)t} \tag{7.8}$$

where C_1 is determined from initial conditions. The homogeneous solution will also be called the **transient** solution when we are dealing with aircraft response.

Example 7.1

Solve the following first-order differential equation

$$\frac{\mathrm{d}x}{\mathrm{d}t} + 2x = 0$$

subject to the following initial condition: $x(0) = 1$

Solution:

$$Px + 2x = 0$$
$$(P + 2)x = 0$$
$$P = -2$$
$$x(t) = C_1 e^{-2t}$$

Using the initial condition $x(0) = 1$ to evaluate C_1

$$1 = C_1$$

and

$$x(t) = e^{-2t}$$

is the solution, or time response.

The solution is graphed in Fig. 7.4. Notice that it starts off at a value of one at time equal to zero and exponentially decays to zero. It is also important to note that a first-order system has a first order or exponential transient response (no oscillations).

Next, we will look at solving a first-order nonhomogeneous differential equation like Eq. (7.6). A forcing function is included with a nonhomogeneous differential equation and the solution is called the nonhomogeneous or particular solution. It is also called the **steady-state** solution when we are dealing with aircraft response. To achieve a solution using differential operators, we must assume a form of the solution based on the form of the forcing function as outlined in Table 7.1.

The first step in solving a nonhomogeneous differential equation involves setting the forcing function to zero and obtaining the homogeneous solution. Next, the appropriate assumed solution is input into the nonhomogeneous differential equation so that the constants A, B, C, (as appropriate) can be determined and the nonhomogeneous solution defined. Finally, the homogeneous and nonhomogeneous are added together to obtain the total solution. Example 7.2 will help clarify these steps.

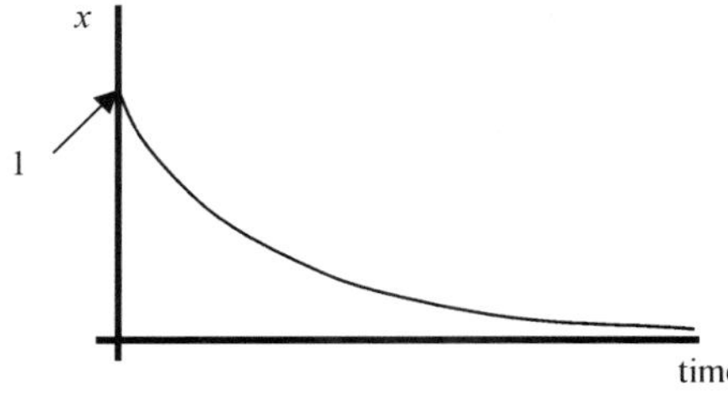

Fig. 7.4 Transient time response of a first-order differential equation.

Table 7.1 Assumed solutions for nonhomogeneous differential equations

Forcing function	Assumed solution
K	A
Kt	$At + B$
Kt^2	$At^2 + Bt + C$
$K \sin wt$	$A \sin wt + B \cos wt$

Example 7.2

Solve the following first-order differential equation:

$$2\frac{dx}{dt} + 3x = 6$$

subject to the following initial condition: $x(0) = 0$

Homogeneous (Transient) Solution:

$$2Px + 3x = 0$$
$$(2P + 3)x = 0$$
$$P = -3/2$$
$$x_h(t) = C_1 e^{(-3/2)t}$$

(Homogeneous Solution)

Nonhomogeneous (Steady-State) Solution:

Assume a steady-state solution of the form $x(t) = A$ because the forcing function is a constant.

Substitute $x_{nh}(t) = A$ into the original differential equation:

$$2(0) + 3A = 6$$
$$A = 2$$

And the nonhomogeneous solution is:

$$x_{nh}(t) = 2$$

The forcing function should be thought of as a constant equal to 2 with the 6 on the right-hand side of the differential equation being equal to the spring constant ($K = 3$) times the forcing function.

The total solution is the combination of the homogeneous and nonhomogeneous solutions:

$$x(t) = x_h(t) + x_{nh}(t) = C_1 e^{(-3/2)t} + 2$$

To evaluate the constant C_1, we use the initial condition $x(0) = 0$

$$0 = C_1 + 2$$
$$C_1 = -2$$

and

$$x(t) = -2e^{(-3/2)t} + 2$$

or

$$x(t) = 2(1 - e^{(-3/2)t})$$

becomes the total solution.

Example 7.2 yields some interesting insights. Evaluating the total solution at $t = 0$ results in $x(t) = 0$ which agrees with our initial condition. At $t = \infty$, the exponential term goes to zero and $x(\infty) = 2$, which is the same as the value of the constant forcing function. The time response for Example 7.2 is presented in Fig. 7.5.

7.1.1.1 Time constant. We will next introduce the important concept of Time Constant (τ). If we return to Eq. (7.6) and solve it in general form for an initial condition of $x(0) = 0$, we obtain

$$x(t) = y(1 - e^{(-K/C)t})$$

Figure 7.6 presents a graph of the time response. Notice that the steady-state value is y, which equates to the value of the displacement of the forcing function.

We will begin referring to a constant forcing function as a step input. Notice also the exponential rise to achieve the steady-state value. The lag time associated with this rise to the steady-state value is an important consideration in determining the acceptability of the response from an aircraft handling qualities standpoint. This lag time is typically quantified with the time constant (τ),

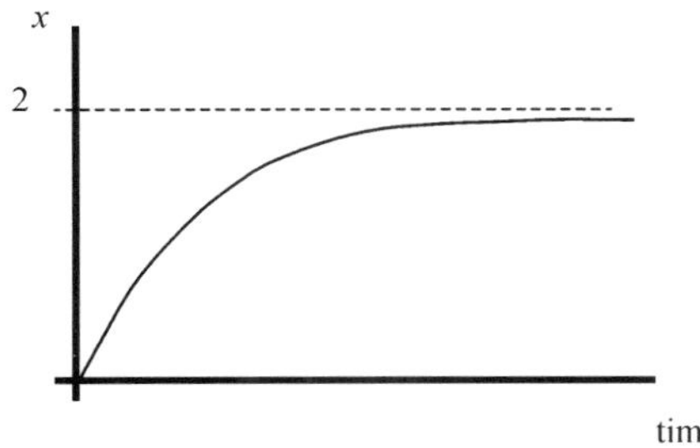

Fig. 7.5 Time response for Example 7.2.

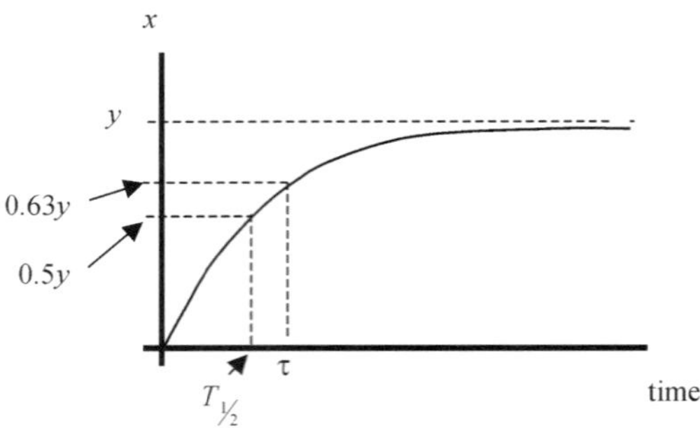

Fig. 7.6 Generalized response of a first-order system.

which is a measure of *the time it takes to achieve 63.2% of the steady-state value*. Why did we pick 63.2%? If we let $t = C/K$, our first-order response to a step input becomes

$$x(t) = y(1 - e^{(-K/C)(C/K)}) = y(1 - e^{-1}) = y(1 - 0.368)$$
$$x(t) = 0.632y \quad (\text{at } t = C/K = \tau)$$

The time constant becomes an easy value to determine because

$$\tau = -\frac{1}{P} \tag{7.9}$$

for a first-order differential equation. In the case of Example 7.2, τ is equal to $2/3$ s. Note also, because τ is equal to C/K for the spring–mass–damper system, that τ will increase (meaning a slower responding system) for an increase in damping constant (C), and that τ will decrease (meaning a faster responding system) for an increase in the spring constant (K). This should make sense when thinking about the physical dynamics of the system.

7.1.1.2 Time to half and double amplitude. Another measure of the lag time associated with a systems response is the time to half amplitude ($T_{1/2}$). Referring to Fig. 7.6, this is simply the time it takes to achieve 50% of the steady-state value. It can be easily shown that

$$T_{1/2} = \tau(\ln 2) = 0.693\tau \tag{7.10}$$

For unstable first order systems ($P > 0$), a measure used as an indication of the instability is the time to double amplitude (T_2). T_2 is the time it takes for the response to achieve twice the amplitude of an input disturbance. It can be found using

$$T_2 = \frac{\ln 2}{P} = \frac{0.693}{P} \tag{7.11}$$

7.1.2 Second-Order Systems

We now return to Eq. (7.5) in its entirety for a spring–mass–damper system where the mass provides significant inertial effects. Equation (7.5) is a second-order differential equation (referring to the highest-order derivative) or simply a second-order system. It becomes

$$M\ddot{x} + C\dot{x} + Kx = Ky$$

or

$$\ddot{x} + \frac{C}{M}\dot{x} + \frac{K}{M}x = \frac{K}{M}y \tag{7.12}$$

To solve Eq. (7.12), we will again use the method of differential operators on the homogeneous differential equation to obtain the transient solution or transient response.

$$(MP^2 + CP + K)x = 0$$

We can then solve for the roots (P) of this equation using the quadratic formula.

$$P_{1,2} = -\frac{C}{2M} \pm \frac{\sqrt{C^2 - 4KM}}{2M}$$

Three cases must be considered based on the sign of the expression under the radical.

Case 1: Two Real Unequal Roots or $C^2 > 4KM$

This results in an **overdamped** system (no oscillations) with a general solution of the form

$$x(t) = C_1 e^{P_1 t} + C_2 e^{P_2 t} \tag{7.13}$$

The constants C_1 and C_2 must be evaluated based on the initial conditions.

Case 2: Two Real Repeated Roots or $C^2 = 4KM$

This results in a **critically damped** system (no oscillations) with a general solution of the form

$$x(t) = C_1 e^{Pt} + C_2 t e^{Pt} \tag{7.14}$$

Again, the constants C_1 and C_2 must be evaluated based on initial conditions.

Case 3: Two Complex Conjugate Roots or $4KM > C^2$
This results in an **underdamped** (with oscillations) system. The roots are

$$P_{1,2} = -\frac{C}{2M} \pm i\frac{\sqrt{4KM - C^2}}{2M} = a \pm ib$$

and the general solution is of the form

$$x(t) = e^{at}[C_1 \sin bt + C_2 \cos bt]$$

or

$$x(t) = C_3 e^{at} \sin(bt + \phi) \tag{7.15}$$

where

$$|C_3| = \sqrt{C_1^2 + C_2^2}$$

and

$$\phi = \tan^{-1}\left(\frac{C_2}{C_1}\right)$$

Notice in Eq. (7.15) that the real part of the root (a) determines the exponential decay (damping) portion of the time response and that the imaginary part of the root (b) is the frequency of the oscillation. The **phase angle** (ϕ) can be thought of for now as a lag between an input and output. Case 3 is typical of three dynamic modes of motion for most aircraft (the short period, the phugoid, and the dutch roll modes), which we will discuss in detail later.

7.1.2.1 Damping ratio and natural frequency. We can recast Eq. (7.12) in terms of two new parameters: damping ratio (ζ) and natural frequency (ω_N). These parameters have physical meaning for Case 3 and lead directly to the time solution for common inputs such as steps and impulses.

$$\ddot{x} + 2\zeta\omega_N\dot{x} + \omega_N^2 x = \omega_N^2 y \tag{7.16}$$

The damping ratio provides an indication of the system damping and will fall between -1 and 1 for Case 3. For stable systems, the damping ratio will be between 0 and 1. For this case, the higher the damping ratio, the more damping is present in the system. Figure 7.7 presents a family of second order responses to a unit step ($y = 1$) input, which show the influence of damping ratio. Notice that the number of overshoots/undershoots varies inversely with the damping ratio.

The natural frequency is the frequency (in rad/s) that the system would oscillate at if there were no damping. It represents the highest frequency that

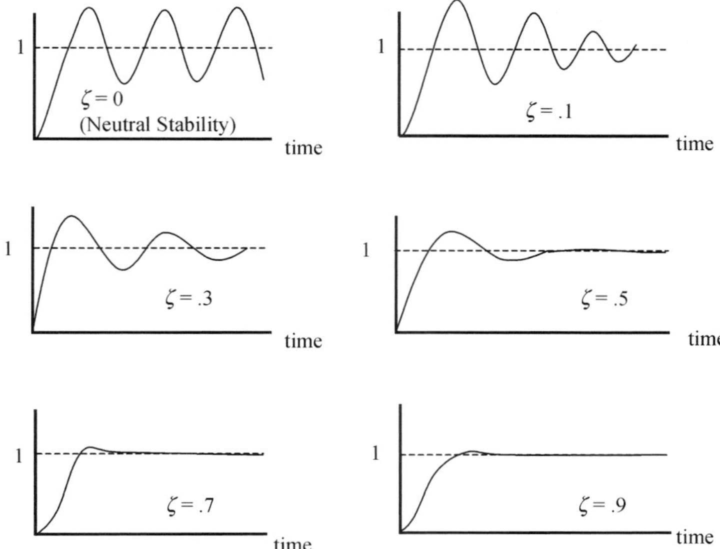

Fig. 7.7 Unit step responses for different damping ratios.

the system is capable of, but it is not the frequency that the system actually oscillates at if damping is present. For the mass–spring–damper system, $\omega_N = \sqrt{K/M}$.

7.1.2.2 Damped frequency. The damped frequency (ω_D) represents the frequency (in rad/s) that the system actually oscillates at with damping present. Returning to Eq. (7.16), we can use the quadratic formula to solve for the roots of the homogeneous form of the equation.

$$P_{1,2} = -\zeta\omega_N \pm i\omega_N\sqrt{1-\zeta^2} = -\zeta\omega_N \pm i\omega_D = a \pm ib \tag{7.17}$$

where

$$\omega_D = \omega_N\sqrt{1-\zeta^2} = \text{Damped Frequency} \tag{7.18}$$

7.1.2.3 Time constant. The time constant (τ) for a second order system can be found by examining the real part of the roots ($-\zeta\omega_n$) in Eq. (7.17) and recalling our discussion in Sec. 7.1.1.1. The time constant (τ) for the generalized case of Eq. (7.16) becomes

$$\tau = \frac{1}{\zeta\omega_N} \tag{7.19}$$

This is similar to the way we computed the time constant for a first-order system ($\tau = -1/P$). Notice that the larger the $\zeta\omega_N$, the smaller τ and the faster the response.

7.1.2.4 Period of oscillation. The period of oscillation (T) for a second-order system is the time it takes between consecutive peaks of an oscillation. The period is inversely proportional to the damped frequency and is defined by

$$T = \frac{2\pi}{\omega_D} \tag{7.20}$$

where ω_D must be in units of radians/second.

The time response for the homogeneous case of Eq. (7.16) is

$$x(t) = C_3 e^{-\zeta\omega_N t} \sin(\omega_D t + \phi) \tag{7.21}$$

Figure 7.8 illustrates this response. The system has been initially disturbed by an impulse input (which may be thought of as a very short duration spike input that excites the system dynamics). The figure also illustrates several of the concepts just discussed.

The steady-state (nonhomogeneous) solution will be defined next for Eq. (7.16). Because the forcing function, $\omega_n^2 y$, is a constant we can assume the form of the solution as

$$x_{SS}(t) = A$$

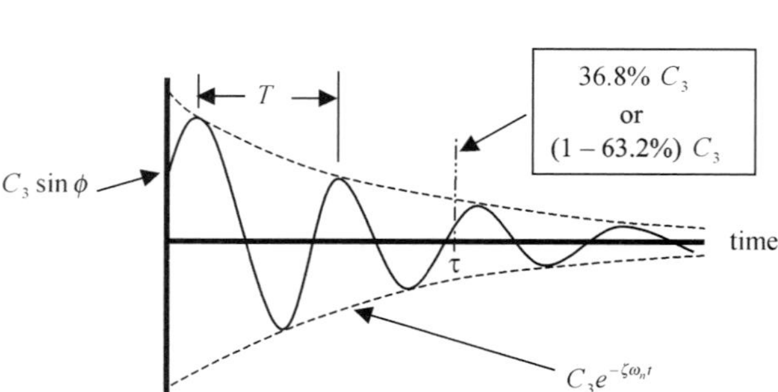

Fig. 7.8 Second-order time response.

Therefore,

$$\frac{d^2A}{dt^2} = 0, \quad \frac{dA}{dt} = 0 \Rightarrow 0 + 0 + \omega_N^2 x = \omega_N^2 y$$

and

$$A = y$$

The particular (steady solution) is

$$x_{SS}(t) = y$$

The total solution then becomes

$$x(t) = x_{\text{transient}}(t) + x_{SS}(t)$$

or

$$x(t) = y(1 + C_4 e^{-\zeta\omega_N t}\sin(\omega_D t + \phi)) \tag{7.22}$$

Please note that the C_4 in Eq. (7.22) will typically not have the same value as the C_3 in Eq. (7.15).

Example 7.3

Find the time solution for the following differential equation:

$$\ddot{x} + 5\dot{x} + 25x = 25$$

First, we apply the generalized form for a second-order differential equation Eq. (7.16):

$$\ddot{x} + 2\zeta\omega_N\dot{x} + \omega_N^2 x = \omega_N^2 y$$

The first thing we notice is the unit step input ($y = 1$). Next, we can determine the values of natural frequency, damping ratio, and damped frequency:

$$\omega_N = \sqrt{25} = 5 \text{ rad/s} \qquad 2\zeta\omega_N = 5 \Rightarrow \zeta = 0.5$$

$$\omega_D = \omega_N\sqrt{1 - \zeta^2} = 4.33 \text{ rad/s}$$

We can then input these values into the generalized solution from Eq. (7.22) for a second-order differential equation with a unit step input:

$$x(t) = 1 + C_4 e^{-2.5t}\sin(4.33t + \phi)$$

We still need to determine C_4 and ϕ, which can be evaluated from initial conditions and a relationship that we will develop in the next section for ϕ.

7.2 Root Representation Using the Complex Plane

The roots of Eq. (7.16) were presented in Eq. (7.17) and are repeated here.

$$P_{1,2} = -\zeta\omega_N \pm i\omega_D = a \pm ib$$

These roots can be represented on a complex plane as shown in Fig. 7.9. The complex plane plots the real part of the root on the horizontal axis and the imaginary part of the root on the vertical axis.

From trigonometry

$$\cos\phi = \frac{|-\zeta\omega_N|}{r} = \frac{|-\zeta\omega_N|}{\omega_N} = \zeta$$

so

$$\zeta = \cos\phi$$

or

$$\phi = \cos^{-1}\zeta \tag{7.23}$$

It is good to keep ϕ in units of radians, as we will soon see. Also

$$\tan\phi = \frac{\omega_N\sqrt{1-\zeta^2}}{|-\zeta_N|} = \frac{\sqrt{1-\zeta^2}}{\zeta}$$

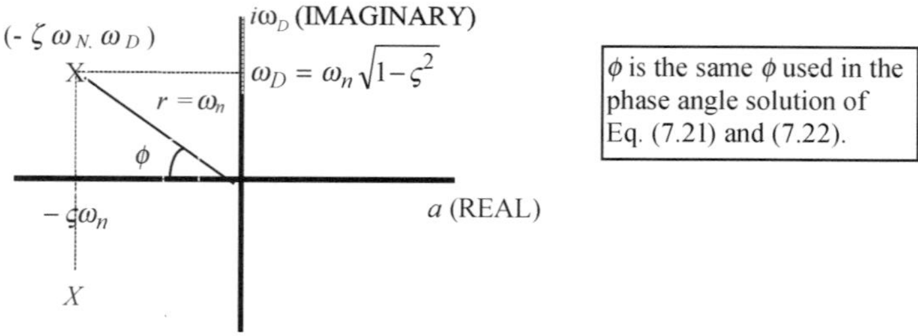

Fig. 7.9 Root representation using the complex plane.

Therefore, if we know ζ we can find ϕ for Eqs. (7.21) and (7.22). This leaves only C_4 to be evaluated. To do this we can assume the initial conditions of $x(0) = \dot{x}(0) = 0$, and use Eq. (7.22).

$$x(0) = y + yC_4(1)\sin\phi = 0$$

to obtain

$$C_4 = -\frac{1}{\sin\phi}$$

Also, we know that

$$\dot{x}(t) = -C_4\zeta\omega_N e^{-\zeta\omega_N t}\sin(\omega_D t + \phi) + C_4\omega_D e^{-\zeta\omega_N t}\cos(\omega_D t + \phi)$$

$$\dot{x}(0) = -C_4\zeta\omega_N \sin\phi + C_4\omega_D\cos\phi = 0$$

$$\frac{\sin\phi}{\cos\phi} = \frac{\omega_D}{\zeta\omega_N} = \frac{\omega_N\sqrt{1-\zeta^2}}{\zeta\omega_N} = \frac{\sqrt{1-\zeta^2}}{\zeta} = \tan\phi$$

This is the same result as found from the complex plane trigonometry, which proves that the ϕs are the same.

Because $C_4 = -1/\sin\phi$ and $\sin\phi = \sqrt{1-\zeta^2}$,

$$C_4 = -\frac{1}{\sqrt{1-\zeta^2}}$$

Therefore, for an underdamped second-order system with a step input of magnitude y, the time response is (assuming 0 I.C.s)

$$x(t) = y\left(1 - \frac{e^{-\zeta\omega_N t}}{\sqrt{1-\zeta^2}}\sin(\omega_D t + \phi)\right) \tag{7.24}$$

This is much nicer than solving for the transient and steady-state solutions and evaluating the constants as in Example 7.2. All we need to determine from the original differential equation is ω_n, ζ, and y. Radians are compatible units for ϕ in Eq. (7.24) because $\omega_D t$ will have units of radians. Use of Eq. (7.24) is illustrated in Example 7.4.

Example 7.4

Find the time response of the following differential equation with zero initial conditions:

$$\ddot{x} + 5\dot{x} + 25x = 25,$$
$$\omega_N = 5 \text{ rad/s},$$
$$\zeta = 0.5,$$
$$y = 1$$
$$\omega_D = 4.33 \text{ rad/s from Example 7.3}$$
$$\phi = \cos^{-1}(\zeta) = \cos^{-1}(0.5) = 60 \text{ deg} = \frac{\pi}{3}$$

substituting into (Eq. 7.24) yields

$$x(t) = 1 - 1.155e^{-2.5t} \sin\left(4.33t + \frac{\pi}{3}\right)$$

Notice the difference in effort required between Example 7.2 and 7.4. Remember that the solution using Eq. (7.24) is for a system with $-1 < \zeta < 1$ and a step input of magnitude y. If $\zeta \geq 1$ or $\zeta \leq -1$, the responses are aperiodic (exponential without oscillations) as discussed in Cases I and II.

The **stability** of a system can be determined directly from looking at the roots of the differential equation. If the root is real and has a negative value, it is stable [for example, $P_1 = -2$ and $x(t) = c_1 e^{-2t}$, which decays to 0 as time goes to infinity]. If the root is real and positive, the response is unstable [$P_1 = 2 \Rightarrow x(t) = C_1 e^{2t}$, which grows without bounds with time]. For complex roots, it is the real part of the root that determines stability. For $P_{1,2} = a \pm ib$, if $a < 0$ the system is stable. In the form $P_{1,2} = -\zeta\omega_N \pm i\omega_D$ this occurs when $\zeta > 0$. If $a > 0$, the system is unstable, which occurs when $\zeta < 0$. Therefore, if the roots occur to the left of the imaginary axis (the left half of the complex plane) the system is stable. Similarly, if the roots are to the right of the imaginary axis ($\zeta = 0$), the system is unstable. If the roots are on the imaginary axis ($\zeta = 0$), the system is neutrally stable (undamped). Time response characteristics for an impulse input are illustrated in Fig. 7.10 for various root locations. Notice the changes in damped frequency and time constant.

Examples 7.5 and 7.6 further illustrate our simplified approach for solving second-order linear differential equations. Remember that the general form of the differential equation is

$$\ddot{x} + 2\zeta\omega_N\dot{x} + \omega_N^2 x = \omega_N^2 y$$

Depending on the value of ζ, the solution will be in the form of Case 1, Case 2, or Case 3.

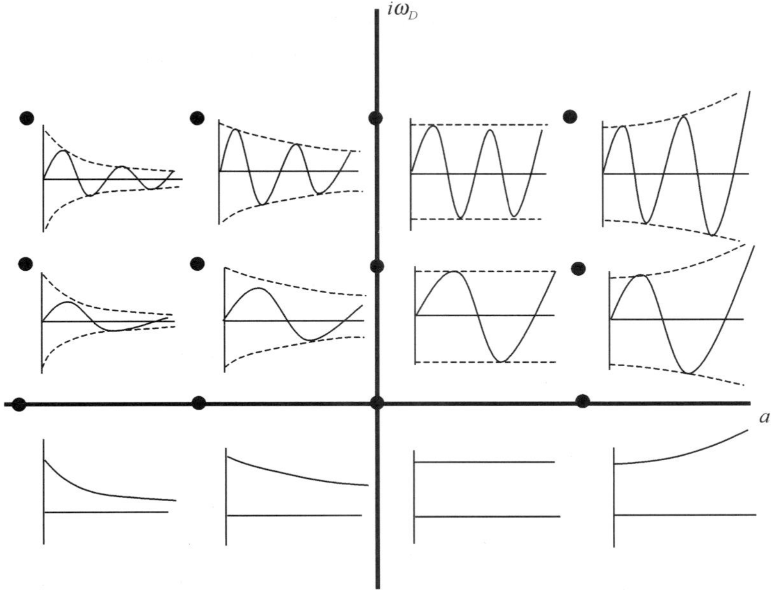

Fig. 7.10 Influence of complex plane root location on the transient response to an impulse input.

Example 7.5

Find the time response for the following system:

$$\ddot{x} + 10\dot{x} + 16x = 32; \quad x(0) = 0; \quad \dot{x}(0) = 0$$

Applying the general form of a second-order differential equation

$$\ddot{x} + 2\zeta\omega_N\dot{x} + \omega_N^2 x = \omega_N^2 y$$

we have

$$\omega_n = \sqrt{16} = 4 \text{ rad/s}$$
$$2\zeta\omega_N = 10 \Rightarrow 2\zeta(4) = 10 \Rightarrow \zeta = 1.25$$

Because $\zeta > 1$ we know that the solution is of the form of Case 1 (2 real unequal roots). Therefore, we need to solve for the roots:

$$P_{1,2} = \frac{-10 \pm \sqrt{10^2 - 4(1)(16)}}{2(1)} = -5 \pm 3 = -2, \ -8$$

The homogeneous (transient solution) is:

$$x_{\text{trans}}(t) = C_1 e^{-2t} + C_2 e^{-8t}$$

To find the steady-state solution, assume $x_{ss}(t) = A$

$$\overset{0}{\cancel{\ddot{A}}} + \overset{0}{\cancel{10\dot{A}}} + 16A = 32 \Rightarrow A = 2$$

We then have

$$x(t) = x_{ss}(t) + x_{\text{trans}}(t) = 2 + C_1 e^{-2t} + C_2 e^{-8t}$$

and

$$\begin{aligned}
&x(0) = 0 = 2 + C_1 + C_2 \\
&\dot{x}(t) = -2C_1 e^{-2t} - 8C_2 e^{-8t} \\
&\dot{x}(0) = 0 = -2C_1 - 8C_2 \Rightarrow C_1 = -4C_2 \\
&2 - 4C_2 + C_2 = 0 \Rightarrow C_2 = \frac{2}{3} \\
&C_1 = -\frac{8}{3}
\end{aligned}$$

Substituting back in, the time response becomes

$$x(t) = 2 - \frac{8}{3} e^{-2t} + \frac{2}{3} e^{-8t}$$

This response has two time constants:

$$\begin{aligned}
\tau_1 &= -\frac{1}{P_1} = -\frac{1}{-2} = 0.5 \text{ s} \\
\tau_2 &= -\frac{1}{P_2} = -\frac{1}{-8} = 0.125 \text{ s}
\end{aligned}$$

Example 7.6

Find the time response for the following system:

$$\ddot{x} + 5\dot{x} + 25x = 75; \quad x(0) = 0, \quad \dot{x}(0) = 0$$

Using the general form of a second-order differential equation, we have

$$\omega_n = \sqrt{25} = 5 \text{ rad/s}; \quad y = 3 \text{ because } \omega_n^2 = 25$$
$$2\zeta\omega_N = 5 \Rightarrow 2\zeta(5) = 5 \Rightarrow \zeta = 0.5$$
$$\omega_D = \omega_n\sqrt{1-\zeta^2} = 5\sqrt{1-(0.5)^2} = 4.333 \text{ rad/s}$$
$$\phi = \cos^{-1}\zeta = \cos^{-1}(0.5) = 60 \text{ deg} = \frac{\pi}{3}$$

Because $-1 < \zeta < 1$ and the input is a step input of magnitude 3 ($y = 3$), the solution is of the form presented in Eq. (7.24)

$$x(t) = y\left[1 - \frac{e^{-\zeta\omega_N t}}{\sqrt{1-\zeta^2}}\sin(\omega_D t + \phi)\right] \qquad \text{(Case 3)}$$

Substituting in the appropriate values

$$x(t) = 3\left[1 - \frac{e^{-(0.5)(5)t}}{\sqrt{1-(0.5)^2}}\sin\left(4.333t + \frac{\pi}{3}\right)\right]$$

and the solution or time response is:

$$x(t) = 3\left[1 - 1.155e^{-2.5t}\sin\left(4.333t + \frac{\pi}{3}\right)\right]$$

Recall that the roots are

$$P_{1,2} = -\zeta\omega_N \pm i\omega_D$$
$$= -2.5 \pm i(4.333)$$

which could easily be plotted on the complex plane. The time constant is:

$$\tau = \frac{1}{\zeta\omega_N} = \frac{1}{2.5} = 0.4 \text{ s}$$

As a reminder, if $\zeta = -1$ or $\zeta = 1$, then the solution would be of the form of Case 2.

7.3 Transforming the Linearized EOM to the Laplace Domain

Another common approach used in solving differential equations is that of Laplace transforms. We will begin with a short review of Laplace transform techniques and then apply these techniques to the six linearized differential equations of motion for the aircraft. The differential operator, P, discussed in Sec. 7.1.1.1, is analogous to the Laplace variable, s. The insight gained with the roots of transformed differential equations obtained using the differential operator will directly transfer to the roots obtained with the Laplace variable, s.

7.3.1 Laplace Transforms

It is assumed that the reader has gained familiarity with solution of differential equations using Laplace transforms from a previous course. It is the intent of this text to simply review highlights of the Laplace method. Simply stated, the Laplace method transforms a linear differential equation from the time domain (the derivatives are with respect to time) into an algebraic equation in the Laplace domain where the variable s is used. We will denote this Laplace transform operation with the symbol L. The methods of algebra are then used in a straightforward manner to solve for the parameter of interest. The resulting equation is transformed back to the time domain, referred to as an inverse Laplace transform operation and denoted with the symbol L^{-1} so that the time response can be obtained. By convention, small letters are used to represent functions of time and upper case letters are used to represent their Laplace transforms.

$$L[f(t)] = F(s)$$

and

$$L^{-1}[F(s)] = f(t)$$

The Laplace transform of the derivative of a function $f(t)$ is given by

$$L[\mathrm{d}f(t)/\mathrm{d}t] = sF(s) - f(0)$$

Where $f(0) = f(t)$ at $t = 0$. This is commonly called an initial condition. A second derivative is given by

$$L[\mathrm{d}^2f(t)/\mathrm{d}t^2] = s^2F(s) - sf(0) - \dot{f}(0)$$

Higher-order derivatives follow in a similar fashion.

7.3.1.1 Standardized inputs. We will be concerned with primarily two types of inputs or forcing functions: the unit impulse $[\delta(t)]$ and the unit step $[1(t)]$. The unit impulse is defined as occuring at $t = 0$, and having zero duration, infinite magnitude, and a strength of unity. When considered as an input to the aircraft, test pilots refer to an impulse as a stick rap. The aircraft is trimmed and the stick is rapidly moved forward or aft from the trimmed position and then returned to the trimmed position. This input can be thought of as basically hitting or rapping the stick and allowing it to return to the trim position. A unit impulse is illustrated in Fig. 7.11.

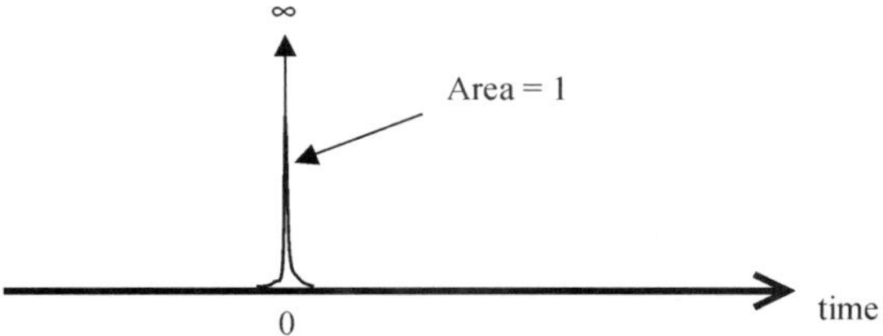

Fig. 7.11 Unit impulse.

A unit step is defined by the following:

$$0 \text{ for } t < 0; \quad 1 \text{ for } t > 0$$

A unit step is illustrated in Fig. 7.12. A test pilot will input a step input by rapidly moving the stick forward or aft from the trimmed postion and holding it. Of course, test pilot inputs only approximate the ideal unit impulse and unit step inputs. These approximations are generally sufficient to excite the dynamics and response of the aircraft, so we have a convenient way to compare the ideal world of the unit impulse and unit step to the practical world of the aircraft in flight.

The Laplace transform of a unit impulse is

$$L[\delta(t)] = 1$$

and for a unit step

$$L[1(t)] = 1/s$$

7.3.1.2 Laplace tables. To simplify transforming expressions from the time domain to the Laplace domain and the taking of the inverse transform to go from the Laplace domain back to the time domain, we will rely on a table of Laplace transforms. Such a table is presented in Appendix E, which contains most of the expressions we will need. More detailed tables are available in a variety of references.

7.3.1.3 Solving differential equations using Laplace transforms. Laplace transforms are used to solve differential equations through a three-step

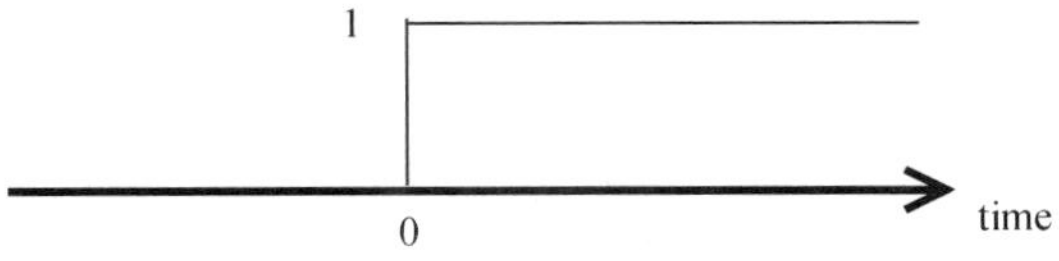

Fig. 7.12 Unit step.

process. First, the Laplace transform of the differential equation is obtained; second, the resulting equation is solved algebraically for the unknown variable/variables; and third, the inverse Laplace transform of each variable is obtained resulting in the desired time solution of the original differential equation. A simple example will illustrate this method.

Example 7.7

Find the time response for the following differential equation:

$$\ddot{x} = 6; \quad x(0) = \dot{x}(0) = 0$$

Taking the Laplace transform,

$$s^2 X(s) = 6/s$$

Solving for $X(s)$,

$$X(s) = 6/s^3$$

and taking the inverse Laplace using #8 in Appendix E,

$$x(t) = 3t^2$$

which is the time solution.

7.3.1.4 Transfer functions and the characteristic equation. A transfer function is defined as the ratio of Laplace transforms of output to input. Outputs for our applications will typically be motion variables such as u, angle of attack, and yaw rate, which describe velocities, angles, or angular rates of the aircraft. Inputs for our applications will be aircraft control surface deflections such as elevator deflection (δ_e) or aileron deflection (δ_a). In simple terms

$$\text{Transfer function} = L\left[\frac{\text{Output}}{\text{Input}}\right]$$

A transfer function will be expressed in Laplace notation and will be obtained from the Laplace form of the aircraft equations of motion. Example 7.8 illustrates how a transfer function is obtained from a differential equation and the utility it has for a variety of inputs.

Example 7.8

Find the transfer function $\phi(s)/\delta_a(s)$ for the following simplified differential equation defining roll angle response. Assume zero initial conditions.

$$\ddot{\phi} + 0.704\dot{\phi} = 0.037\delta_a$$

Taking the Laplace transform

$$s^2\phi(s) + 0.704s\phi(s) = 0.037\delta_a(s)$$

Solving for $\phi(s)$

$$\phi(s) = \frac{0.037\delta_a(s)}{s(s+0.704)}$$

and obtaining the transfer function

$$\frac{\phi(s)}{\delta_a(s)} = \frac{0.037}{s(s+0.704)}$$

For an impulse input, $\delta_a(s) = 1$

$$\phi(s) = \frac{0.037}{s(s+0.704)}$$

and, taking the inverse Laplace transform using the Laplace tables we have the time response

$$\phi(t) = \frac{0.037}{0.704}(1 - e^{-0.704t})$$

For a unit step input, $\delta_a(s) = 1/s$

$$\phi(s) = \frac{0.037}{s^2(s+0.704)}$$

and the time response is

$$\phi(t) = \frac{0.037}{(0.704)^2}(0.704t - 1 + e^{-0.704t})$$

The **characteristic equation** of a transfer function is obtained by setting the polynomial in the denominator of the transfer function equal to zero. For Example 7.8, the characteristic equation is

$$s(s+0.704) = 0 \tag{7.25}$$

The roots of a characteristic equation will define the overall system dynamic characteristics such as time constant (for first-order systems), and damping ratio and natural frequency (for second-order systems) as discussed in Sec. 7.1.2. For Eq. (7.25), the root $s = 0$ leads to a steady state value term in the time solution, and the root $s = -0.704$ leads to a time constant of $1/0.704$ s and a $e^{-0.704t}$ term in the time solution.

7.3.1.5 Partial fraction expansion. A key step in using Laplace transforms to solve differential equations involves using the Laplace transform tables to find inverse transforms. It is impossible to cover every potential case in a table, but the method of partial fractions can aid in breaking up involved transfer functions into pieces, which may be included in a standard table. This text will only review the case of a transfer function with nonrepeated real roots in the characteristic equation. A more detailed coverage of partial fraction expansion techniques is included in most engineering mathematics textbooks such as *Advanced Engineering Mathematics*, by Erwin Kreyszig.

Consider the transfer function:

$$\frac{X(s)}{Y(s)} = \frac{s+2}{s^3+8s^2+19s+12} = \frac{s+2}{(s+3)(s+4)(s+1)}$$

It can be rewritten using a partial fraction expansion as

$$\frac{X(s)}{Y(s)} = \frac{A_1}{s+3} + \frac{A_2}{s+4} + \frac{A_3}{s+1}$$

To evaluate the constants A_1, A_2, and A_3, which are also called residues, we use the following approach:

$$A_1 = \left.\frac{s+2}{(s+4)(s+1)}\right|_{s=-3} = \frac{1}{2}$$

$$A_2 = \left.\frac{s+2}{(s+3)(s+1)}\right|_{s=-4} = -\frac{2}{3}$$

$$A_3 = \left.\frac{s+2}{(s+3)(s+4)}\right|_{s=-1} = \frac{1}{6}$$

The partial fraction representation of the transfer function then becomes

$$\frac{X(s)}{Y(s)} = \frac{\frac{1}{2}}{s+3} + \frac{-\frac{2}{3}}{s+4} + \frac{\frac{1}{6}}{s+1}$$

Each of the partial fraction expressions can be transformed to the time domain using the common first-order transform found in all Laplace tables.

$$L^{-1}\left(\frac{K}{s+a}\right) = Ke^{-at}$$

For this example, we then have

$$\frac{x(t)}{y(t)} = \frac{1}{2}e^{-3t} - \frac{2}{3}e^{-4t} + \frac{1}{6}e^{-t}$$

The residues (A_1, A_2, and A_3) also provide a weighting of the relative magnitude of each component of the response.

Slightly modified partial fraction approaches are defined in mathematics texts for characteristic equation roots, which are repeated real numbers, complex conjugates, and a repeated pair of complex conjugates.

7.3.1.6 Initial and final value theorems. The initial value of a function (at $t = 0$) can be found if its Laplace transform is known using the following theorem:

$$f(t)|_{t\to 0} = \lim|_{s\to\infty} sF(s)$$

For example, consider the function

$$f(t) = e^{-3t}$$
$$L(e^{-3t}) = \frac{1}{s+3}$$

Applying the initial value theorem,

$$\lim|_{t\to 0} e^{-3t} = \lim|_{s\to\infty} s\left(\frac{1}{s+3}\right) = 1 = e^{-3(0)}$$

This theorem is useful in verifying the accuracy of Laplace transforms because the initial conditions are normally known.

The steady-state ($t \to \infty$) value of a time domain function can be found if its Laplace transform is known and if it has a finite steady-state value using the following theorem:

$$\lim|_{t\to\infty} f(t) = \lim|_{s\to 0} sF(s)$$

This theorem does not apply to unstable functions or undamped sinusoidal functions. As an example of application of this theorem, consider the Laplace transform:

$$X(s) = \frac{1}{s(s+1)}$$

To find the steady-state value of x at $t = \infty$, we can apply the final value theorem:

$$\lim|_{t\to\infty} x(t) = \lim|_{s\to 0} s\left(\frac{1}{s(s+1)}\right) = 1 = x(\infty)$$

This result can be checked using

$$L^{-1}\left(\frac{1}{s(s+1)}\right) = 1 - e^{-t} = x(t)$$

and

$$x(\infty) = 1 - e^{-\infty} = 1$$

The final value value theorem provides an easy method to find the steady-state value of a Laplace expression.

7.3.2 Longitudinal Linearized EOM in Laplace Form

We will now use the power of Laplace transforms to recast the aircraft EOM developed in Sec. 6.5. The equations become somewhat long but the concepts are not complex. Equation (6.127), the longitudinal linearized differential EOM for the aircraft, are repeated for reference.

$$\begin{aligned} \dot{u} &= -g\theta\cos\Theta_1 + X_u u + X_{T_u} u + X_\alpha \alpha + X_{\delta_e}\delta_e \\ \dot{w} - U_1 q &= -g\theta\sin\Theta_1 + Z_u u + Z_\alpha \alpha + Z_{\dot{\alpha}}\dot{\alpha} + Z_q q + Z_{\delta_e}\delta_e \\ \dot{q} &= M_u u + M_{T_u} u + M_\alpha \alpha + M_{T_\alpha}\alpha + M_{\dot{\alpha}}\dot{\alpha} + M_q q + M_{\delta_e}\delta_e \end{aligned} \tag{7.26}$$

We will take the Laplace transform of these equations, but first it is important to note that these three EOM have five aircraft motion variables (u, θ, α, w, and q) and δ_e. Because we only have three defining equations, we need to reduce this down to three motion variables and δ_e becomes the input or forcing function for the system We will use the kinematic relations and the approximation for angle of attack, α, to reduce to the three motion variables of α, u, and θ.

From the kinematic equations [Eq. (4.80)] and the assumption of initial trimmed flight with the wings level condition

$$q = \dot{\theta}, \quad \dot{q} = \ddot{\theta}$$

Also, for small perturbations

$$\alpha \approx \frac{w}{U_1} \Rightarrow w = \alpha U_1 \quad \text{and} \quad \dot{w} = \dot{\alpha} U_1$$

Therefore, our aircraft motion variables are reduced to α, u, and θ. These should be thought of as the outputs for our system of differential equations.

With zero initial conditions, the Laplace transform of Eq. (7.26) yields

$$
\begin{aligned}
&su(s) = -g\theta(s)\cos\Theta_1 + X_u u(s) + X_{T_u} u(s) + X_\alpha \alpha(s) + X_{\delta_e}\delta_e(s) \\
&sU_1\alpha(s) - U_1 s\theta(s) = -g\theta(s)\sin\Theta_1 + Z_u u(s) + Z_\alpha \alpha(s) + Z_{\dot\alpha} s\alpha(s) + Z_q s\theta(s) \\
&\qquad + Z_{\delta_e}\delta_e(s) \\
&s^2\theta(s) = M_u u(s) + M_{T_u} u(s) + M_\alpha \alpha(s) + M_{T_\alpha}\alpha(s) + M_{\dot\alpha} s\alpha(s) + M_q s\theta(s) + M_{\delta_e}\delta_e(s)
\end{aligned}
$$

Combining terms yields

$$
\begin{aligned}
&(s - X_u - X_{T_u})u(s) - X_\alpha \alpha(s) + g\cos\Theta_1\theta(s) = X_{\delta_e}\delta_e(s) \\
&- Z_u u(s) + [(U_1 - Z_{\dot\alpha})s - Z_\alpha]\alpha(s) + [-(Z_q - U_1)s + g\sin\Theta_1]\theta(s) = Z_{\delta_e}\delta_e(s) \\
&- (M_u + M_{T_u})u(s) - [M_{\dot\alpha} s + M_\alpha + M_{T_\alpha}]\alpha(s) + (s^2 - M_q s)\theta(s) = M_{\delta_e}\delta_e(s)
\end{aligned}
$$

Notice at this point that we have moved the terms with δ_e (elevator deflection) to the right-hand side of the equal sign because δ_e is the forcing function (or input) for each of the three differential equations. In matrix form this yields

$$
\begin{bmatrix}
(s - X_u - X_{T_u}) & -X_\alpha & g\cos\Theta_1 \\
-Z_u & [s(U_1 - Z_{\dot\alpha}) - Z_\alpha] & [-(Z_q + U_1)s + g\sin\Theta_1] \\
-(M_u + M_{T_u}) & -[M_{\dot\alpha} s + M_\alpha + M_{T_\alpha}] & (s^2 - M_q s)
\end{bmatrix}
\begin{bmatrix} u(s) \\ \alpha(s) \\ \theta(s) \end{bmatrix}
= \begin{bmatrix} X_{\delta_e} \\ Z_{\delta_e} \\ M_{\delta_e} \end{bmatrix}\delta_e(s)
$$

In terms of the transfer functions $\dfrac{u(s)}{\delta_e(s)}, \dfrac{\alpha(s)}{\delta_e(s)}$, and $\dfrac{\theta(s)}{\delta_e(s)}$, we have

$$
\begin{bmatrix}
(s - X_u - X_{T_u}) & -X_\alpha & g\cos\Theta_1 \\
-Z_u & [s(U_1 - Z_{\dot\alpha}) - Z_\alpha] & [-(Z_q + U_1)s + g\sin\Theta_1] \\
-(M_u + M_{T_u}) & -[M_{\dot\alpha} s + M_\alpha + M_{T_\alpha}] & (s^2 - M_q s)
\end{bmatrix}
\begin{bmatrix} \dfrac{u(s)}{\delta_e(s)} \\[2mm] \dfrac{\alpha(s)}{\delta_e(s)} \\[2mm] \dfrac{\theta(s)}{\delta_e(s)} \end{bmatrix}
= \begin{bmatrix} X_{\delta_e} \\ Z_{\delta_e} \\ M_{\delta_e} \end{bmatrix}
\tag{7.27}
$$

and each of the three longitudinal transfer functions can be determined using Cramer's rule as presented in Appendices F and G. It is important at this point not to lose sight of what we have developed. Each of these transfer functions can be represented as the ratio of two polynomials in the Laplace variables.

$$\frac{u(s)}{\delta_e(s)} = \frac{A_u s^3 + B_u s^2 + C_u s + D_u}{Es^4 + Fs^3 + Gs^2 + Hs + I} \tag{7.28}$$

$$\frac{\alpha(s)}{\delta_e(s)} = \frac{A_\alpha s^3 + B_\alpha s^2 + C_\alpha s + D_\alpha}{Es^4 + Fs^3 + Gs^2 + Hs + I} \tag{7.29}$$

$$\frac{\theta(s)}{\delta_e(s)} = \frac{A_\theta s^2 + B_\theta s + C_\theta}{Es^4 + Fs^3 + Gs^2 + Hs + I} \tag{7.30}$$

Notice that all three longitudinal transfer functions have the same input (δ_e) and the same denominator. There is a separate transfer function for each of our three longitudinal motion variables (u, α, and θ). Also, each of these transfer functions has the same characteristic equation:

$$Es^4 + Fs^3 + Gs^2 + Hs + I = 0$$

Recall that *the characteristic equation determines the dynamic stability characteristics* of the response, and therefore all three transfer functions will have the same dynamic characteristics (parameters such as ζ, ω_n, and τ). Notice also that the numerator of each transfer function is different. Each numerator coefficient is designated by an A, B, C, or D with a subscript appropriate to its respective transfer function. The numerator affects the magnitude of the response, and therefore each motion variable will have a different magnitude of response but with the same dynamic characteristics.

7.3.2.1 Three-degree-of-freedom analysis of longitudinal modes of motion. The preceding development included the three motion variables u, α, and θ. This analysis may also be termed a three-degree-of-freedom (3 DOF) determination of the longitudinal transfer functions. Normally, with the help of root solvers such as those available in MATLAB® (a registered trademark of The MathWorks, Inc.), the fourth order characteristic equation for longitudinal motion can be written as the product of two second-order (oscillatory) polynomials.

$$(s^2 + 2\zeta_{SP}\omega_{N_{SP}}s + \omega_{N_{SP}}^2)(s^2 + 2\zeta_{PH}\omega_{N_{PH}}s + \omega_{N_{PH}}^2) = 0 \tag{7.31}$$

The subscript *SP* refers to the short period mode and the subscript *PH* refers to the phugoid mode. All airplanes have these two longitudinal dynamic modes. Each of these polynomials can be thought of as a separate characteristic equation that defines the dynamic characteristics of its respective dynamic mode.

The coefficients (and roots) of each characteristic equation change with flight condition, airplane mass, mass distribution, airplane geometry, and aerodynamic characteristics. These changes translate to changes in ω_n and ζ, but the fundamental presence of the short period and phugoid modes is maintained.

The **short period mode** is characterized by **complex conjugate roots** with a moderate to relatively high damping ratio and relatively high natural frequency and damped frequency (short period). It is easily demonstrated by first trimming the aircraft and then disturbing it from trim with a forward-aft-neutral pitch stick input (commonly called a doublet). The resulting response back to trim may be either first order (exponential decay) or second order (oscillatory). Significant variations in the angle of attack (α), and pitch attitude (θ) longitu-dinal motion variables occur while the airspeed (u) motion variable remains fairly constant. Trim is generally regained in a few seconds, thus the descriptive name short period and the small variation in airspeed. Typical time histories for the short period response of a fighter aircraft to a doublet input are presented in Fig. 7.13.

Notice that the response is second order (oscillatory) and that u remains fairly constant. Oscillations of larger magnitude are observed with α and θ. Notice also that the response is stable.

The **phugoid mode** is characterized by **complex conjugate roots** with a relatively low damping ratio and natural/damped frequency (long period). It is demonstrated by trimming the aircraft in level flight, then inputting aft stick for approximately 2–3 s, bleeding off some airspeed, and then returning the stick to the neutral (trimmed) position. The resulting response is usually oscillatory with significant variations in pitch attitude and airspeed, while angle of attack remains relatively constant. The phugoid has been described as an up and down roller coaster oscillation in the sky that trades off kinetic and potential energy. As the oscillation starts, airspeed decreases while the airplane gains altitude (pitch angle is positive). The aircraft then begins to lose altitude, and airspeed increases while the pitch angle decreases. This is followed by the aircraft pulling up gradually and returning to the climb portion of the phugoid oscillation. The period for the phugoid is typically quite long (somewhere

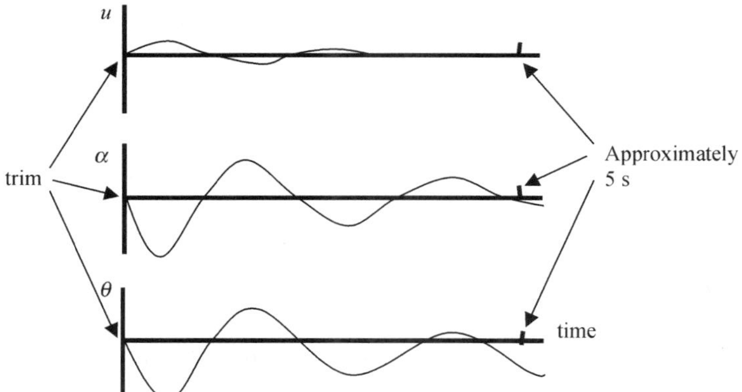

Fig. 7.13 u, α, and θ time history plots illustrating the short period mode.

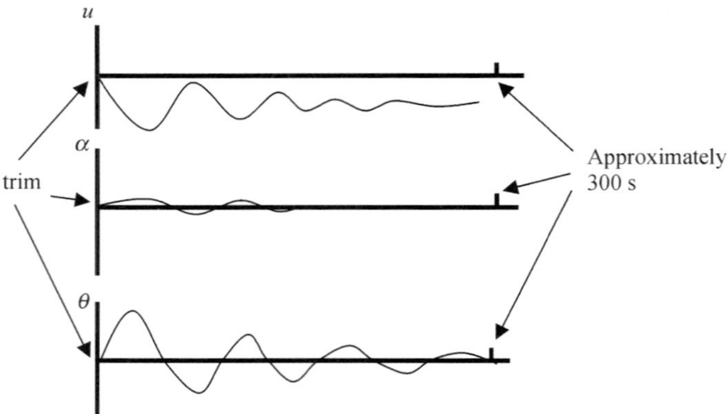

Fig. 7.14 u, α, and θ time history plots for the phugoid mode.

between 30 and 120 s). Typical time histories for the phugoid mode are presented in Fig. 7.14.

Notice that the response is also second order for the phugoid and that angle of attack remains relatively constant. The frequency of the oscillation for the phugoid is much lower than that observed for the short period mode. The phugoid as shown in Fig. 7.14 is stable, but this may not be the case for all flight conditions.

Example 7.9

The α/δ_e transfer function for a T-37 cruising at 30,000 ft and 0.46 Mach follows. Find the natural frequency, damping ratio, damped frequency, and time constant for the short period and phugoid modes.

$$\frac{\alpha}{\delta_e} = -0.0924\frac{(s + 336.1)(s^2 + 0.0105s + 0.0097)}{(s^2 + 4.58s + 21.6)(s^2 + 0.0098s + 0.0087)}$$

We go immediately to the two characteristic equations

$$s^2 + 4.58s + 21.6 = 0$$
$$s^2 + 0.0098s + 0.0087 = 0$$

The natural frequency (ω_n) for the first characteristic equation is

$$\omega_n = \sqrt{21.6} = 4.65 \text{ rad/s}$$

and for the second equation

$$\omega_n = \sqrt{0.0087} = 0.0933\,\text{rad/s}$$

We can identify the first characteristic equation as being for the short period mode because of the higher natural frequency. The phugoid dynamics are contained in the second characteristic equation. Thus

$$\omega_{n_{sp}} = 4.65\ \text{rad/s}$$
$$\omega_{n_{ph}} = 0.0933\ \text{rad/s}$$

and

$$\zeta_{sp} = \frac{4.58}{2\omega_{n_{sp}}} = 0.493$$
$$\zeta_{ph} = \frac{0.0098}{2\omega_{n_{ph}}} = 0.0525$$

Notice that both the short period and phugoid responses will be second order because ζ is less than 1. Both responses are stable because $\zeta\omega_n$ is positive for each mode. The damped frequency is

$$\omega_D = \omega_n\sqrt{1-\zeta^2}$$
$$\omega_{D_{sp}} = \omega_{n_{sp}}\sqrt{1-\zeta_{sp}^2} = 4.65\sqrt{1-(0.493)^2} = 4.046\ \text{rad/s}$$
$$\omega_{D_{ph}} = \omega_{n_{ph}}\sqrt{1-\zeta_{ph}^2} = 0.0933\sqrt{1-(0.0525)^2} = 0.0932\ \text{rad/s}$$

and for the time constant

$$\tau_{sp} = \frac{1}{\zeta_{sp}\omega_{n_{sp}}} = \frac{1}{(0.493)(4.65)} = 0.436\ \text{s}$$
$$\tau_{ph} = \frac{1}{\zeta_{ph}\omega_{n_{ph}}} = \frac{1}{(0.0525)(0.0933)} = 204.2\ \text{s}$$

7.3.2.2 Two-degree-of-freedom short period approximation. To gain insight into the stability parameters and derivatives that influence the dynamic characteristics of the short period mode, we will look at a two-degree-of-freedom (2 DOF) approximation. This is a solution in which the motion is constrained to two motion variables rather than three. Recalling our discussion on the short period mode, we will make the simplifying assumption that ***u* remains near zero** and can be removed from Eq. (7.27). With this assumption and the **elimination of the *x*-force equation** (which is assumed to have a negligible effect if u is approximately constant), we retain the z-force equation and the pitching moment equation along with the motion variables α and θ.

$$\begin{bmatrix} \cancel{(s - X_u - X_{T_u})} & \cancel{-X_\alpha} & \cancel{g \cos\Theta_1} \\ \cancel{-Z_u} & [s(U_1 - Z_{\dot{\alpha}}) - Z_\alpha] & [-(Z_q + U_1)s + g \ \sin\Theta_1] \\ \cancel{-(M_u + M_{T_u})} & -[M_{\dot{\alpha}}s + M_\alpha + M_{T_\alpha}] & (s^2 - M_q s) \end{bmatrix} \begin{bmatrix} \cancel{u(s)} \\ \alpha(s) \\ \theta(s) \end{bmatrix}$$
$$= \begin{bmatrix} \cancel{X_{\delta_e}} \\ Z_{\delta_e} \\ M_{\dot{\delta}_e} \end{bmatrix} \delta_e(s)$$

Equation (7.27) becomes

$$\begin{bmatrix} [s(U_1 - Z_{\dot{\alpha}}) - Z_\alpha] & [-(Z_q + U_1)s + g \sin\Theta_1] \\ -[M_{\dot{\alpha}}s + M_\alpha + M_{T_\alpha}] & (s^2 - M_q s) \end{bmatrix} \begin{bmatrix} \alpha(s) \\ \theta(s) \end{bmatrix} = \begin{bmatrix} Z_{\delta_e} \\ M_{\delta_e} \end{bmatrix} \delta_e(s) \tag{7.32}$$

We will next focus on the dynamic characteristics and the characteristic equation (which is the determinant of the first coefficient matrix). We will look at the short period approximation assuming that $Z_{\dot{\alpha}} = Z_q = \Theta_1 = M_{T_\alpha} = 0$ because these terms are generally small compared to the others.

Equation (7.32) becomes

$$\begin{bmatrix} sU_1 - Z_\alpha & -U_1 s \\ -[M_{\dot{\alpha}}s + M_\alpha] & s^2 - M_q s \end{bmatrix} \begin{bmatrix} \alpha(s) \\ \theta(s) \end{bmatrix} = \begin{bmatrix} Z_{\delta_e} \\ M_{\delta_e} \end{bmatrix} \delta_e(s)$$

The characteristic equation is

$$(sU_1 - Z_\alpha)(s^2 - M_q s) - (-U_1 s)(-[M_{\dot{\alpha}}s + M_\alpha]) = 0$$

or

$$sU_1\left[s^2 - \left(M_q + \frac{Z_\alpha}{U_1} + M\dot{\alpha}\right)s + \left(\frac{Z_\alpha M_q}{U_1} - M_\alpha\right)\right] = 0$$

and in simplified form

$$s^2 - \left(M_q + \frac{Z_\alpha}{U_1} + M\dot{\alpha}\right)s + \left(\frac{Z_\alpha M_q}{U_1} - M_\alpha\right) = 0 \tag{7.33}$$

For this 2-DOF approximation we can find $\alpha(s)/\delta_e(s)$ and $\theta(s)/\delta_e(s)$ using Cramer's rule as before.

$$\frac{\alpha(s)}{\delta_e(s)} = \frac{Z_{\delta_e}s + (M_{\delta_e}U_1 - M_q Z_{\delta_e})}{U_1\left[s^2 - \left(M_q + \frac{Z_\alpha}{U_1} + M_{\dot{\alpha}}\right)s + \left(\frac{Z_\alpha M_q}{U_1} - M_\alpha\right)\right]}$$

$$\frac{\theta(s)}{\delta_e(s)} = \frac{(U_1 M_{\delta_e} + Z_{\delta_e}M_{\dot{\alpha}})s + (M_\alpha Z_{\delta_e} - Z_\alpha M_{\delta_e})}{sU_1\left[s^2 - \left(M_q + \frac{Z_\alpha}{U_1} + M_{\dot{\alpha}}\right)s + \left(\frac{Z_\alpha M_q}{U_1} - M_\alpha\right)\right]} \tag{7.34}$$

An approximation of natural frequency and damping ratio can then be determined using Eq. (7.33).

$$\omega_{n_{SP}} \approx \sqrt{\frac{Z_\alpha M_q}{U_1} - M_\alpha} \tag{7.35}$$

$$\zeta_{SP} \approx \frac{-\left(M_q + \frac{Z_\alpha}{U_1} + M_{\dot{\alpha}}\right)}{2\omega_{n_{SP}}} \tag{7.36}$$

Typically, $-M_\alpha$ is much larger that $Z_\alpha M_q/U_1$ (as long as the c.g. is not too far aft). This results in

$$\omega_{n_{sp}} \approx \sqrt{-M_\alpha} = \sqrt{\frac{-C_{m_\alpha}\bar{q}_1 S\bar{c}}{I_{yy}}} \tag{7.37}$$

The following insights can be observed from Eq. (7.37) for the short period mode natural frequency: 1) $\omega_{n_{SP}}$ will increase as static longitudinal stability ($-C_{m_\alpha}$) increases or as the distance between the c.g. and the aircraft AC increases; 2) $\omega_{n_{SP}}$ will increase as dynamic pressure ($\bar{q}_1$) increases; and 3) $\omega_{n_{SP}}$ will decrease as the pitching moment of inertia (I_{yy}) increases.

Equation (7.36) also leads to important insights for the short period damping ratio: 1) M_q, the pitch damping derivative, is the driving term, because 2) Z_α/U_1 is generally driven by other requirements, and 3) $M_{\dot{\alpha}}$ is generally driven by the same design features (horizontal tail size and the distance from the c.g. to the AC of the tail) as M_q, and $M_{\dot{\alpha}}$ is typically about one-third the value of M_q. One of the limitations of this approximation for damping ratio is that it assumes that ζ is positive (a stable case), which is not always true. It is recommended that unstable cases be analyzed using a 3-DOF solution.

7.3.2.3 Two-degree-of-freedom phugoid approximation. As with the short period approximation, we will look at a 2-DOF approximation for the phugoid mode to gain insight into the parameters that influence dynamic characteristics. For the phugoid approximation, we will assume that **α is constant** with u and θ (or q) as the motion variables. We can eliminate the $\alpha(s)$ terms and the moment equation in Eq. (7.27) to yield two equations with two motion variables.

$$\begin{bmatrix} s - X_u - X_{T_u} & \cancel{-X_\alpha} & g\cos\Theta_1 \\ -Z_u & \cancel{[s(U_1 - Z_{\dot{\alpha}}) - Z_\alpha]} & [-(Z_q + U_1)s + g\sin\Theta_1] \\ \cancel{-(M_u + M_{T_u})} & \cancel{-[M_{\dot{\alpha}}s + M_\alpha + M_{T_\alpha}]} & \cancel{(s^2 - M_q s)} \end{bmatrix} \begin{bmatrix} u(s) \\ \cancel{\alpha(s)} \\ \theta(s) \end{bmatrix} = \begin{bmatrix} X_{\delta_e} \\ Z_{\delta_e} \\ \cancel{M_{\delta_e}} \end{bmatrix} \delta_e(s)$$

or

$$\begin{bmatrix} s - X_u - X_{T_u} & g\cos\Theta_1 \\ -Z_u & [-(Z_q + U_1)s + g\sin\Theta_1] \end{bmatrix} \begin{bmatrix} u(s) \\ \theta(s) \end{bmatrix} = \begin{bmatrix} X_{\delta_e} \\ Z_{\delta_e} \end{bmatrix} \delta_e(s)$$

If we assume $X_{\delta_e} = Z_q = \Theta_1 \approx 0$ we get

$$\begin{bmatrix} s - X_u - X_{T_u} & g \\ -Z_u & -U_1 s \end{bmatrix} \begin{bmatrix} u(s) \\ \theta(s) \end{bmatrix} = \begin{bmatrix} 0 \\ Z_{\delta_e} \end{bmatrix} \delta_e(s) \tag{7.38}$$

The characteristic equation becomes

$$\begin{aligned} (s - X_u - X_{T_u})(-U_1 s) + gZ_u &= 0 \\ -U_1\left[s^2 - (X_u + X_{T_u})s - \frac{Z_u}{U_1}g\right] &= 0 \end{aligned} \tag{7.39}$$

We then have

$$\omega_{n_{ph}} \approx \sqrt{\frac{-Z_u g}{U_1}} = \sqrt{\frac{-g}{U_1}\left[\frac{-(\bar{q}_1 S)(C_{L_u} + 2C_{L_1})}{mU_1}\right]}$$

or

$$\omega_{n_{ph}} \approx \sqrt{\frac{g(\bar{q}_1 S)(C_{L_u} + 2C_{L_1})}{U_1^2 m}}$$

Typically $C_{L_u} \ll 2C_{L_1}$ and $C_{L_1} = mg/\bar{q}_1 S$. With these assumptions we have

$$\omega_{n_{ph}} \approx \sqrt{\frac{g(\bar{q}_1 S)}{U_1^2 m}\left[\frac{2mg}{\bar{q}_1 S}\right]} = \sqrt{\frac{2g^2}{U_1^2}}$$

or

$$\omega_{n_{ph}} \approx \frac{g}{U_1}\sqrt{2} \tag{7.40}$$

Therefore, we can observe that the natural frequency of the phugoid mode is approximately inversely proportional to the forward velocity, U_1.

Returning to the characteristic equation [Eq. (7.39)], we can define an approximation for the phugoid damping ratio:

$$\zeta_{ph} \approx \frac{-(X_u - X_{T_u})}{2\omega_{n_{ph}}}$$

Because

$$X_u = -\frac{(C_{D_u} + 2C_{D_1})\bar{q}_1 S}{mU_1} \qquad \text{and} \qquad X_{T_u} = \frac{(C_{T_{x_u}} + 2C_{T_{x_1}})\bar{q}_1 S}{mU_1}$$

We can substitute these values into the expression for ζ_{ph}

$$\zeta_{ph} \approx \frac{(C_{D_u} + 2C_{D_1} - C_{T_{x_u}} - 2C_{T_{x_1}})\bar{q}_1 S}{2mU_1\omega_{n_{ph}}} \tag{7.41}$$

Equation (7.41) provides us with an approximation for the phugoid damping ratio. To gain a little more insight, we will look at the case of unpowered or gliding flight where

$$C_{T_{x_1}} = C_{T_{x_u}} = 0$$

With this assumption

$$\zeta_{ph} = \frac{(C_{D_u} + 2C_{D_1})\bar{q}_1 S}{2mU_1\omega_{n_p}} = \frac{(C_{D_u} + 2C_{D_1})\bar{q}_1 S U_1}{2mU_1 g\sqrt{2}}$$

$$\zeta_{ph} = \frac{(C_{D_u} + 2C_{D_1})\bar{q}_1 S}{2\sqrt{2}mg} = \frac{(C_{D_u} + 2C_{D_1})}{2\sqrt{2}}\frac{1}{C_{L_1}}$$

At this point we can make one additional assumption for low-speed flight where

$$C_{D_u} \approx 0.$$

With this additional assumption, we have

$$\zeta_{ph} \approx \frac{2C_{D_1}}{2\sqrt{2}C_{L_1}} = \frac{1}{\sqrt{2}}\frac{C_{D_1}}{C_{L_1}} \tag{7.42}$$

Equation (7.42) indicates that the phugoid damping ratio is inversely proportional to the lift to drag ratio (L/D). Of course we must keep in mind all the assumptions we made to obtain this result. It does indicate that airplanes with high values of L/D may have poor phugoid damping. If this is the case, precise control of speed becomes difficult, which can be a problem during the initial phases of a landing pattern. However, after the gear and flaps have been lowered, L/D is reduced and damping of the phugoid improves.

Example 7.10

To illustrate the concepts of transfer functions, characteristic equations, and the modes of motion, we will consider a Lear Jet flying at 0.7 Mach and 40,000 ft. The 3-DOF longitudinal transfer functions are approximated by

$$\frac{u(s)}{\delta_e(s)} = \frac{-(6.312)s^2 + (4927)s + 4302}{(675.9)s^4 + (1371)s^3 + (5459)s^2 + (86.31)s + 44.78}$$

$$\frac{\alpha(s)}{\delta_e(s)} = \frac{-(0.746)s^3 - (208.3)s^2 - (2.665)s - 1.39}{(675.9)s^4 + (1371)s^3 + (5459)s^2 + (86.31)s + 44.78}$$

$$\frac{\theta(s)}{\delta_e(s)} = \frac{-(208.1)s^2 - (136.9)s - 2.380}{(675.9)s^4 + (1371)s^3 + (5459)s^2 + (86.31)s + 44.78}$$

Find the natural frequency, damping ratio, damped frequency, time constant, and period of oscillation for the short period and phugoid modes.

The characteristic equation is found by setting the denominator of the transfer function equal to 0. The characteristic equation for the Lear Jet's longitudinal motion is

$$675.9s^4 + 1371s^3 + 5459s^2 + 86.31s + 44.78 = 0$$

or

$$s^4 + 2.0284s^3 + 8.0766s^2 + 0.1277s + 0.06625 = 0$$

Using a root solver, such as those available in MATLAB, the four roots are found to be

$$s_{1,2} = -\zeta_{SP}\omega_{N_{SP}} \pm i\omega_{D_{SP}} = -1.008 \pm i(2.651)$$
$$s_{3,4} = -\zeta_{PH}\omega_{N_{PH}} \pm i\omega_{D_{PH}} = -0.0069 \pm i(0.0905)$$

The roots with the largest ω_D are obviously associated with the short period mode of motion, while the other roots are associated with the phugoid mode of motion.

$$\zeta_{SP}\omega_{N_{SP}} = 1.008 \quad \omega_{D_{SP}} = 2.651 \text{ rad/s} \quad \text{(short period)}$$
$$\zeta_{PH}\omega_{N_{PH}} = 0.0069 \quad \omega_{D_{PH}} = 0.0905 \text{ rad/s} \quad \text{(phugoid)}$$
$$\omega_{N_{SP}} = \sqrt{(-\zeta\omega_N)^2_{SP} + \omega^2_{D_{SP}}} = \sqrt{(-1.008)^2 + (2.651)^2} = 2.836 \text{ rad/s}$$
$$\omega_{N_{PH}} = \sqrt{(-\zeta\omega_N)^2_{PH} + \omega^2_{D_{PH}}} = \sqrt{(-0.0069)^2 + (0.0905)^2} = 0.091 \text{ rad/s}$$
$$\zeta_{SP} = \frac{\zeta_{SP}\omega_{N_{SP}}}{\omega_{N_{SP}}} = \frac{1.008}{2.836} = 0.355$$
$$\zeta_{PH} = \frac{\zeta_{PH}\omega_{N_{PH}}}{\omega_{N_{PH}}} = \frac{0.0069}{0.091} = 0.076$$
$$\tau_{SP} = \frac{1}{\zeta_{SP}\omega_{N_{SP}}} = \frac{1}{1.008} = 0.992 \text{ s}$$
$$\tau_{PH} = \frac{1}{\zeta_{PH}\omega_{N_{PH}}} = \frac{1}{0.0069} = 144.93 \text{ s}$$

The fourth-order characteristic equation can therefore be written as two second-order (oscillatory) characteristic equations in the form

$$(s^2 + 2\zeta_{SP}\omega_{N_{SP}}s + \omega^2_{N_{SP}})(s^2 + 2\zeta_{PH}\omega_{N_{PH}}s + \omega^2_{N_{PH}}) = 0$$

For the Lear Jet example, this is

$$(s^2 + 2.016s + 8.0429)(s^2 + 0.0138s + 0.00828) = 0$$

Note that the relative magnitudes of the short period and phugoid characteristics are as expected

$$\omega_{N_{SP}} = 2.836 \text{ rad/s} > \omega_{N_{PH}} = 0.091 \text{ rad/s}$$
$$\zeta_{SP} = 0.355 > \zeta_{PH} = 0.076$$
$$\omega_{D_{SP}} = 2.651 \text{ rad/s} > \omega_{D_{PH}} = 0.905 \text{ rad/s}$$
$$\tau_{SP} = 0.992 \text{ s} < \tau_{PH} = 144.93 \text{ s}$$

The period of oscillation can be found using

$$T = \frac{2\pi}{\omega_D}$$

$$T_{sp} = \frac{2\pi}{2.651} \text{ s} = 2.37 \text{ s}$$

$$T_{ph} = \frac{2\pi}{0.0905} \text{ s} = 69.43 \text{ s}$$

It is also worthwhile to plot the short period and phugoid roots from Example 7.10 on the complex plane. This is accomplished in Fig. 7.15. Notice that the short period roots are further out from the origin and have a higher damping ratio than the phugoid roots. The relative location of these roots is typical for most aircraft.

Example 7.11

Use the short period and phugoid 2-DOF approximations to estimate the natural frequency, damping ratio, damped frequency, and period of oscillation for the Lear Jet. Compare the approximation results to those obtained in Example 7.10.

Using the short period approximation Eq. (7.34), we have

$$s^2 + 2.0173s + 8.0777 = 0$$

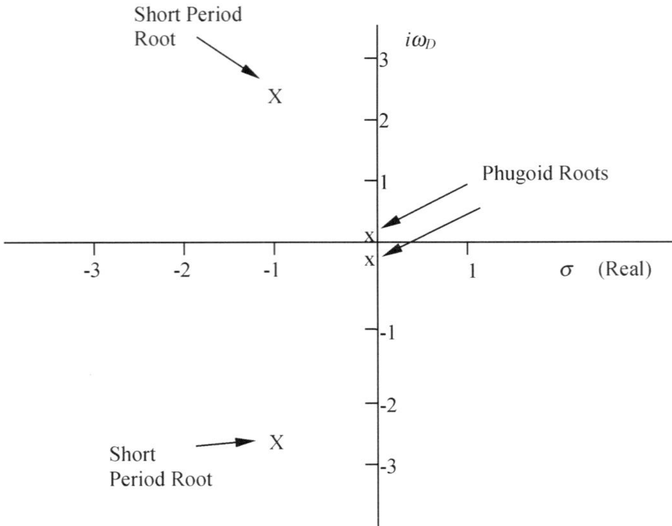

Fig. 7.15 Complex plane plot of longitudinal roots for Example 7.10.

and

$$\omega_{n_{sp}} = \sqrt{8.0777}\ \text{rad/s} = 2.842\ \text{rad/s} \qquad (2.836\ \text{rad/s for the 3-DOF case})$$

Because

$$2\zeta_{sp}\omega_{n_{sp}} = 2.0173 \Rightarrow \zeta_{sp} = 0.355 \qquad (0.355\ \text{for the 3-DOF case})$$

and

$$\omega_{D_{sp}} = \omega_{n_{sp}}\sqrt{1-\zeta_{sp}^2} = 2.657\ \text{rad/s} \qquad (2.651\ \text{rad/s for the 3-DOF case})$$

$$T_{sp} = \frac{2\pi}{\omega_{D_{sp}}} = 2.365\ \text{s} \qquad (2.37\ \text{s for the 3-DOF case})$$

As can be seen, the comparisons with the 3-DOF case are very good.

For the phugoid approximation, we use Eq. (7.39) to obtain the characteristic equation:

$$s^2 + 0.0075s + 0.00663 = 0$$

For natural frequency we have

$$\omega_{n_{PH}} = \sqrt{0.00663}\ \text{rad/s} = 0.0814\ \text{rad/s} \qquad (0.091\ \text{rad/s for the 3-DOF case})$$

To obtain damping ratio,

$$2\zeta_{PH}\omega_{n_{PH}} = 0.0075 \Rightarrow \zeta_{PH} = 0.0461 \qquad (0.076\ \text{for the 3-DOF case})$$

and

$$\omega_{D_{PH}} = \omega_{n_{PH}}\sqrt{1-\zeta_{PH}^2} = 0.0813\ \text{rad/s} \qquad (0.0905\ \text{rad/s for the 3-DOF case})$$

$$T_{PH} = \frac{2\pi}{\omega_{D_{PH}}} = 77.27\ \text{s} \qquad (69.43\ \text{s for the 3-DOF case})$$

In this example, the 2-DOF phugoid approximation provides estimates to approximately 10% accuracy with the exception of damping ratio, which has a 40% error.

7.3.3 Lateral-Directional Linearized EOM in Laplace Form

We next use the same approach developed in Sec. 7.3.2 to transform the linearized lateral-directional EOM developed in Sec. 6.5 to the Laplace domain. Again, the linearized EOM Eq. (6.131) are repeated for reference

$$\begin{aligned} \dot{v} + U_1 r &= g\phi\cos\Theta_1 + Y_\beta\beta + Y_p p + Y_r r + Y_{\delta_a}\delta_a + Y_{\delta_r}\delta_r \\ \dot{p} - \bar{A}_1\dot{r} &= L_\beta\beta + L_p p + L_r r + L_{\delta_a}\delta_a + L_{\delta_r}\delta_r \\ \dot{r} - \bar{B}_1\dot{p} &= N_\beta\beta + N_{T_\beta}\beta + N_r r + N_{\delta_a}\delta_a + N_{\delta_r}\delta_r \end{aligned} \tag{7.43}$$

where

$$\bar{A}_1 = \frac{I_{xz}}{I_{xx}} \qquad \text{and} \qquad \bar{B}_1 = \frac{I_{xz}}{I_{zz}}$$

Notice that the three EOM have five aircraft motion variables (v, p, r, ϕ, and β) along with δ_a and δ_r. δ_a and δ_r are the inputs or forcing functions for the lateral-directional system. Because we have only three equations, we will reduce the number of motion variables to three by using the kinematic equations [Eq. (4.80)] and the assumption of a small pitch attitude angle (Θ_1). Also recall that we previously made the assumption of $\phi = 0$ (wings level in trimmed flight) in Sec. 7.3.2. With these assumptions, we have

$$p = \dot{\phi} \qquad \text{and} \qquad r = \dot{\psi}$$

Also, for small perturbations

$$\beta \approx \frac{v}{U_1} \Rightarrow v \approx \beta U_1 \qquad \text{and} \qquad \dot{v} \approx \dot{\beta} U_1$$

Therefore, we can reduce our aircraft motion variables to β, ϕ, and ψ. These should be thought of as the outputs for our system of lateral-directional differential equations. We could have easily chosen v, p, and r for the output variables but instead chose the three angles. With zero initial conditions, the Laplace transform of Eq. (7.43) becomes

$$\begin{aligned} s\beta(s)U_1 + s\psi(s) &= g\phi(s)\cos\Theta_1 + Y_\beta\beta(s) + Y_p s\phi(s) + Y_r s\psi(s) + Y_{\delta_a}\delta_a(s) \\ &\quad + Y_{\delta_r}\delta_r(s) \\ s^2\phi(s) - \bar{A}_1 s^2\psi(s) &= L_\beta\beta(s) + L_p s\phi(s) + L_r s\psi(s) + L_{\delta_a}\delta_a(s) + L_{\delta_r}\delta_r(s) \\ s^2\psi(s) - \bar{B}_1 s^2\phi(s) &= N_\beta\beta(s) + N_{T_\beta}\beta(s) + N_p s\phi(s) + N_r s\psi(s) + N_{\delta_a}\delta_a(s) \\ &\quad + N_{\delta_r}\delta_r(s) \end{aligned}$$

Combining terms and moving the control inputs to the right-hand side of the equal sign, we have

$$(sU_1 - Y_\beta)\beta(s) - (sY_p + g\cos\Theta_1)\phi(s) + s(U_1 - Y_r)\psi(s) = Y_{\delta_a}\delta_a(s) + Y_{\delta_r}\delta_r(s)$$
$$- L_\beta\beta(s) + (s^2 - L_p s)\phi(s) - (s^2\bar{A}_1 + L_r s)\psi(s) = L_{\delta_a}\delta_a(s) + L_{\delta_r}\delta_r(s)$$
$$- (N_\beta + N_{T_\beta})\beta(s) - (s^2\bar{B}_1 + N_p s)\phi(s) + (s^2 - sN_r)\psi(s) = N_a\delta_a(s) + N_r\delta_r(s)$$

Regrouping in matrix form

$$\begin{bmatrix} (sU_1 - Y_\beta) & -(sY_p + g\cos\Theta_1) & s(U_1 - Y_r) \\ -L_\beta & (s^2 - L_p s) & -(s^2\bar{A}_1 + sL_r) \\ -N_\beta - N_{T_\beta} & -(s^2\bar{B}_1 + N_p s) & (s^2 - sN_r) \end{bmatrix} \begin{bmatrix} \beta(s) \\ \phi(s) \\ \psi(s) \end{bmatrix} = \begin{bmatrix} Y_{\delta_a} & Y_{\delta_r} \\ L_{\delta_a} & L_{\delta_r} \\ N_{\delta_a} & N_{\delta_r} \end{bmatrix} \begin{bmatrix} \delta_a(s) \\ \delta_r(s) \end{bmatrix} \tag{7.44}$$

and each of the six lateral-directional transfer functions can be determined using Cramer's rule as presented in Appendices F and G. The six lateral-directional transfer functions are

$$\beta(s)/\delta_a(s), \quad \beta(s)/\delta_r(s), \quad \phi(s)/\delta_a(s), \quad \phi(s)/\delta_r(s),$$
$$\psi(s)/\delta_a(s), \quad \text{and} \quad \psi(s)/\delta_r(s).$$

Notice that there are six lateral-directional transfer functions vs the three we have for longitudinal motion. This results from the fact that we have two possible control inputs (δ_a and δ_r), each of which can cause changes in the three lateral-directional motion variables. In an attempt to minimize the number of equations needed to represent these transfer functions, we will temporarily drop the subscript on δ_a and δ_r because the transfer functions for each have the same general form (Appendix G). Each of the lateral-directional transfer functions can then be represented as the ratio of two polynominals in the Laplace variables.

$$\frac{\beta(s)}{\delta(s)} = \frac{A_\beta s^3 + B_\beta s^2 + C_\beta s + D_\beta}{E's^4 + F's^3 + G's^2 + H's + I'} \tag{7.45}$$

$$\frac{\phi(s)}{\delta(s)} = \frac{A_\phi s^2 + B_\phi s + C_\phi}{E's^4 + F's^3 + G's^2 + H's + I'} \tag{7.46}$$

$$\frac{\psi(s)}{\delta(s)} = \frac{A_\psi s^3 + B_\psi s^2 + C_\psi s + D_\psi}{s(E's^4 + F's^3 + G's^2 + H's + I')} \tag{7.47}$$

Equations (7.45), (7.46), and (7.47) represent the general form of the six lateral-directional transfer functions. For example, to obtain the $\beta(s)/\delta_a(s)$

transfer function, simply use Eq. (7.45) and use the δ_a derivatives (Y_{δ_a}, L_{δ_a}, and N_{δ_a}) in the determination of A_β, B_β, C_β, and D_β (see Appendix G). In a similar fashion, to obtain the $\beta(s)/\delta_r(s)$ transfer function, Eq. (7.45) is used, and the δ_r derivatives (Y_{δ_r}, L_{δ_r}, and N_{δ_r}) are used in the determination of A_β, B_β, C_β, and D_β. As in the case of the longitudinal transfer functions, the numerator of each lateral-directional transfer function is different. Because the numerator affects the magnitude of the response, each of the three motion variables will have a different magnitude of response.

All six lateral-directional transfer functions have essentially the same denominator, which leads to the same characteristic equation:

$$E's^4 + F's^3 + G's^2 + H's + I' = 0$$

The extra s in the denominator of the $\psi(s)/\delta(s)$ transfer function indicates that the airplane is neutrally stable in heading (that is, it will not return to a trim heading when disturbed). The associated piece of the characteristic equation ($s = 0$) leads to a constant in the time response but is not that interesting from the standpoint of dynamic stability. Notice that E', F', G', H', and I', are not the same as the value of E, F, G, H, and I in Eqs. (7.28–7.30) (the longitudinal transfer functions). Because the characteristic equation determines the dynamic stability characteristics of the response, all six transfer functions will have the same dynamic characteristics (ζ, ω_n, and τ) but a different magnitude of response.

7.3.3.1 Three-degree-of-freedom analysis of the lateral-directional modes of motion. The preceding development of transfer functions for the three lateral-directional motion variables β, ϕ, and ψ leads to a 3-DOF solution for lateral-directional motion. Normally, the fourth-order characteristic equation for lateral-directional motion is written as the product of one second-order (oscillatory) and two first-order (nonoscillatory) polynomials.

$$(s^2 + 2\zeta_{DR}\omega_{n_{DR}} s + \omega_{n_{DR}}^2)\left(s + \frac{1}{\tau_r}\right)\left(s + \frac{1}{\tau_s}\right) = 0 \tag{7.48}$$

The subscript DR refers to the dutch roll mode, the subscript r refers to the roll mode, and the subscript s refers to the spiral mode. All airplanes have these three lateral-directional dynamic modes. Each of these polynominals can be thought of as a separate characteristic equation that defines the dynamic characteristics of its respective mode.

As with the longitudinal case, *the coefficients (and roots) of each characteristic equation change with flight condition, airplane mass, mass distribution, airplane geometry, and aerodynamic characteristics*. These changes translate to changes in $\omega_{n_{DR}}$, ζ_{DR}, τ_r, and τ_s, but the fundamental presence of the dutch roll, roll, and spiral modes is maintained.

The **dutch roll mode** is a second-order response (**complex conjugate roots**) usually characterized by concurrent oscillations in the three lateral-directional motion variables β, ϕ, and ψ. In the discussion on static stability, it was observed that sideslip generates both yawing and rolling moments that lead to a coupled motion between β, ϕ, and ψ. These oscillations may be of high or low frequency and may be lightly or heavily damped. The dutch roll usually begins with a sideslip perturbation followed by oscillations in roll and yaw. The dutch roll motion is something like that of an ice skater's body weaving back and forth as weight shifts from one foot to the other. As the magnitude of C_{l_β} (lateral static stability) becomes larger, more roll coupling is present during dutch roll oscillations, and the dutch roll characteristics typically become more objectionable. Objectionable dutch roll characteristics adversely affect precision tasks like air-to-air and air-to-ground tracking, and formation flying.

The **roll mode** has a **real root** and a first-order (nonoscillatory) response that involves almost a pure rolling motion about the x stability axis. It is usually stable at low and moderate angles of attack but may be unstable at high angles of attack. The roll mode can be excited by a disturbance or an aileron input. It is easiest to characterize the roll mode when discussing response to an aileron input. If a step aileron input (δ_a) is made to the aircraft, there is an exponential rise in roll rate ($\approx \dot{\phi}$) until a steady state roll rate is achieved. We saw this first-order type response in Sec. 7.1.1.1. From the pilot's standpoint, the time taken during the exponential rise to steady state is interpreted as a finite delay, which we usually characterize with the time constant (τ_r). If τ_r is too large, the aircraft is considered sluggish because it may take too long for the commanded roll rate to build up. Likewise, if τ_r is too small, the aircraft may be too responsive to external disturbances such as turbulence.

The **spiral mode** is a first-order response (**real root**) that involves a relatively slow roll and yawing motion of the aircraft. It may be stable or unstable. The spiral is usually initiated by a displacement in roll angle and appears as a descending turn with increasing roll angle if unstable. If the spiral is stable, the aircraft simply returns to wings level after a roll angle displacement. The primary motion variables during the spiral are ϕ and ψ, while β remains close to zero. A high degree of lateral stability (C_{l_β}) will tend to make the spiral stable, while a high degree of directional stability (C_{n_β}) will tend to make the spiral unstable. Fortunately, spiral instability can be tolerated as long as the time to double amplitude (based on the initial roll angle displacement) is gradual (greater than approximately 4 s). Under these conditions, the pilot can return the aircraft to wings level flight with little difficulty using an aileron input. If the spiral mode is unstable, the time to double amplitude (T_2) is calculated with

$$T_2 = \frac{\ln 2}{\text{unstable root}} = \frac{0.693}{\text{unstable root}} \tag{7.49}$$

Spiral stability is usually compromised for good dutch roll characteristics that are typically achieved with relatively high directional stability and relatively low lateral stability.

Example 7.12

The β/δ_a transfer function for a T-37 cruising at 30,000 ft and 0.46 Mach is given next. Find the natural frequency, damping ratio, and damped frequency for the dutch roll mode, and the time constant for the dutch roll, roll, and spiral modes.

$$\frac{\beta}{\delta_a} = 1.41 \frac{(s+2.05)(s+0.0741)}{(s+1.27)(s+0.0037)(s^2+0.227s+5.80)}$$

We go immediately to the three characteristic equations

$$s^2 + 0.227s + 5.8 = 0$$
$$s + 1.27 = 0$$
$$s + 0.0037 = 0$$

The second-order equation is for the dutch roll mode, and we have

$$\omega_{n_{DR}} = \sqrt{5.8} = 2.41 \text{ rad/s}$$

The damping ratio for dutch roll becomes

$$\zeta_{DR} = \frac{0.227}{2\omega_{n_{DR}}} = 0.0471$$

and the damped frequency is

$$\omega_{D_{DR}} = \omega_{n_{DR}}\sqrt{1-\xi_{DR}^2} = 2.41\sqrt{1-(0.0471)^2} = 2.407 \text{ rad/s}$$

The time constant for dutch roll is

$$\tau_{DR} = \frac{1}{\zeta_{DR}\omega_{n_{DR}}} = \frac{1}{(0.0471)(2.41)} = 8.81 \text{ s}$$

At this point, we should comment on the relatively low damping ratio and large time constant for the dutch roll mode. As we will see, these do not pass military specifications. As a result, the T-37 has a yaw damper installed to improve the basic airframe dutch roll characteristics.

The two first-order characteristic equations are for the roll and spiral modes. Referring back to Eq. (7.48) and realizing that the roll mode will have a smaller time constant than the spiral mode, we have

$$\tau_r = \frac{1}{1.27} = 0.787 \text{ s}$$
$$\tau_s = \frac{1}{0.0037} = 270 \text{ s}$$

A few observations are in order. With the roll mode time constant being less than 1 s, we can see that the T-37 roll response is fairly crisp. The spiral mode is stable for this flight condition as indicated by the negative root

($s = -0.0037$). Because it is stable, the time constant indicates the time it takes to return to 36.8% (1–0.632) of the initial displaced roll angle as the aircraft returns to a wings-level attitude. For example, if the initial displaced roll angle is 10 deg, it will take about 4.5 min (270 s) to return to a 3.68-deg roll attitude (assuming the pilot makes no aileron input).

If the characteristic equation for the spiral had been

$$s - 0.0037 = 0$$

then the spiral would be unstable (a positive root at $s = 0.0037$) and the time to double amplitude would have been [using Eq. (7.49)]

$$T_2 = \frac{0.693}{0.0037} = 187 \text{ s}$$

A summary of root and response characteristics for the two longitudinal and three lateral-directional dynamic modes of aircraft motion is presented in Table 7.2.

7.3.3.2 One-degree-of-freedom roll approximation. To gain an understanding of the stability parameters and derivatives that influence the roll mode, we can eliminate two of the three degrees of freedom or motion variables. The roll mode is the simplest of the five dynamic modes. We begin with Eq. (7.44) and retain only the ϕ motion variable, the δ_a control input, and the rolling moment equation. Thus, Eq. (7.44) simplifies to

$$(s^2 - L_p s)\phi(s) = L_{\delta_a}\delta_a(s)$$

The roll approximation transfer function becomes

$$\frac{\phi(s)}{\delta_a(s)} = \frac{L_{\delta_a}}{s(s - L_p)} \tag{7.50}$$

Table 7.2 Root and response characteristics for the aircraft dynamic modes of motion

Mode	Root type	Response
Longitudinal		
Short period	Complex conjugate	Oscillatory
Phugoid	Complex conjugate	Oscillatory
Lateral-directional		
Dutch roll	Complex conjugate	Oscillatory
Roll	Real	Nonoscillatory
Spiral	Real	Nonoscillatory

Equation (7.50) yields two roots for the characteristic equation: 0 and L_p. The root at $s = 0$ is of little interest because it leads to the steady-state value; however, the root $s = L_p$ is of more interest because it leads directly to an estimate of the time constant for the roll mode

$$\tau_r \approx -\frac{1}{L_p} \tag{7.51}$$

Recall that L_p is the roll damping stability parameter that is a direct function of C_{l_p} the roll damping derivative. L_p typically is negative, which makes the roll mode stable. Thus, we can deduce from Eq. (7.51) that the higher the roll damping, the smaller the roll mode time constant. This may seem counter-intuitive at first, but remember that the time constant is only an indicator of the time to a steady-state value. Our intuition tells us that more damping should lead to a lower steady-state value, so we will investigate roll rate response to a step aileron input using our approximation. We will define the magnitude of the step aileron input as δ_a. Thus

$$\delta_a(s) = \frac{\delta_a}{s}$$

and from Eq. (7.50)

$$\phi(s) = \frac{L_{\delta_a}\delta_a}{s^2(s - L_p)}$$

The first step in finding the time response involves partial fractions.

$$\phi(s) = \frac{A}{s^2} + \frac{B}{s} + \frac{C}{(s - L_p)}$$

where

$$A = s^2\phi(s)|_{s=0} = -\frac{L_{\delta_a}\delta_a}{L_p}$$

$$B = \frac{\mathrm{d}}{\mathrm{d}s}[s^2\phi(s)]_{s=0} = -\left.\frac{L_{\delta_a}\delta_a}{(s - L_p)}\right|_{s=0} = -\frac{L_{\delta_a}}{L_p^2}\delta_a$$

$$C = (s - L_p s)\phi(s)|_{s=L_p} = \left.\frac{L_{\delta_a}\delta_a}{s^2}\right|_{s=L_p} = \frac{L_{\delta_a}}{L_p^2}\delta_a$$

Combining and taking the inverse Laplace,

$$\phi(t) = -\frac{L_{\delta_a}}{L_p}\delta_a t - \frac{L_{\delta_a}}{L_p^2}\delta_a + \frac{L_{\delta_a}}{L_p^2}\delta_a e^{L_p t} = -\frac{L_{\delta_a}}{L_p}\delta_a t + \left(\frac{L_{\delta_a}}{L_p^2}\delta_a\right)(e^{L_p t} - 1)$$

Figure 7.16 plots a time response of the previous equation for roll angle to illustrate a graphical method for obtaining the time constant.

To obtain roll rate, we take the time derivative:

$$\dot{\phi}(t) = p = -\frac{L_{\delta_a}}{L_p}\delta_a + L_p\left(\frac{L_{\delta_a}}{L_p^2}\right)\delta_a e^{L_p t}$$

or

$$\dot{\phi} = -\frac{L_{\delta_a}\delta_a}{L_p}(1 - e^{L_p t}) \tag{7.52}$$

We can make several observations based on Eq. (7.52). The steady-state roll rate achieved will be $-L_{\delta_a}\delta_a/L_p$, which indicates that the larger the aileron control power stability parameter (L_{δ_a}) and/or the larger the magnitude of the aileron input (δ_a), the larger the steady-state roll rate will be. We can also see that the magnitude of the steady-state roll rate is inversly proportional to the roll damping stability parameter (L_p). In addition, the time constant predicted by Eq. (7.51) is evident in the exponential term. To illustrate this, let $t = -1/L_p$ in Eq. (7.52).

$$\dot{\phi} = -\frac{L_{\delta_a}\delta_a}{L_p}(1 - e^{L_p(-1/L_p)}) = -\frac{L_{\delta_a}\delta_a}{L_p}(1 - e^{-1}) = 0.632\left(-\frac{L_{\delta_a}\delta_a}{L_p}\right)$$

Thus, we can see that at $t = \tau r$, the roll rate is equal to 63.2% of the steady-state value. Figure 7.17 illustrates these points.

7.3.3.3 Two-degree-of-freedom spiral approximation. Spiral motion is dominated by bank angle, ϕ, and heading angle, ψ, while β is very small. To

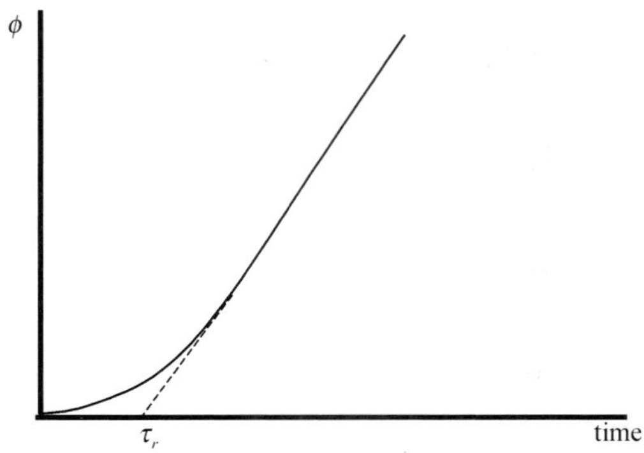

Fig. 7.16 Roll angle response to a step aileron input.

achieve a 2-DOF approximation, we will neglect the roll angle motion variable and the sideforce equation. We neglect ϕ because banking does not induce terms that cause the aircraft to roll out (there are no terms like C_{l_ϕ}). β terms do have an effect on roll out because of C_{l_β} to (lateral static stability). Thus, Eq. (7.44) can be simplified to

$$\begin{bmatrix} -L_\beta & -s(s\bar{A}_1 + L_r) \\ -N_\beta & s(s - N_r) \end{bmatrix} \begin{bmatrix} \beta(s) \\ \psi(s) \end{bmatrix} = \begin{bmatrix} L_{\delta_a} & L_{\delta_r} \\ N_{\delta_a} & N_{\delta_r} \end{bmatrix} \begin{bmatrix} \delta_a(s) \\ \delta_r(s) \end{bmatrix}$$

The characteristic equation becomes:

$$-L_\beta s(s - N_r) - (-N_\beta)\left[-s\left(s\frac{I_{xz}}{I_{xx}} + L_r\right)\right] = 0$$

Close inspection reveals that a common factor in the characteristic equation is s, which can be cancelled. Algebraic manipulation yields the root of the spiral approximation characteristic equation as

$$s \approx \frac{L_\beta N_r - N_\beta L_r}{L_\beta + N_\beta\left(\dfrac{I_{xz}}{I_{xx}}\right)} \tag{7.53}$$

For the spiral to be stable, this root must be negative. The denominator of Eq. (7.53) is normally negative. Because L_β and N_r are negative, and N_β and L_r are positive, we can deduce that the magnitude of L_β (lateral static stability) should be larger than the magnitude of N_β (directional static stability) for a stable spiral mode (assuming N_r and L_r are of approximately equal magnitude). As discussed earlier, the unfavorable impact of this tradeoff on the dutch roll may drive the designers to accept an unstable spiral (more directional stability)

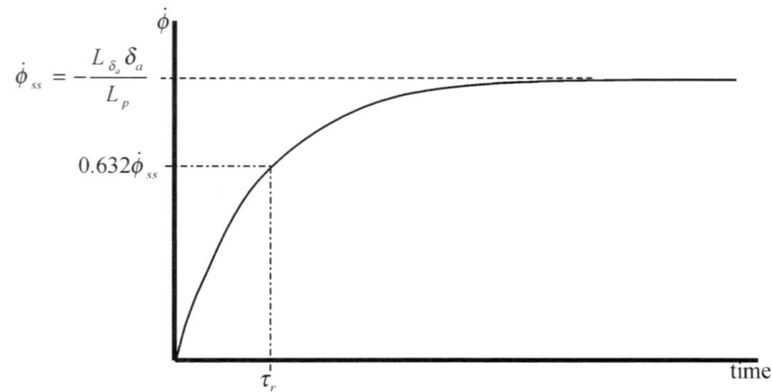

Fig. 7.17 Roll rate response to a step aileron input.

in favor of acceptable dutch roll characteristics, especially at low speeds. From Eq. (7.53), the spiral time constant can be approximated by

$$\tau_s \approx \frac{L_\beta + N_\beta\left(\frac{I_{xz}}{I_{xx}}\right)}{N_\beta L_r - L_\beta N_r} \tag{7.54}$$

Unfortunately, the 2-DOF spiral approximation tends to yield poor results. It does, however, provide insight into stability parameters and design features that affect the spiral mode.

7.3.3.4 Two-degree-of-freedom dutch roll approximations. The dutch roll mode is probably the most difficult aircraft dynamic mode to analyze. Several 2-DOF approximations are possible for the dutch roll based on which simplifying assumptions are made. In many cases, it is best to look at an estimate of the absolute value of the ratio of roll angle to sideslip $|\phi/\beta|$ present during a dutch roll oscillation to help guide dutch roll approximation assumptions. It is basically the ratio of the numerator of the $\phi(s)/\delta(s)$ transfer function to the numerator of the $\beta(s)/\delta(s)$ transfer function evaluated at the specific damping ratio and damped frequency conditions. The ϕ/β ratio tells us if the dutch roll is composed of mostly yawing motion, mostly rolling motion, or approximately equal excursions of each. ϕ/β can be visualized by thinking about the pattern a wing tip light traces as the aircraft goes through a dutch roll oscillation. If the pattern is a horizontal ellipse (major axis horizontal), ϕ/β is less than 1 and the roll angle excursions are low compared to the sideslip excursions. If the wing tip pattern is a circle, ϕ/β is approximately 1. A vertical ellipse pattern indicates ϕ/β greater than 1 and the aircraft is considered "rolly", generally because of a high degree of lateral stability. This last case is usually objectionable for precision tracking tasks. The approximation

$$\left|\frac{\phi}{\beta}\right| \approx \left[\frac{C_{l_\beta}}{C_{n_\beta}}\frac{I_{zz}}{I_{xx}}\frac{1}{\rho U_1}\right] \tag{7.55}$$

will give us an estimate of ϕ/β as a start. We will look at three different approximations for dutch roll dynamic characteristics based on this estimate.

Test Pilot School approximation. This approximation is used by the USAF Test Pilot School and assumes the dutch roll motion is mostly sideslip ($|\phi/\beta|$ very low). It is believed that the approximations for damping ratio and natural frequency are based on experience.

$$\begin{aligned} \zeta_{DR} &\approx \left(\frac{1}{8}\sqrt{2Sb^3}\right)(-C_{n_r})\left(\frac{\rho}{I_{zz}C_{n_\beta}}\right)^{1/2} \\ \omega_{n_{DR}} &\approx \left(\frac{Sb}{2}\right)^{1/2} U_1\left(\frac{C_{n_\beta}\rho}{I_{zz}}\right)^{1/2} \end{aligned} \tag{7.56}$$

Two-degree-of-freedom approximation for low ϕ/β. Here again, the assumption is made that the motion consists primarily of sideslipping and yawing. This is generally the case for aircraft with relatively low lateral stability (C_{l_β}). This approximation eliminates the ϕ motion variable and the rolling moment equation from Eq. (7.44). The result is

$$\begin{bmatrix} (sU_1 - Y_\beta) & s(U_1 - Y_r) \\ -N_\beta & s^2 - sN_r \end{bmatrix}\begin{bmatrix} \beta(s) \\ \psi(s) \end{bmatrix} = \begin{bmatrix} Y_{\delta_a} & Y_{\delta_r} \\ N_{\delta_a} & N_{\delta_r} \end{bmatrix}\begin{bmatrix} \delta_a(s) \\ \delta_r(s) \end{bmatrix}$$

The characteristic equation is obtained from the determinant of the first matrix

$$s\left[s^2 - s\left(N_r + \frac{Y_\beta}{U_1}\right) + \left(\frac{Y_\beta N_r}{U_1} + N_\beta - \frac{N_\beta Y_r}{U_1}\right)\right] = 0$$

and natural frequency and damping ratio are obtained in the usual manner

$$\begin{aligned} \omega_{n_{DR}} &\approx \sqrt{\{N_\beta + \frac{1}{U_1}(Y_\beta N_r - N_\beta Y_r)\}} \\ \zeta_{DR} &= \frac{-\left(N_r + \dfrac{Y_\beta}{U_1}\right)}{2\omega_{n_d}} \end{aligned} \tag{7.57}$$

With the previous approximation equations, the strong influence of directional stability (N_β) and yaw damping (N_r) can be seen.

Two-degree-of-freedom approximation for high ϕ/β. This approximation makes the assumption that the dutch roll consists primarily of rolling motion. This may be the case for aircraft with high lateral stability. The ψ motion variable and the yawing moment equation are eliminated from Eq. (7.44).

$$\begin{bmatrix} sU_1 - Y_\beta & -s(Y_p + g) \\ -L_\beta & s^2 - L_p s \end{bmatrix}\begin{bmatrix} \beta(s) \\ \phi(s) \end{bmatrix} = \begin{bmatrix} Y_{\delta_a} & Y_{\delta_r} \\ L_{\delta_a} & L_{\delta_r} \end{bmatrix}\begin{bmatrix} \delta_a(s) \\ \delta_r(s) \end{bmatrix}$$

and, after a few simplifications (Y_p and $\ddot{\phi}$ negligible), the significant characteristic equation becomes

$$(sU_1 - Y_\beta)(-L_p s) - (-g)(-L_\beta) = 0$$

or

$$s^2 - \frac{Y_\beta}{U_1}s + \frac{g}{U_1}\frac{L_\beta}{L_p} = 0$$

We then have

$$\omega_{n_{DR}} \approx \sqrt{\frac{g}{U_1}\frac{L_\beta}{L_p}}$$
$$\zeta_{DR} \approx -\frac{Y_\beta}{U_1}\sqrt{\frac{U_1}{g}\frac{L_p}{L_\beta}} \tag{7.58}$$

Notice the strong influence of lateral stability (L_β) and roll damping (L_p) in the previous equations.

The three approximations presented for estimating dutch roll dynamic characteristics all have significant limitations because of the highly coupled motion of the dutch roll. A 3-DOF solution is generally preferred when analyzing the dutch roll.

Example 7.13

Using the same aircraft and flight condition as in Example 7.10, the six lateral-directional transfer functions for the Lear Jet are approximated by

$$\frac{\beta(s)}{\delta_a(s)} = \frac{(4.184)s^2 + (5.589)s + 0.363}{(674.9)s^4 + (421.2)s^3 + (1808)s^2 + (897.9)s + 0.903}$$

$$\frac{\phi(s)}{\delta_a(s)} = \frac{(79.59)s^2 + (14.24)s + 189.3}{(674.9)s^4 + (421.2)s^3 + (1808)s^2 + (897.9)s + 0.903}$$

$$\frac{\psi(s)}{\delta_a(s)} = \frac{-(4.189)s^3 + (2.150)s^2 - (0.150)s + 8.991}{s[(674.9)s^4 + (421.1)s^3 + (1808)s^2 + (897.9)s + 0.903]}$$

$$\frac{\beta(s)}{\delta_r(s)} = \frac{(0.185)s^3 + (18.16)s^2 + (8.285)s - 0.0933}{(674.9)s^4 + (421.2)s^3 + (1808)s^2 + (897.9)s + 0.903}$$

$$\frac{\phi(s)}{\delta_r(s)} = \frac{(8.188)s^2 - (2.045)s - 53.85}{(674.9)s^4 + (421.2)s^3 + (1808)s^2 + (897.9)s + 0.903}$$

$$\frac{\psi(s)}{\delta_r(s)} = \frac{-(18.08)s^3 - (8.921)s^2 - (0.4481)s + 2.559}{s[(674.9)s^4 + (421.2)s^3 + (1808)s^2 + (897.9)s + 0.903]}$$

Find the time constant for the roll and spiral modes and the natural frequency, damping ratio, damped frequency, time constant, and period of oscillation for the dutch roll modes.

The characteristic equation is

$$674.9s^4 + 421.2s^3 + 1808s^2 + 897.9s + 0.903 = 0$$

or, in standard form

$$s^4 + 0.6241s^3 + 2.6789s^2 + 1.3304s + 0.001338 = 0$$

Using a root solver such as that available in MATLAB, we have two real roots and one pair of complex conjugates:

$$\begin{aligned} s_1 &= -0.00101 = s_{\text{SPIRAL}} \quad \text{(smaller real root)} \\ s_2 &= -0.507 = s_{\text{ROLL}} \quad \text{(larger real root)} \\ s_{3,4} &= -0.0580 \pm j1.617 = -\zeta_{DR}\omega_{N_{DR}} \pm j\omega_{D_{DR}} \end{aligned}$$

With these roots, the characteristic equation can be written as

$$(s + 0.00101)(s + 0.507)(s^2 + 0.116s + 2.618) = 0$$

For the roll mode, we have:

$$\tau_{\text{ROLL}} = -\frac{1}{-0.507}\ \text{s} = 1.972\ \text{s}$$

For the spiral mode, the root is stable (negative) so we have

$$\tau_{\text{SPIRAL}} = -\frac{1}{-0.00101}\ \text{s} = 991.8\ \text{s}$$

And for the dutch roll, we have:

$$\begin{aligned} \omega_{D_{DR}} &= 1.617\ \text{rad/s} \\ \omega_{N_{DR}} &= 1.618\ \text{rad/s} \\ \zeta_{DR} &= 0.036 \\ \tau_{DR} &= -\frac{1}{-\zeta\omega_N} = -\frac{1}{-0.058} = 17.24\ \text{s} \\ T_{DR} &= \frac{2\pi}{\omega_{D_{DR}}} = \frac{2\pi}{1.617} = 3.88\ \text{s} \end{aligned}$$

Example 7.14

Use the 1-DOF roll approximation, the 2-DOF spiral approximation, and the three dutch roll approximations to approximate the lateral-directional stability characteristics for the Lear Jet. Compare to the 3-DOF values obtained in Example 7.13.

For the roll approximation, we use Eq. (7.50) and recall that $\dot{\phi} = p = s\phi$

$$\frac{p(s)}{\delta_a(s)} = s\frac{\phi(s)}{\delta_a(s)} = s\frac{L_{\delta_a}}{s(s - L_p)}$$

Substituting in the values for the Lear Jet:

$$\frac{p(s)}{\delta_a(s)} = \frac{6.77}{s + 0.437}$$

The time constant becomes:

$$\tau_r = -\frac{1}{-0.437} = 2.29 \text{ s} \quad \text{(1-DOF approximation)}$$

Recalling that the 3-DOF roll time constant was 1.972 s, the approximation compares within approximately 16%.

For the spiral mode, Eq. (7.53) yields: $s = -0.0065$. For the time constant we have

$$\tau_s = -\frac{1}{-0.0065} = 153.8 \text{ s} \quad \text{(2-DOF approximation)}$$

The spiral approximation gives a poor prediction of spiral time constant when compared to the 3-DOF value of 991.8 s.

For the dutch roll mode, we first use Eq. (7.55) to calculate

$$\left|\frac{\phi}{\beta}\right| = 3.66$$

We next use the dutch roll approximations to obtain estimates for natural frequency and damping ratio [using Eqs. (7.56–7.58)]. These results are summarized in Table 7.3.

We can see that the Test Pilot School approximation gives reasonably good values for both damping ratio and natural frequency. The 2-DOF low ϕ/β ratio approximation gives a close value for natural frequency but a poor prediction

Table 7.3 Comparison of dutch roll 3-DOF and approximation solutions

	ω_n	ζ
Exact solution (3-DOF Case)	1.618 rad/s	0.036
1) Test Pilot School Approx.	1.69 rad/s	0.033
2) DOF Approx. Low $\lvert\phi/\beta\rvert$	1.62 rad/s	0.058
3) 2 DOF Approx. High $\lvert\phi/\beta\rvert$	0.68 rad/s	0.062

for damping ratio. The 2-DOF high ϕ/β approximation has poor predictions for both natural frequency and damping ratio (despite this being a relatively high ϕ/β case). The limitations of each approximation are, of course, directly related to the assumptions made.

7.4 Dynamic Stability Guidelines

In designing an aircraft, there are several guidelines available for natural frequency, damping ratio, and/or time constant for each of the dynamic modes. These guidelines are based on approximately the past half century of flying experience. Dynamic stability characteristics directly affect the "flyability" or handling qualities of an aircraft. The interface between the human pilot (with physical and mental limitations) and inherent aircraft response characteristics must allow for accomplishment of mission objectives throughout the flight envelope. Precision tasks such as landing approach, tracking, and formation flying can only be accomplished successfully if the aircraft's dynamic stability characteristics are within acceptable ranges. These ranges are usually presented in terms of the dynamic characteristics we have discussed. Of course, another aspect of acceptable aircraft handling qualities involves sufficient control authority (usually referred to as control power) to trim and maneuver the aircraft throughout the flight envelope. We will focus on acceptable dynamic stability characteristics as presented in a Military Specification (MIL-F-8785C).[1] Although this specification is no longer a requirement for military aircraft, it does provide a good reference for the designer and establishes the concept of flying qualities levels. Another specification is published for civilian aircraft in Federal Aviation Requirement (FAR) documents.

The advent of modern high-performance aircraft with high-authority control augmentation systems (F-15) and fly-by-wire control systems (F-16 and F-22) has resulted in dynamic stability characteristics that do not conform to classic first- and second-order responses. Military Standard 1797 is currently used to address these advances. However, for purposes of this text, MIL-F-8785C provides a good first step in the discussion of acceptable dynamic stability characteristics.

7.4.1 Aircraft Class

Acceptable flying qualities are a function of the size and mission of an aircraft. To account for this, MIL-F-8785C specifies four classes of aircraft as presented in Table 7.4. The determination of aircraft class is usually the first step in utilization of MIL-F-8785C.

7.4.2 Flight Phase Category

MIL-F-8785C dynamic stability requirements also are a function of the flight phase, or mission segment, that an aircraft is engaged in because different demands are placed on the pilot. For example, air-to-ground tracking requires a higher degree of dutch roll damping than cruising flight. The different flight phases may also be associated with different dynamic pressure conditions, which have a direct effect on dynamic stability characteristics. Table 7.5

Table 7.4 MIL-F-8785C aircraft classes

Class	General aircraft types	Specific examples
Class I	Light utility	T-41
small, light	Primary trainer	T-6
airplanes	Light observation	O-1, O-2
Class II	Heavy utility/search and rescue	C-21
medium	Light or medium transport/cargo/tanker	C-130
weight; low-to-	Early warning/ECM/Command & control	E-2
medium	Anti-submarine	S-3A
maneuverability	Assault transport	C-130
airplanes	Reconnaissance	U-2
	Tactical bomber	B-66
	Heavy attack	A-6
	Trainer for Class II	T-1A
Class III	Heavy transport/cargo/tanker	KC-10, C-17
large, heavy,	Heavy bomber	B-52, B-1, B-2
low-to-medium	Patrol/Early warning/ECM/Command & control	P-3, SR-71
maneuverability	Trainer for Class III	TC-135
airplanes		
Class IV	Figher/Intecepter	F-22, F-15, F-16
high-	Attack	F-15E, A-10
maneuverability	Tactical reconnaissance	RF-4
airplanes	Observation	OV-10
	Trainer for Class IV	T-38

presents the three flight phase categories into which MIL-F-8785C divides all mission segments for military aircraft. Terminal refers to the takeoff and landing phases accomplished in a terminal area. Normally, the gear and flaps down configuration is associated with the terminal flight phase (Category C), while the gear and flaps up configuration is associated with Category A and B. When no flight phase category is stated in a dynamic stability requirement, that requirement applies to all three categories.

7.4.3 Flying Quality Levels

An aircraft's compliance to the dynamic stability requirements of MIL-8785C is defined in terms of three flying quality levels. These are summarized in Table 7.6.

Level 1 is the highest level of flying qualities and is the requirement within the operational flight envelope with all aircraft systems in their normal operating state. It is important to define the term operational flight envelope. Flight envelopes are usually defined by boundaries of speed, altitude, load factor, angle of attack, and/or sideslip. The operational flight envelope is the innermost or inside envelope when compared with the boundaries of the service

Table 7.5 MIL-F-8785C flight phase categories

Category A	Those nonterminal flight phases that require rapid maneuvering, precision tracking, or precise flight-path control. Included in this category are: (a) Air-to-air combat (CO) (b) Ground attack (GA) (c) Weapon delivery/launch (WD) (d) Aerial recovery (AR) (e) Reconnaissance (RC) (f) In-flight refueling (receiver) (RR) (g) Terrain following (TF) (h) Antisubmarine search (AS) (i) Close formation flying (FF)
Category B	Those nonterminal flight phases that are normally accomplished using gradual maneuvers and without precision tracking, although accurate flight-path control may be required. Included in this category are: (a) Climb (CL) (b) Cruise (CR) (c) Loiter (LO) (d) In-flight refueling (tanker) (RT) (e) Descent (D) (f) Emergency descent (ED) (g) Emergency deceleration (DE) (h) Aerial delivery (AD)
Category C	Those terminal flight phases that are normally accomplished using gradual maneuvers and which usually require accurate flight-path control. Included in this category are: (a) Takeoff (TO) (b) Catapult takeoff (CT) (c) Approach (PA) (d) Wave-off/go-around (WO) (e) Landing (L)

flight envelope and the permissible flight envelope (to be discussed later in this section). The boundaries of the operational flight envelope are set by **mission requirements**. Expected missions are analyzed to determine what speed, altitude, load factor, angle of attack, and/or sideslip ranges will be needed to accomplish each mission, and this information is used to define the operational flight envelope.

Level 2 implies an increase in pilot workload and/or a degredation in mission effectiveness because of decreased dynamic stability (or control power) characteristics. Level 2 is considered acceptable when the cumulative probability of all failure states that could result in Level 2 flying qualities within the operational flight envelope is less than once every 100 flights. For example, an aircraft has ten failure states that result in Level 2 flying qualities in the opera-

Table 7.6 MIL-F-8785C flying quality levels

Level 1	Flying qualities clearly adequate for the mission flight phase
Level 2	Flying qualities adequate to accomplish the mission flight phase, but some increase in pilot workload or degradation in mission effectiveness, or both, exists
Level 3	Flying qualities such that the airplane can be controlled safely, but pilot workload is excessive or mission effectiveness is inadequate, or both. Category A flight phases can be terminated safely, and Category B and C flight phases can be completed

tional flight envelope, two of which each have a probability of 5×10^{-3}/flight, and the other eight each have probabilities of 1×10^{-3}/flight. Such a situation fails the MIL-F-8785C requirement because the cumulative probability is 1.8×10^{-2}/flight, nearly twice the allowable limit. The reliability of these failure states would have to be improved to meet the cumulative requirement.

Level 3 requires that control of the airplane is maintained but allows excessive pilot workload and/or inadequate mission effectiveness. It is basically a "get home" level and is considered acceptable when the cumulative probability of all failure states that result in Level 3 in the operational flight envelope is less than once every 10,000 flights, and the cumulative probability of all failure states that result in Level 3 in the service flight envelope is less than once every 100 flights. Before defining the service flight envelope, we will define the outermost flight envelope, the permissible flight envelope. The boundaries of the permissible flight envelope are set by aircraft performance and safety limits. The aircraft either is not capable of exceeding these limits, or, if it can exceed these limits, potentially catastrophic failures may occur (such as structural failure or engine failure) when beyond these limits. The boundaries of the service flight envelope are between the operational flight envelope and the permissible flight envelope (that is, its boundaries contain the operational flight envelope but it is contained within the permissible flight envelope). Outside the service flight envelope, but within the permissible flight envelope, the handling qualities with the aircraft systems in the normal state are expected to be at least recoverable. This means that controlled flight may be temporarily lost (as in a stall), but the pilot can safely return to the service flight envelope and regain control. The service flight envelope basically acts as a safety margin between the operational flight envelope and the permissible flight envelope. The handling qualities of an aircraft may degrade as it nears the limits of the permissible flight envelope, but we want the degradation to be gradual, not sudden. Level 2 handling qualities are generally the minimum requirement, with all systems in the normal state, within the service flight envelope but outside the operational flight envelope. This ensures that the pilot has handling qualities good enough to avoid entering or exceeding the permissible flight envelope inadvertently. A level better than the one specified is also considered acceptable for a given failure situation. Section 7.3.5 will discuss the Cooper-Harper rating scale for aircraft flying qualities which has a direct correlation to the MIL-F-8785C levels discussed here. A simplified decision process for using MIL-F-8785C is presented in Fig. 7.18.*

7.4.4 Short Period

Dynamic stability guidelines for the short period mode are covered in two documents. Both will be addressed in this section.

7.4.4.1 MIL-F-8785C. MIL-F-8785C requires that the short period mode meet both a damping ratio and natural frequency requirement. The equivalent damping ratio requirements are presented in Table 7.7 and are a function of

*Private communication with D. Leggett, Dec. 1999.

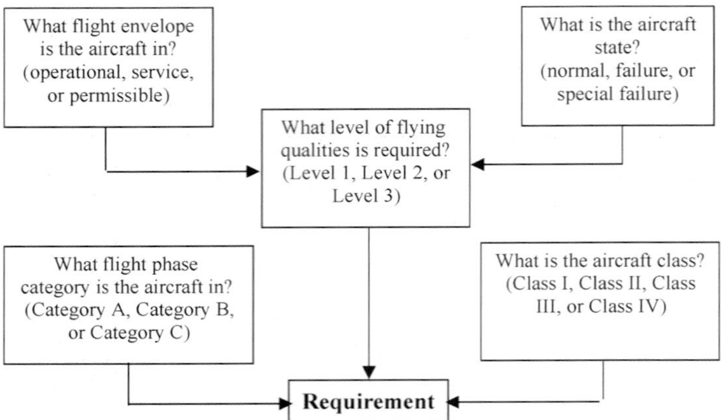

Fig. 7.18 Simplified decision process for using MIL-F-8785C.

category. The term equivalent is used so that aircraft with high authority augmentation systems or fly-by-wire systems can also be included. These aircraft have dynamic response characteristics (by design) that are significantly different from those of the basic airframe.

MIL-F-8785C requires that the short period natural frequency fall within an upper and lower limit as a function of the aircraft's n/α ratio and flight phase category. Figures 7.19–7.21 present the Level 1, 2, and 3 regions for short period natural frequency by category. Notice the logarithmic scale for n/α. This parameter can be estimated for an aircraft using

$$\frac{n}{\alpha} \approx -\frac{Z_\alpha}{g} \tag{7.59}$$

n/α can be thought of as a load factor (n) sensitivity parameter. It increases with increases in C_{L_α} and wing area, and it decreases as weight increases.

The MIL-F-8785C short period requirement will be satisfied if the roots of the short period mode fall within a region on the s plane. In general terms, this region is presented as the crosshatched area in Fig. 7.22.

Table 7.7 Short period damping ratio (ζ_{SP}) limits

	Category A and C flight phases		Category B flight phases	
	Minimum	Maximum	Minimum	Maximum
Level 1	0.35	1.30	0.30	2.00
Level 2	0.25	2.00	0.20	2.00
Level 3	0.15*	no maximum	0.15*	no maximum

* May be reduced at altitudes above 20,000 ft if approved by the procuring activity.

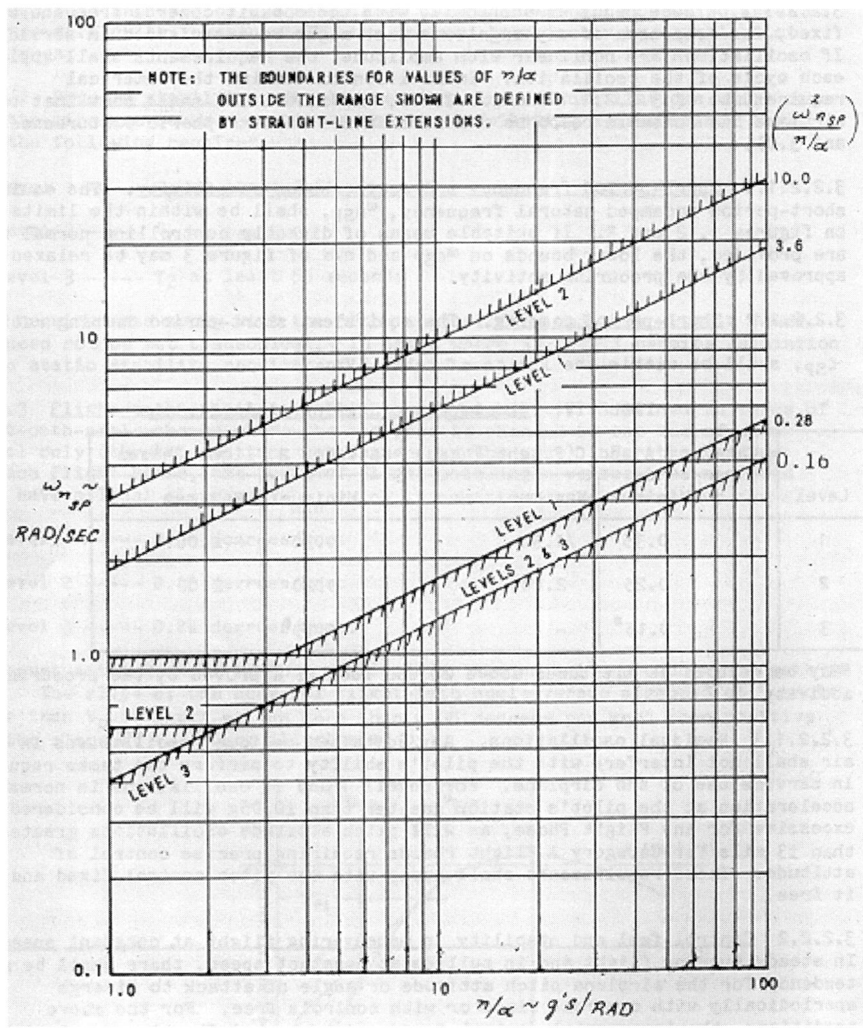

Fig. 7.19 MIL-F-8785C short period natural frequency requirements—Category A flight phases.

7.4.4.2 MIL-STD-1797A. MIL-STD-1797A provides almost identical requirements for the short period mode to those presented in MIL-F-8785C. The terms "equivalent frequency" and "equivalent damping ratio" are used, as with MIL-F-8785C, to account for the dynamics experienced by highly augmented aircraft. The concepts of short period natural frequency and damping ratio are retained when using these guidelines. The MIL-STD-1797A short period requirements are recast in terms of the control anticipation parameter (CAP) over an

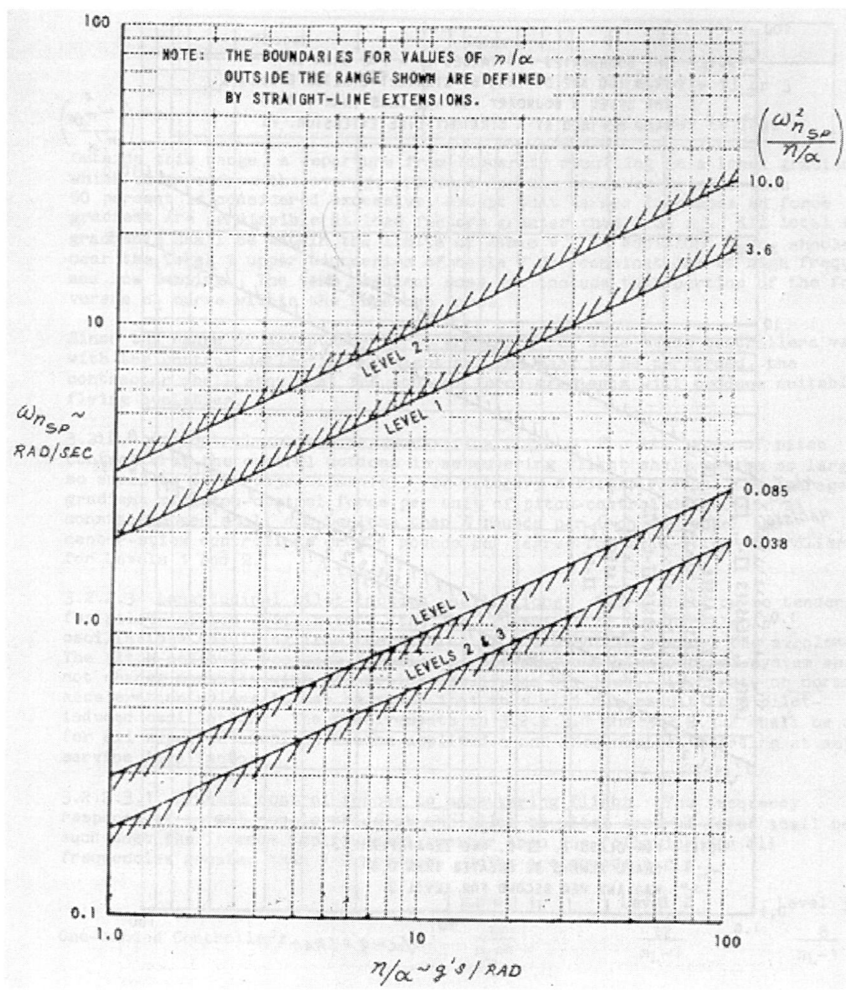

Fig. 7.20 MIL-F-8785C short period natural frequency requirements—Category B flight phases.

acceptable range of damping ratios. The CAP is estimated as

$$\text{CAP} \approx \frac{\omega_{n_{sp}}^2}{n/\alpha} \tag{7.60}$$

The CAP is represented on Figs. 7.19, 7.20, and 7.21 as the sloped boundaries of the Level 1, 2, and 3 regions. MIL-STD-1797A presents short-period compliance regions in terms of the CAP (which is directly proportional to the

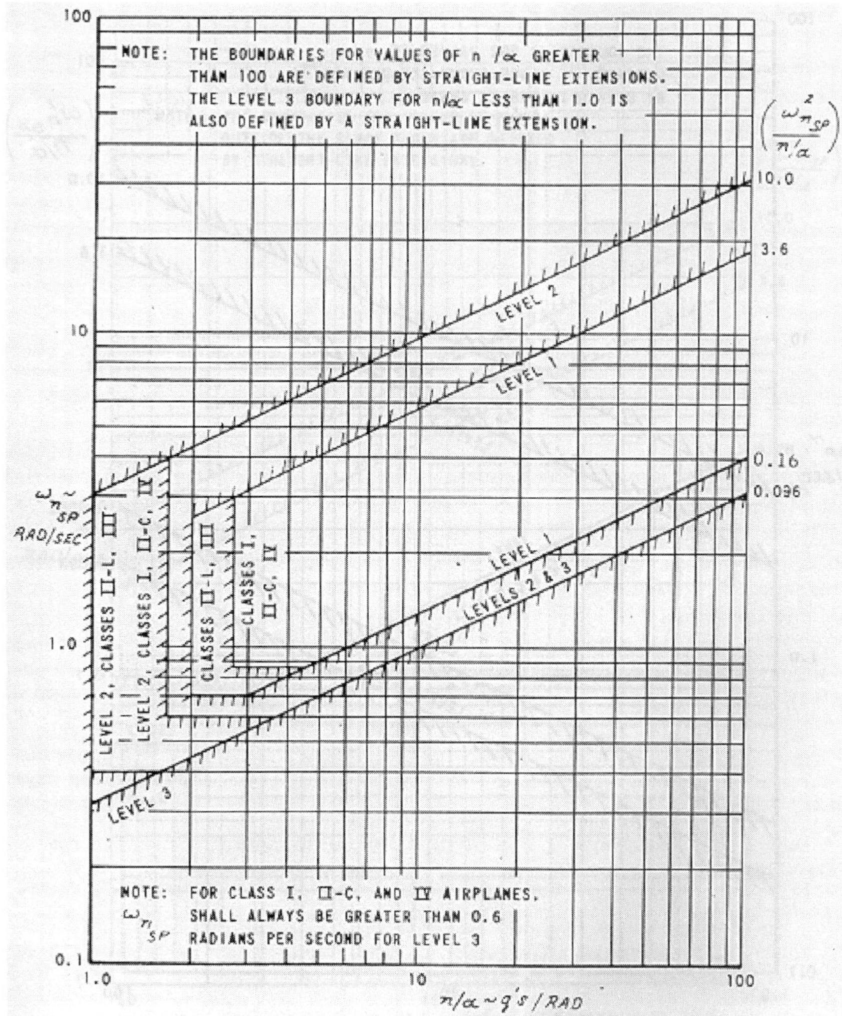

Fig. 7.21 MIL-F-8785C short period natural frequency requirements—Category C flight phases.

square of short period natural frequency) and damping ratio using Fig. 7.23. The only difference between the MIL-F-8785C requirements and those presented in Fig. 7.23 is that MIL-STD-1797A has dropped the lower Level 3 CAP limit. For the short period, these equivalent system requirements work well for highly augmented aircraft as long as the aircraft has a classical-looking response such as with an α-command, q-command, or g-command system (see Chapter 9). They do not work well for nonclassical response types such as

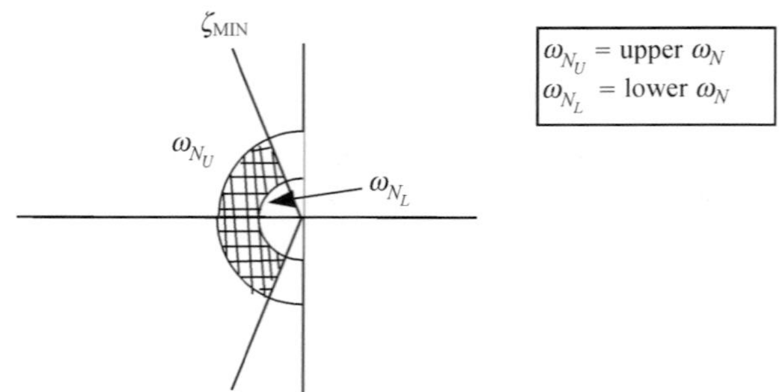

Fig. 7.22 S-plane MIL-F-8785C short period compliance region.

θ-command or flight path command systems. Consequently, MIL-STD-1797A also contains requirements on frequency-response shape that are not defined as a function of the classical dynamic modes discussed in this text.[3]

The level boundaries in Figs. 7.19–7.21 are specified in MIL-STD-1797A using Table 7.8.

In practice, the parameter of equivalent time delay must be considered when applying either MIL-F-8785C or MIL-STD-1797A. Only a snapshot of MIL-STD-1797A has been presented in this section. A detailed reading of this document is necessary for application to an actual aircraft design or flight test evaluation.

7.4.5 Phugoid

MIL-F-8785C has only a requirement on damping ratio for the phugoid mode. This requirement is independent of class and category and is presented in Table 7.9.

For Level 3, the phugoid mode is allowed to be unstable as long as the time to double amplitude is greater than or equal to 55 s. Equation (7.49) can be used to compute T_2. Under Level 3 conditions, a phugoid with neutral or positive stability will always satisfy the Level 3 requirement. The phugoid requirement is not very demanding because the phugoid is a low-frequency mode that generally has little effect on precision tasks. The pilot typically has sufficient time to correct for any undesirable phugoid characteristics. However, it can be important during unattended or divided-attention operation of the aircraft. Figure 7.24 presents the acceptable region for Level 1 phugoid roots on the s plane.

7.4.6 Roll

MIL-F8785C specifies maximum limits on the roll mode time constant (τ_r) which depend on class and category. Table 7.10 presents these requirements.

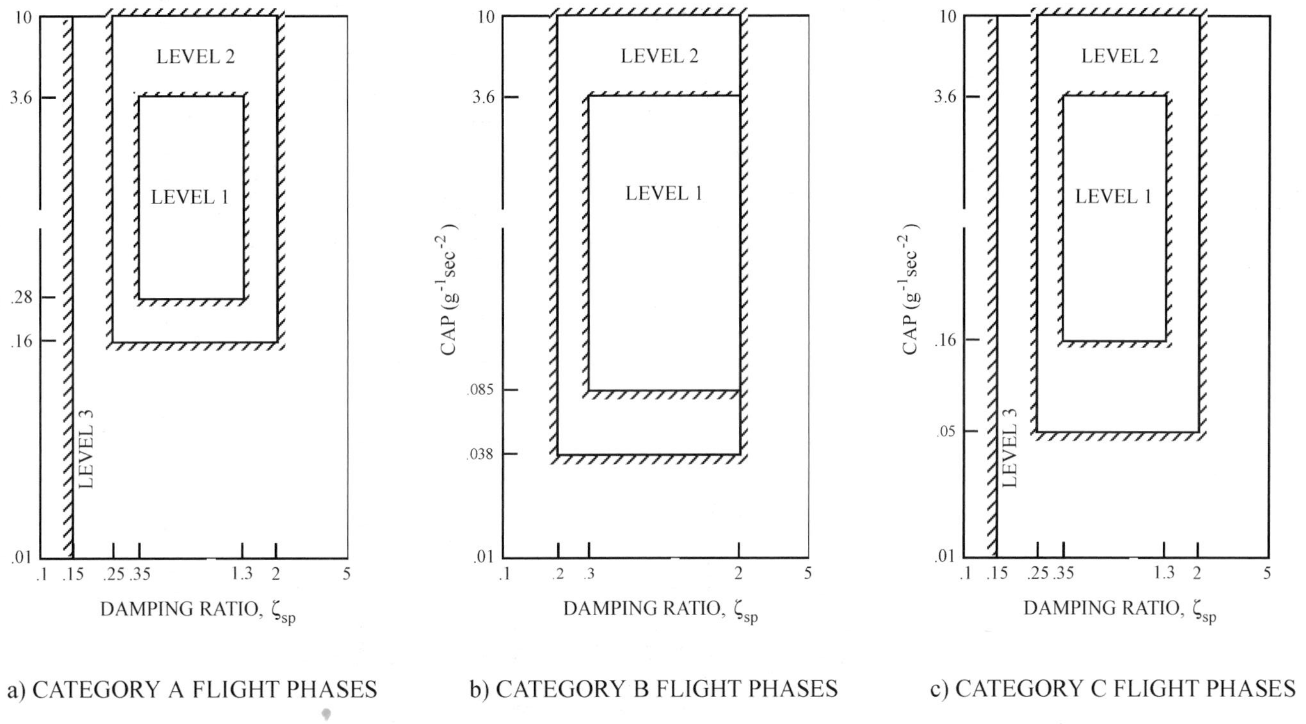

Fig. 7.23 MIL-STD-1797A short-period dynamic requirements.

Table 7.8 Additional short-period dynamic requirements

Flight phase category	Aircraft class	Level 1	Level 2	Level 3
A	All	$\omega_{sp} \geq 1$ rad/s	$\omega_{sp} \geq 0.6$ rad/s	$\zeta_{sp} \geq 0.15$ may be relaxed above 20,000 ft
B	All			T_2, the time to double amplitude based on the unstable root, should be no less than 6 s. In the presence of any other Level 3 flying qualities, ζ_{sp} should be at least 0.05.
C	I, II- C, IV	$\omega_{sp} \geq 0.87$ rad/s $n/\alpha \geq 2.7$ g/rad	$\omega_{sp} \geq 0.6$ rad/s $n/\alpha \geq 1.8$ g/rad	
	II-L, III	$\omega_{sp} \geq 0.7$ rad/s $n/\alpha \geq 2.0$ g/rad	$\omega_{sp} \geq 0.4$ rad/s $n/\alpha \geq 1.0$ rad	

Table 7.9 Phugoid damping requirements

Level 1	$\zeta > 0.04$
Level 2	$\zeta > 0$
Level 3	$T_2 \geq 55$ s

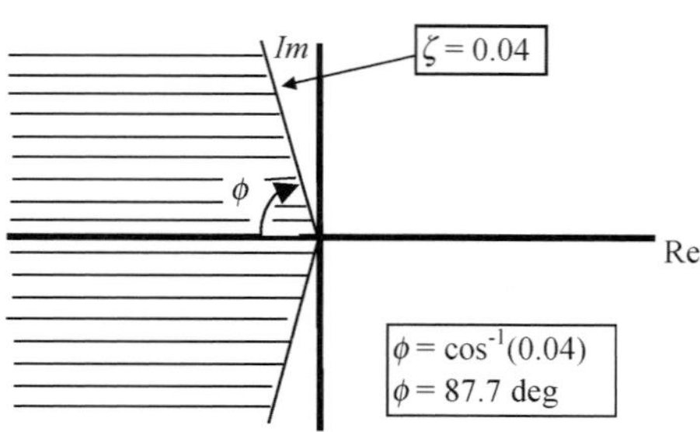

Fig. 7.24 S-plane MIL-F-8785C phugoid level 1 compliance region.

Table 7.10 Maximum roll mode time constant (seconds)

Flight phase category	Class	Level 1	Level 2	Level 3
A	I and IV	1.0	1.4	10
	II and III	1.4	3.0	10
B	All	1.4	3.0	10
C	I, II-C, and IV	1.0	1.4	10
	II-L and III	1.4	3.0	10

Recalling Eq. (7.48), the roll mode root of the characteristic equation is

$$s = -\frac{1}{\tau_r}$$

and the compliance region for the roll mode root becomes

$$s < -\frac{1}{\tau_{r_{\max}}}$$

Because the roll mode is first order, the root will lie on the negative real axis and the compliance region as shown in Fig. 7.25.

By specifying a maximum roll mode time constant, MIL-F-8785C is essentially specifying that the step response of the roll mode must be faster than or equal to the time constant value specified.

7.4.7 Spiral

MIL-F-8785C specifies that the spiral mode meet the requirements for the time to double amplitude of the bank angle as presented in Table 7.11 for bank angle disturbances of up to 20 deg. Therefore, an unstable spiral is acceptable if the time to double amplitude is slow enough. Obviously, any stable spiral (negative root of the characteristic equation) is Level 1. The normal flight test

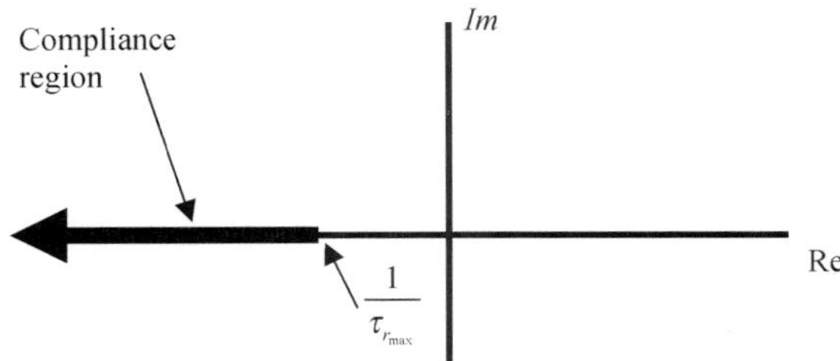

Fig. 7.25 S-plane MIL-F-8785C roll mode compliance region.

Table 7.11 MIL-F-8785C spiral mode minimum time to double amplitude

Flight phase category	Level 1	Level 2	Level 3
A and C	12 s	8 s	4 s
B	20 s	8 s	4 s

approach to evaluate the spiral is to trim the aircraft for wings-level, zero-yaw rate flight and then bank the aircraft to a 20-deg roll angle and neutralize the controls. If the time to 40 deg is less than the time specified in Table 7.11, then the spiral requirements of MIL-F-8785C are not met.

Recalling Eq. (7.49), the spiral mode root of the characteristic for an unstable spiral is equal to:

$$s = \frac{0.693}{T_2}$$

Therefore, the compliance region for the spiral root in the s plane becomes everything on the real axis to the left of:

$$s = \frac{0.693}{T_{2_{\min}}}$$

The spiral compliance region is shown in Fig. 7.26.

7.4.8 Dutch Roll

The MIL-F-8785C requirements for the dutch roll consist of a minimum ζ, a minimum ω_N, and a minimum $\zeta\omega_N$, as shown in Table 7.12. The table must

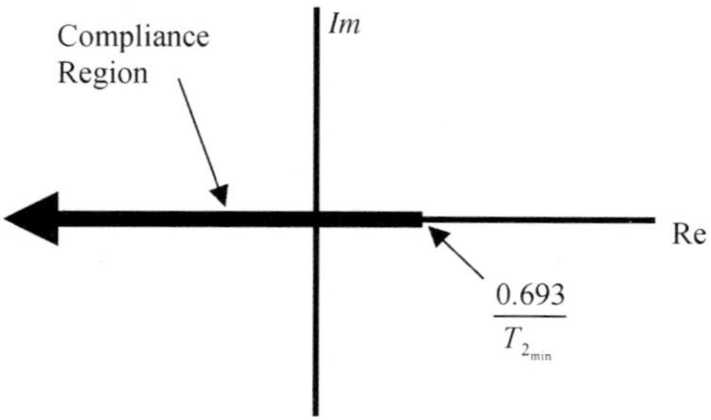

Fig. 7.26 MIL-F-8785C spiral compliance region.

Table 7.12 Minimum dutch roll frequency and damping

Level	Flight phase category	Class	Minimum ζ*	Minimum $\zeta\omega_N$,* rad/s	Minimum ω_N, rad/s
	A (CO, GA, RR, TF, RC, FF, and AS)	I, II, III, and IV	0.4	0.4	1.0
	A	I and IV	0.19	0.35	1.0
1		II and III	0.19	0.35	0.4
	B	All	0.08	0.15	0.4
	C	I, II-C, and IV	0.08	0.15	1.0
		II-L and III	0.08	0.10	0.4
2	All	All	0.02	0.05	0.4
3	All	All	0	—	0.4

* The governing damping requirement is that yielding the larger value of ζ_d, except that a ζ_d of 0.7 is the maximum required for Class III aircraft.

be modified if $\omega_D^2|\phi/\beta|_{DR} > 20\ \text{rad}^2/\text{s}^2$ by increasing the $\zeta\omega_N$ requirement (details are provided in MIL-F-8785C).

A generalized compliance region for dutch roll roots is presented as the shaded area of Fig. 7.27.

Example 7.15

For the Lear Jet in cruise using the same flight condition (40,000 ft and M = 0.7) and data used in Examples 7.10 and 7.12, determine if the aircraft satisfies the dynamic stability requirements of MIL-F-8785C for Level 1.

To start the evaluation, we must first determine the aircraft class and category:

$$\text{Class II, Category B (Cruise)}$$

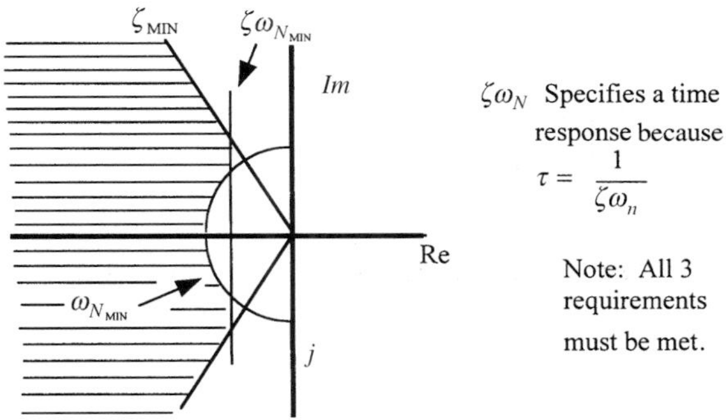

Fig. 7.27 S-plane MIL-F-8785C dutch roll compliance region.

Longitudinal Modes of Motion:

Short Period:

From Example 7.10,

$$\omega_{n_{sp}} = 2.836 \text{ rad/s}$$
$$\zeta_{SP} = 0.335$$

Referring to Table 7.7, because $\zeta_{sp} > 0.3$ and $\zeta_{sp} < 2.0$, the Lear Jet passes the damping ratio requirement at Level 1.

To evaluate $\omega_{n_{sp}}$ for compliance, we need to determine n/α. From Eq. (7.59), and knowing that Z_α is equal to $-451.7\,\text{ft/s}^2$ for this aircraft and flight condition, we have

$$\frac{n}{\alpha} = -\frac{Z_\alpha}{g} = \frac{451.7}{32.2} = 14.03 \text{ g/rad}$$

Using Fig. 7.20 for a Category B Flight Phase, we plot $\omega_{n_{SP}}$ at an $n/\alpha = 14.03$. The point falls within the Level 1 compliance region and therefore passes the Level 1 short period natural frequency requirement. Because the Lear Jet passes both short period dynamic stability requirements at Level 1, the Lear Jet's short period dynamic characteristics are considered Level 1.

Phugoid:

From Example 7.10,

$$\zeta_{PH} = 0.076$$

Referring to Table 7.9, $\zeta_{PH} > 0.04$ and the Level 1 phugoid requirement is passed.

The Lear Jet (at this flight condition), therefore, meets the MIL-F-8785C Level 1 dynamic stability requirements for the longitudinal modes of motion (short period and phugoid).

Lateral-Directional Modes of Motion:

Roll:

From Example 7.12,

$$\tau_r = 1.972 \text{ s}$$

Using Table 7.10 for Category B, $\tau_r > 1.4\,\text{s}$ and it fails the Level 1 requirement. Because $\tau_r < 3\,\text{s}$, it passes Level 2. Therefore, the roll mode is Level 2.

Spiral:

From Example 7.12, the spiral mode is stable ($s = -0.00101$) with a time constant of

$$\tau_s = 991.8 \text{ s}$$

Because the spiral is stable, the time-to double-amplitude requirements of Table 7.11 are automatically met. Therefore the Lear Jet spiral mode passes the Level 1 MIL-F-8785C dynamic stability requirement.

Dutch Roll:

From Example 7.12,

$$\omega_{N_{DR}} = 1.168 \text{ rad/s}$$

$$\zeta_{DR} = 0.036$$

Referring to Table 7.12, we will look at the dutch roll damping ratio requirement first.

$$\zeta_{DR} = 0.036 < 0.08,$$

so it *fails* Level 1. Because

$$\zeta_{DR} = 0.036 > 0.02$$

it passes Level 2.

We will next look at the Table 7.12 requirement for $\zeta_{DR}\omega_{N_{DR}}$. This product must be greater than 0.15 to pass Level 1. For our case,

$$\zeta_{DR}\omega_{N_{DR}} = 0.0582 < 0.15$$

therefore, it *fails* Level 1 for this requirement. Because $\zeta_{DR}\omega_{N_{DR}} > 0.05$ it barely passes Level 2.

Finally, we must look at the dutch roll natural frequency requirement in Table 7.12. We have

$$\omega_{N_{DR}} = 1.618 \text{ rad/s} > 0.4 \text{ rad/s}$$

therefore, the Lear Jet meets the Level 1 dutch roll natural frequency requirement.

The Lear Jet dutch roll mode is rated Level 2 based on failing the Level 1 $\zeta_{DR}\omega_{N_{DR}}$ dynamic stability requirement in MIL-F-8785C.

The Lear Jet lateral-directional mode is rated Level 2 based on both the dutch roll and roll mode characteristics being rated Level 2.

Overall, the Lear Jet meets MIL-F-8785C Level 2 dynamic stability requirements because of the dutch roll and roll mode characteristics.

7.5 Cooper–Harper Ratings

All of the previous discussion on dynamic stability guidelines may imply that good aircraft handling qualities are simply a matter of satisfying published criteria on parameters such as natural frequency, damping ratio, and time constant for each of the dynamic modes. Nothing could be further from the truth. The criteria discussed provides a starting point for designers and a tool for flight testers to evolve the handling qualities of an aircraft. A vital contribution to this evolution is pilot comments obtained from simulations and test flying of the aircraft. A structured rating scale for aircraft handling qualities was developed by NASA in the late 1960s called the Cooper–Harper rating scale. This rating scale applies to specific pilot-in-the-loop tasks such as air-to-air tracking, formation flying, and approach. It does not apply to open-loop aircraft characteristics such as yaw response to a gust. An important part of using the Cooper–Harper rating scale is careful definition of the evaluation task and performance standards for that task. Figure 7.28 presents the Cooper–Harper rating scale.

The Cooper–Harper rating scale is a decision tree for pilots to rate a specific mission task. The pilot begins the decision process at the lower left corner. Aircraft controllability, pilot compensation (workload), and task performance are key factors in the pilot's evaluation. A Cooper–Harper rating of "one" is the highest or best and a rating of "ten" is the worst, indicating the aircraft cannot be controlled during a portion of the task and that improvement is mandatory. It is important to observe that a Cooper–Harper rating of one through three generally corresponds to Level 1 flying qualities, a rating of four through six corresponds to Level 2 flying qualities, and a rating of seven through nine corresponds to Level 3.

It should be understood that two different test pilots may rate the same mission task differently based on different experience levels and aircraft background. This is to be expected, and generally the discussions that evolve based on differences in the ratings help further define the nature of the handling qualities for that aircraft.

With today's highly augmented aircraft, adjustments can be made within the flight control system during the development phases to achieve nearly optimal dynamic stability characteristics. In this process, the pilot comments, using the Cooper–Harper rating scale, play an important role along with MIL-F-8785C and MIL-STD1797A.

7.6 Experimental Determination of Second-Order Parameters

For second-order oscillatory responses, several methods are available to experimentally determine key dynamic stability parameters such as damping ratio and natural frequency. Two such methods will be discussed, both of which require a time history of a key aircraft motion parameter for that mode. The time history is normally obtained in flight test by trimming the aircraft and then exciting the mode with a control doublet (a cyclic control input that perturbs the aircraft on both sides of the trim condition). The time history needed begins immediately after the doublet stops and is often referred to as the **free response** or **transient response**. For example, for the short period mode, the aircraft would be trimmed and a flight test data acquisition system would be activated to record a time history of angle of attack and/or pitch rate. The pilot would input a doublet (a quick pitch stick input: forward–aft–neutral) and the time history needed would begin when the stick is returned to the neutral or trimmed position.

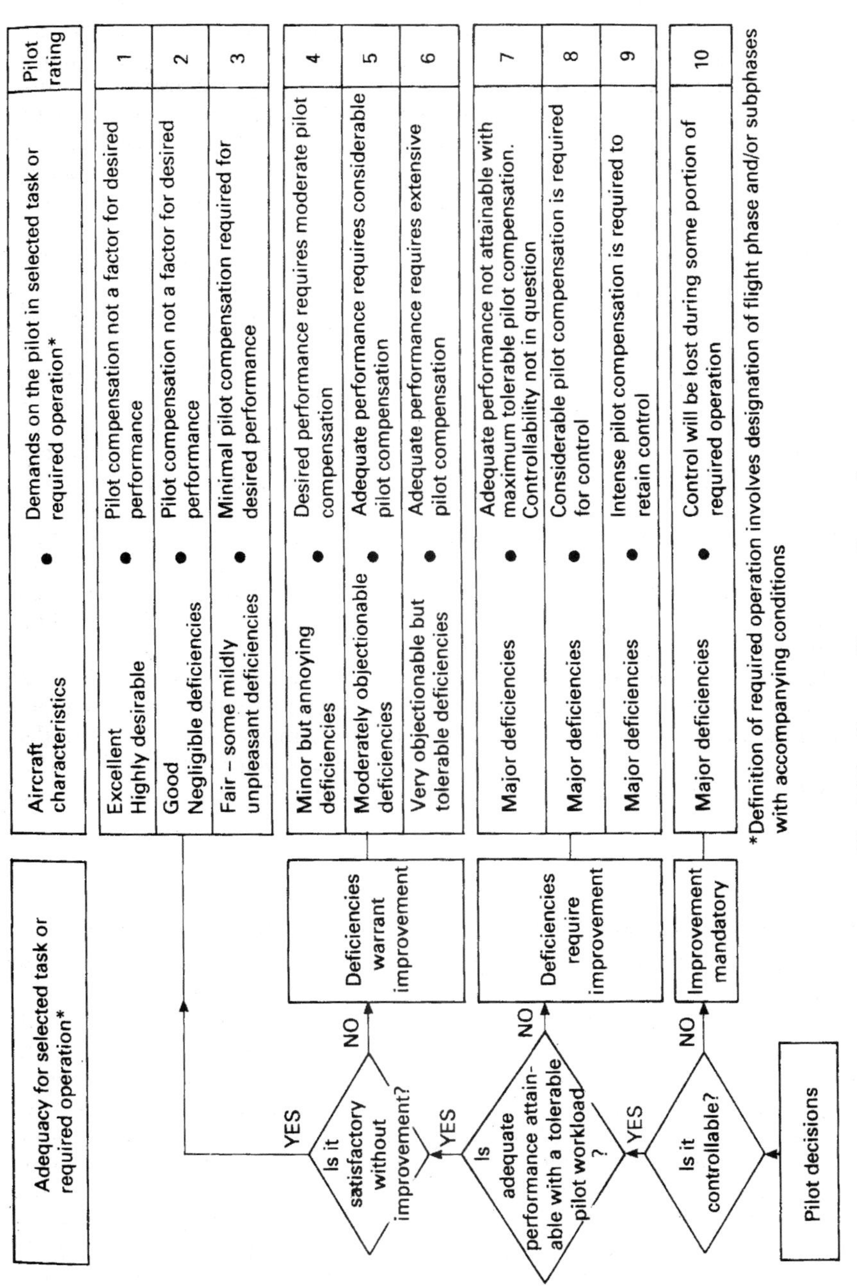

Fig. 7.28 Cooper–Harper rating scale.

7.6.1 Log Decrement Method

If the transient response has **three or more overshoots**, the log decrement method (also called the subsidence ratio or the transient peak ratio method) can be used. The value of each peak deviation from the trim condition (ΔX_0, ΔX_1, ΔX_2, etc.) is measured as shown in Fig. 7.29.

Next, the transient peak ratios, such as $\Delta X_1/\Delta X_0$, $\Delta X_2/\Delta X_1$, $\Delta X_3/\Delta X_2$, are determined, and Fig. 7.30 is used to determine corresponding values for damping ratio (ζ_1, ζ_2, ζ_3, etc.). The average of these values is used to determine the overall ζ.

The period, T, is determined based on a full cycle of an oscillation as shown in Fig. 7.29. The damped frequency can then be found using

$$\omega_D = \frac{2\pi}{T}$$

and the natural frequency can be found using

$$\omega_N = \frac{\omega_D}{\sqrt{1-\zeta^2}}$$

7.6.2 Time Ratio Method

If the transient response has **two or less overshoots**, the time ratio method is appropriate. A well-defined, free-response oscillation is selected and then the amplitude of the peak is used to determine the time to 73.6%, 40.9%, and 19.9% of the peak value as show in Fig. 7.31.

Each of these values (t_1, t_2, and t_3) is then used to form the ratios:

$$\frac{t_2}{t_1}, \quad \frac{t_3}{t_1}, \quad \text{and} \quad \frac{(t_3-t_2)}{(t_2-t_1)}.$$

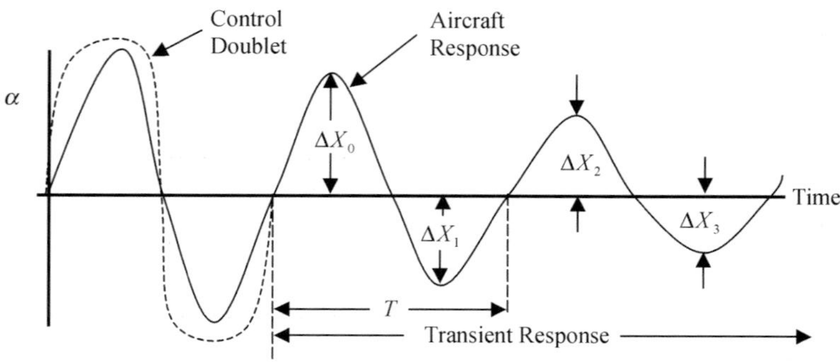

Fig. 7.29 Log decrement method relationships.

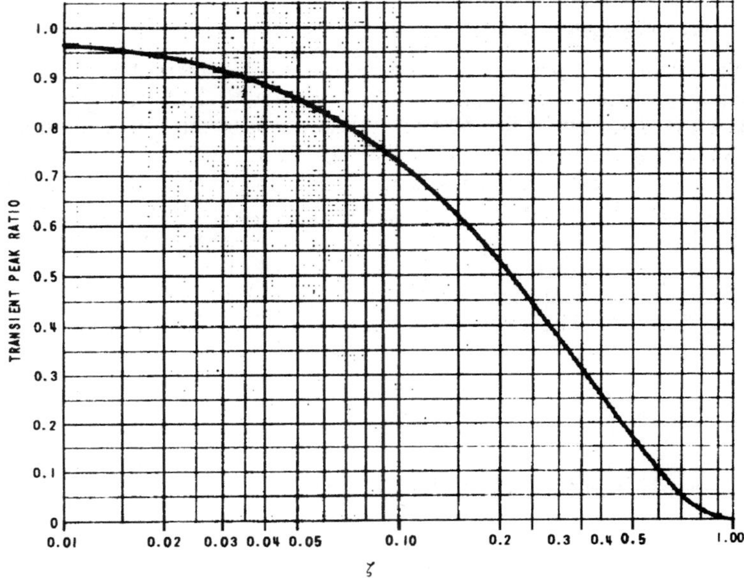

Fig. 7.30 Transient peak ratio vs damping ratio–log decrement method.

Figure 7.32 is then entered on the time ratio (right) side to find the corresponding damping ratio for each time ratio. These values are then averaged to determine the overall ζ. With the overall ζ, Fig. 7.32 is then re-entered to obtain the frequency time products $\omega_N t_1$, $\omega_N t_2$, and $\omega_N t_3$. To determine the natural frequency, compute

$$\omega_{N_1} = \frac{\omega_N t_1}{t_1}, \quad \omega_{N_2} = \frac{\omega_N t_2}{t_2}, \quad \omega_{N_3} = \frac{\omega_N t_3}{t_3}$$

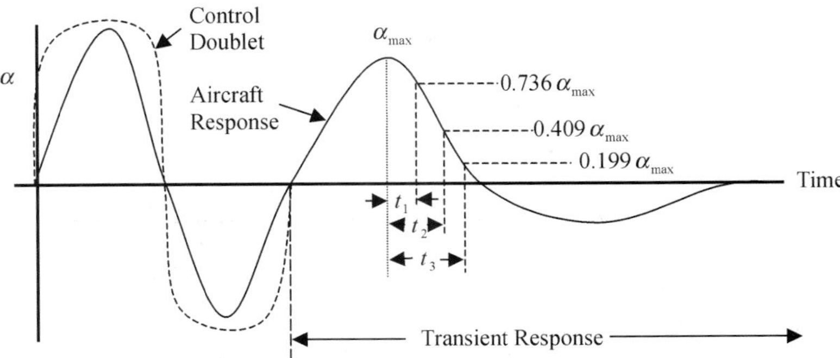

Fig. 7.31 Time ratio method relationships.

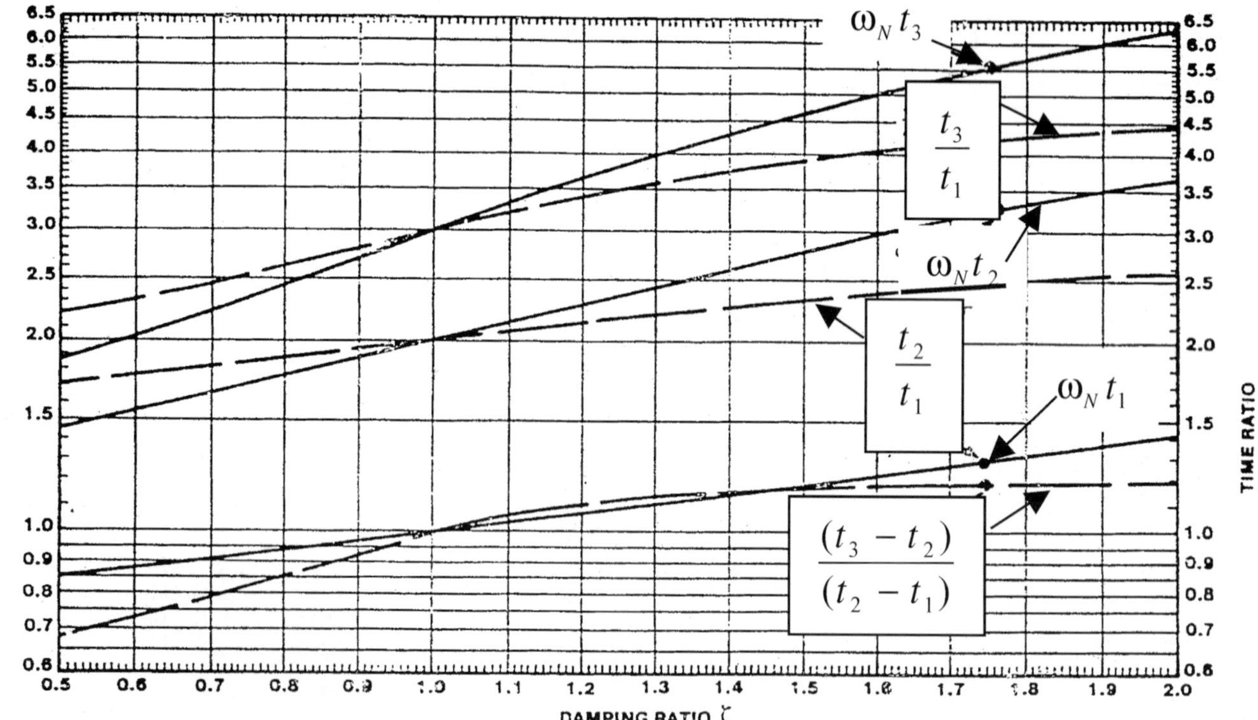

Fig. 7.32 Frequency time product and time ratio vs damping ratio–time ratio method.

ω_{N_1}, ω_{N_2}, and ω_{N_3} are then averaged to determine the overall natural frequency.

7.6.3 Test Pilot Approximation for Damping Ratio

To obtain a rough estimate of the damping ratio for the short period and dutch roll modes, test pilots sometimes use the following approximation which lends itself to easy application in flight:

$$\zeta \approx \frac{7 - (\text{number of over/undershoots})}{10} \tag{7.61}$$

The number of over/undershoots during the free response is counted by the pilot after the doublet is finished, and simple mental arithmetic in flight using the approximation will lead to an approximate value of the damping ratio. **The approximation is only applicable to situations where the number of free response over/undershoots is less than seven**. If seven or more over/undershoots are experienced and the response eventually damps out, the pilot may only conclude that the damping ratio is somewhere between 0 and 0.1, and that the response is lightly damped.

Using the response of Fig. 7.31 as an example, the pilot would count two peaks or over/undershoots. Using Eq. (7.61), the test pilot approximation would yield an estimate of 0.5 for the damping ratio. Of course, more refined methods such as the log decrement method or time ratio method are used after time history plots are available.

7.7 Historical Snapshot—The A-10A Prototype Flight Evaluation

The Air Force conducted a competitive flight evaluation of two prototype close air support aircraft designs in 1972. The overall program was named the "A-X" Competitive Flyoff and the two competing aircraft were the Northrop A-9A and the Fairchild Republic A-10A.[3,4] Figures 7.33 and 7.34 present

Fig. 7.33 A-9A prototype aircraft.

Fig. 7.34 A-10A prototype aircraft.

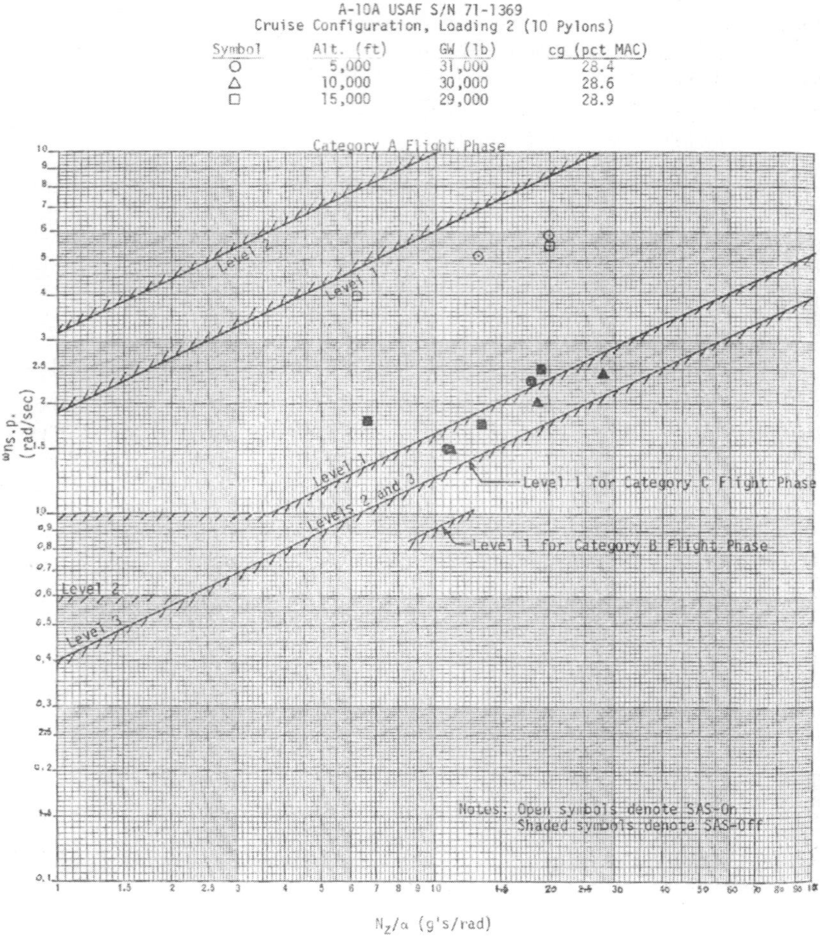

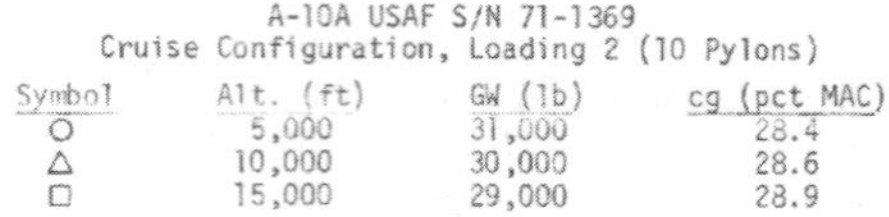

A-10A USAF S/N 71-1369
Cruise Configuration, Loading 2 (10 Pylons)

Symbol	Alt. (ft)	GW (lb)	cg (pct MAC)
○	5,000	31,000	28.4
△	10,000	30,000	28.6
□	15,000	29,000	28.9

Fig. 7.35 A-10A prototype short period dynamic stability characteristics.

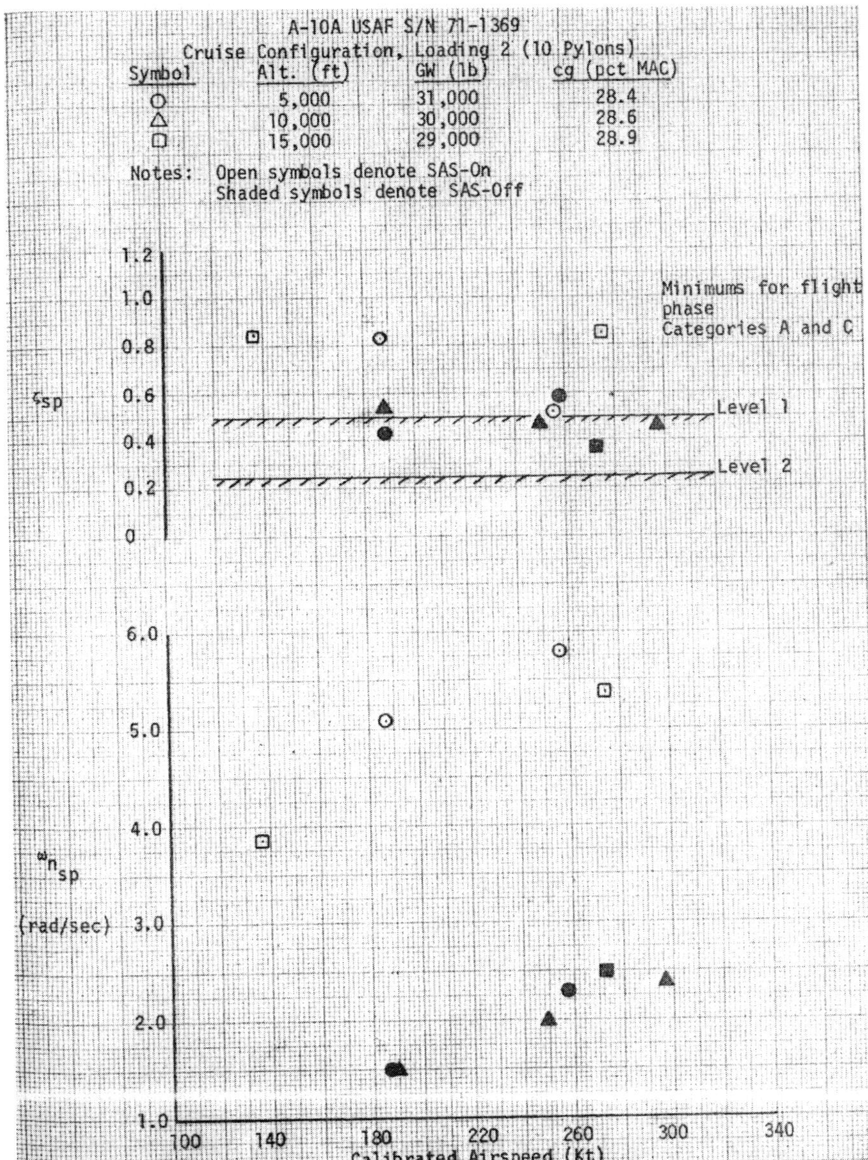

Fig. 7.36 A-10A prototype short period damping ratio and natural frequency characteristics.

pictures of these prototype aircraft, which had significantly different design approaches to satisfy the same mission requirements.

The A-10 was declared the winner in 1973 and went on to full-scale production of 733 aircraft. It also established its reputation for lethality, responsivness, and survivability during the cold war years of the late 1970s through 1980s, and as a key aircraft in the destruction of tanks, artillery, ground vehicles, and missile sites during the Gulf War.

During the A-X Competitive Flyoff, the flying qualities of both aircraft were evaluated using techniques presented in this chapter. To evaluate the dynamic longitudinal stability of the short period mode, elevator doublets were performed at selected airspeed and altitude combinations to determine damping ratio and natural frequency characteristics. Figure 7.35 presents a MIL-F-8785C short period natural frequency compliancy plot for the A-10A.

Notice that the open symbols are for evaluations with the stability augmentation system (SAS) on and the darkened symbols are with the SAS off. The SAS is a system that enhances dynamic stability characteristics and will be discussed in detail in Chapter 9. With the SAS on, the A-10's short period natural frequency characteristics clearly met Level 1 requirements. With the SAS off, several points fell in the Level 2 category; however, this is permitted by the MIL SPEC for a degraded system. Figure 7.36 presents A-10A short period damping ratio characteristics along with natural frequency data.

Note that the A-10 also passed the Level 1 damping ratio requirement with the SAS on, but several points are in the Level 2 category with the SAS off.

References

[1]MIL-F-8785C, *Flying Qualities of Piloted Airplanes*, Military Specification, 1980.

[2]Papa, J. A., Douglas, A. F., Markwardt, J. H., and Fortner, L. D., "A-10A Prototype Task II Performance and Flying Qualities Evaluation," Air Force Flight Test Center TR 73-7, Edwards AFB, CA, March 1973.

[3]Yechout, T. R., Lucero, F. N., and Bridges, R. D., "Air Force Flight Evaluation of the A-10A Prototype Aircraft," Air Force Flight Test Center TR 73-3, Edwards AFB, CA, March 1973.

Problems

7.1 Given the following system, determine the solution of the homogenous equation (which is normally called either the complementary or transient solution) for the given initial conditions:

$$5\dot{X} + 2X = 0$$
$$X(0) = 2$$

7.2 Given the following equation for a mass–spring–damper system

$$M\ddot{X} + C\dot{X} + KX = KY$$

where we assume
(a) the system is massless
(b) the spring has an initial displacement of $Y = 10$ in.
(c) the spring constant equals 5 lb/in.
(d) the damping coefficient equals 0.1 (lb-s/in.).

Given the initial condition, $X(0) = 2$, find
(1) the complementary or transient solution
(2) the particular or steady state solution
(3) the general or total solution

7.3 Using the baseline system of Problem 7.2, find the general or total solution if the forcing function is not a constant (that is, step) but of the form

$$f(t) = K \sin \omega t$$

with $\omega = 2$ rad/s.

7.4 For Problem 7.2 (with the step input) find
(1) time constant
(2) time for the system to reach 98.2% of its final value

7.5 For the following second-order system and initial conditions, find the transient solution:

$$\ddot{X} + 8\dot{X} + 12X = 15$$
$$X(0) = 2$$
$$\dot{X}(0) = 2$$

7.6 Given $M = 2$, $C = 12$, $K = 50$, $Y = 2$
(1) Write out the equation for a mass–spring–damper system where the initial displacement is 2.
(2) Find the transient or complementary solution.
(3) Find the steady state or particular solution.
(4) Find the total or general solution.
(5) Evaluate the two unknown coefficients in the general solution given the following initial conditions:

$$X(0) = \dot{X}(0) = 0$$

7.7 Write out the general solution to Problem 7.6, Part (5), in the phase angle form.

7.8 Sketch the time response to Problem 7.7 by calculating $X(t)$ at the following times:

$$t = 0,\ 0.2,\ 0.5,\ 0.8,\ 1,\ 1.4,\ 1.7$$

What does the steady-state value appear to be? Does this agree with the particular solution?

7.9 For Problem 7.6, calculate the natural frequency (ω_n), damping ratio (ζ), and damped frequency (ω_D). Having done this, do the following

(1) Rewrite the expression for the mass–spring–damper system as a function of damping and natural frequency.
(2) With damping calculated, would you expect an oscillatory type response?
(3) Rewrite the answer to Problem 7.7 (which is in terms of the phase angle) in the same form, but use the damping coefficient, natural and damped frequency expressions.
(4) Because the time constant for a second-order system is $\tau = 1/\zeta\omega_N$, how long would it take for the system to reach 63.2 and 98.2% of its final value?
(5) Looking back at the sketch you made of this system's response, do you agree that at four time constants, the system is within plus or minus 2% of its final value?

7.10 Given the following equation for a mass–spring–damper system

$$\ddot{X} + 1.8\dot{X} + 9X = 9$$

What is the solution at $t = 0.3$ s?

7.11 Given the location of the following roots on the complex plane

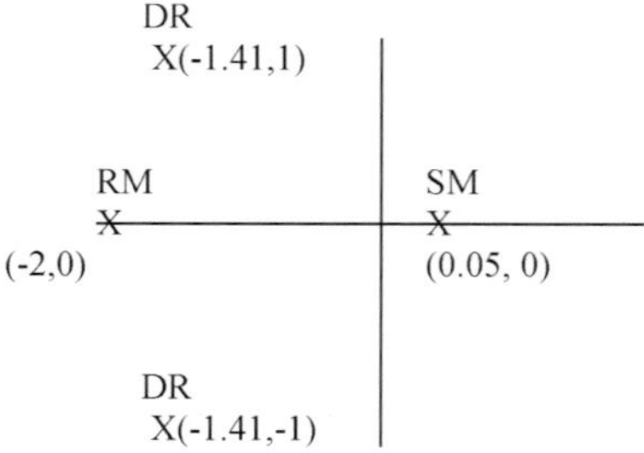

Determine the time constant for the dutch roll, spiral, and roll modes. In addition, determine the natural frequency and damping ratio of the dutch roll mode.

7.12 Match the following root locations on the complex plane:

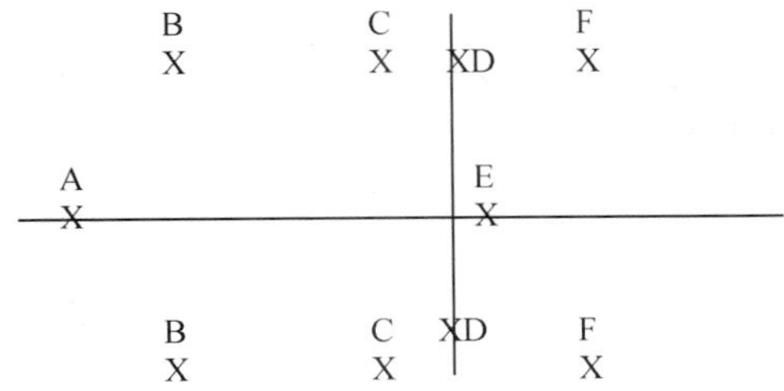

with the following responses

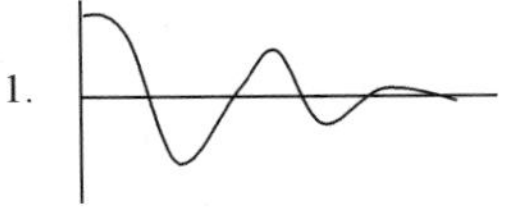

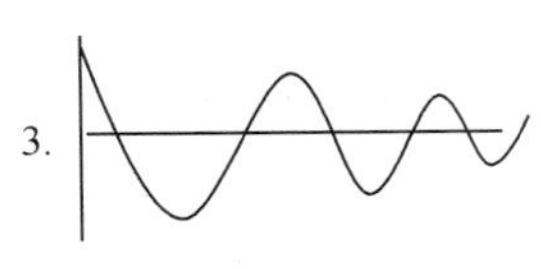

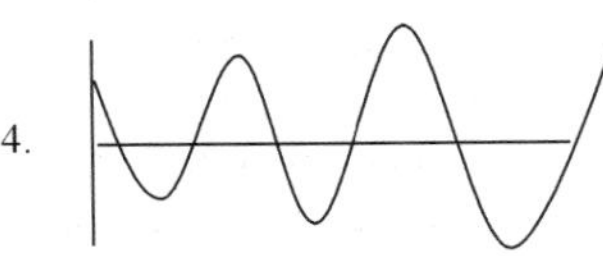

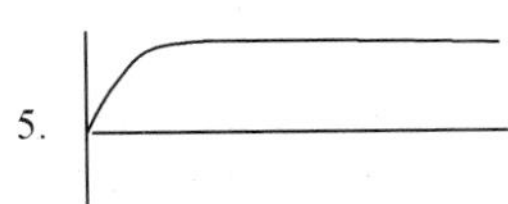

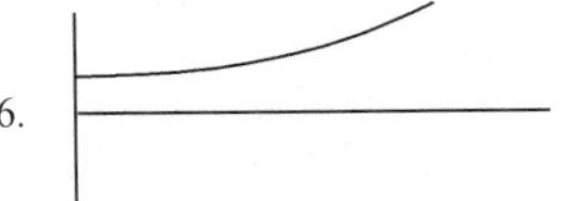

7.13 Determine the Laplace transform of the following functions (using the transform table):

(1) $6e^{-5t}$ (2) $3e^{2t}$

(3) 1 (4) $12\sin 5t$

(5) $\ddot{X} + 2\zeta\omega_n\dot{X} + \omega_n^2 X = 0$ (6) $\ddot{X} + 2\zeta\omega_n\dot{X} + \omega_n^2 X = \omega_n^2 Y$

Assume the initial conditions are zero for parts 5 and 6.

7.14 Using the results of Problem 7.13, Part (6), write (in Laplace form) the ratio of the output to input, $X(s)/Y(s)$.

7.15 Find the inverse Laplace of

$$\frac{2.56}{s(s^2 + 1.6s + 2.56)} \quad \text{at } t = 3 \text{ s}$$

7.16 Determine the partial fraction expansion and inverse transformation of the following function.

$$F(s) = \frac{s+2}{s(s+3)(s+1)}$$

7.17 For the following AOA to elevator transfer function

$$\frac{\alpha(s)}{\delta_e(s)} = \frac{\omega_n^2}{(s^2 + 2\zeta\omega_n s + \omega_n^2)}$$

Determine the steady-state value of AOA for a unit step elevator input using the final value theorem.

7.18 Show how you would go from the pitching moment differential equation presented in Eq. (7.26) to its equivalent in Laplace and matrix form Eq. (7.27).

7.19 Given the 2-DOF short period approximation, Eq. (7.33), determine the $\alpha(s)/\delta_e(s)$ transfer function using Matrix algebra.

7.20 A Marine Corps A-4 flying at 15,000 ft, Mach 0.6 with an alpha trim of 3.4 deg, has the following short period characteristics:

$$\text{short period damping ratio} = 0.304$$
$$\text{short period natural frequency} = 3.69\ \text{rad/s}$$

If the pilot has just made a unit step elevator input, calculate how long it takes for the A-4 to reach the first overshoot, the magnitude of that overshoot, and how long it takes for the A-4 angle of attack to be within $\pm 2\%$ of its final steady-state value. Assume a second-order response.

7.21 A Navy A-4 flying alongside the Marine Corps A-4 has the following phugoid characteristics:

$$\text{phugoid natural frequency} = 0.0635\ \text{rad/s}$$
$$\text{phugoid damping ratio} = 0.0867$$

Does the A-4 have a stable or unstable phugoid? After the Naval Flight Officer has made a unit step elevator input, calculate the time to half amplitude or time to double amplitude as appropriate.

7.22 The following is the longitudinal characteristic equation for an F-89 Scorpion flying at 20,000 ft at Mach 0.638. Determine the short period and phugoid natural frequencies:

$$(s^2 + 4.2102s + 18.2329)(s^2 + 0.00899s + 0.003969) = 0$$

7.23 An aircraft has the following short period approximation:

$$\frac{\alpha}{\delta_e} = \frac{-0.746s - 208.6}{675s^2 + 1361.6s + 5452.45}$$

Find the natural frequency, damping ratio, and the steady-state value of AOA in response to a unit step input using the final value theorem.

7.24 An F-4 flying at 35,000 and 876 ft/s, has the following lateral-directional characteristic equation:

$$(s + 0.01311)(s + 1.339)(s^2 + 0.23137s + 5.7478) = 0$$

(1) Determine the aircraft category, class, and level if you know that all systems are operating properly and the aircraft is in the pre-contact refueling position.
(2) Determine the roll mode time constant.
(3) Determine the dutch roll damping ratio, the damped frequency, and the natural frequency.
(4) Compute the spiral mode $T_{1/2}$ or T_2, as appropriate.
(5) Estimate the ϕ/β ratio.

$$(C_{l_\beta} = -0.08;\ C_{n_\beta} = 0.125;\ I_{zz} = 139{,}800;\ I_{xx} = 25{,}000)$$

7.25 Given the complex plane root locations for a typical business jet, match each location to the appropriate dynamic modes of motion:
(1) Roll mode ________
(2) Dutch roll ________
(3) Phugoid ________
(4) Spiral ________
(5) Short period ________

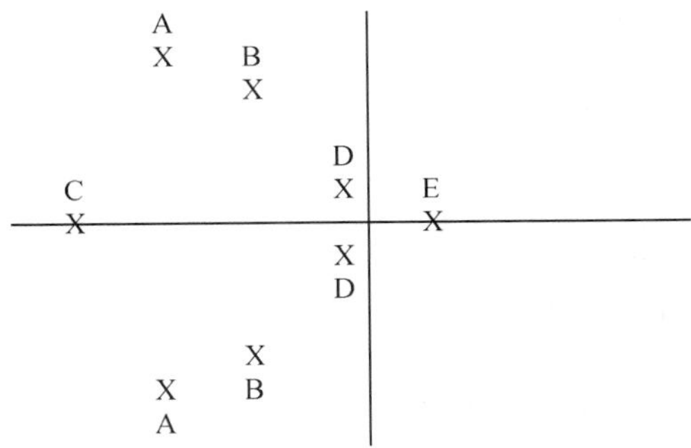

7.26 An F-16 flying at 40,000 ft and Mach 0.8 has the following longitudinal dynamic stability characteristics with all systems up and operating normally:

short period damping: 0.56

short period natural frequency: 2.9 rad/s

$\frac{n}{\alpha}$: 10.8

What is the appropriate class, category, and level and are the longitudinal short period MIL-F-8785C requirements met?

7.27 The C-5A cruising at $M = 0.7$ and 35,000 ft with a c.g. at 41% has the following flight characteristics:

phugoid natural frequency: 0.075 rad/s

phugoid time to half amplitude: 369.6 s.

Determine the C-5A damping ratio for that flight condition and determine if it satisfies MIL-F-8785C.

7.28 Find $y(t)$ for $\ddot{y} + 3\dot{y} + y = 1$, subject to $y(0) = 0$, $\dot{y}(0) = 1$. Check your answer to make sure initial conditions are met.

7.29 Solve

$$\dot{x} = 3x - 4y$$
$$\dot{y} = 2y - 8y$$

subject to $x(0) = 5$; $y(0) = 3$.

7.30 Given

$$U_1 = 700 \text{ ft/s}, \quad Z_\alpha = 400 \text{ ft/s}^2, \quad M_q = -0.9/\text{s},$$
$$M_\alpha = -8/\text{s}^2, \quad M_{\dot{\alpha}} = -0.4/\text{s}$$
$$Z_{\delta_e} = -0.75 \text{ ft/s}^2, \quad M_{\delta_e} = -0.3/\text{s}^3$$

Use the short period approximation to find $\alpha(s)/\delta_e(s)$.

7.31 Starting with Eq. (7.44), show which row and column of the Laplace matrix are eliminated to develop the dutch roll approximation. Which motion parameter is assumed to remain relatively constant in this approximation?

7.32 Find the dutch roll approximation for ω_{ND} and ζ_D given

$$U_1 = 200 \text{ ft/s} \qquad N_r = -0.1/\text{s} \qquad Y_r = 0.8 \text{ ft/s}$$
$$Y_B = -56 \text{ ft/s}^2 \qquad N_B = 2.6/\text{s}^2$$

7.33 A system has the following transfer function:

$$\frac{\theta_0}{\theta_i}\frac{(s)}{(s)} = \frac{1}{(s+10)(s^2+2s+16)}$$

Find the general form of the time response $\theta_0(t)$ if $\theta_i(s)$ is a unit impulse.
Find the general form of the time response $\theta_0(t)$ if $\theta_i(s)$ is a unit step.
What are the two time constants associated with the transient response terms?

7.34 If an aircraft has minor but annoying deficiencies and desired task performance required moderate pilot compensation, the Cooper–Harper rating should be __________ and the flying qualities level would be __________.

7.35 If a hydraulic system failure was estimated to occur every 500 flights on a new fighter aircraft, what flying qualities level should the aircraft be expected to fly at for this failure?

8
Classical Feedback Control

As we discussed in Chapter 7, an aircraft must have dynamic stability characteristics that generally meet specified criteria if the pilot is to consider the handling qualities of the aircraft acceptable. In many aircraft designs, a tradeoff exists between acceptable performance and acceptable dynamic stability characteristics. For example, to meet a performance requirement, aircraft drag may be minimized by decreasing the size of the horizontal tail. This, of course, will decrease the short period damping ratio and may result in handling qualities that do not meet Level 1 requirements. Most modern aircraft incorporate feedback control systems that augment the dynamic stability characteristics of an aircraft so that excellent performance and handling qualities are achievable. An aircraft feedback control system senses key aircraft motion parameters and, through control laws (computer calculations) and control surface actuators, deflects the appropriate control surface to oppose undesired motion (for example, add damping) and/or amplify the pilot's control command. Because of these feedback control systems, most modern aircraft have excellent handling qualities and are rated by test pilots very high on the Cooper–Harper scale. This chapter introduces the fundamentals of feedback control system design needed to tailor the dynamic stability characteristics of an aircraft.

8.1 Open-Loop Systems, Transfer Functions, and Block Diagrams

We are familiar with aircraft transfer functions and how they are used to represent the output/input for the aircraft. Transfer functions are used for a variety of applications and include dynamic behavior because they are based on the differential equations that define the system. Most systems can be thought of as involving a process where an input is converted to an output, as shown in Fig. 8.1.

The system represented in Fig. 8.1 is referred to as an open-loop system because the output does not affect the input. A common toaster is an example of an open-loop system. The first time a piece of bread is put in, the output may range anywhere from the same piece of bread, to nicely toasted bread, to completely burned. Of course, after we observe the output of the first round, we adjust the setting on the toaster to compensate. As soon as we take this step, the system is no longer open-loop because we have adjusted the input

Fig. 8.1 Simple open-loop system.

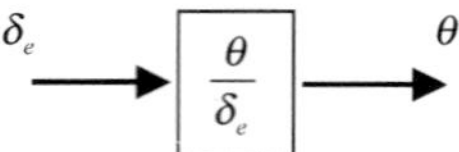

Fig. 8.2 Aircraft open-loop system.

(toaster setting) based on the output. Another example of an open-loop system is the first shot with a rifle when target practicing. The output (where the bullet hits) does not affect the input (how we aim the rifle) on the first shot. After we see where the first shot hits, we typically compensate on the second shot and, at that point, we no longer have an open-loop system. For an aircraft, the transfer functions we developed in Chapter 7 can be used to represent the process. For example, if the θ/δ_e transfer function is the process we are interested in, Fig. 8.1 becomes Fig. 8.2.

In most open-loop systems, a controller is needed to control or activate the process. This is illustrated in Fig. 8.3.

An example of a controller for an aircraft system is a hydraulic actuator used to move an aircraft control surface. A control valve on the actuator is positioned by either a mechanical or electrical input, the control valve ports hydraulic fluid under pressure to the actuator, and the actuator piston moves until the control valve shuts off the hydraulic fluid. A hydraulic actuator is shown in Fig. 8.4.

Clearly, the actuator piston cannot move instantaneously because it takes a finite time for the hydraulic fluid to flow through the ports from the control valve. In response to a step input, the resulting motion (x) of a hydraulic actuator can be modeled as an exponential rise:

$$x(t) = Z(1 - e^{-at}) \tag{8.1}$$

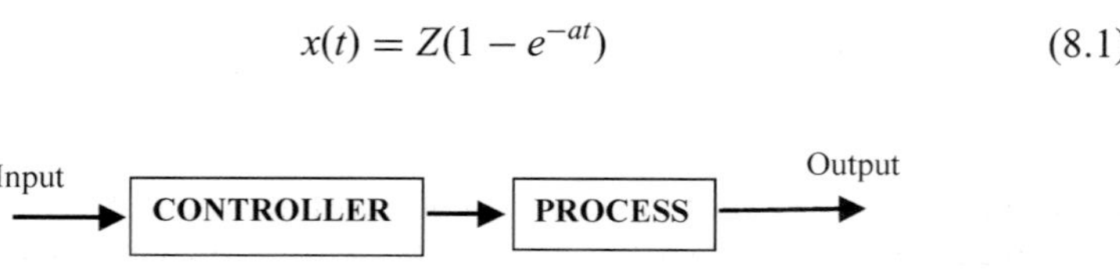

Fig. 8.3 Open-loop system with a controller.

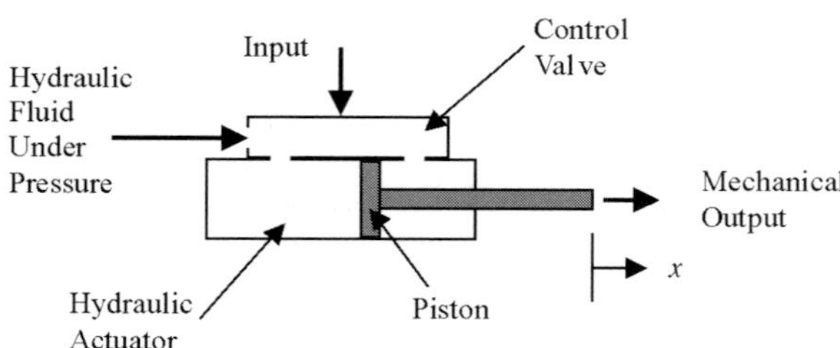

Fig. 8.4 Simplified hydraulic actuator.

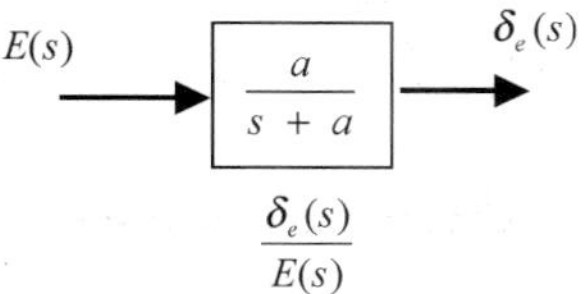

Fig. 8.5 Simplified representation of an aircraft hydraulic actuator in transfer function form.

where Z is the final displacement value of the actuator. Using the techniques developed in Sec. 7.3.1, the generalized transfer function for a hydraulic actuator is then

$$\frac{X(s)}{E(s)} = \frac{a}{s+a} \tag{8.2}$$

where $E(s)$ is the Laplace transform of the input and $X(s)$ is the Laplace transform of the output. For the case of Eq. (8.1), $E(s) = Z/s$. The time constant for the hydraulic actuator represented by Eq. (8.2) is

$$\tau_{\substack{\text{hydraulic}\\ \text{acutator}}} = \frac{1}{a} \tag{8.3}$$

and the final value of $x(t)$ for a unit step input is

$$x(\infty) = \lim_{s \to 0} \left(\frac{1}{s}\right) s \left(\frac{a}{s+a}\right) \Rightarrow 1$$

using the final value theorem. If the output of the hydraulic actuator $x(t)$ is mechanically connected to a control surface to cause a control surface displacement such as δ_e, we then have the open-loop situation shown in Fig. 8.5.

The block diagram representations of the two open-loop systems represented by Figs. 8.2 and 8.5 are normally combined in series for an aircraft, as shown in Fig. 8.6.

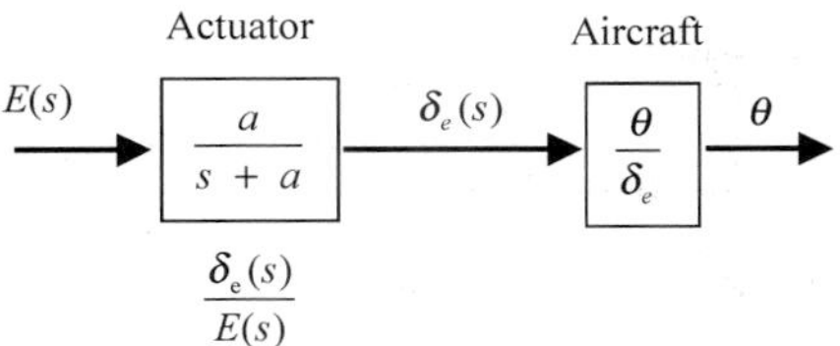

Fig. 8.6 Block diagram representation of open-loop aircraft system.

Example 8.1

Find the open-loop θ/E transfer function for the Lear Jet using the data provided in Example 7.10 and assuming a hydraulic actuator with a 0.1 s time constant.

From Example 7.10, we have

$$\frac{\theta}{\delta_e} = \frac{(208.1)s^2 + (136.9)s + 2.380}{(675.9)s^4 + (1371)s^3 + (5459)s^2 + (86.31)s + 44.78}$$

With a 0.1 s actuator time constant, the a in Eq. (8.3) becomes 10, and we have

$$\frac{\delta_e}{E} = \frac{10}{s + 10}$$

and the θ/E transfer function becomes

$$\frac{\theta}{E} = \frac{10[208.1)s^2 + (136.9)s + 2.380]}{(s + 10)[(675.9)s^4 + (1371)s^3 + (5459)s^2 + (86.31)s + 44.78]}$$

Notice that we simply multiplied two transfer functions, represented by block diagrams in series, to obtain a single transfer function for this open-loop system. We also have started to drop the (s) associated with Laplace variables. For example,

$$\frac{\theta}{E} = \frac{\theta(s)}{E(s)}$$

For this system, we can think of E as being the input (for example, the pilot's stick displacement) and θ as being the output (how the aircraft reacts in terms of an aircraft motion variable). We will begin referring to E as the error signal when we discuss closed-loop systems.

8.2 Closed-Loop Systems

Closed-loop systems are systems where the output is measured and fed back to modify the input that would normally be seen by the open-loop system. A closed-loop system is presented in general form in Fig. 8.7.

The feedback loop makes a measurement of the output (usually using a sensor), multiplies this measurement by an adjustable gain and then provides this feedback signal to a comparator that subtracts it from the input. The comparator is indicated by the circle enclosing X and the subtraction is indicated by negative sign to the lower left of the comparator.

We are surrounded by examples of closed-loop systems in everyday life. A thermostat to control the temperature in a house is a common example. The thermostat is adjusted to the desired temperature (input), a controller activates the furnace if the desired house temperature is below the desired temperature,

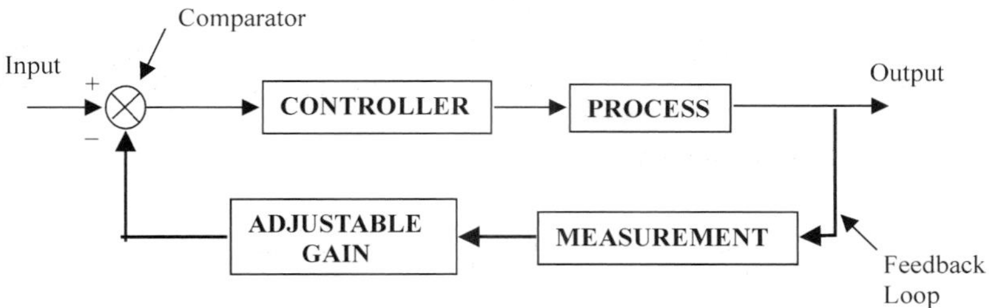

Fig. 8.7 Generalized closed-loop system.

the furnace provides the heating process to increase the temperature of the house (output), and a sensor in the thermostat measures the temperature of the house and feeds this back to the comparator where it is subtracted from the input signal. When the desired temperature and house temperature match, the error signal coming from the comparator is zero and the controller shuts down the furnace. Another example is cruise control on an automobile. The input is the desired speed and the output is the actual speed. The actual speed is measured and fed back so that the throttle can be adjusted automatically to maintain the desired speed.

Closed-loop feedback control systems are used extensively on aircraft to modify dynamic stability characteristics and implement pilot relief (autopilot) functions. As an example of a pilot relief mode, a simplified pitch attitude hold autopilot is presented in Fig. 8.8.

θ_c is the desired pitch attitude commanded by the pilot. The actuator and aircraft transfer functions were discussed in the previous section. The output is the actual pitch attitude of the aircraft (θ), while the vertical gyro (a sensor) measures the pitch attitude and feeds it back to the comparator. An error signal (E) is then generated and input to the actuator. The end result is that the aircraft automatically holds the commanded pitch attitude without the need for continuous pilot inputs.

Figure 8.9 shows the portion of a feedback control system usually accomplished within a flight control computer.

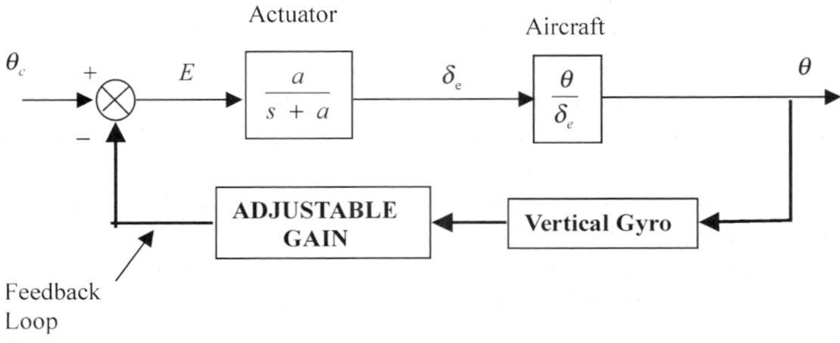

Fig. 8.8 Simplified aircraft closed-loop pitch attitude hold system.

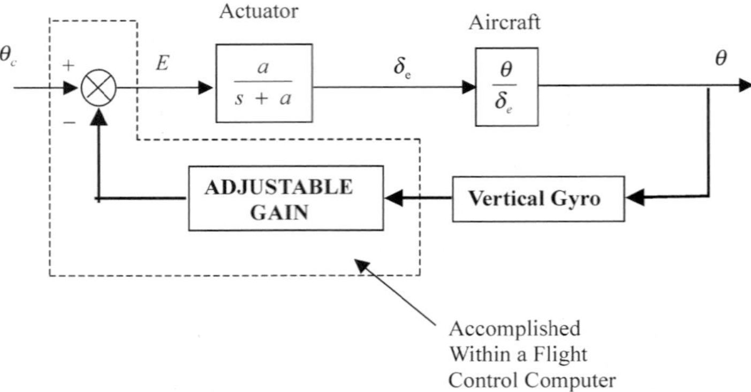

Fig. 8.9 Closed-loop system illustrating functions performed within a flight control computer.

The command signal and vertical gyro signal are input to the computer in the form of voltages or digital signals. Computer software multiplies the vertical gyro signal by the value of the adjustable gain (which is a fixed value for a final configuration), and then performs the comparator subtraction. Finally, the computer outputs the error signal (E) to an electromechanical actuator in the form of a voltage. The electromechanical actuator converts the voltage to a mechanical displacement, which is input into the control valve of the aircraft hydraulic actuator. Many aircraft integrate the electromechanical actuator with the hydraulic actuator as one unit. Thus, Fig. 8.9 presents only one transfer function for the integrated actuator.

8.3 Closed-Loop Analysis of a Second-Order System

We have seen how closed-loop systems can be used for automatic control functions such as a pitch attitude hold autopilot. They are also extremely useful for modifying the dynamic stability characteristics of the aircraft. In Sec. 7.1, we saw that the generalized transfer function for a second-order system was

$$\frac{X(s)}{Y(s)} = \frac{\omega_n^2}{s^2 + 2\zeta\omega_n s + \omega_n^2} \tag{8.4}$$

Think of this transfer function as representative of the short period mode of an aircraft with the natural frequency and damping ratio representing the dynamic characteristics of the basic airframe. The time domain differential equation for zero initial conditions is

$$\ddot{x} + 2\zeta\omega_n\dot{x} + \omega_n^2 x = \omega_n^2 y(t) \tag{8.5}$$

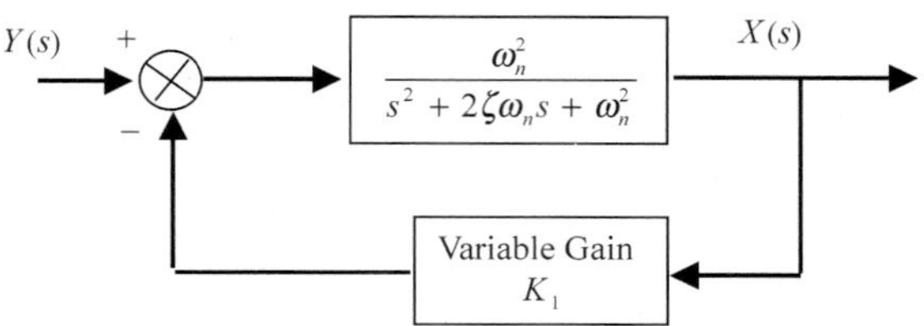

Fig. 8.10 Position feedback system.

We will analyze how these dynamic characteristics might be modified using three types of closed-loop systems.

Figure 8.10 presents a simple closed-loop position feedback system. The term "position" refers to the fact that the output variable (x) is fed back as itself (not as a derivative of x).

Notice that the controller and measurement transfer functions have been omitted for simplicity. The closed-loop differential equation can be seen to be

$$\ddot{x} + 2\zeta\omega_n\dot{x} + \omega_n^2 x = \omega_n^2(y(t) - K_1 x)$$

and the closed-loop transfer function is

$$\frac{X(s)}{Y(s)} = \frac{\omega_n^2}{s^2 + 2\zeta\omega_n s + \omega_n^2(1 + K_1)}$$

The closed-loop characteristic equation for this system is

$$s^2 + 2\zeta\omega_n s + \omega_n^2(1 + K_1) = 0$$

and we can see that

$$\omega_{n_{\substack{\text{position}\\\text{feedback}}}} = \omega_n\sqrt{1 + K_1} \tag{8.6}$$

Likewise, the closed-loop damping ratio has become

$$\zeta_{\substack{\text{position}\\\text{feedback}}} = \frac{\zeta}{\sqrt{1 + K_1}} \tag{8.7}$$

The important point is that a closed-loop position feedback system provides the opportunity to change both the basic airframe natural frequency and damping ratio by adjusting the variable gain K_1. As seen from Eq. (8.6), position feedback allows the designer to increase the natural frequency of the closed-loop system as K_1 is increased positively from zero. Note that when K_1 is equal to zero we have the original open-loop system. It is unfortunate for most applications that the closed-loop damping ratio [Eq. (8.7)] decreases as K_1 is increased. The keen observer will also notice that the closed-loop time

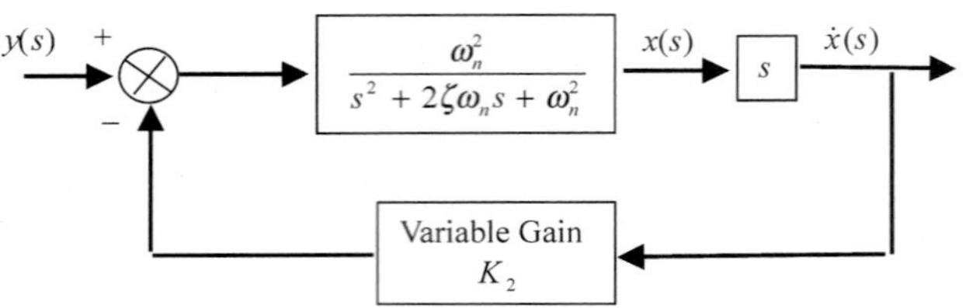

Fig. 8.11 Rate feedback system.

constant $(1/\zeta_{\substack{\text{position}\\\text{feedback}}}\omega_{n_{\substack{\text{position}\\\text{feedback}}}})$ remains unchanged from the open-loop time constant $(1/\theta\omega_n)$. Thus, a position feedback system provides the advantage of automatic control of a motion variable (as in the pitch attitude hold example), but will be accompanied by an increase in natural frequency, a decrease in damping ratio, and no change in time constant.

Figure 8.11 presents a simple closed-loop rate feedback system. Rate refers to the fact that the derivative of the output variable (x) is fed back. Notice that $\dot{x}(s)$ is generated in the block diagram by simply multiplying $x(s)$ by the Laplace operator s. As discussed in Sec. 7.3.1, the derivative in the Laplace domain of a variable such as $x(s)$ is $sx(s)$ with zero initial conditions. The closed-loop differential equation for the rate feedback system becomes

$$\ddot{x} + 2\zeta\omega_n\dot{x} + \omega_n^2 x = \omega_n^2(y(t) - K_2\dot{x})$$

and the closed-loop transfer function is

$$\frac{x(s)}{y(s)} = \frac{\omega_n^2}{s^2 + (2\zeta\omega_n + K_2\omega_n^2)s + \omega_n^2}$$

The closed-loop characteristic equation for the system is

$$s^2 + (2\zeta\omega_n + K_2\omega_n^2)s + \omega_n^2 = 0$$

and we can see that the natural frequency of the open-loop and closed-loop system remains constant and is not affected by the value of K_2. The closed-loop damping ratio becomes

$$\zeta_{\substack{\text{rate}\\\text{feedback}}} = \frac{2\zeta + K_2\omega_n}{2} \tag{8.8}$$

Rate feedback allows the designer to increase damping ratio as K_2 is increased positively from zero. This provides a powerful design tool to tailor the handling qualities of an aircraft and meet dynamic stability damping ratio requirements.

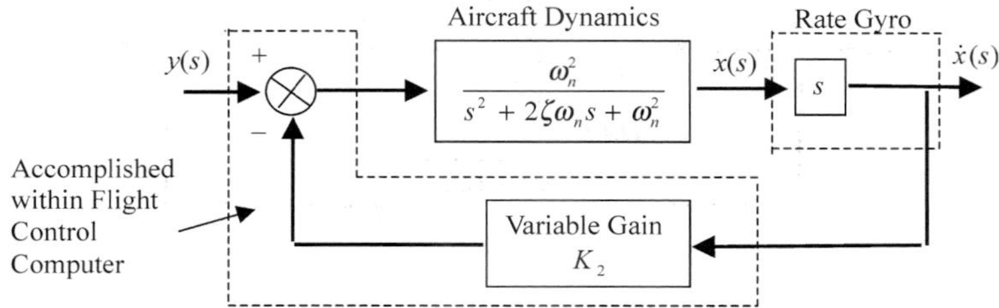

Fig. 8.12 Rate feedback system with component breakouts.

A rate feedback system typically involves adding a rate gyro to the aircraft to provide the $\dot{x}$ measurement and feedback signal shown in Fig. 8.11. Figure 8.12 illustrates where the rate gyro fits into the system.

A rate gyro is a sensor that outputs a voltage proportional to an angular rate. Most highly augmented aircraft have pitch rate (Q), roll rate (P), and yaw rate (R) gyros to tailor dynamic stability and response characteristics for all three rotational degrees of freedom.

Figure 8.13 presents a simple closed-loop acceleration feedback system. Acceleration refers to the fact that the second derivative of the output variable (x) is fed back. Notice that $\ddot{x}(s)$ is generated in the block diagram by simply multiplying $x(s)$ by the Laplace operator s^2. As discussed in Sec. 7.3.1, the second derivative in the Laplace domain of a variable such as $x(s)$ is $s^2\, x(s)$ with zero initial conditions. The closed-loop differential equation for the acceleration feedback system becomes

$$\ddot{x} + 2\zeta\omega_n\dot{x} + \omega_n^2 x = \omega_n^2(y(t) - K_3\ddot{x})$$

and the closed-loop transfer function is

$$\frac{x(s)}{y(s)} = \frac{\omega_n^2}{(1 + K_3\omega_n^2)s^2 + 2\zeta\omega_n s + \omega_n^2}$$

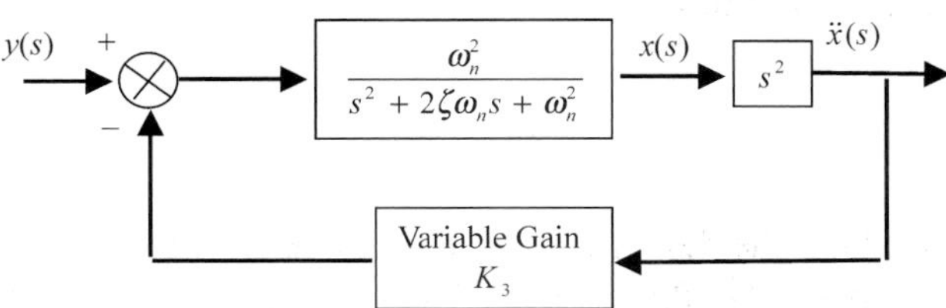

Fig. 8.13 Acceleration feedback system.

The closed-loop characteristic equation for the system is

$$s^2 + \frac{2\zeta\omega_n}{(1 + K_3\omega_n^2)} s + \frac{\omega_n^2}{(1 + K_3\omega_n^2)} = 0$$

The closed-loop natural frequency becomes

$$\omega_{n_{\substack{\text{accleration} \\ \text{feedback}}}} = \frac{\omega_n}{\sqrt{1 + K_3\omega_n^2}} \tag{8.9}$$

and the closed-loop damping ratio is

$$\zeta_{\substack{\text{acceleration} \\ \text{feedback}}} = \frac{\zeta}{\sqrt{1 + K_3\omega_n^2}} \tag{8.10}$$

Equations (8.9) and (8.10) indicate that acceleration feedback decreases both the natural frequency, and damping ratio of the system as K_3 is increased positively from zero. Acceleration feedback involves the addition of an accelerometer in the aircraft to provide the $\ddot{x}$ measurement.

With the position, rate, and acceleration feedback, we have the ability to increase or decrease the natural frequency and damping ratio of an open-loop system. The dynamic stability characteristics and handling qualities of an aircraft can be tailored with these tools, and the roots of the closed-loop characteristic equation can be positioned in the complex-plane to meet stated requirements. In some cases, a combination of position, rate, and/or acceleration feedback is needed to achieve the desired characteristics. A multiloop system using all three types of feedback is presented in Fig. 8.14.

The next section will develop analysis tools to simplify the process of finding closed-loop transfer functions and dynamic stability characteristics.

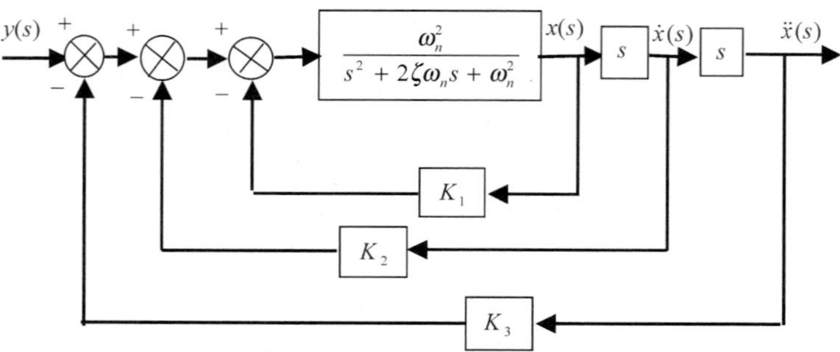

Fig. 8.14 A multiloop system using three feedback loops.

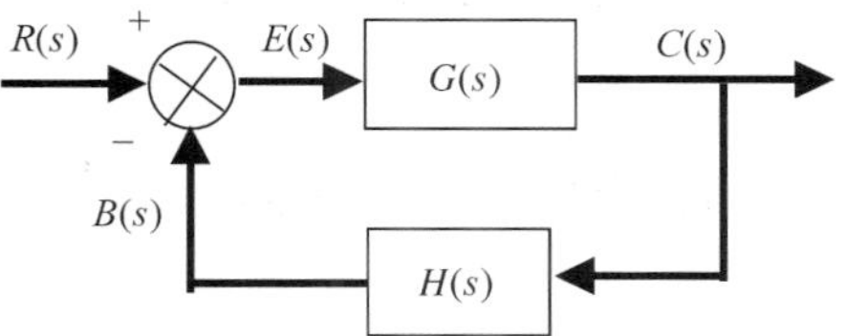

Fig. 8.15 Simple closed-loop system.

8.4 Closed-Loop Transfer Functions

The closed-loop transfer function (CLTF) of any feedback control system can be determined by using a relatively straightforward process. Consider the simple closed-loop system of Fig. 8.15.

We define $R(s)$ as the input, $C(s)$ as the output or control variable, $G(s)$ as the forward path transfer function, $H(s)$ as the feedback path transfer function, $C(s)/R(s)$ as the CLTF, $E(s)$ as the error signal, and $B(s)$ as the feedback signal. $G(s)$ includes all transfer functions and gains in the forward path of the closed-loop system, and $H(s)$ includes all transfer functions and gains in the feedback path. We will now drop the (s) with each of these terms for simplicity, realizing that all transfer functions and signals in a block diagram are expressed in the Laplace domain. To develop an expression for the CLTF, block diagram algebra is used beginning with an expression for the output.

$$C = GE$$

Of course, the error signal is

$$E = R - B$$

Substituting this into the equation for the output, we have

$$C = G(R - B)$$

The feedback signal is

$$B = HC$$

Substituting this into the previous equation, we have

$$C = G(R - HC) = GR - GHC$$

Rearranging,

$$C(1 + GH) = GR$$

$R \rightarrow \boxed{\dfrac{G}{1 + GH}} \rightarrow C$

Fig. 8.16 Open-loop representation of closed-loop system.

and the CLTF becomes:

$$\frac{C}{R} = \frac{G}{1 + GH} = CLTF \tag{8.11}$$

This powerful relationship forms the basis for simplifying closed-loop systems to one transfer function. This process can be thought of as representing a closed-loop system as one transfer function in open-loop form. For example, through the use of Eq. (8.11), the block diagram of Fig. 8.15 is equivalent to the open-loop block diagram of Fig. 8.16.

The closed-loop characteristic equation is easily obtained from Eq. (8.11) as

$$1 + GH = 0 \tag{8.12}$$

Of course, the closed-loop characteristic equation leads directly to determination of closed-loop dynamic stability characteristics such as natural frequency and damping ratio.

Example 8.2

Find the CLTF for the following system.

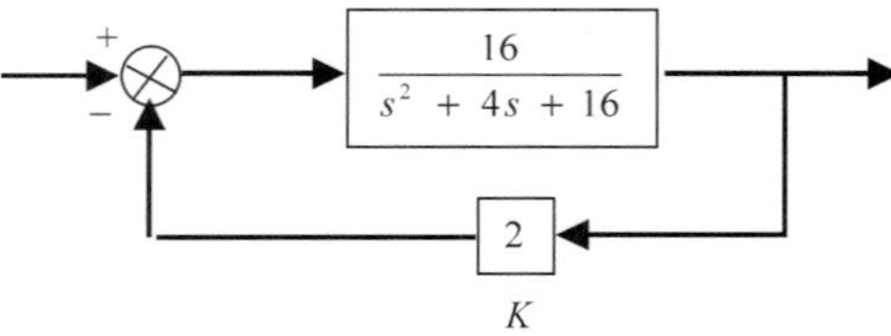

We first notice that the forward path transfer function has a natural frequency of 4 rad/s and a damping ratio of 0.5. By adding the feedback path with a gain of $K = 2$, the CLTF becomes

$$CLTF = \frac{G}{1 + GH} = \frac{\dfrac{16}{s^2 + 4s + 16}}{1 + \dfrac{16(2)}{s^2 + 4s + 16}}$$

Simplifying,

$$CLTF = \frac{16}{s^2 + 4s + 48}$$

Notice that the closed-loop system has a natural frequency of $\sqrt{48}$ or 6.93 rad/s and a damping ratio of 0.289. This is a classic case of position feedback (appropriate to autopilot functions) where the natural frequency is increased and the damping ratio is decreased as addressed in Sec. 8.3.

Example 8.3

Find the CLTF for the following system.

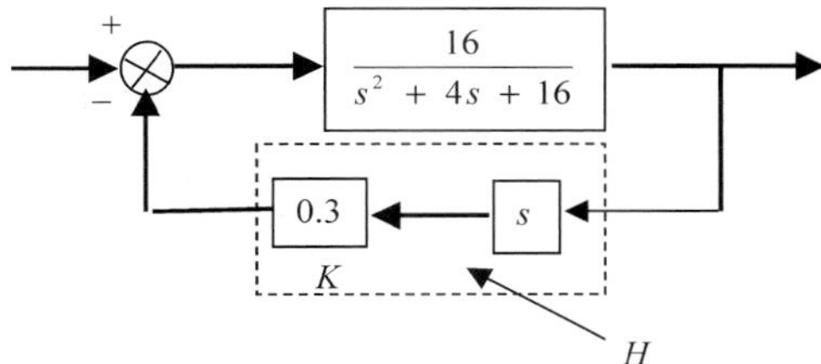

The feedback-path transfer function (H) becomes s times 0.3 or $0.3s$. Notice that this is a case of rate feedback. The only difference from that discussed in Sec. 8.3 is that the derivative is taken in the feedback loop. We expect that the damping ratio will be increased and that the natural frequency will stay constant.

$$CLTF = \frac{G}{1 + GH} = \frac{\dfrac{16}{s^2 + 4s + 16}}{1 + \dfrac{16(0.3s)}{s^2 + 4s + 16}}$$

Simplifying,

$$CLTF = \frac{16}{s^2 + 8.8s + 16}$$

Indeed, the natural frequency stays constant at 4 rad/s but the damping ratio has increased to 1.1, making the closed-loop system behave as a first-order response.

Example 8.4

Find the CLTF for the following system.

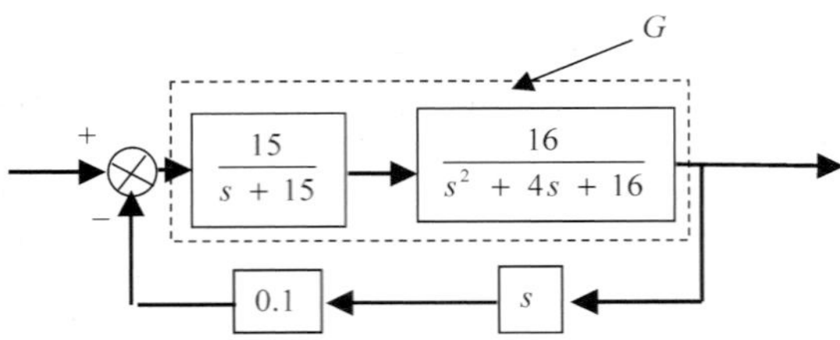

In comparing this system to that of Example 8.3, notice that we have added an actuator in the forward path ($\tau = 1/15$ s) and changed the gain in the feedback path to 0.1. The forward-path transfer function becomes

$$G = \left(\frac{15}{s+15}\right)\left(\frac{16}{s^2+4s+16}\right) = \frac{240}{s^3+19s^2+76s+240}$$

and the CLTF is

$$CLTF = \frac{G}{1+GH} = \frac{\dfrac{240}{s^3+19s^2+76s+240}}{1+\dfrac{240(0.1s)}{s^3+19s^2+76s+240}}$$

Simplifying,

$$CLTF = \frac{240}{s^3+19s^2+100s+240} = \frac{240}{(s+12.56)(s^2+6.44s+19.13)}$$

This example illustrates how to find the CLTF with more than one transfer function in the forward path. It is also interesting to note that closing the loop with rate feedback has appeared to move the actuator root from -15 to -12.56 (indicating a slower response or larger time constant). In addition, the natural frequency of the second-order polynominal has increased from 4 rad/s to $\sqrt{19}$ rad/s (4.36 rad/s) for the closed-loop case, and the damping ratio has increased from 0.5 to 0.739. The increase in damping ratio was expected, but the effects on the actuator root and the natural frequency may be somewhat surprising based on our discussion of rate feedback in Sec. 8.3. However, remember that our discussion there was strictly for a forward-path transfer function with a second-order denominator (and characteristic equation). The effects of closing the loop with higher order forward loop transfer functions are slightly more complex, and additional analysis tools will be developed to assist in understanding these effects.

8.5 Time Response Characteristics

As we have seen, closed-loop systems allow modification of the dynamic stability characteristics of the forward-path transfer function. Dynamic stability characteristics have been defined by familiar parameters such as natural frequency, damping ratio, damped frequency, and time constant. These parameters can be easily identified on a complex plane (s plane) plot of the roots of the characteristic equation, and they provide some insight into the time response characteristics expected of the system. Experience has shown that additional parameters that specifically define time response characteristics are very useful, especially when discussing the flying qualities of an aircraft with test pilots during flight test development of the feedback control system. The additional parameters discussed in this section are based on the step input time response of a system (aircraft), as presented in Fig. 8.17.

The second-order response shown in Fig. 8.17 is representative of the classic second-order transfer function

$$TF = \frac{\omega_n^2}{s^2 + 2\zeta\omega_n s + \omega_n^2}$$

However, the time response characteristics discussed in this section are applicable to nearly all systems. Of course, first-order systems will not have an overshoot; therefore, a few of the parameters lose meaning.

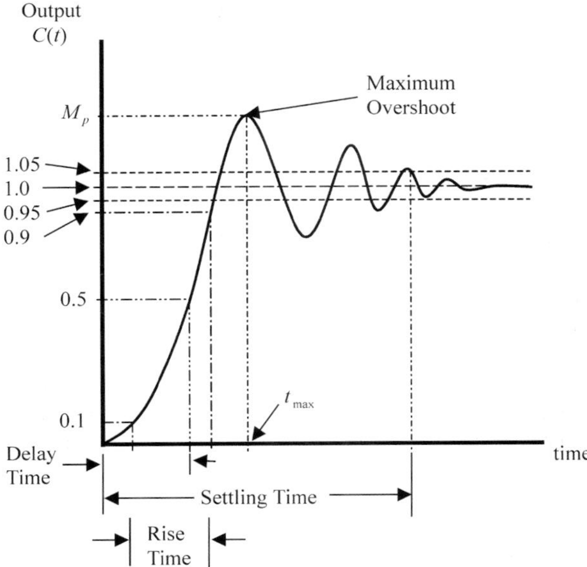

Fig. 8.17 Time response characteristics.

The rise time (t_r) is the time required for a step-input response to rise from 10% to 90% of its steady state (final) value. The delay time (t_d) is the time it takes for the response to reach 50% of the steady-state value. Rise time and delay time indicate how fast the system responds to a step input. For the classic second-order system, these parameters can be estimated in terms of the natural frequency and damping ratio.

$$t_r \approx \frac{1 + 1.1\zeta + 1.4\zeta^2}{\omega_n} \tag{8.13}$$

and

$$t_d \approx \frac{1 + 0.6\zeta + 0.15\zeta^2}{\omega_n} \tag{8.14}$$

The settling time (t_s) is the time required for the response to stay within a specified percentage (usually 2% or 5%) of the steady-state value. Figure 8.17 illustrates a 5% settling time. Again, for the classic second-order system, estimates are available.

$$t_{s_2\%} \approx \frac{4}{\zeta\omega_n} \tag{8.15}$$

$$t_{s_5\%} \approx \frac{3}{\zeta\omega_n} \tag{8.16}$$

It is interesting to note that the settling time estimates are directly based on the properties of the time constant. For an impulse input (rather than a step), $t = 4\tau$ is required to obtain 98% of the steady-state value ($t_{s_2\%}$). Likewise, 95% of the steady-state value will be obtained at $t = 3\tau$, which is the estimate for $t_{s_5\%}$. Thus, the estimates presented in Eqs. (8.15) and (8.16) will be slightly low for a step input response.

The maximum overshoot is the difference between the magnitude of the maximum overshoot and the steady-state value. The magnitude of the peak (M_p) is the value of the response at maximum overshoot. For a second-order system,

$$M_p \approx 1 + e^{\frac{-\zeta\pi}{\sqrt{1-\zeta^2}}} \tag{8.17}$$

Another useful parameter related to the maximum overshoot is the percentage of max overshoot which defines the magnitude of the max overshoot in terms of a percentage of the steady-state value of the response ($M_{\substack{\text{steady}\\\text{state}}}$).

$$\%_{\substack{\text{max}\\\text{overshoot}}} = \frac{M_p - M_{\substack{\text{steady}\\\text{state}}}}{M_{\substack{\text{steady}\\\text{state}}}} \times 100 \tag{8.18}$$

The time to max overshoot (t_{max}) is the time required for the response to rise to the maximum overshoot (which is the first overshoot in the case of the classic second-order system). The time to max overshoot can be estimated by

$$t_{max} \approx \frac{\pi}{\omega_n\sqrt{1-\zeta^2}} = \frac{\pi}{\omega_D} \tag{8.19}$$

This estimate can be seen to be one-half the period of oscillation (T) for a second-order system.

Again, the estimates defined for time response characteristics in Eqs. (8.13–8.19) are based on the classic second-order transfer function. They are also useful for more complex systems in providing insight into the effect of key parameters. For the majority of cases, time response characteristics will be measured directly from the time response in a manner similar to Fig. 8.17.

8.6 Root Locus Analysis

We have seen that in going from an open-loop system (the basic aircraft) to a closed-loop feedback control system, the roots of the characteristic equation can be significantly changed. Of course, these roots directly affect the dynamic stability characteristics of the aircraft. Sect. 7.2 provided an understanding of how the root locations on the complex plane translate to the dynamic stability parameters of natural frequency, damping ratio, damped frequency, and time constant. A correlation was also made between complex plane root location and time response. We will now develop a powerful design tool called the root locus, which graphically presents how the roots of the closed-loop characteristic equation change on the complex plane as the adjustable gain (K) is varied from zero to infinity.

We will begin by considering two locations for the adjustable gain (K). Figure 8.18 presents the two possibilities. Case 1 has the adjustable gain in the feedback path and is the case we have used for our previous development. Case 2 has the adjustable gain in the forward path. Referring back to Fig. 8.9, this case is also easily implemented in the flight control computer. Thus, the designer has both options available.

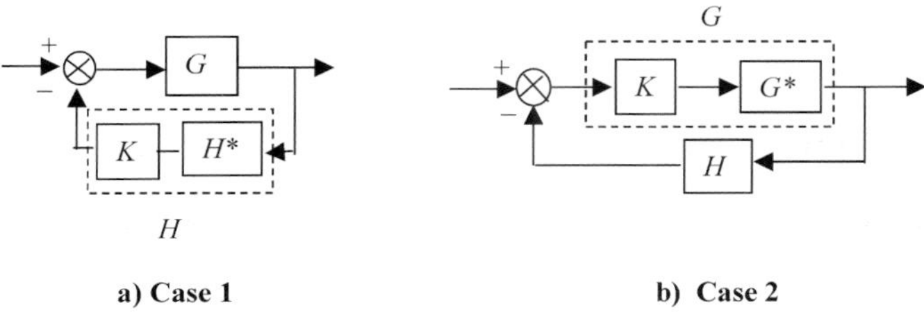

Fig. 8.18 Possible locations for the adjustable gain.

The root locus focuses on the roots of the closed-loop characteristic equation as a function of K. From Eqs. (8.11) and (8.12), the general form of the closed-loop characteristic equation is

$$1 + GH = 0$$

Because K is part of H in case 1 and part of G in case 2, we will break it out in our representation of the characteristic equation. Referring to Fig. 8.18, for case 1 we have

$$H = KH^*$$

and

$$1 + KGH^* = 0$$

or

$$KGH^* = -1 = GH \tag{8.20}$$

Programs that automatically generate a root locus plot will normally use GH^* as the input transfer function for case 1 systems.

For case 2, we have

$$G = KG^*$$
$$1 + KG^*H = 0$$

or

$$KG^*H = -1 = GH \tag{8.21}$$

Programs that automatically generate a root locus plot will normally use G^*H as the input transfer function for case 2 systems.

Notice that Eqs. (8.20) and (8.21) simply have K times the remaining transfer functions in the forward path and feedback path set equal to -1. GH is key to determination of the roots of the characteristic equation. We refer to it as the open-loop transfer function (not to be confused with the forward path transfer function, G). The open-loop transfer function (OLTF) can be expressed as

$$GH = OLTF = \frac{K \prod_{i=1}^{i=m} (s + Z_i)}{\prod_{j=1}^{j=n} (s + P_j)} \tag{8.22}$$

Values of s that make the numerator of Eq. (8.22) go to zero are referred to as open-loop zeros, or roots of the OLTF numerator. The zeros in Eq. (8.22) can be seen to be equal to $-Z_1, -Z_2, \ldots - Z_m$. Values of s that make the denominator of Eq. (8.22) go to zero are referred to as open-loop poles, or roots of the open-loop characteristic equation. The poles in Eq. (8.22) can be seen to be $-P_1, -P_2, \ldots - P_n$. It is believed that the word "pole" was chosen based on the fact that a pole causes the denominator of a transfer function to go to zero, which causes the transfer function to go to infinity. With a stretch of the imagination, this could be viewed as the effect a tent pole has on the roof of a tent (that is, pointing toward infinity). A few examples will help clarify these definitions.

Example 8.5

Find the OLTF poles and zeros for the following system.

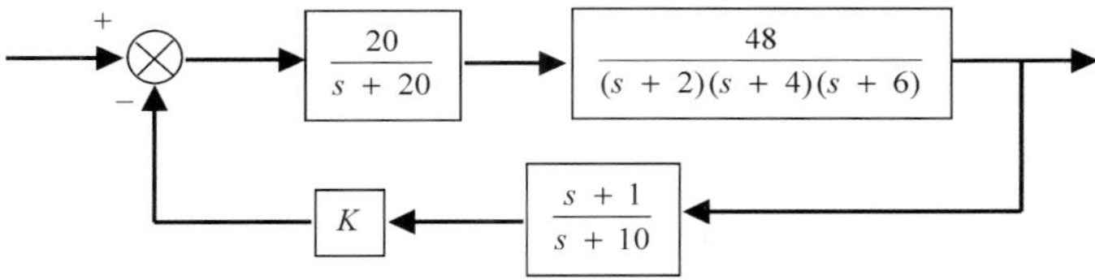

This is a Case 1 closed-loop system. The OLTF is

$$OLTF = GH = \frac{K(20)(48)(s+1)}{(s+20)(s+2)(s+4)(s+6)(s+10)}$$

There is one open-loop zero at $s = -1$. There are five open-loop poles at $s = -2, -4, -6, -10$, and -20.

Example 8.6

Find the open-loop poles and zeros for the following system.

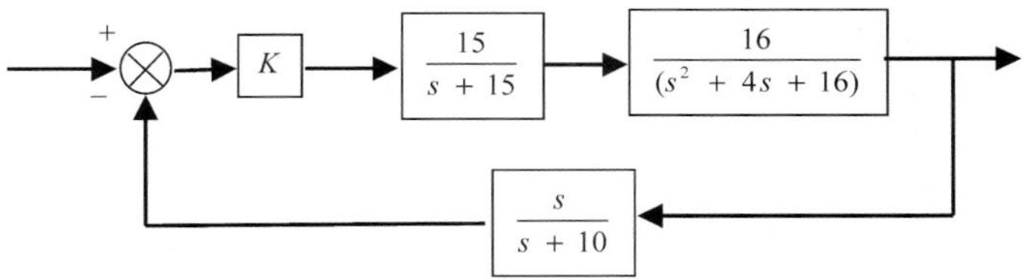

For this Case 2 system, the OLTF is

$$GH = OLTF = \frac{K(15)(16)s}{(s+15)(s+10)(s^2+4s+16)}$$

There is one open-loop zero at $s = 0$. There are four open-loop poles at $s = -15, -10$, and $-2 \pm 3.46j$. Notice that two of the poles are complex conjugates.

8.6.1 Root Locus Fundamentals

The root locus plots the roots of the closed-loop characteristic equation as K is varied from zero to infinity. Returning to Eq. (8.12) and incorporating Eq. (8.22), we can express the closed-loop characteristic equation as

$$1+\frac{K\prod_{i=1}^{i=m}(s+Z_i)}{\prod_{j=1}^{j=n}(s+P_j)}=0$$

Rearranging, the closed-loop characteristic equation becomes

$$\prod_{j=1}^{j=n}(s+P_j)+K\prod_{i=1}^{i=m}(s+Z_i)=0 \tag{8.23}$$

For a value of $K=0$, we can see that the roots of the closed-loop characteristic equation are located at the open-loop poles ($s=-P_1,\ -P_2,\ldots-P_n$). For a value of $K=\infty$ (infinity), the roots of the closed-loop characteristic equation approach the open-loop zeros ($s=-Z_1,\ -Z_2,\ldots-Z_m$). Based on this, we can make an important observation. **The root locus will begin at the open-loop poles for a value of $K=0$, and end at open-loop zeros for a value of $K=\infty$.** We will see that asymptotes may also be involved for the $K=\infty$ case. Consider the closed-loop system of Fig. 8.19.

The OLTF is

$$OLTF=GH=\frac{K(s+2)(s+4)(s+6)}{(s+10)(s^2+4s+16)}$$

The open-loop zeros are at $s=-2$, -4, and -6, and the open-loop poles are at $s=-10$, and $-2\pm 3.46i$. Figure 8.20 presents a plot of the open-loop poles and zeros on the complex plane.

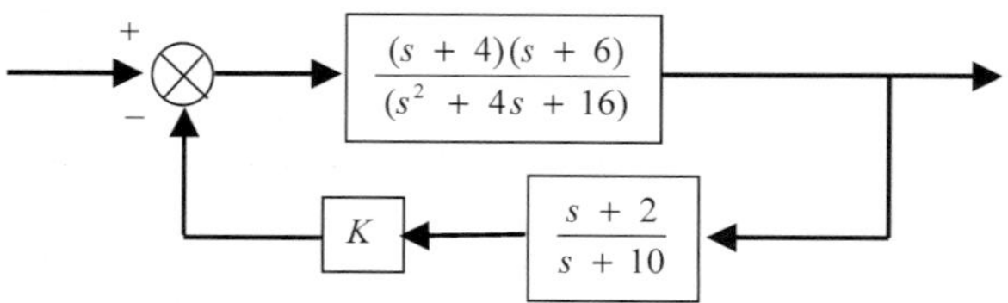

Fig. 8.19 Closed-loop system example.

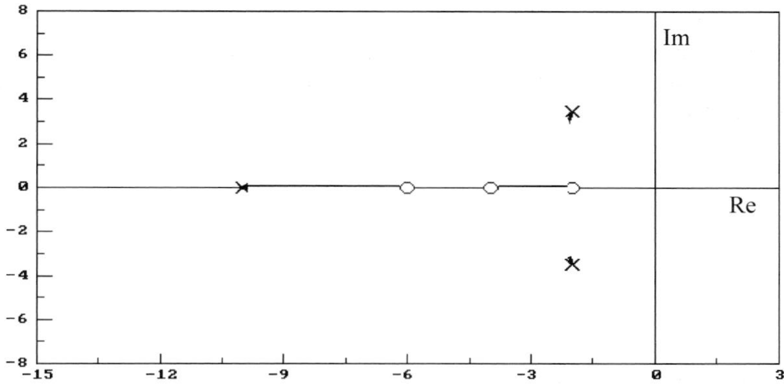

Fig. 8.20 Plot of open-loop poles and zeros for system of Fig. 8.19.

Notice that the open-loop poles are each plotted with an "x", and that the open-loop zeros are each plotted with a "o". Using Eq. (8.23), the closed-loop characteristic equation is

$$(s + 10)(s^2 + 4s + 16) + K(s + 2)(s + 4)(s + 6) = 0$$

We will next find the roots of this characteristic equation for $K = 2$. Using a root-finder, we have

$$s = -8, -2.33 \pm 2.29i \text{ at } K = 2$$

In a similar manner, we will find the roots at $K = 10$ and 40.

$$s = -6.82, -2.68 \pm 1.16i \text{ at } K = 10$$
$$s = -6.29, -3.36, -2.40 \text{ at } K = 40$$

Figure 8.21 adds these roots to the plot of Fig. 8.20.

A close review of Fig. 8.21 shows that one of the closed-loop roots defines a branch of the root locus that starts (for $K = 0$) at the open-loop pole, $s = -10$, and migrates along the real axis as K is increased. It reaches the open-loop zero at $s = -6$ when $K = \infty$. Another branch of the root locus starts at the open-loop pole $s = -2 + 3.46i$ and moves toward the real axis. It meets the final branch of the root locus at a break-in point on the real axis (for $K \approx 30$) and then one closed-loop root moves toward the open-loop zero at $s = -2$ and the other root moves toward the open-loop zero at $s = -4$. The uncluttered plot of the root locus for this system is presented in Fig. 8.22.

A more typical aircraft case has more poles than zeros in the open-loop transfer function. For example, consider Fig. 8.23, which is a rate feedback system similar to that presented in Fig. 8.11.

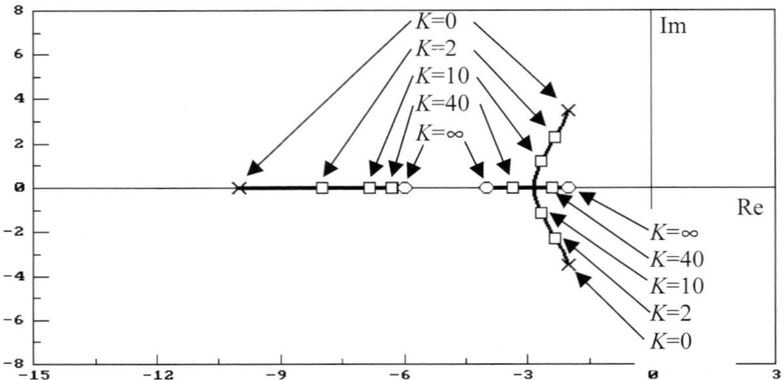

Fig. 8.21 Plot of closed-loop roots at selected gain values for system of Fig. 8.19.

The OLTF is

$$OLTF = \frac{160Ks}{(s+10)(s^2+4s+16)}$$

which has an open-loop zero at $s = 0$ and open-loop poles at $s = -10$, and $-2 \pm 3.46i$. A plot of the root locus is presented in Fig. 8.24.

Notice that we have one open-loop zero and three open-loop poles. The open-loop pole at $s = -10$ migrates to the open-loop zero at $s = 0$ as K is increased from 0 to infinity. As can be seen in Fig. 8.24, the two complex

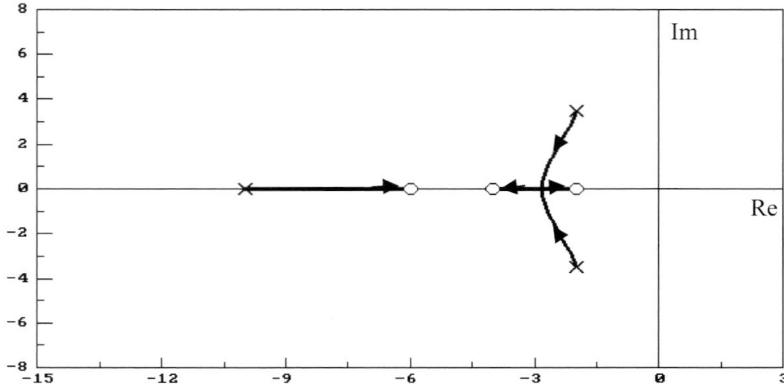

Fig. 8.22 Root locus of the system of Fig. 8.19.

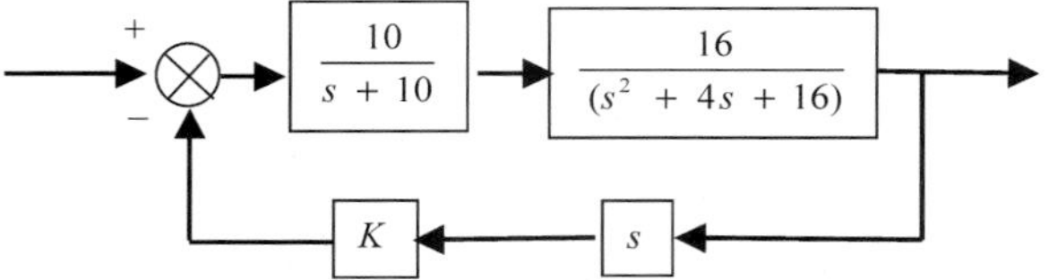

Fig. 8.23 Closed-loop system example for rate feedback.

conjugate poles migrate, as K is increased to infinity, toward asymptotes that go to positive and negative infinity along the imaginary axis. This example illustrates another principle of the root locus: **The number of branches of the root locus that go to infinity along an asymptote is equal to the number of OLTF poles minus the number of OLTF zeros**. Simply stated, each OLTF zero will draw one branch of the root locus toward it, and that branch will terminate at the zero for $K =$ infinity. The remaining branches (originating from an OLTF pole with no zero available for termination) will go to infinity along an asymptote as K goes to infinity.

8.6.2 *Magnitude and Angle Criteria*

As we have seen, each branch of the root locus defines how a root of the closed-loop characteristic equation migrates as K is increased from zero to infinity. The closed-loop roots of the characteristic equation are also called **closed-loop poles** and are expressed as complex numbers(s) using the $s = a + bi$ format. Recalling Eq. (8.12) for the closed-loop characteristic equation, we had

$$1 + GH = 0$$

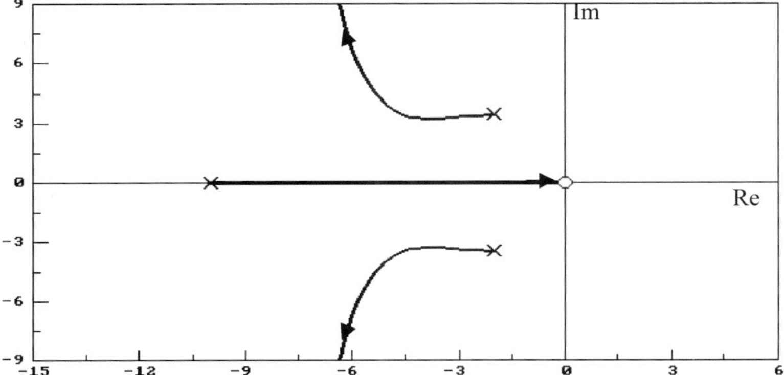

Fig. 8.24 Root locus plot for the system of Fig. 8.23.

or

$$GH = -1 \tag{8.24}$$

Each value of s in the complex plane that is part of a branch of the root locus must therefore satisfy the criteria defined by Eq. (8.24). Because a complex number can also be expressed as a vector with a magnitude and an angle, we can think of the complex number $s = a + bi$ as

$$s = |\sqrt{a^2 + b^2}| \angle \tan^{-1}\left(\frac{b}{a}\right)$$

Using the magnitude and angle representation, the root locus criteria of Eq. (8.24) becomes

$$\begin{aligned} |GH| &= 1 \\ \angle GH &= \pm 180 \text{ deg}, \pm 540 \text{ deg, etc.} \end{aligned} \tag{8.25}$$

Therefore, each point on the root locus must satisfy the requirements of Eq. (8.25) (as a reminder, we are restricting ourselves to the $K > 0$ case). Another way of looking at this is that each root of the closed-loop characteristic equation must satisfy the criteria of Eq. (8.25). Using the general form for the OLTF presented in Eq. (8.22), we have

$$\frac{K \prod_{i=1}^{i=m} |(s + Z_i)|}{\prod_{j=1}^{j=n} |(s + P_j)|} = 1 \tag{8.26}$$

and

$$\begin{aligned} &\angle K + \angle(s + Z_1) + \angle(s + Z_2) + \cdots \angle(s + Z_m) \\ &\quad - \angle(s + P_1) - \angle(s + P_2) - \cdots \angle(s + P_n) = \pm 180 \text{ deg}, \pm 540 \text{ deg, etc.} \end{aligned}$$

when applying the criteria of Eq. (8.25). Normally, the angle criteria is tested first because K can be adjusted to satisfy the magnitude criteria. An example will be used to illustrate application of this criteria.

Example 8.7

For the following system, determine which of the following values of s are on the root locus [satisfy the criteria of Eq. (8.25)].

$$s = -1.5; s = -2.5; s = -3.615 + 1.015i$$

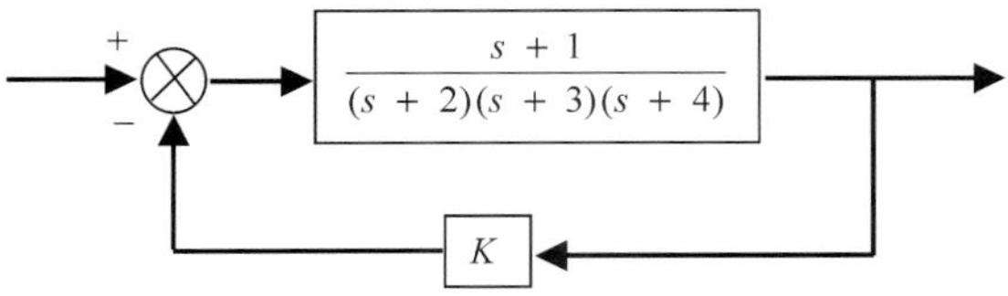

We begin by plotting the OLTF poles and zeros on the complex plane.

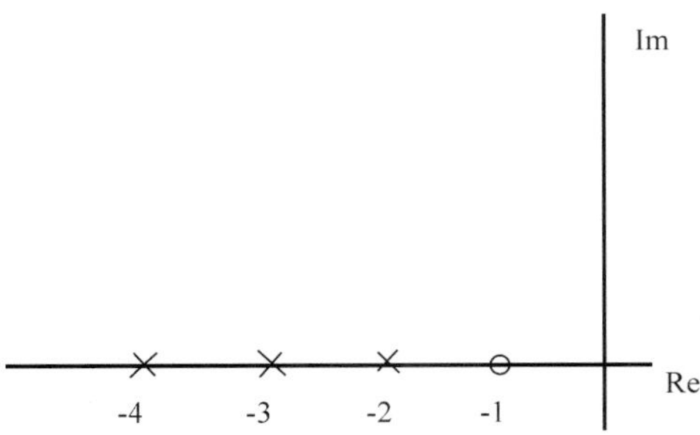

Next, the angle criteria are checked for each of the points by using vectors that originate at each of the open-loop poles and zero. For $s = -1.5$ we have

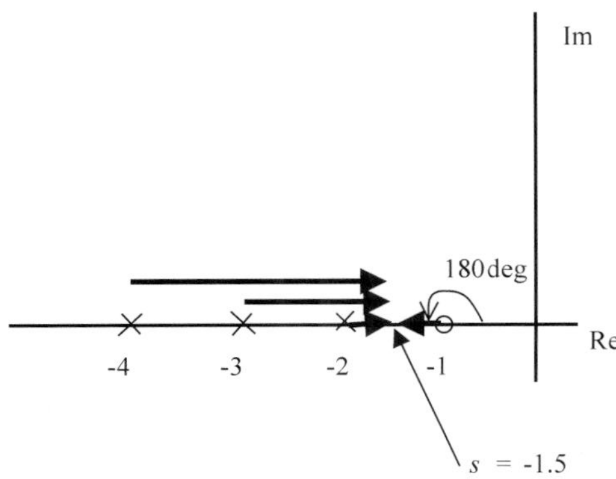

Notice that vectors are drawn from each OLTF pole and zero to the test point. We use a counterclockwise rotation from the real axis to define the angle

associated with each vector. We have

$$\angle(s+1) = 180 \text{ deg}$$
$$\angle(s+2) = 0 \text{ deg}$$
$$\angle(s+3) = 0 \text{ deg}$$
$$\angle(s+4) = 0 \text{ deg}$$

Because the $\angle K = 0$ deg in applying the angle criteria of Eq. (8.26) we add the angle of the zeros and subtract the angle of each pole. We have

$$0 \text{ deg} + 180 \text{ deg} - 0 \text{ deg} - 0 \text{ deg} - 0 \text{ deg} = 180 \text{ deg}$$

and the angle criteria is satisfied for $s = -1.5$. It is a point on the root locus. The value of K at $s = -1.5$ can be determined from the magnitude criteria. Using Eq. (8.26),

$$K = \frac{\prod_{j=1}^{j=n} |(s+P_j)|}{\prod_{i=1}^{i=m} |(s+Z_i)|} = \frac{(0.5)(1.5)(2.5)}{0.5} = 3.75$$

Next, we will test $s = -2.5$ by using the angle criteria.

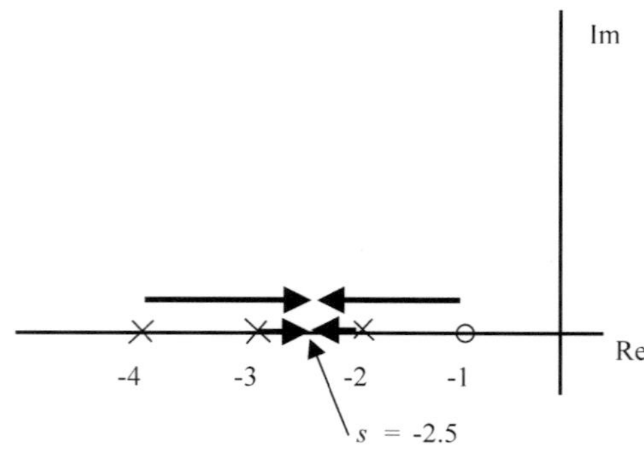

We have

$$\angle(s+1) = 180 \text{ deg}$$
$$\angle(s+2) = 180 \text{ deg}$$
$$\angle(s+3) = 0 \text{ deg}$$
$$\angle(s+4) = 0 \text{ deg}$$

and checking the angle criteria with Eq. (8.26),

$$0 \text{ deg} + 180 \text{ deg} - 180 \text{ deg} - 0 \text{ deg} - 0 \text{ deg} = 0 \text{ deg}$$

which does not equal ±180 deg, ±540 deg. Therefore, $s = -2.5$ is not on the root locus.

Finally, we will check $s = -3.615 + 1.0115i$

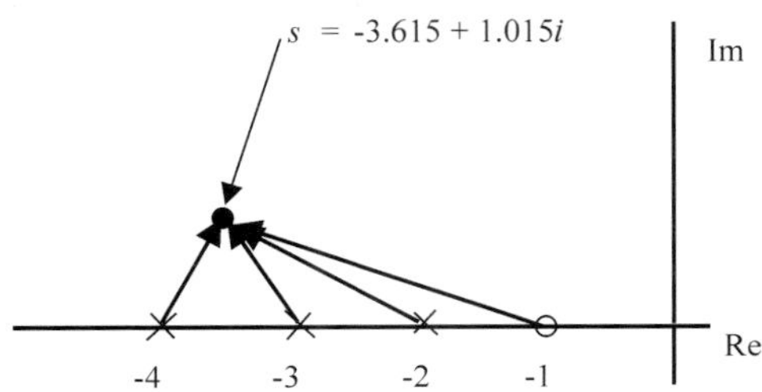

We have

$$\begin{aligned}
\angle(s+1) &= 158.79 \text{ deg} \\
\angle(s+2) &= 147.85 \text{ deg} \\
\angle(s+3) &= 121.21 \text{ deg} \\
\angle(s+4) &= 69.23 \text{ deg}
\end{aligned}$$

and checking the angle criteria, we have

$$\begin{aligned}
0 \text{ deg} + 158.79 \text{ deg} - 147.85 \text{ deg} - 121.21 \text{ deg} - 69.23 \text{ deg} \\
= -179.5 \text{ deg} \approx -180 \text{ deg}
\end{aligned}$$

This is close enough (based on round off errors) to satisfy the angle criteria. The point $s = -3.615 + 1.0115i$ is on the root locus. The value of K at this point can be determined using the magnitude criteria.

$$K = \frac{\prod_{j=1}^{j=n} |(s+P_j)|}{\prod_{i=1}^{i=m} |(s+Z_i)|} = \frac{(1.907)(1.187)(1.086)}{2.805} = 0.876$$

A plot of the complete root locus follows:

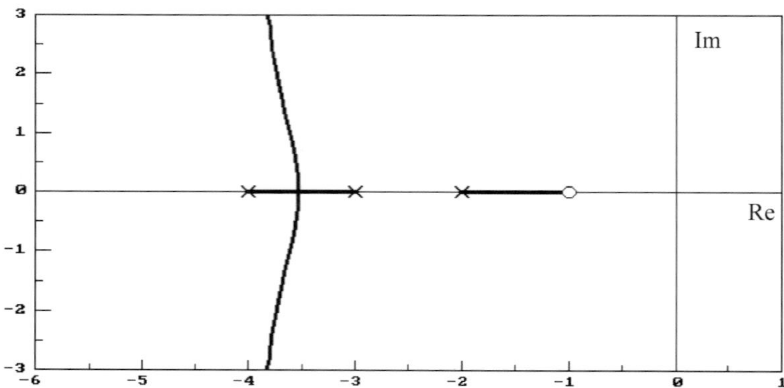

Fortunately, this agonizing process is not used to plot the root locus. Only an understanding of the angle and magnitude criteria are needed so that several plotting rules can be developed. In addition, excellent computer programs are available to generate root locus plots. However, no matter how a root locus plot is obtained, a clear understanding of the information it provides and how it can be used is needed.

8.6.3 Plotting the Root Locus

Nine "rule of thumb" plotting rules are used to obtain the general shape of a root locus. The insight provided by these plotting rules will significantly aid the design process when attempting to design a system to meet specific requirements.

1) *Number of Root Locus Branches:* The number of branches of the root locus is equal to the number of poles of the OLTF (for example, 5 poles will have 5 branches).
2) *Root Migration:* The root locus begins at the open-loop poles (at $K = 0$) and goes to the open-loop zeros or to infinity (at $K = \infty$). The number of branches going to infinity is equal to the number of open-loop poles minus the number of open-loop zeros.
3) *Real Axis Location:* A branch of the root locus is on the real axis if it is to the left of an odd number of poles and zeros. This is illustrated in Fig. 8.25 for a point s_1 on the real axis. Note that s_1 is to the left of an odd number (1) of poles.

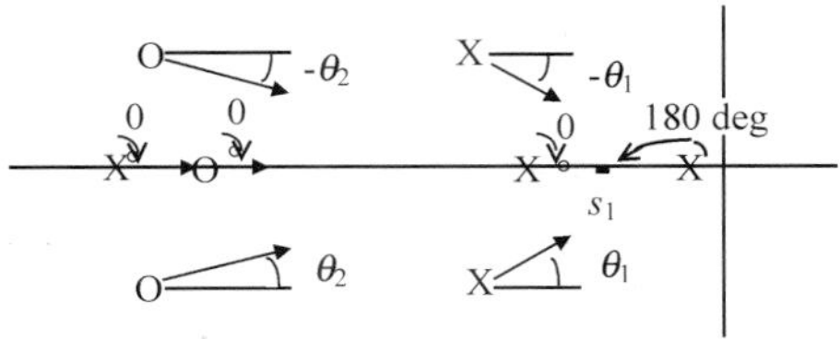

Fig. 8.25 Illustration of root locus real axis location.

A check using the angle criteria shows that the point is on the root locus.

$$\angle GH = \underbrace{(\theta_2 - \theta_2 = 0 \text{ deg})}_{\text{angles of zeros to root at } s_1} - \underbrace{(0 \text{ deg} + \theta_1 - \theta_1 - 0 \text{ deg} + 180 \text{ deg})}_{\text{angle of poles to } s_1}$$
$$= 180 \text{ deg}$$

Notice that the complex conjugate pairs cancel each other. Real axis locations of the root locus for the above example are between the two poles nearest the origin and between the pole and zero to the far left of the plot.

4) *Symmetry:* The root locus is symmetric about the real axis. In other words, branches of the root locus that extend into the upper half of the complex plane will have mirror reflections in the lower half.
5) *Angle of Asymptotes:* As discussed in Sec. 8.6.1, branches of the root locus that go to infinity approach asymptotes. The angle (φ) that the asymptotes of the root locus make with the real axis can be computed by

$$\varphi = \frac{\pm 180 \text{ deg}(1 + 2n)}{\begin{pmatrix}\# \text{ of OLTF} \\ \text{zeros}\end{pmatrix} - \begin{pmatrix}\# \text{ of OLTF} \\ \text{poles}\end{pmatrix}} = \frac{\pm 180 \text{ deg}(1 + 2n)}{\#Z - \#P} \tag{8.27}$$

Equation (8.27) must be evaluated at $n = 0, 1, 2, \ldots,$ until the angles start to repeat.

For example, consider the following OLTF,

$$GH = \frac{K}{(s + 3)(s + 1)}$$

We observe that there are two more poles than zeros telling us that there are two asymptotes. Using Eq. (8.27) to find the angle of these asymptotes with respect to the real axis, we have

$$\varphi = \frac{\pm 180 \text{ deg}}{0-2} = \mp 90 \text{ deg for } n = 0$$
$$\varphi = \frac{\pm 540 \text{ deg}}{0-2} = \mp 270 \text{ deg} = \pm 90 \text{ deg for } n = 1$$

Notice that we obtain two angles (± 90 deg) for $n = 0$ and a check with $n = 1$ equals the same result. These asymptotes are shown in Fig. 8.26.

6) *Asymptote Centroid:* The real axis intercept or centroid (σ) of the root locus asymptotes can be found with

$$\sigma = \frac{\Sigma\text{Poles} - \Sigma\text{Zeros}}{\#\text{Poles} - \#\text{Zeros}} \tag{8.28}$$

Using the previous example, Eq. (8.28) becomes

$$\sigma = \frac{\lfloor -3 + (-1) \rfloor - 0}{2-0} = \frac{-4}{2} = -2$$

The asymptote centroid is also shown in Fig. 8.26.

Example 8.8

Find the real axis location and the angle and centroid of the asymtotes for the following system:

$$GH = \frac{K}{(s+1)(s+2)(s+3)}$$

Because we have three poles and no zeros, the real axis location of the root locus will be between the pole at -1 and -2, and from -3 to negative infinity

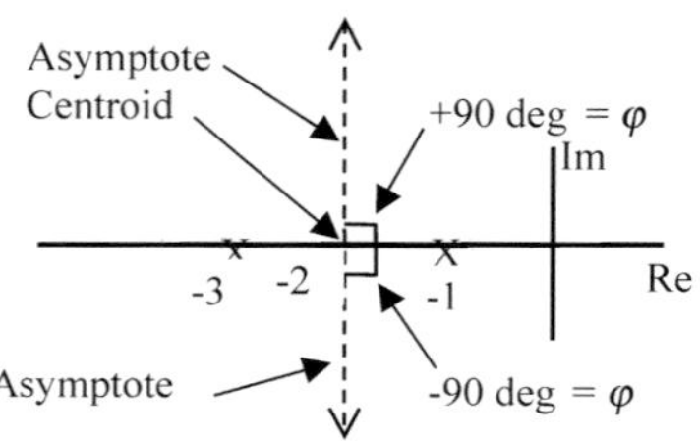

Fig. 8.26 Illustration of angle of asymptotes.

using rule 3. The angle of the asymptotes is next computed using Eq. (8.27).

$$\varphi = \frac{\pm 180\ \text{deg}(1+2n)}{0-3}\left\} \begin{array}{l} n=0 \Rightarrow \varphi = \pm 60\ \text{deg (two asymptotes)} \\ n=1 \Rightarrow \varphi = \pm 180\ \text{deg (one asymptote)} \end{array} \right.$$

As expected, we have three asymptotes. The centroid is computed with Eq. (8.28).

$$\sigma = \frac{\lfloor -1 + (-2) + (-3) \rfloor - 0}{3 - 0} = -2$$

Finally, we sketch the root locus using this information.

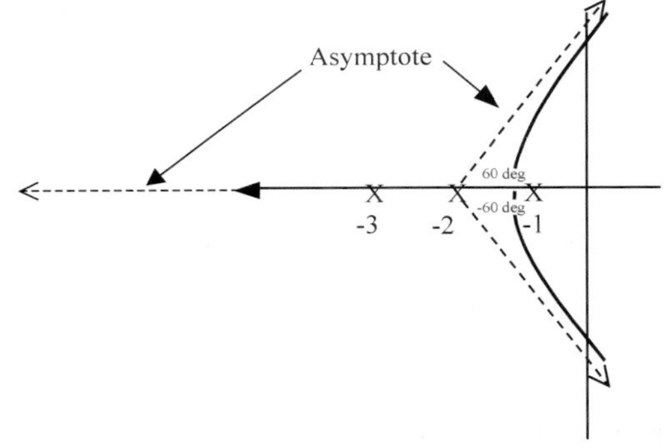

---

7) *Angle of Departure:* The angle of departure of the root locus from complex OLTF pole can be found using the angle criteria. We begin by choosing a root, s_1, very close to the complex pole such that the angles from the remaining poles and zeros to s_1 are essentially the same as the angles to the complex pole. For example, consider the OLTF,

$$GH = \frac{K}{s(s^2 + 2s + 2)}$$

with open-loop poles at $s = 0, -1 \pm i$. A root locus is presented in Fig. 8.27.

Applying the angle criteria to s_1, we have

$$\begin{aligned} \angle GH = \angle \text{Zeros to } s_1 - \angle \text{Poles to } s_1 &= 0 - (90\ \text{deg} + 135\ \text{deg} + \theta_3) \\ &= -225 - \theta_3 = 180\ \text{deg} \\ &\Rightarrow \theta_3 = -405 = -45\ \text{deg} \end{aligned}$$

This result shows that the root locus departs the complex pole at -45 deg. Similarly, the root locus will depart its complex conjugate at

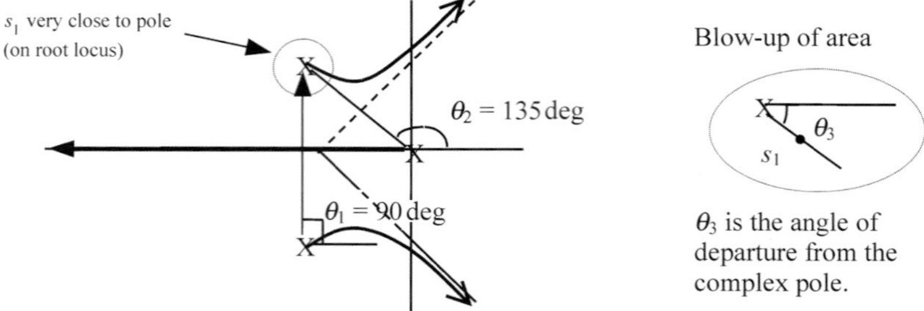

Fig. 8.27 Illustration of angle of departure.

45 deg using the symmetry rule. Determination of the angle of departure from complex poles can help refine the shape of the root locus but is not always necessary.

8) *Imaginary Axis Crossing:* The imaginary axis crossing point of the root locus can also be found. At the imaginary axis crossing points we know that the closed-loop root, s, equals $\pm i\omega$ with a zero real component. Consider the OLTF:

$$GH = \frac{K}{s(s+3)(s+5)}$$

with open-loop poles at $s = 0, -3, -5$. The closed-loop characteristic equation is

$$1 + GH = s^3 + 8s^2 + 15s + K = 0$$

We then substitute in $s = i\omega$

$$(i\omega)^3 + 8(i\omega)^2 + 15i\omega + K = 0$$

Simplifying,

$$-i\omega^3 - 8\omega^2 + 15i\omega + K = 0$$

We then separate out the real and imaginary components of the characteristic equation.

$$\begin{aligned} \text{Real:} \quad & -8\omega^2 + K = 0 \\ \text{Imaginary:} \quad & -i\omega^3 + 15i\omega = 0 \\ & -i\omega(\omega^2 - 15) = 0 \end{aligned}$$

and solve the imaginary part to determine ω.

$$\omega = \pm\sqrt{15}$$

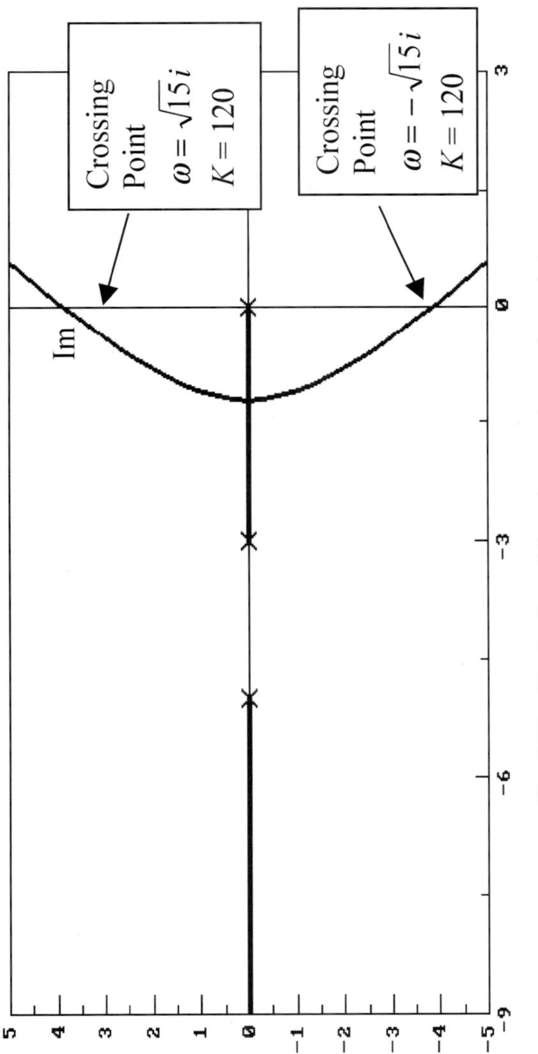

Fig. 8.28 Illustration of imaginary axis crossing points.

We then substitute this result into the real part to determine K.

$$K = 8(\sqrt{15})^2 = 120$$

Therefore, for this example, the root locus crosses the imaginary axis at $\omega = \pm\sqrt{15}i$ with $K = 120$, as illustrated in Fig. 8.28.

The imaginary axis crossing point is important because it tells us the point at which the closed-loop system is about to go unstable. In the previous example, the closed-loop system is stable for $K < 120$, it has neutral stability for $K = 120$, and it is unstable for $K > 120$.

9) *Breakaway Points:* Breakaway points are points on the real axis where the root locus "breaks away" and the closed-loop roots become complex. Breakaway points can be found by solving for the gain (K) in the closed-loop characteristic equation, taking the derivative of the gain with respect to s, and setting it equal to 0. For example, consider the OLTF

$$GH = \frac{K}{(s+1)(s+3)}$$

with open-loop poles at $s = -1, -3$. The characteristic equation is

$$1 + GH = s^2 + 4s + 3 + K = 0$$

Solving for K we have

$$K = -(s^2 + 4s + 3)$$

and taking the derivative and setting it equal to zero:

$$\frac{\partial K}{\partial s} = -2s - 4 = 0$$

We then simply solve the resulting relationship for s to determine the breakaway point.

$$s = -2$$

Figure 8.29 presents the root locus for this example.

In this simple example, the asymptote centroid and the breakaway point are the same. This is not normally the case.

Example 8.9

Find the asymptote centroid and breakaway point for the following OLTF:

$$GH = \frac{K}{s^3 + 8s^2 + 15s}$$

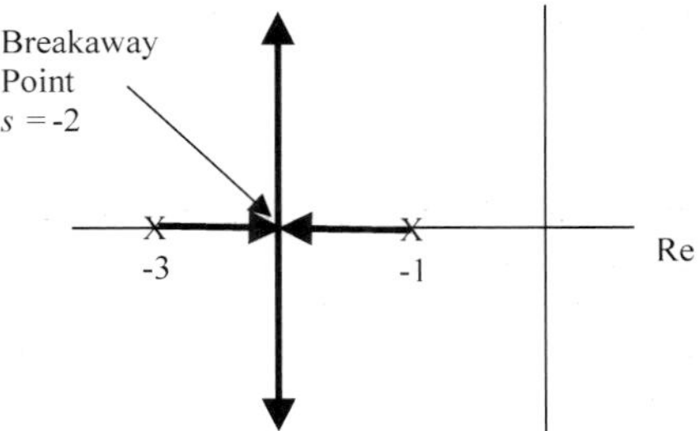

Fig. 8.29 Illustration of breakaway point.

The open-loop poles are located at $s = 0, -3, -5$. Because there are three more poles than zeros, the angle of the asymptotes is

$$\varphi = \pm 180 \text{ deg}, \pm 60 \text{ deg}$$

The asymptote centroid is

$$\sigma = \frac{-3 + (-5)}{3} = -\frac{8}{3}$$

To compute the breakaway point, we form the closed-loop characteristic equation

$$1 + GH = s^3 + 8s^2 + 15s + K = 0$$

Solving for K,

$$K = -s^3 - 8s^2 - 15s$$

Taking the derivative with respect to s and setting it equal to zero,

$$\frac{dK}{ds} = -3s^2 - 16s - 15 = 0$$

We then solve the above equation for s.

$$s = -4.12, -1.21$$

A plot of the root locus, presented next, will aid in determining if one or both of these breakaway points are valid.

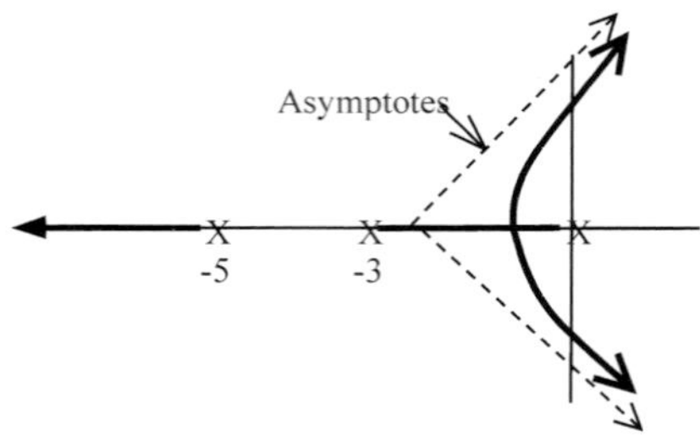

Because $s = -1.12$ is part of the root locus real axis location, it is a breakaway point. The point $s = -4.12$ is not part of the root locus real axis location and therefore cannot be a breakaway point. It is discarded. Often, when there are 2 roots, one will correspond to the break-in point (illustrated in Example 8.10). Notice that in this example, the asymptote centroid and the breakaway point are not the same.

Example 8.10

Sketch the root locus for the following OLTF using the nine steps discussed in this section:

$$GH = \frac{K(s+5)}{(s+2)(s+3)} \qquad Z = -5 s = -2, -3$$

We first observe that the system has one open-loop zero at $s = -5$ and two open-loop poles at $s = -2, -3$.

Step 1: 2 poles $\Rightarrow$ 2 branches of the root locus
Step 2: Root locus begins at open loop poles and goes to zeros or to ∞.
Step 3: A branch is on the real axis if it is to the left of an odd number of poles and zeros.

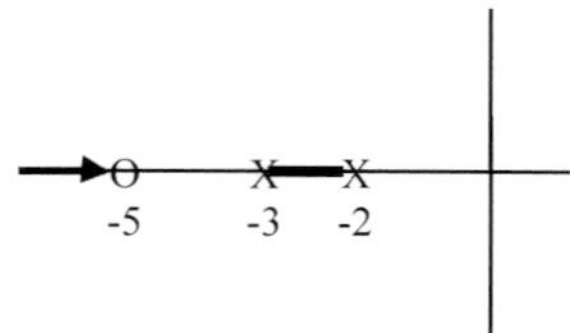

Step 4: Root locus is symmetric about the real axis.
Step 5: The angle of the asymptotes is

$$\varphi = \frac{\pm 180 \text{ deg}}{1-2} = \pm 180 \text{ deg}$$

because we have one more pole than zero.

Step 6: The centroid of the asymptotes is

$$\sigma = \frac{-5 - (-5)}{2 - 1} = 0$$

For the ± 180 deg asymptote case, we know the asymptote is the negative real axis and the centroid has little meaning.

Step 7: The departure angle is not needed because we do not have complex open-loop poles.

Step 8: We do not have an imaginary axis crossing because we do not have asymptotes that cross the imaginary axis (that is ± 60 deg). This will be the case if the difference between OLTF poles and zeros is two or fewer.

Step 9: To find the breakaway point, we determine the roots of $\mathrm{d}K/\mathrm{d}s = 0$.

$$1 + GH = s^2 + 5s + 6 + Ks + 5K = 0$$

$$K = \frac{-s^2 - 5s - 6}{(s + 5)} = \frac{u}{v}$$

and using the chain rule

$$\mathrm{d}\left(\frac{u}{v}\right) = \frac{v\mathrm{d}u - u\mathrm{d}v}{v^2}$$

$$\frac{\mathrm{d}K}{\mathrm{d}s} = \frac{(s + 5)(-2s - 5) - (-s^2 - 5s - 6)(1)}{(s + 5)^2} = 0$$

or

$$-s^2 - 10s - 19 = 0 \Rightarrow s = -7.45, -2.55$$

Thus, potential locations for breakaway points are $s = -7.45, -2.55$. A plot of the root locus is presented next. For the locus between the poles -2 and -3, there is a breakaway point at -2.55. The point at -7.45 is a break in point as shown. Notice that, after the break-in point, one branch of the root locus goes to the zero at $s = -5$ and the other branch goes to the 180-deg asymptote.

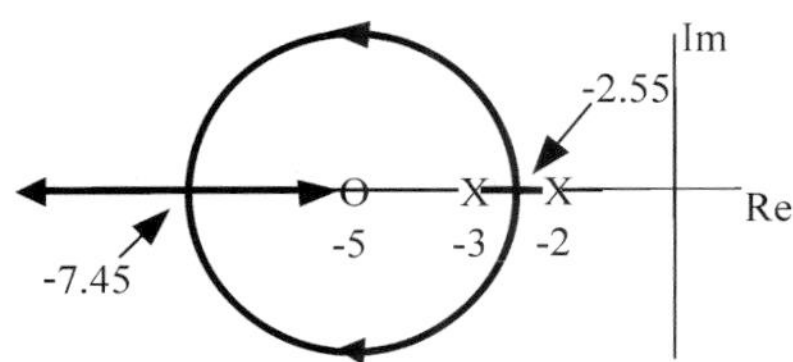

Example 8.11

Sketch the root locus for the following OLTF using the nine steps discussed in this section

$$GH = \frac{K}{(s+1)(s+2)(s+3)}$$

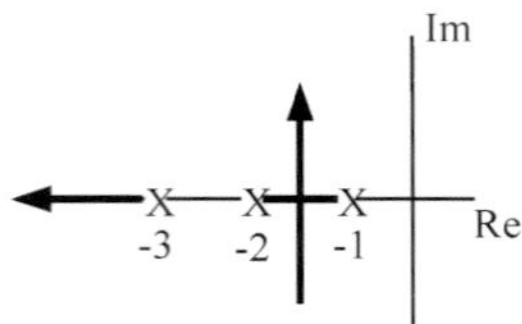

Step 1: 3 poles $\Rightarrow$ 3 branches
Step 2: Root locus begins at open loop poles and goes to zero or ∞ (3 branches go to ∞ because we have no zeros).
Step 3: A branch is on the real axis if it is to the left of an odd number of poles and zeros. Real axis root locus locations are shown in the sketch.
Step 4: Root locus is symmetric about the real axis.
Step 5: The angle of the asymptotes is

$$\varphi = \frac{\pm 180\ \text{deg}(1+2n)}{0-3}\left\} \begin{array}{l} n=0 \Rightarrow \varphi = \pm 180\ \text{deg (one asymptote)} \\ n=1 \Rightarrow \varphi = \pm 60\ \text{deg (two asymptotes)} \end{array}\right.$$

We have three asymptotes because we have three more poles than zeros.
Step 6: The centroid of the asymptotes is

$$\sigma = \frac{-1+(-2)+(-3)}{3-0} = -2$$

Step 7: The departure angle is not needed because we do not have complex open-loop poles.
Step 8: The imaginary axis crossing is determined by substituting $s = i\omega$ into the closed-loop characteristic equation.

$$\begin{aligned} 1 + GH = 0 &= s^3 + 6s^2 + 11s + 6 + K \\ -K &= s^3 + 6s^2 + 11s + 6 \\ -K &= (i\omega)^3 + 6(i\omega)^2 + 11(i\omega) + 6 \\ -K &= -i\omega^3 - 6\omega^2 + 11i\omega + 6 \end{aligned}$$

Separating into the imaginary and real parts, we have

$$-i\omega(\omega^2 - 11) = 0 \Rightarrow \omega = \sqrt{11} = 3.3166 \text{ rad/s}$$
$$K = 6\omega^2 - 6 = 6(3.3166)^2 - 6 = 60$$

Thus, the imaginary axis crossing is at $s = \pm i3.3166$ for a value of $K = 60$.
Step 9: To find the breakaway point, we determine the roots of $\mathrm{d}K/\mathrm{d}s = 0$

$$-K = s^3 + 6s^2 + 11s + 6$$
$$\frac{\mathrm{d}K}{\mathrm{d}s} = -3s^2 - 12s - 11 = 0 \Rightarrow s = -1.4226, -2.577$$

The point $s = -1.4226$ is on the real axis root locus, therefore it is a breakaway point. The other point, $s = -2.577$, is not. A sketch of the root locus plot is provided in Example 8.8.

8.7 Historical Snapshot—The C-1 Autopilot

One of the first closed-loop systems applied to aircraft was the C-1 autopilot installed on the B-17E bomber during World War II (see Fig. 8.30).

The C-1 became operational in 1943 and was primarily intended to stabilize the aircraft during bombing. In preparation for a bomb run, the pilot would trim the aircraft in straight and level flight and then engage the C-1 autopilot. The autopilot's job was to provide a stable platform for the bombardier so that he could acquire the target and make small corrections to achieve the proper release point. It also eliminated undesirable aircraft motion during the bomb release, which contributed to improved accuracy. The C-1 was also used to relieve pilot fatigue on long flights.

Operation of the C-1 began with potentiometers on the vertical and directional gyros to sense changes in pitch, roll, and yaw attitude. The electronic signal from the potentiometers was amplified and input to solenoids and servomotors, which were attached to the flight control surfaces through cable drives.

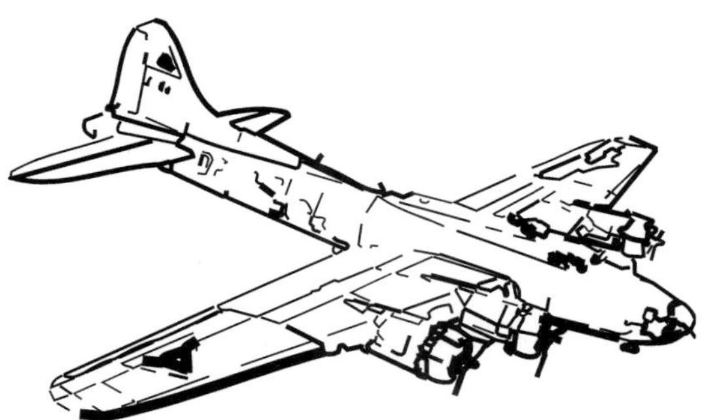

Fig. 8.30 Boeing B-17 Bomber.

If a potentiometer sensed a nose-up pitch attitude change, the result was a trailing edge down elevator deflection to correct for the deviation from trim. Similar feedback loops operated the ailerons and the rudder. The C-1 was built by Honeywell, Inc. Other 1940s era autopilots used on military aircraft included the E-4 and the A-12, both of which were built by the Sperry Corporation.

Problems

8.1 You are given the following block diagram:

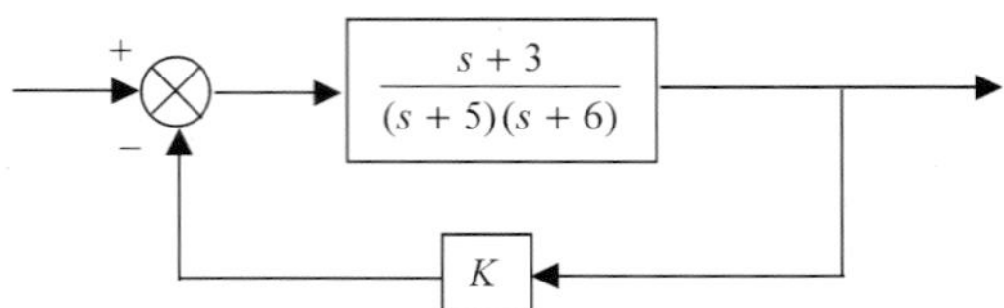

(1) What is the natural frequency and damping ratio of the open-loop system?
(2) Form the CLTF.
(3) At $K = 0$, what is the characteristic equation?
(4) What are the poles and zeros of the open-loop system?
(5) Sketch the open-loop poles and zeros on the complex plane.

8.2 Using the results from Problem 8.1, determine
(1) The closed-loop characteristic equation and roots (using the quadratic formula) at $K = 0$, 1, 5, 10 and 100. Form a table.
(2) On the sketch from Problem 8.1, Part (5), plot the movement of the roots of the CLTF characteristic equation as the gain is varied from 0 to 100.
(3) Where does the branch associated with the OLTF pole at $s = -6$ appear to be going? How about the branch associated with the OLTF pole at $s = -5$? What role does the open loop area seem to be playing? Summarize these results in your own words.

8.3 The following is a model of pitch displacement autopilot with rate feedback:

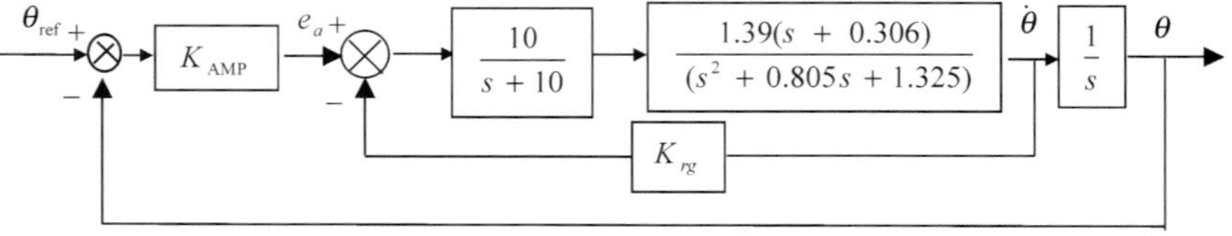

Determine the CLTF for $\theta/\theta_{\text{ref}}$. Note, you will have to determine the CLTF for the inner loop first. K_{AMP} and K_{rg} represent the gain of the amplifier and the rate gyro, respectively.

8.4 If a gain, K, is part of the "G" term, you would expect which of the following to happen?
(a) The input to the root locus program would be GH^*.
(b) The CLTF characteristic equation would not be dependent on the value of K.
(c) The CLTF poles would change as a function of K.
(d) None of the above.

8.5 A pole of the CLTF will be found to be a zero of the $1 + GH$ term.
(a) True
(b) False

8.6 Which of the following depicts the transfer function for a servo actuator with a time constant of 0.2 s.

(a) $\dfrac{0.2}{s+0.2}$

(b) $\dfrac{5}{s+5}$

(c) $\dfrac{0.2}{5s+1}$

(d) $\dfrac{5}{s+0.2}$

8.7 For the following system, find $\dfrac{C(s)}{R(s)}$

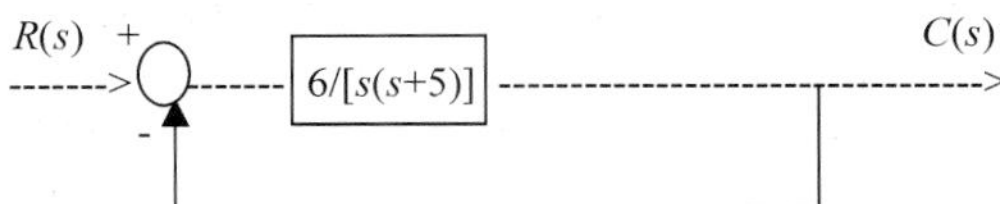

For a step input, find $c(t)$ using a partial fraction expansion.

8.8 Find T_p, M_p, and max overshoot for

$$C(s) = \frac{16}{s(s^2 + 4s + 16)}$$

8.9 Is the following system stable?

$$TF = \frac{5}{s^2 + 8s + 7}$$

8.10 Find the OLTF and CLTF for the following system:

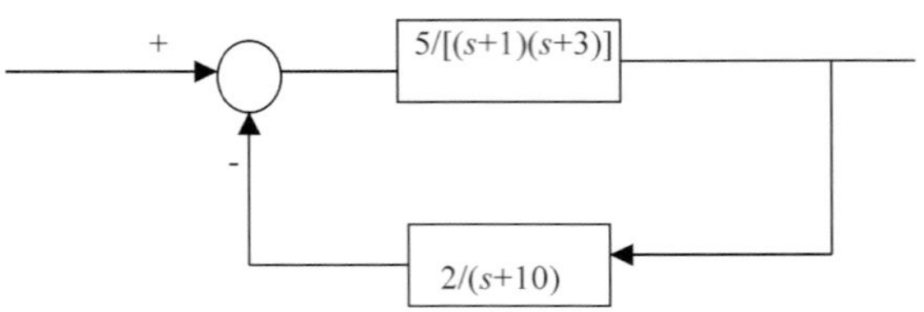

8.11 Hand plot the root locus for the following system:

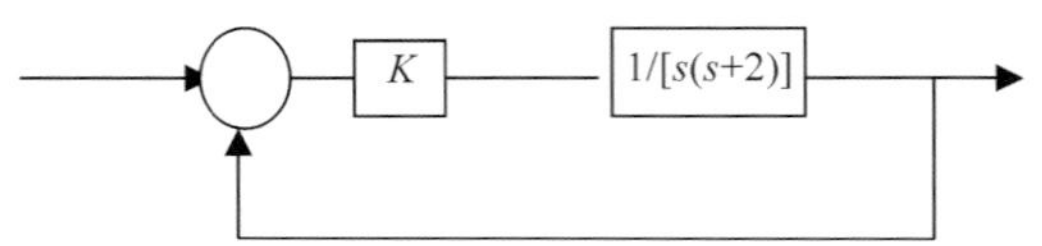

Using $K = 0, \frac{1}{2}, 1, 2, 3$, and ∞

CLOSED-LOOP ROOT LOCATIONS AS A FUNCTION OF *K*

K	s_1	s_2
0	$0 + j0$	$-2 - j0$ (open-loop poles)
	$-0.3 + j0$	$-1.7 - j0$
1	$-1 + j0$	$-1 - j0$
2	$-1 + j1$	$-1 - j1$
3	$-1 + j\sqrt{2}$	$-1 - j\sqrt{2}$
∞	$-1 + j\infty$	$-1 - j_{\infty}$

8.12 Sketch the root locus for

$$GH = \frac{K}{s(s + 25)(s^2 + 50s + 2600)}$$

Calculate the number of branches, real axis loci, location, θ, and γ for the asymptotes, breakaway points, and departure angles.

8.13 Which set of poles has a higher magnitude of overshoot for $\xi =$ constant?

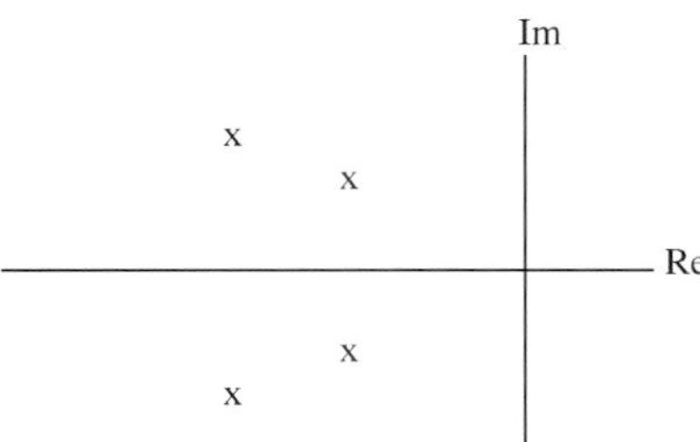

Which set of poles has a higher magnitude overshoot for $\omega_N =$ constant?

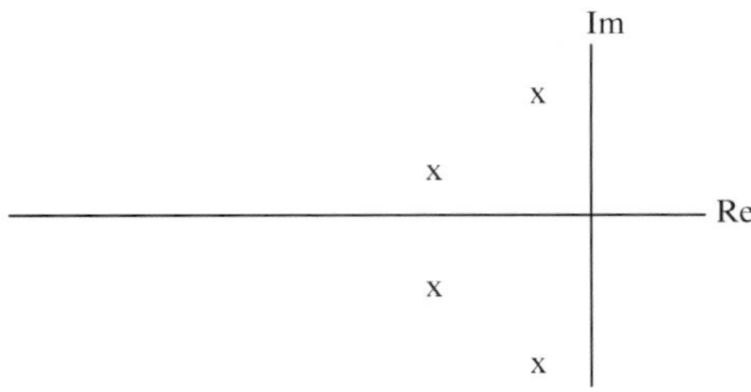

8.14 Select K_1 and K_2 for the following system to achieve an ω_N of 4 and a ξ of 0.25.

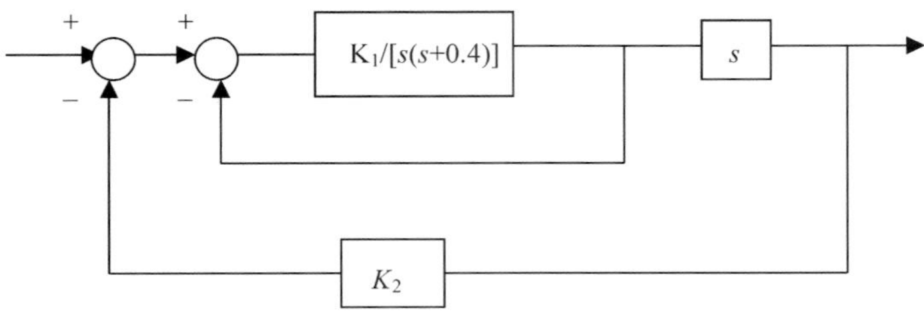

8.15 Plot the root locus for the following system:

$$G(s) = \frac{K}{s(s+1)(s+5)}$$

Plot the modified root locus with the following lead compensators added and discuss.

(a) $G_c = \dfrac{s+0.75}{s+7.5}$

(b) $G_c = \dfrac{s+1}{s+10}$

(c) $G_c = \dfrac{s+1.5}{s+15}$

8.16 Why shouldn't a washout filter be used in cascade?

9
Aircraft Stability and Control Augmentation

The classical feedback control techniques developed in Chapter 8 provide the foundation for the design of aircraft feedback control systems. The root locus will be a primary design tool used in this text, but the reader should be aware that additional analysis methods, such as Bode plots, are also used extensively by designers. Fortunately, several computer programs have been developed (such as MATLAB® (a registered trademark of The MathWorks, Inc.) and XMATH) that provide a user-friendly generation of root locus plots and associated open-loop and closed-loop time response characteristics. Two such programs, Program CC developed by Systems Technology, Inc., and MATLAB will be used for the generation of root locus plots throughout this chapter. The detailed discussion of root locus plotting techniques in the last chapter should provide design insight in developing strategies for improving dynamic stability characteristics. The reader is encouraged to review the complex plane time response relationships presented in Sec. 7.2 and the closed-loop design concepts presented in Chapter 8.

9.1 Inner-Loop Stability and Control

The dynamic stability characteristics of aircraft have been improved during the past 50 years using a variety of inner-loop feedback control systems. Inner-loop simply refers to the fact that these systems are represented as the inner loop in a block diagram representation when married with outer-loop autopilot modes such as attitude hold. Although not exact, inner-loop feedback control systems can be grouped into three broad categories: stability augmentation systems, control augmentation systems, and fly-by-wire systems.

9.1.1 Stability Augmentation Systems

Stability augmentation systems (SAS) were generally the first feedback control system designs intended to improve dynamic stability characteristics. They were also referred to as dampers, stabilizers, and stability augmenters. Aircraft such as the F-104, T-37, T-38, and F-4 had SAS. These systems generally fed back an aircraft motion parameter, such as pitch rate, to provide a control deflection that opposed the motion and increased damping characteristics. The SAS had to be integrated with the primary mechanical control system of the aircraft consisting of the stick, pushrods, cables, and bellcranks leading to the control surface or the hydraulic actuator that activated the control surface. The control authority (percentage of full surface deflection available)

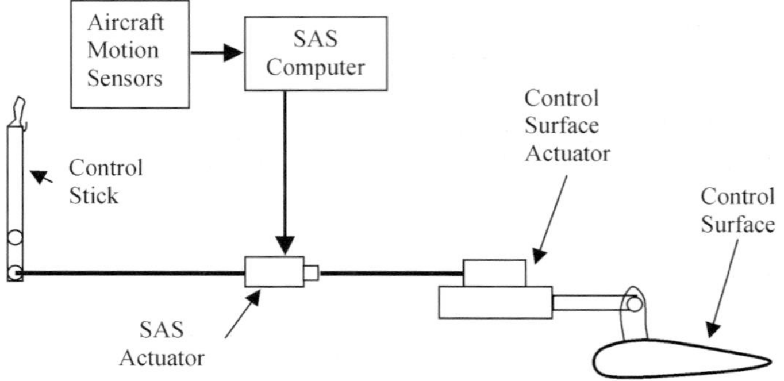

Fig. 9.1 Simplified SAS.

of SAS was generally limited to about 10%. Fig. 9.1 presents a simplified SAS.

One problem with SAS was the fact that the feedback loop provided a command that opposed pilot control inputs. As a result, the aircraft became less responsive for a given stick input. This was typically addressed with the addition of a washout filter in the feedback loop that attenuated the feedback signal for constant values of the aircraft motion parameter. A block diagram of a typical SAS is presented in Fig. 9.2.

Another concern was the limited authority of the SAS actuator that was necessitated by safety-of-flight requirements. SAS sensors and computers were normally nonredundant or dual redundant and thus did not approach the system reliability of the mechanical flight control system. Despite these concerns, SAS was effective in improving aircraft flying qualities.

9.1.2 Control Augmentation Systems

The next step in the evolution of aircraft feedback control was control augmentation systems (CAS). CAS added a pilot command input into the flight control computer. A force sensor on the control stick was usually used to provide this command input. With a CAS, a pilot stick input is provided to the flight control system in two ways—through the mechanical system and through the CAS electrical path. The CAS design eliminated the SAS problem of pilot

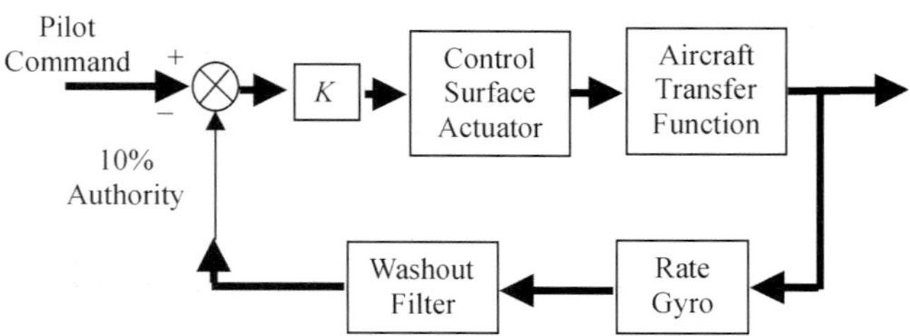

Fig. 9.2 Typical SAS block diagram.

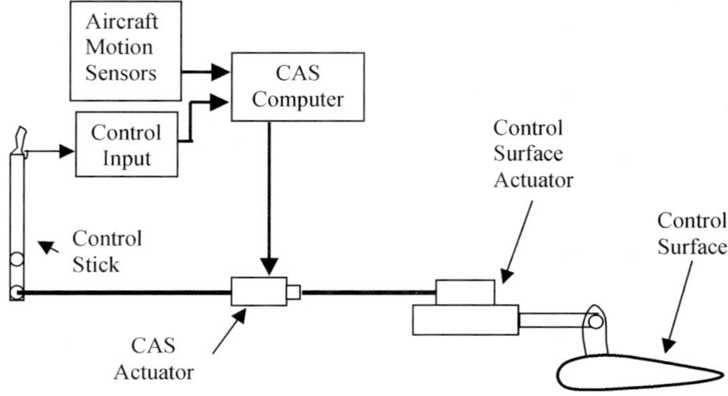

Fig. 9.3 Simplified CAS.

inputs being opposed by the feedback loop. Aircraft such as the A-7, F-111, F-14, and F-15 have CAS. Figure 9.3 presents a simplified CAS.

Additional reliability was designed into CAS so that the control authority could be increased (to approximately 50%). With a CAS, the aircraft dynamic response is typically well-damped, and control response is scheduled with the control system gains to maintain desirable characteristics throughout the flight envelope. A block diagram for a typical CAS is presented in Fig. 9.4.

CAS provided dramatic improvements in aircraft handling qualities. Both dynamic stability and control response characteristics could be tailored and optimized to the mission of the aircraft. One pioneering exploratory development program in the early 1970s, the A-7D Digital Multimode Flight Control System Program, developed specific feedback flight control designs using a CAS and an A-7D aircraft, which tailored the aircraft's handling qualities to specific mission tasks such as air-to-air tracking and air-to-ground gunnery.

9.1.3 Fly-By-Wire Systems

Based on the excellent handling qualities achieved with CAS, the next logical step in feedback control systems development was to remove the mechani-

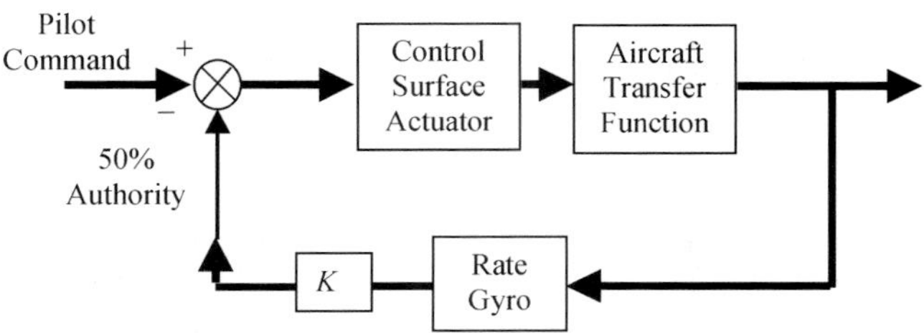

Fig. 9.4 Typical CAS block diagram.

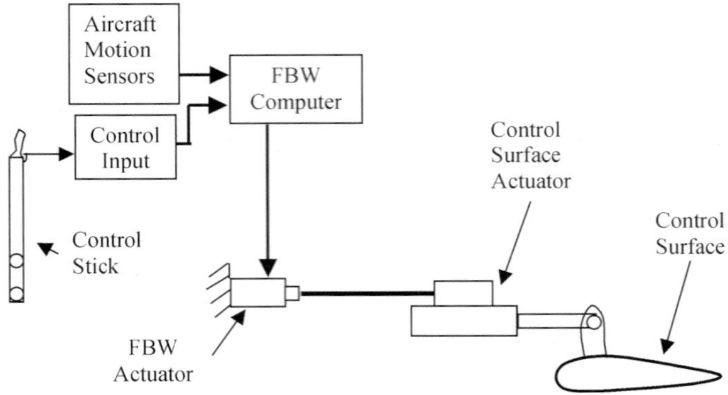

Fig. 9.5 Simplified FBW system.

cal control system and provide the CAS full authority. Such systems are known as fly-by-wire (FBW) systems. This major step involved proving that the reliability of the FBW system, composed of mostly electrical components, was equal to or better than the trusted mechanical system. To achieve this reliability, triple and quad redundancy in system components, along with self-test software is used. Aircraft such as the F-16, C-17, and F-22 have FBW systems. In the case of the F-22, another term, fly-by-light, is sometimes used to indicate that fiberoptic links are used rather than wire. The full authority provided by FBW allows very significant tailoring of stability and control characteristics. This ability has led to FBW systems with several feedback parameters and weighting of feedback gains based on flight condition and other parameters. Figure 9.5 presents a simplified FBW system.

Block diagrams for FBW systems can become complex because of the number of feedback sensors involved. Figure 9.6 presents a simplified block diagram for the F-16 longitudinal FBW system.

9.1.4 Typical Inner-Loop Systems

Several inner-loop systems will be discussed and analyzed in this section. The general analysis techniques developed in Chapter 8 will be used extensively.

9.1.4.1 Yaw damper. A yaw damper is primarily used to improve dutch roll characteristics. Good dutch roll characteristics are essential to tight tracking tasks such as air-to-air refueling, formation flying, approach, air-to-air tracking, and air-to-ground tracking. To illustrate the design of a yaw damper, we will start with an approximation of the $\dot{\psi}/\delta_r$ transfer function[1] for the C-5 at 0.22 Mach, sea level, in the power approach configuration (gear and flaps down)

$$\frac{\dot{\psi}}{\delta_r} = -\frac{0.213(s+1.2)(s^2+0.6s+0.1525)}{(s+0.028)(s+1.13)(s^2+0.24s+0.2848)}$$

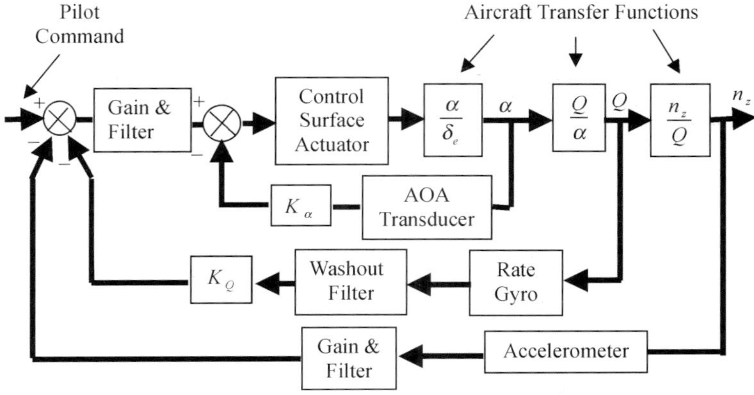

Fig. 9.6 Simplified F-16 longitudinal FBW block diagram.

Figure 9.7 presents a yaw damper configuration based on the SAS approach presented in Fig. 9.2, except that the gain is located in the feedback path.

The washout filter in Fig. 9.7 will drive the feedback signal to zero after a steady-state or relatively constant yaw rate is achieved. This prevents the yaw damper from fighting a pilot command. The washout can be thought of in general terms as

$$\frac{\tau s}{\tau s + 1}$$

where τ is the time constant that determines the pace at which the feedback signal is driven to zero. Figure 9.8 presents the time response of a washout filter to a unit step input.

In this figure, the unit step occurs at time equal to zero, and the full signal is immediately passed through the washout filter. Because the unit step input stays constant for time greater than zero, the attenuation provided by the washout is faster for smaller values of time constant.

A root locus plot for the yaw damper of Fig. 9.7 is presented in Fig. 9.9. The plot focuses on the dutch roll roots that have an open-loop damping ratio

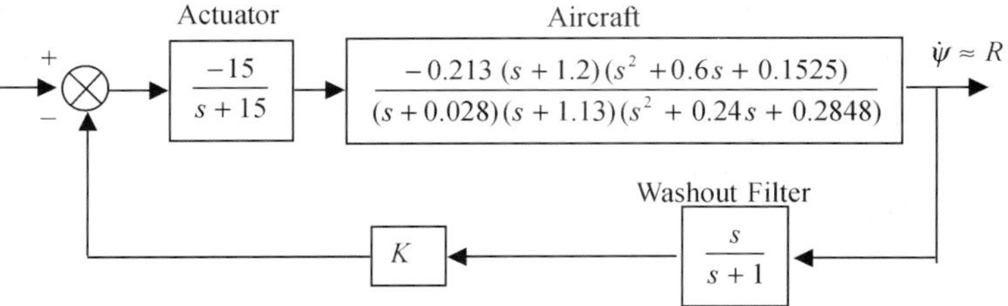

Fig. 9.7 Yaw damper block diagram.

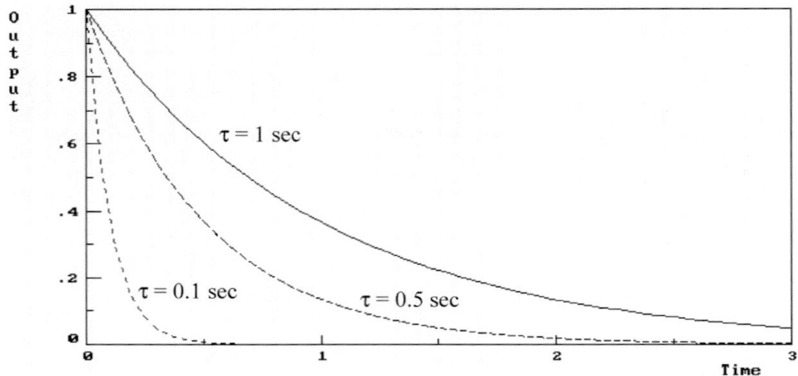

Fig. 9.8 Unit step time response of a washout filter with three different time constants.

of 0.2227 and a natural frequency of 0.532 rad/s. By closing the loop with a gain (K) of 4.277, we can achieve a damping ratio of 0.5 and a natural frequency of 0.5011 rad/s. The roots of the closed-loop system are also shown in Fig. 9.9.

Notice that even higher values for dutch roll damping ratio could be obtained with higher values of K. Also notice that the dutch roll natural frequency remains fairly constant throughout the gain range as discussed for the case of rate feedback in Sec. 8.3. Figure 9.10 presents the open-loop and closed-loop time response to a unit impulse input.

The open-loop response clearly shows the oscillations associated with a 0.2227 damping ratio. As expected, the closed-loop response is more heavily

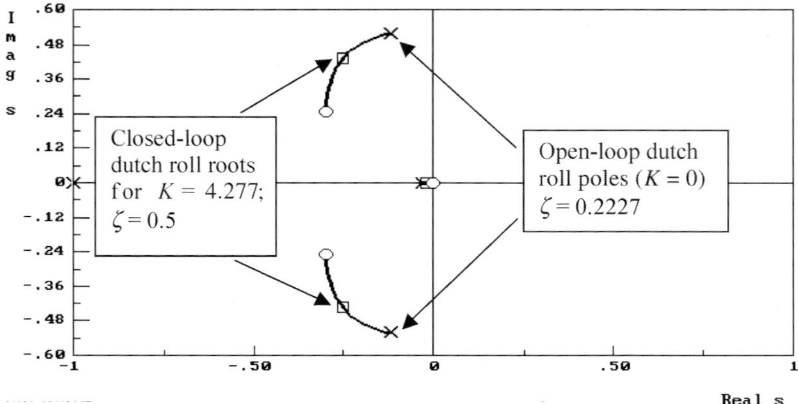

Fig. 9.9 Yaw damper root locus plot focusing on dutch roll roots.

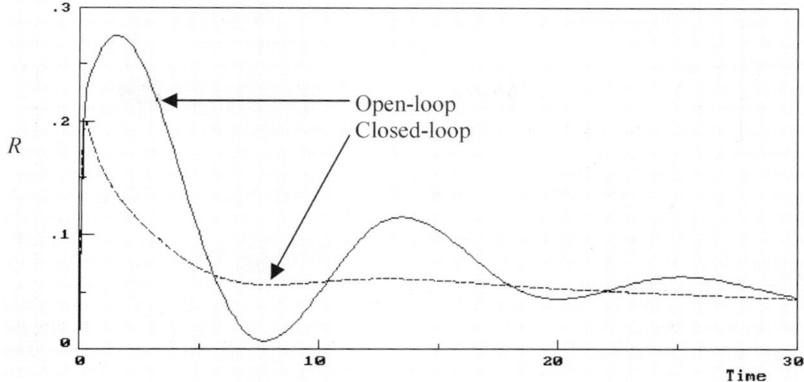

Fig. 9.10 Yaw damper time response to an impulse input.

damped. The full root locus plot for the yaw damper is presented in Fig. 9.11 so that the migration of the other poles can be observed. Note the uneven scales.

Closing the loop moves the actuator pole from −15 to approximately −14 (a slightly slower time constant). The spiral root at −0.028 moves to −0.0187 (still stable but a slower time constant), the roll mode root moves from −1.13 to −1.65 (a faster time constant), and the root added by the washout filter moves from −1 to −1.24.

To make the yaw damper more effective throughout the aircraft flight envelope, the gain, K, may be scheduled with dynamic pressure or other flight parameters. Definition of the gain schedule involves analysis of transfer functions representative of the entire aircraft flight envelope and selection of a gain value

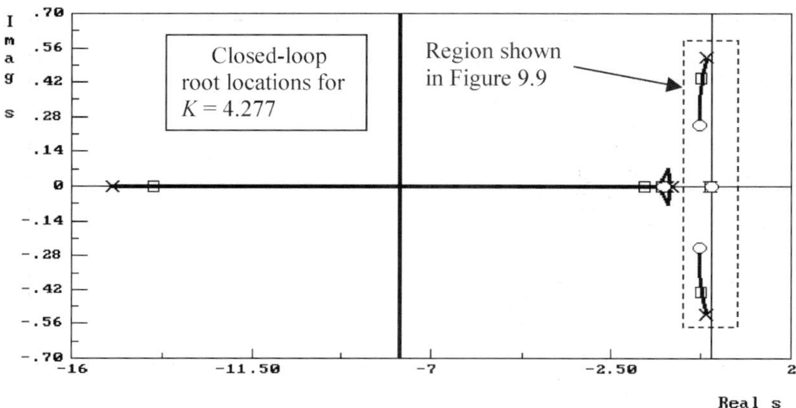

Fig. 9.11 Complete yaw damper root locus.

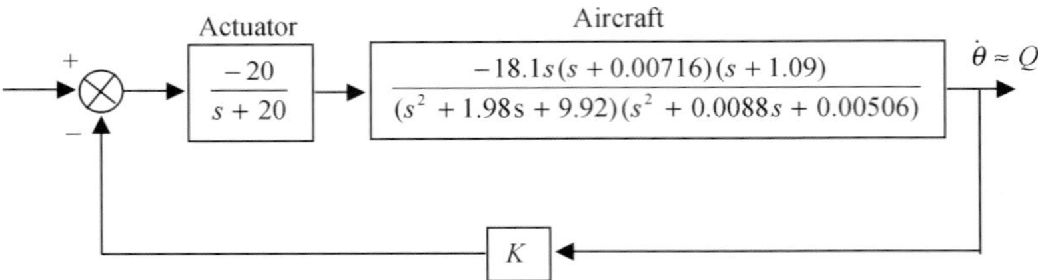

Fig. 9.12 Pitch damper block diagram.

appropriate to each. A schedule is then developed that defines the gain as a function of flight condition. In addition, selection of the washout filter time constant will affect the value of the gain needed to obtain the desired stability characteristics.

9.1.4.2 Pitch damper. A pitch damper is primarily used to improve short period characteristics. Good short period characteristics are critical to the same tight tracking tasks as for dutch roll. To illustrate the design of a pitch damper, we start with an approximation[1] of the $\dot{\theta}/\delta_e$ transfer function for the A-7D at 0.6 Mach, 15,000 ft, in the cruise configuration (gear and flaps up).

$$\frac{\dot{\theta}}{\delta_e} = \frac{-18.1s(s+0.00716)(s+1.09)}{(s^2+1.98s+9.92)(s^2+0.0088s+0.00506)}$$

Figure 9.12 presents a pitch damper configuration based on the CAS approach. Notice that a washout filter is not included.

The pitch damper root locus is presented in Fig. 9.13. It focuses on the short period roots that have an open-loop damping ratio of 0.3184 and a natural frequency of 3.14 rad/s. By closing the loop with a gain of 0.1592, we

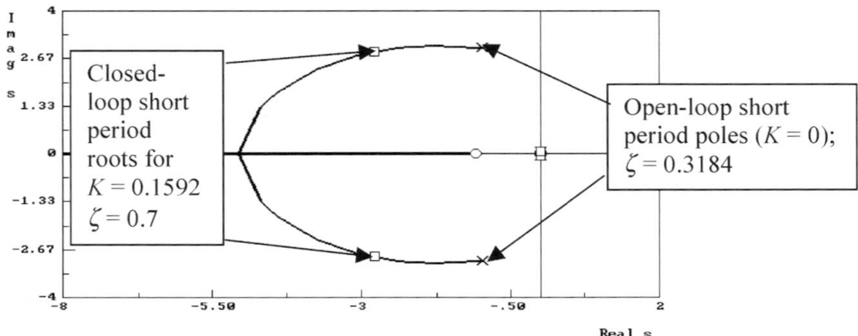

Fig. 9.13 Pitch damper root locus plot focusing on short period roots.

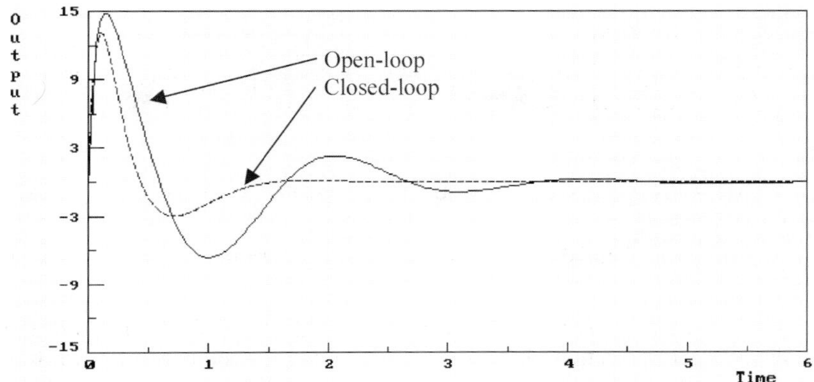

Fig. 9.14 Pitch damper time response to an impulse input.

achieve a damping ratio of 0.7 and a natural frequency of 3.98 rad/s. The roots of the closed-loop system are also shown in Fig. 9.13.

Here again, even higher values of short period damping could be achieved with higher values of K. The short period natural frequency does increase because this system is more complex than the simple second-order example used in Sec. 8.3. Figure 9.14 presents the open-loop and closed-loop time response to a unit step input.

As expected, the closed-loop response is more heavily damped. The full root locus plot for the pitch damper is presented in Fig. 9.15 so that migration of the actuator pole can be observed.

Notice that the actuator pole moves from -20 to approximately -16.5 (a slightly longer time constant). Migration of the phugoid roots is difficult to

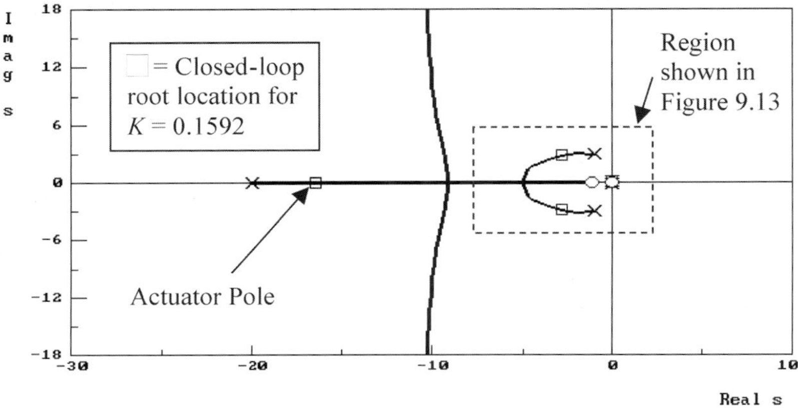

Fig. 9.15 Complete pitch damper root locus.

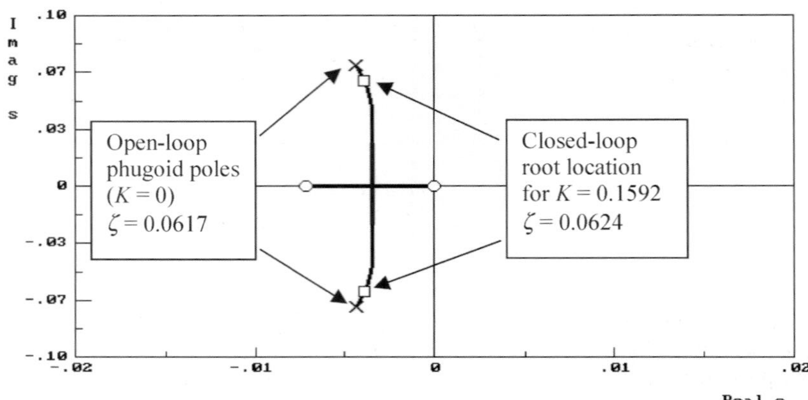

Fig. 9.16 Pitch damper root locus focusing on phugoid roots.

observe in Figs. 9.13 and 9.15 because of the relatively large scale. Figure 9.16 presents a focused view of the phugoid roots that are concentrated around the origin.

The phugoid roots only move slightly with a very small increase in damping ratio and decrease in natural frequency.

9.1.4.3 Angle of attack feedback. Angle of attack (AOA) feedback can be used to increase short period natural frequency and the static stability of the aircraft (C_{m_α}). Modern aircraft with relaxed static stability in the basic airframe design will use AOA feedback to return static stability characteristics to the closed-loop aircraft (see Fig. 9.6 for the F-16). To illustrate the design of an AOA feedback system, we will start with an approximation (see Ref. 1) of the α/δ_e transfer function for the F-16 at 135 kn true airspeed and sea level in the power approach configuration.

$$\frac{\alpha}{\delta_e} = \frac{-0.083(s^2 + 0.086s + 0.0594)(s + 35.4)}{(s^2 + 0168s + 0.0832)(s + 1.75)(s - 0.4825)}$$

Notice that one of the roots is positive, making the aircraft statically unstable. This is typical for aircraft with the c.g. aft of the aerodynamic center. Advantages of designing an aircraft with basic airframe static instability include reduced horizontal stabilizer size and quicker longitudinal response. Of course, a closed-loop system must be added to provide apparent static stability. Figure 9.17 presents an AOA feedback configuration.

A simplified root locus for the AOA feedback system is presented in Fig. 9.18. Notice that closing the loop causes the unstable root to move into the stable region of the complex plane. The two open-loop oscillatory roots move to the real axis, one moving toward the origin and the other moving to the left. We then have two breakaway points that form branches for the more recogniz-

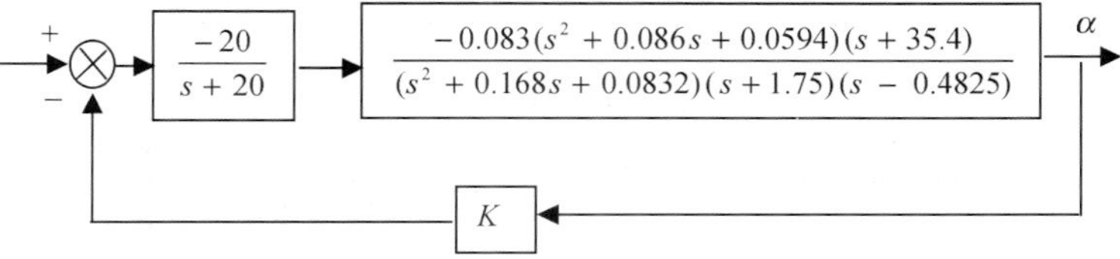

Fig. 9.17 AOA feedback block diagram.

able short period (roots away from the origin) and phugoid (roots close to the origin) second-order roots. For illustration, we select a gain of $K = 0.415$, which provides a short-period damping ratio of 0.9 and a natural frequency of 0.715 rad/s. The closed-loop phugoid mode has a damping ratio of 0.942 and a natural frequency of 0.0643 rad/s.

Even larger values of gain will give the aircraft more recognizable short-period and phugoid roots. AOA feedback requires an accurate AOA measurement, a sometimes difficult parameter to measure with high fidelity.

9.1.4.4 Load factor feedback. Another method that may increase short-period damping ratio is load factor feedback. It has the added benefit of helping to linearize the stick force gradient (stick force/g) of the aircraft. To develop the n/δ_e transfer function, we approximate the perturbed load factor as

$$n = \frac{U_1 \dot{\gamma}}{g}$$

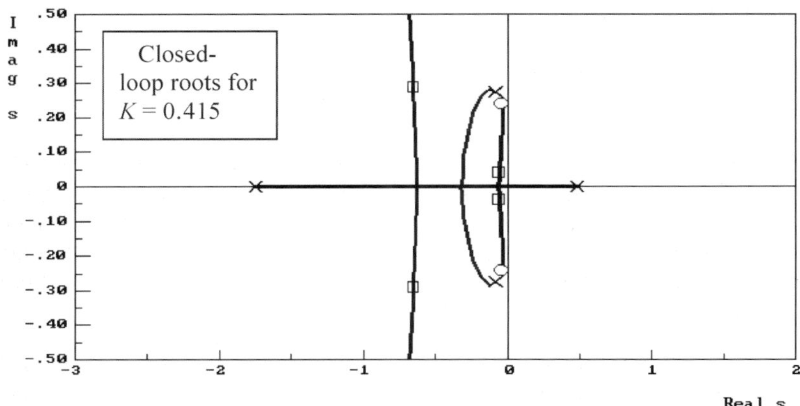

Fig. 9.18 Simplified AOA feedback root locus.

where $U_1\dot{\gamma}$ is the centripetal acceleration acting on the aircraft because of the rotation of the velocity vector and g is $32.2\,\mathrm{ft/sec^2}$ in the English system. Recall that the flight path angle (γ) is the angle between the aircraft velocity vector and the horizon. Because $\gamma = \theta - \alpha$,

$$\frac{n}{\delta_e} = \frac{U_1\dot{\gamma}(s)}{32.2\delta_e(s)} = \frac{U_1 s}{32.2}\left(\frac{\theta(s)}{\delta_e(s)} - \frac{\alpha(s)}{\delta_e(s)}\right) \tag{9.1}$$

For the A-7D at 0.6 Mach and 15,000 ft (see Ref. 1), the n/δ_e transfer function is approximated by

$$\frac{n}{\delta_e} = \frac{19.7s(-0.157s^3 - 0.8977s^2 + 19.77s + 0.1122)}{(s^2 + 1.98s + 9.92)(s^2 + 0.0088s + 0.00506)}$$

Figure 9.19 presents a load factor feedback configuration.

The load factor feedback root locus scaled to show the phugoid roots is presented in Fig. 9.20. Notice that the short period roots become more stable as K is increased. However, the phugoid roots quickly become unstable with an increase in K, as shown in Fig. 9.21. As a result, we select a value of gain that will increase the damping of the short-period mode while keeping the phugoid stable. A gain of 0.016 is selected that results in a short-period damping ratio of 0.658 and a natural frequency of 1.91 rad/s. With $K = 0.016$, the phugoid has a damping ratio of 0.037 and a natural frequency of 0.114 rad/s. A multiloop system that incorporates both rate and load factor feedback are common on modern high-performance aircraft.

9.2 Outer-Loop Autopilot/Navigation Control

Most airplanes are equipped with pilot-relief or autopilot functions. These systems take control of the aircraft and perform holding or navigation functions. Since the pilot is not directly in the loop as with inner-loop stability and control, the autopilot modes are referred to as outer-loop functions. Four selected autopilot functions will be presented in this section. Several others are needed for a complete autopilot. More detailed texts provide an analysis of most autopilot functions. The intent of this text is to familiarize the reader with design approaches and not to provide a complete discussion of all possible configurations.

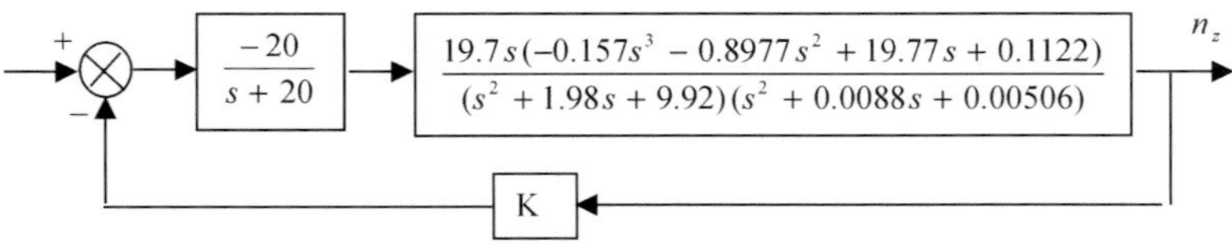

Fig. 9.19 Load factor feedback block diagram.

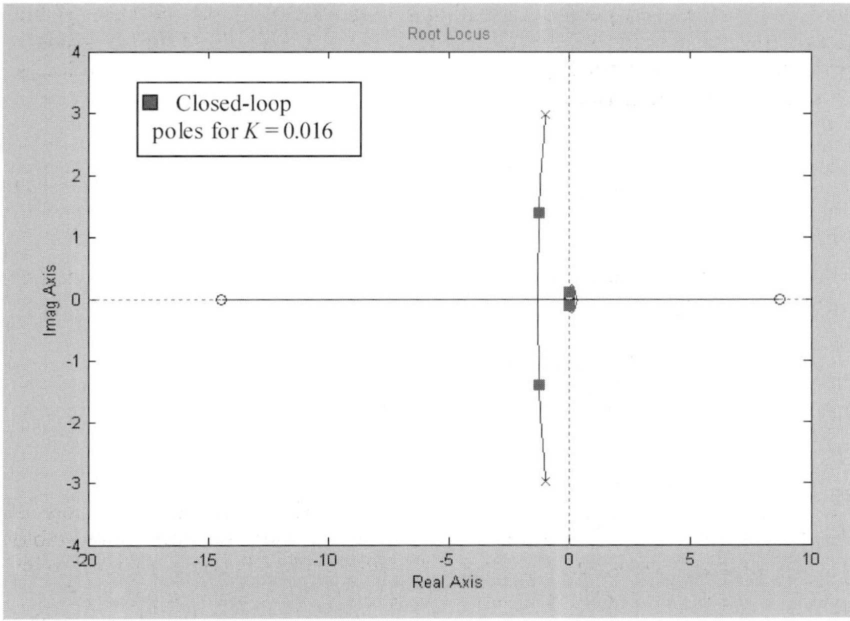

Fig. 9.20 Load factor feedback root locus plot.

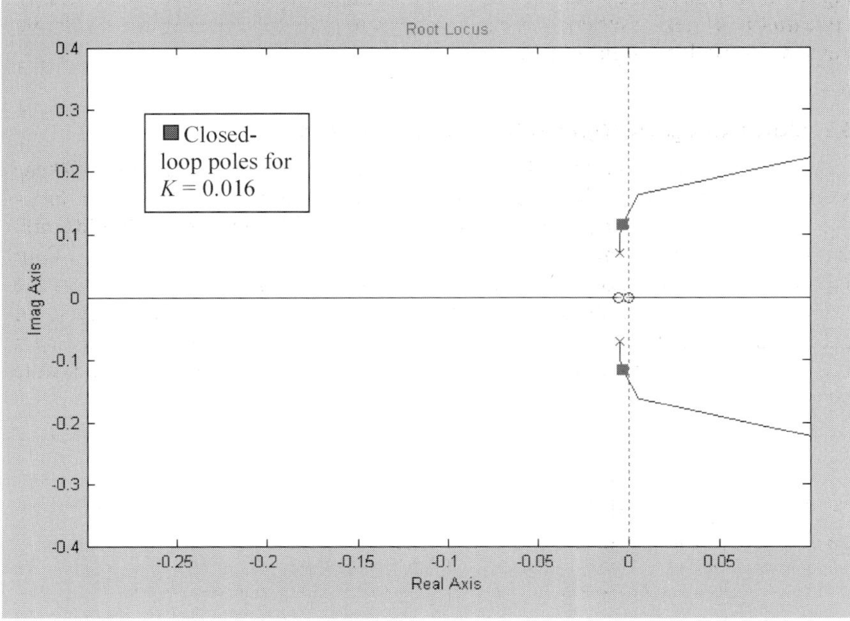

Fig. 9.21 Load factor feedback root locus showing closed-loop phugoid poles.

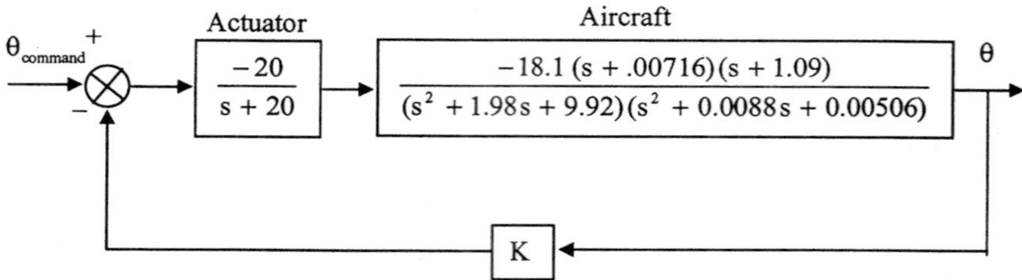

Fig. 9.22 Pitch attitude hold block diagram.

9.2.1 *Pitch Attitude Hold*

A pitch attitude hold mode provides automatic control of pitch attitude with a closed-loop feedback system. It is a mode which can significantly reduce pilot workload in turbulent air. To illustrate the design of a pitch attitude hold system, we return to the A-7D pitch damper block diagram presented in Fig. 9.12 with the output modified to pitch attitude (Fig. 9.22).

The attitude hold root locus for the previous system is presented in Fig. 9.23. Notice that as K is increased, the damping ratio decreases and the natural frequency increases. Figure 9.24 presents an expanded view of this root locus focusing on the short period roots.

If $K = 0.215$ is selected, the short-period damping ratio decreases from 0.313 to 0.2. The short-period natural frequency increases from 3.16 rad/s to 3.6 rad/s. Of course, we must not lose sight of our original purpose for closing the attitude hold loop, incorporation of the automatic pilot relief mode. Pitch attitude feedback has the added advantage of increasing the damping ratio of the phugoid mode.

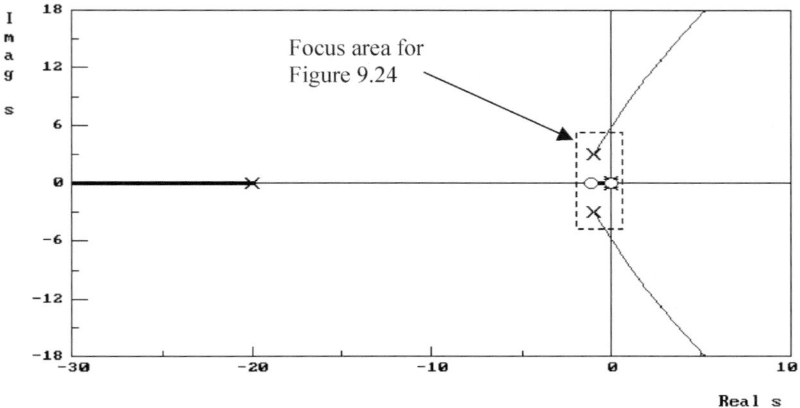

Fig. 9.23 Pitch attitude hold root locus plot.

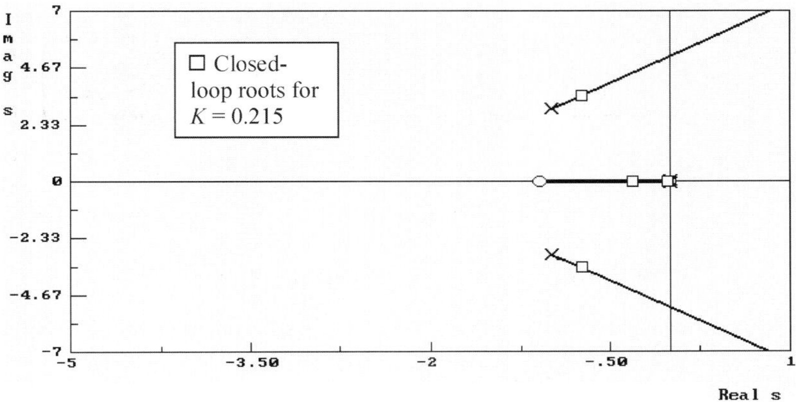

Fig. 9.24 Pitch attitude hold root locus focused on short period roots.

9.2.2 *Altitude Hold*

Altitude hold is another important outer-loop autopilot mode. Simply stated, a feedback loop is used to keep altitude (h) constant using the elevator as the controller. We begin by deriving the h/δ_e transfer function. In Sec. 3.6 we found that the rate of climb (ROC) was

$$\mathrm{ROC} = \dot{h} = U_1 \sin\gamma \approx U_1\gamma$$

Taking the Laplace transform,

$$sh(s) = U_1\gamma(s)$$

and forming the desired transfer function,

$$\frac{h(s)}{\delta_e(s)} = \frac{U_1}{s}\left(\frac{\gamma(s)}{\delta_e(s)}\right)$$

Because $\gamma = \theta - \alpha$, the altitude to elevator transfer function becomes

$$\frac{h(s)}{\delta_e(s)} = \frac{U_1}{s}\left(\frac{\theta(s)}{\delta_e(s)} - \frac{\alpha(s)}{\delta_e(s)}\right) \tag{9.2}$$

For the A-7D at 0.6 Mach and 15,000 ft, the h/δ_e transfer function is approximated by Ref. 1.

$$\frac{h}{\delta_e} = \frac{634(-0.157s^3 - 0.8977s^2 + 19.77s + 0.1122)}{s(s^2 + 1.98s + 9.92)(s^2 + 0.0088s + 0.00506)}$$

Figure 9.25 presents an altitude hold configuration.

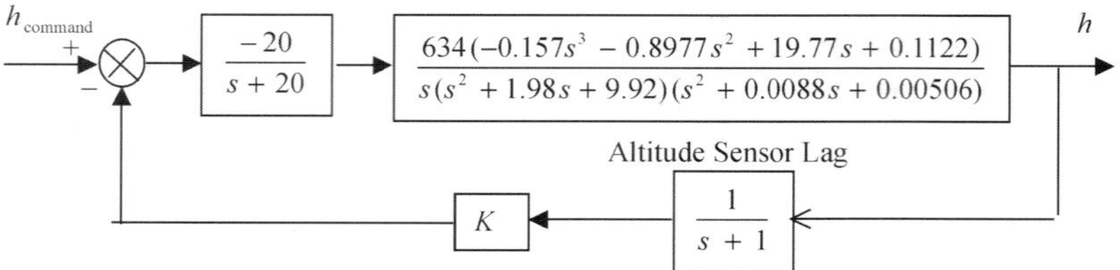

Fig. 9.25 Altitude hold block diagram.

Notice that a lag with a time constant of 1 s has been added to the feedback loop to simulate the lag associated with barometric altimeters. The altitude hold root locus is presented in Fig. 9.26.

Notice that with a low value of gain ($K = 0.00218$) the short-period roots can be kept stable. However, the zero in the right-half plane causes one of the closed-loop roots to be unstable. To solve this problem, two approaches can be used. A pitch attitude feedback can be added (as in Sec. 9.2.1), or a compensation filter can be designed and incorporated into the block diagram that will alter the root locus branches so that a gain can be selected, which keeps all seven roots stable (compensation is discussed in Sec. 9.3).

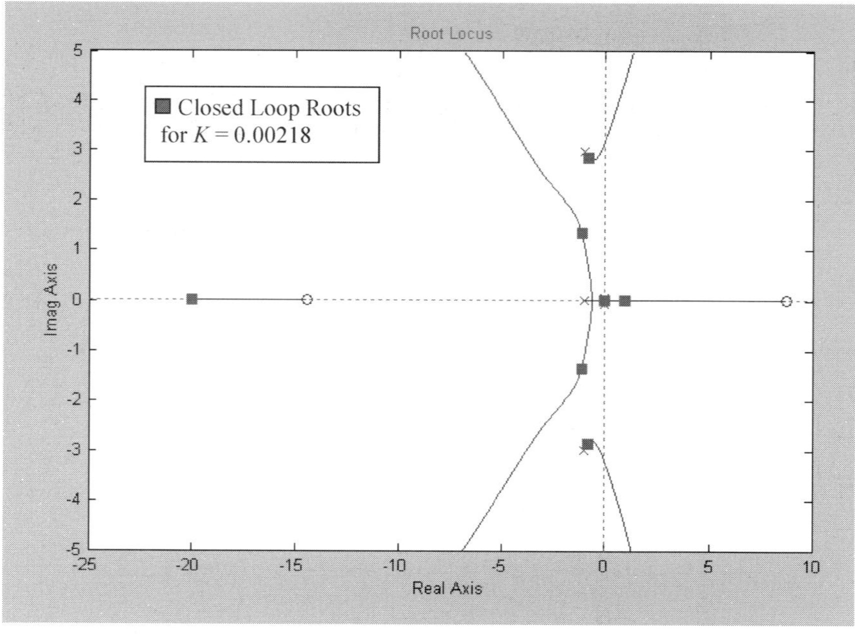

Fig. 9.26 Altitude hold root locus.

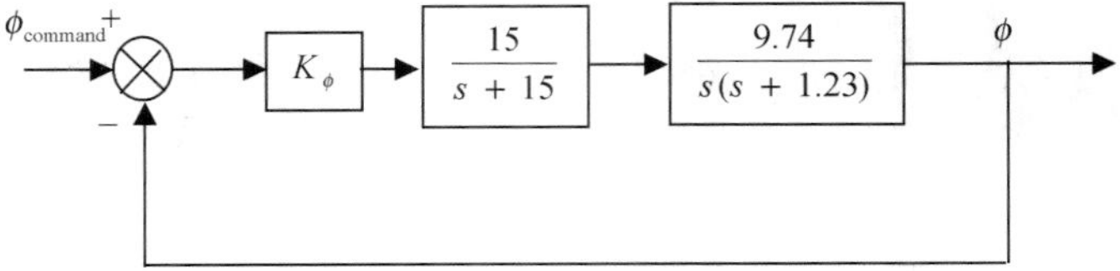

Fig. 9.27 Bank angle hold block diagram.

9.2.3 Bank Angle Hold

A complete autopilot is made up of several feedback loops to take care of the many functions that come naturally to a pilot. Another autopilot loop is bank angle hold, which is sometimes called the wing leveller, when the commanded bank angle is zero (typical of cruising flight). To illustrate the design of a bank angle hold system, we will use the θ/δ_a transfer function using the one-degree-of-freedom (DOF) roll approximation (Sec. 7.3.3.2) for the F-4 at 0.9 Mach and 35,000 ft (Ref. 1).

$$\frac{\phi}{\delta_a} = \frac{9.74}{s(s + 1.23)}$$

Figure 9.27 presents a bank angle hold configuration.

The bank angle hold root locus is presented in Fig. 9.28. Notice that the two poles in the ϕ/δ_a transfer function become complex conjugates for the closed-loop case with a value of $K = 0.077$. For this case, the damping ratio decreases from 1 to 0.66. Also, as seen from the root locus, the gain cannot be increased to large values or the system will go unstable.

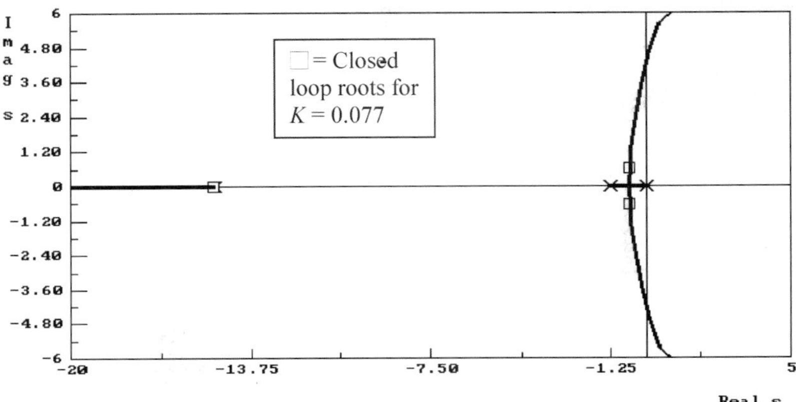

Fig. 9.28 Bank angle hold root locus.

9.2.4 Heading Hold

Autopilot control of aircraft heading is another outer-loop function that can reduce pilot workload. Because heading corrections are usually made by banking (and turning) the aircraft, a heading hold system usually involves a multiloop approach with a bank angle hold system used as the inner loop. A bank angle gyro and a heading angle gyro are needed in this type of system. To draw the block diagram, a Laplace relationship between heading angle and bank angle is needed. This is presented next and was derived from the level turn relationships in Sec. 3.9.2.

$$\frac{\psi(s)}{\phi(s)} = \frac{g}{U_1 s)} \tag{9.3}$$

For the F-4 at 0.9 Mach and 35,000 ft (the same conditions as used in the bank angle hold example), this becomes $\psi/\phi = 0.0368/s$. We can now develop the block diagram for the heading hold mode. The inner-loop bank angle hold system can be thought of as the controller that provides corrections to the heading angle through a level turn. Figure 9.29 presents a heading hold configuration using the 1-DOF roll approximation.

Using the gain selected for the inner-loop in the previous example ($K_\phi = 0.077$) and Eq. (8.11), the inner-loop closed-loop transfer function (CLTF) becomes:

$$\frac{\phi}{\phi_{\text{command}}} = \frac{11.25}{s^3 + 16.23s^2 + 18.45s + 11.25}$$

Multiplying this by Eq. (9.3) (with values for the F-4), we have

$$\frac{\psi}{\phi_{\text{command}}} = \frac{0.414}{s(s^3 + 16.23s^2 + 18.45s + 11.25)}$$

which is the basis for generating the root locus for the outer loop. The outer loop root locus is presented in Fig. 9.30. Notice that the open-loop poles are precisely the same as the inner-loop (closed-loop) poles defined in the previous example. A K_ψ of 7.86 was selected for the outer loop, which provided a damping ratio of 0.5 for the complex conjugate roots.

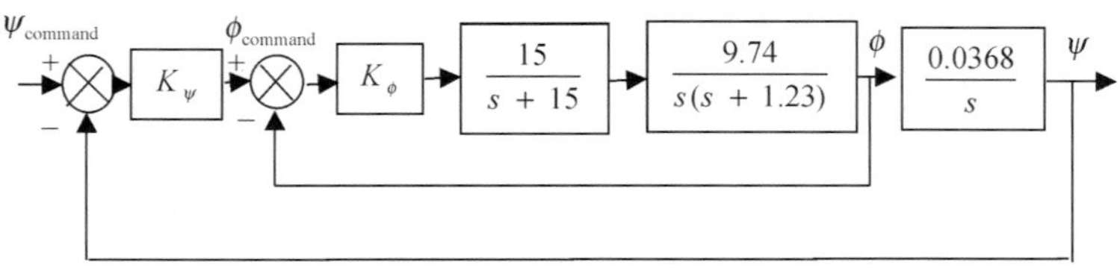

Fig. 9.29 Heading hold block diagram.

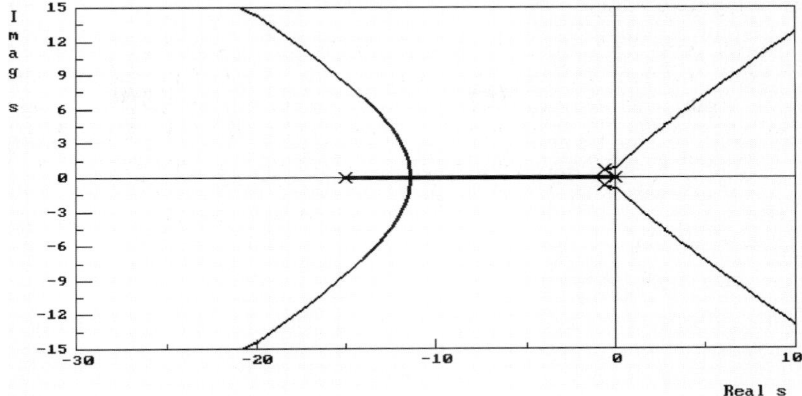

Fig. 9.30 Outer-loop heading hold root locus.

9.3 Compensation Filters

Another powerful tool available to the control system designer is a compensation filter. Compensation filters can take a variety of forms and are very effective in tailoring the aircraft response. We will look at three types of compensation filters, but several more are in common use.

In a generic feedback control system, a compensation filter can be located in generally three possible positions, as shown in Fig. 9.31.

Prefilter compensation modifies the CLTF directly, while forward path and feedback compensation modify the forward path and feedback path transfer functions, which are inputs for the root locus.

The compensation filters that will be discussed add an equal combination of poles and zeros. Prefilter compensators generally use the principle of cancellation of an undesirable closed-loop pole with a prefilter zero, and cancellation of a closed-loop zero with a prefilter pole. Having worked with several root locus plots at this point, a few general observations can be made: open-loop zeros attract branches of the root locus, open-loop zeros in the right half of the complex plane will draw closed-loop roots into the unstable region for certain

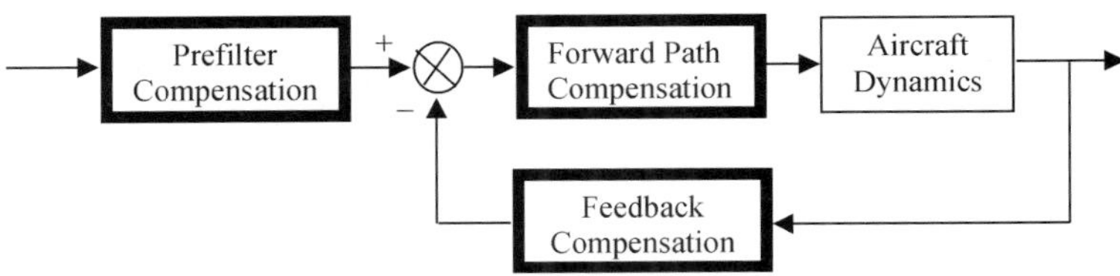

Fig. 9.31 Possible locations of compensation filters.

values of K, and closed-loop roots near the origin of the complex plane provide a slow component (large time constant) of the time response. Forward path and feedback path compensation filters allow modification of the root locus through the addition of poles and zeros in desirable locations.

The filters are generally implemented in the flight control computer and can be thought of as added software for digital systems and additional circuitry for analog systems. A more detailed discussion of compensation filter implementation is available in several texts.

9.3.1 Lead Compensator

Lead compensators are generally used to quicken the system response by increasing natural frequency and/or decreasing time constant. Lead compensators also increase the overall stability of the system. A lead compensator has the general form

$$TF_{\substack{\text{lead}\\\text{compensator}}} = \frac{b(s+a)}{a(s+b)} \qquad a < b \tag{9.4}$$

The b/a in Eq (9.4) simply keeps the steady-state value of the compensator as one. The practical limit in choosing the pole and zero for the lead compensator is $b < 10a$. A common application of lead compensators is to cancel a pole at $s = -a$, which is slowing the time response or causing the system to be unstable. A washout filter, as discussed with rate feedback, is a special case of a lead compensator.

Example 9.1

Design a prefilter lead compensator to decrease the time constant of the following system to less than 0.2 s.

$$\longrightarrow \boxed{\frac{4(s+5)}{(s+1)(s+20)}} \longrightarrow$$

The time response of the system will be composed of two components, each directly dependent on the characteristics of the two poles. Notice that the pole at $s = -1$ has a time constant of approximately 1 s. The pole at $s = -20$ has a time constant of 1/20th of a second and is not a problem. A simple lead compensator can be used to cancel the problem pole.

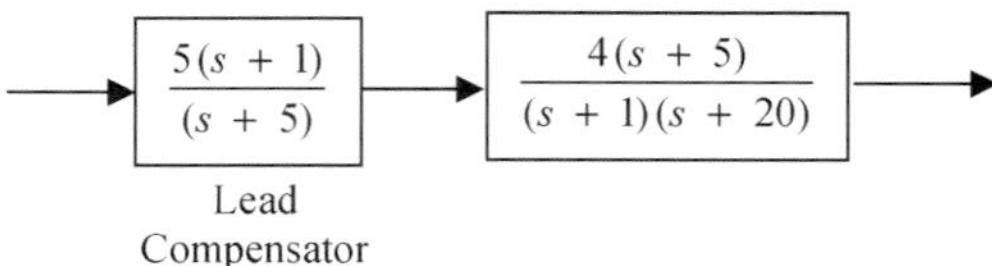

Notice also that we have cancelled the zero at $s = -5$ with the pole on the lead compensator. The next figure shows the time response characteristics with and without the lead compensator.

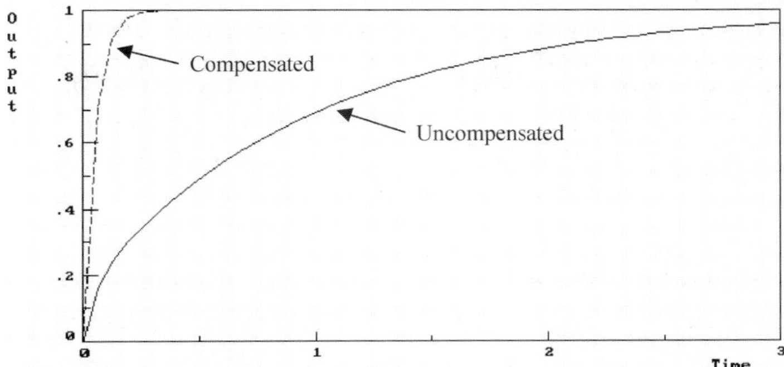

Forward-path or feedback-lead compensation can also be used to shift the root locus to the left.

Example 9.2

Design a forward-path lead compensator for the following system that will shift the root locus to the left.

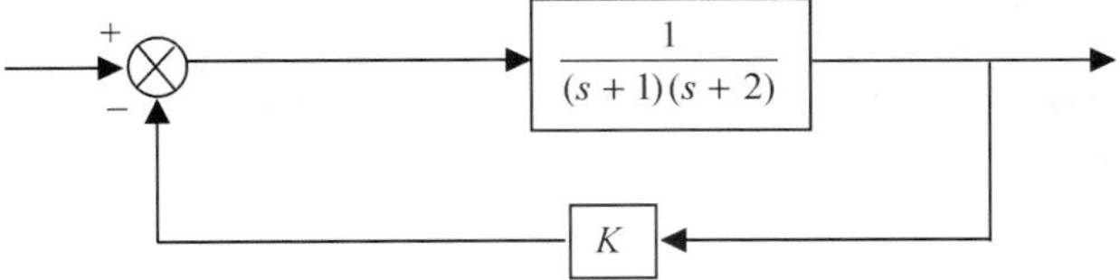

The root locus for the uncompensated system follows:

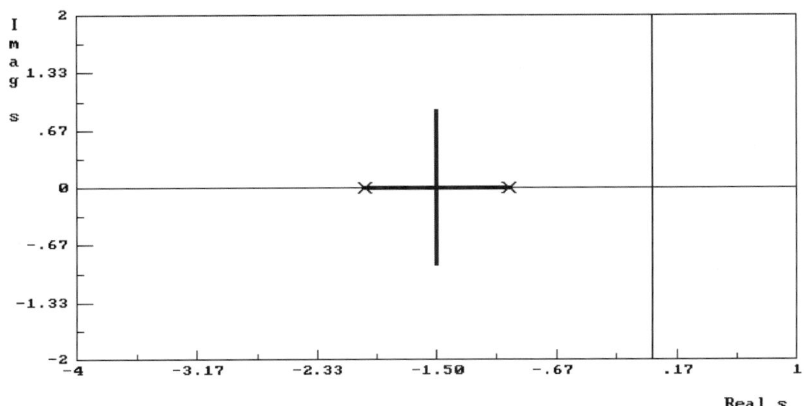

We add a forward-path compensator that will place a pole and zero to the left of the two open-loop poles. The root locus is attracted to the compensator zero at $s = -5$.

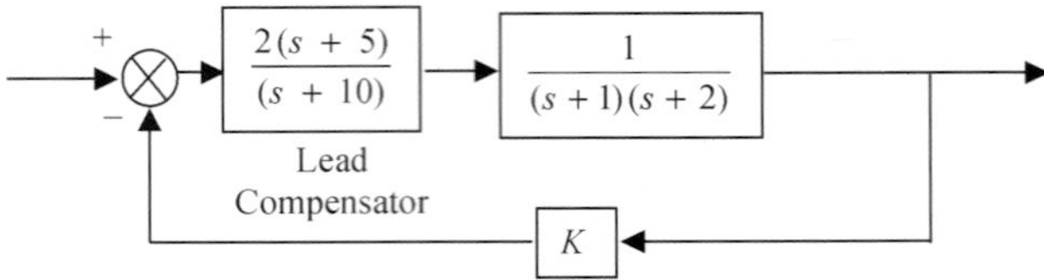

The compensated root locus is presented next. Notice that the vertical branches have shifted to the left.

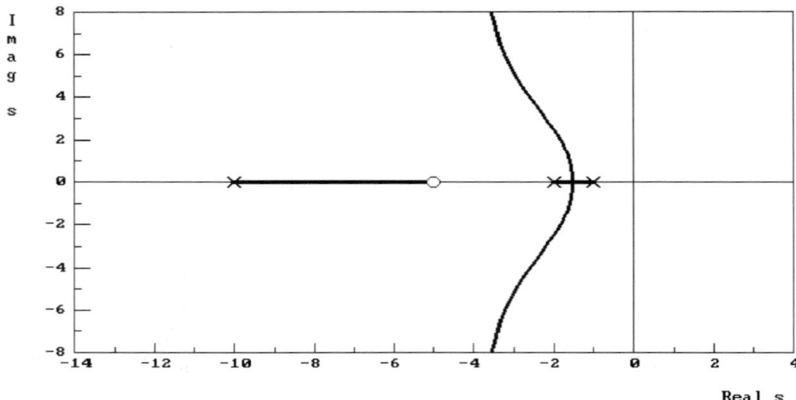

9.3.2 Lag Compensator

Lag compensators are generally used to slow the system response by decreasing natural frequency and/or increasing time constant. They also tend to decrease the overall stability of the system. A lag compensator has the general form

$$TF_{\substack{\text{lag}\\\text{compensator}}} = \frac{b(s+a)}{a(s+b)} \qquad a > b \tag{9.5}$$

With lag compensation, a pole is added to the right of a zero. The pole may be used to cancel a zero, or it may be used to shift the root locus to the right. Lag compensation may also reduce the steady-state error of a system, a topic discussed in Sec. 10.1.

Example 9.3

For the same system defined in Example 9.2, design a feedback path lag compensator that will shift the root locus to the right.

The lag compensator pole is placed at $s = -3$ to repel the root locus to the right.

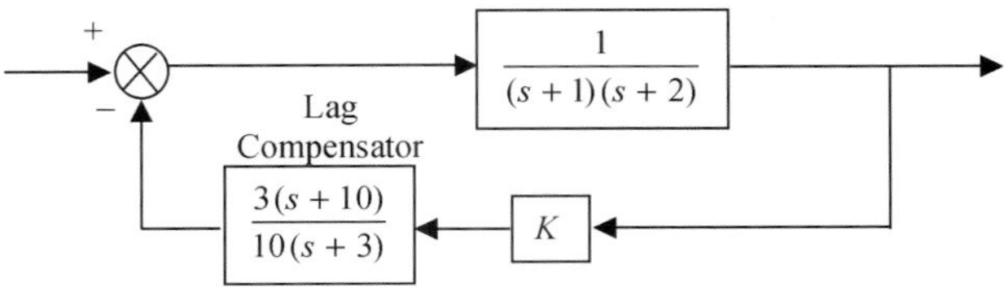

The root locus for the compensated system is presented next. Notice that the root locus branches have shifted to the right.

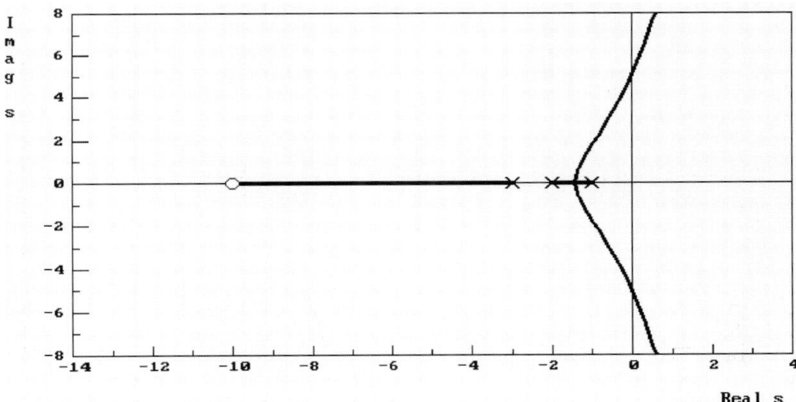

9.3.3 Lead-Lag Compensator

The combined benefits of a lead compensator and a lag compensator may be realized using lead-lag compensation. A lead-lag compensator has the general form

$$TF_{\substack{\text{lead-lag}\\\text{compensator}}} = \frac{bd(s+a)(s+c)}{AC(s+b)(s+d)} \qquad a > b;\ a < c;\ c < d \tag{9.6}$$

The $(s + a)/(s + b)$ component represents the lag filter, and the $(s + c)/(s + d)$ component represents the lead filter. Each component will have the same type of effect on a root locus as discussed in the previous two sections. A lead-lag compensator normally adds two zeros that are fairly close together and that provide a powerful attraction for root locus branches. In many cases, the useful gain range (before a system goes unstable) can be increased using a lead-lag compensator.

Another common use of lead-lag compensators is the attenuation of a specific frequency range (sometimes called a notch filter). For example, an aircraft structural resonant frequency can be filtered out with a lead-lag compensator if a feedback sensor is erroneously affected by that frequency. Design of such filters is easily done using Bode plots, which are discussed in Chapter 10.

Example 9.4

Starting with the following system, design a lead-lag feedback path compensator that will provide for stable roots at higher values of K.

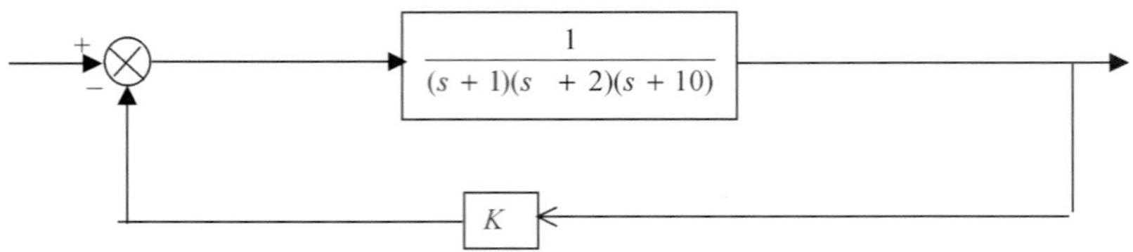

We begin by reviewing the root locus for the previous system.

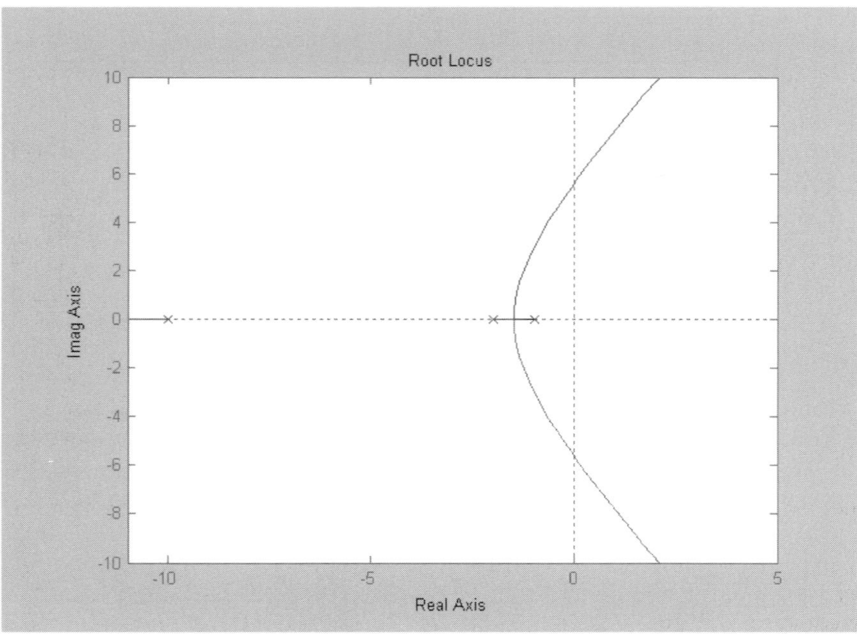

The two complex branches of the root locus go unstable for values of K greater than 391. To allow for a larger range of stable gain values, we add the feedback path lead-lag compensator

$$\frac{1.5(s+4.5)(s+4)}{(s+3)(s+9)}$$

that places two zeros to the left of the pole at −2.

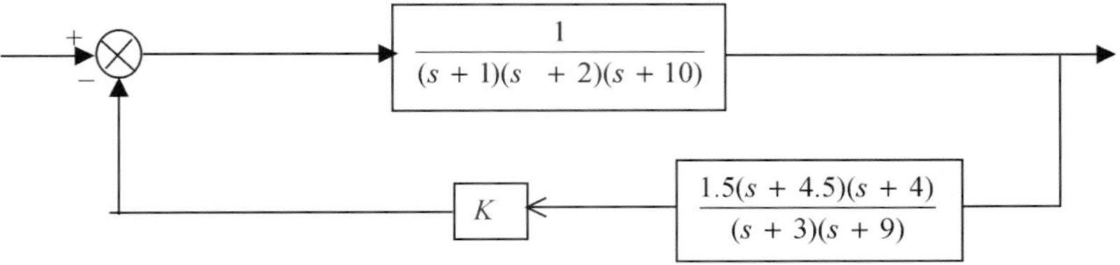

The resulting root locus is:

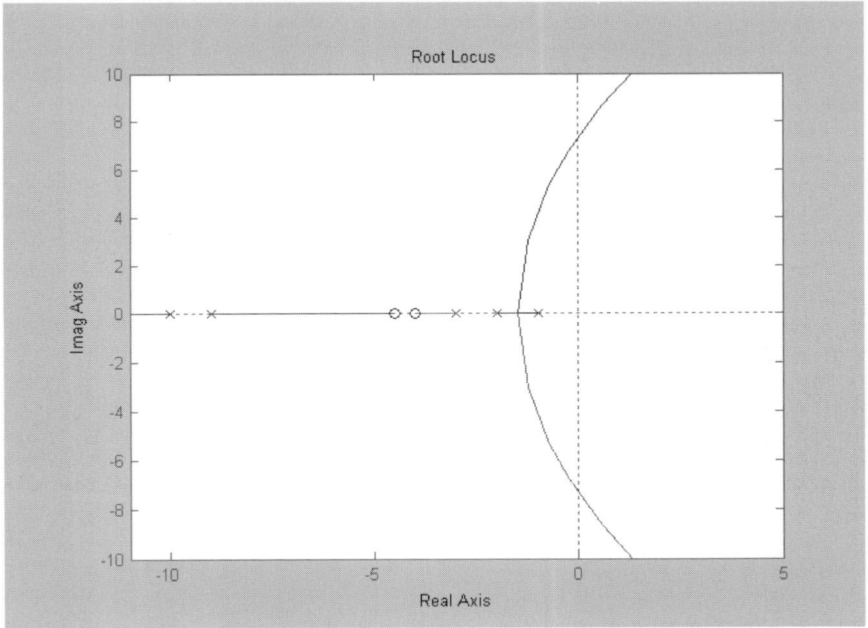

The two complex branches of the compensated root locus now go unstable for K values greater than 588. Thus, we have gained a larger range of stable values of K. As discussed in Chapter 10, this will have a positive effect on the steady-state error of the system. Of course, additional analysis is necessary to determine if this design is acceptable. The example is intended to simply illustrate the effect that a lead-lag compensator can have on the root locus.

9.4 Combined Systems

Modern aircraft combine many or all of the functions discussed in Secs. 9.1 and 9.2, along with additional functions not discussed in this text. The inner-loop stability/control functions are normally designed first, followed by the outer-loop autopilot modes. In many cases, autopilot functions tend to have an adverse effect on dynamic stability characteristics, and the design of inner and outer loops must be accomplished concurrently. This is best understood with an example.

Example 9.5

Develop an autopilot for the DC-8 that will maintain aircraft pitch angle, θ, and will have an overall damping ratio of the short period mode of $\zeta = 0.55 \pm 0.01$. Use an actuator with a 0.1-s time constant and the following DC-8 transfer function based on the short period approximation.

$$\frac{\theta}{\delta_e} = \frac{-1.39(s + 0.306)}{s(s^2 + 0.805s + 1.325)}$$

First, we look at a displacement autopilot for the DC-8 using only θ feedback. Our initial autopilot is

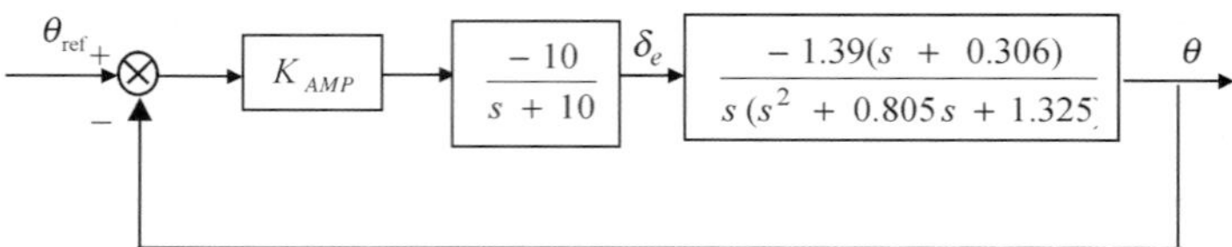

The transfer function for input to the root locus is

$$G^*H = \frac{13.9(s + 0.306)}{(s + 10)s(s^2 + 0.805s + 1.325)}$$

The root locus is

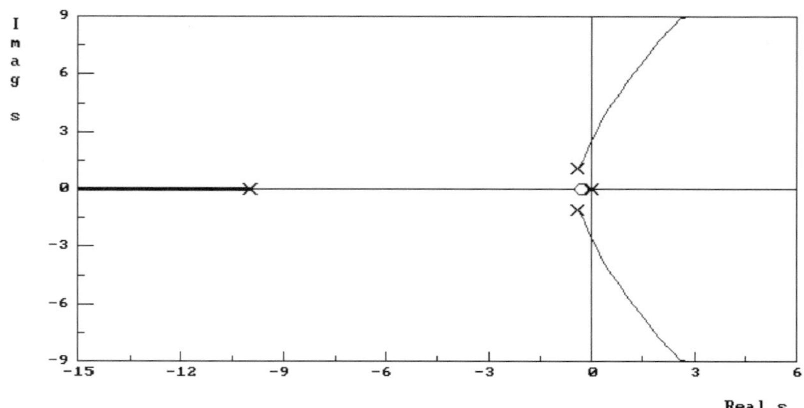

While this autopilot has stable short-period roots for a limited range of K_{AMP} (0 to 3.65), it fails our requirement of $\zeta = 0.55$ because the maximum short-period damping ratio is 0.35, which occurs at the open-loop poles. As seen from the root locus, the damping ratio decreases as K_{AMP} is increased and a damping ratio of 0.55 cannot be achieved with this feedback configuration. Therefore, this system is unsatisfactory. It is important to note the migration tendency of the short-period root locus with pitch attitude feedback: as K_{AMP} is increased, the damping ratio decreases and eventually the system goes unstable.

To meet the short-period damping ratio requirement, we will add an inner-loop utilizing pitch rate $\dot{\theta}$ feedback and a rate gyro. Recall

$$\frac{\dot{\theta}(s)}{\delta_e(s)} = s\frac{\theta(s)}{\delta(s)} = \frac{-1.39(s + 0.306)}{s^2 + 0.805s + 1.325}$$

The new autopilot with $\dot{\theta}$ and θ inner and outer loops is

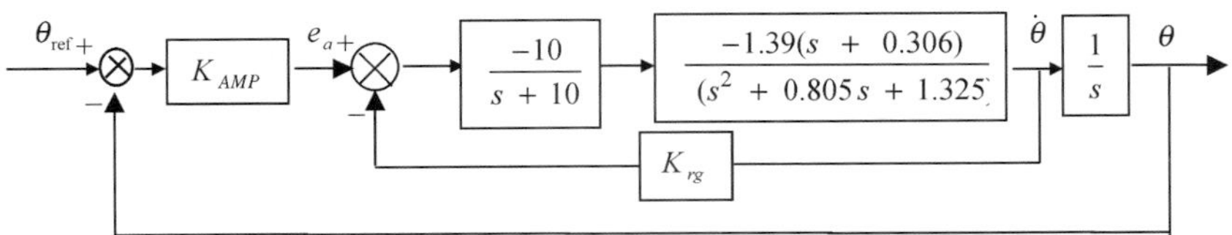

The root locus transfer function for the inner loop is

$$GH^* = \frac{13.9(s + 0.306)}{(s + 10)(s^2 + 0.805s + 1.325)}$$

The inner-loop root locus is

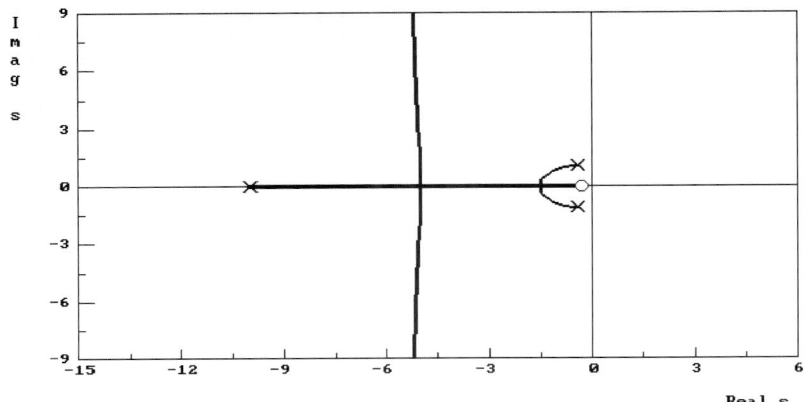

With rate feedback, notice that as K_{rg} is increased, the short period damping ratio is increased from the open-loop value of 0.35. To obtain an overall system response of $\zeta = 0.55$, we must select a damping ratio for the inner loop larger than 0.55 because we know that closure of the outer loop will decrease as the damping ratio as K_{AMP} is increased. The choice of the inner loop ζ (and the K_{rg} needed to obtain it) is typically a compromise between the desirability of rapid response and the desire to reduce excessive overshoot. For our inner loop we select a $\zeta = 0.7$, which is obtained with $K_{rg} = 0.6577$. The closed-loop system for the inner loop is

$$\frac{\dot{\theta}}{e_a} = \frac{\dfrac{13.9(s + 0.306)}{(s + 10)(s^2 + 0.805s + 1.325)}}{1 + \dfrac{(0.6577)(13.9)(s + 0.306)}{(s + 10)(s^2 + 0.805s + 1.325)}} = \frac{G}{1 + GH}$$

or, in simplified form,

$$\frac{\dot{\theta}}{e_a} = \frac{13.9(s + 0.306)}{s^3 + 10.805s^2 + 18.52s + 16.05}$$

The block diagram for the system with $K_{rg} = 0.6577$ can be represented as

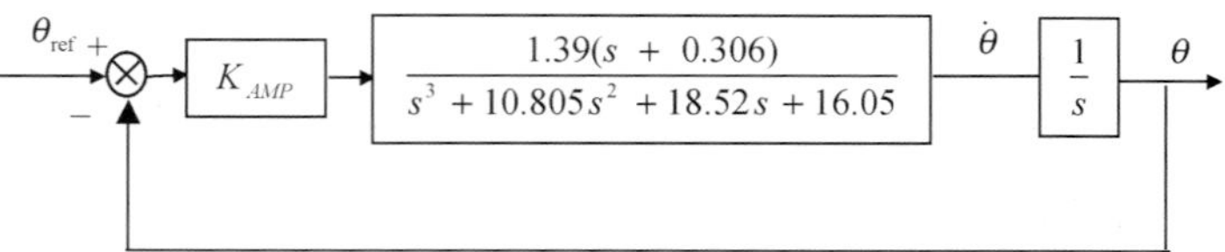

The next step is to find a K_{AMP} such that the overall damping ratio is 0.55. The root locus transfer function for the outer loop is

$$G^*H = \frac{13.9(s + 0.306)}{s(s^3 + 10.805s^2 + 18.52s + 16.05)}$$

The outer-loop root locus is

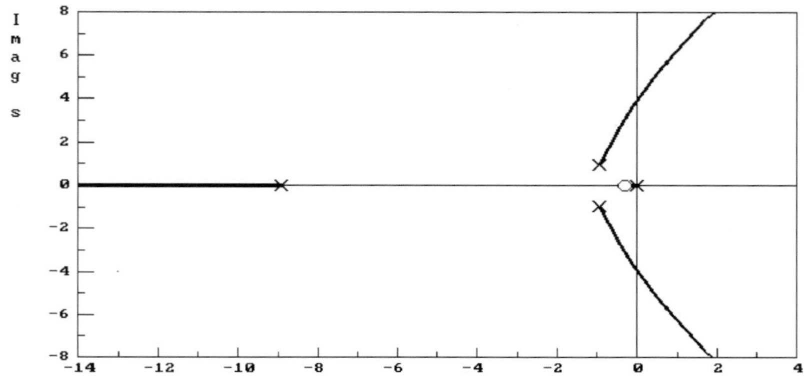

This looks like the root locus for the autopilot with θ feedback; however, the open-loop poles are now at $\zeta = 0.7$ because of the inner-loop utilizing rate ($\dot{\theta}$) feedback. By selecting $K_{AMP} = 0.4688$, we can obtain an overall system damping ratio that meets our requirement of $\zeta = 0.55 \pm 0.01$.

The CLTF for the autopilot is

$$\frac{\theta}{\theta_{\text{REF}}} = \frac{G}{1+GH} = \frac{\dfrac{(0.4688)(13.9)(s+0.306)}{s(s^3+10.805s^2+18.52s+16.05)}}{\dfrac{(0.4688)(13.9)(s+0.306)}{s(s^3+10.805s^2+18.52s+16.05)}}$$

Notice that K_{AMP} is in the forward path and appears as part of G when forming the CLTF. This simplifies to

$$\frac{\theta}{\theta_{\text{REF}}} = \frac{6.52(s+0.306)}{s^4+10.805s^3+18.52s^2+22.56s+1.99}$$

or

$$\frac{\theta}{\theta_{\text{REF}}} = \frac{6.52(s+0.306)}{(s+0.0954)(s^2+1.68s+2.31)(s+9.028)}$$

The pitch displacement autopilot with $K_{rg} = 0.6577$ and $K_{AMP} = 0.4688$ can be represented as

$$\theta_{\text{REF}} \longrightarrow \boxed{\frac{6.52\,(s+0.306)}{(s+0.0954)\,(s^2+1.68s+2.31)\,(s+9.028)}} \longrightarrow \theta$$

If we input a unit step, $\theta_{\text{REF}} = 1/s$, the time response of θ is

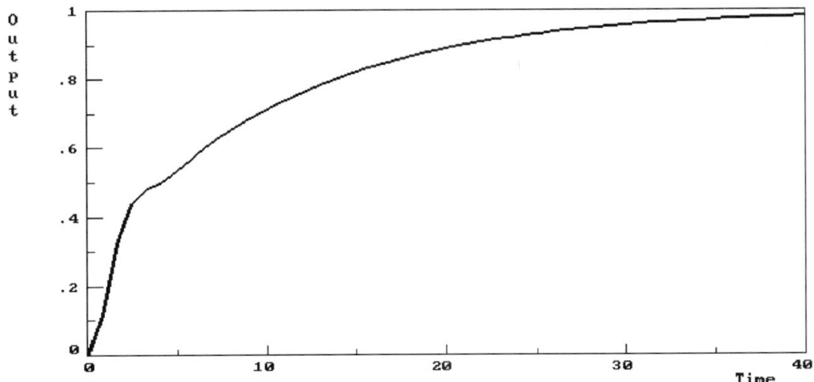

While the response has a sufficient damping ratio ($\zeta = 0.55$), it takes more than 40 s to reach steady state. This is because of the root near the origin at

$s = -0.0954$. This root has a large time constant that slows the time response. Although not specifically requested, we can significantly improve the time response with the addition of a prefilter lead compensator to eliminate the pole and zero near the origin. The lead compensator to do this is

$$TF\text{-}_{\substack{\text{lead}\\ \text{compensator}}} = \frac{0.306(s + 0.0954)}{0.0954(s + 0.306)}$$

This prefilter compensator will eliminate the pole at $s = -0.0954$ and zero at $s = -0.306$ and hence should improve the time response. The system with the lead compensator is

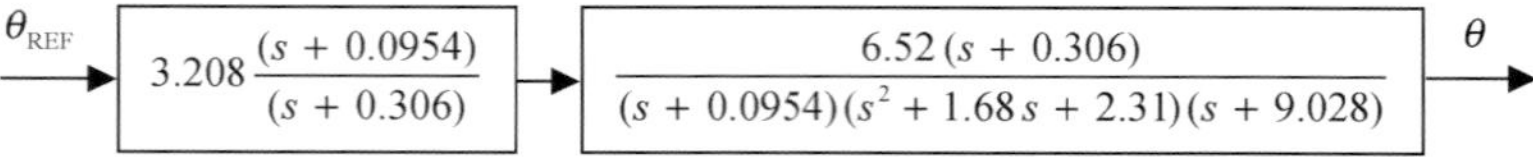

or equivalently,

$$\theta_{REF} \longrightarrow \boxed{\frac{20.9}{(s^2 + 1.68s + 2.31)(s + 9.028)}} \longrightarrow \theta$$

The equivalent block diagram represents a system where the slow pole and zero have been eliminated. The time response of the compensated autopilot to a unit step input, is now

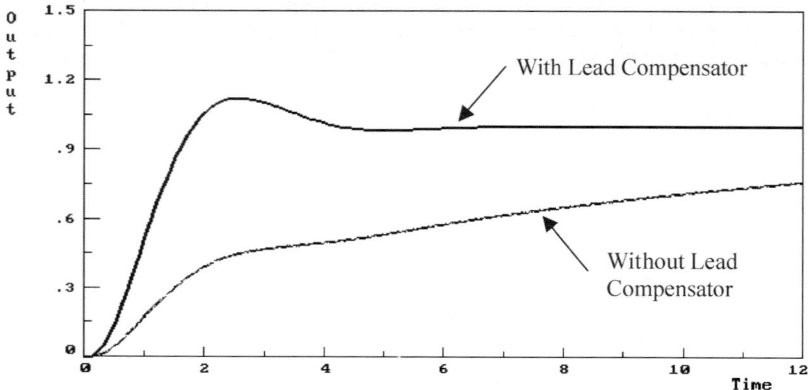

Notice that the time response has been significantly improved with the lead compensator. Recalling our discussion of time-response characteristics in Sec. 8.5, the compensated pitch attitude autopilot design has a rise time of 1.176 s, a time to peak of 2.59 s, and a 5% settling time of 3.593 s. A complete block

diagram of our final system is

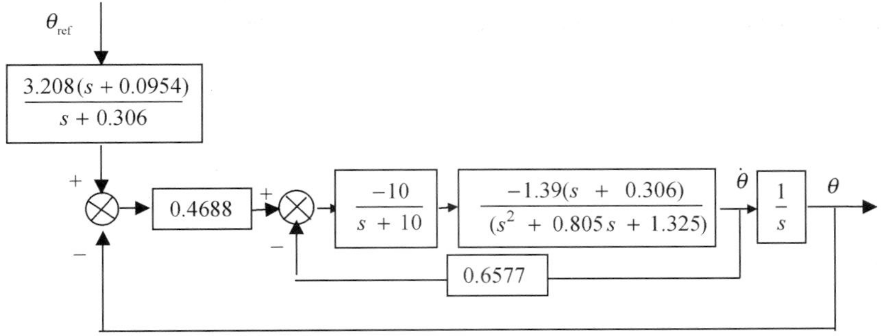

9.5 Historical Snapshot—The A-7D DIGITAC Digital Multimode Flight Control System Program

During the mid-1970s, the Air Force Flight Dynamics Laboratory, the Air Force Flight Test Center, and Honeywell, Inc., developed and flight tested the first digital and multimode control augmentation system in history.[2,3] This program provided an important, sequential step in the evolution of flight control systems from simple analog stability augmentation to much more capable digital system mechanizations with control laws specifically tailored to individual mission tasks. The test aircraft for the program was the second prototype A-7D shown in Fig. 9.31.

The system also addressed the difficult problem of increasing system reliability and providing passive system failures. Fail-operational, fail-safe capability was achieved for failures of the dual-redundant system through the use of

Fig. 9.32 A-7D DIGITAC test aircraft.

in-line, self-test software that ran continuously during system operation. This approach led directly to the triple channel, 'fail-op, fail-op, fail-safe' systems now in modern aircraft. Figure 9.33 presents the redundancy management mechanization.

The multimode control laws consisted primarily of a flight path (FP) mode and a precision attitude (PA) mode. Both were implemented via software in the digital computers using a CAS approach. The FP mode was designed to enhance control of aircraft flight path at the expense of aircraft attitude. Related mission tasks included large combat maneuvers and fine flight-path

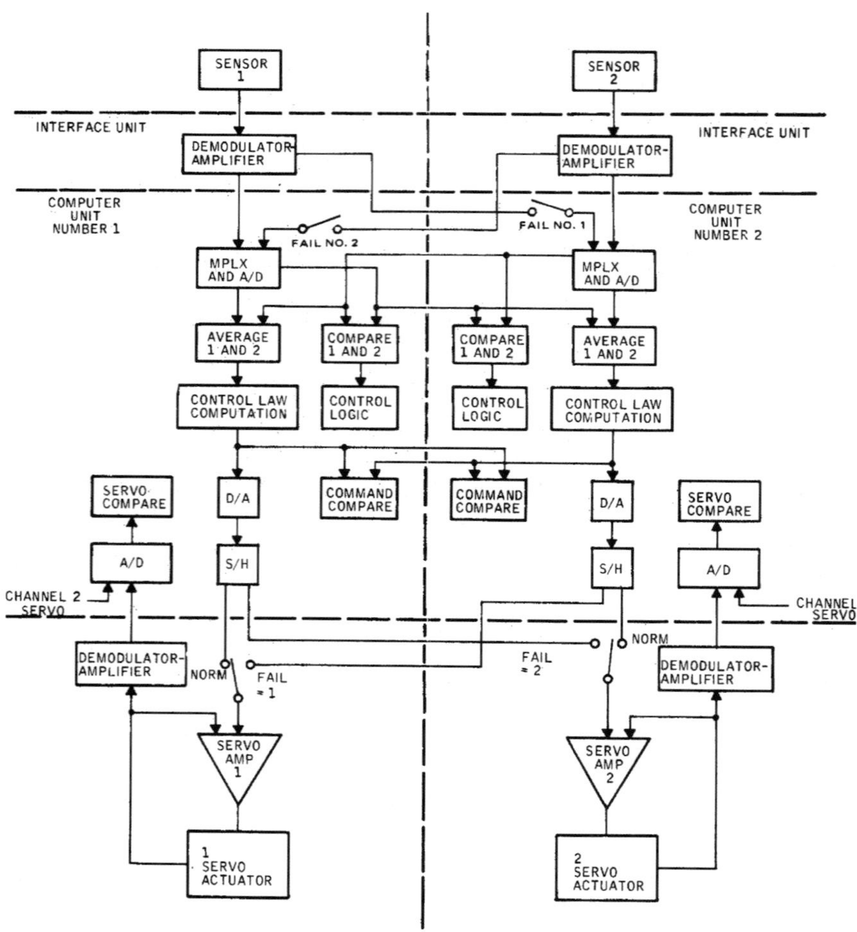

Fig. 9.33 A-7D DIGITAC redundancy management approach.

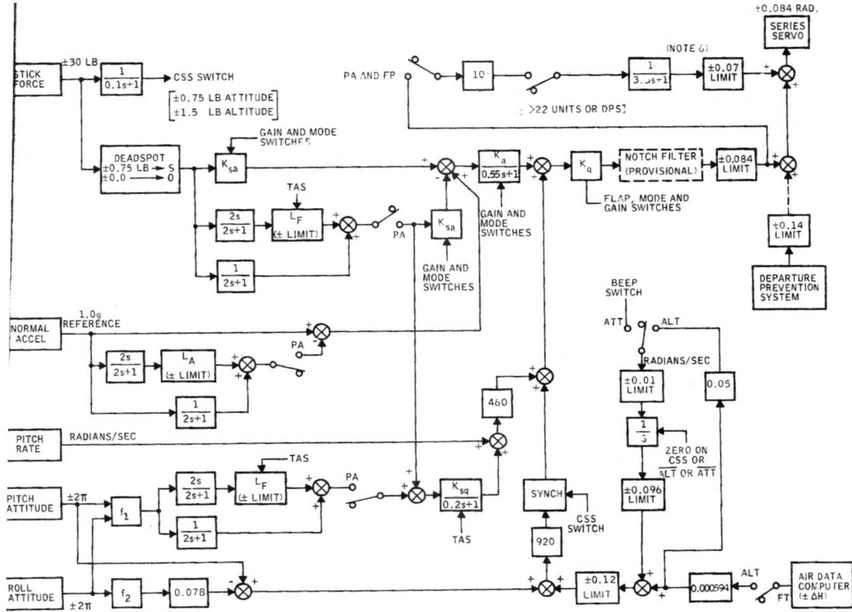

Fig. 9.34 DIGITAC pitch axis control laws.

tracking. This mode provided rapid yet well damped normal acceleration and roll rate response to control stick inputs. The PA mode was designed to enhance control of aircraft attitude. Appropriate mission tasks included gunnery and air-to-air tracking. This mode provided rapid yet well damped pitch rate and roll-rate response. Turn coordination was task related. A selectable gunnery submode of PA forced the aircraft to roll about the bullet stream during roll corrections to reduce the pendulum effect associated with terminal tracking. Figures 9.34–9.36, present the pitch, roll, and yaw axis control laws.

During the flight-test program, quantitative tracking data were obtained to evaluate the effectiveness of each mode as compared to the standard A-7D flight control system. Comparison time history data of gunsite pipper position are presented in Fig. 9.37 for a typical gunnery pass.

Overall, a 38% average reduction in air-to-air and air-to-ground tracking error was demonstrated during the program with the PA mode. Few advantages were found with the FP mode. The program also proved the feasibility of digital implementation of flight control computations, paving the way for similar systems in the Space Shuttle, F-15, F-16, and F-22.

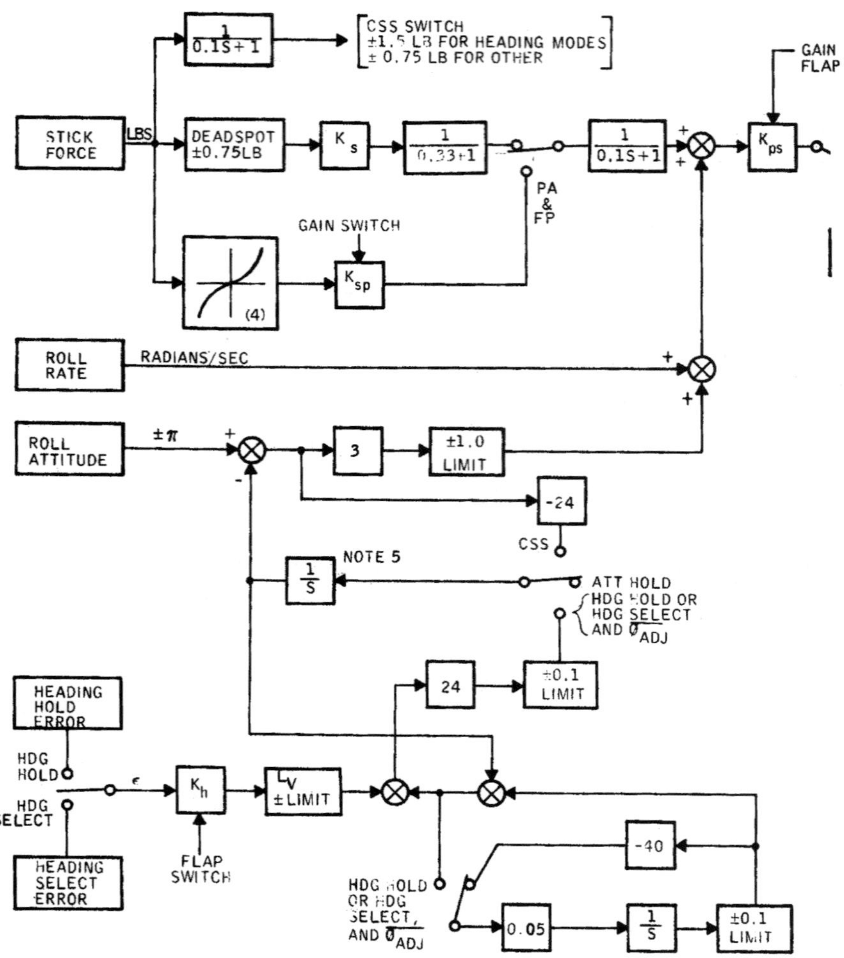

Fig. 9.35 DIGITAC roll axis control laws.

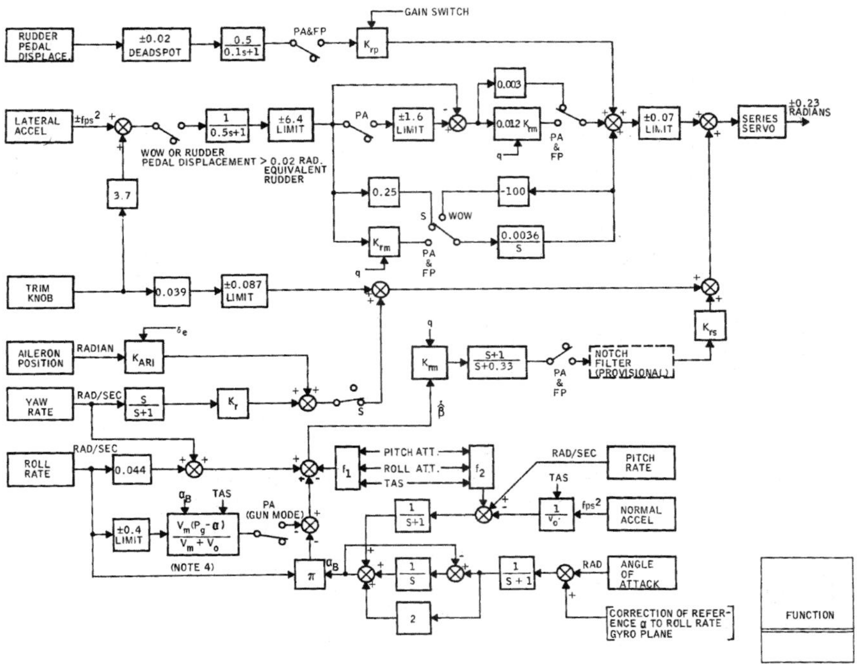

Fig. 9.36 DIGITAC yaw axis control laws.

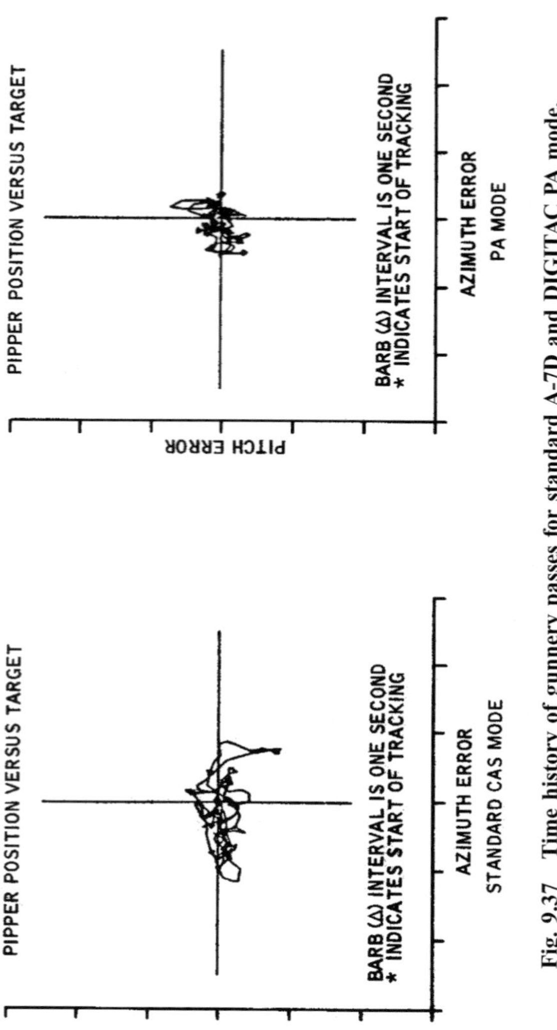

Fig. 9.37 Time history of gunnery passes for standard A-7D and DIGITAC PA mode.

References

[1]USAF Test Pilot Scholl, *Flight Control Systems*, chapter 14, Edwards AFB, CA, 1988.

[2]Yechout, T. R., and Oelschlaeger, D. R., "Flight Test Evaluation of a Digital Multimode Flight Control System in an A-7D Aircraft," AIAA TP 76-1913, AIAA Guidance and Control Conference, San Diego, CA, Aug. 1976

[3]Yechout, T. R., and Oelschlaeger, D. R., "DIGITAC Multimode Flight Control System," AIAA TP 75-1085, AIAA Guidance and Control Conference, Boston, Aug. 1975.

Problems

9.1 The following is a block diagram for a roll control autopilot for a jet fighter:

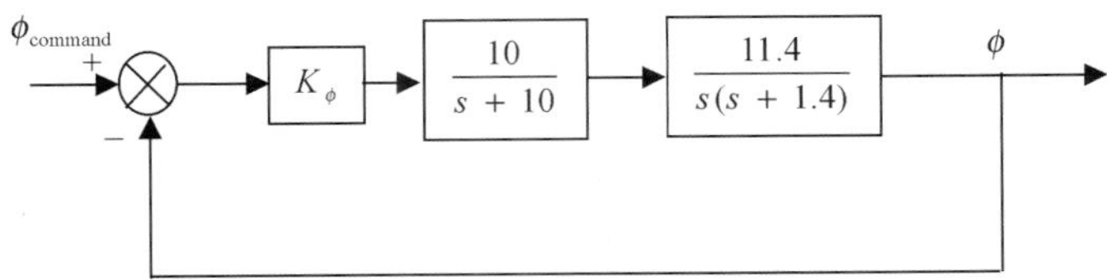

Determine
(1) the OLTF and the CLTF.
(2) the poles and zeros of the OLTF and CLTF.

Sketch the root locus for this autopilot as the amplifier gain is varied from 0 to infinity. Be sure to determine
(a) the number of branches required.
(b) where the root locus exists on the real axis.
(c) where it crosses the imaginary axis.
(d) where it breaks away from the real axis.
(e) the gain where it crosses the imaginary axis.

9.2 For a pitch attitude hold autopilot, the short period damping ratio tends to ________ as K is increased.
(1) increase
(2) decrease
(3) stay the same

9.3 If an "inner-loop" rate feedback is added to a pitch attitude hold autopilot, you should select an inner-loop, short-period damping ratio that is ________ than that desired for the outer loop.
(1) larger
(2) smaller
(3) the same

9.4 For the following OLTF, select the forward path compensator that will shift the vertical branches of the root locus to the left.

$$OLTF = \frac{10K}{(s+1)(s+2)}$$

(1) $\dfrac{0.1(s+30)}{s+3}$

(2) $\dfrac{0.5(s+1)}{s+0.5}$

(3) $\dfrac{10(s+3)}{s+30}$

(4) $\dfrac{4(s+2)}{s+0.5}$

9.5 After selecting a gain for the outer loop of our pitch attitude hold autopilot, what did we do to improve the time response characteristics? Which prefilter compensator would solve a similar problem for the following CLTF?

$$CLTF = \frac{25(s+4)(s+0.05)}{(s+0.01)(s+5)(s+30)}$$

(1) $\dfrac{0.8(s+5)}{s+4}$

(2) $\dfrac{400(s+0.01)}{s+4}$

(3) $\dfrac{100(s+5)}{s+0.05}$

(4) $\dfrac{5(s+0.01)}{s+0.05}$

9.6 Which of the following is an example of a lag filter?

(1) $\dfrac{(s+1)}{5(s+0.2)}$

(2) $\dfrac{s+0.05}{s+0.005}$

(3) $\dfrac{(s+1)}{10(s+10)}$

(4) None of these

9.7 For a yaw damper system, you want to select a washout circuit time constant that is really fast to eliminate any pilot inputs from interfering with system operation.
(1) True
(2) False

9.8 We are trying to increase the natural frequency of the short-period mode. All but which of the following should do this?
(1) pitch attitude feedback
(2) pitch rate feedback
(3) angle of attack feedback
(4) load factor feedback
For the short period mode, what parameter would you feedback to increase damping Ratio? To increase ω_N?

9.9 What effect does β feedback have on
(a) N_β?
(b) roll mode?
(c) spiral mode?
(d) dutch roll damping?

9.10 Why are an inner and an outer loop needed for a pitch attitude hold system?

9.11 Given the following unstable high α case, find the closed-loop impulse response for a $K_{rg} = 0.3$ and $K_{1G} = 1.75$. Show the inner-loop and outer-root locus. Is the response stable?

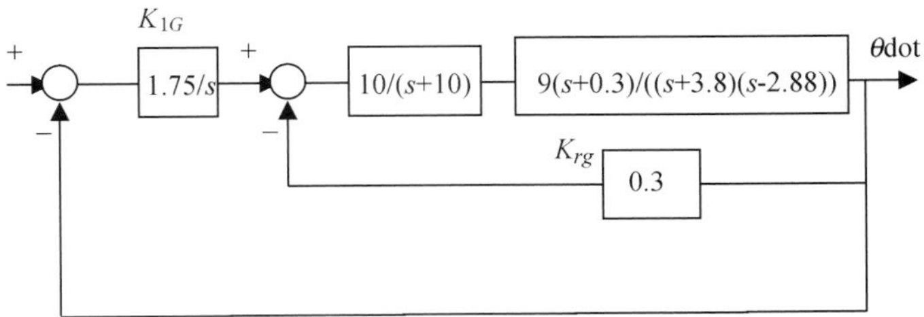

10
Special Topics

This chapter will discuss special topics that amplify and add depth to the material covered in Chapters 7–9. This material primarily focuses on additional analysis techniques for feedback control systems and various types of aircraft flight control systems.

10.1 System Type and Steady-State Error

When analyzing a unity feedback system such as the one shown in Fig. 10.1, one can deduce much about the system from the number of "free integrators" in the denominator of the forward-path transfer function [$G(s)$]. The number of free integrators, or free s terms, in the denominator of $G(s)$ determines the system type.

$G(s)$ can be expressed in general terms as shown in Eq. (10.1). The value of n in the free s term defines the system type, where $n = 0, 1, 2, \ldots$.

$$G(s) = \frac{K_n(\tau_1 s + 1)(\tau_2 s + 1)\cdots}{s^n(\tau_a s + 1)(\tau_b s + 1)\cdots} \tag{10.1}$$

The system type has a direct impact on the steady-state error of a system. Consider the following development for analyzing the error signal $E(s)$ in Fig. 10.1.

$$E(s) = R(s) = C(s)$$

where $R(s)$ is the input and $C(s)$ is the output. Next, relate the output to the error signal.

$$C(s) = E(s)G(s)$$

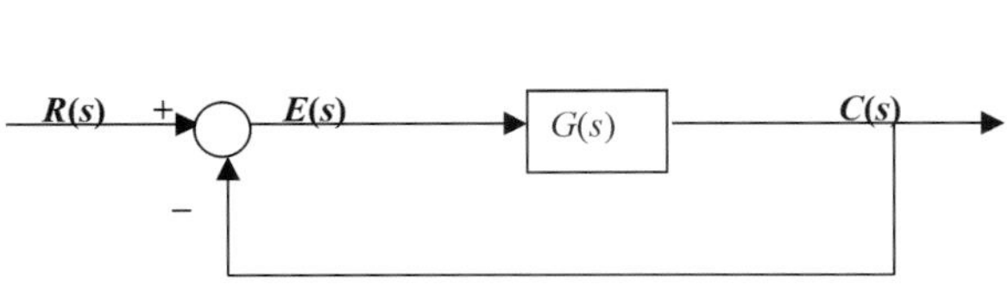

Fig. 10.1 Unity feedback system.

Substituting $C(s)$ into the previous equation yields

$$E(s) = R(s) - E(s)G(s)$$

Simplifying,

$$E(s)[1 + G(s)] = R(s)$$

Finally, the transfer function for the error signal to the input becomes

$$\frac{E(s)}{R(s)} = \frac{1}{1 + G(s)}$$

Multiplying both sides by the input yields an expression for the error signal, as shown in Eq. (10.2).

$$E(s) = \frac{R(s)}{1 + G(s)} \tag{10.2}$$

Evaluating the steady-state error involves use of the final value theorem, presented in Sec. 7.3.1.6. When applied to steady-state error, Eq. (10.3) results.

$$e(t)_{ss} = \lim_{s \to 0} sE(s) = \lim_{s \to 0} s\left[\frac{R(s)}{1 + G(s)}\right] \tag{10.3}$$

The three most common types of inputs $[R(s)]$ are step, ramp, and parabolic. Each will be analyzed in the following sections for Type 0, Type 1, and Type 2 systems so that the effect of system type on steady-state error can be clearly seen.

10.1.1 Step Input

The concept of a step input was first introduced in Sec. 7.3.1.2. In this case, the step input magnitude is represented by the magnitude R as shown in Fig. 10.2. A physical approximation of a step input occurs when the pilot *quickly* moves the stick from the neutral position to another position.

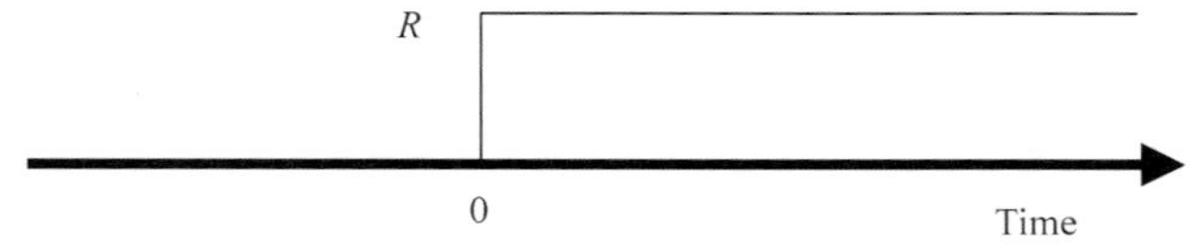

Fig. 10.2 Step input of magnitude *R*.

The time domain representation is $y(t) = R^*\delta(t)$, where $y(t)$ is the output and $\delta(t)$ is the delta function defined in Sec. 7.3.1.2. The Laplace transform of a step input with a magnitude of R is

$$R(s) = \frac{R}{s}$$

Applying this input to Eq. (10.3) results in a steady-state error equation for step inputs as shown in the following equation:

$$e(t)_{ss} = \lim_{s\to 0} s\left\{\left[\frac{R}{1+G(s)}\right]\frac{1}{s}\right\}$$

which simplifies to Eq. (10.4)

$$e(t)_{ss} = \frac{R}{1 + \lim\limits_{s\to 0} G(s)} \tag{10.4}$$

This leads to the definition of the **position error constant, K_p**. It is related to the steady-state error between the input and output when the **input is a step function**, and provides a means to compare system types. Equation (10.5) presents the mathematical definition for K_p.

$$K_p = \lim_{s\to 0} G(s) \tag{10.5}$$

Combining Eqs. (10.4) and (10.5) yields a compact notation for steady-state error for a step input.

$$e(t)_{ss} = \frac{R}{1+K_p} \tag{10.6}$$

This formulation applies regardless of system type. The system type affects K_p, which in turm affects the steady-state error. The specific system types and how they affect K_p and the steady-state error for a step input will be addressed next.

10.1.1.1 Type 0 system. A Type 0 ($n = 0$) system has no free integrators, and may be represented by the following transfer function:

$$G(s) = \frac{K_0(\tau_1 s+1)(\tau_2 s+1)\cdots}{(\tau_a s+1)(\tau_b s+1)\cdots}$$

Placing $G(s)$ into Eq. (10.5) yields

$$K_p = \lim_{s\to 0}\left[\frac{K_0(\tau_1 s+1)(\tau_2 s+1)\cdots}{(\tau_a s+1)(\tau_b s+1)\cdots}\right] = K_0 \tag{10.7}$$

It is important to realize that the $G(s)$ transfer function must be put in the previous form to discern the K_0 gain by inspection. A common mistake is to use the gain with roots in the form of $(s + \text{root})$. In that case, the transfer function is of the form

$$G(s) = \frac{K_{new}(s + z_1)(s + z_2)\cdots}{(s + p_1)(s + p_2)\cdots}$$

which results in a different value for K_p.

$$K_p = \frac{K_{new}(z1)(z2)}{(p1)(p2)}$$

Thus, the form of the transfer function affects the gain that appears in the numerator. It is important to note that the steady-state error will be the same regardless of the form of the transfer function. However, the gain K_0 will be different. The end result is that the steady-state error for a Type 0 system with a step input is a constant, as shown in Eq. (10.8).

$$e(t)_{ss} = \frac{R}{1 + K_p} = \frac{R}{1 + K_0} \tag{10.8}$$

Inspection reveals that higher values of K_0 result in lower values of steady-state error. The input can be compared to the actual closed unity feedback output presented in Chapter 8, Eq. (8.11). It is repeated here for convenience.

$$\frac{C(s)}{R(s)} = \frac{G(s)}{1 + G(s)}$$

This concept is best reinforced with an example.

Example 10.1

Find the steady-state error for the transfer function $G(s)$ with unity feedback when given a unit step input $R(s) = 1/s$.

$$G(s) = \frac{2}{(s + 3)} = \frac{0.667}{(0.333s + 1)}$$

$G(s)$ is a Type 0 system because it has no free integrators, so a constant steady-state error is expected for a step input.

$$K_p = \lim_{s \to 0}\left[\frac{0.667}{(0.333s + 1)}\right] = 0.667$$

As expected, K_p is equal to K_0. Therefore, the steady-state error is

$$e(t)_{ss} = \frac{R}{1+K_0} = \frac{1}{1+0.667} = 0.6$$

Figure 10.3 shows graphically the physical significance of the steady-state error for unity feedback using the $G(s)$ transfer function.

10.1.1.2 Type 1 system. A Type 1 ($n = 1$) system has one free integrator, so the value for K_p will be infinity, as the following equation shows:

$$K_p = \lim_{s\to 0}\left[\frac{K_1(\tau_1 s+1)(\tau_2 s+1)\cdots}{s(\tau_a s+1)(\tau_b s+1)\cdots}\right] = \infty$$
$$e(t)_{ss} = \frac{R}{1+K_p} = \frac{R}{1+\infty} = 0 \tag{10.9}$$

Thus, the steady-state error is zero and the output will match the input.

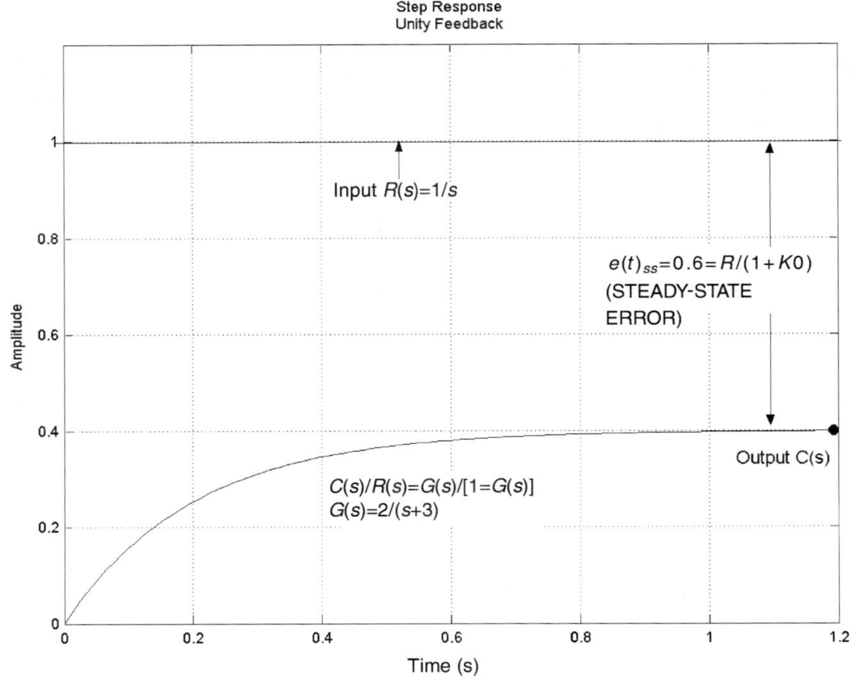

Fig. 10.3 Type 0 system response to a unit step input (unity feedback).

Example 10.2

Find the steady-state error for the transfer function $G(s)$ with unity feedback when given a unit step input $[(R(s) = 1/s]$.

$$G(s) = \frac{2}{s(s+3)} = \frac{0.667}{s(0.333s+1)}$$

$G(s)$ is a Type 1 system because it has one free integrator; therefore, zero steady-state error is expected for a step input.

$$K_p = \lim_{s \to 0}\left[\frac{0.667}{s(0.333s+1)}\right] = \infty$$

As expected, K_p is equal to infinity. Therefore, the steady-state error is

$$e(t)_{ss} = \frac{R}{1+K_p} = \frac{1}{1+\infty} = 0$$

Figure 10.4 shows graphically the physical significance of zero steady-state error for unity feedback using the $G(s)$ transfer function.

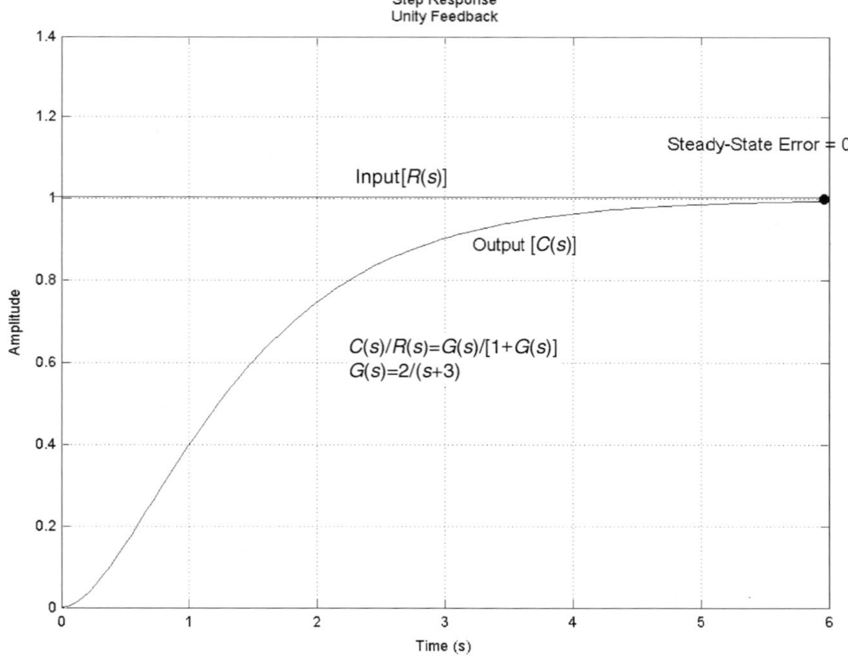

Fig. 10.4 Type 1 system response to a unit step input (unity feedback).

Table 10.1 Steady-state error summary for a step input

Type	K_p	$e(t)_{ss}$
0	K_0	$R/(1 + K_0)$
1	∞	0
2	∞	0

10.1.1.3 Type 2 system. A Type 2 ($n = 2$) system has two free integrators; therefore K_p will be infinity, as is the case with a Type 1 system for a step input.

$$K_p = \lim_{s \to 0} \left[\frac{K_2(\tau_1 s + 1)(\tau_2 s + 1) \cdots}{s^2(\tau_a s + 1)(\tau_b s + 1) \cdots} \right] = \infty \tag{10.10}$$

Thus, the steady-state error is

$$e(t)_{ss} = \frac{R}{1 + K_p} = \frac{R}{1 + \infty} = 0$$

10.1.1.4 Step input summary. Table 10.1 summarizes the steady-state error relationship to system type for a step input. Simply stated, if zero steady-state error is desired for a unity feedback system that is given a step input, then the system type must be Type 1 or greater. Zero steady-state error is also implied when K_p is equal to infinity.

10.1.2 Ramp Input

A ramp input is shown in Fig. 10.5. A physical example of a ramp input occurs when the pilot gradually moves the stick from the center position to another position in a uniform manner. In this simple example, the magnitude R is the slope of the ramp that begins at $t = 0$ s.

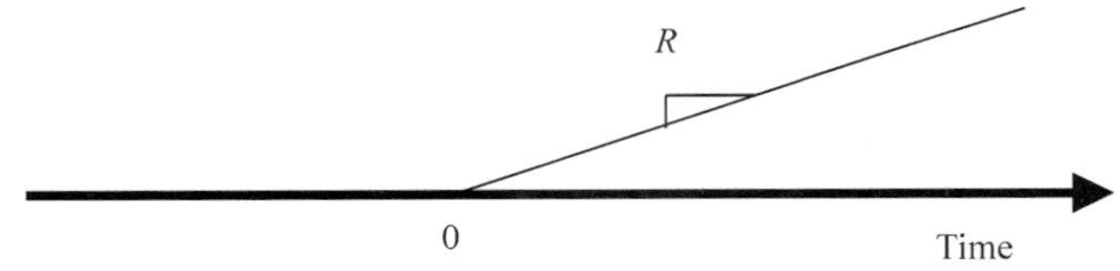

Fig. 10.5 Ramp input of magnitude *R*.

The time domain representation of the ramp input is $y(t) = Rt$, where $y(t)$ is the output and t is time. The Laplace transform of a ramp input with magnitude (slope) R is

$$R(s) = \frac{R}{s^2}$$

Applying this input to Eq. (10.3) results in a steady-state error equation for ramp inputs as shown in the following form:

$$e(t)_{ss} = \lim_{s\to 0} s\left\{\left[\frac{R}{1+G(s)}\right]\frac{1}{s^2}\right\}$$

which simplifies to

$$e(t)_{ss} = \lim_{s\to 0}\left\{\frac{R}{[s+sG(s)]}\right\}$$

and finally to Eq. (10.11).

$$e(t)_{ss} = \frac{R}{\lim\limits_{s\to 0} sG(s)} \tag{10.11}$$

This allows the definition of the **velocity error constant, K_v** which is always **associated with a ramp input**. The mathematical definition of K_v is shown in Eq. (10.12).

$$K_v = \lim_{s\to 0} sG(s) \tag{10.12}$$

This leads to a concise definition of the steady-state error for a ramp input, independent of system type. It is common for students to confuse the mathematical definition of K_v and the final value theorem described in Sec. 7.3.1.6. In fact, the right-hand side of both equations are exactly the same. What is different is the context of the equations. The final value theorem applies to finding the final value for any stable system. The K_v equation applies to finding the velocity error constant for a system when given a ramp input. Combining Eqs. (10.11) and (10.12) yields

$$e(t)_{ss} = \frac{R}{K_v} \tag{10.13}$$

As in the case of the step input, this equation applies to all system types for a ramp input. Application of this equation to three different system types follows.

10.1.2.1 Type 0 system. From Eq. (10.12), the velocity error coefficient for a Type 0 system (no free integrators) is

$$K_v = \lim_{s\to 0} s\left[\frac{K_0(\tau_1 s + 1)(\tau_2 s + 1)\cdots}{(\tau_a s + 1)(\tau_b s + 1)\cdots}\right] = 0 \tag{10.14}$$

This results in a steady-state error of

$$e(t)_{ss} = \frac{R}{K_v} = \frac{R}{0} = \infty \tag{10.15}$$

Simply stated, a Type 0 system cannot track a ramp input. Example 10.3 illustrates this case.

Example 10.3

What is the steady-state error for the transfer function $G(s)$ with unity feedback when given a unit ramp input $[R(s) = 1/s^2]$?

$$G(s) = \frac{2}{(s+3)} = \frac{0.667}{(0.333s + 1)}$$

$G(s)$ is a Type 0 function; therefore, an infinite steady-state error is expected for a ramp input.

$$K_v = \lim_{s\to 0} sG(s) = \lim_{s\to 0} s\left[\frac{2}{(s+3)}\right] = 0$$

Therefore, the steady-state error is

$$e(t)_{ss} = \frac{1}{0} = \infty$$

Figure 10.6 shows the time response for a unit ramp input to this system. It is easy to see from the response that the error approaches infinity as time increases.

10.1.2.2 Type 1 system. A Type 1 system has one free integrator; therefore, the steady-state error is a constant. In fact, the steady-state error is directly related to the gain of a Type 1 system, K_1. This is shown in Eq. (10.16).

$$K_v = \lim_{s\to 0} s\left[\frac{K_1(\tau_1 s + 1)(\tau_2 s + 1)\cdots}{s(\tau_a s + 1)(\tau_b s + 1)\cdots}\right] = K_1 \tag{10.16}$$

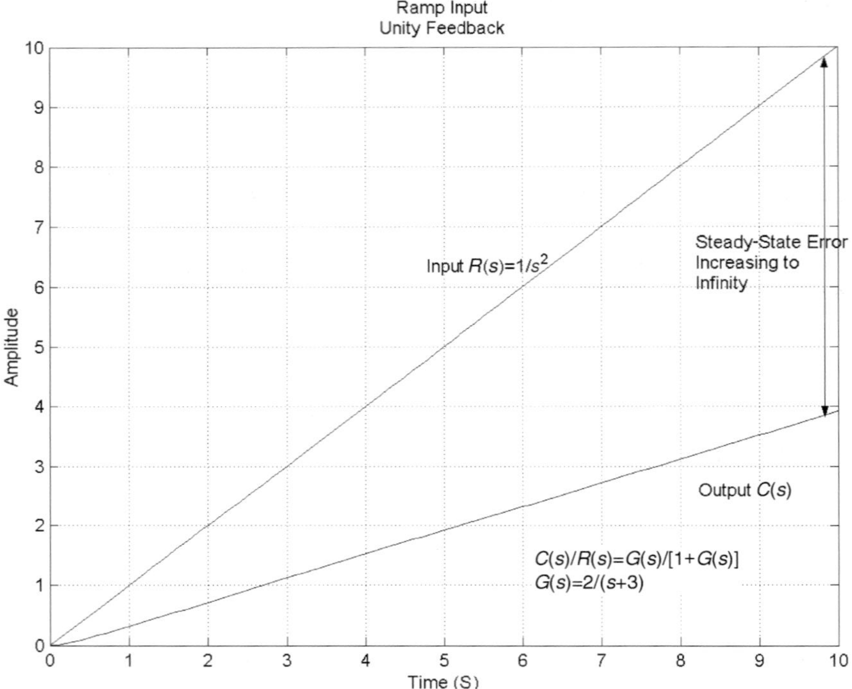

Fig. 10.6 Type 0 system response to a ramp input (unity feedback).

Thus, the steady-state error is a constant.

$$e(t)_{ss} = \frac{R}{K_v} = \frac{R}{K_1} \tag{10.17}$$

Example 10.4 illustrates this graphically.

Example 10.4

What is the steady-state error for the transfer function $G(s)$ with unity feedback when given a unit ramp input $[R(s) = 1/s^2]$?

$$G(s) = \frac{2}{s(s+3)} = \frac{0.667}{s(0.333s+1)}$$

$G(s)$ is a Type 1 system because it has one free integrator; therefore, a constant steady-state error is expected for a ramp input.

$$K_v = \lim_{s \to 0} s\left[\frac{2}{s(s+3)}\right] = 0.667$$

For a unit ramp input, the steady-state error is

$$e(t)_{ss} = \frac{R}{K_v} = \frac{1}{0.667} = 1.5$$

Figure 10.7 shows the time response of $G(s)$ to a unit ramp input. Clearly, the time response produces a constant steady-state error.

10.1.2.3 Type 2 system. A Type 2 system has two free integrators; therefore zero steady-state error is expected.

$$K_v = \lim_{s \to 0} s\left[\frac{K_2(\tau_1 s + 1)(\tau_2 s + 1)\cdots}{s^2(\tau_a s + 1)(\tau_b s + 1)\cdots}\right] = \infty \tag{10.18}$$

Thus, the steady-state error is

$$e(t)_{ss} = \frac{R}{K_v} = \frac{R}{\infty} = 0$$

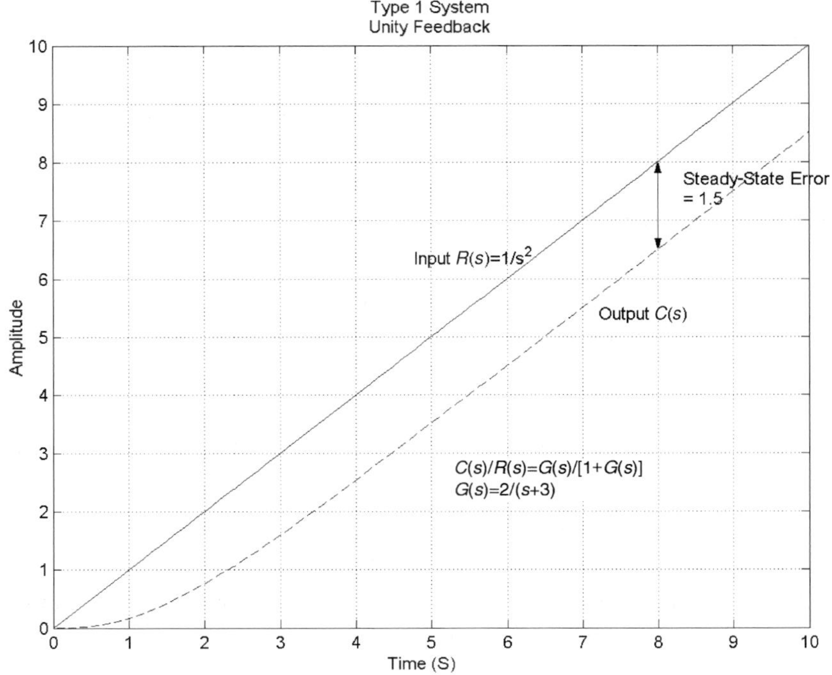

Fig. 10.7 Type 1 system response to a unit ramp input (unity feedback).

Example 10.5 illustrates zero steady-state error for a ramp input and a Type 2 system with unity feedback.

Example 10.5

What is the steady-state error for the transfer function $G(s)$ with unity feedback when given a unit ramp input $[R(s) = 1/s^2]$?

$$G(s) = \frac{2(s+2)}{s^2(s+3)} = \frac{1.333(0.5s+1)}{s^2(0.333s+1)}$$

Because $G(s)$ is a Type 2 system, a zero steady-state error is expected for a ramp input.

$$K_v = \lim_{s \to 0} s\left[\frac{2(s+2)}{s^2(s+3)}\right] = \infty$$

Thus, the steady-state error is

$$e(t)_{ss} = \frac{1}{\infty} = 0$$

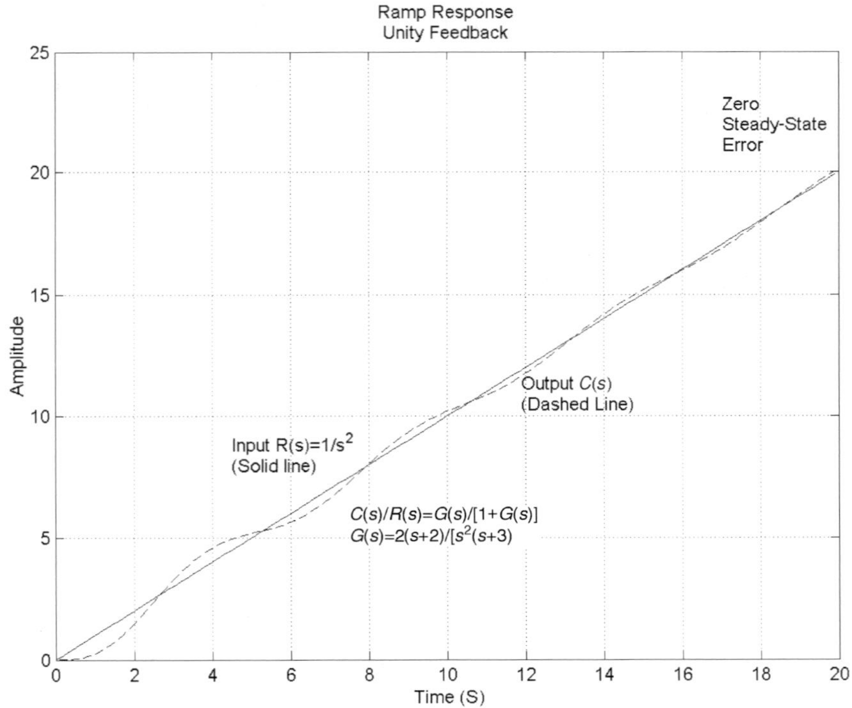

Fig. 10.8 Type 2 system response to a ramp input (unity feedback).

Table 10.2 Steady-state error summary for a ramp input

Type	K_v	$e(t)_{ss}$
0	0	∞
1	K_1	R/K_1
2	∞	0

Figure 10.8 illustrates the response of a Type 2 system to a ramp input. Notice that there is zero steady-state error.

10.1.2.4 Ramp input summary. Table 10.2 summarizes the steady-state error relationships for each system type. For a ramp input, the Type 0 system will produce infinite steady-state error, a Type 1 system will produce a finite constant steady-state error, and the Type 2 system will produce no steady-state error.

10.1.3 Parabolic Input

The final type of input considered is a parabolic input. Figure 10.9 shows a drawing of a parabolic input. A physical example of a parabolic input occurs when the pilot moves the stick away from the center slowly at first, and then gradually increases the rate of the stick movement.

The time domain representation of a parabola is $y(t) = t^2$. When including the scale factor R, the relation becomes $y(t) = Rt^2$. The Laplace transform for a parabola with a magnitude R is

$$R(s) = \frac{R}{s^3}$$

Applying this input to Eq. (10.3) results in a steady-state error equation for parabolic inputs shown by the following equations:

$$e(t)_{ss} = \lim_{s \to 0} s\left\{\left[\frac{R}{1 + G(s)}\right]\frac{1}{s^3}\right\}$$

which simplifies to

$$e(t)_{ss} = \lim_{s \to 0}\left\{\frac{R}{[s^2 + s^2 G(s)]}\right\}$$

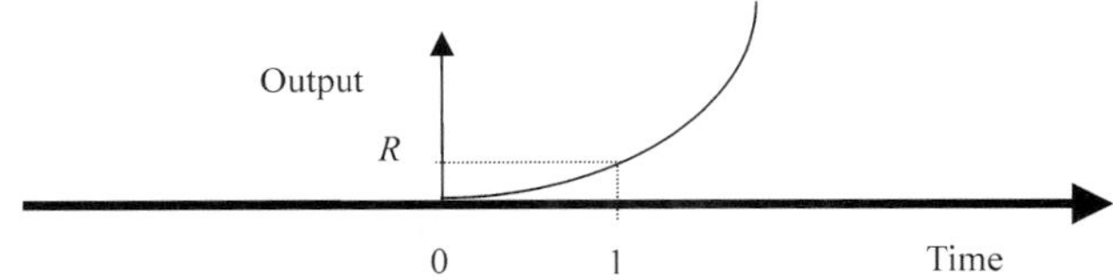

Fig. 10.9 Parabolic input of magnitude *R*.

and finally to Eq. (10.19).

$$e(t)_{ss} = \frac{R}{\lim_{s\to 0} s^2 G(s)} \tag{10.19}$$

This allows the definition of the **acceleration error constant, K_a**. It allows a compact notation for steady-state error **for a parabolic input**. It also provide a means to compare system types.

$$K_a = \lim_{s\to 0} s^2 G(s) \tag{10.20}$$

which leads to

$$e(t)_{ss} = \frac{R}{K_a} \tag{10.21}$$

The steady-state error for all three system types for a parabolic input will be presented next.

10.1.3.1 Type 0 system. The K_a for a Type 0 system is calculated using Eq. (10.22).

$$K_a = \lim_{s\to 0} s^2 \left[\frac{K_0(\tau_2 s + 1)(\tau_2 s + 1)\cdots}{(\tau_a s + 1)(\tau_b s + 1)\cdots}\right] = 0 \tag{10.22}$$

Thus, the steady-state error is

$$e(t)_{ss} = \frac{R}{K_a} = \infty$$

This means that a Type 0 system is not capable of tracking a parabolic input.

10.1.3.2 Type 1 system. The K_a for a Type 1 system is shown in Eq. (10.23).

$$K_a = \lim_{s\to 0} s^2 \left[\frac{K_1(\tau_1 s + 1)(\tau_2 s + 1)\cdots}{s(\tau_a s + 1)(\tau_b s + 1)\cdots}\right] = 0 \tag{10.23}$$

Thus, the steady-state error is

$$e(t)_{ss} = \frac{R}{K_a} = \infty$$

This means that a Type 1 system is also unable to track a parabolic input. Example 10.6 shows this error for a parabolic input to a Type 1 system.

Example 10.6

What is the steady-state error for the transfer function $G(s)$ with unity feedback when given a unit parabolic input $[R(s) = 1/s^3]$?

$$G(s) = \frac{2}{s(s+3)} = \frac{0.667}{s(0.333s+1)}$$

Because this is a Type 1 system with a parabolic input, infinite steady-state error is expected.

$$K_a = \lim_{s \to 0} s^2 \left[\frac{2}{s(s+3)} \right] = 0$$

Thus, the steady-state error is

$$e(t)_{ss} = \frac{1}{0} = \infty$$

The resulting time response shown in Fig. 10.10 displays the increasing steady-state error as time increases.

10.1.3.3 Type 2 system. Finally, the K_a for a Type 2 system given a parabolic input is shown in Eq. (10.24).

$$K_a = \lim_{s \to 0} s^2 \left[\frac{K_2(\tau_1 s+1)(\tau_2 s+1)\cdots}{s^2(\tau_a s+1)(\tau_b s+1)\cdots} \right] = K_2 \tag{10.24}$$

Thus

$$e(t)_{ss} = \frac{R}{K_a} = \frac{R}{K_2}$$

When the unity feedback is applied, the Type 2 system will track a parabolic input with a constant steady-state error. This is shown in Example 10.7.

Example 10.7

What is the steady-state error for the transfer function $G(s)$ with unity feedback when given a unit parabolic input $[R(s) = 1/s^3]$?

$$G(s) = \frac{2(s+2)}{s^2(s+3)} = \frac{1.333(0.5s+1)}{s^2(0.333s+1)}$$

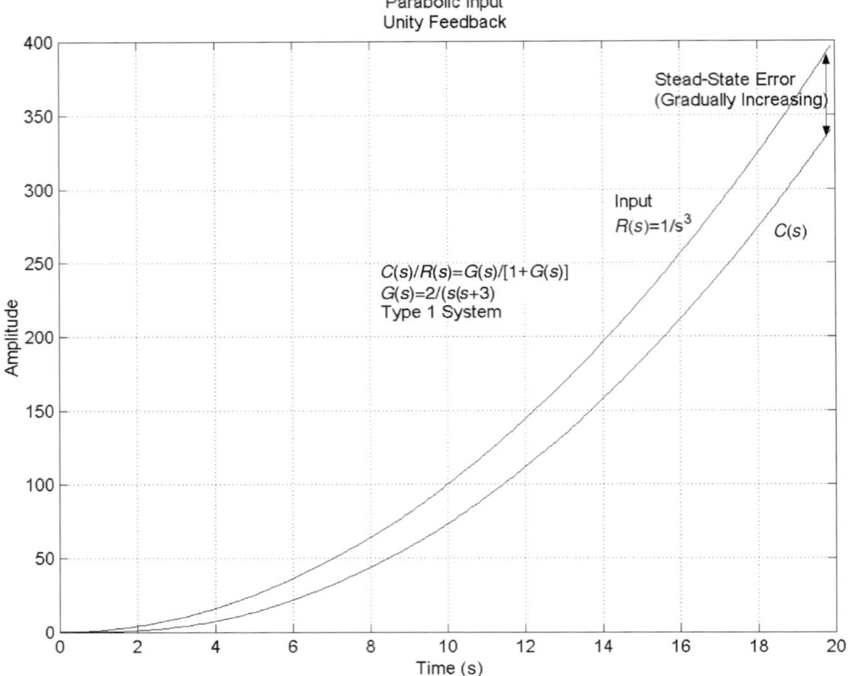

Fig. 10.10 Type 1 system response to a parabolic input (unity feedback).

Because $G(s)$ is a Type 2 system, a constant steady-state error is expected for a ramp input.

$$K_a = \lim_{s\to 0} s^2 \left[\frac{2(s+2)}{s^2(s+3)} \right] = 1.333$$

Thus, the steady-state error is

$$e(t)_{ss} = \frac{1}{1.333} = 0.75$$

Figure 10.11 illustrates the response of a Type 2 system to a parabolic input. Notice that there is a finite steady-state error of 0.75.

10.1.3.4 Parabolic input summary. Table 10.3 summarizes the steady-state error relationship for a parabolic input. Type 0 and Type 1 systems cannot track a parabolic input; while a Type 2 system can track a parabolic input with a constant steady-state error.

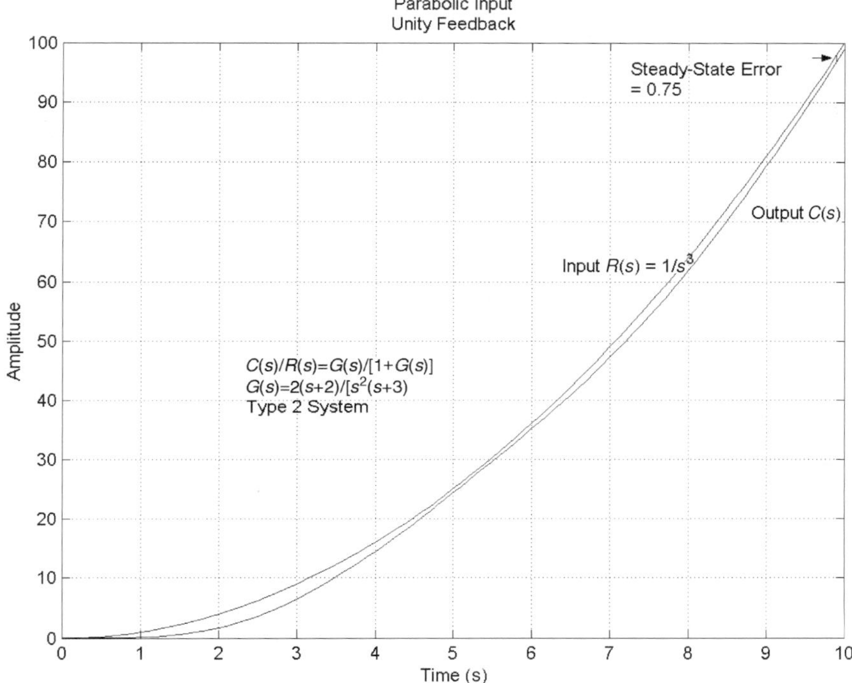

Fig. 10.11 Type 2 system response to a parabolic input.

10.1.4 Overall Steady-State Error Summary

Table 10.4 summarizes the overall steady-state error for Type 0, Type 1, and Type 2 systems for step, ramp, and parabolic inputs. Notice that as the system type increases, the steady-state error performance increases (steady-state error decreases) for the same type of input. This table only applies for the case of stable unity feedback systems.

Table 10.3 Steady-state error summary for a parabolic input

Type	K_a	$e(t)_{ss}$
0	0	∞
1	0	∞
2	K_2	R/K_2

Table 10.4 Steady-state error summary

	Error constants			Steady-state error		
System type	Step	Ramp	Parabolic	Step	Ramp	Parabolic
	$K_p =$	$K_v =$	$K_a =$	$R/(1+K_p)$	R/K_v	R/K_a
0	K_0	0	0	$R/(1+K_0)$	∞	∞
1	∞	K_1	0	0	R/K_1	∞
2	∞	∞	K_2	0	0	R/K_2

10.1.5 Error Constants for Stable Nonunity Feedback Systems

It is possible to assess error constants for stable nonunity feedback systems. Figure 10.12 shows the block diagram of an ideal system represented by a transfer function $T_d(s)$, and an actual transfer function of $C(s)/R(s)$ [which is an approximation of $T_d(s)$].

R is the input to both systems, and E is the difference (error) between the desired output and the actual output. The three nonunity feedback or general error constants are defined based on this representation.

The general **step error constant K_s** is defined as:

$$K_s \equiv \frac{1}{\lim_{s\to 0}\left[T_d(s) - \dfrac{C(s)}{R(s)}\right]} \tag{10.25}$$

The steady-state error for a general system when the input is a unit step function is related to K_s by

$$e(t)_{ss} = \frac{1}{K_s} \tag{10.26}$$

The general ramp error constant K_r is defined as

$$K_r \equiv \frac{1}{\lim_{s\to 0}\dfrac{1}{s}\left[T_d(s) - \dfrac{C(s)}{R(s)}\right]} \tag{10.27}$$

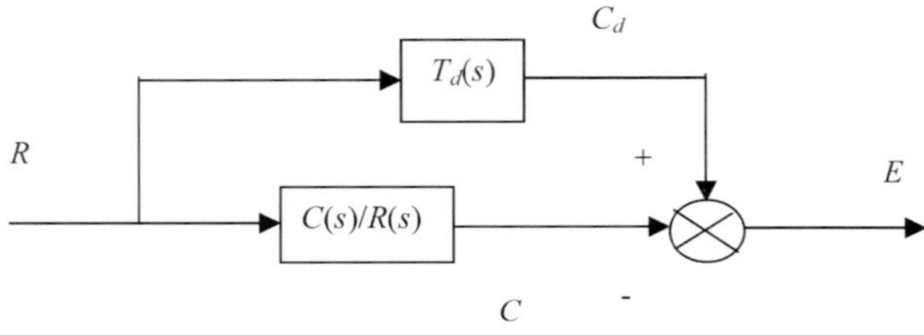

Fig. 10.12 Nonunity feedback system representation.

The steady-state error for a general system when the input is a unit ramp function is related to K_r by

$$e(t)_{ss} = \frac{1}{K_r} \tag{10.28}$$

The general parabolic error constant K_{pa} is defined as

$$K_{pa} \equiv \frac{1}{\lim\limits_{s\to 0} \frac{1}{s^2}\left[T_d(s) - \frac{C(s)}{R(s)}\right]} \tag{10.29}$$

The steady-state error for a general system when the input is a unit parabolic function is related to K_{pa} by

$$e(t)_{ss} = \frac{1}{K_{pa}} \tag{10.30}$$

For an aircraft, this becomes a realistic scenario because sensors that measure parameters have dynamics associated with them. Thus, it may not be possible to obtain unity feedback. One case of a nonunity feedback system is presented in Example 10.8.

Example 10.8

For the following nonunity feedback system, find the general error constants if the desired transfer function is $\frac{1}{2}$.

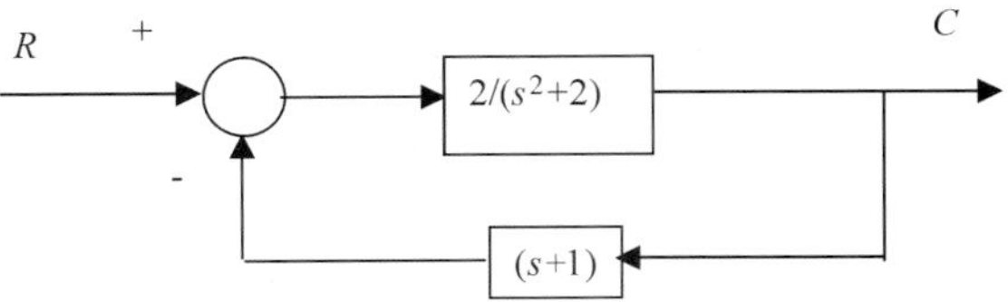

The first step in solving this problem is to form the $C(s)/R(s)$ ratio. This is simply the closed-loop transfer functions with $G(s) = 2/(s^2+2)$ and $H(s) = (s+1)$. The resulting equation is

$$\frac{C(s)}{R(s)} = \frac{2}{s^2+2s+4}$$

Next, the difference or error between the actual and desired transfer functions is formed.

$$T_d(s) - \frac{C(s)}{R(s)} = 0.5 - \frac{2}{s^2+2s+4} = \frac{s(s+2)}{2(s^2+2s+4)}$$

Therefore

$$K_s = \frac{1}{\lim_{s\to 0}\left[\frac{s(s+2)}{2(s^2+2s+4)}\right]} = \infty$$

$$K_r = \frac{1}{\lim_{s\to 0}\frac{1}{s}\left[\frac{s(s+2)}{2(s^2+2s+4)}\right]} = 4$$

$$K_{pa} = \frac{1}{\lim_{s\to 0}\frac{1}{s^2}\left[\frac{s(s+2)}{2(s^2+2s+4)}\right]} = 0$$

10.2 Frequency Response

Time domain response of a system has already been covered in great detail in Sec. 8.5. In the time domain, the system response is measured in terms of such things as rise time and settling time. However, another important aspect of system performance is the frequency domain. This section explains the background on frequency response, discusses frequency response curves and how to construct them, shows frequency domain analysis and specifications, and ends with several subsections detailing frequency domain compensation. If logarithmic plots are used, frequency domain analysis is sometimes called Bode analysis.

10.2.1 Background

Frequency response analysis is centered on the concept of the sinusoidal response of a system. When a sine wave is provided as the input to a system, as shown in Fig. 10.13, the steady-state output will be a sinusoid at the same frequency but at a different magnitude and phase. Notice that the input amplitude is A_1 and the output is some different amplitude, A_2. Also, notice that the peak amplitude occurs later in time. This time delay is known as "phase lag" in the frequency domain.

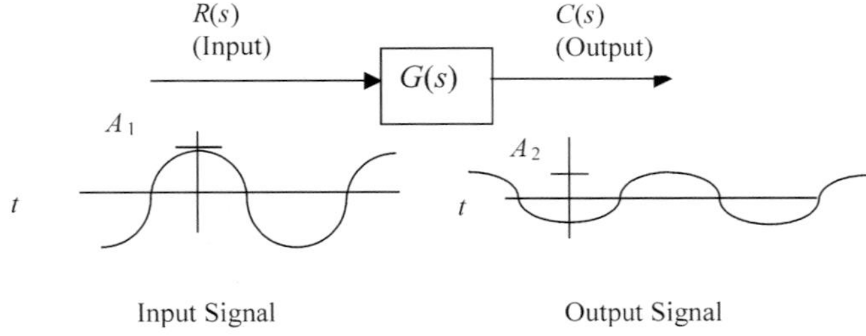

Fig. 10.13 Input/output relationship for a transfer function.

Before proceeding, it is important to relate the time domain to the frequency domain. The variable s, as described in Secs. 7.2 and 7.3, is a complex variable of the form:

$$s = \sigma \pm i\omega_d = \zeta\omega_n \pm i\omega_d = a \pm ib \tag{10.31}$$

where σ is the transient part of the response and ω_d is the steady-state part of the response. The time response can be studied knowing the values of ζ, ω_n, and ω_d (see Sec. 8.5). Because the observed frequency is usually the damped frequency (except in the case of a pure, undamped sinusoid), the frequency (ω) discussed for frequency analysis will be the damped frequency (ω_d). Also, note that for frequency analysis the **steady-state response is used**. Thus, s simplifies to

$$s = i\omega_d \equiv i\omega \tag{10.32}$$

The frequency response is thus easy to relate to the aircraft transfer functions developed in previous sections. For example, take an actuator transfer function

$$G(s) = \frac{10}{(s + 10)}$$

This can now be expressed in frequency domain terms as

$$G(i\omega) = \frac{10}{(i\omega + 10)}$$

The same transformation from the s domain to the frequency domain is simply made by substituting Eq. (10.32) for s into any $G(s)$ transfer function.

10.2.1.1 Input/output relationship. To understand the frequency domain fully, it is best to go back to the roots of the time domain. As mentioned earlier, the output of a system with a sinusoidal input will have the same frequency as the input, but at a different amplitude (magnitude) and phase angle. Given the input

$$r(t) = A_1 \sin(\omega t) \tag{10.33}$$

the output will be

$$c(t) = A_2 \sin(\omega t + \phi) \tag{10.34}$$

where A_1 is the input amplitude, A_2 is the output amplitude, and ϕ is the phase shift angle. The amplitudes are read from the peak values of the input sinusoid for A_1 and the output sinusoid for A_2 in the frequency response. It is also quite easy to find the phase lag from a frequency response. Because the input and output sinusoids have the same frequency, from the basic definition of frequency response, they have the same period. As stated earlier, the time domain equivalent of phase lag is time delay, which is easily obtained from a frequency domain plot. The time delay between the peak of the input and output is the phase lag. Because one period contains 360 deg, the phase lag can be obtained from the following ratio:

$$\frac{t_{del}(\mathrm{s})}{T(\mathrm{s})} = \frac{\phi(\mathrm{deg})}{360(\mathrm{deg})} \tag{10.35}$$

where t_{del} is the delay between the input and output peak amplitudes (in seconds), T is the period of oscillation (in seconds), and ϕ is the phase lag (in degrees). Figure 10.14 shows the input/output relationship for a sinusoidal input to a typical transfer function.

Using the Type 0 transfer function from the previous section $[G(s) = 2/(s+3)]$ and an input of $r(t) = 1\sin(4t)$ results in an output amplitude of 0.4 and a phase shift ϕ of approximately 53 deg. In this case, the period $T = 2\pi/\omega = 2\pi/4 = 1.57$ sec, and the delay is approximately 0.23 sec. Thus

$$\phi = 360\ \mathrm{deg}\frac{0.23\ \mathrm{s}}{1.57\ \mathrm{s}} = 53\ \mathrm{deg}$$

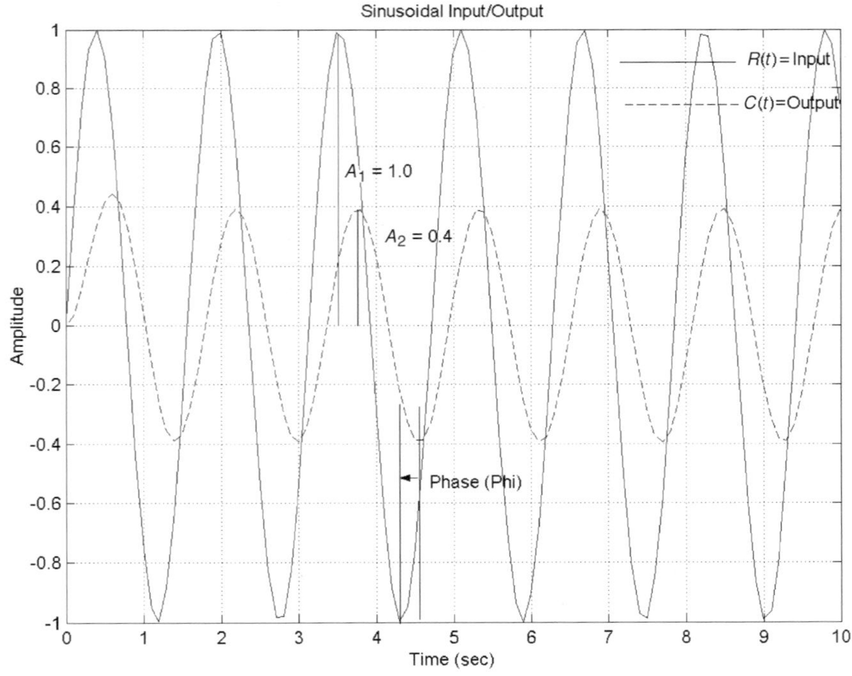

Fig. 10.14 Sinusoidal input/output relationship.

This can be related to Eq. (10.34) because $\omega = 4$ rad/s, the output amplitude $A_2 = 0.4$, and the phase angle ϕ is 53 deg. Thus, $c(t) = 0.4\sin(4t + 53 \text{ deg})$ is the resulting output sinusoid.

Frequency response analysis simply relates the output to the input by examining the variation of a sinusoid between the entry and exit to a system. To fully characterize a system, a sine sweep feeds sinusoids of varying frequencies into a system and then measures the output. More about this will be discussed in the experimental frequency response section. It should be noted that at times the transient dynamics are important. The disadvantage of frequency analysis is the lack of transient response characteristics that may be needed for time domain specifications.

10.2.1.2 Transient analysis. To understand why the transient part is not used in frequency analysis, consider that a sinusoid is a continuous input. Thus, any initial transients are not major contributors to the overall response. Consider a sinusoidal input of $r(t) = \sin(4t)$ into the following transfer function:

$$G(s) = \frac{4}{s^2 + 2s + 4}$$

Figure 10.15 shows the transient portion associated with the output. Notice the variation in the amplitude for the first three cycles of the output. This is the transient response. All cycles after that are the same amplitude, and the response has thus reached steady-state. For a sinusoidal input, which is the basis of frequency analysis, the majority of the cycles are associated with steady-state. Thus, it should be apparent from Fig. 10.15 that ignoring the transient portion of the response makes sense for frequency analysis.

10.2.1.3 Aircraft application. Why is frequency analysis important to aircraft applications? The pilot input can actually be thought of as sinusoidal stick movement as the pilot makes corrections to keep the aircraft at a specified trim condition. The pilot can move the stick forward and back slowly in calm flight conditions, but may have to move the stick rapidly forward and back to account for turbulence. If the pilot is trying to maintain a pitch angle, then the aircraft response is an output sinusoid in terms of nose position, for example. The input is the pilot's stick input, and the aircraft dynamics are the transfer function through which the sinusoid is applied.

10.2.2 Frequency Response Curves

To perform analysis in the frequency domain, it is most convenient to plot the magnitude and phase of the output vs frequency. At any individual frequency, there is a relationship between the input and output similar to Fig. 10.14. As the frequency is varied, the relationship between the input and output amplitude and phase varies. The ratio between the input and output forms the magnitude plot, and the variation of phase angle (ϕ) is plotted as the phase plot.

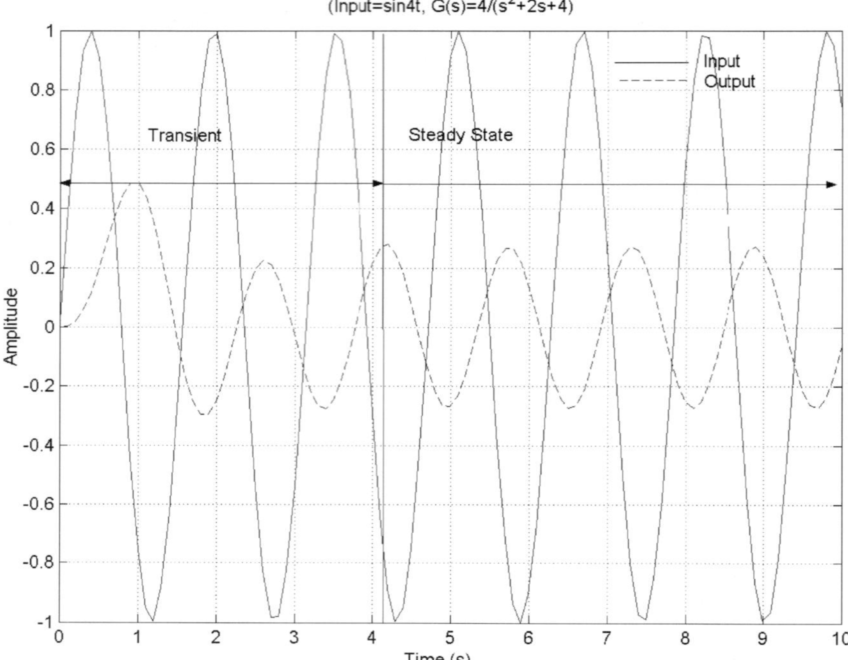

Fig. 10.15 Transient associated with sinusoidal input.

Because the frequency response curves possess both a magnitude and angle, it is often easier to express the quantities in phasor form. Given a generic complex number $\bar{z}$

$$\bar{z} = \mathrm{Re} + iIm \tag{10.36}$$

The phasor form of a complex number $\bar{z}$ is shown in Eq. (10.37)

$$|\bar{z}| < \phi \tag{10.37}$$

where the magnitude $|z|$ is written as

$$|\bar{z}| = \sqrt{(\mathrm{Re}^2 + Im^2)} \tag{10.38}$$

and the phase angle ϕ is written as

$$\angle\phi = \tan^{-1}\left(\frac{Im}{\mathrm{Re}}\right) \tag{10.39}$$

Thus, the complex notation for a transfer function $\bar{G}(i\omega)$, written in terms of the magnitude $|G(i\omega)|$ and the phase angle ϕ, is

$$\bar{G}(i\omega) = |G(i\omega)\angle\phi \tag{10.40}$$

This means that at a particular frequency ω, the transfer function becomes one complex number. As the frequency varies, the value of the complex number varies. The compilation of all of these complex numbers for different frequencies becomes the frequency response for the transfer function.

Example 10.9

To illustrate frequency response plots, a typical transfer function $\bar{G}(i\omega)$ is defined as

$$\bar{G}(i\omega) = \frac{2}{(i\omega + 3)}$$

The frequency response for $\bar{G}(i\omega)$ is shown in Fig. 10.16 as the magnitude vs frequency plot and in Fig. 10.17 as the phase vs frequency plot. Note that frequency response plots involve both magnitude and phase plots vs frequency.

Notice that one frequency point for this plot is shown in Fig. 10.14. Examining the response at 4 rad/s yielded an amplitude ratio of 0.4 and phase lag of 53 deg. This can be seen as a single point on the magnitude and frequency plots that follow. The rest of the plot is generated identically using different frequencies.

10.2.3 Bode Plots

The most common method to display frequency response plots is on a logarithmic scale. This allows presentation and simplifies the analysis of a large range of frequencies in a compact plot. In addition, notice that in the case of Figs. 10.16 and 10.17 most of the change occurs at low frequency. The logarithmic scale allows expanded display of the lower frequencies. Another advantage is that when logarithms are used, multiplications and divisions become simple additions and subtractions. Finally, using plots of this type allows for graphical analysis of transfer functions. This type of logarithmic frequency plot is called a Bode plot. The name is in honor of H.W. Bode, who used them for the study of feedback amplifiers.

Before proceeding, it is necessary to define a logarithm mathematically as it pertains to Bode plots. This is best illustrated by first putting Eq. (10.40) into exponential form.

$$\bar{G}(i\omega) = |G(i\omega)|e^{i\phi(\omega)} \tag{10.41}$$

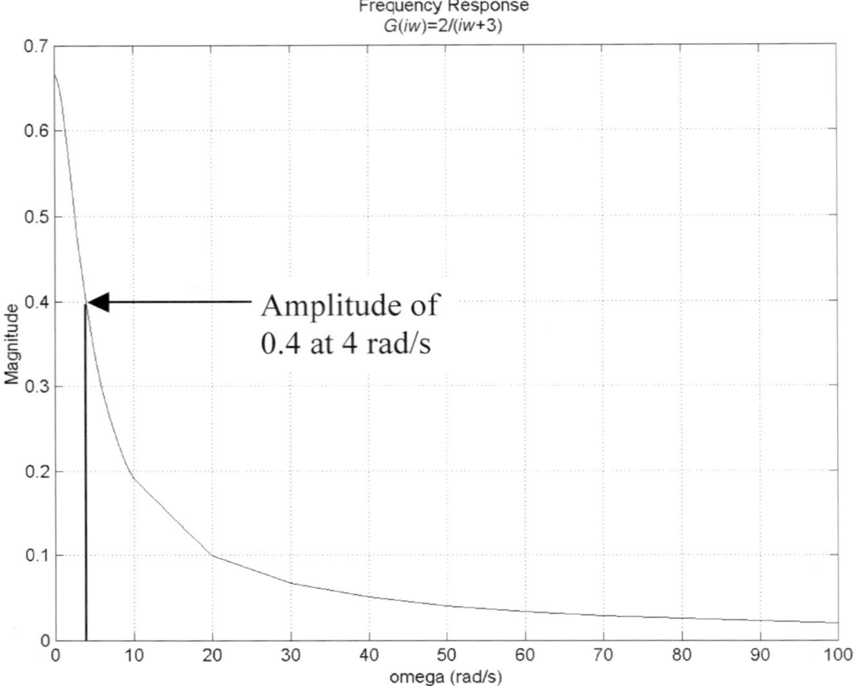

Fig. 10.16 Frequency response curve—magnitude.

Taking the logarithm of both sides results in

$$\log \bar{G}(i\omega) = \log|G(i\omega)| + \log e^{i\phi(\omega)} = \underbrace{\log|G(i\omega)|}_{\text{Magnitude}} + \underbrace{i0.434\phi(\omega)}_{\text{Phase}}$$

The real part is the logarithm of the magnitude and the imaginary part is a scale factor of 0.434 times the phase angle in radians. It is standard procedure when discussing Bode plots to omit the 0.434 scale factor and use only the angle $\phi(\omega)$.

It is also common in feedback control systems to use the decibel scale for the log magnitude. This allows display of a large range of magnitude numbers on a graph. The logarithm for the magnitude of a transfer function $\bar{G}(i\omega)$ with units of decibels is represented as

$$20\log|G(i\omega)|$$

with units of dB for decibels. Sometimes this is expressed as the log magnitude of $\bar{G}(i\omega)$, or Lm $\bar{G}(i\omega)$ for short.

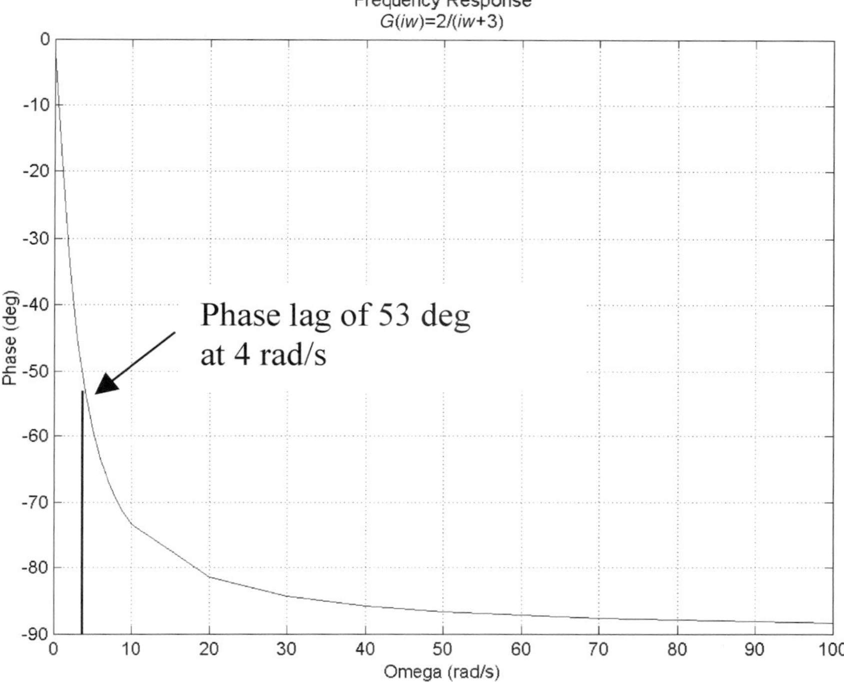

Fig. 10.17 Frequency response curve—phase.

10.2.3.1 Bode form. To properly evaluate transfer functions with Bode plots, it is convenient to rearrange the equation such that all non-$i\omega$ terms are equal to one.

$$\bar{G}(i\omega) = \frac{K_n(i\omega)^m(1+i\tau_1\omega)(1+i\tau_2\omega)\left(1+\left(\frac{i\omega}{\omega_{n_1}}\right)^2+i\left(\frac{2\zeta\omega}{\omega_{n_1}}\right)\right)}{(i\omega)^n(1+i\tau_3\omega)(1+i\tau_4\omega)\left(1+\left(\frac{i\omega}{\omega_{n_2}}\right)^2+i\left(\frac{2\zeta\omega}{\omega_{n_2}}\right)\right)} \tag{10.42}$$

This is called the Bode form of a transfer function, and the gain K_n is called the Bode gain.

Example 10.10

Place the following transfer function into Bode form and identify the Bode gain.

$$\bar{G}(iw) = \frac{4(i\omega+2)(i\omega+5)((i\omega)^2+2i\omega+6)}{(i\omega)(i\omega+3)(i\omega+7)((i\omega)^2+4i\omega+8)}$$

To put this transfer function into Bode form, the numerator and denominator poles/zeros must have the non-$i\omega$ terms (2,5,6 for the numerator and 3,7,8 for the denominator) factored out so that they have a value of 1.

$$\bar{G}(i\omega) = \frac{4(2)(5)(6)\left(\dfrac{i\omega}{2}+1\right)\left(\dfrac{i\omega}{5}+1\right)\left(\dfrac{(i\omega)^2}{6}+\dfrac{2i\omega}{6}+1\right)}{(3)(7)(8)(i\omega)\left(\dfrac{i\omega}{3}+1\right)\left(\dfrac{i\omega}{7}+1\right)\left(\dfrac{(i\omega)^2}{8}+\dfrac{4i\omega}{8}+1\right)}$$

This simplifies to the Bode form

$$\bar{G}(i\omega) = \frac{1.4286\left(\dfrac{i\omega}{2}+1\right)\left(\dfrac{i\omega}{5}+1\right)\left(\dfrac{(i\omega)^2}{6}+\dfrac{i\omega}{3}+1\right)}{(i\omega)\left(\dfrac{i\omega}{3}+1\right)\left(\dfrac{i\omega}{7}+1\right)\left(\dfrac{(i\omega)^2}{8}+\dfrac{i\omega}{2}+1\right)}$$

The gain in Bode form is the Bode gain, which is 1.4286 in this case.

10.2.3.2 Additive/subtractive nature of Bode plots. The power of Bode plots is their simplicity. Because the Bode plots are based on logarithms, terms in the transfer function $\bar{G}(i\omega)$, which are multiplied together, may be added logarithmically. In the same manner, terms in $\bar{G}(i\omega)$ that are divided are subtracted logarithmically. In other words, numerator terms (constant and zeros) are added and denominator terms (poles) are subtracted. This is best illustrated by an example using the general transfer function of Eq. (10.42). Taking the log magnitude in decibels results in the following representation for magnitude.

Remember that with Eq. (10.38) the magnitude of a complex number is obtained, and that the phase for a complex number is obtained with Eq. (10.39). At any particular frequency each term is simply a complex number with magnitude and phase. By using the logarithmic approach, each of these individual terms are simply added or subtracted to form the overall response.

$$\begin{aligned} 20\log|G(i\omega)| &= 20\log K_n + 20m\log i\omega + 20\log|1+i\tau_1\omega| + 20\log|1+i\tau_2\omega| \\ &+ 20\log\left|1-\frac{\omega^2}{\omega_{n_1}^2}+i\frac{2\zeta\omega}{\omega_{n_1}}\right| - 20n\log i\omega - 20\log|1+i\tau_3\omega| - 20\log|1+i\tau_4\omega| \\ &- 20\log\left|1-\frac{\omega^2}{\omega_{n_2}^2}+i\frac{2\zeta\omega}{\omega_{n_2}}\right| \end{aligned} \quad (10.43)$$

The phase is represented as

$$\angle G(i\omega) = \angle K_n + m90\ \deg + \tan^{-1}\omega\tau_1 + \tan^{-1}\omega\tau_2 + \tan^{-1}\left(\frac{\dfrac{2\zeta\omega}{\omega_{n_1}}}{1-\dfrac{\omega^2}{\omega_{n_2}^2}}\right)$$

$$- n90\ \deg - \tan^{-1}\omega\tau_3 - \tan^{-1}\omega\tau_4 - \tan^{-1}\left(\frac{\dfrac{2\zeta\omega}{\omega_{n_2}}}{1-\dfrac{\omega^2}{\omega_{n_2}^2}}\right) \tag{10.44}$$

Each term can be treated as a separate entity. Thus, Bode analysis allows each contributing pole or zero to be analyzed separately. There are four types of terms in a transfer function, and therefore in Bode plots.

1) Constant term, K_n
2) Pole or zero at the origin, $(i\omega)^{\pm n}$ ($+n$ for a zero, $-n$ for a pole)
3) First-order real pole or zero, $(1 + i\omega\tau)^{\pm n}$
4) Second-order quadratic (complex conjugate) poles or zeros, $(1 - \omega^2/\omega_n^2 + i2\zeta\omega/\omega_n)^{\pm n}$

It is instructive to show an example of how to manually calculate a Bode plot at a single frequency. The computer iteratively performs these calculations for different frequencies to form the Bode plot. Therefore, understanding how to perform the calculation for a single frequency allows understanding of how the entire Bode plot is calculated.

Example 10.11

Given the following transfer function $\bar{G}(s)$, calculate the log magnitude and phase at a frequency $\omega = 10$ rad/s.

$$\bar{G}(s) = \frac{5(s+2)(s^2+2s+4)}{s(s+3)(s^2+4s+6)}$$

The first step is to transform $\bar{G}(s)$ into $\bar{G}(i\omega)$ by substituting $s = i\omega$. Next, put the resulting transfer function into Bode form.

$$G(i\omega) = \frac{5(2)(4)\left(\dfrac{i\omega}{2}+1\right)\left(\dfrac{(i\omega)^2}{4}+\dfrac{2i\omega}{4}+1\right)}{3(6)(i\omega\left(\dfrac{i\omega}{3}+1\right)\left(\dfrac{(i\omega)^2}{6}+\dfrac{4i\omega}{6}+1\right)}$$

which simplifies to a Bode form

$$G(i\omega) = \frac{2.22\left(\frac{i\omega}{2}+1\right)\left(1-\frac{\omega^2}{4}+\frac{i\omega}{2}\right)}{(i\omega)\left(\frac{i\omega}{3}+1\right)\left(1-\frac{\omega^2}{6}+\frac{i2\omega}{3}\right)}$$

After the transfer function is in Bode form, it is easy to calculate the log magnitude.

$$20\log|G(i10)| = 20\log 2.22 + 20\log|1+i(10)/2| + 20\log\left|1-\frac{10^2}{4}+i(10)/2\right| - 20\log|i10| - 20\log|1+i(10)/3| - 20\log\left|1-\frac{10^2}{6}+i\frac{2(10)}{4}\right|$$

Simplifying,

$$20\log|G(i10)| = 6.93 + 14.15 + 27.79 - 20 - 10.82 - 24.62 = -6.57\text{ dB}$$

Similarly, for the phase:

$$\phi(10) = 0\text{ deg} + \tan^{-1}\frac{10}{2} + \tan^{-1}\frac{\frac{10}{2}}{1-\frac{10^2}{4}} - 90\text{ deg} - \tan^{-1}\frac{10}{3} - \tan^{-1}\frac{\frac{2(10)}{3}}{1-\frac{10^2}{6}}$$

This simplifies to

$$\begin{aligned}\phi(10) &= 0\text{ deg} + 78.76\text{ deg} + 168.23\text{ deg} - 90\text{ deg} - 73.31\text{ deg} - 156.94\text{ deg}\\ &= -73.42\text{ deg}\end{aligned}$$

Figure 10.18 shows the excellent correlation between the Bode plot at $\omega = 10$ rad/s and the previous calculations. It is the accumulation of the log magnitude and phase calculations that produce the Bode plot, after all. Thus, there should be a one-to-one correlation at each frequency point calculated.

10.2.3.3 Bode plot construction. With the advent of computers and engineering software, it is very easy to have calculated Bode plots at one's fingertips (such as the one shown in Fig. 10.17). This allows an engineer to quickly analyze multiple transfer functions over a wide range of frequencies. However, the importance of understanding the basics of Bode plot construction cannot be overstated. A tremendous amount of insight is gained from recognizing how the curves in a Bode plot are generated. Ultimately, a Bode plot is the superposition of all the individual terms (poles, zeros, and gain) in the transfer function. Each of the four types of terms described in the previous section can be plotted one by one.

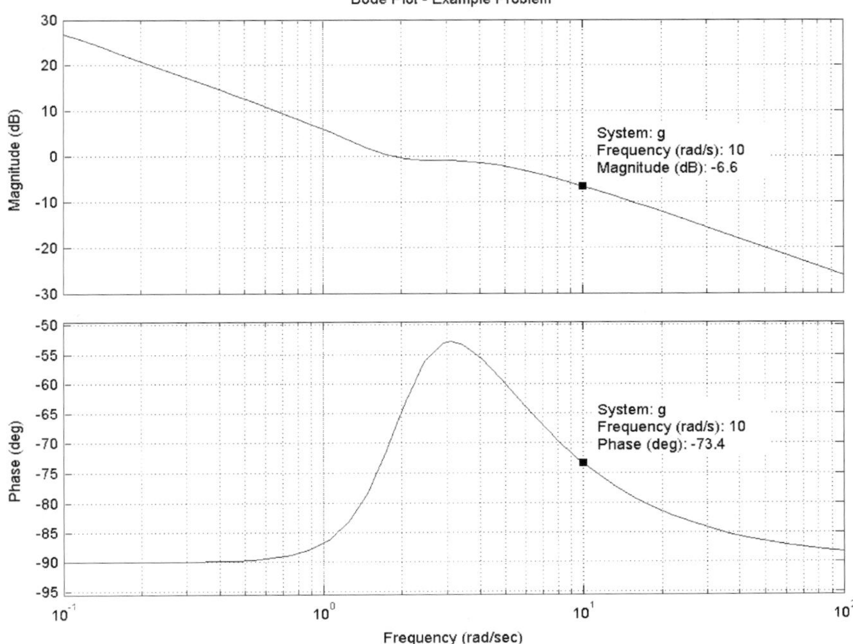

Fig. 10.18 Bode plot for Example 10.11 with ω = 10 rad/s marked.

Hand-drawn Bode plots start with straight-line asymptotic approximations. They are called asymptotic approximations because the actual Bode plots approach these lines asymptotically. The maximum error for these approximations occurs at the break point of a pole or zero. A break point is at the frequency that corresponds to the pole or zero location. For example, the break point for a pole at $(s + 2)$ [or $2(i\omega/2 + 1)$ in Bode form] occurs at 2 rad/s. For a second-order (or quadratic) term, the break point is at the natural frequency, ω_n. Sometimes, the break point is called the corner frequency. The following sections will describe how to draw the asymptotic approximations for the four different types of terms in a Bode plot.

10.2.3.4 Constant terms. If a transfer function is in Bode form, then the constant present is the Bode gain. As expected, the magnitude and phase contributions to a Bode plot from a constant are constants. The gain contribution is exact, and thus not an approximation. Stated another way, the asymptotic approximation for a constant equals the actual Bode plot contribution. The magnitude contribution from a constant K_n is simply:

$$20 \log |K_n| = \text{const} \tag{10.45}$$

The phase depends strictly on the sign.

$$+K_n = 0 \text{ deg}$$

or

$$-K_n = +/- 180 \text{ deg} \tag{10.46}$$

Figure 10.19 shows the Bode plot contribution for both $K_n = 10$ and $K_n = -10$. Notice that the magnitude plot is the same for both, but the phase is different.

10.2.3.5 Poles or zeros at origin. The second type of term in a Bode plot is a pole or zero at the origin. If there is a pole at the origin, this is called an integrator (recall the discussion on system type in Sec. 10.1). A zero at the origin is called a differentiator. These terms are represented mathematically by

$$s^{\pm n} = (i\omega)^{\pm n}$$

where n is the number of terms present. Poles correspond to $-n$ while zeros correspond to $+n$.

All first-order terms have a slope of either positive or negative 20 dB/decade. A decade is a factor of 10 change in frequency. For example, going from $\omega = 1$ to $\omega = 10$ or from $\omega = 10$ to $\omega = 100$ is a decade change in frequency. First-order poles have a −20 dB/decade slope, and first-order zeros

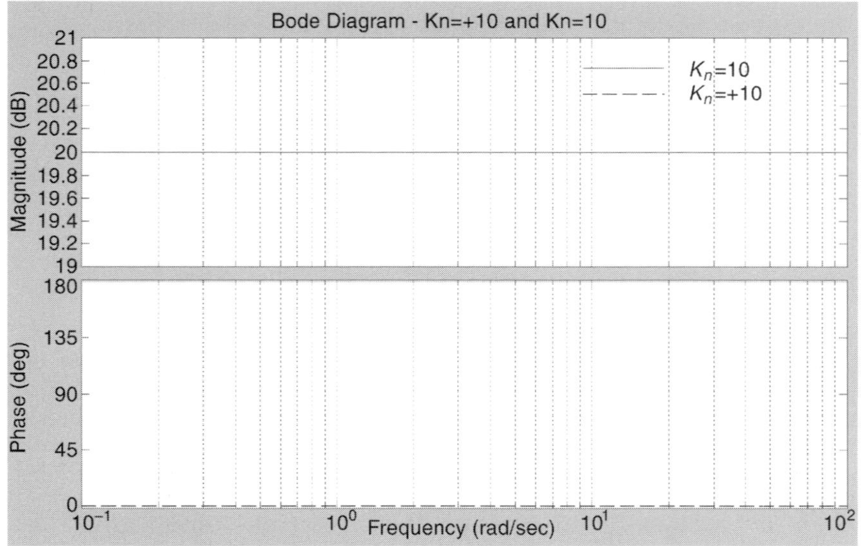

Fig. 10.19 Bode plot for positive and negative constant $K_n = 10$.

have +20 dB/decade slope. The value of +/− 20 dB/decade is because of the fact that an order of 10 increase (or decrease) in magnitude (ω in this case) corresponds to an increase (or decrease) of 20 dB, as seen in Eq. (10.47). Poles and zeros at the origin are simply a special case of first-order poles and zeros. Because there is no break point for these terms, the asymptotic approximation is also the exact Bode plot contribution. The magnitude is represented by the following equation:

$$20 \log |i\omega|^{\pm n} = \pm 20(n) \log |i\omega| \tag{10.47}$$

An interesting magnitude characteristic of poles and zeros at the origin is that they have a value of 1, or 0 dB, at $\omega = 1$ rad/s. Thus, all terms at the origin cross through 0 dB on the Bode plot at $\omega = 1$ rad/s.

The general equation for slope for terms that pass through the origin is:

$$\text{slope} = \pm 20(n)\text{dB/decade} \tag{10.48}$$

Thus, if two integrators are present, they would be drawn as a line with a slope of −40 db/decade, which crosses the 0 dB line at 1 rad/s.

The second aspect of the Bode plot asymptotic approximation is phase. All first-order terms add either +90 deg of phase (for a zero) or −90 deg of phase (for a pole). Here, too, terms that pass through the origin possess an interesting characteristic. They provide all of their phase contribution immediately. Thus, they provide a constant phase regardless of frequency. Mathematically, this is represented as

$$\text{phase} = \pm 90(n) \tag{10.49}$$

This can be seen from examination of the complex plane. For Bode analysis, $s = i\omega$ has already been defined. Figure 10.20 shows that s is essentially a vector along the imaginary axis of magnitude ω and phase 90 deg.

Thus, a for a numerator s term, $s = |\omega|\angle 90$ deg. For a denominator s term, or $1/s$,

$$\frac{1}{s} = \frac{1}{|\omega|\angle 90\ \text{deg}} = \frac{\angle -90\ \text{deg}}{|\omega|}$$

Figure 10.21 shows cases for both one and two poles at the origin as well as one and two zeros at the origin. Note that there is change in slope and phase contribution, but all pass through 0 dB at 1 rad/s.

10.2.3.6 First-order poles and zeros. One of the most common terms seen in Bode plots are first-order poles and zeros. They correspond to first-order roots in the numerator or denominator of the transfer functions. These are sometimes called simple poles and zeros. Here, straight line representations of the frequency response are approximations. The general form for a first-order

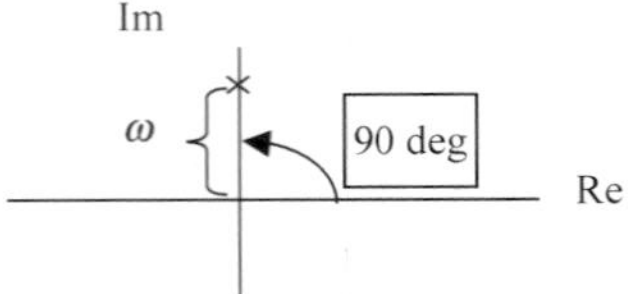

Fig. 10.20 Complex plane representation of s.

term is

$$\left(s + \frac{1}{\tau}\right)^{\pm n} \Rightarrow (1 + i\omega\tau)^{\pm n} \tag{10.50}$$

As with the previous section, n corresponds to the number of poles or zeros. If there is one root, then $n = 1$. Zeros have $a + n$ power, while poles have $a - n$ power. The log magnitude for first-order terms is expressed as

$$20 \log |1 + i\omega\tau|^{\pm n} = \pm 20(n) \log |1 + i\omega\tau| \tag{10.51}$$

Each single pole has a magnitude slope of -20 dB/decade, while each single zero has a contribution of $+20$ dB/decade. However, the slope does not occur until after the corner frequency or break point, ω_c, which occurs at the negative value of the root. For example, an $(s + 3)$ term would have a corner frequency of 3 rad/s. The reason for the $+/-$ 20 dB/decade slope is the same

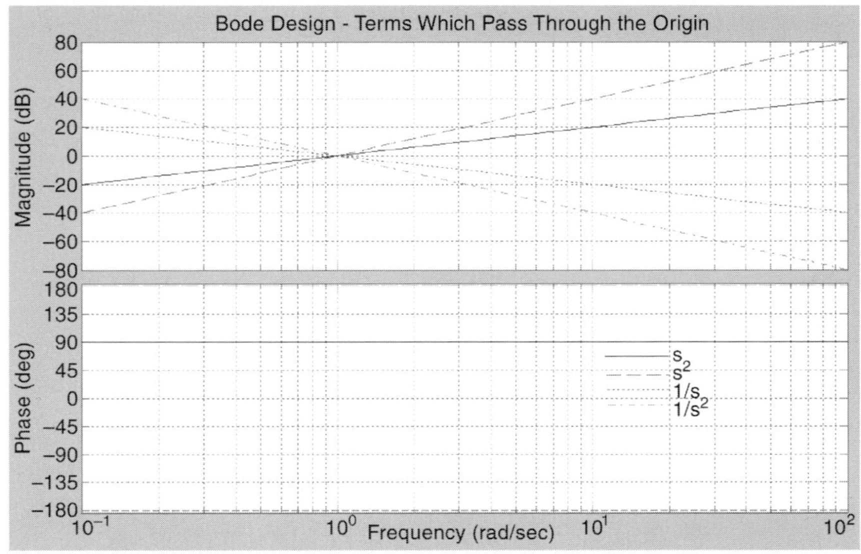

Fig. 10.21 Bode plot and asymptotic sketch for poles/zeros at the origin.

as for the pole or zero at the origin—the magnitude increases primarily because of ω at frequencies well past the break frequency. Therefore, the magnitude increases by an order of magnitude for every decade change in frequency. The corner frequency is usually abbreviated ω_c. The magnitude thus varies in relation to the corner frequency

$$\text{For } \omega \ll \omega_c, \text{Mag} = 0 \text{ dB and slope} = 0 \text{ dB/decade}$$
$$\text{For } \omega \gg \omega_c, \text{ slope} = \pm 20(n)\text{dB/decade} \tag{10.52}$$

If $\omega \approx \omega_c$, then an exact solution is necessary for the term in question (Example 10.11).

The asymptotic straight line approximation for a first-order term is quite simple. Below the corner frequency, the magnitude response is a straight line along the 0 dB line. At the corner frequency, the slope becomes $+/-20(n)$ dB/decade. Most of the error in the approximation is near the corner frequency. This is shown in the following log magnitude plot (Fig. 10.22) for the term $(1 + i\omega/2)$, where the corner frequency is 2 rad/s. The error at the corner frequency is approximately 3 dB, while the error at an octave (two times or one half the frequency) is approximately 1 dB. The error at 2 rad/s is 3 dB and the error at 1 rad/s and 4 rad/s is approximately 1 dB for a first-order pole or zero.

There is also a straight-line approximation for the phase associated with a first-order pole or zero. The actual phase is

$$\text{Angle}[1 + i\omega\tau]^{\pm n} = \pm(n)\tan^{-1}(\omega\tau) \tag{10.53}$$

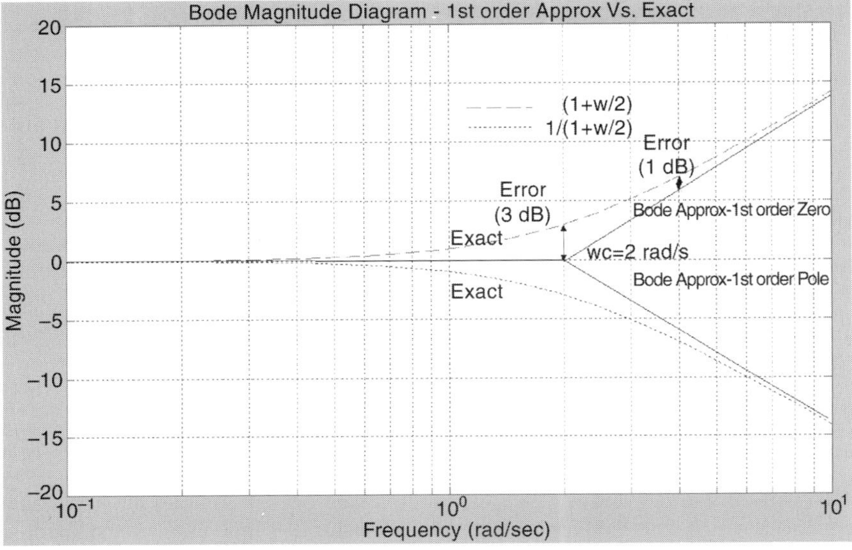

Fig. 10.22 Bode log magnitude plot—actual vs approximation.

The straight-line approximation for phase starts at a decade below the corner frequency at 0 deg and ends a decade above the corner frequency at $+/-(n)$ deg. The phase at the corner frequency is $+/-(n)(45)$ deg. This approximation is shown graphically in Fig. 10.23 for the first-order pole previously discussed with a corner frequency of 2 rad/s. This is summarized in Table 10.5. The poles have a negative phase contribution, and the zeros have a positive phase contribution.

10.2.3.7 Second-order complex poles and zeros. The last type of terms in Bode plots are second-order (or quadratic) poles and zeros. These are terms that are complex conjugate roots and can have much variation from the straight line asymptotic approximation. The primary accuracy consideration between the actual frequency response and the approximation is the damping. Light damping results in large frequency peaks at the resonant or natural frequency of the roots. The general form for second-order roots is

$$(s^2 + 2\zeta\omega_n s + \omega_n^2)^{\pm n} \Rightarrow \left(1 - \frac{\omega^2}{\omega_n^2} + i\frac{2\zeta\omega}{\omega_n}\right)^{\pm n} \tag{10.54}$$

The actual magnitude equation for second-order poles/zeros is

$$\log \text{Magnitude}\left(1 - \frac{\omega^2}{\omega_n^2} + i\frac{2\zeta\omega}{\omega_n}\right)^{\pm n} = \pm 20(n)\log\sqrt{\left[1 - \left(\frac{\omega}{\omega_n}\right)^2\right] + \left[2\zeta\left(\frac{\omega}{\omega_n}\right)\right]^2} \tag{10.55}$$

while the actual phase equation is

$$\text{Angle}\left(1 - \frac{\omega^2}{\omega_n^2} + i\frac{2\zeta\omega}{\omega_n}\right)^{\pm n} = \pm(n)\tan^{-1}\left[\frac{2\zeta\dfrac{\omega}{\omega_n}}{1 - \left(\dfrac{\omega}{\omega_n}\right)^2}\right] \tag{10.56}$$

Because there are two poles or zeros associated with second-order roots, the asymptotic magnitude slope is twice that of first-order poles/zeros. In a similar manner to first-order poles and zeros, for frequencies below the natural frequency (or corner frequency for first-order terms), the magnitude contribution is zero. After the natural frequency, the slope becomes $+/-(n)40$ dB/decade.

$$\begin{gathered}\text{For } \omega \ll \omega_n, \text{ Magnitude} = 0 \text{ dB and slope} = 0 \text{ dB/decade} \\ \text{For } \omega \gg \omega_n, \text{ slope} = +/-(n)40 \text{ dB/decade}\end{gathered} \tag{10.57}$$

As with first-order terms, the maximum error occurs around the resonant frequency, which corresponds to the damped natural frequency. As the damping decreases, the resonant peak occurs closer to the undamped natural frequency. For poles with damping greater than approximately 0.5, the actual

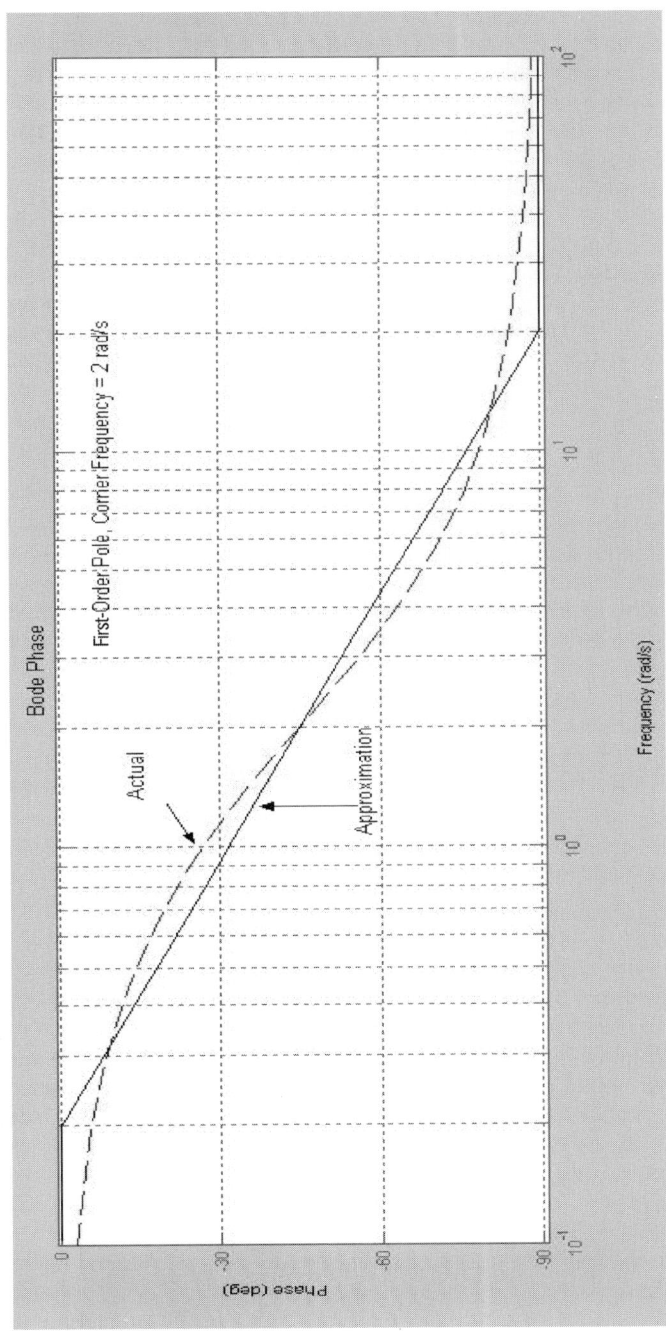

Fig. 10.23 Bode phase plot—actual vs approximation.

Table 10.5 First-order pole frequency approximation

$\frac{\omega}{\omega_c}$	0.1	1.0	10.0
Angle	0 deg	$\pm(n)45$ deg	$\pm(n)90$ deg

response falls below the approximation. For poles with damping below 0.5, the actual response lies above the approximation. For zeros, the exact opposite is true. This is shown in Fig. 10.24. Notice that for lightly damped systems, the approximation is very poor around the natural (or corner) frequency.

The phase approximation of second-order poles/zeros is similar to that of first-order poles/zeros in the sense that the phase contribution starts one decade below the natural frequency and ends one decade above. Half of the phase contribution occurs at the natural frequency. Because there are two poles/zeros involved, the overall phase contribution is +/− 180 deg instead of +/− 90 deg. Once again, the most dramatic variation from the approximation occurs for lightly damped poles/zeros. The phase variation for poles with varied damping is shown in Fig. 10.25. For second-order zeros, the plot would be inverted. The approximation for second-order poles/zeros is summarized in Table 10.6. For poles the phase angle is negative, and for zeros the phase angle is positive.

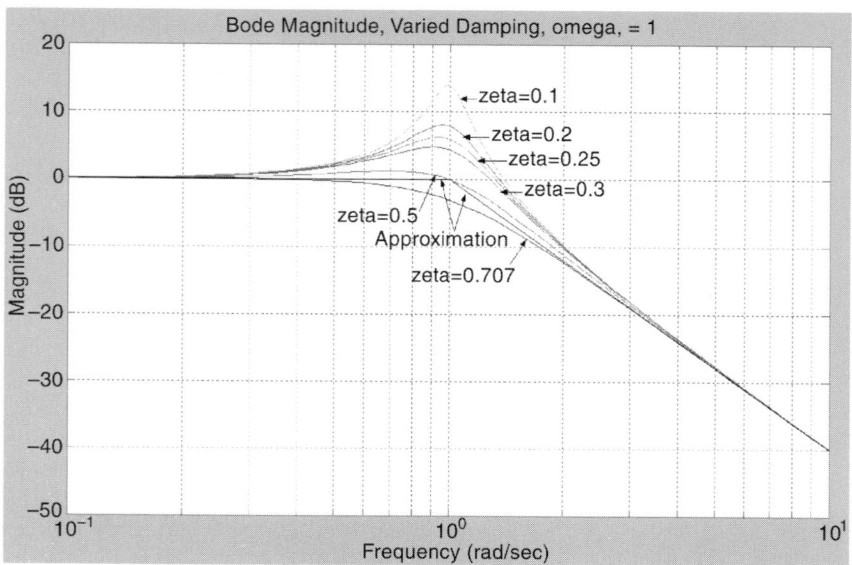

Fig. 10.24 Log magnitude curve for second-order poles.

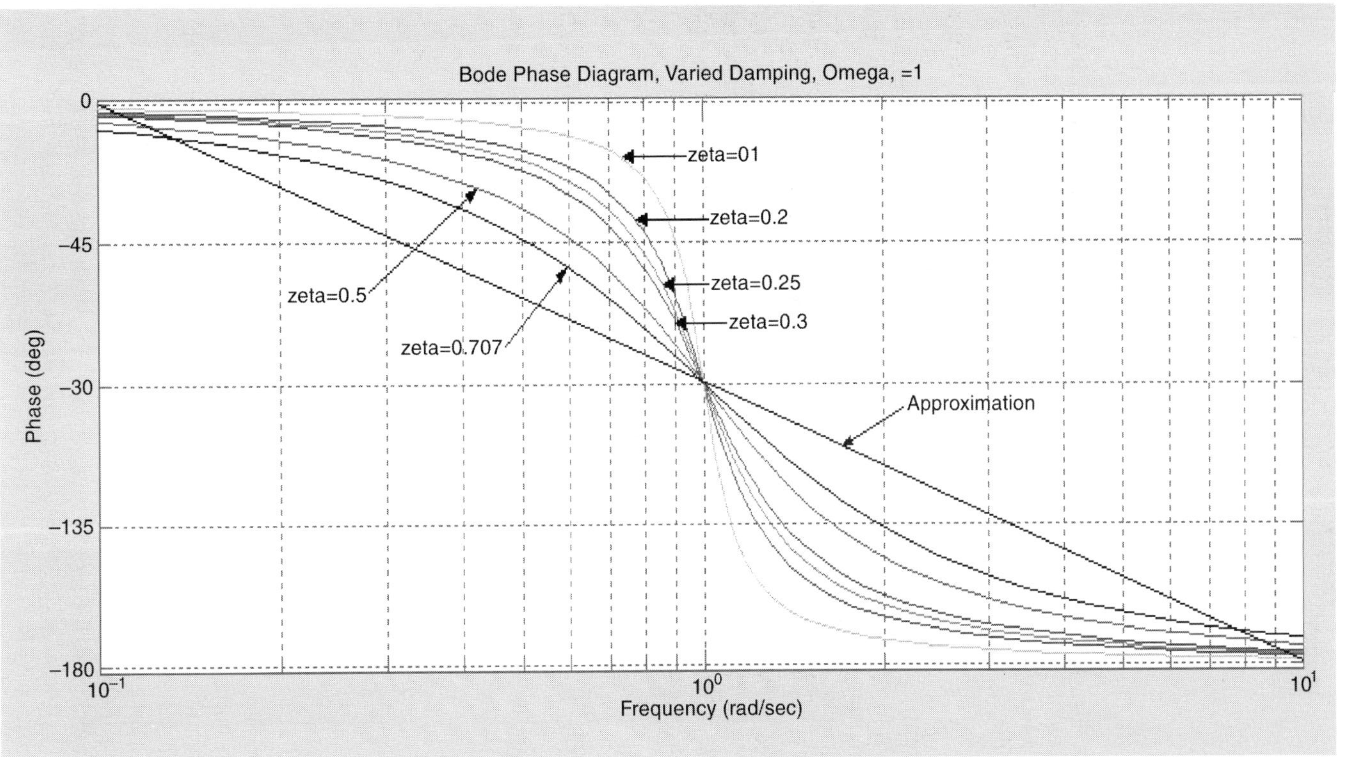

Fig. 10.25 Phase curve for second-order poles.

Table 10.6 Second-order pole frequency approximation

$\frac{\omega}{\omega_c}$	0.1	1.0	10.0
Angle	0 deg	$\pm(n)90$ deg	$\pm(n)180$ deg

10.2.3.8 Bode approximation example. It is easiest to understand how to construct a Bode plot approximation after looking at a complete example. The easiest way to construct an approximation is to plot each term individually, then add all the terms together. This is sometimes called superposition.

The procedure is relatively simple:

1) Put the transfer function into Bode form.
2) Draw the asymptotic approximations for each individual term in the transfer function.
3) Starting at the lowest frequency (left of the plot), add all the magnitudes at a specific frequency together to get the overall point. For example, if the Bode gain is 10 dB and an integrator is present, then at $\omega = 0.1$ rad/s the integrator contribution is -20 dB. Therefore, the overall magnitude at $\omega = 0.1$ rad/s is 10 dB $-$ 20 dB $= -10$ dB.

Do this for various frequencies, and the overall Bode approximation is obtained.

Example 10.12

Using the transfer function of Example 10.11, construct an asymptotic approximation for the Bode plot.

$$G(i\omega) = \frac{2.22\left(\frac{i\omega}{2}+1\right)\left(1-\frac{\omega^2}{4}+\frac{i\omega}{2}\right)}{(i\omega)\left(\frac{i\omega}{3}+1\right)\left(1-\frac{\omega^2}{6}+\frac{i2\omega}{3}\right)}$$

Before constructing the plot, note that the gain is 20 log 2.22 = 6.93 dB. Also, note the corner frequency for the first-order zero is at 1 rad/s and the natural frequency for the second-order zero is 2 rad/s. There is a first-order integrator pole and a first-order pole with a corner frequency of 3 rad/s. Finally, there is a second-order pole with a natural frequency of 2.45 rad/s. Figure 10.26 shows the comparison between the actual Bode magnitude plot and the asymptotic plot. As expected, the plot is very accurate far above and below the corner and natural frequencies. However, near the corner and natural frequencies is approximately 5 dB of error. Knowing the damping for the second-order poles/zero and the corrections for the first-order pole/zero, and accounting for these

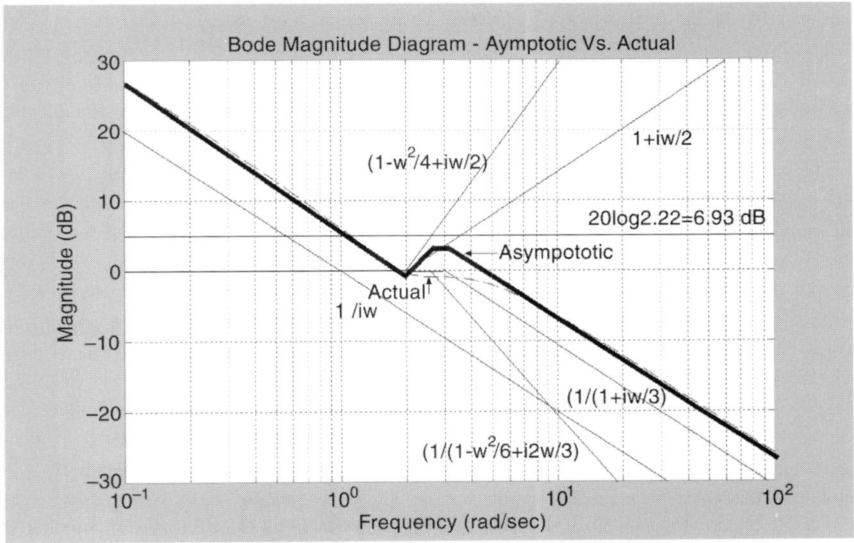

Fig. 10.26 Asymptotic vs actual Bode magnitude plot.

on the sketch would give much greater accuracy. The magnitude corrections for the second-order terms when the damping is known is shown in Fig. 10.24 and for the first-order terms in Fig. 10.22.

Similarly, the asymptotic phase plot vs the actual plot is shown in Fig. 10.27. The same trends as for the magnitude are noticed. Namely, the asymptotic plots are accurate far above and below the corner and natural frequencies but not as accurate near them. Once again, accounting for corrections in phase can improve the approximation sketch by using Fig. 10.25 for the second-order terms and Fig. 10.23 for the first-order terms.

10.2.4 Bode Stability

For controls practitioners, Bode plots can provide a very valuable insight into system stability. Chapter 8 discusses how closing the loop on an open-loop transfer function (OLTF) can provide varied closed-loop responses depending on the gain. As the root locus shows, the closed-loop poles move from the open-loop poles to the open-loop zeros as the gain increases. Thus, the OLTF has a large impact on the closed-loop response.

It is beneficial to look at a closed-loop block diagram in Fig. 10.28 before proceeding. Recall that the OLTF is $KG(s)H(s)$ and that the closed-loop transfer function (CLTF) is:

$$CLTF(s) = \frac{C(s)}{R(s)} = \frac{KG(s)}{1 + KG(s)H(s)} \tag{10.58}$$

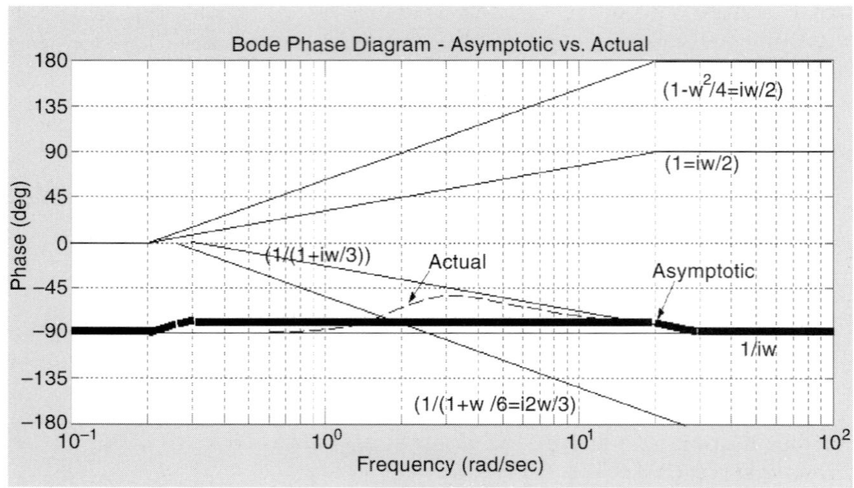

Fig. 10.27 Asymptotic vs actual Bode phase plot.

The closed-loop stability is determined from the characteristic equation of Eq. (10.58), or

$$1 + KG(s)H(s) = 0 \tag{10.59}$$

which can also be expressed as

$$KG(s)H(s) = -1 \tag{10.60}$$

From a Bode plot perspective, this means that OLTF Bode plot must have a log magnitude of less than 0 dB before the phase reaches -180 if the system is to be closed-loop stable. The magnitude of $KG(s)H(s) = 1$ and the phase $= -180$ deg at the point where the closed-loop system transitions from a stable response to an unstable response. This is the equivalent of the roots moving from the left-half plane on a root locus plot (stable) to the right-half plane (unstable). When the roots are on the imaginary axis, the response is neutrally stable. From examination of Fig. 10.28, it is clear that if the original

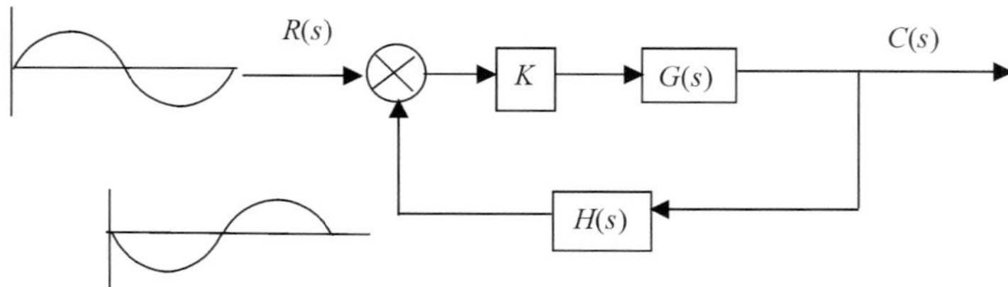

Fig. 10.28 Closed-loop block diagram.

signal is phase shifted 180 deg, then inverting this signal and adding it to the original signal will result in a doubling of the amplitude, which also leads to instability.

There are two measures of relative stability from the OLTF Bode plot. The first is the **gain margin** (GM), which is the additional gain in dB that can be added to the system before it goes unstable (i.e., reaches 0 dB at a phase angle of −180 deg). This is clearly analogous to the root locus limit of gain, which can be added before the closed-loop roots go into the right-hand plane. A stable system has a positive GM, while an unstable system has a negative GM. The GM may also be defined as the reciprocal of the gain at which the phase angle reaches −180 deg. This can be expressed mathematically as

$$GM = 20\log\frac{1}{[KGH(i\omega)]} \tag{10.61}$$

Another measure of relative system stability from an OLTF Bode plot is the **phase margin** (PM). This is the amount of phase shift or lag (in degrees) allowed before the system goes unstable (i.e., 0 dB at a phase angle of −180 deg). A stable system has a positive PM, while an unstable system has a negative PM. The PM is expressed mathematically as:

$$PM = 180\text{ deg} + \phi \tag{10.62}$$

where ϕ is the phase angle at the frequency where the magnitude crosses 0 dB.

When discussing Bode stability, two additional terms are used. The **gain crossover point** is the frequency where the magnitude curve crosses 0 dB. In addition, the **phase crossover point** is the frequency where the phase curve crosses −180 deg. The key Bode stability parameters are shown in Fig. 10.29.

One important note is that if a system will never go unstable using root locus analysis, then the same system will not go unstable using Bode analysis. It is therefore possible to have infinite gain margin (if the phase never crosses −180 deg) or infinite phase margin (if the gain never crosses 0 dB).

Example 10.13

Find the gain margin and phase margin for the following OLTF. Is the closed-loop system stable?

$$kGH(i\omega = \frac{1.25\left(\frac{i\omega}{2}+1\right)}{\left[(i\omega)\left(\frac{i\omega}{4}+1\right)\left(\frac{i\omega}{8}+1\right)\left(\frac{i\omega}{10}+1\right)\right]}$$

Figure 10.30 shows the Bode plot and the resulting Bode stability margins. The gain margin is 20.7 dB and the phase margin is 87.8 deg. Because both the gain margin and phase margin are positive for the Bode plot of the OLTF, the resulting closed-loop response is stable.

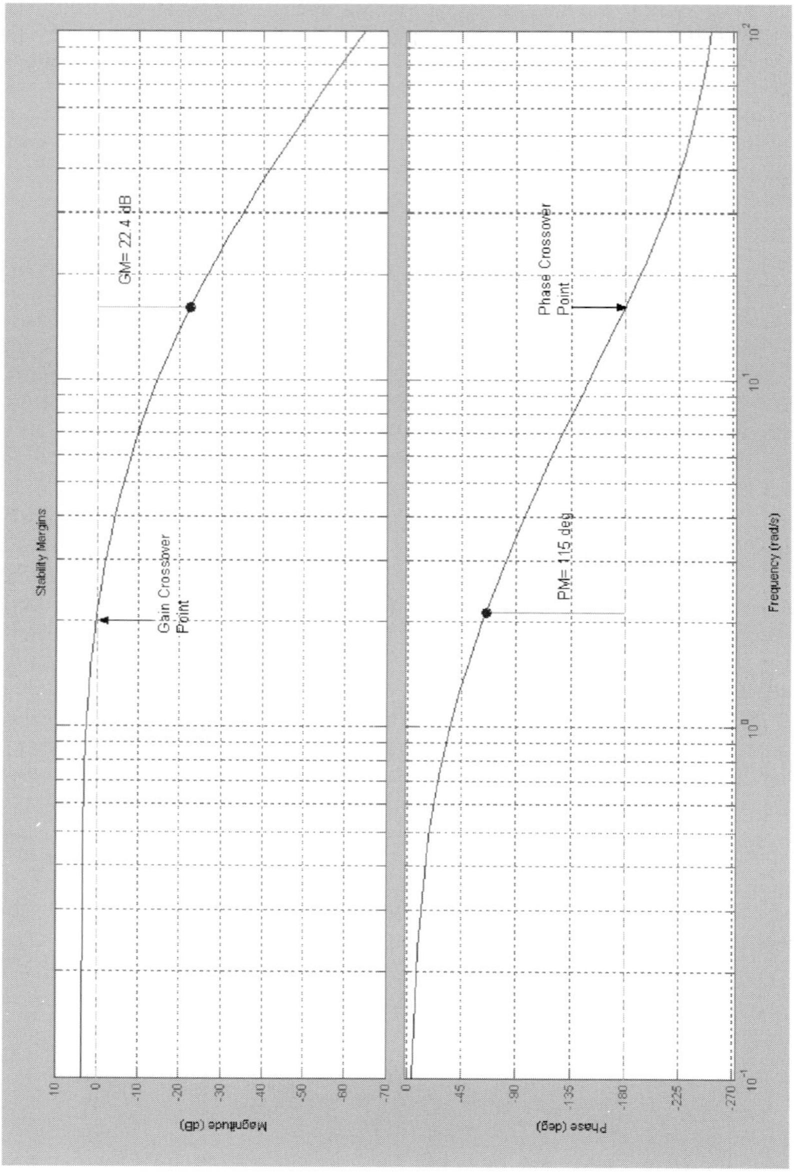

Fig. 10.29 Bode stability parameters.

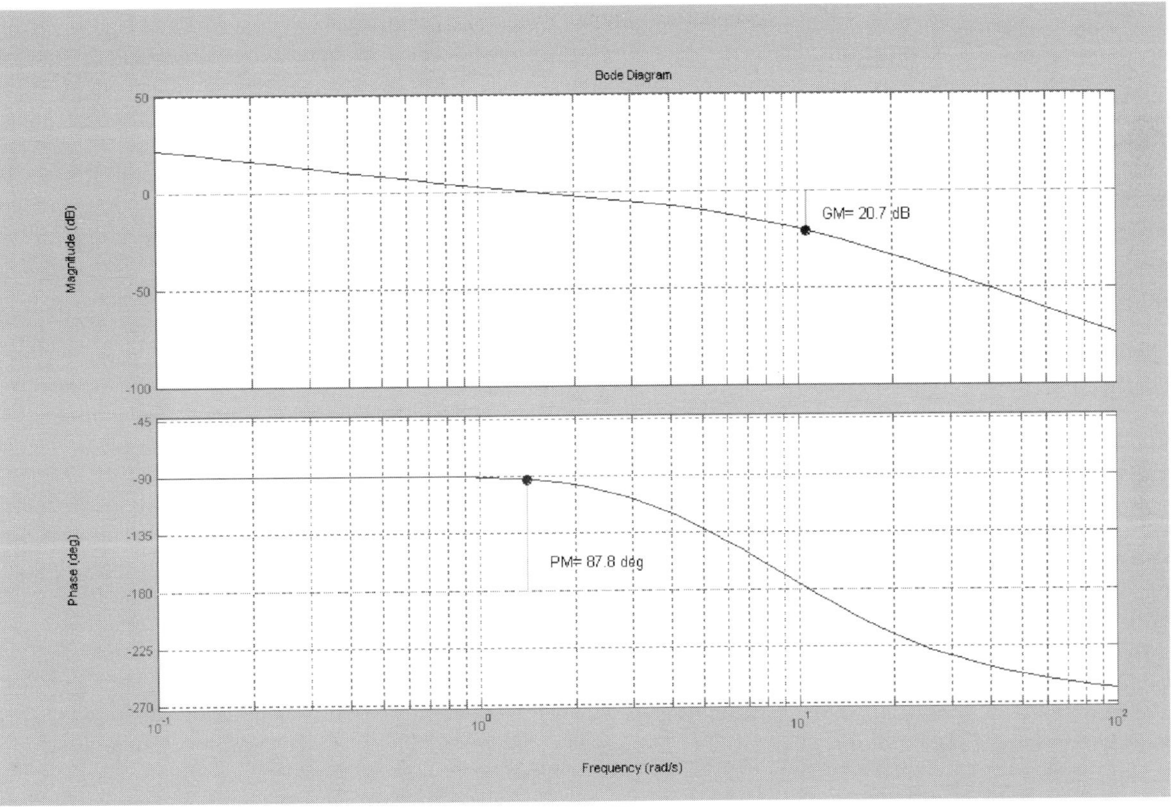

Fig. 10.30 Example 10.13 Bode plot.

10.2.5 Frequency Domain Specifications

Just as system performance can be discussed in terms of specifications in the time domain (as was done in Chapter 8), frequency domain specifications allow quantification of system performance. There are three primary frequency domain specifications—**bandwidth, resonant peak**, and **cutoff rate**.

There are several definitions for bandwidth, but the one used for this text is the frequency range when the log magnitude of the frequency response is above -3dB. This corresponds to when the output to input amplitude is 0.707, and frequency at which it occurs is called the cutoff frequency (ω_{co}). When the output magnitude drops below that threshold, the system is not as responsive. Thus, it does not have the ability to adequately track a signal whose frequency results in an output magnitude being less than 0.707 of the input. It is possible to have a bandwidth up to a certain frequency, as shown in Fig. 10.31, or between frequencies, as shown in Fig. 10.32.

Figure 10.31 also shows the resonant peak. This results from a system with second-order roots. If the roots are lightly damped, the resonant peak will be higher. The peak occurs at the natural frequency for a second-order root. Sometimes, the frequency at which the peak occurs is called the resonant frequency, which is shown as ω_r in Fig. 10.31.

Finally, the third specification is called the cutoff rate. This is the rate at which the magnitude decreases after the cutoff frequency, and is measured in dB/dec. It is directly related to the excess of poles over zeros because the slope $-20(n)$ dB/dec, where n is the number of excess poles in the transfer function. If a high cutoff rate is present, high-frequency noise is more readily attenuated, which can improve system performance.

10.2.6 Experimental Frequency Response Determination of System Transfer Functions

At times, real-world systems can be difficult to model mathematically. Fortunately, there is a convenient frequency response approach that allows experimental determination of the system transfer function. From a practical standpoint, frequency response can be determined by inputting a sinusoidal input at varying frequency into a system. The output magnitude and frequency, which will also be sinusoidal, are then measured. The relationship between the input and output sinusoid at each sinusoidal frequency is then compared to produce a magnitude and phase at each frequency. Then, just as for the analytical case when a mathematical model is present, a Bode plot can be constructed. Figure 10.33 shows a block diagram of this experimental setup.

Before proceeding, it is necessary to discuss the difference between a minimum phase system and a nonminimum phase system. A minimum phase system has no poles or zeros in the right-half plane, while a nonminimum phase system has at least one pole or zero in the right-half plane. This affects the phase of a system. If a system is known to be minimum phase, the system transfer function can be obtained from the magnitude plot alone. If it is not known in advance, then both magnitude and phase information are needed.

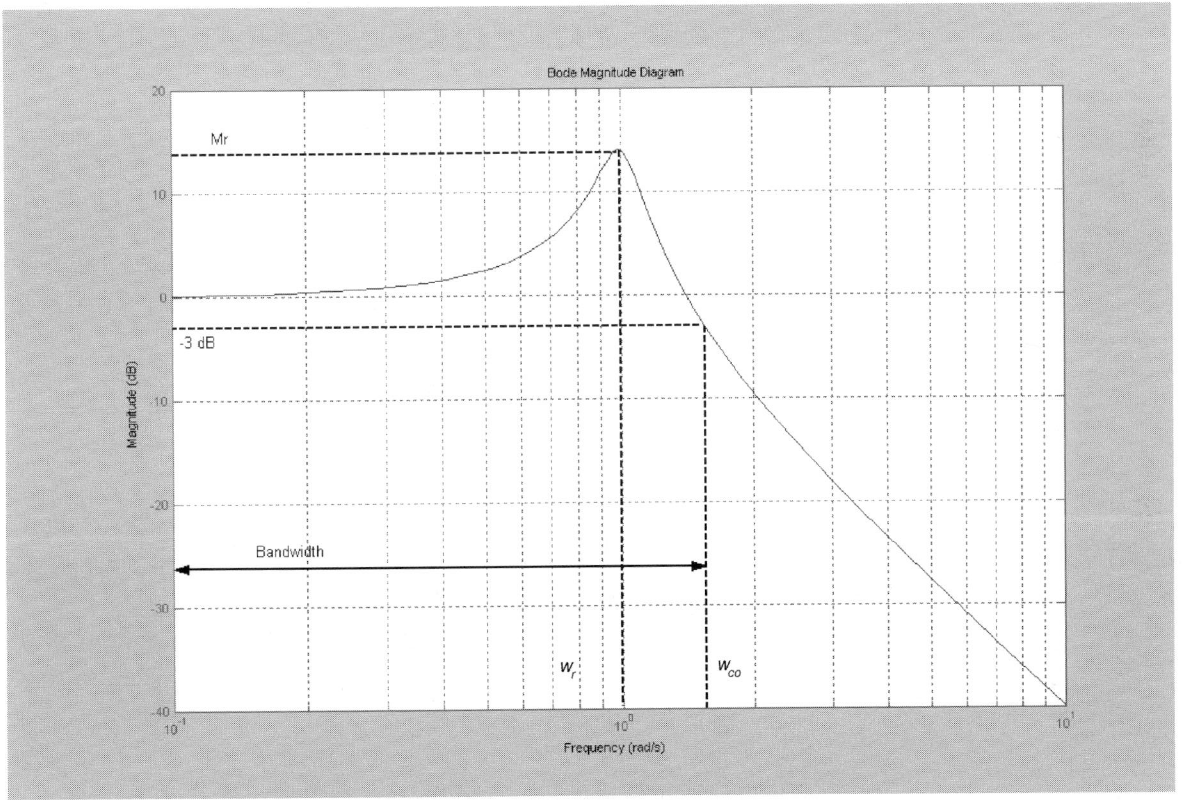

Fig. 10.31 Frequency domain characteristics—resonant peak.

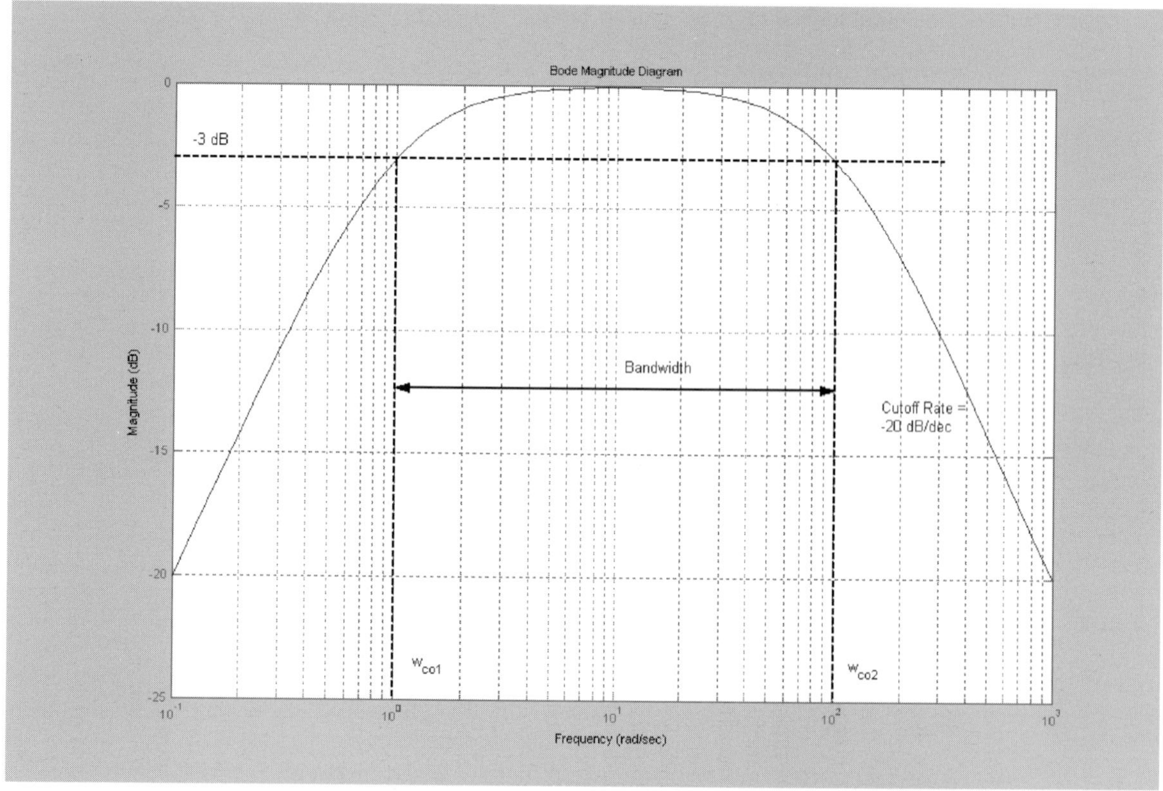

Fig. 10.32 Frequency domain characteristics—bandwidth range.

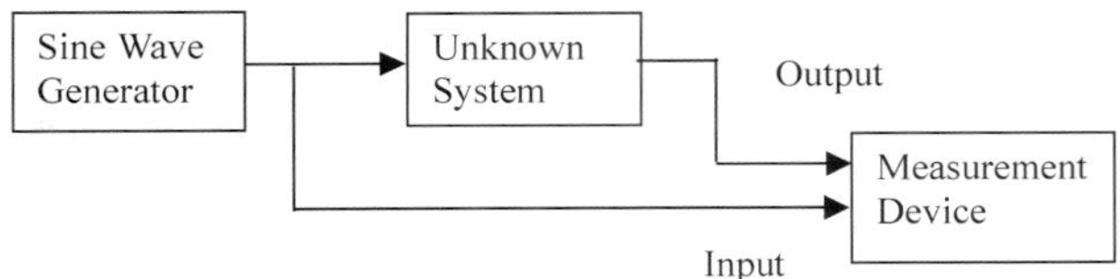

Fig. 10.33 Experimental frequency response setup.

After the Bode plot is formed from the sets of experimental magnitude and phase data at different frequencies, the system transfer can be obtained. The procedure is to fit asymptotic Bode approximation lines at the corner frequencies to determine pole/zero locations. One important fact to point out is that two real poles/zeros that are close together have a frequency response very similar to a second-order pole/zero pair with high damping (>0.707). Thus, the experimental method gives you only approximate system poles and zeros.

The general procedure for finding the system transfer function given an experimental frequency response is as follows:

1) Find all single poles (−20 dB/decade changes).
2) Find all single zeros (+20 dB/decade changes).
3) Find all double real poles (−40 dB/decade changes with no resonant peak).
4) Find all double real zeros (+40 dB/decade changes with no resonant undershoot).
5) Find complex pole pairs (−40 dB/decade changes with a resonant peak).
6) Find complex zero pairs (+40 dB/decade changes with a resonant undershoot).
7) Find value for the Bode gain K ($K = 1$ or 0 dB before any poles/zeros; If differentiators or integrators are present, look at the $\omega = 1$ point where both have values of 0 dB).

Example 10.14

Given the experimental frequency response shown in Fig. 10.34, determine the system transfer function. Assume the system is minimum phase.

The asymptotic Bode solution is shown in Fig. 10.35. Because the system has a slope of −20 dB/decade at low frequencies, a free integrator is present ($1/i\omega$). Because the magnitude for a pure integrator at $\omega = 0.1$ rad/s is normally 20 dB, it is clear that a Bode gain of 20 dB (that corresponds to a normal gain of 10) is present because the value of 40 dB at $\omega = 0.1$ rad/s. The slope decreases (becomes more positive) around 0.5 rad/s, which is indicative of a first-order zero $(1 + \omega/0.5)$. The slope starts to become more negative around 5 rad/s, which relates to a pole at $1/(1 + \omega/5)$. Finally, there is a resonant peak at 20 rad/s. The approximate size of the peak is 6 dB, which corresponds to a zeta of 0.2 from Fig. 10.24. Thus, the second-order pole has a natural frequency ω_n of 20 rad/s and a damping z of 0.2. This corresponds to

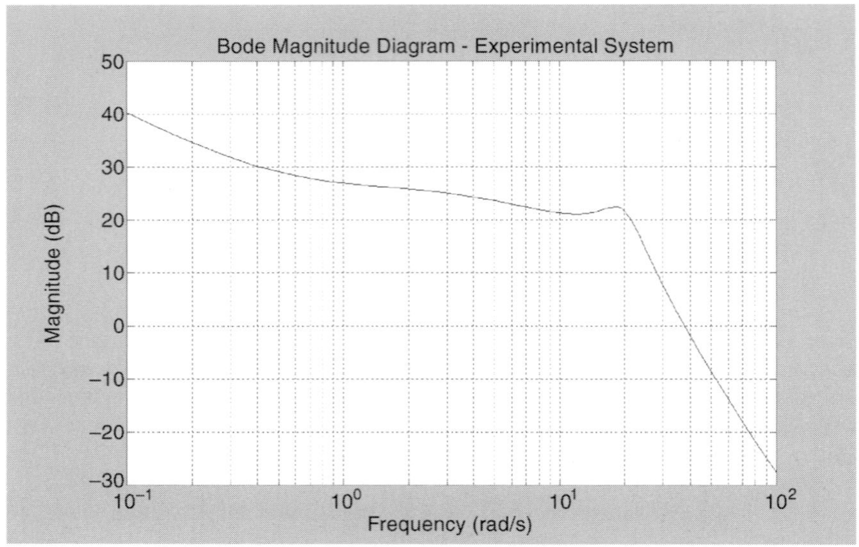

Fig. 10.34 Experimental frequency response.

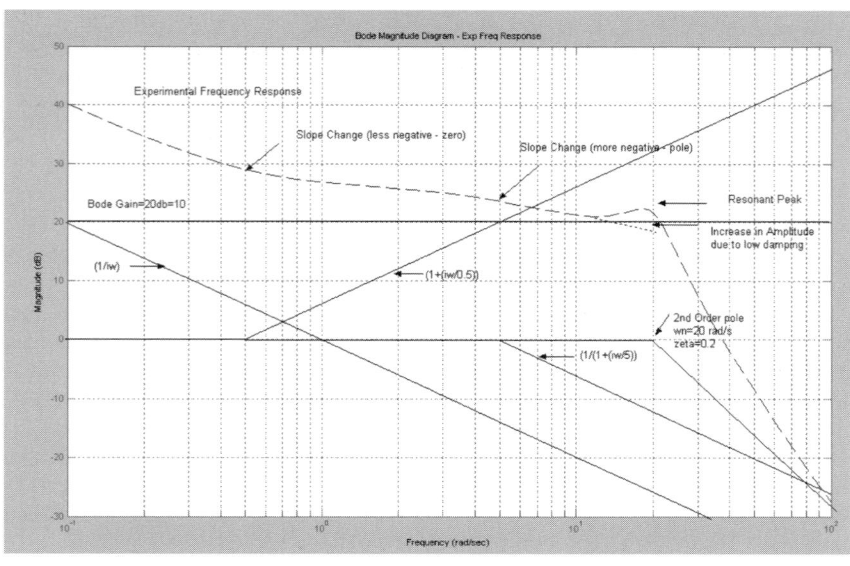

Fig. 10.35 Experimental frequency response solution.

a second-order pole of

$$\frac{1}{\left(1-\left(\frac{\omega}{\omega_n}\right)^2+i\frac{2\zeta\omega}{\omega_n}\right)}=\frac{1}{\left(1-\left(\frac{\omega}{20}\right)^2+i\frac{0.4\omega}{20}\right)}$$

The overall transfer function is the combination of all the individual parts.

$$G(i\omega)=\frac{10\left(1+\frac{i\omega}{0.5}\right)}{\left((i\omega)\left(1+\frac{i\omega}{5}\right)\left(1-\left(\frac{\omega}{20}\right)^2+i\frac{0.4\omega}{20}\right)\right)}$$

10.2.7 Frequency Compensation Devices

The goal of frequency compensators is to augment the performance of the response so that it falls within desired specifications. This can be done by placing a transfer function in various locations either inside or outside of the feedback loop. Figure 10.36 shows the three compensator locations—prefilter, forward path (cascade), and feedback. Note, there are parallels between the compensators described in Sec. 9.3 for s-domain root locus applications. In many control systems, the compensating device is an electrical circuit. Other forms of compensators may include mechanical, hydraulic, and pneumatic devices.

At each location, different types of compensators are used. The different types are high-pass filters, low-pass filters, or a combination of both called high-low pass or low-high pass filters. The high pass filters are lead compensators described in Sec. 9.3.1, the low-pass filters are the lag compensators described in Sec. 9.3.2, and the high-low pass filters are the lead-lag compensators described in Sec. 9.3.3. The following sections will describe each of the compensators in terms of the frequency domain as opposed to the root-locus domain.

10.2.7.1 High-pass filters. The purpose of a high-pass filter compensator is to attenuate low-frequency signals and amplify high-frequency signals. Attenuation occurs when the output is of lower amplitude than the input, and is the opposite of amplification. The net effect is that when a high-pass filter is used, the system response is quickened.

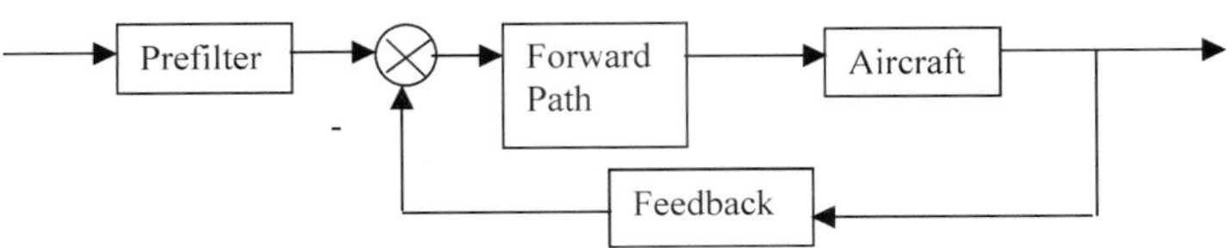

Fig. 10.36 Compensator locations.

10.2.7.2 Differentiator. The ideal high-pass filter is a pure differentiator, and has the form

$$G_c(s) = K_d s \tag{10.63}$$

or

$$G_c(\omega) = iK_d\omega$$

where K_d is the derivative gain.

Much insight can be gained from looking at the Bode plot for this pure differentiator, which is shown in Fig. 10.37. Notice that for frequencies less than 1 rad/s, the output is <0 dB, which means it is attenuated. For frequencies >1 rad/sec, the output is amplified. Thus, a differentiator is the simplest form of a high-pass filter. Also, notice the phase is always +90 deg, so it adds phase lead to a system. Unfortunately, pure differentiators have a serious drawback that limits their practical application—namely, amplification of high-frequency noise. Notice that at frequencies greater than 100 rad/s, the signal is amplified by 40 db. Said another way, the output is 100 times the original noise.

10.2.7.3 Proportional plus derivative filter. A more usable type of high-pass filter is a proportional plus derivative (PD) high-pass filter. The block diagram is shown in Fig. 10.38. In the case, the output of the signal is proportional to the magnitude and rate of the input signal.

This results in a transfer function of

$$G_c(s) = 1 + K_d s \tag{10.64}$$

or

$$G_c(\omega) = 1 + iK_d\omega = 1 + i\frac{\omega}{\omega_c}$$

where

$$\omega_c = \frac{1}{K_d} \tag{10.65}$$

Notice the Bode plot of the PD filter in Fig. 10.39 (with $K_d = 1$). Once again, notice the +90 deg phase contribution, or phase lead. The difference is that the phase is not added until a decade below the corner frequency defined by ω_c. In this case, the lower frequencies are not attenuated, but are passed through at the same amplitude. Higher frequencies are amplified after the corner frequency. The PD filter has the same problem of amplifying high-frequency signals, where noise usually occurs.

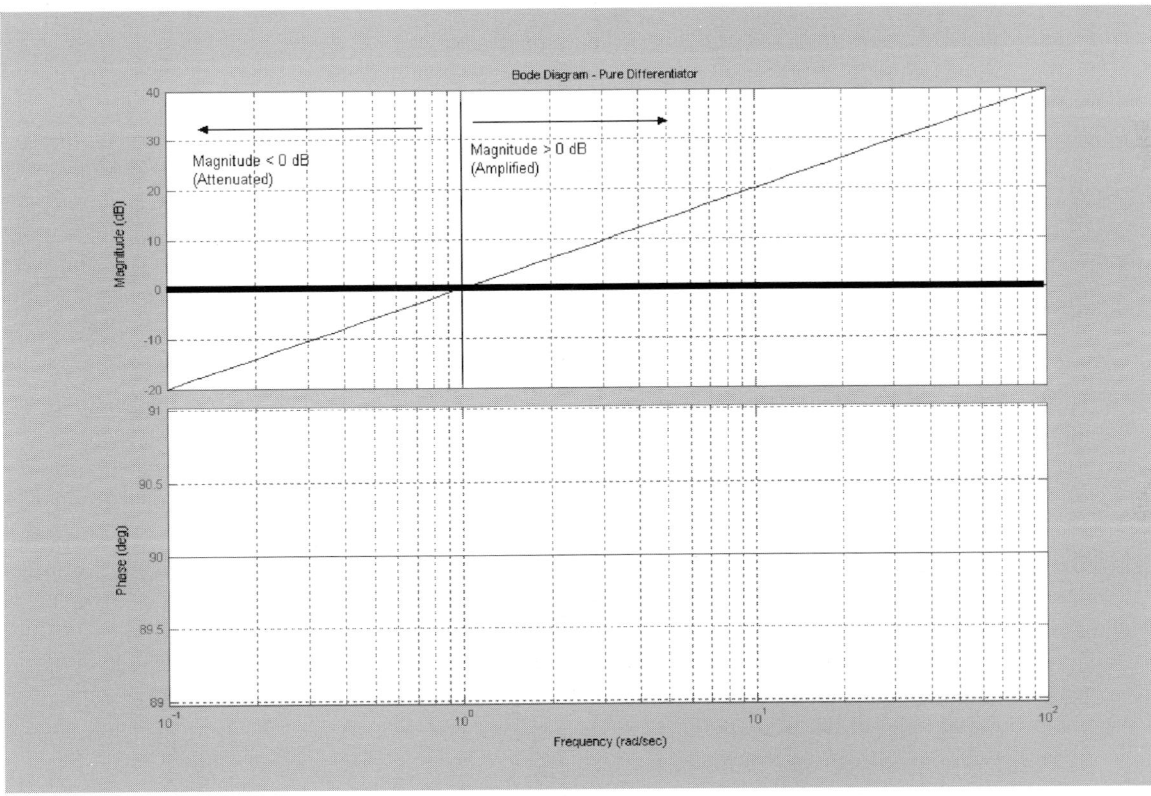

Fig. 10.37 Bode plot of an ideal high-pass filter (differentiator).

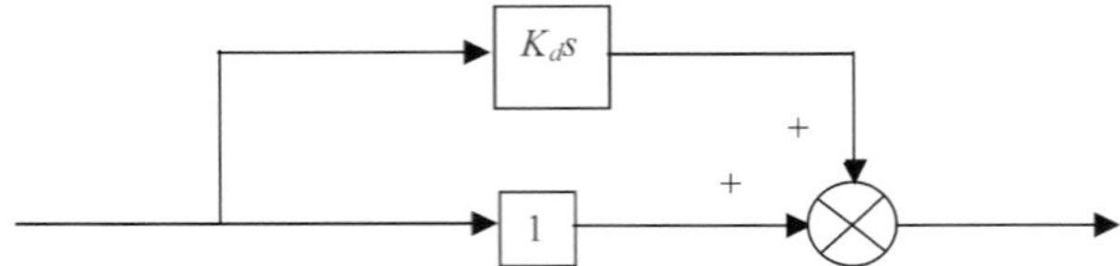

Fig. 10.38 Proportional plus derivative filter.

10.2.7.4 Lead compensator. Probably the most-used high pass filter is a lead compensator. This is defined as Eq. (9.4) in the previous chapter, but is now put in frequency domain form.

$$G_c(s) = \frac{b(s+a)}{a(s+b)} \tag{10.66}$$

or

$$G_c(\omega) = \frac{\left(\dfrac{i\omega}{a} + 1\right)}{\left(\dfrac{i\omega}{b} + 1\right)}; a < b$$

A Bode plot for a sample lead compensator of $G_c = 10(s+1)/(s+10)$ is shown in Fig. 10.40. Here, below frequency a (the zero in the filter) the signal is not amplified. Between a and b, the amplitude gradually increases at a rate of 20 dB/dec. After b, the amplitude is a positive gain that depends on the separation between a and b. Thus, high frequencies are amplified (passed) and low frequencies are not.

The phase starts at zero, and is positive a decade below a until a decade above b, when it returns to zero. The amount of phase depends on the ratio between a and b. The phase lead is added only around a specified region, which can be very useful when designing a compensator. Notice that a finite gain is applied to higher frequencies, which makes this filter much more realistic to implement physically.

Figure 10.41 is a useful design aid that shows how the gain and phase are affected by the ratio between a and b. One can control how much phase is added and how much the final gain is applied to high frequencies by controlling this ratio. This plot is generated by placing the zero (the a value) at 1 rad/s, and moving the pole (the b value) further away. Notice that for a ratio of 2, the final gain is approximately 6 dB, while for the ratio of 20 the final gain is approximately 26 dB. This makes sense because the variation is 20 dB/dec, therefore delaying the pole for a decade results in an additional gain of 20 dB.

The phase exhibits similar characteristics. When the ratio is 2, a maximum phase of approximately +20 deg is added. When the ratio reaches 20, approxi-

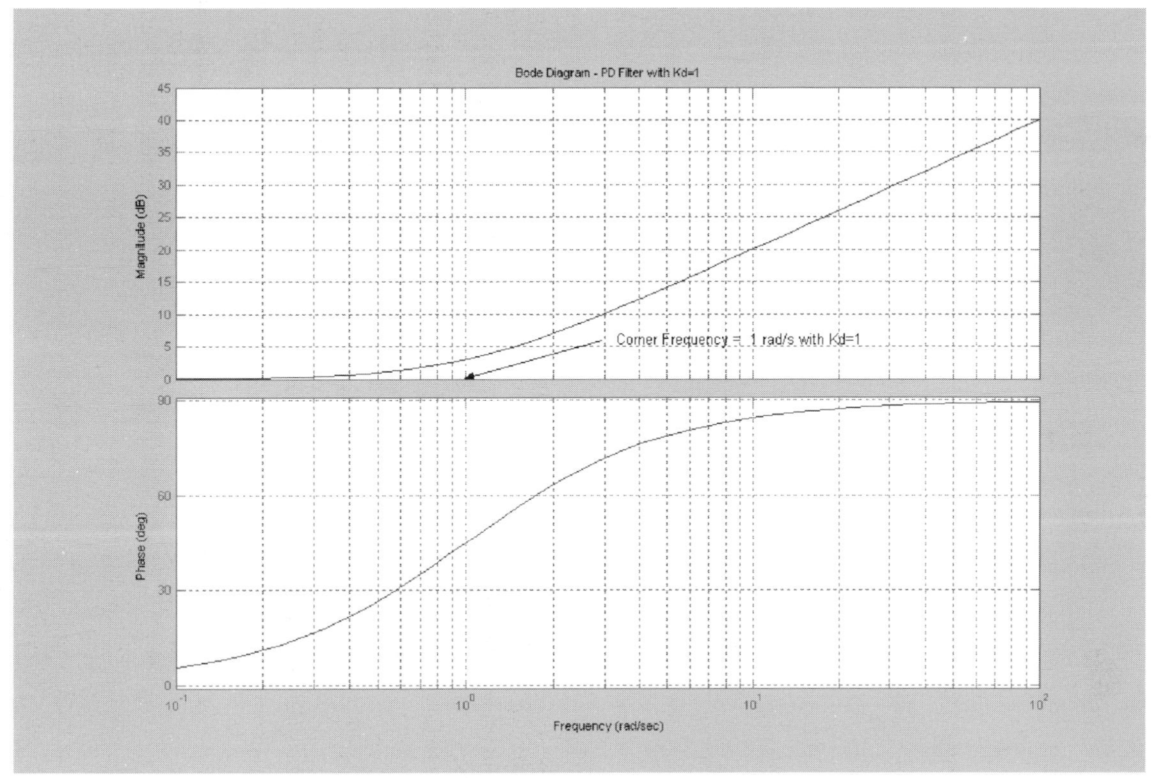

Fig. 10.39 Bode plot of PD filter with $K_d = 1$.

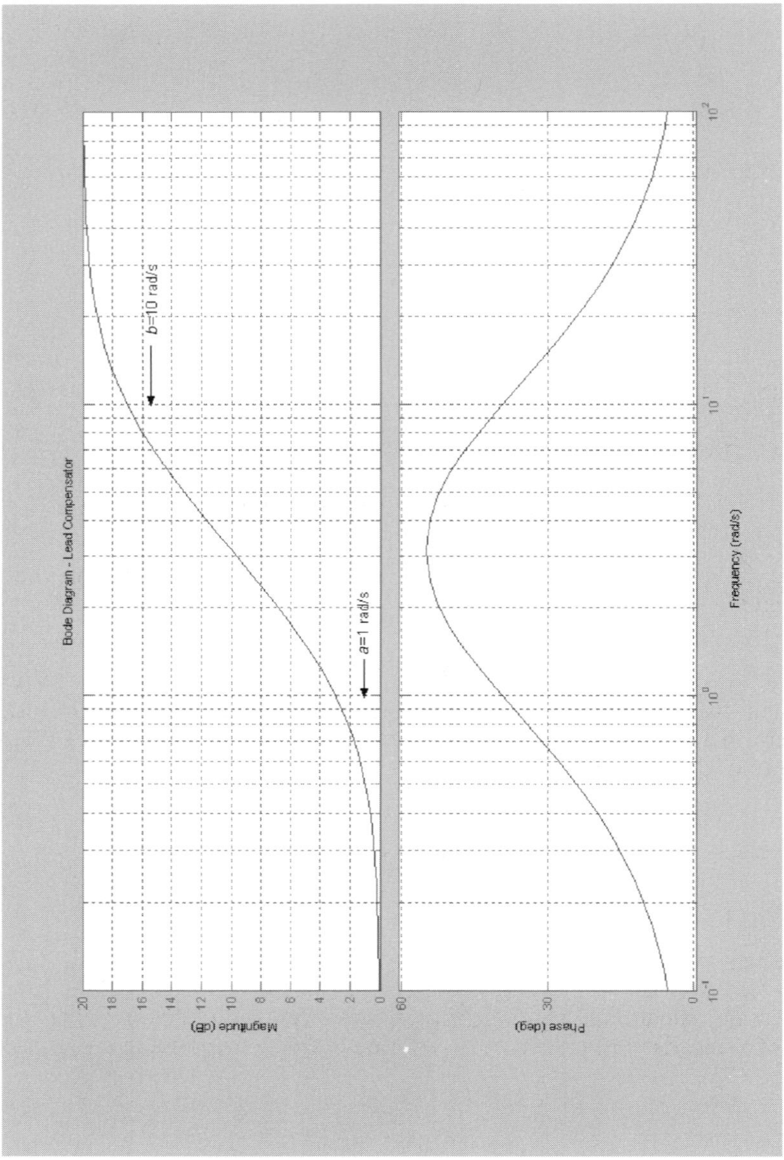

Fig. 10.40 Bode plot—lead compensator.

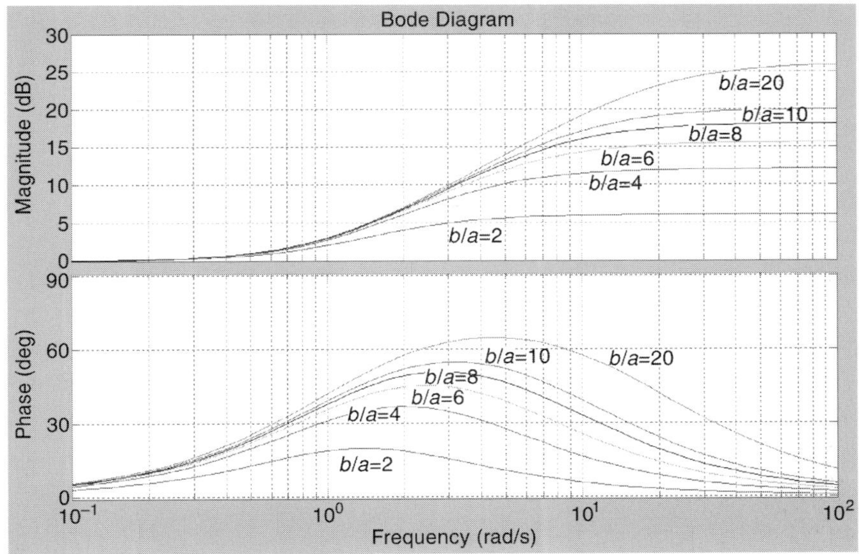

Fig. 10.41 Pole/zero ratio variation affect on lead compensators.

mately +65 deg of phase is added. The maximum phase contribution occurs between the pole and zero of the compensator.

Example 10.14

Using the transfer function of the root-locus lead compensator design problem in Example 9.2, design a prefilter lead compensator to speed up the response of the system.

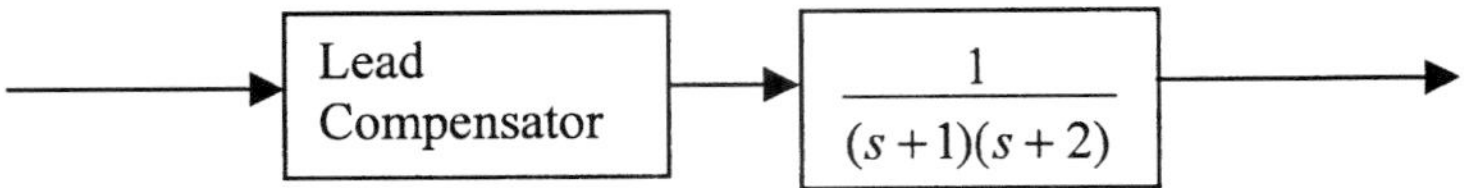

The key in designing a lead compensator in the frequency domain is to add phase lead near the first-break frequency for the basic system. Thus, a lead compensator should add phase at approximately 1 rad/s in this example. As a basis of comparison, use the lead compensator designed in Example 9.2.

$$G_c = \frac{2(s+5)}{(s+10)}$$

Figure 10.42 shows the Bode plot for the basic system, the lead compensator, and the combined system. Notice that the combined system has less phase lag because of the lead compensator, so the expected time response is faster.

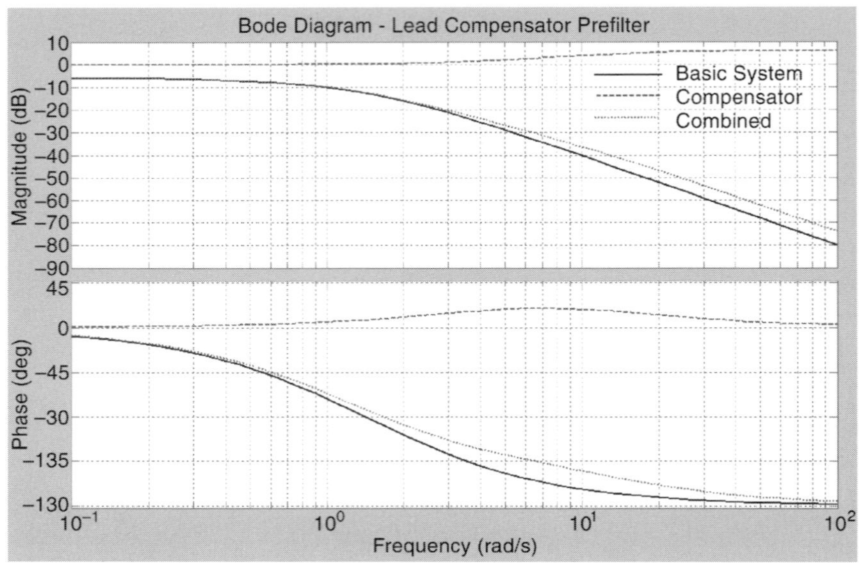

Fig. 10.42 Bode plot of basic, compensator, and combined systems.

Indeed, the time response shown in Fig. 10.43 shows an increase in the speed of response for the compensated system over the basic system.

10.2.7.5 Washout filter. Another type of high-pass filter is used commonly in aircraft stability augmentation systems—a washout filter. It is a special case of the lead compensator where the zero is actually a differentiator. It has the form

$$Gc(s) = \frac{K_{wo}s}{(s+b)} \tag{10.67}$$

or

$$Gc(\omega) = \frac{\left(\dfrac{K_{wo}}{b}\right)i\omega}{\left(1+\dfrac{i\omega}{b}\right)}$$

The Bode plot for a washout filter with a pole at 2 rad/s and a gain of 20 is shown in Fig. 10.44. Notice that low-frequency signals are attenuated, or washed out. Only changes in the input are passed through. This is valuable for aircraft feedback control because feeding back a parameter such as roll rate with a washout filter will not affect the steady-state roll rate. Without a washout filter, the stability augmentation system would constantly oppose the roll

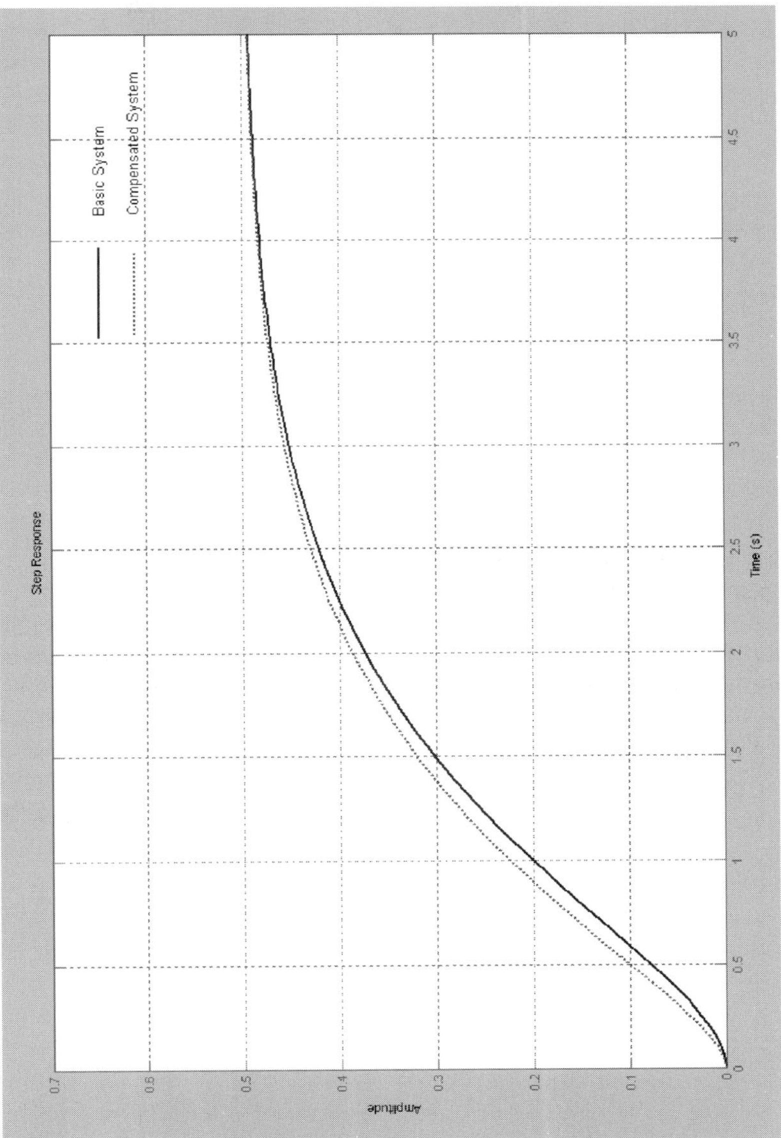

Fig. 10.43 Time response with and without lead compensator.

rate and decrease the aircraft performance. The gain for high frequencies is determined by the corner frequency and the washout filter gain K_{wo}. Additionally, the phase lead is added at lower frequencies.

10.2.7.6 Low-pass filters. A low-pass filter amplifies the lower-frequency signals while attenuating high-frequency signals. In addition, they have the added benefit of reducing steady-state error by increasing filter gain. They tend to add phase lag to a system, which slows down the response. Typically, they are used to eliminate noise, which is usually high frequency.

10.2.7.7 Integrator. The ideal low-pass compensator is an integrator, and has the form

$$G_c(s) = \frac{K_i}{s} \tag{10.68}$$

or

$$G_c(\omega) = \frac{K_i}{i\omega}$$

where K_i is the integral gain.

The Bode plot of a pure integrator $1/s$ is shown in Fig. 10.45. Notice that low frequencies are amplified (passed), while high frequencies are attenuated. The use of pure integral control has the disadvantage of excessive lag and no direct path between a pilot's command and the control surface. In addition, note that the phase is always −90 deg, which is a phase lag. This tends to slow down the response.

10.2.7.8 Proportional plus integral filter. Most applications of integral control involve a direct linkage of the control input to the output. This is known as a proportional plus integral (PI) filter, as shown in Fig. 10.46. The resulting transfer function is

$$G_c(s) = \frac{(s + \omega_c)}{s} \tag{10.69}$$

or

$$G_c(\omega) = \frac{\omega_c\left(1 + \dfrac{\omega}{\omega_c}\right)}{s}$$

where

$$\omega_c = \frac{1}{K_i} \tag{10.70}$$

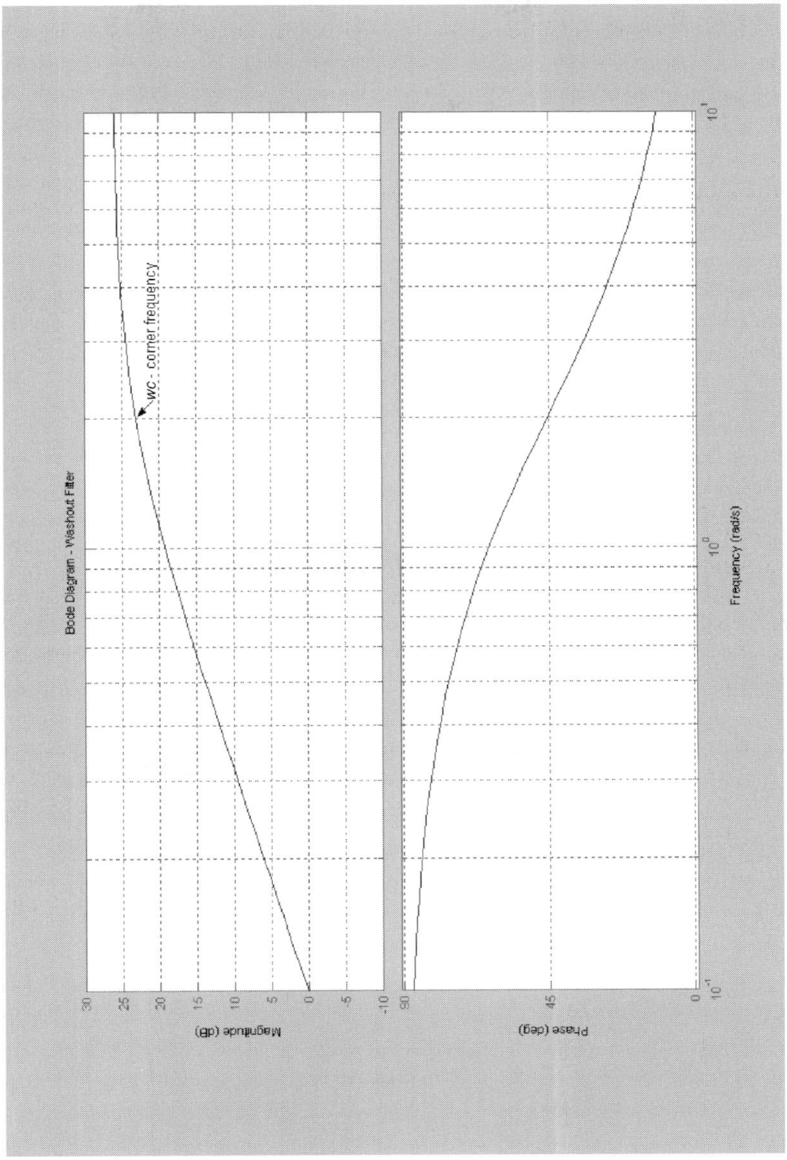

Fig. 10.44 Washout filter Bode plot.

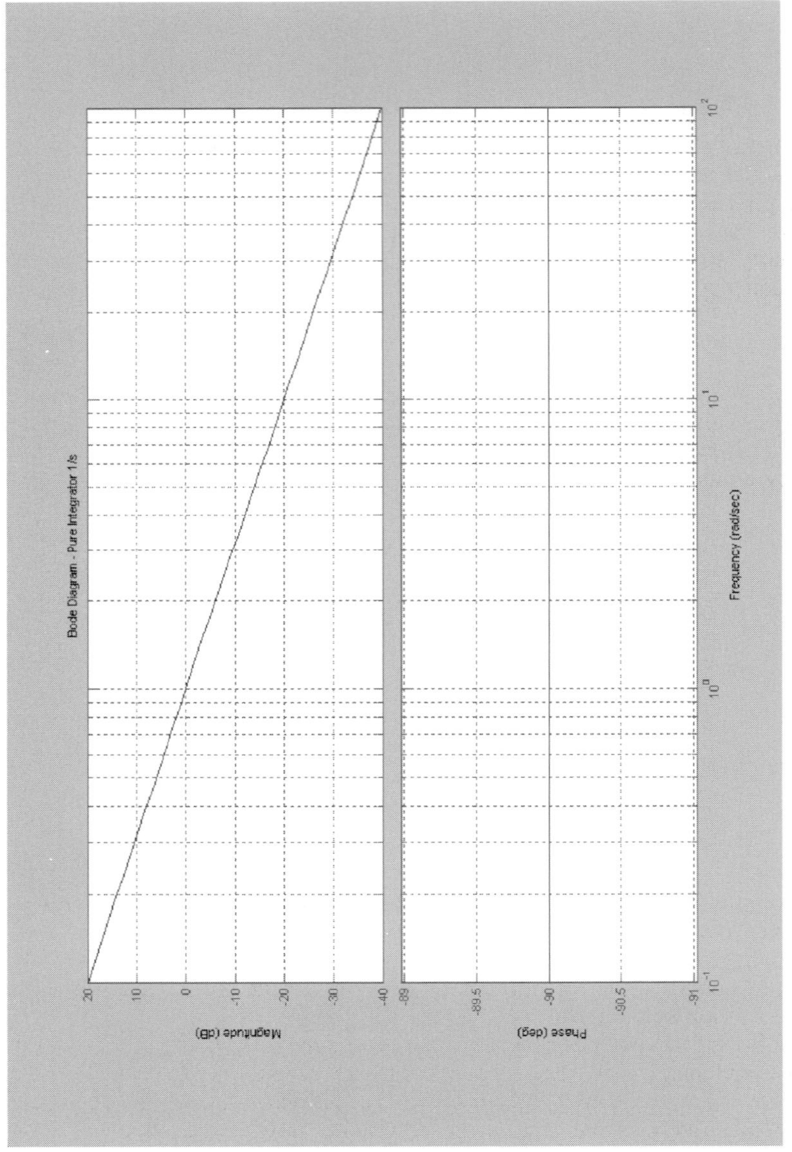

Fig. 10.45 Bode plot for a pure integrator (1/s).

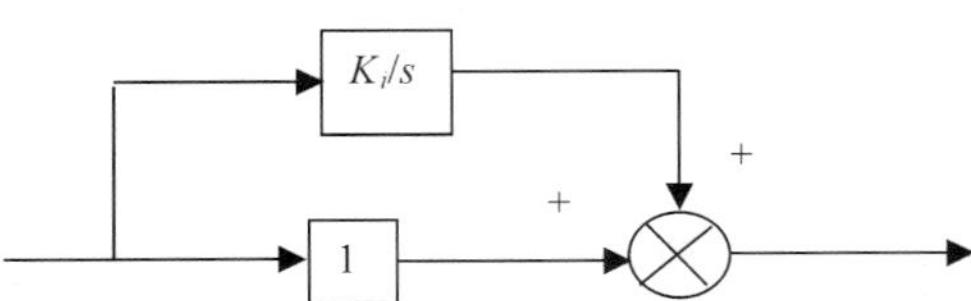

Fig. 10.46 PI filter.

The Bode plot for a PI filter or controller with $K_i = 1$ is shown in Fig. 10.47. Notice that at lower frequencies the amplitude is still positive. As the frequency moves past the cutoff frequency (ω_c), the gain becomes 0 dB (or 1). Thus, the lower frequencies are passed with a higher amplitude. Also, the phase starts at -90 deg and ends up at almost 0 deg at a decade above the break frequency. Thus, the overall contribution is a negative phase, or a phase lag.

10.2.7.9 Lag compensator. The most frequently used low-pass filter is a lag compensator. The lag compensator was described in terms of the s domain in Sec. 9.3.2. However, the lag compensator will now be defined in terms of the frequency domain.

$$G_c(s) = \frac{b(s + a)}{a(s + b)} \tag{10.71}$$

or

$$G_c(\omega) = \frac{\left(1 + \dfrac{i\omega}{a}\right)}{\left(1 + \dfrac{i\omega}{b}\right)}; a > b$$

A Bode plot for a lag compensator of $G_c = 0.1(s + 10)/(s + 1)$ is shown in Fig. 10.48. Because the pole occurs before the zero in a lag compensator, negative phase is initially added, and magnitude loss occurs. As was the case with a lead compensator, the amount of phase and magnitude change depends on the separation between the pole and zero.

As with the lead compensator, it is useful to provide a figure to show how the phase and magnitude vary with the ratio of separation between the pole and the zero. Figure 10.49 shows the variation of phase and magnitude for a pole at $\omega = 1$ rad/s as the zero is moved further away. Remember, for a lag compensator the pole always occurs before the zero. The same trend is noticed for lag compensators as for lead compensators. Also, because the magnitude drops by 20 db/decade, each decade away from the pole that the zero is placed results in a magnitude decrease of 20 dB.

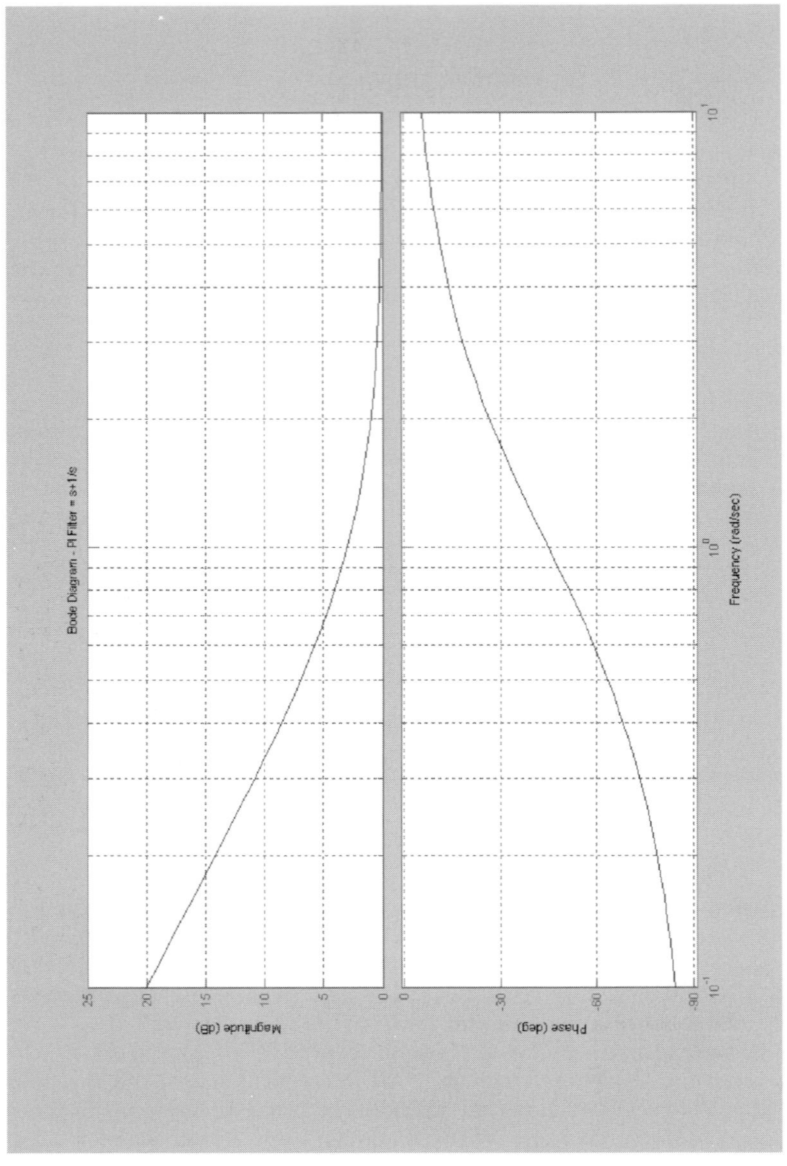

Fig. 10.47 Bode plot of a PI filter $(s+1)/s$.

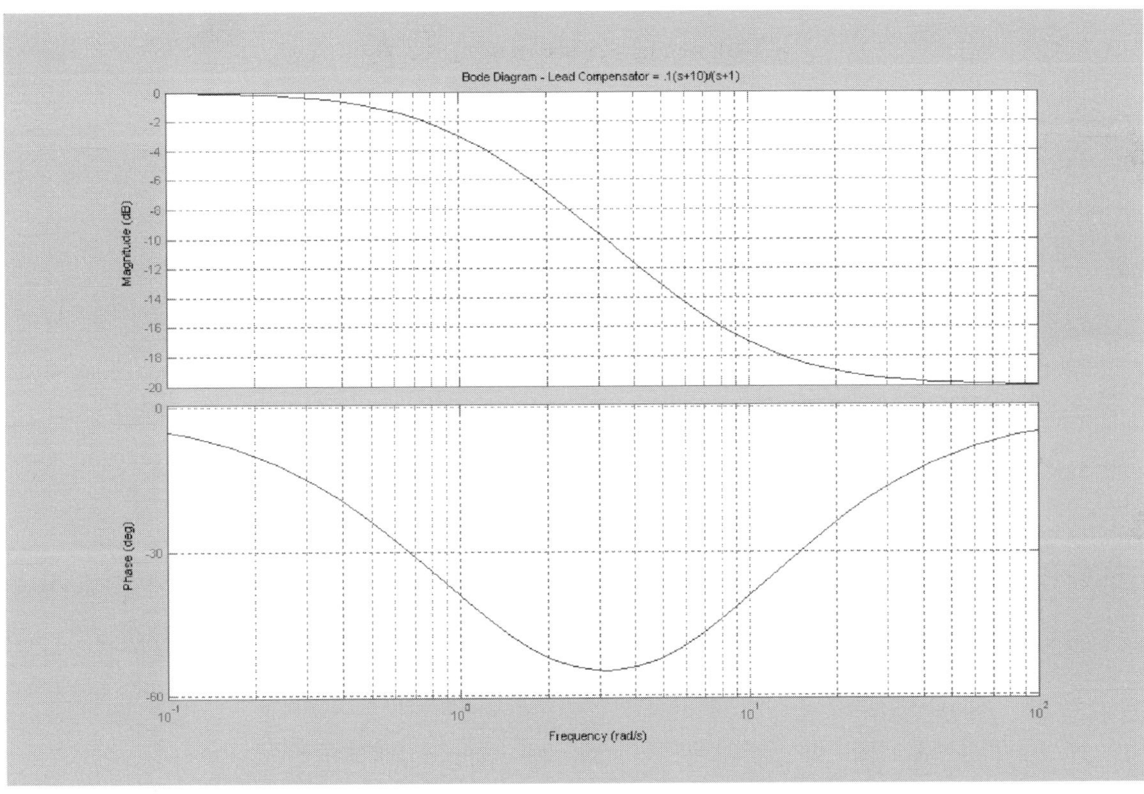

Fig. 10.48 Bode plot of a lag compensator—$G_c(s)$−0.1(s+10)/s + 1).

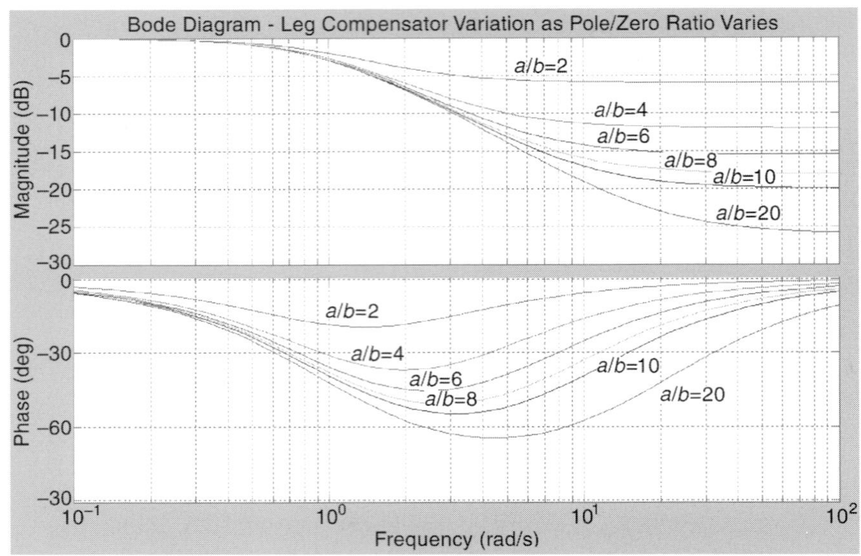

Fig. 10.49 Pole/zero variation affect on lag compensators.

Example 10.15

Using the same transfer function as Example 10.14 and Example 9.3, design a prefilter lag compensator to slow down the response of the basic system.

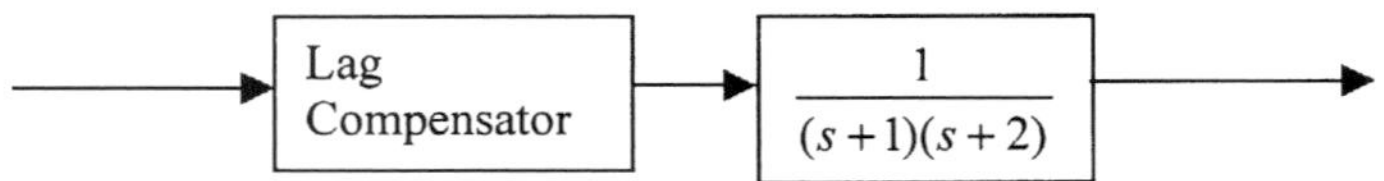

As with the lead compensator design, the process starts with an examination of the basic system Bode plot. In this case, because the desired objective is to slow down the response, phase lag is needed around the first-break frequency for the system. The lag compensator used in Example 9.3 will be used for comparison purposes.

$$G_c(s) = \frac{3(s+10)}{10(s+3)}$$

The Bode plot of the basic system, lag compensator, and combined system is shown in Fig. 10.50. Notice that in this case there is more phase lag in the combined system, therefore, the expected response is slower. The time response, shown in Fig. 10.51, validates the assertion that the time response is slower.

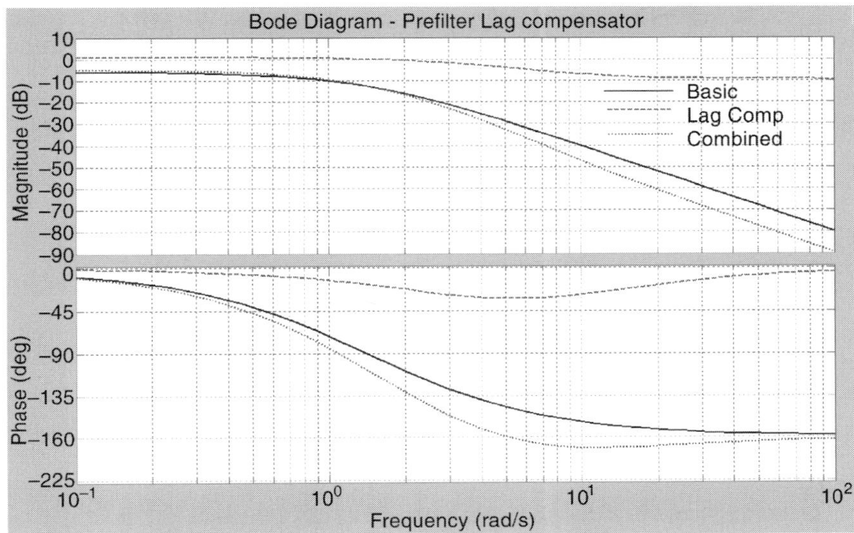

Fig. 10.50 Bode plot for basic, lag compensator, and combined systems.

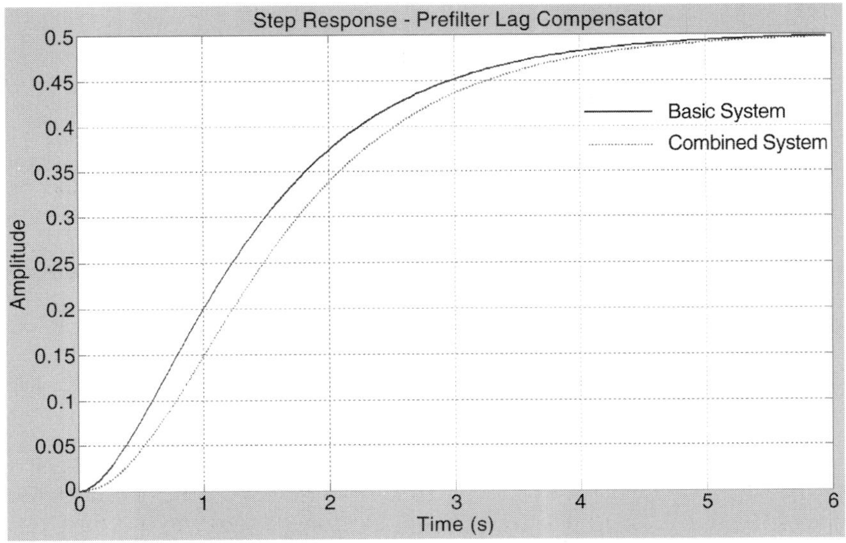

Fig. 10.51 Step response for basic system and lag compensated system.

10.2.7.10 Noise filter. The final type of low-pass filter discussed is a noise filter. This is a special case of the lag compensator where there is no zero term. Thus, the amplitude for low frequencies is passed until the corner frequency, when the magnitude changes at a rate of -20 dB/decade. This allows high frequencies, where noise is usually present, to be attenuated while the frequencies of interest are passed. A noise filter has the following transfer function:

$$G_c(s) = \frac{K}{(s+b)} \tag{10.72}$$

or

$$G_c(i\omega) = \frac{\frac{K}{b}}{\left(1 + \frac{i\omega}{b}\right)}$$

The corner frequency is located at b rad/s. The placement of the pole determines which frequencies are passed and which are attenuated. Figure 10.52 shows a Bode plot of a noise filter with a pole at 10 rad/s and a value of $K = 10$.

10.2.7.11 High–low-pass filters. It is possible to use the advantages of both high-pass and low-pass filters in the same filter, called a high–low-pass filter. Low-pass filters can produce an increase in gain, which reduces steady-state error. High-pass filters can quicken the response, thus increasing ω_n. The resulting transfer function, which combines both filters into one, is

$$Gc(s) = K\frac{(s+a)(s+c)}{(s+b)(s+d)} \tag{10.73}$$

or

$$G_c(i\omega) = \frac{K\frac{ac}{bd}\left(1 + \frac{i\omega}{a}\right)\left(1 + \frac{i\omega}{c}\right)}{\left(1 + \frac{i\omega}{b}\right)\left(1 + \frac{i\omega}{d}\right)}$$

where

$$a > b, a < c, c < d$$

This is best observed by looking at a pole zero map shown in Fig. 10.53. Typically, the low-pass filter is closer to the origin and the high-pass filter is further out.

The frequency response of a typical high–low-pass filter is shown in the Bode plot of Fig. 10.54. In the case, a high–low-pass filter $Gc(s) =$

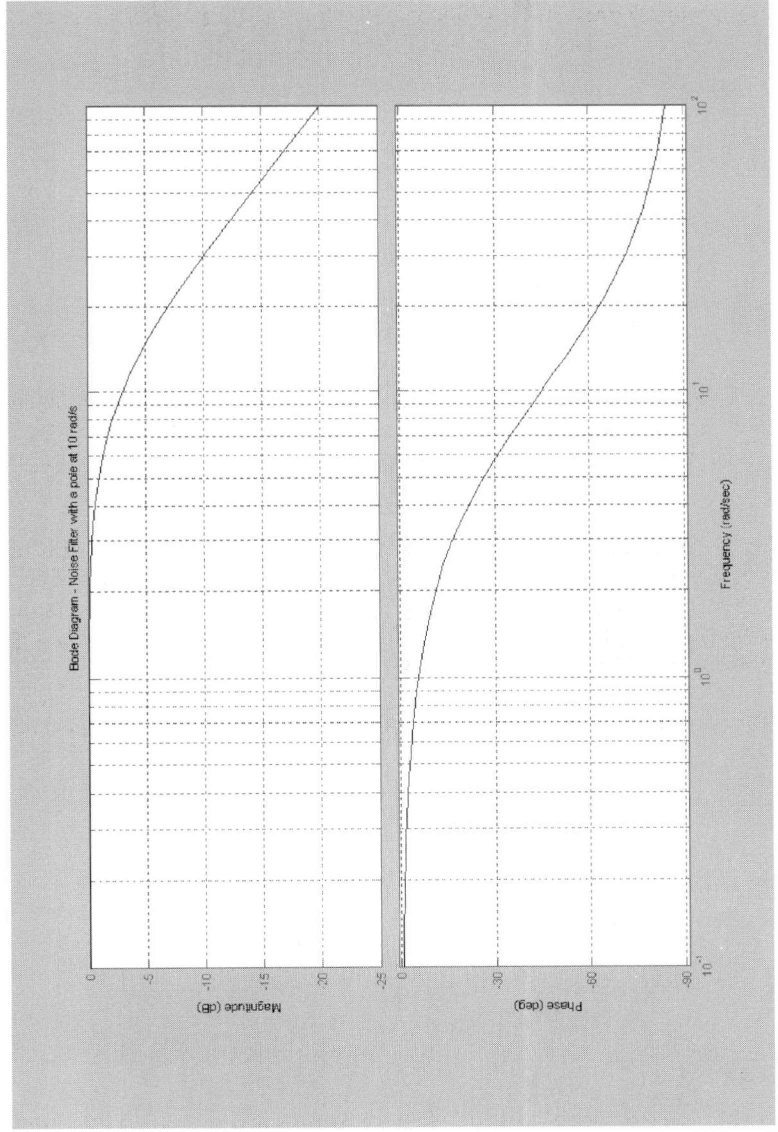

Fig. 10.52 Bode plot of a noise filter with a pole at 10 rad/s.

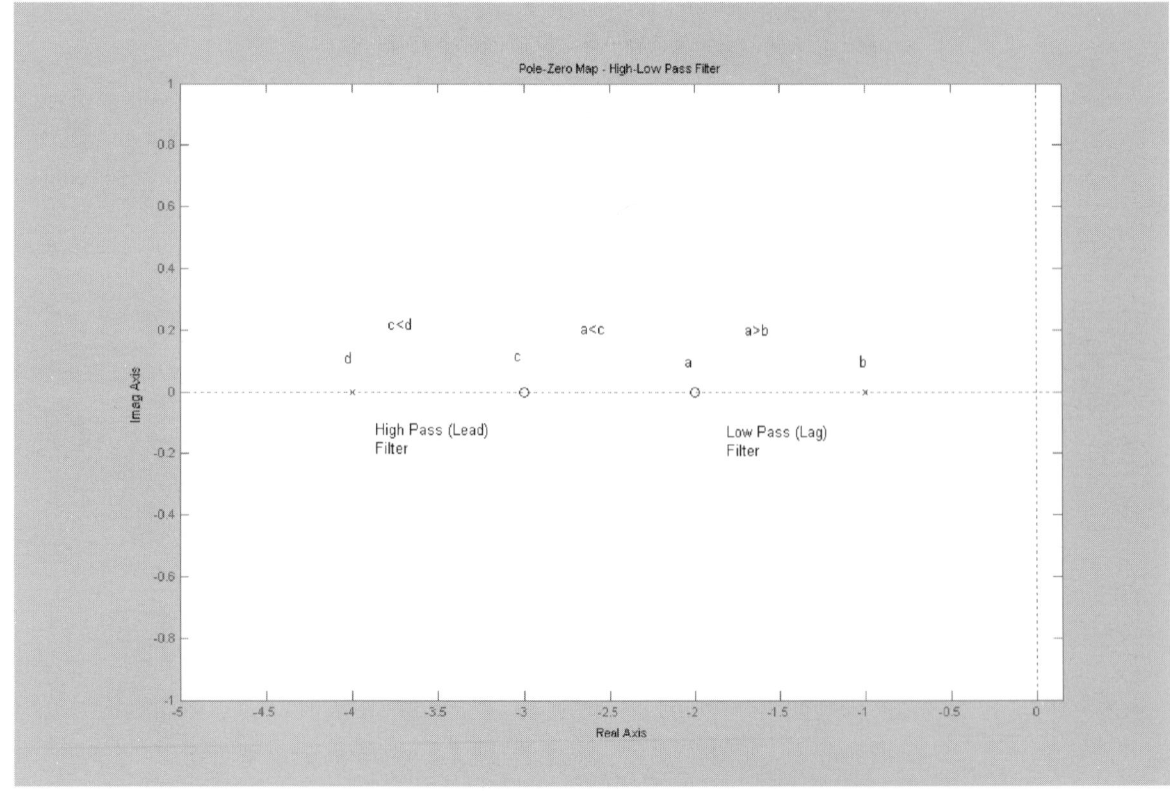

Fig. 10.53 Pole-zero map of a high–low-pass filter.

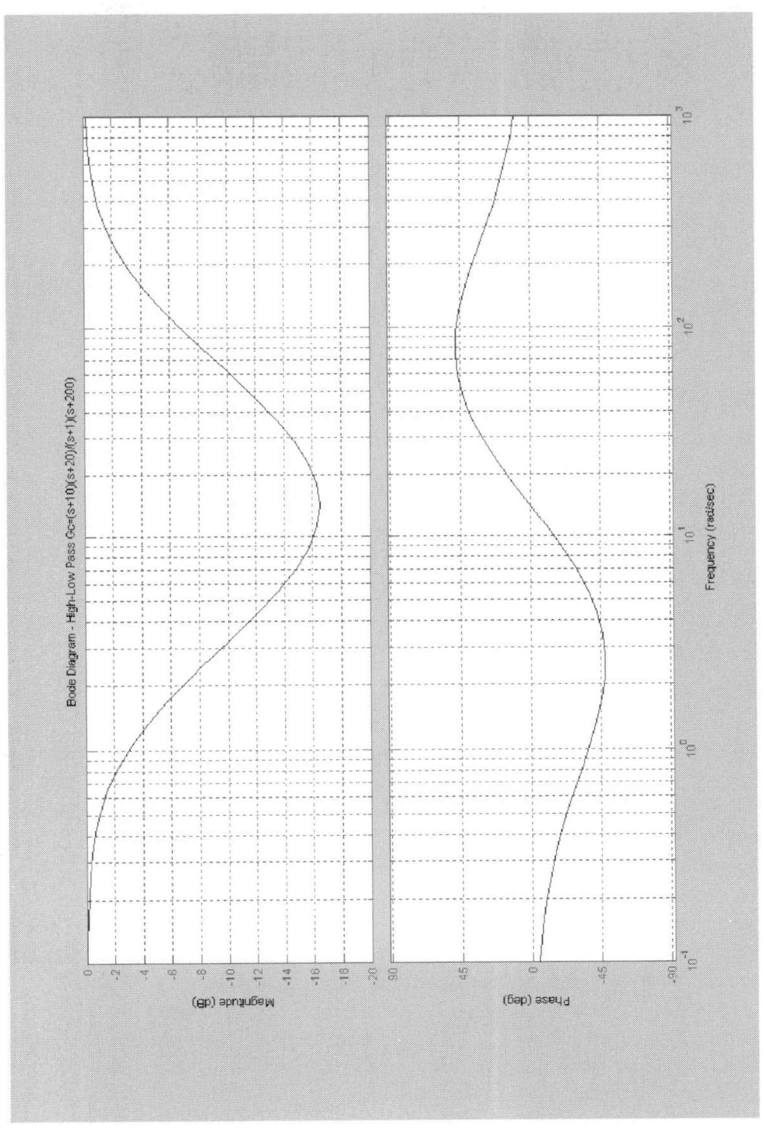

Fig. 10.54 Bode plot of a typical high–low-pass filter.

$(s+10)(s+20)/(s+1)(s+200)$ is shown. Notice the phase initially goes negative, but then goes positive before ending up at 0 deg. The magnitude starts at 0 dB, goes negative, and then returns to 0 dB.

A special application of the high–low-pass filter that attenuates a very small frequency range is called a notch filter. Typically, these filters are used to take out frequencies that may cause excitation of different aircraft dynamic modes.

10.2.7.12 Bandpass filters. A special case of the high–low-pass filter, but in a slightly different form, is often used in many applications. By putting the lead compensator before the lag compensator in Eq. (10.67), a region of positive gain is available. This type of high–low-pass filter is called a bandpass filter because it applies gain to a specified frequency range. In this case, $a < b$, $b < d$, and $c > d$. For a sample bandpass filter of $Gc = [(s+1)(s+200)/(s+10)(s+20)]$ the Bode plot is shown in Fig. 10.55. Notice that the frequency response is exactly the opposite of that shown in Fig. 10.54, and that a region of frequency is amplified.

10.3 Digital Control

Any discussion of modern flight control must include digital control effects because most new aircraft that have come on line, such as the Boeing 777 and the F-22 Raptor, have digital flight control systems (as opposed to analog systems). The goal of this section is a top-level overview of some of the key concepts for digital control systems. There are many advantages to digital flight control, but there are also several disadvantages.

10.3.1 Digital Control Advantages

There are many advantages to digital flight control systems:

1) They are more versatile than analog because they can be easily reprogrammed without changing hardware.
2) It is easy to implement gain scheduling to vary flight control gains as the aircraft dynamics change with flight condition.
3) Digital components in the form of electronic parts, transducers, and encoders are often more reliable, more rugged, and more compact than analog equivalents.
4) Multimode and more complex digital control laws can be implemented because of fast, light, and economical microprocessors.
5) It is possible to design "robust" controllers that can control the aircraft for various flight conditions, including some mechanical failures.
6) Improved sensitivity with sensitive control elements that require relatively low-energy signals.

10.3.2 Digital Control Disadvantages

As with most engineering applications, there are positive and negative implications for the use of digital control. The disadvantages are listed next:

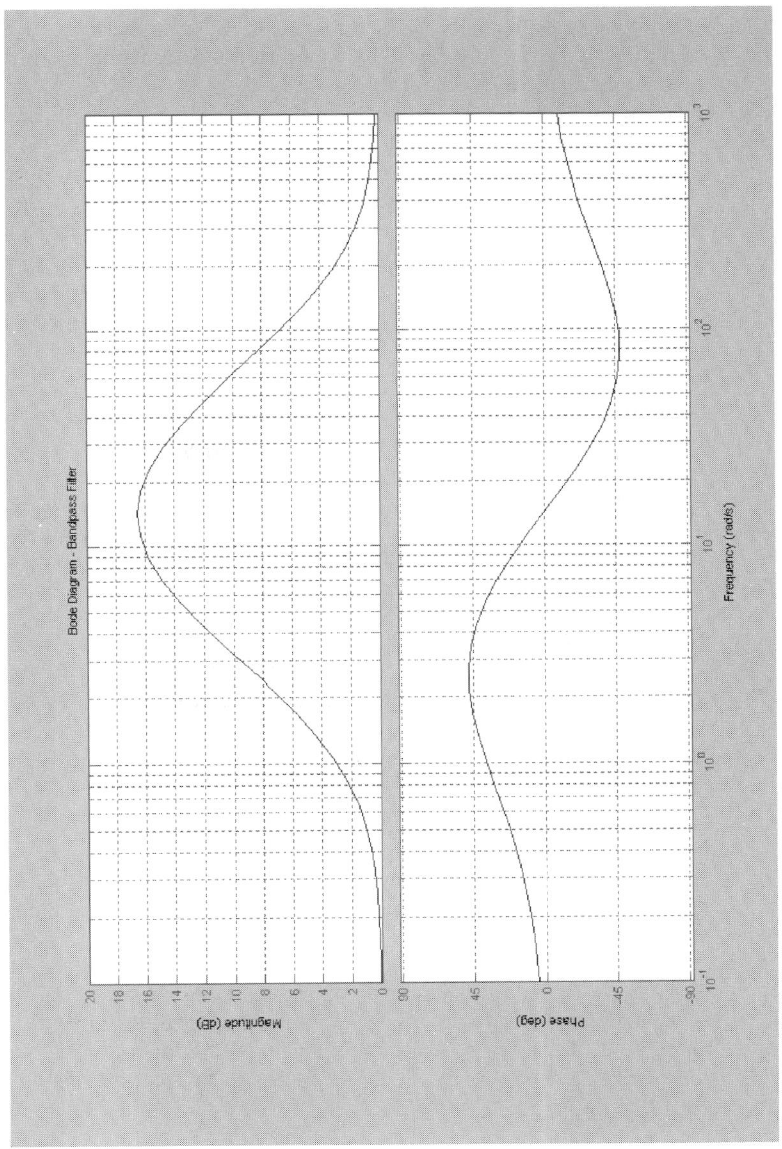

Fig. 10.55 Bode plot of a typical bandpass filter.

1) The lag associated with the sampling process reduces the system stability.
2) The mathematical analysis and system design of a sampled-data system is sometimes more complex than an analog system.
3) The signal information may be lost because it must be digitally reconstructed from an analog signal.
4) The complexity of the control process is in the software-implemented control algorithms that may contain errors.
5) Software verification becomes critical because of the safety of flight issues—software errors can cause the aircraft to crash.

Some of these disadvantages become less stringent because of rapid increases in computer and microprocessor technology that allow high sampling rates. However, the biggest issue is getting all the software "bugs" out before flight.

10.3.3 Digital Control Overview

A digital controller must take an analog signal, sample it with an analog-to-digital converter (A/D), process the information in the digital domain, and then convert the signal to analog with a digital-to-analog (D/A) converter. The key here is to provide redundant paths in the event of a hardware failure. An overall digital flight control block diagram is shown in Fig. 10.56. Here, the signal comes from a sensing device, such as a gyro. Next, it is fed in parallel along multiple paths (three in this case) to an A/D. After the signal is in digital form, the flight control computers execute the control algorithms. The output from the flight control computers is then fed to D/A converters, which in turn operate an actuator. For example, a pitch attitude signal may be fed back for a pitch attitude hold system. In this case, the signal travels from the

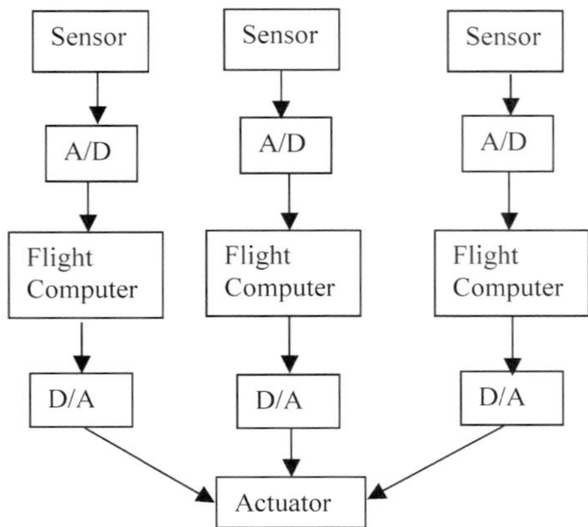

Fig. 10.56 Overall digital flight control system block diagram.

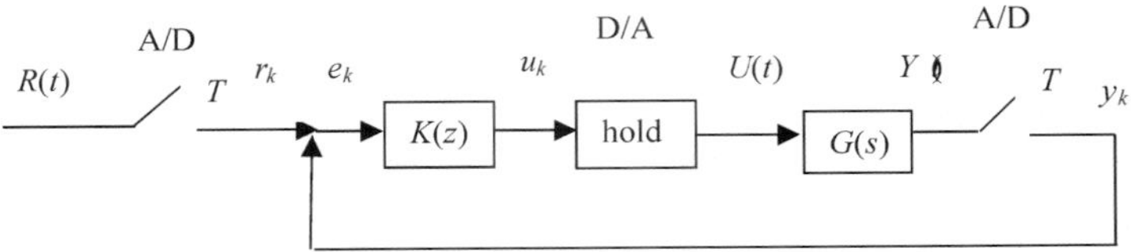

Fig. 10.57 Typical digital control block diagram.

gyro through the A/D, where it is then compared to a reference signal. After the control algorithm calculates the necessary control signal, the output is routed through a D/A converter and fed to the elevator to maintain pitch attitude.

The processing must consider sampling delays when working on the control information. A block diagram of a simple digital controller is shown in Fig. 10.57. In this figure, the capital T denotes the sampling time associated with the sampling process. It is the time interval between each sampling of the signal. The continuous time signal are functions of time (t), while the discrete signals are functions of the specific sample number k. The basic processes such as A/D and D/A conversion are presented next.

10.3.4 Analog-to-Digital Conversion

The A/D converter is a sampling device that converts an analog signal to a digital signal by sampling at a discrete time interval T. It is possible to sample at different rates; however, normally all samples are taken at the same rate for simplicity in analysis and design. Figure 10.58 shows the block diagram of a simple A/D converter.

The fidelity of the A/D conversion depends upon the sampling rate and the highest frequency in the signal. The sample signal is actually a train of impulses defined by

$$r^*(t) = r(t)\delta_t(t)$$

Figure 10.59 shows a sampled signal compared to an analog signal. Notice that the sampled signal is available at discrete points that attempt to recreate the signal. The best representation of the signal is analog, which can never realistically be achieved with a digital sampling of the signal. If the sampling

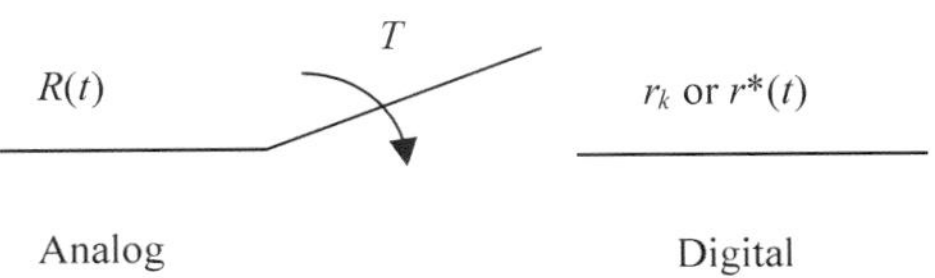

Fig. 10.58 A/D converter block diagram.

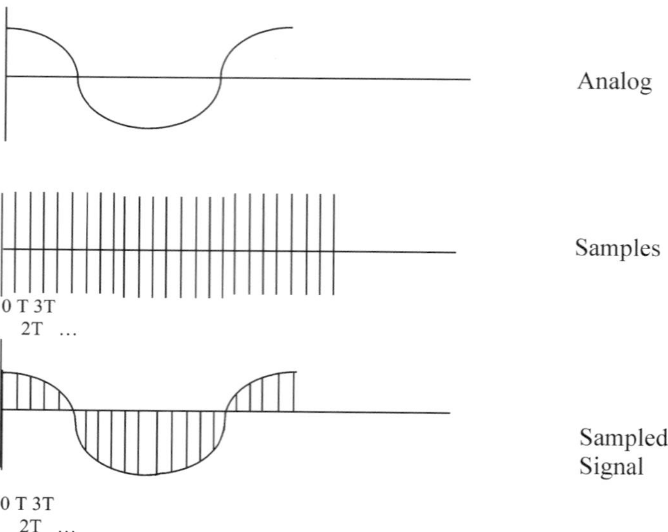

Fig. 10.59 Sampled signal example.

rate is low compared to the signal frequency, inaccurate data will result. Figure 10.60 shows the digital representation of the same signal with a much lower sampling rate.

Unfortunately, it is not practical to instantaneously sample a signal and feed it into a computer. The practical application of an A/D converter is a sample and hold (S/H) device that samples the signals at discrete intervals and holds that value until the next sample. Figure 10.61 shows an example S/H signal.

10.3.5 Digital-to-Analog Conversion

After the signal has been converted to digital and processed by the computer, the signal must then be converted to analog to run devices like control actuators. This process is called D/A conversion. The most common type of D/A device is a zero-order hold (ZOH). It essentially works the opposite of an A/D sample and hold device in that a digital signal at a discrete value is held for a sample period until the next discrete value is used. Figure 10.62 illustrates what a ZOH D/A signal looks like.

One concern with early digital flight control systems were quantization effects associated with the digital nature of the D/A ZOH output. Pilots were concerned that the discrete nature of the command signal to the flight control actuators would translate to abrupt discrete aircraft motion. This was shown in early flight tests to be easily overcome with sample rates above 40 per s.

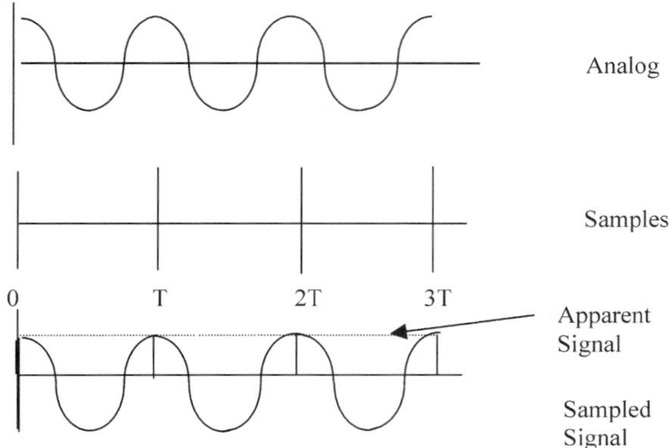

Fig. 10.60 Incorrect sampling time for a signal.

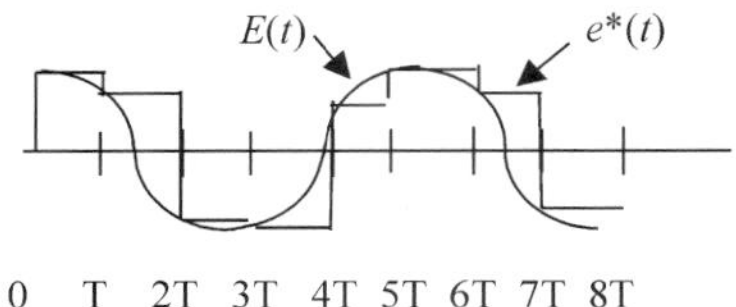

Fig. 10.61 SH signal.

10.4 Advanced Control Algorithms

Although the purpose of this text is to introduce students to classical control techniques and flight control systems, a short discussion of more advanced flight control systems and techniques is included. Students in advanced flight control jobs will use many additional tools, and this section mentions some of them so the student has an idea of other techniques that are available.

10.4.1 Integrated Flight Control Systems

Traditionally, aircraft flight controls, propulsion, avionics, and sensors have been separate components of an aircraft that only interact through control by

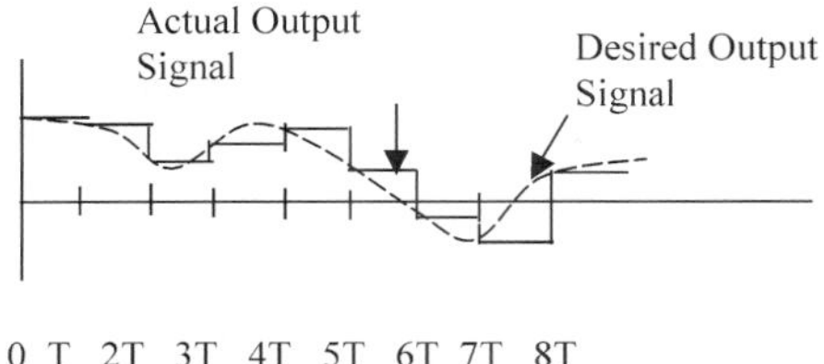

Fig. 10.62 D/A ZOH output example.

the pilot. With fly-by-wire systems, integration of these important aspects of an aircraft is possible.

Integration of flight controls with avionics and sensors gives the capability of automated air-to-surface and air-to-air weapons deliveries with significant reduction in pilot workload. It also provides the capability for automated collision avoidance systems that significantly enhance safety. One of the most recent advances that can easily be integrated into an overall flight control system is the Global Positioning System for navigation purposes. Also, integration of flight controls with the propulsion system promises increased capability for range–payload performance and maneuverability.

One early example (mid 1980s) of an aircraft that considered many aspects in the flight control system design is the Advanced Fighter Technology Insertion (AFTI) F16. This system sought to develop an automated maneuvering attack system (AMAS) for both air-to-air and air-to-surface attack. The system used automated maneuvering using inertial navigation and a conformally mounted surface tracker that contained an infrared tracker, laser range finder, and target state estimator (also known as a Kalman filter). Automated maneuvering consisted of ingress steering, curvilinear weapons delivery, and a target egress. A block diagram of this system is shown in Fig. 10.63.

10.4.2 Robust Control

Another area that shows great potential for flight applications is robust control. Robust control, from an aircraft feedback control standpoint, is the ability of a single controller to operate under multiple flight conditions. Also, robust controllers have the potential to operate the aircraft despite failures of some noncatastrophic aircraft systems. Currently, the changes in flight conditions are handled through gain scheduling based on flight condition. However, problems can occur if there is sensor noise on the parameters for which the gain is scheduled. Also, the gain changes may produce sudden changes in the aircraft's handling characteristics, which cause the pilot's workload to increase.

Typically, robust controllers will design for several plant conditions for a desired "performance envelope." Although a robust controller may not be the optimum solution at any specific flight condition, it will provide acceptable

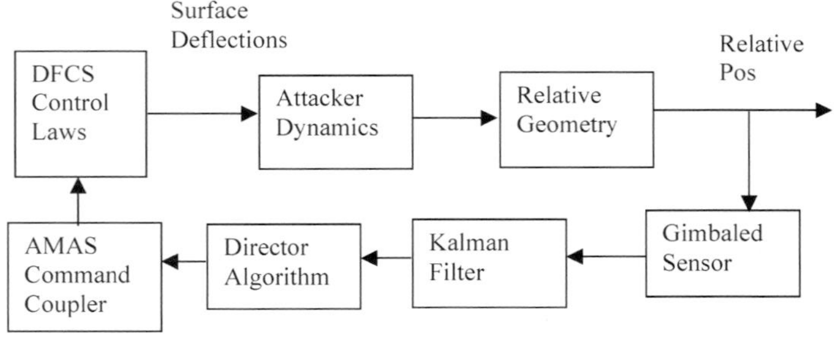

Fig. 10.63 AMAS closed-loop control.

flight performance for all flight conditions within its designed envelope. There are many different robust control techniques that have been applied to aircraft applications. These include H-infinity, quantitative feedback theory, fuzzy logic, direct reduced-order optimization, neural networks, and adaptive control. All have shown some degree of success for flight applications.

10.5 Reversible and Irreversible Flight Control Systems

The system that connects the pilot's stick (or yoke) and rudder pedals to the aircraft flight control surfaces can affect the stability and control characteristics of the aircraft. Two broad categories of flight control system connections will be discussed to illustrate the important differences.

10.5.1 Reversible Flight Control Systems

In a reversible flight control system, the cockpit controls are directly connected to the aircraft flight control surfaces through mechanical linkages such as cables, pushrods, and bellcranks. There is no hydraulic actuator is this path and the muscle to move the control surfaces is provided directly by the pilot. Figure 10.64 provides an illustration of a reversible flight control system.

With no hydraulic or electrical power on the aircraft, a reversible flight control system will have the following characteristics: movement of the stick and rudders will move the respective control surface, and hand movement of each control surface will result in movement of the respective cockpit control—hence the name "reversible." A quick check to determine if a system is reversible can be accomplished by simply walking up to an aircraft and moving the trailing edge of the elevator up and down by hand. If the pilot's stick moves back and forth, the system is reversible. Of course, the ailerons and rudder should be checked in the same manner.

Reversible flight control systems are normally used on light general aviation aircraft such as those produced by Cessna, Beachcraft, and Piper. They have the advantage of being relatively simple and "pilot feel" is provided directly by the airloads on each control surface being transferred to the stick or rudder pedals. They have the disadvantage of increasing stick and rudder forces as the airspeed of the aircraft increases. As a result, the control forces present may exceed the pilot's muscular capabilities if the aircraft is designed to fly at high speed. Of course, the definition of "high speed" depends on the size of the

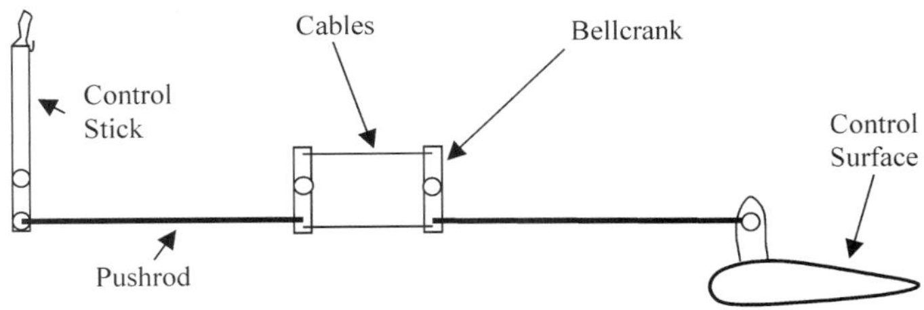

Fig. 10.64 Example of a reversible flight control system.

control surface and the size of the aircraft because these also directly effect pilot control forces. When high pilot control forces become a problem, it is generally time to incorporate a hydraulically boosted or irreversible flight control system into the design of the airplane.

Two types of static stability must be considered with reversible flight control systems. Stick-fixed stability implies that the control surfaces are held in a fixed position by the pilot during a perturbation. Stick-fixed stability was assumed for all the static-stability derivative development in Chapter 5 and subsequent chapters. Stick-fixed stability provides the largest magnitude for derivatives such as C_{m_α} and C_{n_β}. Stick-free stability implies that the stick and rudder pedals are not held in a fixed position by the pilot but rather left to seek their own position during a flight perturbation. In other words, this is the situation where the pilot lets go of the stick (and pedals). For example, a positive perturbation from the trim angle of attack will result in the elevator floating up (or a negative change in δ_e) for the stick-free condition because of the resulting aero loads on the elevator. This situation provides less restoring moment back toward trim and reduced static stability, as illustrated in Fig. 10.65.

The conclusion is that stick-free stability is lower in magnitude than stick-fixed stability. This also directly affects the concepts of neutral point and maneuver point as presented in Secs. 5.4.3 and 5.4.4. The stick-fixed situation was assumed in the presentation of neutral point and maneuver point in these sections. The stick-free neutral point will be forward of the stick-fixed neutral point, which indicates that neutral stability will occur for the stick-free case at a c.g. location that still has positive static stability for the stick-fixed case. As a result, the stick-free case is the most critical and the one to be considered when determining the aft c.g. limits for an aircraft with a reversible flight control system. Another conclusion is that the pilot can increase the static stability of the aircraft by holding the stick fixed rather than letting it go. This also has dynamic stability implications because of the static stability derivative contribution to short-period and dutch-roll characteristics as discussed in Secs. 7.3.2.2 and 7.3.3.4. A similar effect is present for maneuver point where the stick-free maneuver is located forward of the stick-fixed maneuver point. Maneuvering may appear to be a situation where stick free would not be applicable. However, it should be kept in mind that the aircraft can be trimmed in a flight condition where the load factor is greater than one. Figure 10.66

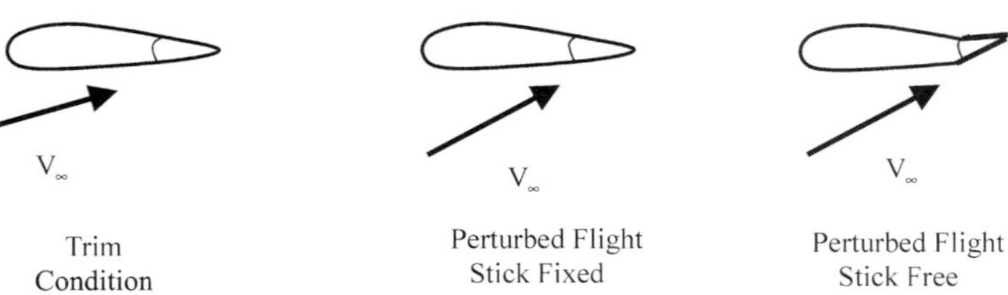

Fig. 10.65 Illustration of stick-fixed and stick-free elevator travel during an angle of attack perturbation.

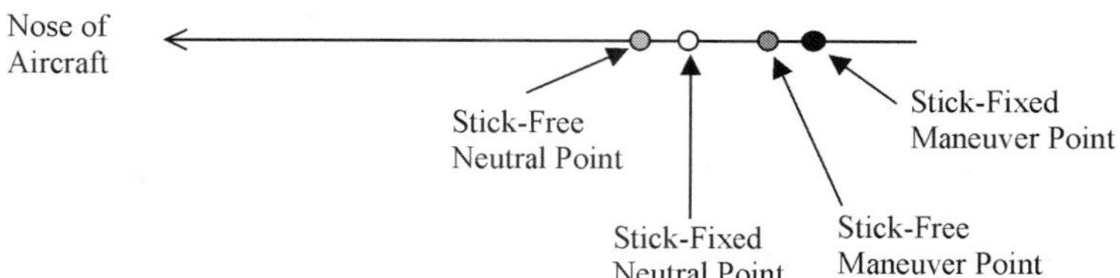

Fig. 10.66 Relative location of stick-fixed and stick-free neutral point and maneuver point.

illustrates the relative locations of the stick-fixed and stick-free neutral point and maneuver point.

Methods to determine the location of the stick-free neutral point and maneuver point involve analysis of the hinge moments acting on each surface and gearing ratios appropriate to the flight control system design. Reference 1 contains methods for doing this.

10.5.2 Irreversible Flight Control Systems

In an irreversible flight control system, the cockpit controls are either directly or indirectly connected to a controller that transforms the pilot's input into a commanded position for a hydraulic or electromechanical actuator. The most common form of an irreversible flight control system connects the pilot's displacement or force command from the stick or rudder pedals to a control valve on a hydraulic actuator. The control valve positions the hydraulic actuator that, in turn, moves the flight control surface. Irreversible systems were first implemented to provide the hydraulic muscle needed to move flight control surfaces at high speeds. Nearly all high speed aircraft flying today have irreversible flight control systems. These generally include all aircraft outside the general aviation category. Figure 10.67 illustrates an irreversible flight control system.

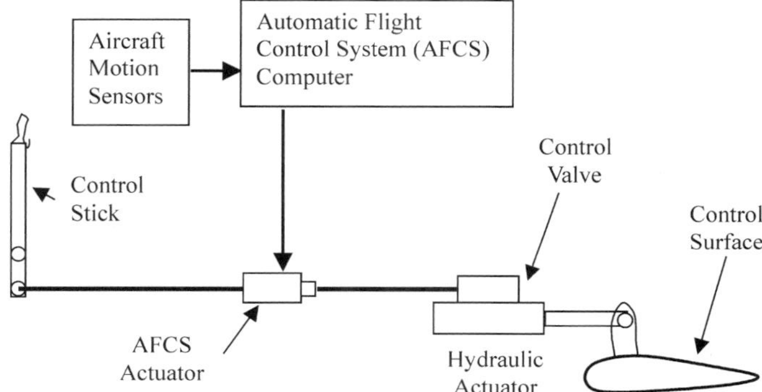

Fig. 10.67 Example of an irreversible flight control system.

Such a system is called irreversible because manual movement of a control surface (only possible when the hydraulic power is off) will not be transferred to movement of the stick or rudder pedals. Movement of the control surface will move the hydraulic actuator but will not be transferred back through the control valve. Irreversible flight control systems behave as essentially a stick-fixed system when the aircraft undergoes a perturbation because the hydraulic actuator holds the control surface in the commanded position—even if the pilot has let go of the stick. Thus, the stick-free concepts discussed with reversible flight control systems are not applicable to irreversible systems. Irreversible flight control systems are also ideal for incorporation of automatic flight control system (AFCS) functions such as inner-loop stability and outer-loop autopilot modes (Reference Chapter 9) because the AFCS actuator only has to reposition the actuator control valve and not the heavily air-loaded control surface. A disadvantage of irreversible flight control systems is that artificial pilot feel must be designed into the stick and rudder pedals because the air loads on the control surfaces are not transmitted back. Artificial pilot feel is normally accomplished with a combination of springs and a bobweight (for longitudinal feel) on the stick, and springs on the rudder pedals. Flight test development is normally required to optimize the design of any artificial feel system.

10.6 Spins

Spins are perhaps the most complex of the many coupled motions that an aircraft can perform. The beginning of a spin may follow a stall if the aircraft is allowed to continue the pitching, rolling, and/or yawing motions occurring during the post-stall phase without corrective action. Because of the post-stall rolling and yawing motions, an imbalance in angle of attack may occur between wings. One wing may experience an increase in lift and a decrease in drag compared to the other wing, which can result in moments that amplify the rolling and yawing, and place the aircraft in an auto-rotation. In a spin, the aircraft descends rapidly toward the Earth in a helical motion about a vertical axis at an angle of attack greater than the stall angle of attack. Departures from controlled flight may be momentary, benign events or may be violent and lead to an undesirable trimmed flight condition—the spin. Early high-performance airplanes such as the F-86 had benign stall characteristics, with buffet followed by a g break resulting in a symmetrical nose down motion. As wings became more swept back with sharper leading edges and stronger vortical flows, the angle of attack at stall and resulting maximum lift increased significantly. This led to improved turn performance because turn rate increases with lift coefficient (Sec. 3.9). It also led to a different type of stall, called yaw divergence, because directional stability was usually lost near the increased stall angle of attack. Yaw divergence can result in a departure from controlled flight and may cause the aircraft to enter a spin from which it may or may not recover depending on the effectiveness of the aerodynamic controls. Adequate lateral and directional stability (C_{l_β}, C_{n_β}), either natural or augmented by the flight control system, became essential to prevent departure. In addition to a buildup of sideslip because of inadequate directional stability, a departure can also result from an asymmetric orientation of vortices on the forebody, resulting in a large side force and yawing moment. This phenomenon is known as "nose slice." The A-7 and F-4 both suffered from nose slice type departures. More than 60 F-4s were lost in out-of-control situations resulting from high angle of attack stall/spin problems, including three during Air Force and Navy spin tests.

A spin is often characterized as a steady, helical motion about the vertical axis with a rapid rate of vertical descent as shown in Fig. 10.68.

This is done in the interest of simplicity, as many spins are oscillatory and therefore unsteady in nature. For the simplified steady condition with the spin angular rate about the spin axis represented by Ω, the aircraft drag balances the weight while the lift is balanced by the centrifugal force (CF):

$$W = \text{mg} = 1/2\rho V^2 SC_D \tag{10.74}$$

$$CF = mR_s\Omega^2 = 1/2\rho V^2 SC_L \tag{10.75}$$

The spin radius (R_s) may be solved from these as

$$R_s = g(C_L/C_D)/\Omega^2 \tag{10.76}$$

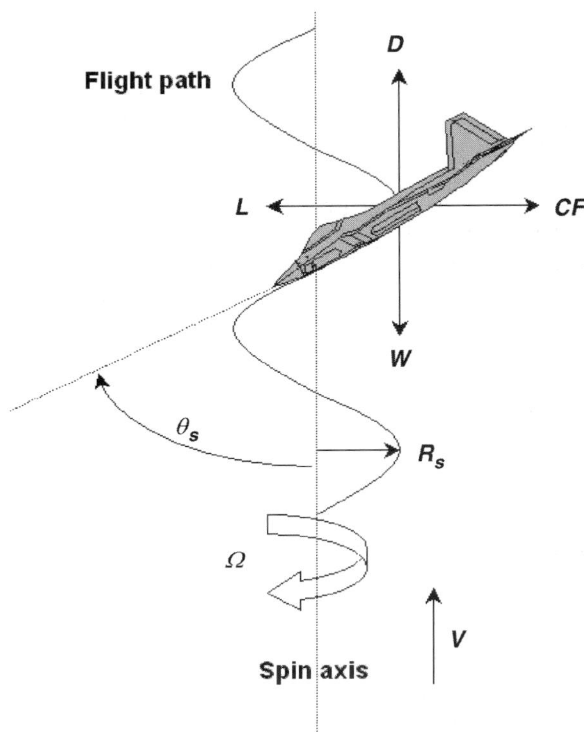

Fig. 10.68 Fundamental spin characteristics.

The radius is typically a fraction of the wing span. For a spin to be maintained, the aerodynamic and inertial moments because of the motion must be balanced. Neglecting the product of inertia (I_{xz}) and moments because of thrust, the moment equations of motion [see Eq. (4.69)] may be written as

$$L_A = (1/2)\rho V^2 SbC_l = \dot{P}I_{xx} + QR(I_{zz} - I_{yy}) \tag{10.77}$$

$$M_A = (1/2)\rho V^2 ScC_m = \dot{Q}I_{yy} + PR(I_{xx} - I_{zz}) \tag{10.78}$$

$$N_A = (1/2)\rho V^2 SbC_n = \dot{R}I_{zz} + PQ(I_{yy} - I_{xx}) \tag{10.79}$$

The steady nature of the spin gives

$$\dot{P} = \dot{Q} = \dot{R} = \dot{\theta} = \dot{\phi} = 0 \tag{10.80}$$

Realizing that the spin rate, Ω, may be denoted as $\mathrm{d}\psi/\mathrm{d}t$, the kinematic equations [Eq. (4.80)] can be reduced to

$$P = -\dot{\psi}\sin\theta + \dot{\phi} = -\Omega\sin\theta \tag{10.81}$$

$$Q = \dot{\psi}\sin\phi\cos\theta + \dot{\theta}\cos\phi = \Omega\cos\theta\sin\phi \tag{10.82}$$

$$R = \dot{\psi}\cos\phi\cos\theta - \dot{\theta}\sin\phi = \Omega\cos\theta\cos\phi \tag{10.83}$$

Because the velocity vector in a spin is near vertical, it is conventional to define the spin pitch angle (θ_s) from this axis, namely

$$\theta_s = 90 + \theta \tag{10.84}$$

Spins are generally categorized as steep or flat, and can be entered in either an upright or inverted orientation. In a steep spin, the nose is pointed down with spin pitch angles ranging from approximately 30 to 50 deg. There is a significant spiral motion with a marked radius. A flat spin looks like it sounds, with the vehicle nearly horizontal. Flat spin pitch angles range from approximately 70 to 90 deg and the spin radius is very small. The rate of rotation in a flat spin is approximately double that of a steep spin, and recovery is more difficult.

The spin rate is typically expressed in terms of the nondimensional quantity ($\Omega b/2V$). Substituting Eqs. (10.81–10.84) into Eqs. (10.77–10.79) results in the following:

$$C_l = \left(\frac{\Omega b}{2V}\right)^2 \frac{4}{\rho Sb^3}(I_{zz} - I_{yy})\sin^2\theta_s \sin 2\phi \tag{10.85}$$

$$C_m = \left(\frac{\Omega b}{2V}\right)^2 \frac{4}{\rho Scb^2}(I_{xx} - I_{zz})\sin 2\theta_s \cos\phi \tag{10.86}$$

$$C_n = \left(\frac{\Omega b}{2V}\right)^2 \frac{4}{\rho Sb^3}(I_{yy} - I_{xx})\sin 2\theta_s \sin\phi \tag{10.87}$$

These are the equations used to analyze a simplified steady spin. The aerodynamic moments appear on the left with the inertial moments on the right. For typical aircraft, the concentration of mass along the fuselage results in a yaw inertia much larger than the roll inertia ($I_{ZZ} \gg I_{XX}$). This indicates that a negative (nose down) pitching moment is required to maintain a spin. As the spin becomes flatter, the size of the required moment reduces. The required aerodynamic roll and yaw moments are usually small because the roll angle in a spin (ϕ) is small. The rudder is the primary control used to recover from a spin. It provides a yawing moment that reduces the rotational velocity, Ω, by deflecting it in opposition to the motion. Trailing edge down elevator is sometimes used after application of the rudder to reduce the angle of attack to below stall. However, there is not one general rule (which covers all aircraft) for the elevator in spin recovery. For modern fighter aircraft, I_{yy} is generally greater than I_{xx}. For this case, the aircraft my be excited in pitch and roll to generate a pitch rate (Q) and roll rate (P) so that an inertial yaw acceleration opposite to the spin direction will result in accordance with Eq. (10.79) to arrest the spin. Another technique is to pump the stick fore and aft in an attempt to induce pitching oscillations that can be used to reduce the angle of attack and initiate a spin recovery.

Spin characteristics can be measured in a wind tunnel by rotating an aircraft model about an axis aligned with the freestream velocity. By testing at a variety of pitch and roll orientation angles and various rotation rates, a database can be developed that is canvassed for points satisfying Eqs. (10.85–10.87). The variation in yawing moment with rotation rate for the F-18 and X-29 with neutral controls at $\theta_s = 90$ deg are shown in Fig. 10.69.

Here, the aerodynamic pitching and yawing moment required to sustain a spin are zero. The F-18 shows a stable slope, with restoring yaw moments

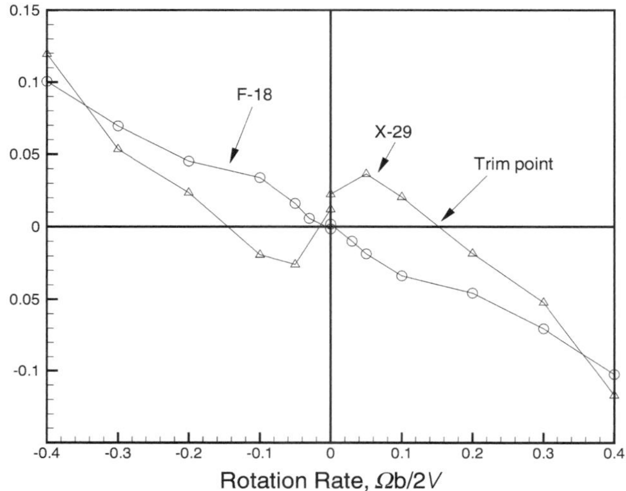

Fig. 10.69 Wind tunnel spin characteristics for the X-29 and F-18.

found with rotation. The X-29 is unstable in yaw, with a negative slope at the higher rotation rates. It intersects the x axis (zero yawing moment) at $\Omega b/2V = 0.14$, satisfying Eq. (10.87). The slope of the yawing moment curve must be negative (stable) at this intersection to maintain a spin. With a stable slope, an increase in spin rate results in a negative yawing moment that slows the spin, while a decrease in spin rate results in a positive yawing moment, thereby increasing the spin.

An example will help clarify how the wind tunnel data may be used. For a forward c.g. location, Eqs. (10.85–10.87) were all satisfied at $\theta_s = 89$ deg, indicating a flat spin mode for the X-29. To compute the spin properties for this case, the aerodynamic data from $\theta_s = 90$ deg will be used assuming an aircraft weight of 15,000 lb, a 27.2-ft wing span, a 185 sq ft wing area, and flight at 30,000 ft ($\rho = 0.000889$ sl/ft^3). The lift and drag coefficients were measured as 0.14 and 2.64, respectively. Equation (10.74) may be used to solve for the velocity, which in a spin is the descent rate. A value of 262 ft/s is obtained. The rotation rate is computed from $\Omega b/2V = 0.14$ as 2.7 rad/s. The time to complete one turn in the spin is $2\pi/\Omega = 2.3$ s, which is indicative of a fast spin. The spin radius is computed from Eq. (10.76) as 0.3 ft.

A design procedure does not exist for a spin proof airplane, but guidelines have been developed over the years. The relative location of the horizontal and vertical tail is known to be important, with T-tail or aft-mounted horizontal tail locations preferred. Spin recovery is contingent on having available aerodynamic control, primarily rudder control, at extreme angles of attack. For many years, a guideline known as the tail damping power factor (TDPF) was used to estimate spin recovery characteristics. At high angles of attack, some (or all) of the vertical tail and rudder may become blanketed by the horizontal tail. The TDPF attempted to quantify this effect by assuming that only the portion of the vertical tail and rudder that were not blanketed by horizontal tail would contribute to spin recovery. It also considered the portion of the aft fuselage that was directly beneath the horizontal tail. A landmark test completed in 1989 measured the pressure distribution on the fuselage, horizontal, and vertical tails of a proposed trainer aircraft. This test showed that the assumptions behind the TDPF were incorrect. Large pressure differences were found across the vertical tail whether or not a horizontal tail was present. Significant pressure differences were also found from the portion of the aft fuselage forward of the horizontal tail. The study concluded that the TDPF should not be used as a guide.

10.7 Historical Snapshot—The F-16 Fly-by-Wire System

The F-16 was the first production aircraft to incorporate a fly-by-wire control system as discussed in Sec. 9.1.3. The prototype F-16 made its first flight in the mid-1970s and, since then, fly-by-wire has been fairly common in high-performance aircraft designs. A simplified version of the F-16 multiloop flight control system will be analyzed in this section. Keep in mind that although the block diagrams are shown in the s domain, the actual implementation is through A/D and D/A sensors that interface with a flight control computer on the aircraft. The other assumption not explicitly stated is that the sample rate is fast enough so that the sampling lag that might adversely affect stability is negligible. The F-16 iteration rates are on the order of 80/s, which is fast enough to overcome sampling lag problems. If there is a case where this is not true, digital design techniques exist to allow for sampling delays directly. The most common of these is the z transform, which will not be discussed because it is beyond the scope of this book.

For the longitudinal F-16 flight control scheme, there are three pertinent transfer functions—α/δ_e, q/δ_e, n_z/δ_e. For a Mach 0.6 F-16 at sea level, the transfer functions are

$$\frac{\alpha}{\delta_e} = \frac{0.203(s + 0.0087 \pm 0.067i)(s + 106.47)}{(s - 0.087)(s + 2.373)(s + 0.098 \pm 0.104i)} \frac{\text{deg}}{\text{deg}}$$

$$\frac{q}{\delta_e} = \frac{21.516s(s + 0.0189)(s + 1.5)}{(s - 0.087)(s + 2.373)(s + 0.098 \pm 0.104i)} \frac{\text{deg/s}}{\text{deg}}$$

$$\frac{n_z}{\delta_e} = \frac{0.0889s(s + 0.0158)(s + 1.165 \pm 11.437i)}{(s - 0.087)(s + 2.373)(s + 0.098 \pm 0.104i)} \frac{g}{\text{deg}}$$

The n_z transfer function is for the accelerometer location. The simplified F-16 longitudinal flight control system for Mach 0.6 at sea level is shown in Fig. 10.70.

The gains F2 and F3 are scheduled based on flight condition. Gain F2 is a function of dynamic pressure divided by static pressure, while gain F3 is a function of static pressure alone. The system is shown in a clearer representation with feedback loops in Fig. 10.71.

For analysis, the F-16A block must be represented in terms of the three transfer functions previously defined. This is shown in Fig. 10.72.

The q/α transfer function is obtained by dividing the q/δ_e transfer function by the α/δ_e transfer function. Similarly, the n_z/q transfer function is obtained by dividing the n_z/δ_e transfer function by the q/δ_e transfer function. Please note that in the following development any transfer function with GH is an OLTF. Also, any transfer function with a "_cl" represents a CLTF.

The inner loop is closed at a gain of $K_\alpha = 0.5$. This results in an innermost loop CLTF of:

$$G\_\alpha\_cl(s) = \frac{4.1(s + 0.0087 \pm 0.067i)(s + 106.47)(s + 10)}{(s + 0.0083 \pm 0.0643i)(s + 0.478 \pm 3.03i)(s + 12.1)(s + 19.6)}$$

The angle of attack feedback has stabilized the short-period roots (complex roots furthest from the imaginary axis), but the damping is low for the short period −0.156. The natural frequency is 3.07 rad/s.

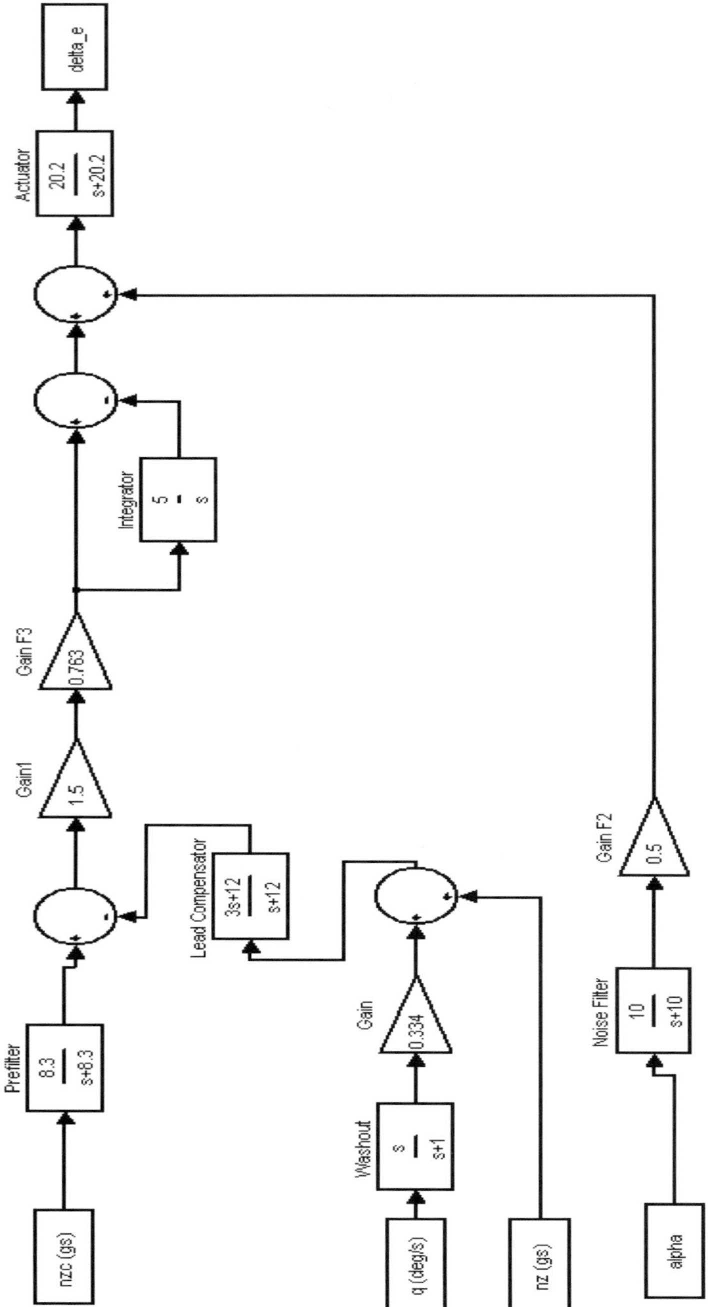

Fig. 10.70 Simplified F-16 longitudinal flight control system, Mach = 0.6 at sea level.

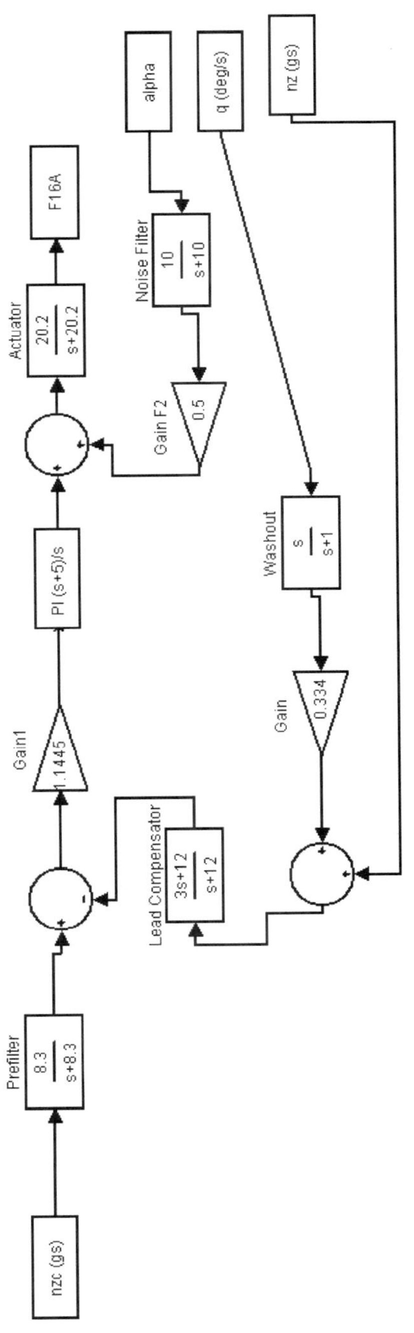

Fig. 10.71 F-16 feedback loop view.

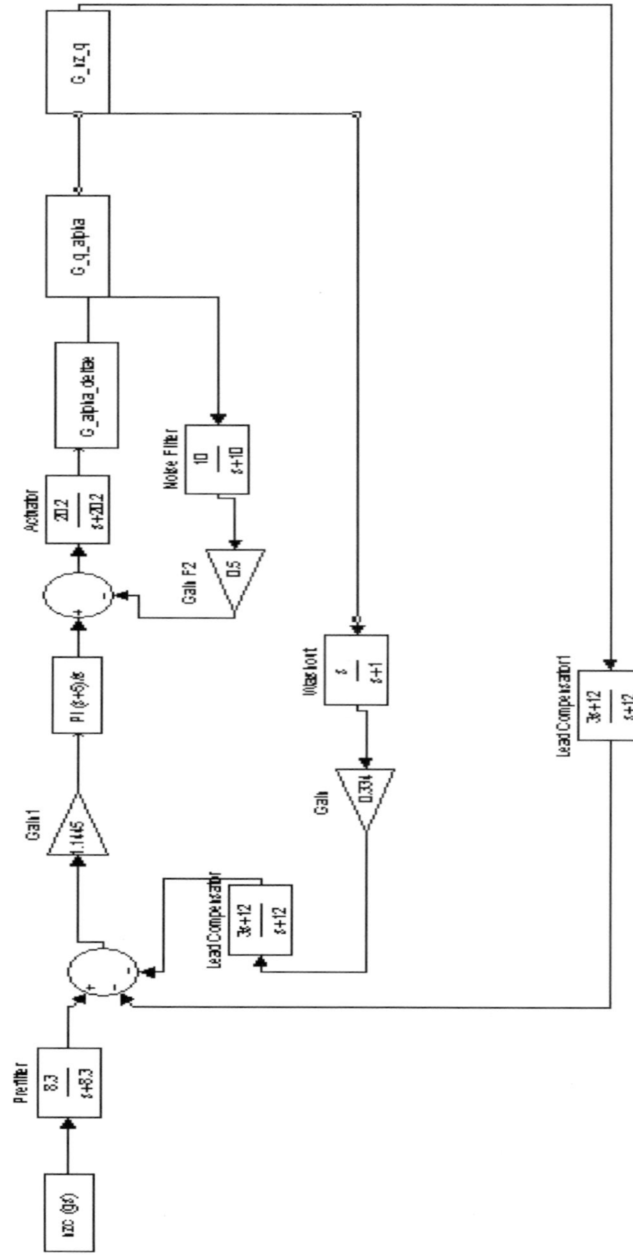

Fig. 10.72 Simplified F-16 longitudinal flight control system for analysis.

The next step is to improve (increase) the damping with pitch rate (q) feedback. The block diagram for this step is shown in Fig. 10.73. Thus, the OLTF for the pitch loop is

$$\text{G\_oltf\_q} = \text{Gain1*PI*G\_}\alpha\text{\_cl*Gq/}\alpha * \text{GLead*gain*washout}$$

or

$$GHq(s) = \frac{498.42(s+0.0189)(s+1.5)(s+4)(s+5)}{(s+0.0083 \pm 0.0643i)(s-0.478 \pm 3.03)(s+12.1)(s+19.6)(s+1)(s+12)}$$

Closing the pitch rate loop with "static" gain of 498.42 results in a pitch rate CLTF of

$$G\_q\_\text{alpha}\_cl(s) = \frac{497.43(s+0.0189)(s+1.5)(s+10)(s+5)(s+1)(s+12)}{(s+0.0093 \pm 0.023i)(s+1.33)(s+3.78 \pm i2.68) \times (s+10.2)(s+13.3 \pm 20.2i)}$$

This results in a new short-period damping of $\zeta_{sp} = 0.816$ and a new natural frequency of $\omega_{nsp} = 4.63$ rad/s.

The CLTF from the inner two loops now becomes part of the outer-loop transfer function for the outer n_z loop. This is shown in Fig. 10.74.

Thus, the new outer-loop, open-loop transfer function becomes

$$GHn_z(s) = \frac{2.055(s+0.0058)(s+1.165 \pm 11.437i)(s+10)(s+5)(s+1)}{(s+0.0093 \pm 0.023i)(s+1.33)(s+3.78 \pm 2.68i)(s+10.2)} \times \frac{(s+12)3(s+4)}{(s+13.3 \pm 20.2i)(s+12)}$$

Closing this with a gain of $Kn_z = 6.165$ results in an overall CLTF of

$$GHn_{z_{cl}}(s) = \frac{2.055(s+0.0158)(s+1.165 \pm 11.437i)(s+10)}{(s+0.0164)(s+1.74)(s+3.86 \pm 3.32i)(s+0.637)} \times \frac{(s+5)(s+1)(s+12)}{(s+10.3)(s+15.7 \pm 17.6i} \times \frac{8.3}{(s+8.3)}$$

The last term in the previous equation is the prefilter. Notice that there are poles at -1.74 and -0.637. These are close to the short-period roots at $-3.86 \pm 3.32i$, and will have some effect on the short-period response. These two roots will tend to slow down the short-period response. The aircraft response to a unit load factor input is shown in Fig. 10.75.

Notice that the prefilter slows the response down slightly. The response is dominated by the two real poles, rather than the short-period poles.

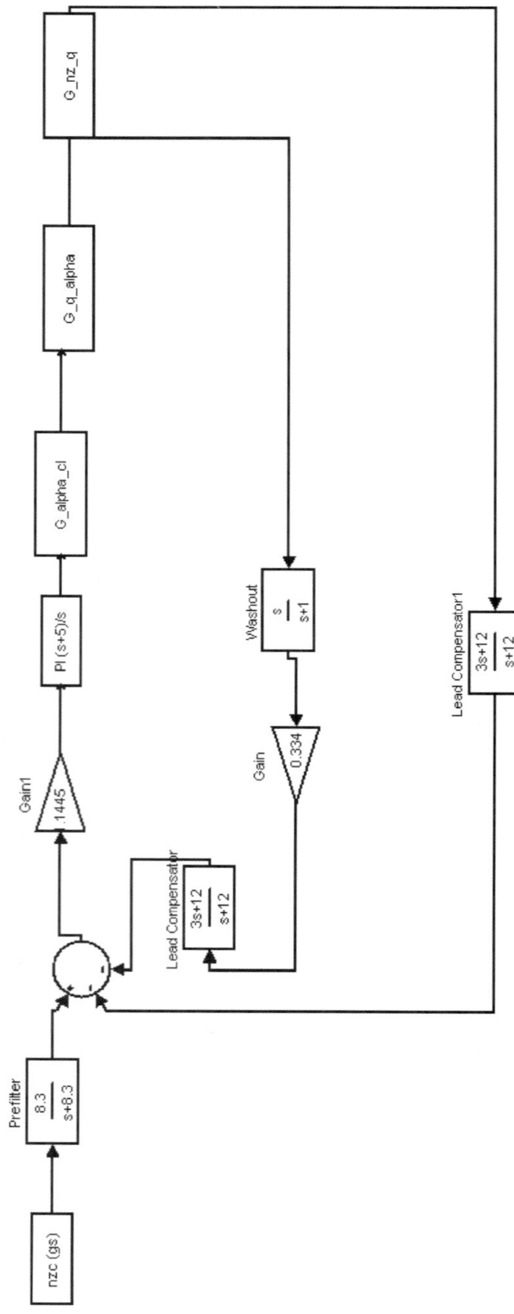

Fig. 10.73 F-16 longitudinal flight control system with $K_\alpha = 20.5$.

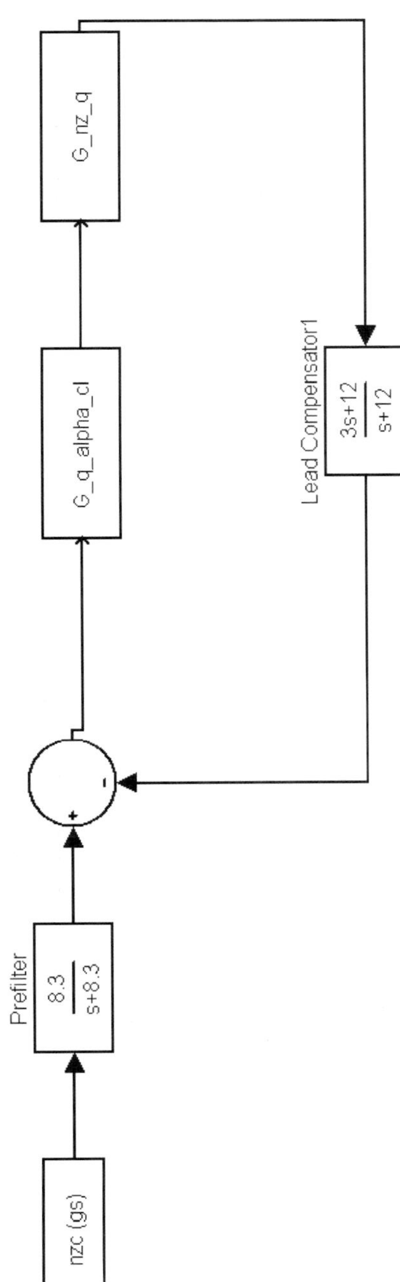

Fig. 10.74 Simplified F-16 longitudinal flight control system with 2 inner loops closed.

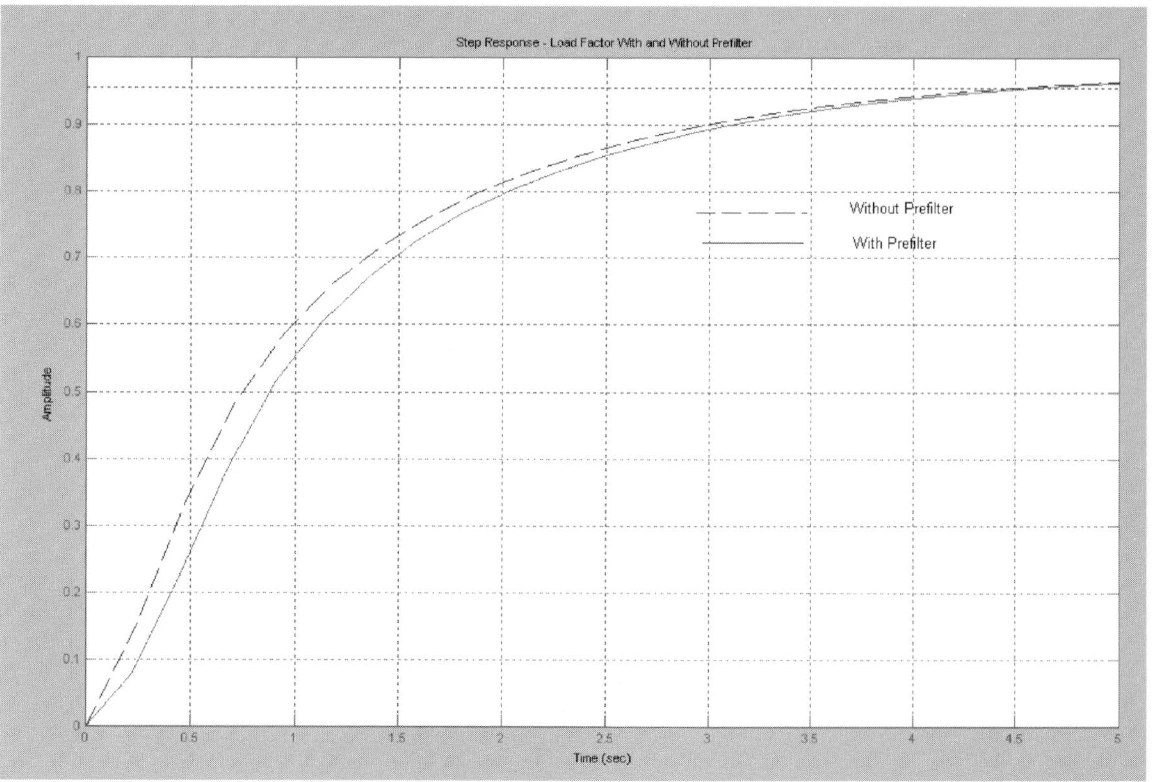

Fig. 10.75 Load factor response with and without prefilter.

Problems

10.1 For the following system unity feedback system, determine the system type, K_p, K_v, and K_a.

$$G(s) = \frac{4(s+2)}{s(s+4)(s+8)}$$

10.2 What is the steady-state error of the previous system for unit step, unit ramp, and unit parabolic inputs?

10.3 Express the complex number $\bar{z} = 2 + 4j$ in terms of magnitude and phase angle.

10.4 Convert the following transfer function to Bode form and identify the Bode gain.

$$G(s) = \frac{3(s+1)(s+4)}{s(s+3)(s^2+2s+8)}$$

10.5 Calculate the log magnitude and phase of the transfer function in Problem 4 for a frequency of 5 rad/s and 10 rad/s. What is the difference in the magnitude and phase for the two frequencies?

10.6 Construct a magnitude and phase angle Bode plot using asymptotic approximations for

$$G(i\omega) = \frac{10(1+i\omega)}{(i\omega)^2\left[1 + \frac{i\omega}{4} - \left(\frac{\omega}{4}\right)^2\right]}$$

10.7 Find the gain margin, phase margin, gain crossover frequency, phase crossover frequency, and bandwidth for the following system using a Bode plot generated by any applicable computer software package.

$$G(s) = \frac{10(s+2)}{s(s+1)(s^2+4s+16)}$$

10.8 Find the transfer function $G(i\omega)$ and $G(s)$ for the following Bode magnitude plot. Assume a minimum phase system.

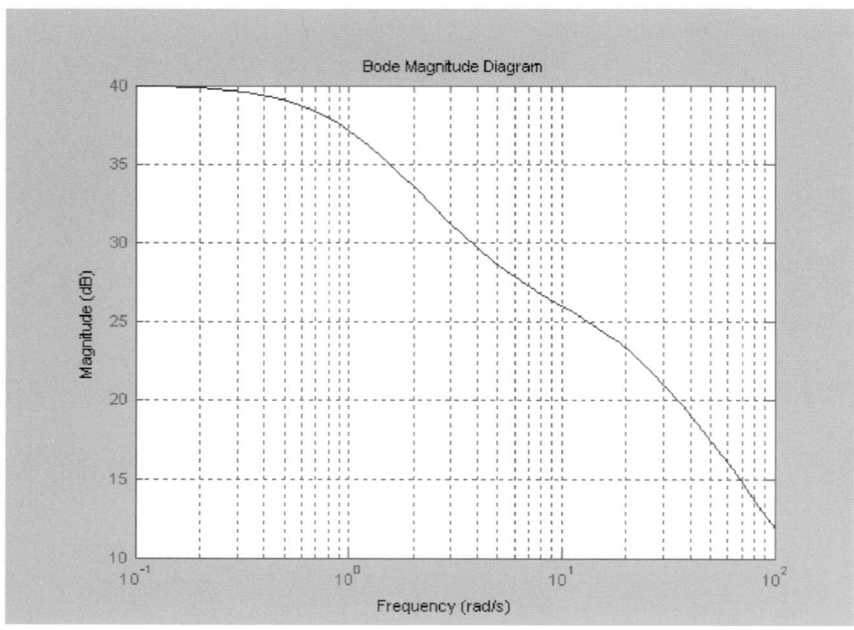

10.9 Estimate the damping ratio and natural frequency for the following second order pole from the following plot.

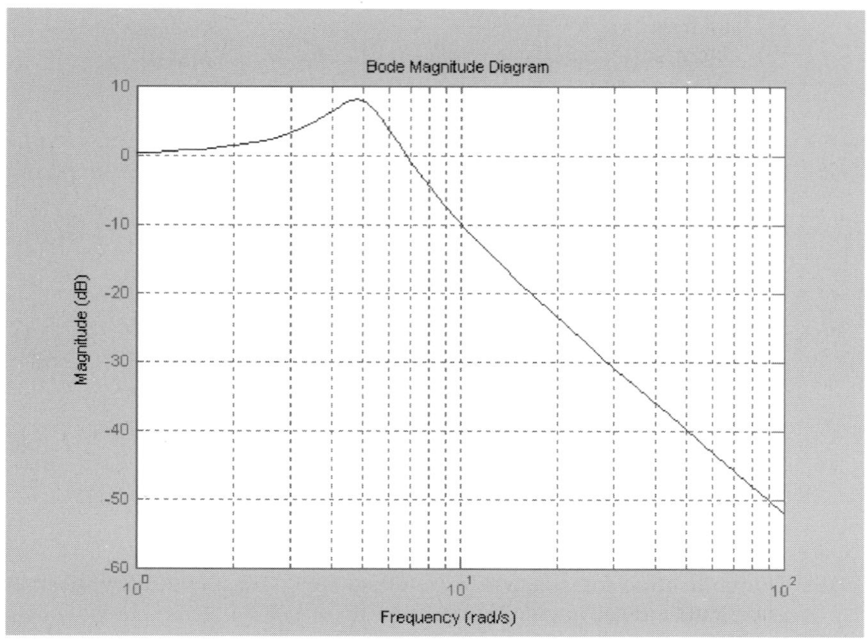

10.10 Match the following:

System Problem/Objective	*Filter to Use*
Pass a range of frequencies _____	(a) lag
Too much SS error _____	(b) lead
System response too slow _____	(c) lag lead
Eliminate a specific frequency _____	(d) noise (lag)
Improve SS error and quicken response _____	(e) notch
Eliminate unwanted high frequency signals_____	(f) band pass filter

10.11 For the following system, is there any value or range of K that can satisfy the following specifications?

$$K_v \geq 2$$

$$\text{gain margin} \geq 4$$

$$G = \frac{K}{s(s+2)(s+4)}$$

If the specifications cannot be met, design a lead compensator to meet them.

10.12 The primary disadvantage of a pure integrator $(1/s)$ is
(a) Increases ω_N
(b) Increases ω_d
(c) Adds phase lag
(d) Increases steady-state error

10.13 The following system has a gain margin of 14.7 dB and a phase margin of 60.4 deg. Add a lead compensator so that the phase margin is between 90 and 110 deg.

$$G = \frac{40(s+1)}{s(s+0.5)(s+5)(s^2+2s+20)}$$

10.14 An aircraft is experiencing problems with 400-Hz power noise. Design a lag-lead compensator in the forward path that will provide at least 10 dB attenuation at 400 Hz.

10.15 Design a noise filter in the forward path that will provide between 10 and 20 dB attenuation at 100 Hz.

10.16 The load factor response for an aircraft is too fast. Design a lag pre-filter to slow the rise time down from 3.33 s to between 4 and 4.5 s using frequency domain techniques. The system transfer function can be approximated as

$$G(s) = \frac{1.31}{(s + 0.75)(s + 1.75)}$$

10.17 A digital controller is experiencing problems with a system whose dominant mode is at 20 Hz. If the sample period is 100 ms, what is the probable cause of the problem?

10.18 What are some of the problems associated with digital control?

Appendix A
Conversions

1. Length:

1 foot (ft) = 0.3048 meters (m)

1 statute mile (sm) = 5280 ft = 0.8690 nautical miles (nm) = 1.609 kilometers (km)

2. Velocity:

1 foot/s (ft/s) = 0.5921 nm/h (kn) = 0.6818 m/h (mph) = 1.097 km/h

1 kt = 1.689 ft/s = 1.151 sm/h = 1.852 km/h

1 mph = 1.467 ft/s = 0.8684 kn = 1.609 km/h

3. Pressure:

1 lb/ft^2 (psf) = 0.006944 lb/in^2 (psi) = 47.88 newtons/m^2 (nt/m^2)

1 psi = 144 psf = 6895 nt/m^2

4. Temperature:

$$^\circ F = \tfrac{9}{5}(^\circ C) + 32$$

$$^\circ C = \tfrac{5}{9}(^\circ F - 32)$$

$$^\circ R = ^\circ F + 459.69$$

$$K = ^\circ C + 273.16$$

5. Frequency/Rotation:

1 radian/s (rad/s) = 57.296 deg/s = 0.1592 revolutions/sec = 9.549 revolutions/min (rpm)

Appendix B
Properties of the U.S. Standard Atmosphere

In this appendix the properties of the 1962 U.S. Standard Atmosphere are tabulated in accordance with Eqs. (1.5–1.17) and (1.21–1.22) of Chapter 1. Two tables are given: Table B1 in English units and Table B2 in Metric units.

Table B1 U.S. Standard atmosphere in English units

Altitude	Temperature	Temperature ratio	Pressure	Pressure ratio	Density	Density ratio	Coefficient of viscosity	Speed of sound
h, ft	T, °R	θ	P, psf	δ	ρ $\frac{\text{slugs}}{\text{ft}^3}$ $\times 10^{-3}$	σ	μ $\frac{\text{lb}\cdot\text{sec}}{\text{ft}^2}$ $\times 10^{-7}$	a, ft/s
0	518.7	1.000	2116	1.000	2.377	1.000	3.737	1,116.4
1,000	515.1	0.9932	2041	0.9644	2.3081	0.97106	3.717	1,112.6
2,000	511.5	0.9863	1963	0.9298	2.2409	0.94277	3.697	1,108.7
3,000	508.0	0.9794	1879	0.8962	2.1751	0.91512	3.677	1,104.0
4,000	504.4	0.9725	1828	0.8637	2.1109	0.88809	3.657	1,101.0
5,000	500.8	0.9657	1761	0.8320	2.0481	0.86167	3.636	1,097.1
6,000	497.3	0.9588	1696	0.8014	1.9868	0.83586	3.616	1,093.2
7,000	493.7	0.9519	1633	0.7716	1.9268	0.81064	3.596	1,089.2
8,000	490.1	0.9459	1572	0.7428	1.8683	0.78602	3.575	1,085.3
9,000	486.6	0.9382	1513	0.7148	1.8111	0.76196	3.555	1,081.4
10,000	483.0	0.9313	1456	0.6877	1.7533	0.73848	3.534	1,077.4
11,000	479.4	0.9244	1400	0.6614	1.7008	0.71555	3.513	1,073.4
12,000	475.9	0.9175	1345	0.6360	1.6476	0.69317	3.493	1,069.4
13,000	472.3	0.9107	1293	0.6113	1.5957	0.67133	3.472	1,065.4
14,000	468.7	0.9038	1243	0.5857	1.5451	0.65003	3.451	1,061.4
15,000	465.2	0.8969	1194	0.5643	1.4956	0.62924	3.430	1,057.3
16,000	461.6	0.8900	1147	0.5420	1.4474	0.60896	3.409	1,053.2
17,000	458.0	0.8831	1101	0.5203	1.4004	0.58919	3.388	1,049.2
18,000	454.5	0.8763	1057	0.4994	1.3546	0.56991	3.366	1,045.1
19,000	450.9	0.8694	1014	0.4791	1.3100	0.55112	3.345	1,041.0
20,000	447.3	0.8625	972	0.4595	1.2664	0.53281	3.324	1,036.8
21,000	443.8	0.8556	932	0.4406	1.2240	0.51497	3.302	1,032.7
22,000	440.2	0.8488	894	0.4223	1.1827	0.49758	3.281	1,028.5
23,000	436.6	0.8419	856	0.4046	1.1425	0.48065	3.259	1,024.4
24,000	433.1	0.8350	820	0.3876	1.1033	0.46417	3.238	1,020.2
25,000	429.5	0.8281	785	0.3711	1.0651	0.44812	3.216	1,016.1
26,000	426.0	0.8213	752	0.3522	1.0280	0.43250	3.194	1,011.7

(continued)

Table B1 U.S. Standard atmosphere in English units (continued)

Altitude	Temperature	Temperature ratio	Pressure	Pressure ratio	Density	Density ratio	Coefficient of viscosity	Speed of sound
h,ft	T,° R	θ	P, psf	δ	ρ $\frac{\text{slugs}}{\text{ft}^3}$ $\times 10^{-3}$	σ	μ $\frac{\text{lb} \cdot \text{sec}}{\text{ft}^2}$ $\times 10^{-7}$	a, ft/s
27,000	422.4	0.8144	719	0.3398	0.9919	0.41730	3.172	1,007.5
28,000	418.8	0.8075	688	0.3250	0.9567	0.40251	3.150	1,003.2
29,000	415.3	0.8006	657	0.3107	0.9225	0.38812	3.128	999.0
30,000	411.7	0.7938	628	0.2970	0.8893	0.37413	3.106	994.7
31,000	408.1	0.7869	600	0.2837	0.8569	0.36053	3.084	990.3
32,000	404.6	0.7800	573	0.2709	0.8255	0.34731	3.061	986.0
33,000	401.0	0.7731	547	0.2586	0.7950	0.33447	3.039	981.6
34,000	397.4	0.7663	522	0.2467	0.7653	0.32199	3.016	977.3
35,000	393.3	0.7594	497	0.2353	0.7365	0.30987	2.994	972.9
36,000	390.3	0.7525	474	0.2243	0.7086	0.29811	2.971	968.5
36,089	390.0	0.7519	471	0.2234	0.7062	0.29710	2.969	968.1
37,000	390.0	0.7519	452	0.2138	0.6759	0.28435	2.969	968.1
38,000	390.0	0.7519	431	0.2038	0.6442	0.27101	2.969	968.1
39,000	390.0	0.7519	410	0.1942	0.6139	0.25829	2.969	968.1
40,000	390.0	0.7519	391	0.1851	0.5851	0.24617	2.969	968.1
41,000	390.0	0.7519	373	0.1764	0.5577	0.23462	2.969	968.1
42,000	390.0	0.7519	355	0.1681	0.5315	0.22361	2.969	968.1
43,000	390.0	0.7519	339	0.1602	0.5065	0.21311	2.969	968.1
44,000	390.0	0.7519	323	0.1527	0.4828	0.20311	2.969	968.1
45,000	390.0	0.7519	308	0.1455	0.4601	0.19358	2.969	968.1
46,000	390.0	0.7519	293	0.1387	0.4385	0.18450	2.969	968.1
47,000	390.0	0.7519	279	0.1322	0.4180	0.17584	2.969	968.1
48,000	390.0	0.7519	266	0.1260	0.3983	0.16759	2.969	968.1
49,000	390.0	0.7519	254	0.1201	0.3796	0.15972	2.969	968.1
50,000	390.0	0.7519	242	0.1145	0.3618	0.15223	2.969	968.1
51,000	390.0	0.7519	230	0.1091	0.3449	0.14509	2.969	968.1
52,000	390.0	0.7519	219	0.1040	0.3287	0.13828	2.969	968.1
53,000	390.0	0.7519	209	0.09909	0.3133	0.13179	2.969	968.1
54,000	390.0	0.7519	199	0.09444	0.2985	0.12560	2.969	968.1
55,000	390.0	0.7519	190	0.09001	0.2845	0.11971	2.969	968.1
56,000	390.0	0.7519	181	0.08578	0.2712	0.11409	2.969	968.1
57,000	390.0	0.7519	172	0.08176	0.2585	0.10874	2.969	968.1
58,000	390.0	0.7519	164	0.07792	0.2463	0.10364	2.969	968.1
59,000	390.0	0.7519	156	0.07426	0.2348	0.098772	2.969	968.1
60,000	390.0	0.7519	150	0.07078	0.2238	0.094137	2.969	968.1
61,000	390.0	0.7519	143	0.06746	0.2133	0.089720	2.969	968.1
62,000	390.0	0.7519	136	0.06429	0.2032	0.085509	2.969	968.1
63,000	390.0	0.7519	130	0.06127	0.1937	0.081497	2.969	968.1
64,000	390.0	0.7519	124	0.05840	0.1846	0.077672	2.969	968.1
65,000	390.0	0.7519	118	0.05566	0.1760	0.074027	2.969	968.1

Table B2 U.S. Standard atmosphere in metric units

Altitude	Temperature	Temperature ratio	Pressure	Pressure ratio	Density	Density ratio	Coefficient of viscosity	Speed of sound
h, m Geop.	T, °K	θ	P, N/m^2	δ	ρ, Kg/m^3	σ	μ, $\frac{N-sec}{m^2}$ $\times 10^{-5}$	a, m/sec
0	288.2	1.0000	101,325	1.0000	1.2250	1.000	1.789	340.3
500	284.9	0.9888	95,460	0.9421	1.1673	0.9529	1.774	338.4
1,000	281.7	0.9775	89,874	0.8870	1.1116	0.9075	1.758	336.4
1,500	278.4	0.9662	84555	0.8345	1.0581	0.8637	1.742	334.5
2,000	275.2	0.9549	79495	0.7846	1.0065	0.8216	1.726	332.5
2,500	271.9	0.9436	74682	0.7371	0.95686	0.7811	1.710	330.6
3,000	268.7	0.9324	70108	0.6919	0.90912	0.7421	1.694	328.6
3,500	265.4	0.9211	65764	0.6490	0.86323	0.7047	1.678	326.6
4,000	262.2	0.9098	61640	0.6083	0.81913	0.6687	1.661	324.6
4,500	258.9	0.8985	57728	0.5697	0.77677	0.6341	1.645	332.6
5,000	255.7	0.8872	54019	0.5331	0.73612	0.6009	1.628	320.5
5,500	252.4	0.8760	50506	0.4985	0.69711	0.5691	1.612	318.5
6,000	249.2	0.8647	47181	0.4656	0.65970	0.5385	1.595	316.4
6,500	245.9	0.8534	44034	0.4346	0.62384	0.5093	1.578	314.4
7,000	242.7	0.8421	41060	0.4052	0.58950	0.4812	1.561	312.4
7,500	239.4	0.8390	38251	0.3775	0.55662	0.4544	1.544	310.2
8,000	236.2	0.8196	35599	0.3513	0.52517	0.4287	1.527	308.1
8,500	232.9	0.8083	33099	0.3267	0.49509	0.4042	1.510	305.9
9,000	229.7	0.7970	30742	0.3034	0.46635	0.3807	1.492	303.8
9,500	226.4	0.7857	28523	0.2815	0.43890	0.3583	1.475	301.6
10,000	223.2	0.7745	26436	0.2609	0.41271	0.3369	1.457	229.5
10,500	219.9	0.7632	24474	0.2415	0.38773	0.3165	1.439	297.3
11,000	216.7	0.7519	22632	0.2234	0.36392	0.2971	1.422	295.1
11,500	216.7	0.7519	20916	0.2064	0.33633	0.2746	1.422	295.1
12,000	216.7	0.7519	19330	0.1908	0.31083	0.2537	1.422	295.1
12,500	216.7	0.7519	17864	0.1763	0.28726	0.2345	1.422	295.1
13,000	216.7	0.7519	16510	0.1629	0.26548	0.2167	1.422	295.1
13,500	216.7	0.7519	15218	0.1506	0.24536	0.2003	1.422	295.1
14,000	216.7	0.7519	14101	0.1392	0.22675	0.1851	1.422	295.1
14,500	216.7	0.7519	13032	0.1286	0.20956	0.1711	1.422	295.1
15,000	216.7	0.7519	12044	0.1189	0.19367	0.1581	1.422	295.1
15,500	216.7	0.7519	11131	0.1099	0.17899	0.1461	1.422	295.1
16,000	216.7	0.7519	10287	0.1015	0.16542	0.1350	1.422	295.1
16,500	216.7	0.7519	9507	0.09383	0.15288	0.1248	1.422	295.1
17,000	216.7	0.7519	8787	0.08672	0.14129	0.1153	1.422	295.1
17,500	216.7	0.7519	8121	0.08014	0.13058	0.1066	1.411	295.1
18,000	216.7	0.7519	7505	0.07407	0.12068	0.09851	1.422	295.1
18,500	216.7	0.7519	6936	0.06845	0.11153	0.09104	1.422	295.1
19,000	216.7	0.7519	6410	0.06326	0.10307	0.08414	1.422	295.1
19,500	216.7	0.7519	5924	0.05847	0.09525	0.07776	1.422	295.1
20,000	216.7	0.7519	5475	0.05403	0.08803	0.07187	1.422	295.1

Appendix C
Airfoil Data

NACA 0006

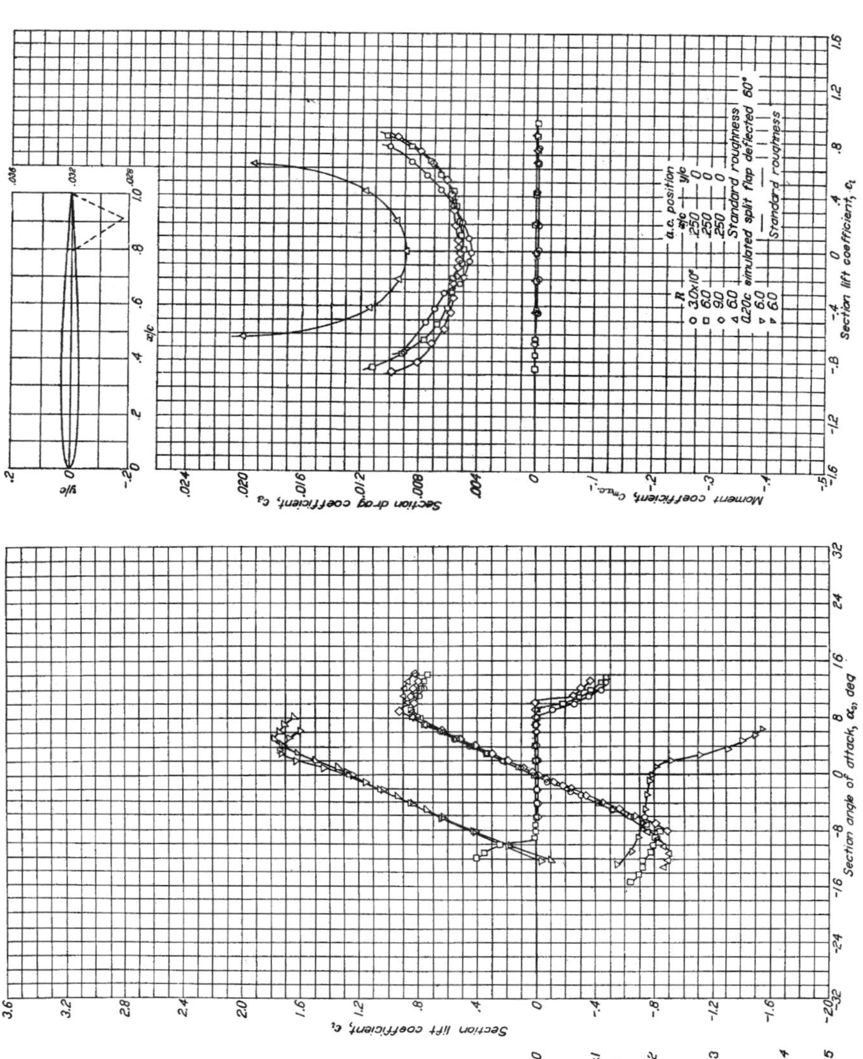

NACA 0009

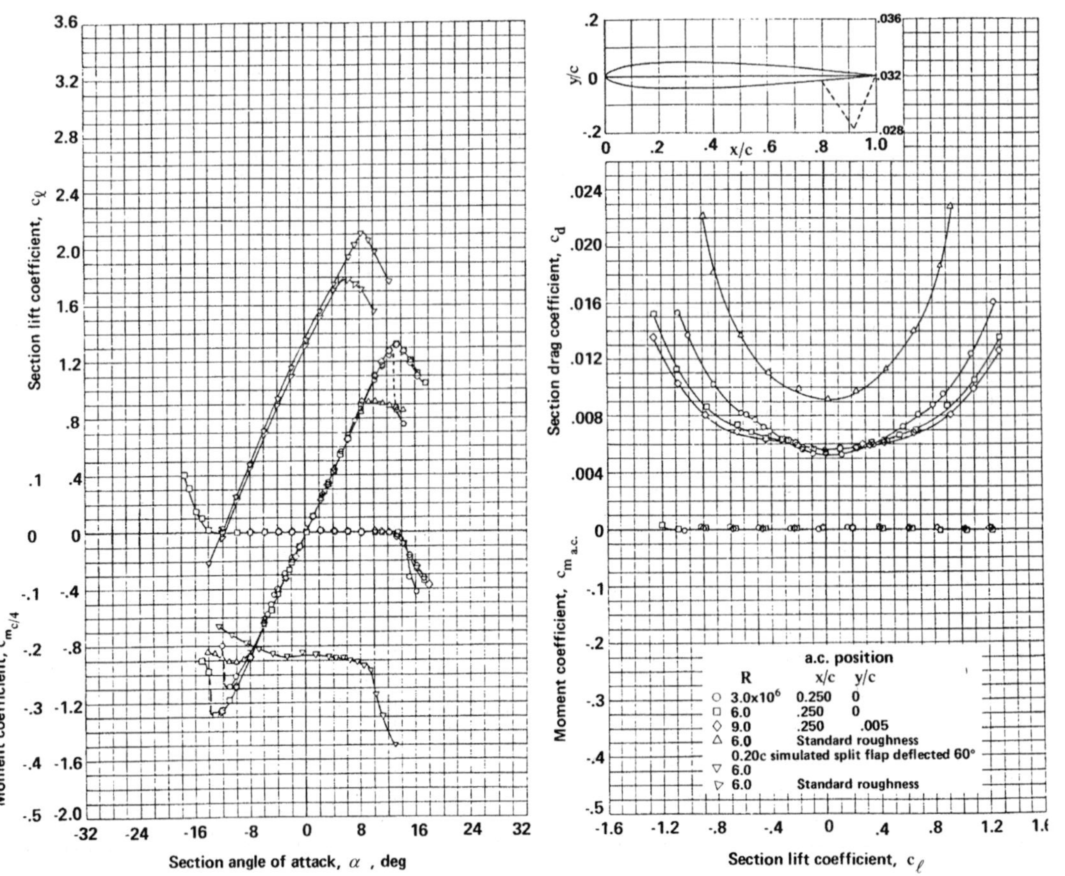

Section lift coefficient, c_ℓ
Moment coefficient, c_m c/4
Section angle of attack, α, deg
y/c
x/c
Section drag coefficient, c_d
Moment coefficient, c_m a.c.
Section lift coefficient, c_ℓ
a.c. position
R x/c y/c
3.0x10^6 0.250 0
6.0 .250 0
9.0 .250 .005
6.0 Standard roughness
0.20c simulated split flap deflected 60°
6.0
6.0 Standard roughness

NACA 0012

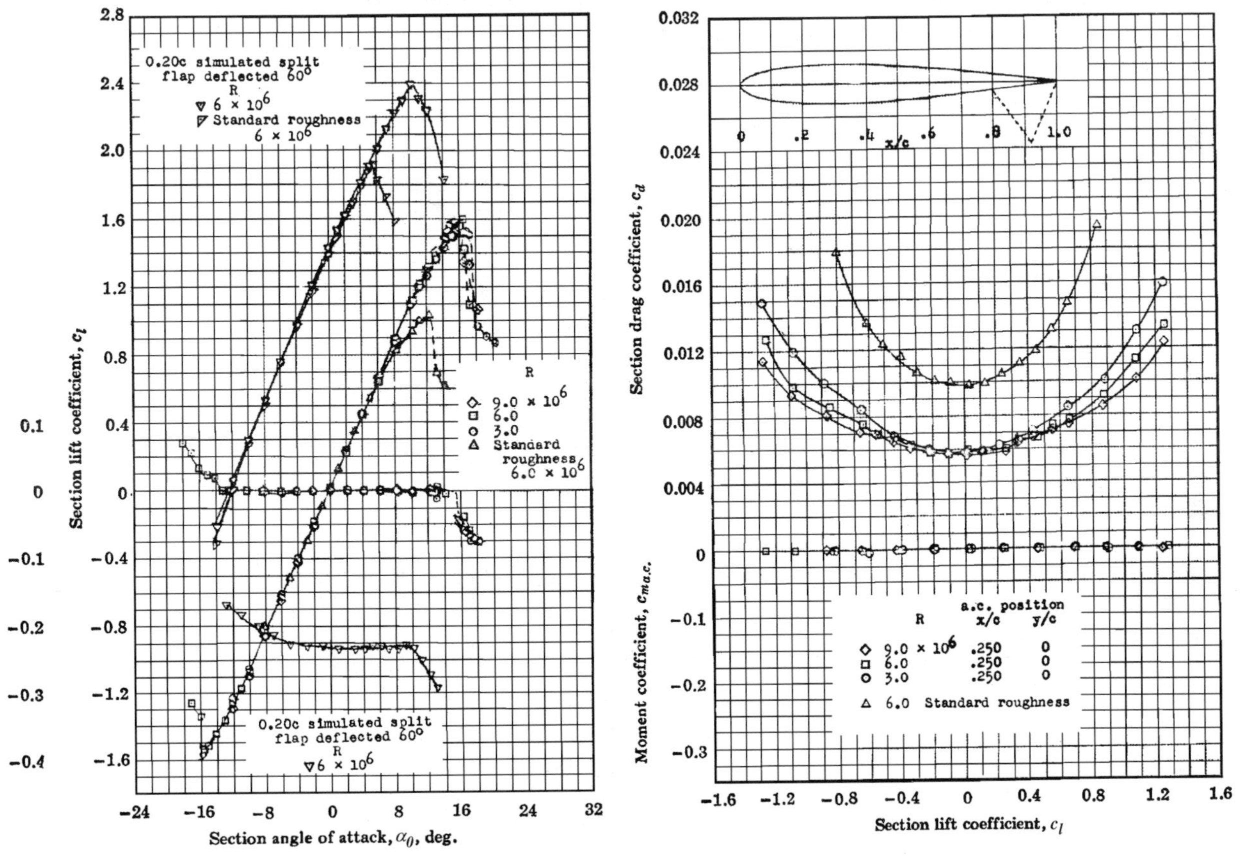

0.20c simulated split flap deflected 60°
R
▽ 6 × 10^6
▽ Standard roughness 6 × 10^6
R
◇ 9.0 × 10^6
□ 6.0
○ 3.0
△ Standard roughness 6.0 × 10^6
0.20c simulated split flap deflected 60°
R
▽ 6 × 10^6
Section lift coefficient, c_l
Moment coefficient, $c_{m_{c/4}}$
Section angle of attack, α_0, deg.
x/c
Section drag coefficient, c_d
Moment coefficient, $c_{m_{a.c.}}$
a.c. position
R x/c y/c
◇ 9.0 × 10^6 .250 0
□ 6.0 .250 0
○ 3.0 .250 0
△ 6.0 Standard roughness
Section lift coefficient, c_l

NACA 1408

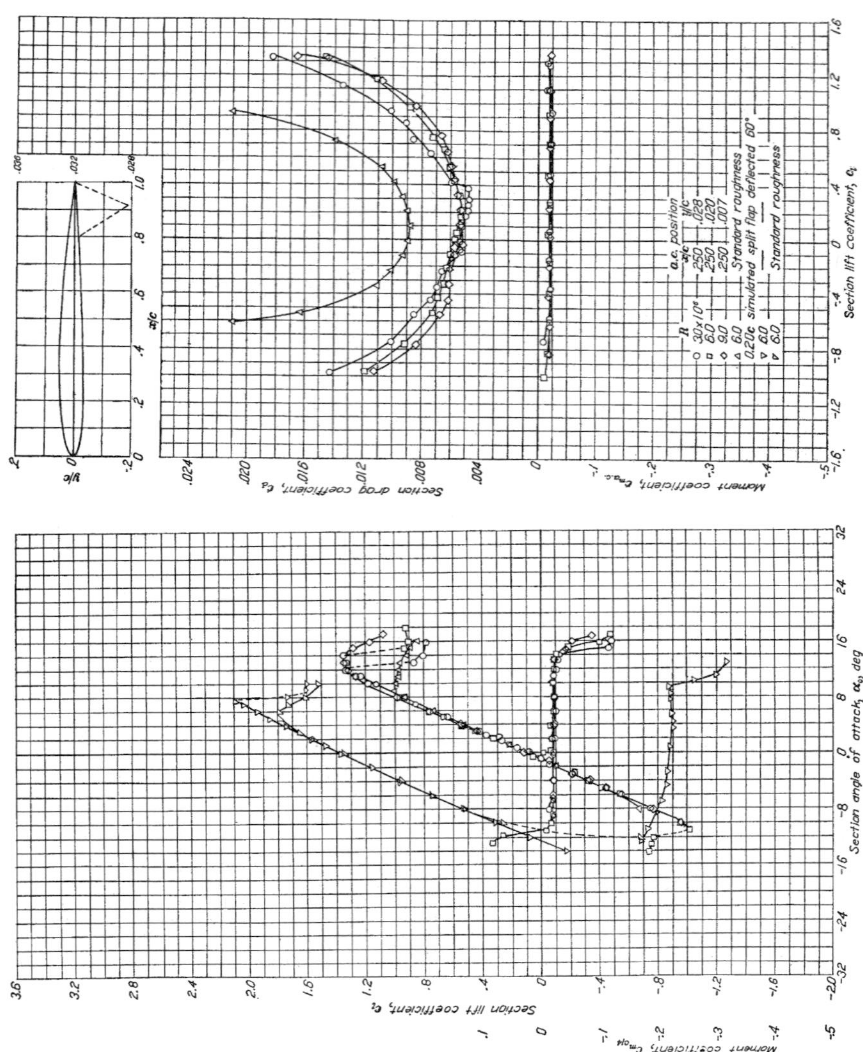
Section lift coefficient, c_l
Section drag coefficient, c_d
Moment coefficient, $c_{m_{a.c.}}$
Section angle of attack, α_0, deg
Moment coefficient, $c_{m_{c/4}}$
x/c
y/c
R
a.c. position
30×10⁶
6.0
9.0
6.0 Standard roughness
0.20c simulated split flap deflected 60°
6.0
6.0 Standard roughness

NACA 1412

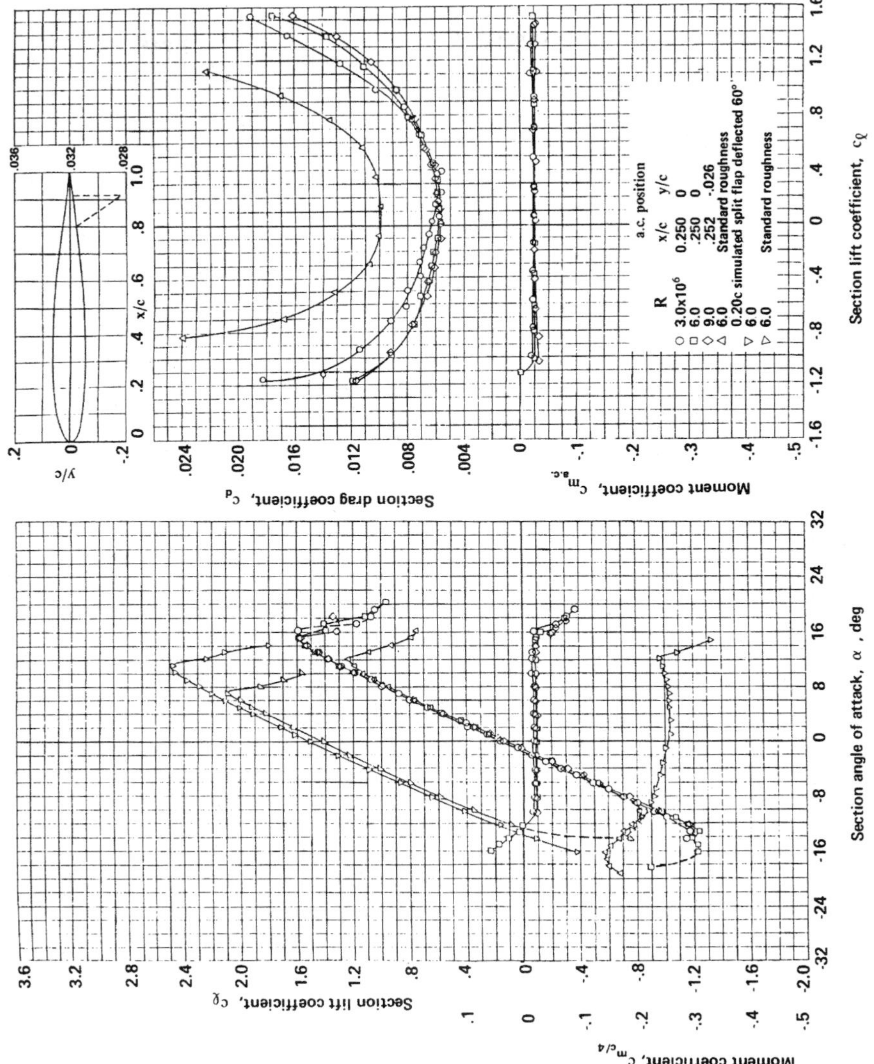
Section lift coefficient, c_ℓ
Section drag coefficient, c_d
Moment coefficient, $c_{m_{a.c.}}$
Moment coefficient, $c_{m_{c/4}}$
Section angle of attack, α, deg
y/c
x/c
a.c. position
R x/c y/c
○ 3.0×10⁶ 0.250 0
□ 6.0 .250 0
◇ 9.0 .252 -.026
△ 6.0 Standard roughness
▽ 6.0 0.20c simulated split flap deflected 60°
▷ 6.0 Standard roughness

NACA 2412

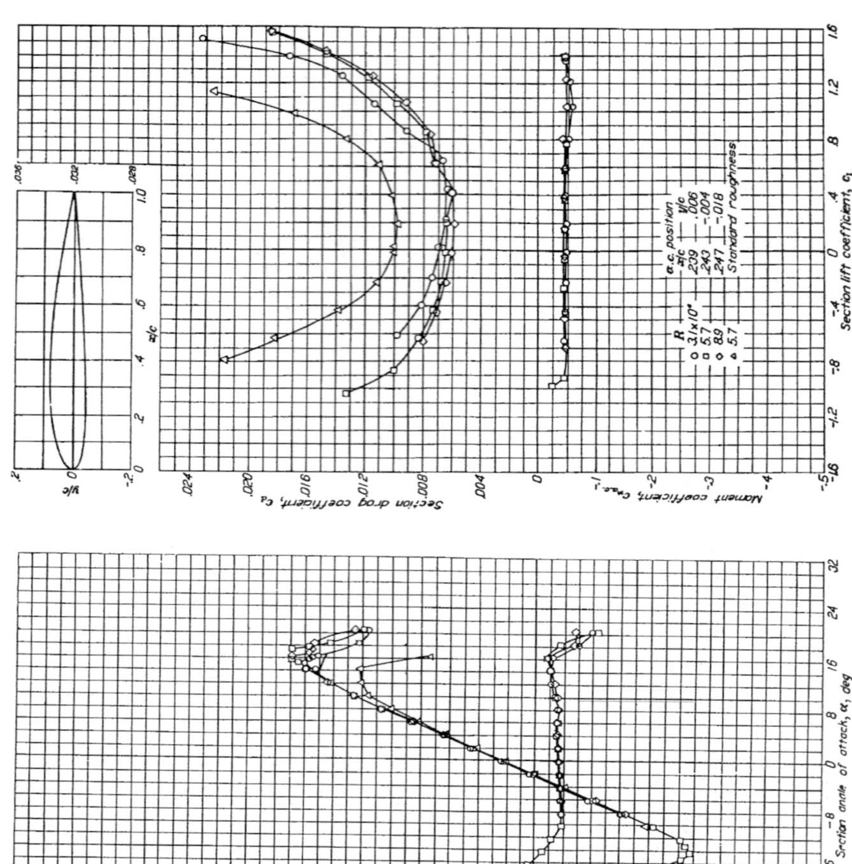
Section lift coefficient, c_l
Section drag coefficient, c_d
Moment coefficient, $c_{m_{a.c.}}$
Section angle of attack, α, deg
Moment coefficient, $c_{m_{c/4}}$
x/c
y/c
R
a.c. position
x/c
y/c
○ 3.1×10⁶ .239 -.006
□ 5.7 .243 -.004
◇ 8.9 .247 -.018
△ 5.7 Standard roughness

NACA 2415

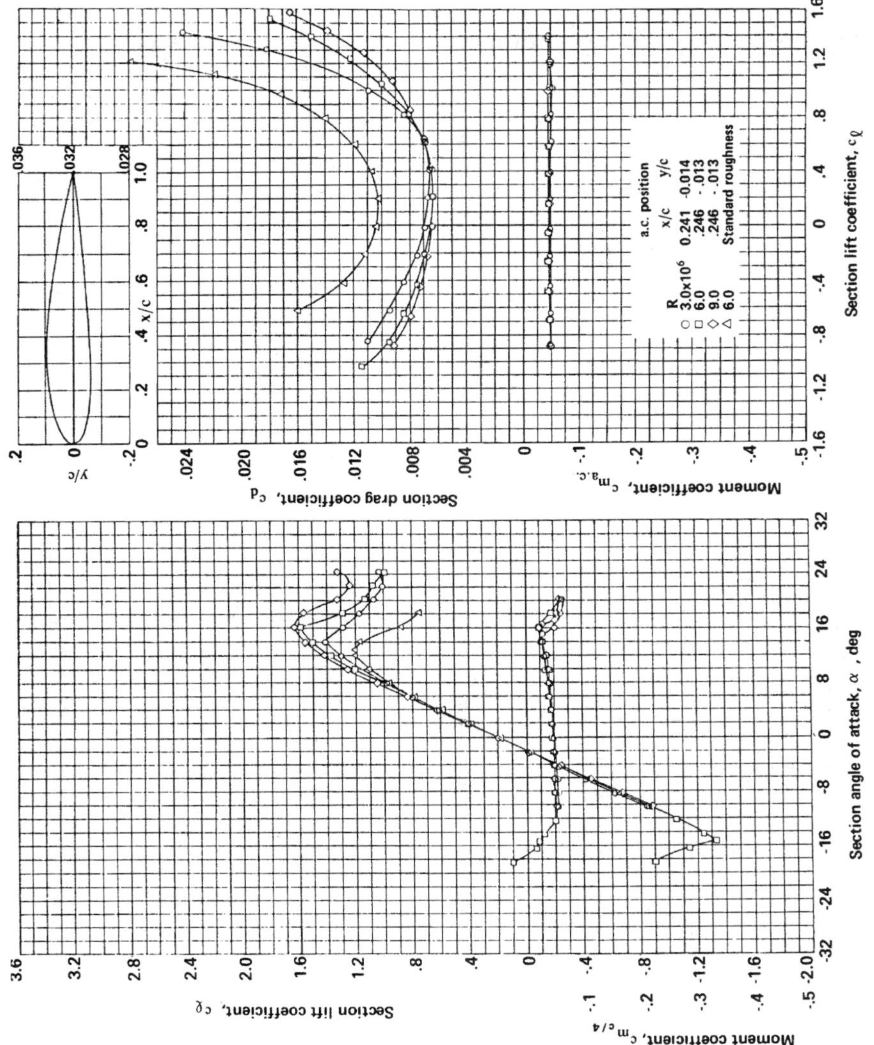
y/c
x/c
Section drag coefficient, c_d
Moment coefficient, $c_{m_{a.c.}}$
Section lift coefficient, c_ℓ
a.c. position
R x/c y/c
○ 3.0×10^6 0.241 -0.014
□ 6.0 .246 -.013
◇ 9.0 .246 -.013
△ 6.0 Standard roughness
Section lift coefficient, c_ℓ
Moment coefficient, $c_{m_{c/4}}$
Section angle of attack, α, deg

NACA 2418

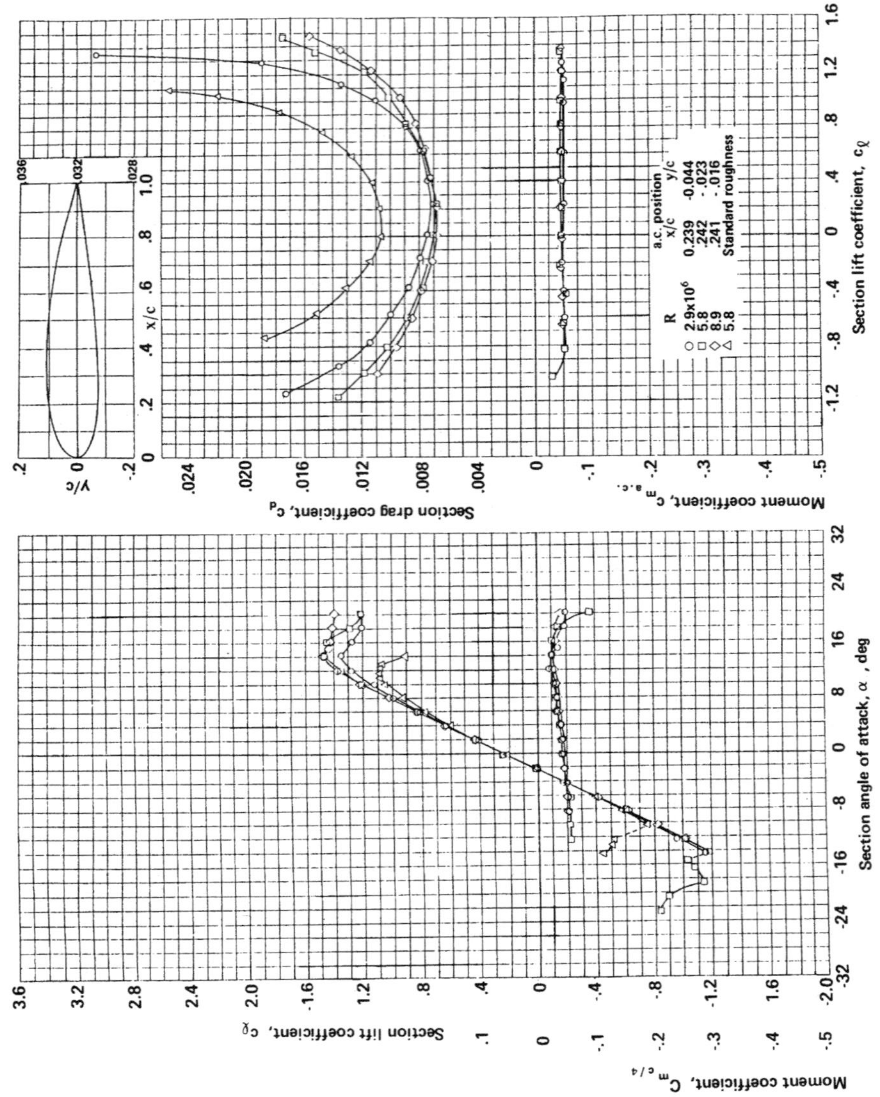
y/c
x/c
Section drag coefficient, c_d
Moment coefficient, $c_{m_{a.c.}}$
Section lift coefficient, c_ℓ
R
2.9x10^6
5.8
8.9
5.8
a.c. position
x/c
0.239
.242
.241
y/c
-0.044
-.023
-.016
Standard roughness
Section angle of attack, α, deg
Moment coefficient, $c_{m_{c/4}}$

NACA 2421

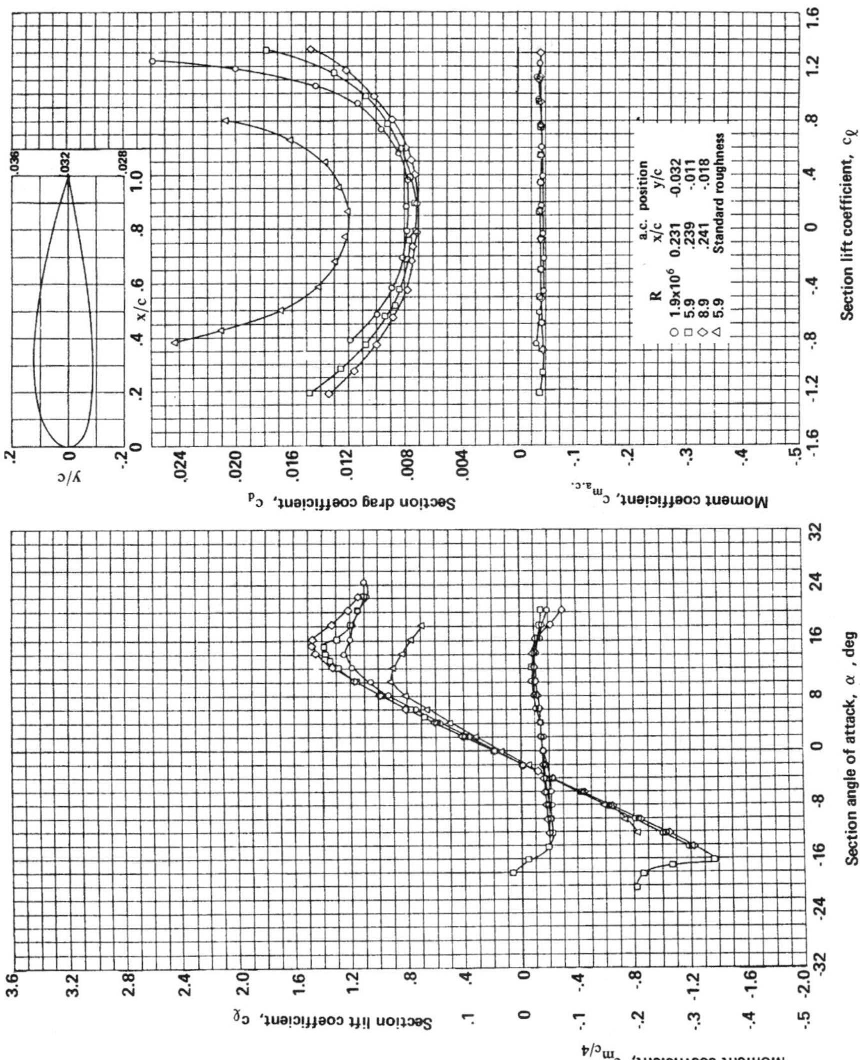
Section lift coefficient, c_l
Section drag coefficient, c_d
Moment coefficient, $c_{m_{a.c.}}$
Moment coefficient, $c_{m_{c/4}}$
Section angle of attack, α , deg
x/c
y/c
R
a.c. position
x/c
y/c
○ 1.9x10[6] 0.231 -0.032
□ 5.9 .239 -.011
◇ 8.9 .241 -.018
△ 5.9 Standard roughness

NACA 4412

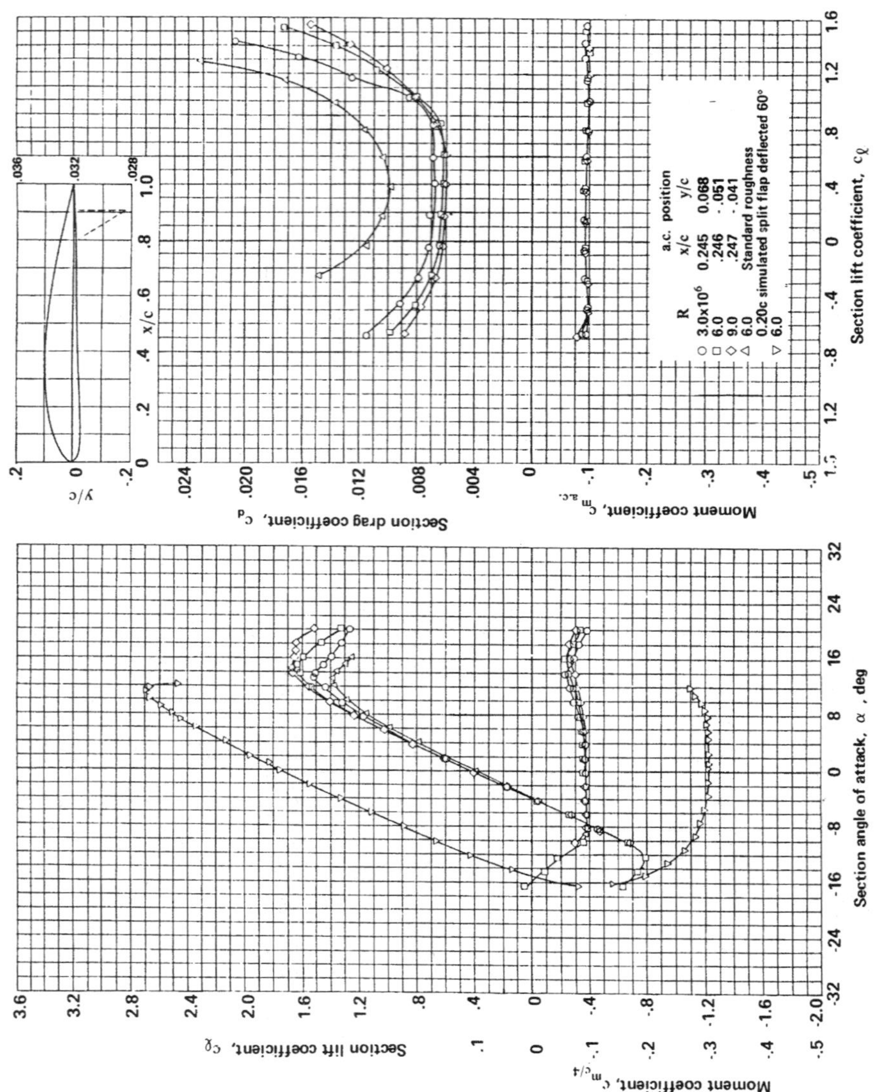
Section lift coefficient, c_ℓ
Moment coefficient, $c_{m_{c/4}}$
Section angle of attack, α, deg
Section drag coefficient, c_d
Moment coefficient, $c_{m_{a.c.}}$
Section lift coefficient, c_ℓ
y/c
x/c
a.c. position
R x/c y/c
3.0x10[6] 0.245 0.068
6.0 .246 -.051
9.0 .247 -.041
6.0 Standard roughness
0.20c simulated split flap deflected 60°
6.0

NACA 4415

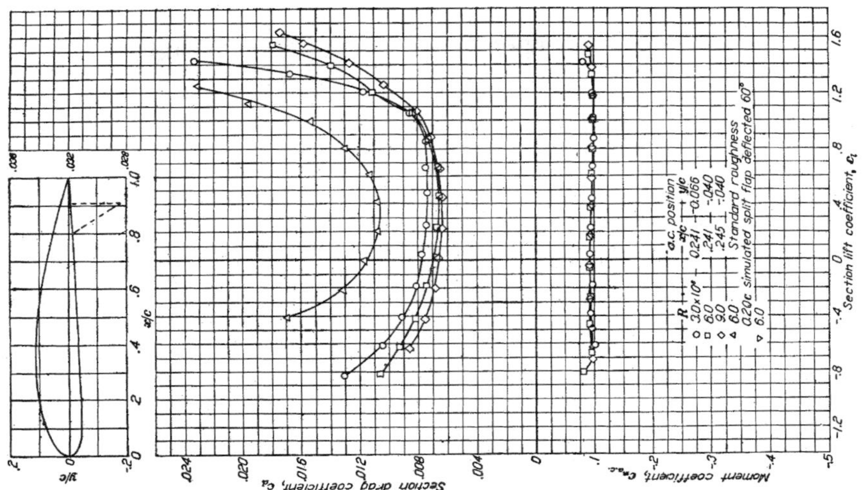
Section drag coefficient, c_d
Moment coefficient, $c_{m_{a.c.}}$
Section lift coefficient, c_l
x/c
y/c
R
a.c. position
x/c
y/c
3.0×10^6 0.241 -0.066
6.0 .241 -.040
9.0 .245 -.040
6.0 Standard roughness
0.20c simulated split flap deflected 60°
6.0

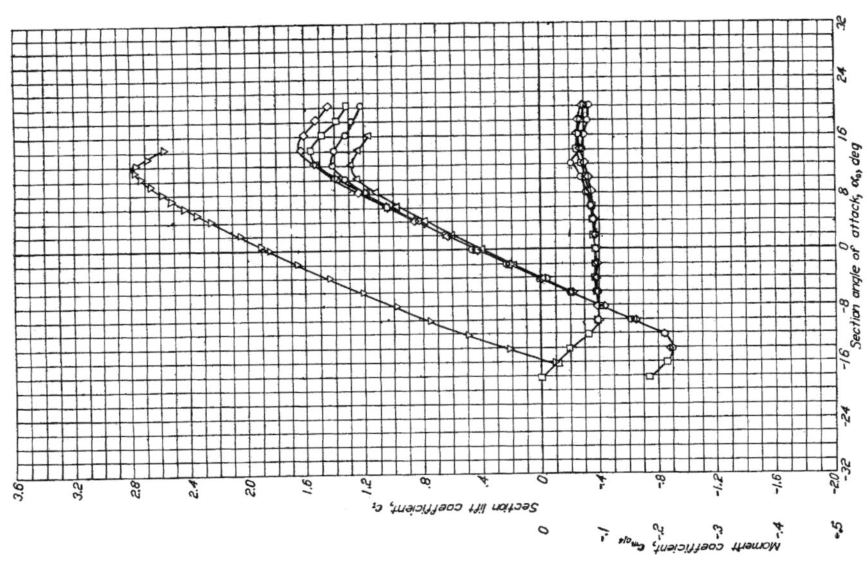
Section lift coefficient, c_l
Moment coefficient, $c_{m_{c/4}}$
Section angle of attack, α_0, deg

NACA 64A210

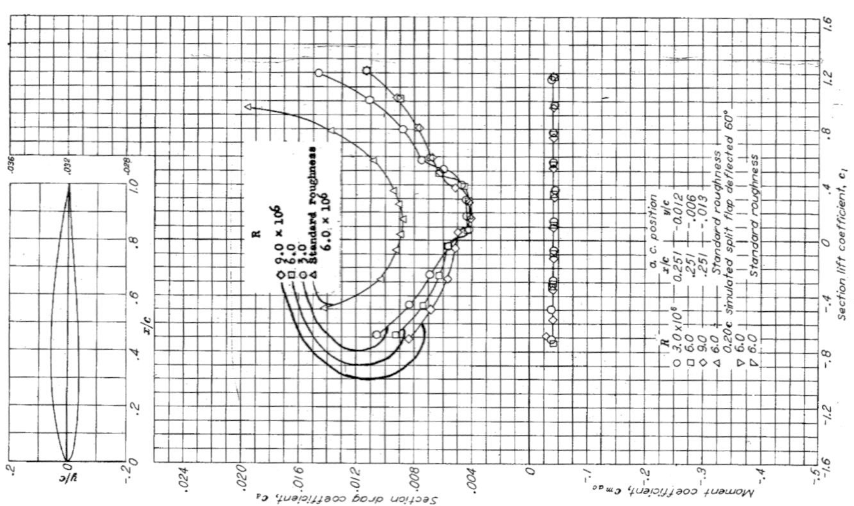

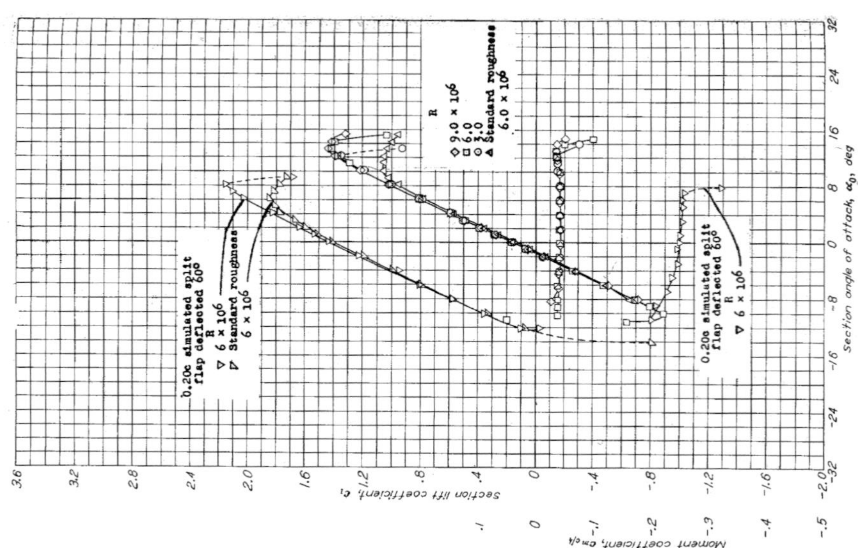

Appendix D
T-38 Performance Data

Dimensions	
Wing	
Total Area	170 ft^2
Span	25 ft 3 in.
Aspect Ratio	3.75
Taper Ratio	0.20
Sweepback (quarter chord)	24 deg
Airfoil Section	NACA 65A004.8
Mean Aerodynamic Chord	92.76 in.
Dihedral	0 deg
Span/Thickness Ratio	51.1
Horizontal Tail	
Total Area	59.0 ft^2
Exposed Area	33.34 ft^2
Aspect Ratio (exposed)	2.82
Taper Ratio (exposed)	0.33
Sweepback (quarter chord)	25 deg
Airfoil Section	NACA 65A004
Span/Thickness Ratio (exposed)	58.6
Vertical Tail	
Total Area	41.42 ft^2
Exposed Area	41.07 ft^2
Apect Ratio (exposed)	1.21
Taper Ratio (exposed)	0.25
Sweepback (quarter chord)	25 deg
Airfoil Section	NACA 65A004
Span/Thickness Ratio	42.2
Airplane	
Height	12 ft 11 in.
Length	43 ft 1 in.
Tread	10 ft 9 in.

T-38 Powerplant Characteristics

Description	
Number	2

Model	J85-GE-5
Manufacturer	General Electric
Type	Turbojet
Augmentation	Afterburning
Compressor	Axial Flow
Exhaust Nozzle	Variable Area
Length (overall)	107.4 in.
Maximum Diameter (afterburner tailpipe)	20.2 in.
Dry Weight	477 lb
Fuel Grade	JP-4
Fuel Specific Weight	6.2 to 6.9 lb/gal

Ratings[1]			
Power Setting	Normal *Power*	Military *Power*	Maximum *Power*
Augmentation	None	None	Afterburner
Engine Speed[2]	96.4	100	100
Thrust per engine—lb			
No losses	2140	2455	3660
Installed	1770	1935	2840
Specific fuel consumption[3]			
Installed	1.09	1.14	2.64
Operating Limitations			
Power Setting	*Normal*	*Military*	*Maximum*
Turbine Discharge			
Total Temp (°F)	1050	1220	1220

Notes
[1]Sea level static ICAO standard conditions with a fuel-specific weight of 6.5 lb/gal.
[2]Units are % rpm where 100% = 16,500 rpm.
[3]Units are lb/hr per lb thrust.

T-38A "Lift Curve"
Coefficient of Lift vs Angle of Attack
(Rigid Wing-and-Body Model, Mach = 0.4
Out of Ground Effect)

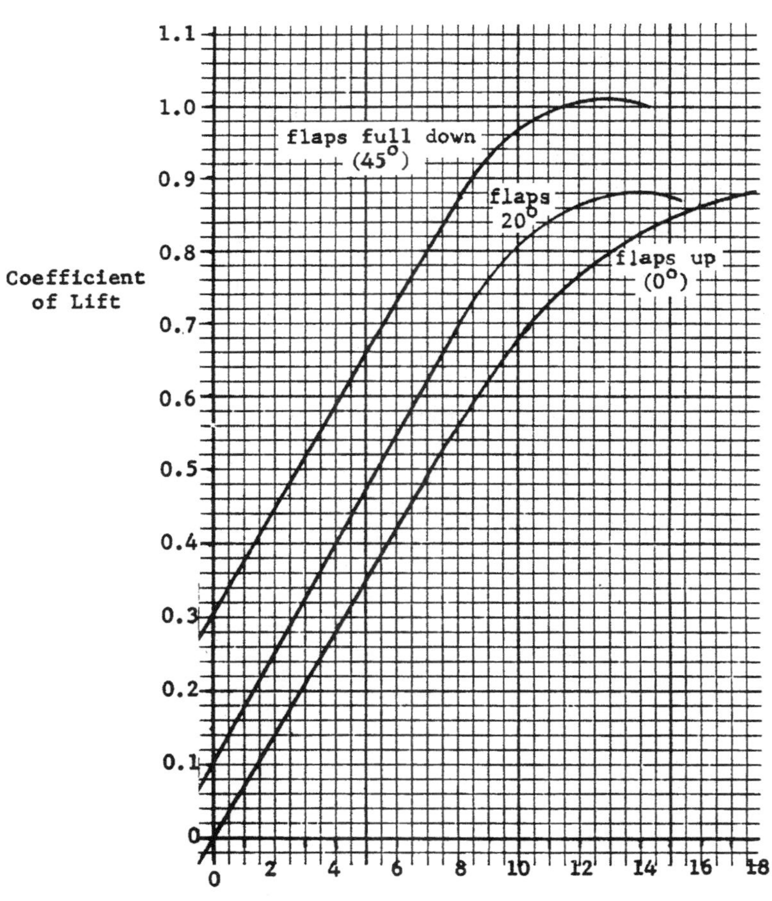

T-38A Zero-Lift Drag Coefficient
Variation with Mach Number
(Full-Scale Model - Clean Configuration)

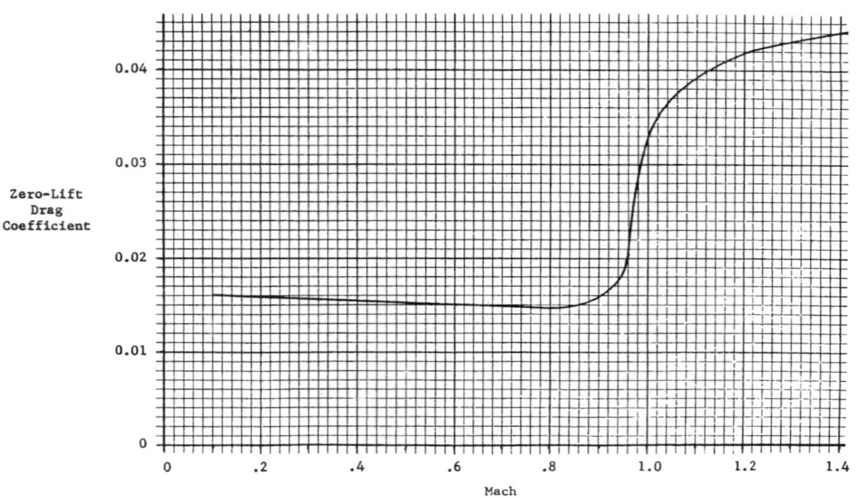

T-38A Drag-Due-to-Lift Factor
Variations with Mach Number
For Five Values of Lift Coefficient
(Full-Scale Model - Clean Configuration)

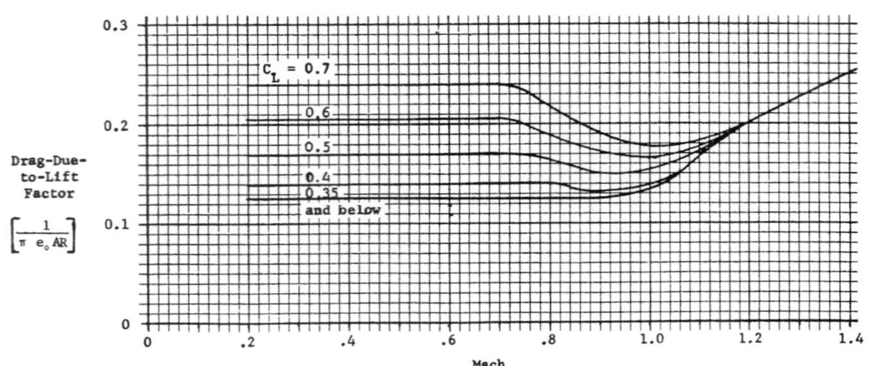

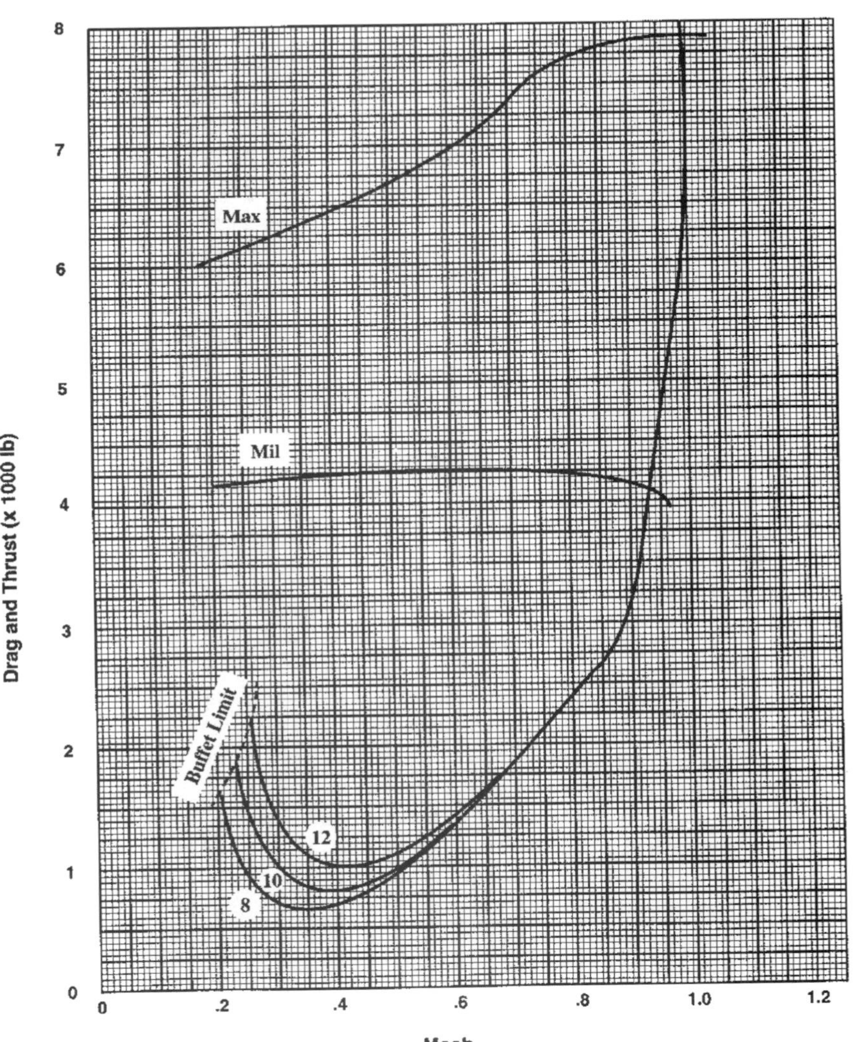
Thrust Required and Thrust Available
(2) J85-GE-5A Engines
Aircraft Weights of 12,000, 10,000 and 8000 lb
at Sea Level
Drag and Thrust (x 1000 lb)
Mach
0
1
2
3
4
5
6
7
8
0
.2
.4
.6
.8
1.0
1.2
Max
Mil
Buffet Limit
12
10
8

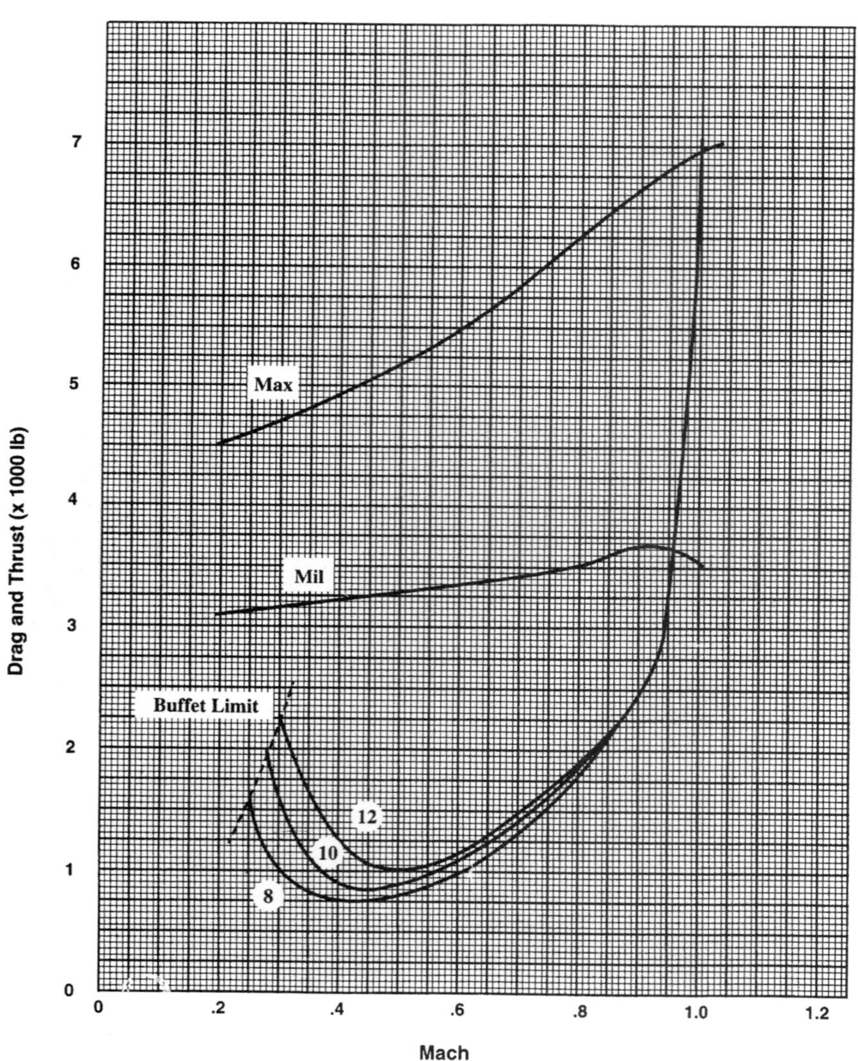
Thrust Required and Thrust Available
(2) J85-GE-5A Engines
Aircraft Weights of 12,000, 10,000 and 8000 lb
at an Altitude of 10,000 ft
Drag and Thrust (x 1000 lb)
0
1
2
3
4
5
6
7
Max
Mil
Buffet Limit
12
10
8
0
.2
.4
.6
.8
1.0
1.2
Mach

Thrust Required and Thrust Available
(2) J85-GE-5A Engines
Aircraft Weights of 12,000, 10,000 and 8000 lb
at an Altitude of 20,000 ft

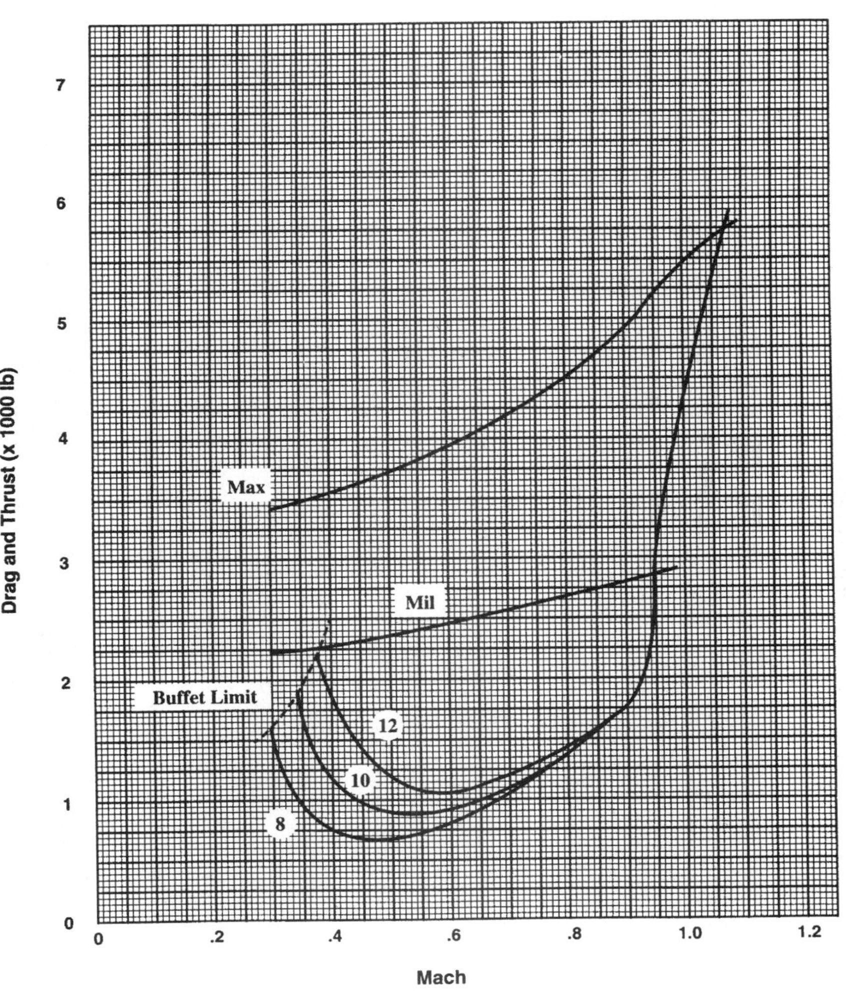

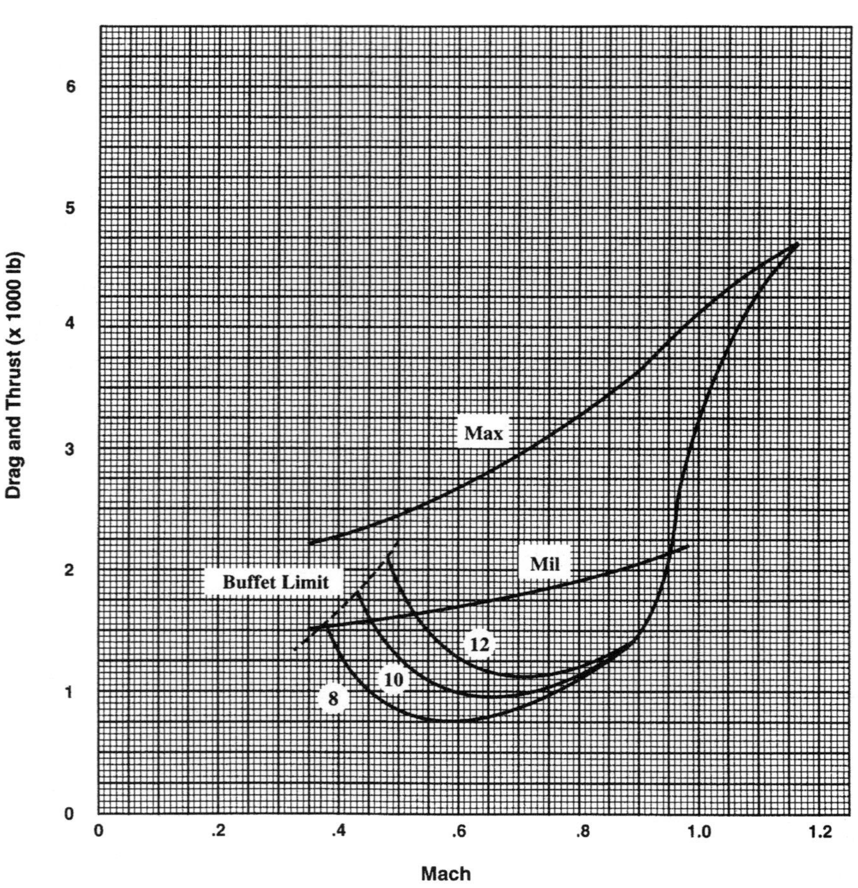
Thrust Required and Thrust Available
(2) J85-GE-5A Engines
Aircraft Weights of 12,000, 10,000 and 8000 lb
at an Altitude of 30,000 ft
Drag and Thrust (x 1000 lb)
0
1
2
3
4
5
6
Max
Mil
Buffet Limit
12
10
8
0
.2
.4
.6
.8
1.0
1.2
Mach

Thrust Required and Thrust Available
(2) J85-GE-5A Engines
Aircraft Weights of 12,000, 10,000 and 8000 lb
at an Altitude of 40,000 ft

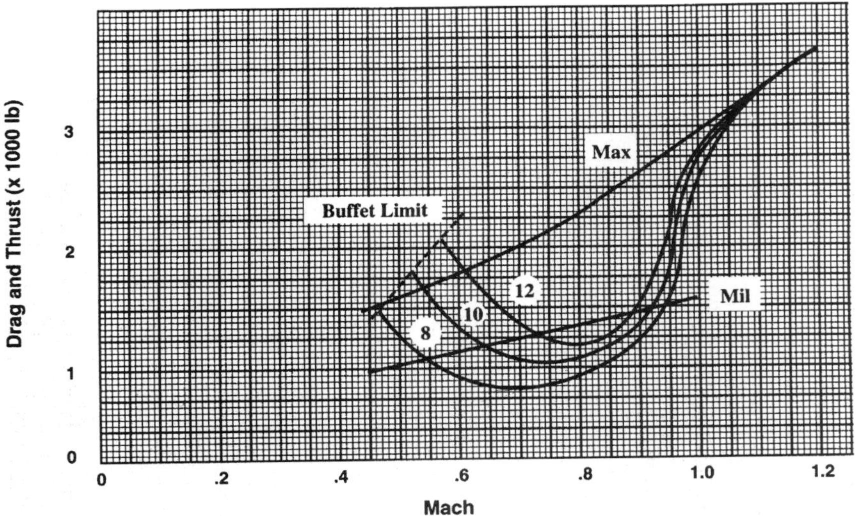

Thrust Required and Thrust Available
(2) J85-GE-5A Engines
Aircraft Weights of 12,000, 10,000 and 8000 lb
at an Altitude of 50,000 ft

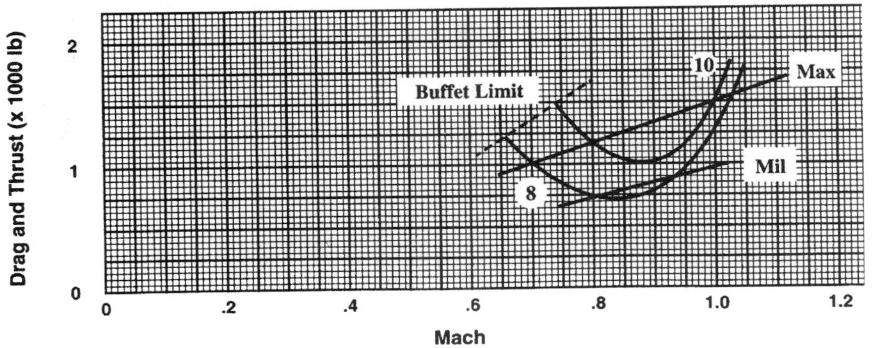

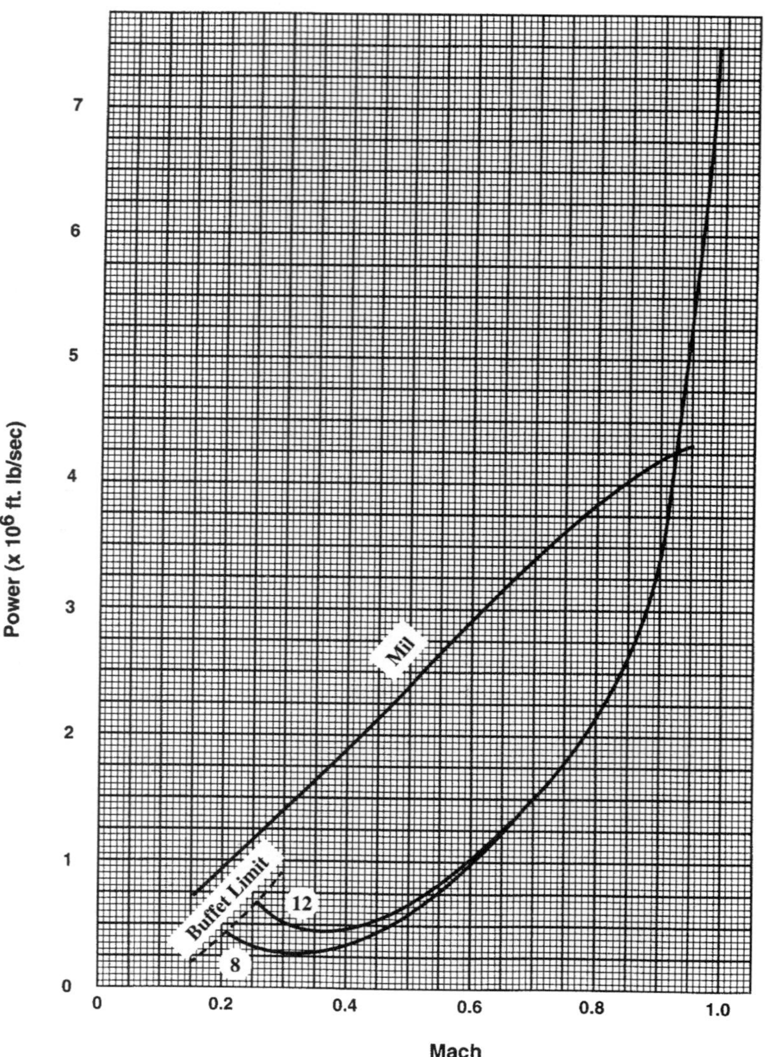
Power Required and Power Available
(2) J85-GE-5A Engines
Aircraft Weights of 8000 and 12,000 lbs
at Sea Level
Power (x 10^6 ft. lb/sec)
Mach
Mil
Buffet Limit
12
8
0
1
2
3
4
5
6
7
0
0.2
0.4
0.6
0.8
1.0

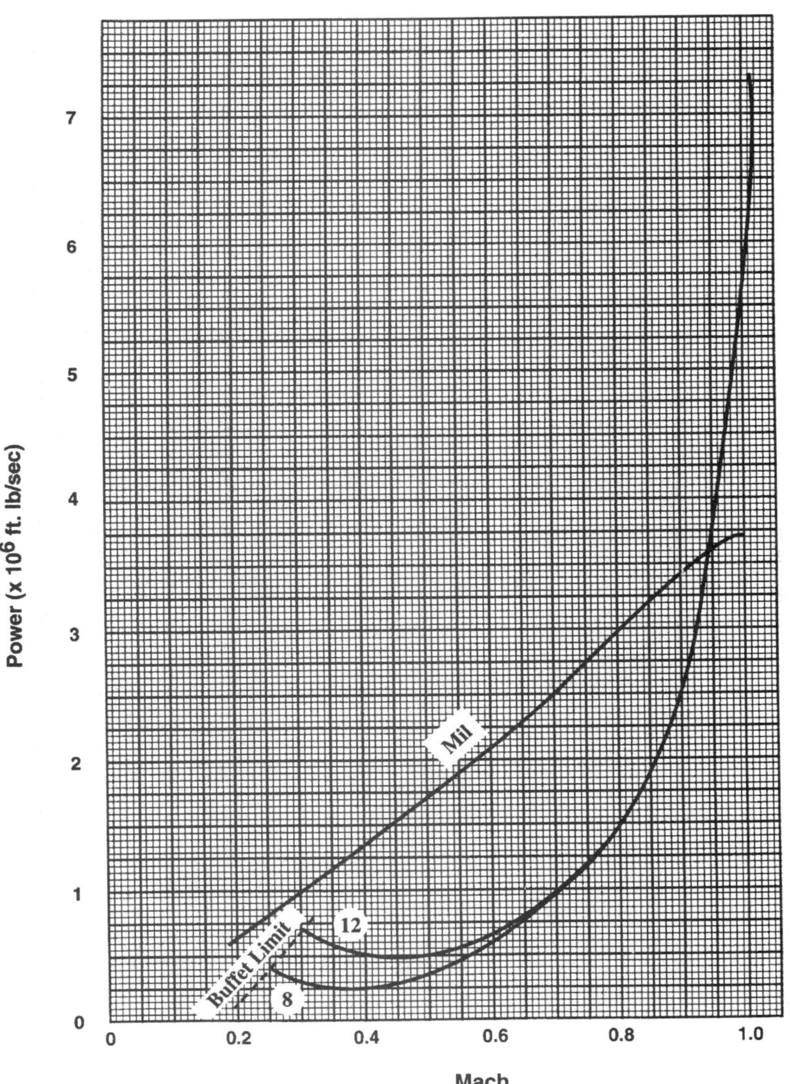
Power Required and Power Available
(2) J85-GE-5A Engines
Aircraft Weights of 8000 and 12,000 lbs
at an Altitude of 10,000 ft
Power (x 10^6 ft. lb/sec)
0
1
2
3
4
5
6
7
Mil
12
8
Buffet Limit
0
0.2
0.4
0.6
0.8
1.0
Mach

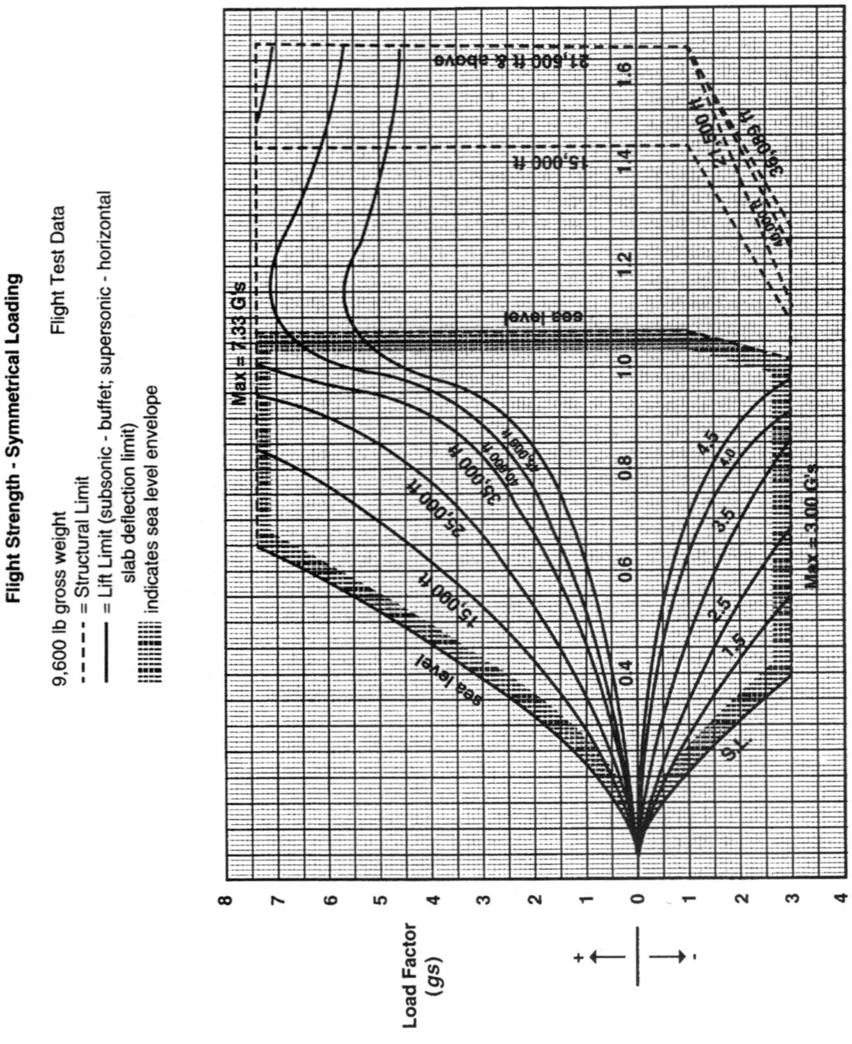

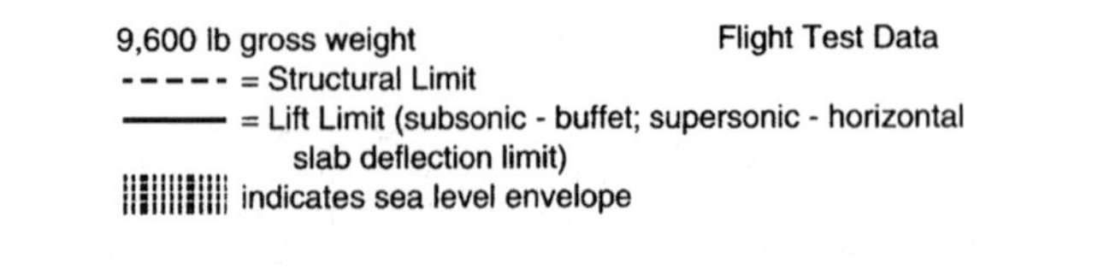
Flight Strength - Symmetrical Loading
9,600 lb gross weight
Flight Test Data
= Structural Limit
= Lift Limit (subsonic - buffet; supersonic - horizontal slab deflection limit)
indicates sea level envelope
Load Factor (gs)
+
-
8
7
6
5
4
3
2
1
0
1
2
3
4
0.4
0.6
0.8
1.0
1.2
1.4
1.6
Max = 7.33 G's
Max = 3.00 G's
sea level
15,000 ft
25,000 ft
35,000 ft
40,000 ft
45,000 ft
sea level
15,000 ft
21,500 ft & above
21,500 ft
36,089 ft
40,000 ft
S.L.
1.5
2.5
3.5
4.0
4.5

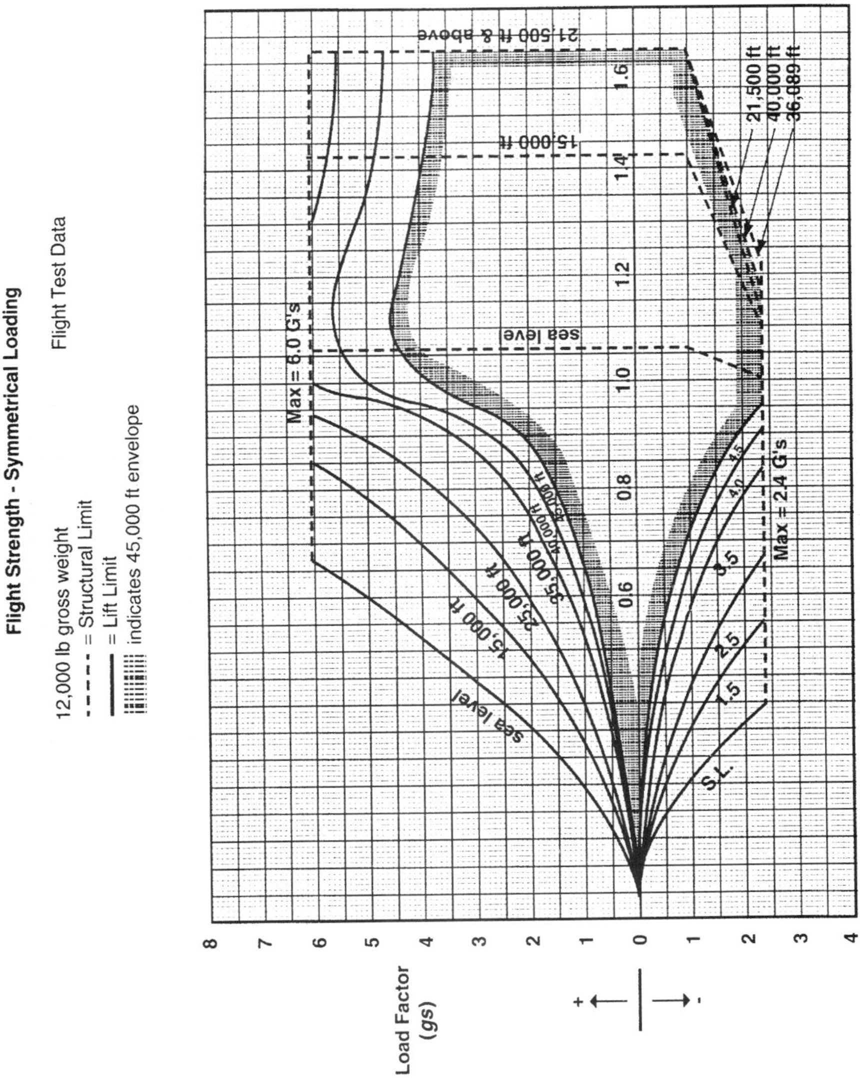

Flight Strength - Symmetrical Loading

Flight Test Data

12,000 lb gross weight

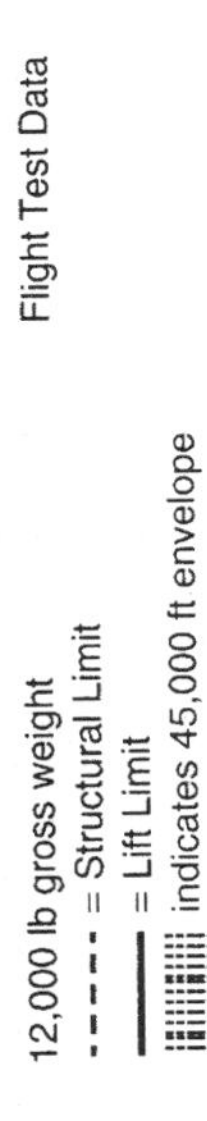

Load Factor (gs)

+ ← → -

8 7 6 5 4 3 2 1 0 1 2 3 4

Max = 6.0 G's

Max = 2.4 G's

21,500 ft & above

15,000 ft

sea level

sea level

15,000 ft

25,000 ft

35,000 ft

40,000 ft

45,000 ft

21,500 ft

40,000 ft

36,089 ft

S.L.

1.5

2.5

3.5

4.0

4.5

0.6 0.8 1.0 1.2 1.4 1.6

Appendix E
Selected Laplace Transforms

	$F(s)$	$f(t)\ t > 0$
1	1	$\delta(t)$
2	$e - Ts$	$\delta(t - T)$
3	$\dfrac{1}{s+a}$	e^{-at}
4	$\dfrac{1}{(s+a)^n}$	$\dfrac{1}{(n-1)!}t^{n-1}e^{-at} \quad n = 1, 2, 3, \ldots$
5	$\dfrac{1}{(s+a)(s+b)}$	$\dfrac{1}{b-a}(e^{-at} - e^{-bt})$
6	$\dfrac{s}{(s+a)(s+b)}$	$\dfrac{1}{a-b}(ae^{-at} - be^{-bt})$
7	$\dfrac{s+z}{(s+a)(s+b)}$	$\dfrac{1}{b-a}[(z-a)e^{-at} - (z-b)e^{-bt}]$
8	$\dfrac{1}{(s+a)(s+b)(s+c)}$	$\dfrac{e^{-at}}{(b-a)(c-a)} + \dfrac{e^{-bt}}{(c-b)(a-b)} + \dfrac{e^{-ct}}{(a-c)(b-c)}$
9	$\dfrac{s+z}{(s+a)(s+b)(s+c)}$	$\dfrac{(z-a)e^{-at}}{(b-a)(c-a)} + \dfrac{(z-b)e^{-bt}}{(c-b)(a-b)} + \dfrac{(z-c)e^{-ct}}{(\alpha-c)(b-c)}$
10	$\dfrac{\omega}{s^2+\omega^2}$	$\sin \omega t$
11	$\dfrac{s}{s^2+\omega^2}$	$\cos \omega t$
12	$\dfrac{s+z}{s^2+\omega^2}$	$\sqrt{\dfrac{z^2+\omega^2}{\omega^2}}\sin(\omega t + \phi) \quad \phi \equiv \tan^{-1}(\omega/z)$
13	$\dfrac{s \sin\phi + \omega\cos\phi}{s^2+\omega^2}$	$\sin(\omega t + \phi)$
14	$\dfrac{1}{(s+a)^2+\omega^2}$	$\dfrac{1}{\omega}e^{-at}\sin \omega t$
15	$\dfrac{1}{s^2+2\zeta\omega_n s+\omega_n^2}$	$\dfrac{1}{\omega_d}e^{-\zeta\omega_n t}\sin \omega_d t \qquad \omega_d \equiv \omega_n\sqrt{1-\zeta^2}$
16	$\dfrac{s+a}{(s+a)^2+\omega^2}$	$e^{-at}\cos \omega t$

(continued)

	$F(s)$	$f(t)\ t > 0$
17	$\dfrac{s+z}{(s+a)^2+\omega^2}$	$\sqrt{\dfrac{(z-a)^2+\omega^2}{\omega^2}}e^{-at}\sin(\omega t+\phi) \quad \phi \equiv \tan^{-1}\left(\dfrac{\omega}{z-a}\right)$
18	$\dfrac{1}{s}$	$u(t)$ or 1 unit step
19	$\dfrac{1}{s}e^{-Ts}$	$u(t-T)$ delayed step
20	$\dfrac{1}{s}(1-e^{-Ts})$	$u(t)-u(t-T)$ rectangular pulse
21	$\dfrac{1}{s(s+a)}$	$\dfrac{1}{a}(1-e^{-at})$
22	$\dfrac{1}{s(s+a)(s+b)}$	$\dfrac{1}{ab}\left(1-\dfrac{be^{-at}}{b-a}+\dfrac{ae^{-bt}}{b-a}\right)$
23	$\dfrac{s+z}{s(s+a)(s+b)}$	$\dfrac{1}{ab}\left(z-\dfrac{b(z-a)e^{-at}}{b-a}+\dfrac{a(z-b)e^{-bt}}{b-a}\right)$
24	$\dfrac{1}{s(s^2+\omega^2)}$	$\dfrac{1}{\omega^2}(1-\cos\omega t)$
25	$\dfrac{s+z}{s(s^2+\omega^2)}$	$\dfrac{z}{\omega^2}-\sqrt{\dfrac{z^2+\omega^2}{\omega^4}}\cos(\omega t+\phi) \quad \phi \equiv \tan^{-1}(\omega/z)$
26	$\dfrac{1}{s(s^2+2\zeta\omega_n s+\omega_n^2)}$	$\dfrac{1}{\omega_n^2}-\dfrac{1}{\omega_n\omega_d}e^{-\zeta\omega_n t}\sin(\omega_d t+\phi)$ $\omega_d \equiv \omega_n\sqrt{1-\zeta^2} \qquad \phi \equiv \cos^{-1}\zeta$
27	$\dfrac{1}{s(s+a)^2}$	$\dfrac{1}{\alpha^2}(1-e^{-at}-ate^{-at})$
28	$\dfrac{s+z}{s(s+a)^2}$	$\dfrac{1}{a^2}[z-ze^{-at}+a(a-z)te^{-at}]$
29	$\dfrac{1}{s^2}$	t unit ramp
30	$\dfrac{1}{s^2(s+a)}$	$\dfrac{1}{a^2}(at-1+e^{-at})$
31	$\dfrac{1}{s^n} \quad n=1,2,3,\ldots$	$\dfrac{t^{n-1}}{(n-1)!} \qquad 0!=1$

Appendix F
Cramer's Rule

This appendix will present a brief review of Cramer's rule for solving systems of simultaneous equations, such as those encountered with the aircraft equations of motion. Most engineering mathematics textbooks provide a more detailed coverage of this subject.

Consider the following set of simultaneous equations:

$$3x - 4y = -6$$
$$2x + 5y = 19$$

To use Cramer's rule to solve these equations for x and y, we first recast the equations in matrix format.

$$\underbrace{\begin{bmatrix} 3 & -4 \\ 2 & 5 \end{bmatrix}}_{\text{Coefficient Matrix}} \begin{bmatrix} x \\ y \end{bmatrix} = \underbrace{\begin{bmatrix} -6 \\ 19 \end{bmatrix}}_{\text{Input Matrix}}$$

We next find the determinant of the coefficient matrix

$$D = \begin{vmatrix} 3 & -4 \\ 2 & 5 \end{vmatrix} = 23$$

The values of x and y can then be found using

$$x = \frac{D_1}{D}, y = \frac{D_2}{D}$$

where the determinants D_1 and D_2 are obtained from the coefficient matrix with one of the appropriate columns replaced with the input matrix column as illustrated:

$$D_1 = \begin{vmatrix} -6 & -4 \\ 19 & 5 \end{vmatrix} = 46$$

$$D_2 = \begin{vmatrix} 3 & -6 \\ 2 & 19 \end{vmatrix} = 69$$

The solution then becomes

$$x = \frac{D_1}{D} = \frac{46}{23} = 2$$
$$y = \frac{D_2}{D} = \frac{69}{23} = 3$$

The same approach is applicable for higher-order sets of simultaneous equations. In the case of developing transfer functions for the longitudinal and lateral–directional equations of motion, Cramer's rule is applied for the case of three equations and three unknowns. A simplified example of this is presented

$$\begin{aligned} 2u - \alpha + 2\theta &= 2\delta_e \\ u + 10\alpha - 3\theta &= 5\delta_e \\ -u + \alpha + \theta &= -3\delta_e \end{aligned}$$

Recasting in matrix form

$$\begin{bmatrix} 2 & -1 & 2 \\ 1 & 10 & -3 \\ -1 & 1 & 1 \end{bmatrix} \begin{bmatrix} u \\ \alpha \\ \theta \end{bmatrix} = \begin{bmatrix} 2 \\ 5 \\ -3 \end{bmatrix} \delta_e$$

and finding the determinant of the coefficient matrix

$$\begin{vmatrix} 2 & -1 & 2 \\ 1 & 10 & -3 \\ -1 & 1 & 1 \end{vmatrix} = 46$$

The solution then becomes

$$u = \frac{\begin{vmatrix} 2 & -1 & 2 \\ 5 & 10 & -3 \\ -3 & 1 & 1 \end{vmatrix}}{D} = \frac{92}{46} = 2\delta_e$$

$$\alpha = \frac{\begin{vmatrix} 2 & 2 & 2 \\ 1 & 5 & -3 \\ -1 & -3 & 1 \end{vmatrix}}{D} = \frac{0}{46} = 0\delta_e$$

$$\theta = \frac{\begin{vmatrix} 2 & -1 & 2 \\ 1 & 10 & 5 \\ -1 & 1 & -3 \end{vmatrix}}{D} = \frac{-46}{46} = -1\delta_e$$

or, in transfer function form

$$\frac{u}{\delta_e} = 2$$
$$\frac{\alpha}{\delta_e} = 0$$
$$\frac{\theta}{\delta_e} = -1$$

Appendix G
Development of Longitudinal and Lateral–Directional Transfer Functions

Table G.1 Longitudinal airplane transfer functions

$$\frac{\begin{vmatrix} X_{\delta_e} & -X_\alpha & g\cos\theta_1 \\ Z_{\delta_e} & \{s(U_1 - Z_{\dot\alpha}) - Z_\alpha\} & \{-(Zq + U_1)s + g\sin\theta_1\} \\ M_{\delta_e} & -\{M_{\dot\alpha}s + M_\alpha + M_{T_\alpha}\} & (s^2 - \mathrm{M_q s}) \end{vmatrix}}{\begin{vmatrix} (s - X_u - X_{T_u}) & -X_\alpha & g\cos\theta_1 \\ -Z_u & \{s(U_1 - Z_{\dot\alpha}) - Z_\alpha\} & \{-(Z_q + U_1)s + g\sin\theta_1\} \\ -(M_u + M_{T_u}) & -\{M_{\dot\alpha}s + M_\alpha + M_{T_\alpha}\} & (s^2 - \mathrm{M_q s}) \end{vmatrix}} = \frac{u(s)}{\delta_e(s)} = \frac{N_u}{D_1}$$

$\bar{D}_1 = Es^4 + Fs^3 + Gs^2 + Hs + I$, where:

$$E = U_1 - Z_{\dot\alpha}$$

$$F = -(U_1 - Z_{\dot\alpha})(X_u + X_{T_u} + M_q) - Z_\alpha - M_{\dot\alpha}(U_1 + Z_q)$$

$$G = (X_u + X_{T_u})\{M_q(U_1 - Z_{\dot\alpha}) + Z_\alpha + M_{\dot\alpha}(U_1 + Z_q)\} + M_q Z_\alpha - Z_u X_\alpha + M_{\dot\alpha} g\sin\theta_1 + \\ - (M_\alpha + M_{T_\alpha})(U_1 + Z_q)$$

$$H = g\sin\theta_1\{M_\alpha + M_{T_u} - M_{\dot\alpha}(X_u + X_{T_u})\} + g\cos\theta_1\{Z_u M_{\dot\alpha} + (M_u + M_{T_u})(U_1 - Z_{\dot\alpha})\} \\ + (M_u + M_{T_u})\{-X_\alpha(U_1 + Z_q)\} + Z_u X_\alpha M_q \\ + (X_u + X_{T_u})\{(M_\alpha + M_{T_u})(U_1 + Z_q) - M_q Z_\alpha\}$$

$$I = g\cos\theta_1\{(M_\alpha + M_{T_\alpha})Z_u - Z_\alpha(M_u + M_{T_u})\} \\ + g\sin\theta_1\{M_u + M_{T_u})X_\alpha - (X_u + X_{T_u})(M_\alpha + M_{T_\alpha})\}$$

$N_u = A_u s^3 + B_u s^2 + C_u s + D_u$, where:

$$A_u = X_{\delta_e}(U_1 - Z_{\dot\alpha})$$

$$B_u = -X_{\delta_e}\{(U_1 - Z_{\dot\alpha})M_q + Z_\alpha + M_{\dot\alpha}(U_1 + Z_q)\} + Z_{\delta_e}X_\alpha$$

$$C_u = X_{\delta_e}\{M_q Z_\alpha + M_{\dot\alpha} g\sin\theta_1 - (M_\alpha M_{T_\alpha})(U_1 + Z_q)\} \\ + Z_{\delta_e}\{M_{\dot\alpha} g\cos\theta_1 - X_\alpha M_q\} + M_{\delta_e}\{X_\alpha(U_1 + Z_q) - (U_1 - Z_{\dot\alpha})g\cos\theta_1\}$$

$$D_u = X_{\delta_e}(M_\alpha + M_{T_\alpha})g\sin\theta_1 - Z_{\delta_e}M_\alpha g\cos\theta_1 M_{\delta_e}(Z_\alpha g\cos\theta_1 - X_\alpha g\sin\theta_1)$$

$$\frac{\begin{vmatrix} (s - X_u - X_{T_u}) & X_{\delta_e} & g\cos\theta_1 \\ -Z_u & Z_{\delta_e} & \{-(Z_q + U_1)s + g\sin\theta_1\} \\ -(M_u + M_{T_u}) & M_{\delta_e} & (s^2 - \mathrm{M_q s}) \end{vmatrix}}{\bar{D}_1} = \frac{\alpha(s)}{\delta_e(s)} = \frac{N_\alpha}{D_1}$$

$N_\alpha = A_\alpha s^3 + B_\alpha s^2 + C_\alpha s + D_\alpha$, where:

$$A_\alpha = Z_{\delta_e}$$

$$B_\alpha = X_{\delta_e} Z_u + Z_{\delta_e}\{-M_q - (X_u + X_{T_u})\} + M_{\delta_e}(U_1 + Z_q)$$

$$C_\alpha = X_{\delta_e}\{(U_1 + Z_q)(M_u + M_{T_u}) - M_q Z_u\} + Z_{\delta_e} M_q (X_u + X_{T_u}) + M_{\delta_e}\{-g\sin\theta_1 - (U_1 + Z_q)(X_u + X_{T_u})\}$$

$$D_\alpha = -X_{\delta_e}(M_u + M_{T_u})g\sin\theta_1 + Z_{\delta_e}(M_u + M_{T_u})g\cos\theta_1 + M_{\delta_e}\{(X_u + X_{T_u})g\sin\theta_1 - Z_u g\cos\theta_1\}$$

$$\frac{\theta(s)}{\delta_e(s)} = \frac{\begin{vmatrix} (s - X_u - X_{T_u}) & -X_\alpha & X_{\delta_e} \\ -Z_u & \{s(U_1 - Z_{\dot\alpha}) - Z_\alpha\} & Z_{\delta_e} \\ -(M_u + M_{T_u}) & -\{M_{\dot\alpha} s + M_\alpha + M_{T_\alpha}\} & M_{\delta_e} \end{vmatrix}}{\bar{D}_1} = \frac{N_\theta}{\bar{D}_1}$$

$N_\theta = A_\theta s^2 + B_\theta s + C_\theta$, where

$$A_\theta = Z_{\delta_e} M_{\dot\alpha} + M_{\delta_e}(U_1 - Z_{\dot\alpha})$$

$$B_\theta = X_{\delta_e}\{Z_u M_{\dot\alpha} + (U_1 - Z_{\dot\alpha})(M_u + M_{T_u})\} + Z_{\delta_e}\{(M_\alpha + M_{T_\alpha}) - M_{\dot\alpha}(X_u + X_{T_u})\} + M_{\delta_e}\{-Z_\alpha - (U_1 - Z_{\dot\alpha})(X_u + X_{T_u})\}$$

$$C_\theta = X_{\delta_e}\{(M_\alpha + M_{T_u})Z_u - Z_\alpha(M_u + M_{t_u})\} + Z_{\delta_e}\{-(M_\alpha + M_{T_u})(X_u + X_{T_u}) + X_\alpha(M_u + M_{T_u})\} + M_{\delta_e}\{Z_\alpha(X_u + X_{T_u}) - X_\alpha Z_u\}$$

Table G.2 Lateral–directional airplane transfer functions

$$\frac{\beta(s)}{\delta(s)} = \frac{\begin{vmatrix} Y_\delta & -(sY_p + g\cos\theta_1) & s(U_1 - Y_r) \\ L_\delta & (s^2 - L_p s) & -(s^2\bar{A}_1 + sL_r) \\ N_\delta & -(s^2\bar{B}_1 + N_p s) & (s^2 - sN_r) \end{vmatrix}}{\begin{vmatrix} (sU_1 - Y_\beta) & -(sY_p + g\cos\theta_1) & s(U_1 - Y_r) \\ -L_\beta & (s^2 - \mathrm{L_p s}) & -(s^2 A_1 + sL_r) \\ -(N_\beta + N_{T_\beta}) & -(s^2\bar{B}_1 + \mathrm{N_p s}) & (s^2 - sN_r) \end{vmatrix}} = \frac{N_\beta}{\bar{D}_2}$$

$$\bar{D}_2 = s(E's^4 + F's^3 + G's^2 + H's + I')$$

$$E' = U_1(1 - \bar{A}_1\bar{B}_1)$$

$$F' = -Y_\beta - (1 - \bar{A}_1\bar{B}_1) - U_1(L_p + N_r + \bar{A}_1 N_p + \bar{B}_1 L_r)$$

$$G' = U_1(L_p N_r - L_r N_p) + Y_\beta(N_r + L_p + \bar{A}_1 N_p + \bar{B}_1 L_r) - Y_p(L_\beta + N_\beta \bar{A}_1 + N_{T_\beta}\bar{A}_1) + U_1(L_\beta \bar{B}_1 + N_\beta + N_{T_\beta}) - Y_r(L_\beta \bar{B}_1 + N_\beta + N_{T_\beta})$$

$$H' = -Y_\beta(L_p N_r - L_r N_p) + Y_p(L_\beta N_r - N_\beta L_r - N_{T_\beta} L_r) - g\cos\theta_1(L_\beta + N_\beta \bar{A}_1 + N_{T_\beta}\bar{A}_1) + U_1(L_\beta N_p - N_\beta L_p - N_{T_\beta} L_p) - Y_r(L_\beta N_p - N_\beta L_p - N_{T_\beta} L_p)$$

$$I' = g\cos\theta_1(L_\beta N_r - N_\beta L_r - N_{T_\beta} L_r)$$

$N_\beta = s(A_\beta s^3 + B_\beta s^2 + C_\beta s + D_\beta)$, where:

$A_\beta = Y_\delta(1 - \bar{A}_1\bar{B}_1)$

$B_\beta = Y_\delta(N_r + L_p + \bar{A}_1 N_p + \bar{B}_1 L_r) + Y_p(L_\delta + N_\delta \bar{A}_1) + Y_r(L_\delta \bar{B}_1 + N_\delta) +$
$\quad - U_1(L_\delta \bar{B}_1 + N_\delta)$

$C_\beta = Y_\delta(L_p N_r - N_p L_r) + Y_p(N_\delta L_r - L_\delta N_r) + g\cos\theta_1(L_\delta + N_\delta \bar{A}_1) + Y_r(L_\delta N_p - N_\delta L_p) +$
$\quad - U_1(L_\delta N_p - N_\delta L_p)$

$D_\beta = g\cos\theta_1(N_\delta L_r - L_\delta N_r)$

$$\frac{\phi(s)}{\delta(s)} = \frac{\begin{vmatrix} (sU_1 - Y_\beta) & Y_\delta & s(U_1 - Y_r) \\ -L_\beta & L_\delta & -(s^2 A_1 + sL_r) \\ -(N_\beta + N_{T_\beta}) & N_\delta & (s^2 - sN_r) \end{vmatrix}}{\bar{D}_2} = \frac{N_\phi}{\bar{D}_2}$$

$N_\phi = s(A_\phi s^2 + B_\phi s + C_\phi)$, where :

$A_\phi = U_1(L_\delta + N_\delta A_1)$

$B_\phi = U_1(N_\delta L_r - L_\delta N_r) - Y_\beta(L_\delta + N_\delta \bar{A}_1) + Y_\delta(L_\beta + N_\beta \bar{A}_1 + N_{T_\beta}\bar{A}_1)$

$C_\phi = -Y_\beta(N_\delta L_r - L_\delta N_r) + Y_\delta(L_r N_\beta + L_r N_{T_\beta} - N_r L_\beta)$
$\quad + (U_1 - Y_r)(N_\beta L_\delta + N_{T_\beta} L_\delta - L_\beta N_\delta)$

$$\frac{\psi(s)}{\delta(s)} = \frac{\begin{vmatrix} (sU_1 - Y_\beta) & -(sY_p + g\cos\theta_1) & Y_\delta \\ -L_\beta & (s^2 - L_p s) & L_\delta \\ -(N_\beta + N_{T_\beta}) & -(s^2\bar{B}_1 + N_p s) & N_\delta \end{vmatrix}}{\bar{D}_2} = \frac{N_\psi}{\bar{D}_2}$$

$N_\psi = (A_\psi s^3 + B_\psi s^2 + C_\psi s + D_\psi)$, where:

$A_\psi = U_1(N_\delta + L_\delta \bar{B}_1)$

$B_\psi = U_1(L_\delta N_p - N_\delta L_p) - Y_\beta(N_\delta + L_\delta \bar{B}_1) + Y_\delta(L_\beta \bar{B}_1 + N_\beta + N_{T_\beta})$

$C_\psi = -Y_\beta(L_\delta N_p - N_\delta L_p) + Y_p(N_\beta L_\delta + N_{T_\beta} L_\delta - L_\beta N_\delta)$
$\quad + Y_\delta(L_\beta N_p - N_\beta L_p - N_{T_\beta} L_p)$

$D_\psi = g\cos\theta_1(N_\beta L_\delta + N_{T_\beta} L_\delta - L_\beta N_\delta)$

Appendix H
Stability Characteristics of Selected Aircraft

Lockheed C-5A

Aircraft Parameter	C-5A	C-5A	C-5A
Altitude (ft)	S.L.	20,000	40,000
Mach	0.25	0.7	0.7
True Airspeed (ft/s)	279	725	678
Dyn Pressure (lb/ft^2)			
Weight (lb)	728,000	728,000	728,000
Wing Area—S—(ft^2)	6,200	6,200	6,200
Wing Span—b—(ft)	219.17	219.17	219.17
Wing Chord—$\bar{c}$—(ft)	30.93	30.93	30.93
C.G. (x $\bar{c}$)	0.25	0.25	0.25
Trim AOA (deg)	11.41	0.68	7.66
I_{xx_S} (slug − ft^2)	3.25×10^7	3.23×10^7	3.22×10^7
I_{yy_S} (slug − ft^2)	3.28×10^7	3.28×10^7	3.28×10^7
I_{zz_S} (slug − ft^2)	6.15×10^7	6.18×10^7	6.19×10^7
I_{xz_S} (slug − ft^2)	-3.34×10^6	2.2×10^6	-1.42×10^6
Longitudinal Derivatives			
X_u (1/s)	−0.0111	−0.00283	−0.00339
X_α (ft/s^2)	−5.58	19.1	8.43
Z_u (1/s)	−0.107	−0.042	−0.06
Z_α (ft/s^2)	−124	−450	−204
M_u (1/ft · s)	−0.00007	−0.00004	0.00005
M_α (1/s^2)	−0.76	−2.2	−1.07
$M_{\dot{\alpha}}$ (1/s)	−0.169	−0.21	−0.113
M_q (1/s)	−0.67	−0.898	−0.426
X_{δ_e} (ft/s^2)	0	0	0
Z_{δ_e} (ft/s^2)	−11.5	−27.7	−13.2
M_{δ_e} (1/s^2)	−1.049	−2.48	−1.17
Lat-Dir Derivatives			
Y_β (ft/s^2)	−21.2	−73.2	−29.4
L_β (1/s)^2	−0.585	−1.516	−0.752
L_p (1/s)	−0.329	−0.432	−0.217
L_r (1/s)	0.256	0.181	0.125
N_β (1/s^2)	0.167	0.449	0.200
N_p (1/s)	−0.0184	−0.0248	−0.0151
N_r (1/s)	−0.12	−0.158	−0.07
Y_{δ_r} (ft/s^2)	4.64	16.7	6.70

L_{δ_r} $(1/s^2)$	0.00974	0.168	0.0533
N_{δ_r} $(1/s^2)$	−0.141	−0.504	−0.203
Y_{δ_a} (ft/s^2)	0	0	0
L_{δ_a} $(1/s^2)$	0.264	0.389	0.304
N_{δ_a} $(1/s^2)$	0.0138	0	0

Boeing 747

Aircraft Parameter	747	747	747
Altitude (ft)	S.L.	20,000	40,000
Mach	0.198	0.650	0.900
True Airspeed (ft/s)	221	673	871
Dyn Pressure (lb/ft^2)	58.0	287.2	222.8
Weight (lb)	564,000	636,636	636,636
Wing Area—S—(ft^2)	5,500	5,500	5,500
Wing Span—b—(ft)	196	196	196
Wing Chord—$\bar{c}$—(ft)	27.3	27.3	27.3
C.G. (x $\bar{c}$)	0.25	0.25	0.25
Trim AOA (deg)	8.5	2.5	2.4
I_{xx_S} $(slug - ft^2)$	1.41×10^7	1.82×10^7	1.82×10^7
I_{yy_S} $(slug - ft^2)$	3.05×10^7	3.31×10^7	3.31×10^7
I_{zz_S} $(slug - ft^2)$	4.27×10^7	4.97×10^7	4.97×10^7
I_{xz_S} $(slug - ft^2)$	-3.49×10^6	-4.05×10^5	-3.50×10^5
Longitudinal Derivatives			
X_u $(1/s)$	−0.0433	−0.0059	−0.0218
X_α (ft/s^2)	11.4738	15.9787	1.2227
Z_u $(1/s)$	−0.2720	−0.1104	−0.0569
Z_α (ft/s^2)	−108.0542	−353.52	−339.0036
M_u $(1/ft \cdot s)$	0.0001	0.0000	−0.0001
M_α $(1/s^2)$	−0.4140	−1.3028	−1.6165
$M_{\dot{\alpha}}$ $(1/s)$	−0.0582	−0.1057	−0.1425
M_q $(1/s)$	−0.3774	−0.5417	−0.4038
X_{δ_e} (ft/s^2)	0.0000	0.0000	0.0000
Z_{δ_e} (ft/s^2)	−6.5565	−25.5659	−18.3410
M_{δ_e} $(1/s^2)$	−0.3997	−1.6937	−1.2124
Lat-Dir Derivatives			
Y_β (ft/s^2)	−19.6694	−71.9142	−55.7808
L_β $(1/s^2)$	−1.2461	−2.7255	−1.2555
L_p $(1/s)$	−0.9871	−0.8434	−0.4758
L_r $(1/s)$	0.3834	0.3224	0.2974
N_β $(1/s^2)$	0.2694	0.9962	1.0143
N_p $(1/s)$	−0.1441	−0.0236	0.0109
N_r $(1/s)$	−0.2338	−0.2539	−0.1793
Y_{δ_r} $(ft/s)^2$	3.2600	9.5872	3.7187

L_{δ_r} $(1/s^2)$	0.0000	0.1363	0.2974
N_{δ_r} $(1/s^2)$	−0.1655	−0.6226	−0.4589
Y_{δ_a} (ft/s^2)	0.0000	1.0386	0.0000
L_{δ_a} $(1/s^2)$	0.2350	0.2214	0.1850
N_{δ_a} $(1/s^2)$	0.0122	0.0112	−0.0135

McDonnell Douglass DC-8

Aircraft Parameter Assume $\alpha_{L=0} = -2^\circ$	DC-8	DC-8	DC-8
Altitude (ft)	S/L	15,000	33,000
Mach	0.218	0.443	0.88
True Airspeed (ft/s)	243	468	868
Dyn Pressure (lb/ft^2)	70.4	164	300
Weight (lb)	190,000	190,000	230,000
Wing Area—S—(ft^2)	2,600.0	2,600.0	2,600.0
Wing Span—b—(ft)	142.3	142.3	142.3
Wing Chord—$\bar{c}$—(ft)	23.0	23.0	23.0
C.G. (x $\bar{c}$)	0.15	0.15	0.15
Trim AOA (deg)	10.36	3.23	0.45
I_{xx_S} $(slug - ft^2)$	3.16×10^6	3.13×10^6	3.77×10^6
I_{yy_S} $(slug - ft^2)$	2.94×10^6	2.94×10^6	3.56×10^6
I_{zz_S} $(slug - ft^2)$	5.51×10^6	5.86×10^6	7.13×10^6
I_{xz_S} $(slug - ft^2)$	-4.14×10^5	-2.20×10^5	2.73×10^5
Longitudinal Derivatives			
X_u $(1/s)$	−0.0291	−0.0071	−0.0463
X_α (ft/s^2)	15.32	15.03	−22.36
Z_u $(1/s)$	−0.251	−0.133	0.062
Z_α (ft/s^2)	−152.8	−354	−746.9
M_u $(1/ft \cdot s)$	0.0000	0.0000	−0.0025
M_α $(1/s^2)$	−2.118	−5.010	−12.003
$M_{\dot{\alpha}}$ $(1/s)$	−0.260	−0.337	−0.449
M_q $(1/s)$	−0.792	−0.991	−1.008
X_{δ_e} (ft/s^2)	0.0	0.0	0.0
Z_{δ_e} (ft/s^2)	−10.17	−23.70	−39.06
M_{δ_e} $(1/s^2)$	−1.351	−3.241	−5.120
Lat-Dir Derivatives			
Y_β (ft/s^2)	−27.05	−47.19	−81.28
L_β $(1/s^2)$	−1.334	−2.684	−5.111
L_p $(1/s)$	−0.949	−1.234	−1.299
L_r $(1/s)$	0.611	0.391	0.352
N_β $(1/s^2)$	0.762	1.272	2.497
N_p $(1/s)$	−0.119	−0.048	−0.008
N_r $(1/s)$	−0.268	−0.253	−0.254

Y_{δ_r} (ft/s)^2	5.782	13.47	20.35
L_{δ_r} (1/s^2)	0.185	0.375	0.639
N_{δ_r} (1/s^2)	−0.389	−0.861	−1.298
Y_{δ_a} (ft/s^2)	0.0	0.0	0.0
L_{δ_a} (1/s^2)	0.725	1.622	2.329
N_{δ_a} (1/s^2)	0.050	0.036	0.062

Learjet C-21

Aircraft Parameter	C-21	C-21	C-21
Altitude (ft)	S/L	40,000	40,000
Mach	0.152	0.7	0.7
True Airspeed (ft/s)	170	677	677
Dyn Pressure (lb/ft^2)	34.3	134.6	134.6
Weight (lb)	13,000	13,000	9,000
Wing Area—S—(ft^2)	230	230	230
Wing Span—b—(ft)	34.0	34.0	34.0
Wing Chord—$\bar{c}$—(ft)	7.0	7.0	7.0
C.G. (x $\bar{c}$)	0.32	0.32	0.32
Trim AOA (deg)	5.0	2.7	1.5
I_{xx_S} (slug − ft^2)	2.79×10^4	2.79×10^4	5.94×10^3
I_{yy_S} (slug − ft^2)	1.88×10^4	1.88×10^4	1.78×10^4
I_{zz_S} (slug − ft^2)	4.71×10^4	4.71×10^4	2.51×10^4
I_{xz_S} (slug − ft^2)	-3.60×10^2	4.03×10^2	9.02×10^2
Longitudinal Derivatives			
X_u (1/s)	−0.0589	−0.0194	−0.0261
X_α (ft/s^2)	11.3335	8.4349	6.6457
Z_u (1/s)	−0.3816	−0.1382	−0.1374
Z_α (ft/s^2)	−103.4862	−450.3834	−649.9336
M_u (1/ft · s)	−0.0002	0.0009	0.0013
M_α (1/s^2)	−1.9387	−7.3772	−7.7917
$M_{\dot{\alpha}}$ (1/s)	−0.3024	−0.3993	−0.4217
M_q (1/s)	−0.8164	−0.9237	−0.9756
X_{δ_e} (ft/s^2)	0.0000	0.0000	0.0000
Z_{δ_e} (ft/s^2)	−7.8162	−35.2731	−50.9500
M_{δ_e} (1/s^2)	−2.8786	−14.2934	−15.0964
Lat-Dir Derivatives			
Y_β (ft/s^2)	−14.2645	−55.9768	−80.8554
L_β (1/s^2)	−1.6621	−4.1470	−17.7208
L_p (1/s)	−0.3747	−0.4260	−2.0024
L_r (1/s)	0.4323	0.1515	0.6230

N_β $(1/s^2)$	0.8546	2.8393	5.2082
N_p $(1/s)$	−0.0741	−0.0045	−0.0232
N_r $(1/s)$	−0.1481	−0.1123	−0.2109
Y_{δ_r} $(ft/s)^2$	2.7357	10.7353	15.5065
L_{δ_r} $(1/s^2)$	0.1345	0.7163	3.7214
N_{δ_r} $(1/s^2)$	−0.4216	−1.6544	−3.1081
Y_{δ_a} (ft/s^2)	0.0000	0.0000	0.0000
L_{δ_a} $(1/s^2)$	1.4315	6.7106	31.5431
N_{δ_a} $(1/s^2)$	−0.2849	−0.4471	−0.8400

Lockheed F-104

Aircraft Parameter	F-104	F-104
Altitude (ft)	S/L	55,000
Mach	0.257	1.800
True Airspeed (ft/s)	287	1,742
Dyn Pressure (lb/ft^2)	97.8	434.5
Weight (lb)	16,300	16,300
Wing Area—S—(ft^2)	196	196
Wing Span—b—(ft)	21.9	21.9
Wing Chord—$\bar{c}$—(ft)	9.6	9.6
C.G. (x $\bar{c}$)	0.07	0.07
Trim AOA (deg)	10	2
I_{xx_S} (slug − ft^2)	5.30×10^3	3.67×10^3
I_{yy_S} (slug − ft^2)	5.90×10^4	5.90×10^4
I_{zz_S} (slug − ft^2)	5.83×10^4	5.99×10^4
I_{xz_S} (slug − ft^2)	-9.64×10^3	-1.97×10^3
Longitudinal Derivatives		
X_u $(1/s)$	−0.0695	−0.0049
X_α (ft/s^2)	14.9575	−32.4692
Z_u $(1/s)$	−0.2243	−0.0176
Z_α (ft/s^2)	−140.2374	−346.6128
M_u $(1/ft \cdot s)$	0.0000	0.0000
M_α $(1/s^2)$	−2.0086	−18.1248
$M_{\dot{\alpha}}$ $(1/s)$	−0.0855	−0.0783
M_q $(1/s)$	−0.3046	−0.1844
X_{δ_e} (ft/s^2)	0.0000	0.0000
Z_{δ_e} (ft/s^2)	−25.9012	−87.9865
M_{δ_e} $(1/s^2)$	−4.9904	−18.1525
Lat-Dir Derivatives		
Y_β (ft/s^2)	−44.6833	−175.8047
L_β $(1/s^2)$	−13.8595	−47.2783
L_p $(1/s)$	−0.8612	−0.8692

L_r (1/s)	0.8007	0.4921
N_β (1/s^2)	3.6508	7.5310
N_p (1/s)	−0.0396	−0.0182
N_r (1/s)	−0.2069	−0.1270
Y_{δ_r} (ft/s)^2	12.4583	14.6364
L_{δ_r} (1/s^2)	3.5480	4.0161
N_{δ_r} (1/s^2)	−1.1845	−1.3537
Y_{δ_a} (ft/s^2)	0.0000	0.0000
L_{δ_a} (1/s^2)	3.1045	8.7948
N_{δ_a} (1/s^2)	0.0302	0.0778

McDonnell Douglass F-4C

Aircraft Parameter	F-4C	F-4C	F-4C
Altitude (ft)	S/L	35,000	35,000
Mach	0.8	0.6	1.2
True Airspeed (ft/s)	893	584	1,167
Dyn Pressure (lb/ft^2)	948	126	503
Weight (lb)	38924	38924	38924
Wing Area—S—(ft^2)	530	530	530
Wing Span—b—(ft)	38.67	38.67	38.67
Wing Chord—$\bar{c}$—(ft)	16.0	16.0	16.0
C.G. (x $\bar{c}$)	0.289	0.289	0.289
Trim AOA (deg)	0.3	9.4	1.6
I_{xx_S} (slug − ft^2)	2.50×10^4	3.56×10^4	2.51×10^4
I_{yy_S} (slug − ft^2)	1.22×10^5	1.22×10^5	1.22×10^5
I_{zz_S} (slug − ft^2)	1.40×10^5	1.29×10^5	1.40×10^5
I_{xz_S} (slug − ft^2)	9.75×10^2	-3.37×10^4	-4.23×10^3
Longitudinal Derivatives			
X_u (1/s)	−0.0162	−0.0176	−0.0136
X_α (ft/s^2)	−0.820	−24.25	−16.53
Z_u (1/s)	−0.073	−0.116	−0.010
Z_α (ft/s^2)	−1374.9	−162.2	−848.3
M_u (1/ft · s)	−0.0017	−0.0001	0.0022
M_α (1/s^2)	−17.76	−1.927	−29.03
$M_{\dot{\alpha}}$ (1/s)	−0.592	−0.144	−0.288
M_q (1/s)	−1.360	−0.307	−0.746
X_{δ_e} (ft/s^2)	0.00	−0.01	−0.01
Z_{δ_e} (ft/s^2)	−141.0	−20.98	−90.44
M_{δ_e} (1/s^2)	−32.30	−4.90	−20.70
Lat-Dir Derivatives			
Y_β (ft/s^2)	−299.2	−33.05	−176.2
L_β (1/s^2)	−27.21	−8.252	−13.23

L_p (1/s)	−3.034	−0.687	−1.372
L_r (1/s)	0.876	0.281	0.329
N_β (1/s^2)	16.03	2.155	12.59
N_p (1/s)	0.009	−0.006	−0.021
N_r (1/s)	−0.753	−0.149	−0.403
Y_{δ_r} (ft/s)^2	39.47	6.600	15.40
L_{δ_r} (1/s^2)	7.774	−0.346	2.765
N_{δ_r} (1/s^2)	−7.920	−1.403	−3.251
Y_{δ_a} (ft/s^2)	−6.644	−0.882	−3.524
L_{δ_a} (1/s^2)	22.15	4.243	10.93
N_{δ_a} (1/s^2)	0.556	−0.124	0.443

LTV Corsair A-7A

Aircraft Parameter	A-7A	A-7A	A-7A
Altitude (ft)	S/L	35,000	35,000
Mach	0.6	0.6	0.9
True Airspeed (ft/s)	670	584	876
Dyn Pressure (lb/ft^2)	534	126	283
Weight (lb)	21,889	21,889	21,889
Wing Area—S—(ft^2)	375	375	375
Wing Span—b—(ft)	38.7	38.7	38.7
Wing Chord—$\bar{c}$—(ft)	10.8	10.8	10.8
C.G. (x $\bar{c}$)	0.30	0.30	0.30
Trim AOA (deg)	2.9	7.5	3.8
I_{xx_S} (slug − ft^2)	1.36×10^4	1.58×10^4	1.38×10^4
I_{yy_S} (slug − ft^2)	5.90×10^4	5.90×10^4	5.90×10^4
I_{zz_S} (slug − ft^2)	6.76×10^4	6.54×10^4	6.74×10^4
I_{xz_S} (slug − ft^2)	-2.55×10^3	-1.10×10^4	-4.24×10^3
Longitudinal Derivatives			
X_u (1/s)	−0.0156	−0.0093	−0.0230
X_α (ft/s^2)	−26.58	−33.35	−29.74
Z_u (1/s)	−0.124	−0.112	−0.091
Z_α (ft/s^2)	−1284.0	−316.1	−881.3
M_u (1/ft · s)	0.0002	0.0002	−0.0026
M_α (1/s^2)	−15.57	−4.183	−13.02
$M_{\dot{\alpha}}$ (1/s)	−0.207	−0.065	−0.143
M_q (1/s)	−1.110	−0.330	−0.539
X_{δ_e} (ft/s^2)	−0.02	0.01	0.01
Z_{δ_e} (ft/s^2)	−165.2	−43.57	−99.62
M_{δ_e} (1/s^2)	−30.60	−8.19	−20.20
Lat-Dir Derivatives			
Y_β (ft/s^2)	−210.4	−49.46	−127.0
L_β (1/s^2)	−44.57	−13.60	−29.89

L_p (1/s)	−4.401	−1.295	−2.956
L_r (1/s)	1.341	0.639	0.705
N_β (1/s^2)	8.126	2.416	6.444
N_p (1/s)	0.022	−0.020	0.026
N_r (1/s)	−0.968	−0.288	−0.490
Y_{δ_r} (ft/s)^2	51.52	15.59	30.39
L_{δ_r} (1/s^2)	11.08	1.854	5.927
N_{δ_r} (1/s^2)	−9.207	−2.754	−5.473
Y_{δ_a} (ft/s^2)	−7.034	−1.559	−3.740
L_{δ_a} (1/s^2)	28.46	7.859	14.24
N_{δ_a} (1/s^2)	0.559	0.098	0.205

Cessna T-37

Aircraft Parameter	T-37	T-37	T-37
Altitude (ft)	S/L	30,000	S/L
Mach	0.313	0.459	0.143
True Airspeed (ft/s)	349	456	160
Dyn Pressure (lb/ft^2)	144.9	92.7	30.4
Weight (lb)	6,360	6,360	6,360
Wing Area—S—(ft^2)	182	182	182
Wing Span—b—(ft)	33.8	33.8	33.8
Wing Chord—$\bar{c}$—(ft)	5.47	5.47	5.47
C.G. (x $\bar{c}$)	27.0	27.0	27.0
Trim AOA (deg)	0.7	2	4.2
I_{xx_S} (slug − ft^2)	7.99×10^3	7.99×10^3	8.00×10^3
I_{yy_S} (slug − ft^2)	3.33×10^3	3.33×10^3	3.33×10^3
I_{zz_S} (slug − ft^2)	1.12×10^4	1.12×10^4	1.12×10^4
I_{xz_S} (slug − ft^2)	-3.91×10^1	-1.12×10^2	-2.34×10^2
Longitudinal Derivatives			
X_u (1/s)	−0.0168	−0.0112	−0.0553
X_α (ft/s^2)	14.8205	10.9335	13.1096
Z_u (1/s)	−0.1844	−0.1416	−0.4027
Z_α (ft/s^2)	−645.1571	−442.4657	−134.4015
M_u (1/ft · s)	0.0000	0.0000	0.0000
M_α (1/s^2)	−28.9722	−19.4229	−5.7417
$M_{\dot{\alpha}}$ (1/s)	−2.2569	−1.1566	−1.0639
M_q (1/s)	−4.8604	−2.4797	−2.1776
X_{δ_e} (ft/s^2)	0.0000	0.0000	0.0000
Z_{δ_e} (ft/s^2)	−53.4070	−42.7090	−11.2048
M_{δ_e} (1/s^2)	−46.4075	−31.0767	−9.5543
Lat-Dir Derivatives			
Y_β (ft/s^2)	−48.1999	−29.5547	−8.4876
L_β (1/s^2)	−9.4992	−6.7383	−1.9210

L_p (1/s)	−2.3783	−1.1693	−1.1305
L_r (1/s)	0.3189	0.2450	0.6270
N_β (1/s^2)	8.3856	5.6418	1.8339
N_p (1/s)	−0.0594	−0.0459	−0.1359
N_r (1/s)	−0.5531	−0.2628	−0.2853
Y_{δ_r} (ft/s)^2	26.7035	17.0836	5.6024
L_{δ_r} (1/s^2)	1.6744	1.0707	0.3505
N_{δ_r} (1/s^2)	−2.9094	−1.8619	−0.6113
Y_{δ_a} (ft/s^2)	0.0000	0.0000	00.000
L_{δ_a} (1/s^2)	19.9583	12.9199	4.1785
N_{δ_a} (1/s^2)	−1.2754	−1.2957	−1.2729

Bibliography

Blake, W. B., Air Force Research Laboratory, Flight Vehicles Directorate, private communication, Jan. 2002.

Bossert, D., and Cohen, K., "Design of Fuzzy Pitch Attitude Hold Systems for a Fighter Jet," *Proceedings of the 2001 AIAA Guidance, Navigation, and Control Conference*.

Bowman, J. S., Jr., Hultberg, R. S., and Martin, C. A., "Measurements of Pressures on the Tail and Aft Fuselage of an Airplane Model During Rotary Motions at Spin Attitudes," NASA TP-2939, Nov. 1989.

Brandt, S. A., Stiles, R. J., Bertin, J. J., and Whitford, R., *Introduction to Aeronautics: A Design Perspective*, AIAA Education Series, AIAA, Reston, VA, 1997.

D'Azzo, J., and Houpis, C., *Linear Control System Analysis and Design, Conventional and Modern*, McGraw-Hill, New York, 1988.

DiStefano, J., Stubberud, A., and Williams, I., *Feedback and Control Systems*, Schaum's Outline Series, McGraw-Hill, New York, 1967.

Houpis, C., and Lamont, G., *Digital Control Systems: Theory, Hardware, Software*, McGraw-Hill, New York, 1985.

Hultberg, R., "Low Speed Rotary Aerodynamic of F-18 Configuration for 0 to 90 Angle of Attack—Test Results and Analysis," NASA CR-3608, Aug. 1984.

Mattingly, J. D., *Elements of Gas Turbine Propulsion*, McGraw-Hill, New York, 1996.

Mattingly, J. D., Heiser, W. H., and Daley, D. H., *Aircraft Engine Design*, AIAA Education Series, AIAA, Reston, VA, 1987.

Nelson, R. C., *Flight Stability and Automatic Control*, McGraw-Hill, New York, 1989.

Ralston, J. N., "Rotary Balance Data and Analysis for the X-29A Airplane for an Angle-of-Attack Range of 0 to 90," NASA CR-3747, Aug. 1984.

Roskam, J., *Airplane Flight Dynamics and Automatic Flight Controls*, *Part I and II*, Design, Analysis, and Research Corporation, Lawrence, KS, 1995.

USAF Test Pilot School, *Feedback Control Theory*, Edwards AFB, CA, 1988, Chaps. 13, 14.

Index

TEXTS PUBLISHED IN THE AIAA EDUCATION SERIES

Introduction to Aircraft Flight Mechanics
Thomas R. Yechout with Steven L. Morris, David E. Bossert, and Wayne F. Hallgren 2003
ISBN 1-56347-577-4

Analytical Mechanics of Space Systems
Hanspeter Schaub and John L. Junkins 2003
ISBN 1-56347-563-4

Flight Testing of Fixed-Wing Aircraft
Ralph D. Kimberlin 2003
ISBN 1-56347-564-2

Aircraft Design Projects for Engineering Students
Lloyd Jenkinson and James Marchman 2003
ISBN 1-56347-619-3

Elements of Spacecraft Design
Charles D. Brown 2002
ISBN 1-56347-524-3

Civil Avionics Systems
Ian Moir and Allan Seabridge 2002
ISBN 1-56347-589-8

Helicopter Test and Evaluation
Alastair K. Cooke and Eric W. H. Fitzpatrick 2002
ISBN 1-56347-578-2

Aircraft Engine Design, Second Edition
Jack D. Mattingly, William H. Heiser, and David T. Pratt 2002
ISBN 1-56347-538-3

Dynamics, Control, and Flying Qualities of V/STOL Aircraft
James A. Franklin 2002
ISBN 1-56347-575-8

Orbital Mechanics, Third Edition
Vladimir A. Chobotov, Editor 2002
ISBN 1-56347-537-5

Basic Helicopter Aerodynamics, Second Edition
John Seddon and Simon Newman 2001
ISBN 1-56347-510-3

Aircraft Systems: Mechanical, Electrical, and Avionics Subsystems Integration
Ian Moir and Allan Seabridge 2001
ISBN 1-56347-506-5

Design Methodologies for Space Transportation Systems
Walter E. Hammond 2001
ISBN 1-56347-472-7

Tactical Missile Design
Eugene L. Fleeman 2001
ISBN 1-56347-494-8

Flight Vehicle Performance and Aerodynamic Control
Frederick O. Smetana 2001
ISBN 1-56347-463-8

Modeling and Simulation of Aerospace Vehicle Dynamics
Peter H. Zipfel 2000
ISBN 1-56347-456-6

Applied Mathematics in Integrated Navigation Systems
Robert M. Rogers 2000
ISBN 1-56347-445-X

Mathematical Methods in Defense Analyses, Third Edition
J. S. Przemieniecki 2000
ISBN 1-56347-396-6

Finite Element Multidisciplinary Analysis
Kajal K. Gupta and John L. Meek 2000
ISBN 1-56347-393-3

Aircraft Performance: Theory and Practice
M. E. Eshelby 1999
ISBN 1-56347-398-4

Space Transportation: A Systems Approach to Analysis and Design
Walter E. Hammond *1999*
ISBN 1-56347-032-2

Civil Jet Aircraft Design
Lloyd R. Jenkinson, Paul Simpkin, and Darren Rhodes *1999*
ISBN 1-56347-350-X

Structural Dynamics in Aeronautical Engineering
Maher N. Bismarck–Nasr *1999*
ISBN 1-56347-323-2

Intake Aerodynamics, Second Edition
E. L. Goldsmith and J. Seddon *1999*
ISBN 1-56347-361-5

Integrated Navigation and Guidance Systems
Daniel J. Biezad *1999*
ISBN 1-56347-291-0

Aircraft Handling Qualities
John Hodgkinson *1999*
ISBN 1-56347-331-3

Performance, Stability, Dynamics, and Control of Airplanes
Bandu N. Pamadi *1998*
ISBN 1-56347-222-8

Spacecraft Mission Design, Second Edition
Charles D. Brown *1998*
ISBN 1-56347-262-7

Computational Flight Dynamics
Malcolm J. Abzug *1998*
ISBN 1-56347-259-7

Space Vehicle Dynamics and Control
Bong Wie *1998*
ISBN 1-56347-261-9

Introduction to Aircraft Flight Dynamics
Louis V. Schmidt *1998*
ISBN 1-56347-226-0

Aerothermodynamics of Gas Turbine and Rocket Propulsion, Third Edition
Gordon C. Oates *1997*
ISBN 1-56347-241-4

Advanced Dynamics
Shuh-Jing Ying *1997*
ISBN 1-56347-224-4

Introduction to Aeronautics: A Design Perspective
Steven A. Brandt, Randall J. Stiles, John J. Bertin, and Ray Whitford *1997*
ISBN 1-56347-250-3

Introductory Aerodynamics and Hydrodynamics of Wings and Bodies: A Software-Based Approach
Frederick O. Smetana *1997*
ISBN 1-56347-242-2

An Introduction to Aircraft Performance
Mario Asselin *1997*
ISBN 1-56347-221-X

Orbital Mechanics, Second Edition
Vladimir A. Chobotov, Editor *1996*
ISBN 1-56347-179-5

Thermal Structures for Aerospace Applications
Earl A. Thornton *1996*
ISBN 1-56347-190-6

Structural Loads Analysis for Commercial Transport Aircraft: Theory and Practice
Ted L. Lomax *1996*
ISBN 1-56347-114-0

Spacecraft Propulsion
Charles D. Brown *1996*
ISBN 1-56347-128-0

Helicopter Flight Dynamics: The Theory and Application of Flying Qualities and Simulation Modeling
Gareth D. Padfield *1996*
ISBN 1-56347-205-8

Flying Qualities and Flight Testing of the Airplane
Darrol Stinton 1996
ISBN 1-56347-117-5

Flight Performance of Aircraft
S. K. Ojha 1995
ISBN 1-56347-113-2

Operations Research Analysis in Test and Evaluation
Donald L. Giadrosich 1995
ISBN 1-56347-112-4

Radar and Laser Cross Section Engineering
David C. Jenn 1995
ISBN 1-56347-105-1

Introduction to the Control of Dynamic Systems
Frederick O. Smetana 1994
ISBN 1-56347-083-7

Tailless Aircraft in Theory and Practice
Karl Nickel and Michael Wohlfahrt 1994
ISBN 1-56347-094-2

Mathematical Methods in Defense Analyses, Second Edition
J. S. Przemieniecki 1994
ISBN 1-56347-092-6

Hypersonic Aerothermodynamics
John J. Bertin 1994
ISBN 1-56347-036-5

Hypersonic Airbreathing Propulsion
William H. Heiser and David T. Pratt 1994
ISBN 1-56347-035-7

Practical Intake Aerodynamic Design
E. L. Goldsmith and J. Seddon 1993
ISBN 1-56347-064-0

Acquisition of Defense Systems
J. S. Przemieniecki, Editor 1993
ISBN 1-56347-069-1

Dynamics of Atmospheric Re-Entry
Frank J. Regan and Satya M. Anandakrishnan 1993
ISBN 1-56347-048-9

Introduction to Dynamics and Control of Flexible Structures
John L. Junkins and Youdan Kim 1993
ISBN 1-56347-054-3

Spacecraft Mission Design
Charles D. Brown 1992
ISBN 1-56347-041-1

Rotary Wing Structural Dynamics and Aeroelasticity
Richard L. Bielawa 1992
ISBN 1-56347-031-4

Aircraft Design: A Conceptual Approach, Second Edition
Daniel P. Raymer 1992
ISBN 0-930403-51-7

Optimization of Observation and Control Processes
Veniamin V. Malyshev, Mihkail N. Krasilshikov, and Valeri I. Karlov 1992
ISBN 1-56347-040-3

Nonlinear Analysis of Shell Structures
Anthony N. Palazotto and Scott T. Dennis 1992
ISBN 1-56347-033-0

Orbital Mechanics
Vladimir A. Chobotov, Editor 1991
ISBN 1-56347-007-1

Critical Technologies for National Defense
Air Force Institute of Technology 1991
ISBN 1-56347-009-8

Space Vehicle Design
Michael D. Griffin and James R. French 1991
ISBN 0-930403-90-8

Defense Analyses Software
J. S. Przemieniecki 1990
ISBN 0-930403-91-6

Inlets for Supersonic Missiles
John J. Mahoney *1990*
ISBN 0-930403-79-7

Introduction to Mathematical Methods in Defense Analyses
J. S. Przemieniecki *1990*
ISBN 0-930403-71-1

Basic Helicopter Aerodynamics
J. Seddon *1990*
ISBN 0-930403-67-3

Aircraft Propulsion Systems Technology and Design
Gordon C. Oates, Editor *1989*
ISBN 0-930403-24-X

Boundary Layers
A. D. Young *1989*
ISBN 0-930403-57-6

Aircraft Design: A Conceptual Approach
Daniel P. Raymer *1989*
ISBN 0-930403-51-7

Gust Loads on Aircraft: Concepts and Applications
Frederic M. Hoblit *1988*
ISBN 0-930403-45-2

Aircraft Landing Gear Design: Principles and Practices
Norman S. Currey *1988*
ISBN 0-930403-41-X

Mechanical Reliability: Theory, Models and Applications
B. S. Dhillon *1988*
ISBN 0-930403-38-X

Re-Entry Aerodynamics
Wilbur L. Hankey *1988*
ISBN 0-930403-33-9

Aerothermodynamics of Gas Turbine and Rocket Propulsion, Revised and Enlarged
Gordon C. Oates *1988*
ISBN 0-930403-34-7

Advanced Classical Thermodynamics
George Emanuel *1987*
ISBN 0-930403-28-2

Radar Electronic Warfare
August Golden Jr. *1987*
ISBN 0-930403-22-3

An Introduction to the Mathematics and Methods of Astrodynamics
Richard H. Battin *1987*
ISBN 0-930403-25-8

Aircraft Engine Design
Jack D. Mattingly, William H. Heiser, and Daniel H. Daley *1987*
ISBN 0-930403-23-1

Gasdynamics: Theory and Applications
George Emanuel *1986*
ISBN 0-930403-12-6

Composite Materials for Aircraft Structures
Brian C. Hoskin and Alan A. Baker, Editors *1986*
ISBN 0-930403-11-8

Intake Aerodynamics
J. Seddon and E. L. Goldsmith *1985*
ISBN 0-930403-03-7

The Fundamentals of Aircraft Combat Survivability Analysis and Design
Robert E. Ball *1985*
ISBN 0-930403-02-9

Aerothermodynamics of Aircraft Engine Components
Gordon C. Oates, Editor *1985*
ISBN 0-915928-97-3

Aerothermodynamics of Gas Turbine and Rocket Propulsion
Gordon C. Oates *1984*
ISBN 0-915928-87-6

Re-Entry Vehicle Dynamics
Frank J. Regan *1984*
ISBN 0-915928-78-7